Biomarkers of Oxidative Stress

Silvana Andreescu • Ralf Henkel
Abderrezak Khelfi
Editors

Biomarkers of Oxidative Stress

Basics and Measurement of Oxidative Stress

Springer

Editors
Silvana Andreescu
Bioanalytical Chemistry
Clarkson University
Potsdam, NY, USA

Abderrezak Khelfi
University of Algiers
Algiers, Algeria

Clarkson University
Potsdam, NY, USA

Ralf Henkel
LogixX Pharma Ltd.
Theale, Berkshire, UK

Department of Medical Bioscience
University of the Western Cape
Bellville, South Africa

Department of Metabolism
Digestion and Reproduction
Imperial College
London, UK

ISBN 978-3-031-60737-0 ISBN 978-3-031-60738-7 (eBook)
https://doi.org/10.1007/978-3-031-60738-7

This Springer imprint is published by the registered company Springer Nature Switzerland AG
The registered company address is: Gewerbestrasse 11, 6330 Cham, Switzerland

Preface

This book pinpoints one of the fastest growing subjects in chemistry and medical science and is therefore much more than just a reference book as the dangers of redox stress to human health and its chemistry behind it needs to be explained.

While previous books rather focused on the chemistry or the involvement of free radicals in diseases and oxidative stress, the content of this book reveals how reactive oxygen and nitrogen species, other free radicals, and antioxidants are involved not only in almost every human disease but also in essential physiological processes; this book discusses their importance, relevance, physiology, and pathology. The book expands on the concept of oxidative stress and explains the physiological relevance of redox homeostasis, i.e., the balance between oxidative and reductive stress. Readers will find novel information, including new radical chemistries and the latest discoveries of how radicals and reactive species react with biomolecules resulting in new biomarkers of oxidative stress.

The book provides a solid background on the chemistry behind the generation of reactive species for novice learners while also offering researchers and physicians, already involved in this field, rich information covering the most recent developments in the study of oxidative stress such as the role of radical and reactive species, novel antioxidant therapies, and methods for assessing free radicals and oxidative stress. The authors present concise, yet thorough, descriptions of the state-of-the-art methods and techniques that any investigator working in the field needs to access. The current methodologies including the development of sensors and biosensors for the detection of ROS/RNS/RHS and of biomarkers of redox stress are thoroughly discussed.

Several leading experts in the field of oxidative stress contributed to this book. They have been involved in oxidative stress-related research for many years and are internationally recognized for developing new procedures and methods. Their expertise ranges from diagnosis to therapeutic aspects of oxidative stress-associated diseases.

This book is addressed to all scientists interested in starting to work in redox stress research. It will be a significant resource for researchers in molecular biology and chemistry. Healthcare professionals in different medical fields who want to understand the molecular mechanisms underlying the redox stress-related diseases might also have interest in this book.

Although primarily a comprehensive source for research, this book also provides easy-to-follow directions for academic use as a manual for graduate students in basic and medical sciences. Each chapter will be both pedagogical and critical with an intended audience of new students and researchers entering the area. This book provides a rich list of literature for instructors to teach the subject while providing students with a wealth of information to gain deeper insight into the subject.

Algiers, Algeria — Abderrezak Khelfi
Theale, Berkshire, UK — Ralf Henkel
Potsdam, NY, USA — Silvana Andreescu
March 2024

Contents

Abbreviations

4-HNE	4-Hydroxyl-2-nonenal
5-ASA	Mesalazine
aAb	Auto-antibodies
AD	Alzheimer's disease
AGEs	Advanced glycation end-products
AH_2	Ascorbic acid
ALEs	Advanced peroxidation end-products
ALS	Amyotrophic lateral sclerosis
AO	Aldehyde oxidase
AOPP	Advanced oxidation protein products
APCI	Atmospheric pressure chemical ionization
APX	Ascorbate peroxidase
AUC	Area under the curve
BAP	Biological antioxidant potential
BOSS	Biomarkers of Oxidative Stress Study
CAT	Catalase
CD	Crohn's disease
CEL	*Nε*-(carboxyethyl)lysine
CF	Cystic fibrosis
CMA	*Nω*-(carboxymethyl)arginine
CML	*Nε*-(carboxymethyl)lysine
COPD	Chronic obstructive pulmonary disease
CoQ	Ubiquinone
$CoQH_2$	Ubiquinol
COX	Cyclooxygenase
CUPRAC	Cupric reducing antioxidant capacity
CVD	Cardiovascular disease
CYP450	Cytochrome P450-dependent monooxygenase
DM	Diabetes mellitus
DNP	Dinitrophenyl
DNPH	2,4-Dinitrophenylhydrazine

DOLD	Deoxyglucosone lysine dimer
EBC	Exhaled breath condensate
EC	Endothelial cells
eNOS	Endothelial nitric oxide synthase
EPO	Eosinoperoxidase
ER	Endoplasmic reticulum
ESCODD	European Standards Committee on Oxidative DNA Damage
ESCULA	European Standards Committee on Urinary Lesion Analysis
ESI	Electrospray ionization
ETC	Electron transport chain
FAPY	Formamidopyrimidine
FapyGua	2,6-Diamino-4-hydroxy-5-formamidopyrimidine
FMO	Flavin monooxygenases
FPG	Formamidopyrimidine DNA glycosylase
FR	Free radicals
FRAP	Ferric reducing antioxidant power
G6PD	Glucose-6-phosphate dehydrogenase
GOLD	Glyoxal-lysine dimer
GPO	Gastric peroxidase
GPx	Glutathione peroxidase
GR	Glutathione reductase
GSH	Glutathione
GSSG	Glutathione disulfide
GST	Glutathione-S-transferase
HAT	Hydrogen atom transfer
HD	Huntington's disease
HETE	Hydroxyeicosatetraenoic acid
HO-1	Heme oxygenase-1
HpETE	Hydroperoxyeicosatetraenoic acid
HPLC	High-performance liquid chromatography
HPLC-EC	High-performance liquid chromatography-electrochemical detection
HPLC-FL	High-performance liquid chromatography-fluorescence detection
HPLC-UV	High-performance liquid chromatography-ultraviolet detection
HRP	Horseradish peroxidase
IBD	Inflammatory bowel disease
IEC	Intestinal epithelial cell
iNOS	Inducible nitric oxide synthase
IsoPs	Isoprostanes
LOD	Limit of detection
LOQ	Limit of quantification
LOX	Lipoxygenase

LPO	Lactoperoxidase
mAb	Monoclonal antibody
MALD	Matrix-assisted laser desorption/ionization
MAPK	Mitogen-activated protein kinase
MDA	Malondialdehyde
MMPs	Matrix metalloproteinases
MOLD	Methylglyoxal-lysine dimer
MPO	Myeloperoxidase
MS	Mass spectrometry
mtDNA	mitochondrial DNA
NAC	N-acetyl cysteine
NAPQI	N-acetyl-p-benzoquinone imine
NAPSQI	N-acetyl-p-benzosemiquinone imine
nDNA	Nuclear DNA
NEC	Necrotizing enterocolitis
NF-κB	Nuclear factor-κB
nNOS	Neuronal nitric oxide synthase
NOS	Nitric oxide synthase
NOX	NAD(P)H oxidase
Nrf-2	Nuclear factor erythroid 2-related factor-2
OGG-1	8-Oxoguanine DNA glycosylase 1
ORAC	Oxygen radical antioxidant capacity
ORP	Oxidation-reduction potential
OS	Oxidative stress
PD	Parkinson's disease
PLA_2s	Phospholipases A_2s
PMNs	Polynuclear leukocytes
Prx	Peroxiredoxins
RA	Rheumatoid arthritis
RCS	Reactive carbon species
RHS	Reactive halide species
RNS	Reactive nitrogen species
ROS	Reactive oxygen species
RS	Reactive species
RSS	Reactive sulfur species
SET	Single electron transfer
SOD	Superoxide dismutase
SPE	Solid phase extraction
SPO	Salivary peroxidase
SR	Sarcoplasmic reticulum
SVHYs	Snake venom hyaluronidases
SVMPs	Snake venom metalloproteinases
T2DM	Type 2 diabetes mellitus
TAC	Total antioxidant capacity

TAP	Total antioxidant performance
TAS	Total antioxidant status
TEAC	Trolox equivalent antioxidant capacity
TOF	Time of flight
TOS	Total oxidant status
TPO	Thyroid peroxidase
TRAP	Total radical antioxidant potential
Trx	Thioredoxin
TrxR	Thioredoxin reductase
UC	Ulcerative colitis
VSMC	Vascular smooth muscle cells
XO	Xanthine oxidase

Chapter 1
Introduction: Physicochemical Properties and Physiological Roles of ROS/Oxidative Eustress and Distress

A. Khelfi and R. Henkel

Abstract Reactive species (RS) have long been known as undesired compounds to be avoided at all costs. This view has changed with the realization that RS, which are constantly produced in all cells, are important actors in physiology. RS can regulate several cellular pathways, having a key role in the homeostasis of all organisms.

Cells contain multiple signal transduction pathways coupling the redox status to specific regulations of gene expression. Excessive RS production is considered to cause multiple changes in cells that can evolve into pathological conditions. In turn, these changes accelerate the generation of RS and increase oxidative modifications of lipids, DNA, and proteins in various tissues.

This chapter covers medical terminology and definitions needed for the comprehension of the processes involving RS. It provides a comprehensive review on the basic functions of RS and their effects on biological molecules and physiological events. In addition, the chapter deals with general reactions involving RS. The chemistry of free radicals (FR), a matter of contemporary interest, is also addressed. The origin of these species and their connection with the dysfunction of cell constituents (oxidative eustress and distress) is described. Other aspects covered in this chapter are the physicochemical properties of RS and FR and their toxic effects.

Keywords Physicochemical properties · Reactive species (RS) · Eustress · Distress · Physiology

A. Khelfi (✉)
Faculty of Pharmacy, University of Algiers1, Algiers, Algeria

National Centre of Toxicology, Algiers, Algeria

Department of Chemistry and Biomolecular Science, Clarkson University, Potsdam, NY, USA
e-mail: ab.khelfi@univ-alger.dz; akhelfi@clarkson.edu

R. Henkel
LogixX Pharma Ltd., Theale, Berkshire, UK

Department of Medical Bioscience, University of the Western Cape, Bellville, South Africa

Department of Metabolism, Digestion & Reproduction, Imperial College, London, UK
e-mail: Ralf.henkel@logixxpharma.com; rhenkel@uwc.ac.za; r.henkel@imperial.ac.uk

S. Andreescu et al. (eds.), *Biomarkers of Oxidative Stress*,
https://doi.org/10.1007/978-3-031-60738-7_1

1.1 Definitions

Oxidative stress (OS) is a relatively new concept, which requires well-established definitions. In this review, chemical terms such as "reactive species (RS)," "free radicals (FRs)," and "oxidants" are used in efforts to discuss complex pathways and processes occurring in biological systems. Terms such as "lesion," "damage," and "injury" are often used in a pathophysiological context to convey a description regarding the status of cells and tissues. Without a clear understanding of the meanings of these terms, mechanisms cannot be discussed properly. To help achieve that end, the following definitions are provided.

RS are generally short-lived, highly reactive molecules that are formed by complete one or more electron reductions of certain atoms, usually oxygen. They can be divided into two distinct groups: radicals and non-radicals.

Radicals are molecules or atoms with one or more unpaired electrons in a single orbit. They are highly unstable and could actively react with other molecules with half-lifetimes in the millisecond (10^{-3}s) to nanosecond (10^{-9}s) range. They can be either charged or uncharged. In general, when two radicals share their unpaired electrons, non-radical forms are created.

What is called "free radicals" are molecules that moved out or diffused from the enzymatic active site or restricted solvent sphere in which they were generated and are no longer "controlled" by that environment. Besides these molecules, caged radicals are those retained within their sites of generation. The distinction between different kinds of radicals is important because radicals generated by the catalytic activity of some enzymes are controlled and the loss of this control has significant toxicological implications [1].

Non-radicals do not necessarily interact with biomolecules through radical reactions and are not necessarily less reactive than radicals. However, because of the reaction of non-radicals with radicals generating other radicals, they have the capacity to undergo chain reactions.

Initially, most of the identified RS were derived from two elements: oxygen and nitrogen; then, with the advances made in technology, the term "RS" has been expanded to cover reactive halide and sulfur species, thus creating reactive oxygen species (ROS), reactive nitrogen species (RNS), reactive halide species (RHS), and reactive sulfur species (RSS). Alternative terms are sometimes used, including "reactive oxidants" and "oxygen-derived species." Although there are reactive carbon species (RCS or carbon-centered radicals) that do not strictly fit into this classification, their interaction frequently involves oxygen and links them to the overall RS network. Besides, xenobiotic radicals may also occur in biological systems and can result in the initiation and propagation of various types of injuries.

The so-called OS is simply the main result of the multiple actions of these species. Cumulatively, "OS" is a broad term used to describe the outcome of RS-mediated reactions, leading to oxidative damage and generation of secondary RS. The term "OS" was initially defined as "a disturbance in the pro-oxidant/antioxidant balance in favor of the former in cells and tissues" [2]. In other words, this definition takes into account possible OS resulting from diminished antioxidant enzymes or antioxidant levels or defects in antioxidant machinery in expressing antioxidant genes. However, this definition does not address the detrimental consequences of such a change on the function of biomolecules. It was later extended to include the detrimental effects of OS as "a disturbance in the pro-oxidant/antioxidant balance in favor of the former, leading to potential damage" [3]. Because of recent discoveries on OS markers and redox signalling pathways, the definition has been once again modified to "an imbalance between oxidants and antioxidants in favor of the oxidants, leading to a disruption of redox signalling and control and/or molecular damage" [4]. Other definitions and caveats have been proposed, such as the widely used definition that leaves open the possibilities of the disturbance resulting from increases in oxidant production, decreases in tissue reductive capacity, or a combination of these processes.

The term "nitrosative stress" emphasizes a close relationship with OS, where among other targets, metals and thiols are modified by oxidation mediated by RNS.

At the point when the level of the RS exceeds the endogenous antioxidant defense mechanisms, the cell is said to be in a state of OS and disturbances in a normal redox state occur [5]. This could lead to damage to all components of the cell.

A number of definitions of biomarkers have been provided according to their applications. The definition of the Food and Drug Administration (FDA) will be followed in this work with emphasis on molecular biomarkers: "Biomarker is a defined characteristic that is measured as an indicator of normal biological processes, pathogenic processes or responses to an exposure or intervention" [6]. Classical biomarkers to characterize OS in biological systems involve increased levels of the RS (radicals and non-radicals), an imbalance in the cellular redox status (decreased levels of antioxidants and increased levels of pro-oxidants), disruption of transcription factors and repair systems, and oxidative damage to cellular components such as lipids, proteins, carbohydrates, and DNA. Global indicators are not classified as biomarkers and are explained later in this work.

1.2 Reactive Species

1.2.1 Physicochemical Properties

All RS have similar electron configurations but exhibit significant differences in structure, stability, solubility, diffusion, reactivity, and other physicochemical properties. They exhibit different biological activities as well and vary in their membrane permeability and in their ability to be scavenged by individual antioxidants. Therefore, the biological impact of the RS will depend on the type of species produced and where. The picture is complex and dynamic because many of these species are derived from one another and many of the reactions are reversible (Fig. 1.1).

Reactivity of the RS is defined as a measure of the rate at which these substances are chemically altered. Reactivity thus depends on the intrinsic properties of the

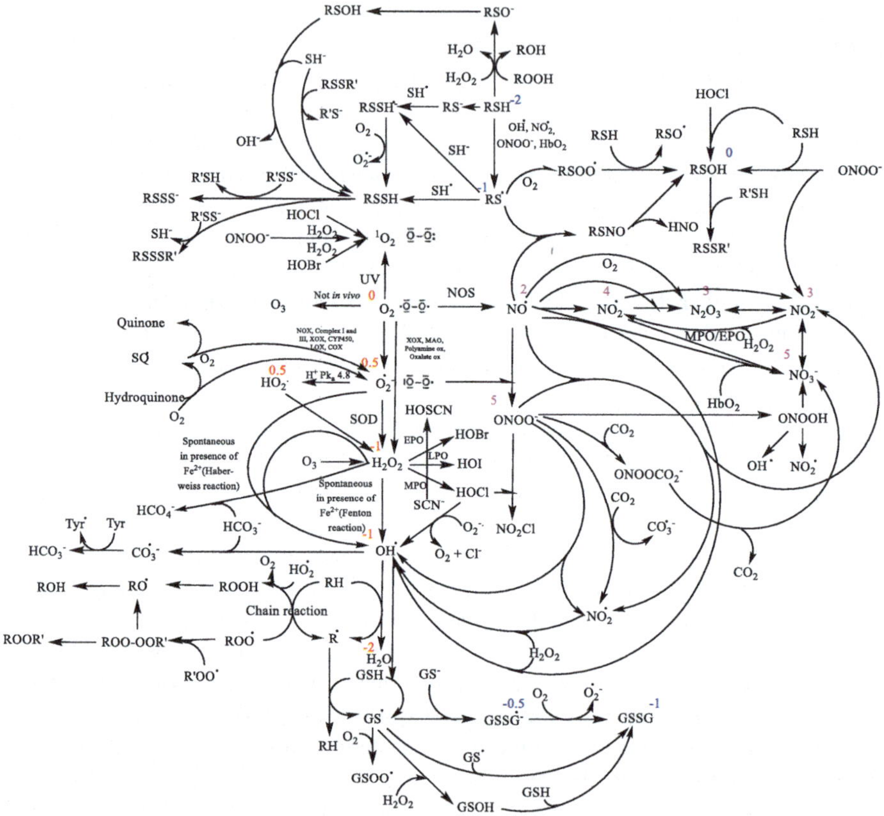

Fig. 1.1 An interplay of physiologically relevant reactive species

The oxidation states of oxygen are presented in red and purple. The oxidation states of sulfur are presented in blue

The scheme covers most RS and selected routes of generation and interconversion. Mitochondria can release either $O_2^{•-}$ or H_2O_2. In the cytosol, $O_2^{•-}$ is converted to H_2O_2 by SOD. NADPH oxidases can also generate $O_2^{•-}$ in the cytosol. Then, $O_2^{•-}$ is converted into various ROS forms, including H_2O_2 and hydroxyl radical (OH•) by the Fenton reaction. SOD, superoxide dismutase; NADPH, nicotinamide adenine dinucleotide phosphate

RS. However, rates of RS decomposition in solution also depend on the environment of the RS. For example, one of the most stable species such as hydrogen peroxide (H_2O_2) is not stable in the presence of some oxido-reductants.

The presence of unpaired electrons is not a prerequisite for reactivity. Some are radicals and undergo one-electron reactions, whereas others are non-radical species that cause two-electron oxidation. The family of non-radical ROS such as H_2O_2, singlet oxygen ($^1O_2$), or peroxynitrite ($ONOO^-$) is chemically reactive in biological systems. For this reason, ROS, as an example, have been defined more broadly than radicals as entities containing one or more oxygen atoms that meet the criteria of being chemically reactive [7].

RS can also be categorized according to their stability as stable, persistent, or unstable. They can be classified as strong or weak oxidants as well. For the most part, RS act as strong oxidants. However, because the direction of any redox reaction depends on the reduction potentials of the couples involved, a species ($O_2^{\bullet -}$, hydroperoxyl radical, semiquinone radicals, glutathione disulfide anion radical, and hydropersulfide) can be an oxidant in some situations and act as a reductant in others.

Radical-based ROS are usually unstable, thus making their identification in solution extremely difficult. Therefore, their detection is commonly accomplished by detecting secondary products arising from their redox reaction with a reagent or by assessing their effect on some biochemical entities. By far, the hydroxyl radical ($OH^{\bullet}$) is the most oxidizing agent, and radicals seem to be more reactive than non-radicals, as depicted in Table 1.1. As reactant and product concentrations also influence the direction of a reaction, under some conditions (e.g., if products are rapidly removed), the same radical can act as a stronger oxidant.

Thermodynamics can help know whether a reaction is feasible and what are the possible factors that can be involved in such a reaction. In this review, we will consider reduction potential as being a measure of the oxidizing strength of a radical. One-electron reduction potential is a reasonable reflection of radical reactivity because the activation energy for radical reactions is low (Table 1.1).

1.2.2 General Reactions for Radical Generation

Free and caged radicals can be formed from neutral molecules through uptake or release of an electron. Another possibility of radical formation lies in permutation or hemolytic cleavage of a covalent bond, a reaction that requires a large amount of energy. A FR can therefore be anionic, cationic, or neutral species, carrying negative, positive, or any charge (Eqs. 1.1, 1.2, 1.3 and 1.4).

$$A \rightarrow \overset{\text{Cation radical}}{A^{+}} + e^{-} \quad \text{Oxidation} \tag{1.1}$$

$$A + e^{-} \rightarrow \overset{\text{Anion radical}}{A^{-}} \quad \text{Reduction} \tag{1.2}$$

$$A{-}B \rightarrow A^{\bullet} + \overset{\text{Neutral radical}}{B^{\bullet}} \quad \text{Scission} \tag{1.3}$$

$$A{-}B + \overset{\text{Neutral radical}}{A^{\bullet}} \rightarrow A{-}A + \overset{\text{Neutral radical}}{B^{\bullet}} \quad \text{Permutation} \tag{1.4}$$

Table 1.1 An overview on RS [54, 62–82]

Family	Class	Reactive species	Properties	Stability and reactivity	Half-life at 37 °C (in seconds)	Formation	Standard reduction potential (V) pH 7 at 25 °C	Physiological roles	Toxic effects	Marker of disease	Neutralizing antioxidant
Reactive oxygen species	Non-radicals	Singlet oxygen ($^1O_2$)	Excited state of molecular oxygen by a peripheral antiparallel spin electron	Highly unstable form and extremely reactive toward molecules rich in electrons	10^{-6}	Mainly formed during photochemical processes (UVA): Generated by the energy transfer between a photosensitizer and the triplet state of O_2 Myeloperoxidase in phagocytes Spontaneous dismutation of $O_2^{\bullet-}$ Disproportionation of H_2O_2 with $ONOO^-$ or hypohalites	–	–	Reactivity mainly toward unsaturated structures (membranes) Considered as the main agent responsible for the harmful effects of UVA radiation	Skin disorders	Vitamin A, APX, vitamin C, vitamin E, uric acid
		Ozone (O_3)	Unstable Oxidizing agent	Extremely reactive	–	Not produced in vivo	–	–	Cytotoxic against the respiratory system and will react with cellular macromolecules in its immediate vicinity (proteins, carbohydrates, DNA, and lipids) React with other organic compounds (nitrites, amines, nitriles, alkanes, alkenes, etc.) Ozonization: Rupture of alkenes at the level of a double bond and generation of aldehyde species	–	–

		Hydrogen peroxide (H_2O_2)	Oxidizing and reducing properties Diffuse quickly through cell membranes	Relatively stable and nonreactive molecule in the absence of peroxidases and transition metals Poorly reactive with most biomolecules	10 (10^{-3}–10^{-5} in the presence of transition metals)	Formation in the mitochondria, RE, and peroxisomes by: -dielectric reduction of molecular oxygen Monoelectric reduction of $O_2^{\bullet-}$ by spontaneous disproportionation or by SOD Possible generation from ozone	$O_2,2H^+/H_2O_2 = 0.36$ $H_2O_2/OH^\bullet = 0.32$ $H_2O_2,2H^+/H_2O = 1.77$	A signalling molecule in various physiological processes, particularly autophagy Both a vasodilator and a vasoconstrictor	May interact with some biological substrates (inactivation of enzymes, oxidation of DNA, and keto acids) Fairly reacts with [Fe–S] clusters	Inflammation, IBD, ischemic stroke, rheumatoid arthritis, diabetes, some cancers, Alzheimer's disease, atherosclerosis in early stage, infertility, COPD, asthma, sepsis, acute leukemia	CAT, vitamin A, APX, vitamin C, vitamin E, GPx, GR, GST, GSH, Trx, TrxR, Prx, ubiquinol, flavonoids, lipoic acid
		Organic hydroperoxides (ROOH)	Diffuse quickly through cell membranes Chemically related to H_2O_2	Unstable in the presence of transition metals	>100	Produced from organic peroxyls	–	–	Involved in lipid peroxidation Readily react with thiols	Rheumatoid arthritis, beta-thalassemia, Alzheimer's disease	Vitamin A, vitamin E, APX, vitamin C, GPx, GR, GST, GSH, uric acid, Prx, bilirubin, albumin, ubiquinol, flavonoids
	Radicals	Superoxide anion ($O_2^{\bullet-}$)	A moderate oxidizing agent and a strong reductant nucleophilic character (proton acceptor) Limited diffusion ability	Unstable Highly reactive	10^{-6}	Monovalent reduction of molecular oxygen during mitochondrial respiration Generation from redox cycling of semiquinones Enzymatic generation by NOX, XO, COX, and LOX	$O_2/O_2^{\bullet-} = -0.16$ $O_2^{\bullet-},2H^+/H_2O_2 = 0.94$	A weak signalling molecule Capacitation of sperm	Relatively unreactive toward most biomolecules Reacts extremely slowly with biological molecules (enzymes, unsaturated fatty acids, and nucleic acids) Hydrolysis of fatty acid esters Inactivate the mitochondrial respiratory chain	Atherosclerosis in early stage, ischemic stroke, rheumatoid arthritis, diabetes, Alzheimer's disease, IBD, infertility, COPD, asthma, sepsis	SOD, APX, vitamin C, vitamin E, vitamin A, GPx, GR, GST, GSH, flavonoids, uric acid
		Hydroperoxyl radical ($HO_2^\bullet$)	Weak acid Stronger oxidizer than $O_2^{\bullet-}$ Easy diffusion in membranes	Unstable and highly reactive	–	Formed from $O_2^{\bullet-}$ in acidic conditions Dehydrogenation of H_2O_2 by $OH^\bullet$ Two-step oxidation of H_2O_2 by Fe(III)	$O_2,H^+/HO_2^\bullet = -0.46$ $HO_2^\bullet,H^+/H_2O_2 = 1.06$	A weak signalling molecule	Slow reactions with organic compounds Preferred target is the fatty acids of the lipid bilayer	Same as the superoxide anion	APX, vitamin C, GPx, GR, GST, GSH, flavonoids, SOD

(continued)

Table 1.1 (continued)

Family	Class	Reactive species	Properties	Stability and reactivity	Half-life at 37 °C (in seconds)	Formation	Standard reduction potential (V) pH 7 at 25 °C	Physiological roles	Toxic effects	Marker of disease	Neutralizing antioxidant
		Hydroxyl radical ($OH^{\bullet}$)	The most powerful oxidant Hardly diffuses within biological media	Unstable Extremely reactive	10^{-9}	Fenton reaction: Decomposition of H_2O_2 in the presence of metals Haber–Weiss reaction: Decomposition of H_2O_2 in the presence of metals Reaction of H_2O_2 with RNS Decomposition of $ONOO^-$ MPO action on HOCl	$OH^{\bullet},H^+/H_2O = 2.31$	Messenger at extremely low concentrations by quickly reacting with signalling molecules	Reaction with a large spectrum of biomolecules at close proximity Instant oxidation of sugars, proteins, lipids, and nucleic acids	Cancerogenesis, ischemic stroke, Alzheimer's disease, rheumatoid arthritis, beta-thalassemia, rhabdomyolysis	APX, vitamin C, vitamin E, GPx, GR, GST, GSH, flavonoids, lipoic acid, uric acid, Sequestration of metals: Ferritin, transferrin, haptoglobin, lactoferrin, hemopexin, ceruloplasmin, albumin, metallothionein
		Peroxyl radicals ($ROO^{\bullet}$)	Moderate oxidizing agents Penetrate biological membranes at moderate distances	Highly reactive More stable than either the carbon-centered radical or the related alkoxyl radicals	7 (10^{-2} $\rightarrow 10^{-1}$ in the presence of transition metals)	Formed from organic hydroperoxide by hydrogen abstraction Addition of molecular oxygen on carbon-centered radicals during lipid oxidation chain reactions	$ROO^{\bullet},H^+/ROOH = 0.77–1.44$	Formation of numerous other important ROS to sustain physiological functions	Reaction with numerous biological substrates (lipids, DNA, and proteins) Strongly involved in lipid peroxidation Generation of different toxic aldehydes	Ionizing radiation, ischemia, aging	Vitamin A, vitamin E, APX, vitamin C (if accessible), GPx, GR, GST, GSH, uric acid bilirubin, ubiquinol, flavonoids
		Alkoxyl radicals ($RO^{\bullet}$)	Strong oxidant species	Unstable and extremely reactive	10^{-6}	Formed from organic hydroperoxides and peroxyls following Fenton-type reaction	$RO^{\bullet},H^+/ROH = 1.60$	Formation of numerous other important ROS to sustain physiological functions	Strongly involved in lipid peroxidation Fragmentation reactions yield toxic aldehydes and ketones	–	Vitamin A, vitamin E, APX, vitamin C (if accessible), GPx, GR, GST, GSH, uric acid, bilirubin, ubiquinol

Reactive nitrogen species	Radicals	Nitric oxide ($NO^•$)	Easy diffusion through cell and subcellular membranes Oxidation to nitrosonium ion NO^+ and reduction to nitroxyl ion NO^-	Unstable Highly reactive	$10^{-3} \rightarrow 1$ (depending on the considered forms of NO)	Formed by NOS in almost all mammalian cells	$NO/NO^- = -0.81$	Regulation of blood pressure and vasodilator Regulation of oxygen consumption Neurotransmission Neuronal plasticity and memory function Endocrine gland regulation Immune regulation Intra- and intercellular signalling in defense mechanisms, apoptosis, and relaxation of smooth muscles Maintenance of gastrointestinal mucosal integrity Resorption of the clot Antiviral, antibacterial, and antiparasitic activities	Fixation on amino acids such as tyrosine to generate nitrotyrosine and amino acids containing thiol groups by S-nitrosylation Vascular decomposition Induction of synthesis of numerous inflammatory cytokines Glutamate excitotoxicity At extremely high concentrations, genotoxic effects and lipid peroxidation	Inflammation and inflammatory diseases: IBD, atherosclerosis in early stage, some cancers, Alzheimer's disease, ischemic stroke, Parkinson's disease, asthma, sepsis	–
		Nitrogen dioxide ($NO_2^•$)	A moderately strong one-electron oxidant One-electron reductant	Unstable Highly reactive	$<10^{-5}$	Formed following the interaction of $NO^•$ with molecular oxygen One-electron reduction of $ONOO^-$ by ferric heme peroxidases Decomposition of nitrosoperoxycarbonate Decomposition of either organic peroxynitrites or	$NO_2^•/NO_2^- = 1.04$	–	Lipid peroxidation Nitration of PUFAs Oxidation of amino acids Formation of nitrotyrosine	–	Lycopene

(continued)

Table 1.1 (continued)

Family	Class	Reactive species	Properties	Stability and reactivity	Half-life at 37 °C (in seconds)	Formation	Standard reduction potential (V) pH 7 at 25 °C	Physiological roles	Toxic effects	Marker of disease	Neutralizing antioxidant
						peroxynitrous acid Formation from H_2O_2 and NO_2^- in the presence of MPO					
	Non-radicals	Peroxynitrite ($ONOO^-$)	A strong oxidant Diffuse through cell membranes by anion channels	Highly reactive	10^{-2}	Formed from the reaction between $NO^•$ and $O_2^{•-}$	$ONOO^-/NO_2^• =$ 1.6 $ONOO^-/NO_2^- =$ 1.3	Intracellular messenger for the production of IL-8 Low-concentration vasoconstriction, but at a higher concentration causes vasodilatation	DNA base nitration Nitration of tyrosine to 3-nitrotyrosine Formation of S-nitrosothiols Lipid oxidation Degradation of hyaluronan Genotoxic at high concentrations	Alzheimer's disease, IBD, atherosclerosis in early stage, ischemic stroke, sepsis	GPx, GR, GST, GSH, Prx, lycopene
		Nitrite (NO_2^-) and nitrate (NO_3^-)	Strong oxidants Taken up by red blood cells via ionic channels	Highly reactive	–	Mainly formed by the conversion of $NO^•$, $NO_2^•$, and $ONOO^-$	$NO_3^-,4H^+/NO = 0.96$ $NO_3^-, 3H^+/HNO_2 = 0.93$ $NO_2^-/2HO^- =$ 0.01	–	Formation of methemoglobin Formation of toxic nitrosamines	Nitrite/nitrate water poisoning Inflammatory diseases, urolithiasis, cancerogenesis	–
		Dinitrogen trioxide (N_2O_3)	A moderate oxidant	Highly reactive	–	Formed from $NO^•$, NO_2^-, or $NO_2^•$	$N_2O_3/NO_2^• = 0.7$	–	S-nitrosylation and formation of S-nitrosothiols Nitrosate amines to form nitrosamines Deamination of nucleobases	Cancerogenesis	–
Carbon reactive species	Radicals	Carbonate radical ($CO_3^{•-}$)	A strong oxidant	Low reactivity toward most extracellular components	–	Reaction of $OH^•$ or $ONOO^-$ with CO_2 Oxidation of bicarbonate by $OH^•$	$CO_3^{•-},H^+/HCO_3^- = 1.78$	–	Reaction with several alcohols and tetrahydrofuran Highest reactivity	–	–

								with indole heterocyclic compounds		
		Semiquinone radicals ($SQ^{\bullet}$)	Reductants	Reactive during the redox-cycling process	Half-lives up to days >100 s	Redox-cycling process	Catechol $SQ^{\bullet}$, H^{+}/catechol = 0.53 Quercetin $SQ^{\bullet}$, H^{+}/quercetin = 0.33	–	Generation of substantial quantities of $O_2^{\bullet -}$	–	–
		Organic radicals ($R^{\bullet}$)	Oxidants and reductants	React extremely quickly with molecular oxygen	Short-lived species	Hydrogen abstraction from unsaturated bonds in fatty acids	$PUFA^{\bullet}$, H^{+}/PUFA-H = 0.60	–	Involved in lipid peroxidation	–	–
Halide reactive species	Non-radicals	Hypochlorous acid (HOCl)	A strong oxidant	High reactivity, particularly with amino and thiol groups	–	Produced in the lysosomes of immune cells containing myeloperoxidases	HOCl/H_2O = 1.28 HOCl/cl^{-} = 1.50	Microbicide activity	Alteration of protein structure and function Generates 3-chlorotyrosine from tyrosine Generation of tyrosyl radical Generation of chloramines Attacks on unsaturated lipids to form chlorohydrins Generation of chlorinated bases Generation of DNA adducts	Inflammation, atherosclerosis in early stage, carcinogenesis and allergic reactions, rheumatoid arthritis, IBD, sepsis	–
		Hypobromous acid (HOBr)	A strong oxidant	High reactivity, particularly with amino and thiol groups	–	Produced in the lysosomes of eosinophils containing eosinoperoxidases	HOBr/H_2O = 1.13	Microbicide activity	Reacts with a wide variety of compounds Generation of bromamines and brominated tyrosine	Inflammation and allergic reactions	–

(continued)

Table 1.1 (continued)

Family	Class	Reactive species	Properties	Stability and reactivity	Half-life at 37 °C (in seconds)	Formation	Standard reduction potential (V) pH 7 at 25 °C	Physiological roles	Toxic effects	Marker of disease	Neutralizing antioxidant
		Hypoiodous acid (HOI)	An oxidant	Reactivity, particularly with amino and thiol groups	–	Oxidation of iodide by heme peroxidases	HOI/ H_2O = 0.78	Involved in the production of thyroglobulin	Generation of iodinated tyrosine	–	–
Pseudohalide reactive species	Non-radicals	Hypothiocyanous acid	A weak oxidant	–	–	Oxidation of thiocyanate (SCN^-) via the action of peroxidase enzymes	HOSCN/SCN- = 0.56	–	Reaction with free and protein-bound sulfhydryls with the generation of disulfide bond Reaction with selenium-containing amino acid residues and Trp residues Protein carbamylation	Atherosclerosis	–
	Radicals	Thiyl radicals ($RS^•$)	Strong oxidants	Unstable Highly reactive More electrophilic character than their chemically analogous alkoxyl radicals	Order of microseconds	One-electron oxidation of thiols Catalyzed by high-valence transition metal ions, $OH^•$, $NO_2^•$, $ONOO^-$, and oxyhemoglobin	$GS^•/GSH$ = 0.92 $RS^•$/ RSH = 0.92	–	Abstraction of hydrogen from proteins and initiation of lipid peroxidation Generation of disulfide bonds and inactivation of proteins Interactions with $NO^•$ and generation of S-nitrosothiols	–	Lycopene
		Peroxylsulfenyl radicals ($RSOO^•$)	–	–	–	Formed upon the interaction of thiyl radicals with molecular oxygen	–	–	Generation of toxic radicals (sulfenic acid Sulfinyl radical Sulfonyl radicals)	–	–

		Disulfide anion radical ($RSSR^{\bullet-}$)	–	–	–	Reaction of the thiyl radicals of glutathione with the surrounding molecules	GSSG/ $GSSG^{\bullet-}$ = –1.5 RSSR/ $RSSR^{\bullet-}$ = –1.5	–	–	–	–
	Non-radicals	Sulfenic acid (RSOH)	–	Electrophile compound Heteroatomic version of the analogous organic hydroperoxides	Short-lived species	Two-electron oxidation of thiol moieties with H_2O_2, ROOH, $ONOO^-$, HOCl, or chloramines	$RSO^{\bullet}$/ RSOH = 0.74	–	Combination with another thiol to form disulfide compounds and quick transformation to sulfenyl radicals	–	–
		Hydropersulfides (RSSH) and polysulfides (RSxH)	–	Highly reactive and unstable due to their nucleophilic and electrophilic character Readily reduced back to RSH	–	Oxidation of thiol moieties mediated by several oxidants	–	–	–	–	–

UVA ultraviolet A; *SOD* superoxide dismutase, *XO* xanthine oxidase, *COX* cyclooxygenase, *LOX* lipoxygenase, *MPO* myeloperoxidase, *PUFA* polyunsaturated fatty acid, *IL-8* interleukin 8, *NA* not applicable, *IBD* inflammatory bowel disease, *COPD* chronic obstructive pulmonary disease

1.2.3 Reactions Involving Radicals

The chemistry of radicals is complex, and this complexity is further emphasized by their interactions with close biomolecules, which is one of the reasons why there are difficulties in conclusively linking radicals with injury and diseases. As a general rule, the free nature of the electrons in radicals allows them to combine easily with other unpaired electrons within small and biological molecules in order to achieve a stable paired electron status.

There are several basic reactions involving radicals (Eqs. 1.5, 1.6, 1.7, 1.8, 1.9, 1.10, 1.11, 1.12, 1.13, 1.14 and 1.15) and three general types of radical reactions that can be defined:

- Initiation reactions in which the number of radical species increases.
- Propagation reactions in which the number of radicals does not change.
- Termination reactions in which the products contain no radical species.

$$A^{\bullet} \rightarrow A^{+} + e^{-} \quad \text{Oxidation} \tag{1.5}$$

$$A^{\bullet} \rightarrow e^{-} + A^{-} \quad \text{Reduction} \tag{1.6}$$

$$A^{\bullet} + A^{\bullet} \rightarrow A\text{—}A \quad \text{Recombination} \tag{1.7}$$

$$CH_3\dot{C}H_2 + CH_3\dot{C}H_2 \rightarrow CH_3CH_3 + H_2C{=}CH_2 \quad \text{Disproportionation} \tag{1.8}$$

$$A^{\bullet} + B^{\bullet} \rightarrow A\text{—}B \quad \text{Addition} \tag{1.9}$$

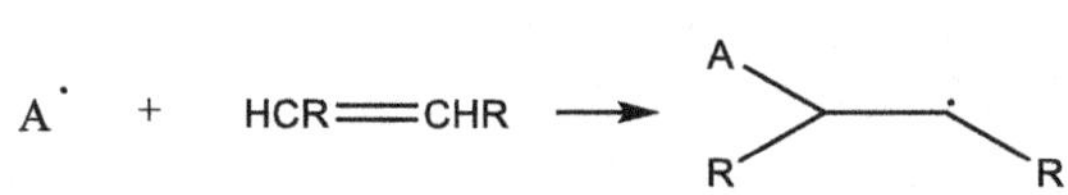

Addition (1.10)

$$R^{\bullet} + O_2 \rightarrow ROO^{\bullet} \quad \text{Addition} \tag{1.11}$$

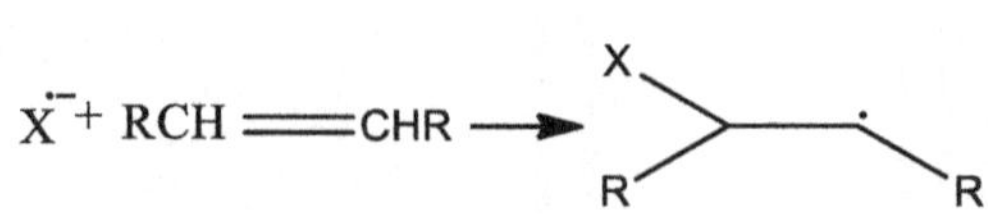

Addition (1.12)

$$H_3C\text{—}\overset{H_2}{C}\text{—}O^{\bullet} \rightarrow \dot{C}H_3 + H_2C{=}O \qquad \text{Scission} \tag{1.13}$$

$$A^{\bullet -} + B \rightarrow A + B^{\bullet -} \qquad \text{Electron transfer} \tag{1.14}$$

$$A^{\bullet} + RH \rightarrow AH + R^{\bullet} \qquad \text{Hydrogen abstraction} \tag{1.15}$$

1.2.4 Physiological Roles of ROS

The atmosphere of the early earth about 3.8 billion years ago had only an extremely low oxygen concentration [8], a gas that is per se a diradical and highly reactive. Oxygen is also a toxic molecular by-product of anaerobic respiration [9]. At a time, after Cyanobacteria and other organisms had produced large amounts of oxygen via chemosynthesis and photosynthesis, its atmospheric concentration increased significantly about 2.3 billion years ago [10]. Following this Great Oxygenation Event (GOE), atmospheric oxygen concentration finally rose again to its current levels approximately 500 million years ago [11].

Given its rising atmospheric levels and the high reactivity of molecular oxygen, this posed a serious dilemma for early life. Consequently, as an adaptation to these high oxygen levels, aerobic respiration and protective mechanisms had to be developed. On the one hand, there are highly effective enzymatic and nonenzymatic antioxidant systems that scavenge and thereby detoxify excessive amounts of oxidants. On the other hand, life adapted to the high oxygen levels by using it to produce energy. All known eukaryotes have mitochondria [12], in which the enzymatically controlled oxidative phosphorylation and oxidation of nicotinamide adenine dinucleotide (NADH) as an electron donator takes place. The electron transfer in the mitochondrial electron transfer chain, even under physiological conditions, however, is not 100% as about 1–5% of the consumed oxygen is converted into FRs [13, 14]. Among the FRs produced is $O_2^{\bullet -}$, which is a highly reactive intermediary metabolite that can enzymatically dismutase into H_2O_2, which has been shown to be crucial for cellular redox regulation, signalling [15], and triggering essential physiological functions such as gene regulation, cellular activities, or synaptic plasticity [16–19]. Therefore, from a physiological point of view, H_2O_2 and $O_2^{\bullet -}$ are not regarded as harmful agents but as essential triggers to cellular functions [20]. Hence, in order to regulate these critical life functions, it is crucial that the balance between oxidation, reduction, and homeostasis remains and that the relevant physiological redox triggers, particularly via H_2O_2, for specific cellular functions are set and recognized.

ROS play a crucial role in molecular signalling, cell proliferation, and survival as a whole. Among the specific actions of ROS, a plethora of genes and signal transduction pathways are regulated [21–23]. ROS produced in cells are involved in the regulation of cellular growth, apoptosis, and necrosis [24, 25]. Examples of genes or proteins that are activated are activator protein (AP)-1, basic fibroblast growth factor (bFGF), collagen, collagenase, *c-fos*, *Fas*, various interleukins, mitogen-activated protein (MAP) kinases, or thioredoxin (Trx) (for review, see Allen and Tresini [16], Ray et al. [26]).

As examples of bodily systems that are ROS-regulated, I want to briefly mention the reproductive and immune systems. In the female reproductive system, a small amount of ROS is required for all essential steps and processes of female reproduction, including steroidogenesis in the ovaries, oocyte development and maturation, follicle growth, ovulation, formation of the corpus luteum and luteolysis, implantation of the embryo, pregnancy, and giving birth [27, 28]. In the male reproductive system, ROS are not only involved in regulating spermatogenesis and sperm maturation in the testes and epididymis but also in sperm functions such as capacitation, acrosome reaction, and sperm–oocyte fusion [27–30].

In the immune system, ROS also play a pivotal role as they have a strong influence on immune cells. Depending on the site and the amount of production, ROS trigger different metabolic pathways such as glycolysis or lipid metabolism in a different way, thus regulating different immune cells, innate and adaptive, inflammatory and noninflammatory [31]. In the adaptive immune response, NADPH oxidase (NOX) and mitochondrial ROS play an essential role in the activation of the immune response [32]. On the other hand, ROS are also involved in innate immune functions by potentiating phagocyte activity [32]. In the new therapeutic concept of "redox medicine," it depends on the type of disease for the patient to benefit most as ROS and RNS functions may differ with different targets and antioxidant treatments may either succeed or fail depending on the individual circumstances of patients in terms of nutrition, lifestyle, or exercise [33]. According to this concept, it might be better to administer specific ROS inhibitors than scavenging ROS and counteracting oxidative stress overall, thus improving the physiological reactions stimulated by ROS [34] and thereby strengthening immunity. Hence, it is important to understand that redox levels may differ in different individuals depending on different stages of possible diseases.

1.3 Oxidative Eustress and Distress

The concept of oxidative stress, which was formulated by Helmut Sies in 1985 [35], is based on the principle of the "general adaptation syndrome" developed by Hans Selye in 1936 [36]. The term "oxidative stress" itself was then coined by Sies in 1985 and was defined as a "disturbance in the pro-oxidant-antioxidant balance in favor of the former, leading to potential damage" [35] and refined in 1997 as "an imbalance between oxidants and antioxidants in favor of the oxidants, potentially leading to damage, is termed 'oxidative stress' [37]. Since under certain circumstances such as aging or excessive exercise, antioxidants can trigger an increased production of ROS, Jones [38] redefined the condition as "a disruption of redox signalling and control" [38]. Since ROS will not accumulate because they have extremely reactive and therefore short half-lifetimes, one can only detect their steady-state levels. Furthermore, ROS do not only cause damage to cells but are also essential triggers for physiological regulatory processes. Therefore, Lushchak [39] proposed that "oxidative stress is a situation when steady-state ROS concentration is transiently or chronically elevated, disturbing cellular metabolism and its regulation and damaging cellular constituents" [39]. This definition was then further modified to: "Oxidative stress is a transient or long-term increase of steady-state ROS levels, disturbing cellular metabolic and signalling pathways, particularly ROS-based ones, and leading to oxidative modifications of an organism's macromolecules that, if not counterbalanced, may culminate in cell death via necrosis or apoptosis" [40].

Based on the fact that a certain limited amount of ROS is necessary to trigger regulatory processes in cells to maintain normal physiological functions, but an excessive amount leads to cellular damage including cell death, Sies [41] further developed the stress theory of Selye [42] and defined "oxidative eustress" as "physiological oxidative stress," whereas "oxidative distress" describes severe "oxidative stress causing oxidative damage" to biomolecules and cells [41, 42]. On the other hand, an inadequate quantity of ROS (stress), either caused by excessive amounts of antioxidants scavenging the essential ROS or too little production of ROS, also impairs normal physiological functions or triggers pathological conditions, including cardiomyopathy, Alzheimer's disease, dysfunctions of the blood–brain barrier, or infertility [43–48]. This condition is then called "sustress" or alternatively "reductive stress" and has been described as harmful as oxidative distress [49]. Hence, one has to realize that redox stress follows a biphasic response to the exposure of oxidants, which is defined as "hormesis" [50].

Hydrogen peroxide has been identified as the major metabolite in cellular redox regulation, in signalling [15], and, together with Ca^{2+} and adenosine triphosphate (ATP), as one of the most important molecules for transcription-independent signalling molecules [51, 52]. Hence, H_2O_2 appears in all mammalian cells as an essential metabolite [53, 54], with physiological concentrations indicating oxidative eustress between 1 and 10 nM [14, 55]. In different cell compartments and different organs, however, the concentrations are different. A concentration of 80 pM was estimated in the cytosol [56], in the mitochondrial matrix 5–20 nM, and in the lumen of the endoplasmic reticulum (ER) 700 nM [57], whereas the concentration in blood plasma was reported at 1–5 μM [58]. Hence, it is clear that the reported physiological concentrations of 1–10 nM H_2O_2 are only an estimate and can vary markedly depending on different conditions such as age, illness, exercise, or lifestyle [59] but can also vary between different cells, all depending on the relevant functional state that the cells are in. Yet, the suggested physiological range can be quite wide, and higher limits of up to 100 nM have also been reported [14]. It is clear, however, that values above 100 nM H_2O_2 are toxic and have to be regarded as "oxidative distress" [60]. Meng et al. [61] defined this variable point of transition between oxidative eustress and oxidative distress as the "redox-stress signalling threshold" (RST) [61].

References

1. Kehrer JP, Robertson JD, Smith CV, editors. Free Radicals and Reactive Oxygen Species 2010.
2. H. S. Oxidative stress. In: Press A, editor. London 1985. p. 1–8.
3. Sies H. Oxidative stress. Oxidants and antioxidants. San diego CA: Academic Press; 1991. p. XV–XXII.
4. Sies H, Jones D. Oxidative Stress*. In: Fink G, editor. Encyclopedia of Stress (Second Edition). New York: Academic Press; 2007. p. 45–8.
5. Sharma P, Jha A, Dubey R, Pessarakli M. Reactive Oxygen Species, Oxidative Damage, and Antioxidative Defense Mechanism in Plants under Stressful Conditions. Journal of Botany. 2012;2012.
6. Califf RM. Biomarker definitions and their applications. Experimental biology and medicine (Maywood, NJ). 2018;243(3):213–21.
7. Kehrer JP, Klotz LO. Free radicals and related reactive species as mediators of tissue injury and disease: implications for Health. Critical reviews in toxicology. 2015;45(9):765–98.
8. Crowe SA, Døssing LN, Beukes NJ, Bau M, Kruger SJ, Frei R, et al. Atmospheric oxygenation three billion years ago. Nature. 2013;501(7468):535–8.
9. Des Marais DJ. Earth's early biosphere. Gravit Space Biol Bull. 1998;11(2):23–30.
10. Luo G, Ono S, Beukes NJ, Wang DT, Xie S, Summons RE. Rapid oxygenation of Earth's atmosphere 2.33 billion years ago. Sci Adv. 2016;2(5):e1600134.
11. Hamilton TL, Bryant DA, Macalady JL. The role of biology in planetary evolution: cyanobacterial primary production in low-oxygen Proterozoic oceans. Environ Microbiol. 2016;18(2):325–40.
12. Brochier-Armanet C, Talla E, Gribaldo S. The multiple evolutionary histories of dioxygen reductases: Implications for the origin and evolution of aerobic respiration. Mol Biol Evol. 2009;26(2):285–97.
13. Boveris A, Chance B. The mitochondrial generation of hydrogen peroxide. General properties and effect of hyperbaric oxygen. The Biochemical journal. 1973;134(3):707–16.
14. Chance B, Sies H, Boveris A. Hydroperoxide metabolism in mammalian organs. Physiological reviews. 1979;59(3):527–605.
15. Marinho HS, Real C, Cyrne L, Soares H, Antunes F. Hydrogen peroxide sensing, signaling and regulation of transcription factors. Redox Biol. 2014;2:535–62.
16. Allen RG, Tresini M. Oxidative stress and gene regulation. Free Radic Biol Med. 2000;28(3): 463–99.
17. Banerjee Mustafi S, Chakraborty PK, Dey RS, Raha S. Heat stress upregulates chaperone heat shock protein 70 and antioxidant manganese superoxide dismutase through reactive oxygen species (ROS), p38MAPK, and Akt. Cell Stress Chaperones. 2009;14(6):579–89.
18. Massaad CA, Klann E. Reactive oxygen species in the regulation of synaptic plasticity and memory. Antioxid Redox Signal. 2011;14(10):2013–54.
19. Tschopp J. Mitochondria: Sovereign of inflammation? European journal of immunology. 2011;41(5):1196–202.
20. Sies H, Belousov VV, Chandel NS, Davies MJ, Jones DP, Mann GE, et al. Defining roles of specific reactive oxygen species (ROS) in cell biology and physiology. Nature reviews Molecular cell biology. 2022;23(7):499–515.
21. Monteiro HP, Stern A. Redox modulation of tyrosine phosphorylation-dependent signal transduction pathways. Free Radic Biol Med. 1996;21(3):323–33.
22. Sen CK, Packer L. Antioxidant and redox regulation of gene transcription. FASEB journal: official publication of the Federation of American Societies for Experimental Biology. 1996;10 (7):709–20.
23. Allen RG. Oxidative stress and superoxide dismutase in development, aging and gene regulation. Age (Omaha). 1998;21(2):47–76.

24. Bushell AJ, Klenerman L, Taylor S, Davies H, Grierson I, Helliwell TR, et al. Ischaemic preconditioning of skeletal muscle. 1. Protection against the structural changes induced by ischaemia/reperfusion injury. J Bone Joint Surg Br. 2002;84(8):1184–8.
25. Yu T, Jhun BS, Yoon Y. High-glucose stimulation increases reactive oxygen species production through the calcium and mitogen-activated protein kinase-mediated activation of mitochondrial fission. Antioxid Redox Signal. 2011;14(3):425–37.
26. Ray PD, Huang BW, Tsuji Y. Reactive oxygen species (ROS) homeostasis and redox regulation in cellular signaling. Cell Signal. 2012;24(5):981–90.
27. du Plessis SS, Harlev A, Mohamed MI, Habib E, Kothandaraman N, Cakar Z. Physiological Roles of Reactive Oxygen Species (ROS) in the Reproductive System. In: Agarwal A, Sharma R, Gupta S, Harlev A, Ahmad G, du Plessis SS, et al., editors. Oxidative Stress in Human Reproduction: Shedding Light on a Complicated Phenomenon. Cham: Springer International Publishing; 2017. p. 47–64.
28. Das A, Roychoudhury S. Reactive Oxygen Species in the Reproductive System: Sources and Physiological Roles. Advances in experimental medicine and biology. 2022;1358:9–40.
29. Dutta S, Henkel R, Sengupta P, Agarwal A. Physiological Role of ROS in Sperm Function. In: Parekattil SJ, Esteves SC, Agarwal A, editors. Male Infertility: Contemporary Clinical Approaches, Andrology, ART and Antioxidants. Cham: Springer International Publishing; 2020. p. 337–45.
30. Juárez-Rojas L, Casillas F, López A, Betancourt M, Ommati MM, Retana-Márquez S. Physiological role of reactive oxygen species in testis and epididymal spermatozoa. Andrologia. 2022;54(4):e14367.
31. Banerjee S, Ghosh S, Mandal A, Ghosh N, Sil PC. ROS-associated immune response and metabolism: a mechanistic approach with implication of various diseases. Arch Toxicol. 2020;94(7):2293–317.
32. Bassoy EY, Walch M, Martinvalet D. Reactive Oxygen Species: Do They Play a Role in Adaptive Immunity? Front Immunol. 2021;12:755856.
33. Meng J, Lv Z, Zhang Y, Wang Y, Qiao X, Sun C, et al. Precision Redox: The Key for Antioxidant Pharmacology. Antioxid Redox Signal. 2021;34(14):1069–82.
34. Dumas A, Knaus UG. Raising the 'Good' Oxidants for Immune Protection. Front Immunol. 2021;12:698042.
35. Sies H. 1 – Oxidative Stress: Introductory Remarks. In: Sies H, editor. Oxidative Stress. London: Academic Press; 1985. p. 1–8.
36. Selye H. A Syndrome produced by Diverse Nocuous Agents. Nature. 1936;138(3479):32-.
37. Sies H. Oxidative stress: oxidants and antioxidants. Exp Physiol. 1997;82(2):291–5.
38. Jones DP. Redefining oxidative stress. Antioxid Redox Signal. 2006;8(9–10):1865–79.
39. Lushchak VI. Environmentally induced oxidative stress in aquatic animals. Aquat Toxicol. 2011;101(1):13–30.
40. Lushchak VI, Storey KB. Oxidative stress concept updated: Definitions, classifications, and regulatory pathways implicated. EXCLI journal. 2021;20:956–67.
41. Sies H. Hydrogen peroxide as a central redox signaling molecule in physiological oxidative stress: Oxidative eustress. Redox Biol. 2017;11:613–9.
42. Selye H. Stress without Distress. In: Serban G, editor. Psychopathology of Human Adaptation. Boston, MA: Springer US; 1976. p. 137–46.
43. Finkel T, Holbrook NJ. Oxidants, oxidative stress and the biology of ageing. Nature. 2000;408 (6809):239–47.
44. Chrousos GP. Stress and disorders of the stress system. Nat Rev Endocrinol. 2009;5(7):374–81.
45. Lloret A, Fuchsberger T, Giraldo E, Vina J. Reductive Stress: A New Concept in Alzheimer's Disease. Curr Alzheimer Res. 2016;13(2):206–11.
46. Mentor S, Fisher D. Aggressive Antioxidant Reductive Stress Impairs Brain Endothelial Cell Angiogenesis and Blood Brain Barrier Function. Current neurovascular research. 2017;14(1): 71–81.

47. Henkel R, Sandhu IS, Agarwal A. The excessive use of antioxidant therapy: A possible cause of male infertility? Andrologia. 2019;51(1):e13162.
48. Ma WX, Li CY, Tao R, Wang XP, Yan LJ. Reductive Stress-Induced Mitochondrial Dysfunction and Cardiomyopathy. Oxid Med Cell Longev. 2020;2020:5136957.
49. Castagné V, Lefèvre K, Natero R, Clarke PG, Bedker DA. An optimal redox status for the survival of axotomized ganglion cells in the developing retina. Neuroscience. 1999;93(1): 313–20.
50. Calabrese EJ, Baldwin LA. Defining hormesis. Human & experimental toxicology. 2002;21(2): 91–7.
51. Cordeiro JV, Jacinto A. The role of transcription-independent damage signals in the initiation of epithelial wound healing. Nature reviews Molecular cell biology. 2013;14(4):249–62.
52. van der Vliet A, Janssen-Heininger YM. Hydrogen peroxide as a damage signal in tissue injury and inflammation: murderer, mediator, or messenger? J Cell Biochem. 2014;115(3):427–35.
53. Sies H, Chance B. The steady state level of catalase compound I in isolated hemoglobin-free perfused rat liver. FEBS Lett. 1970;11(3):172–6.
54. Sies H, Berndt C, Jones DP. Oxidative Stress. Annual review of biochemistry. 2017;86:715–48.
55. Lyublinskaya O, Antunes F. Measuring intracellular concentration of hydrogen peroxide with the use of genetically encoded H(2)O(2) biosensor HyPer. Redox Biol. 2019;24:101200.
56. Lim JB, Huang BK, Deen WM, Sikes HD. Analysis of the lifetime and spatial localization of hydrogen peroxide generated in the cytosol using a reduced kinetic model. Free Radic Biol Med. 2015;89:47–53.
57. Gao C, Tian Y, Zhang R, Jing J, Zhang X. Endoplasmic Reticulum-Directed Ratiometric Fluorescent Probe for Quantitive Detection of Basal H(2)O(2). Anal Chem. 2017;89(23): 12945–50.
58. Forman HJ, Bernardo A, Davies KJA. What is the concentration of hydrogen peroxide in blood and plasma? Archives of Biochemistry and Biophysics. 2016;603:48–53.
59. Sies H. Oxidative eustress: the physiological role of oxidants. Science China Life Sciences. 2023;66(8):1947–8.
60. Wilson C, Muñoz-Palma E, González-Billault C. From birth to death: A role for reactive oxygen species in neuronal development. Seminars in Cell & Developmental Biology. 2018;80:43–9.
61. Meng J, Lv Z, Wang Y, Chen C. Identification of the redox-stress signaling threshold (RST): Increased RST helps to delay aging in C. elegans. Free Radic Biol Med. 2022;178:54–8.
62. Maulucci G, Bačić G, Bridal L, Schmidt HH, Tavitian B, Viel T, et al. Imaging Reactive Oxygen Species-Induced Modifications in Living Systems. Antioxid Redox Signal. 2016;24 (16):939–58.
63. Ozougwu J. The Role of Reactive Oxygen Species and Antioxidants in Oxidative Stress. International Journal of Pharmacy and Biosciences. 2016;3:1–8.
64. Di Meo S, Reed TT, Venditti P, Victor VM. Role of ROS and RNS Sources in Physiological and Pathological Conditions. Oxidative Medicine and Cellular Longevity. 2016;2016:1245049.
65. Tsalamandris S, Vogiatzi G, Antonopoulos A, Tousoulis D. The Role of Oxidative Stress. 2018. p. 95–100.
66. Fransen M, Nordgren M, Wang B, Apanasets O. Role of peroxisomes in ROS/RNS-metabolism: Implications for human disease. Biochimica et Biophysica Acta (BBA) – Molecular Basis of Disease. 2012;1822(9):1363–73.
67. Cortese-Krott MM, Santolini J, Wootton SA, Jackson AA, Feelisch M. The reactive species interactome: Academic Press Ltd-Elsevier Science Ltd; 2020 2020.
68. Huang M-Z, Li J-Y. Physiological regulation of reactive oxygen species in organisms based on their physicochemical properties. Acta Physiologica. 2020;228(1):e13351.
69. Mesa-Garcia M, Plaza-Díaz J, Gomez-Llorente C. Molecular Basis of Oxidative Stress and Inflammation. 2018. p. 41–62.

70. Bagchi M, Zafra-Stone S, Bagchi D, Patel S. Oxidative stress and chemical toxicity. Veterinary Toxicology. 2012:426–39.
71. Ye ZW, Zhang J, Townsend DM, Tew KD. Oxidative stress, redox regulation and diseases of cellular differentiation. Biochim Biophys Acta. 2015;1850(8):1607–21.
72. Rahman T, Hosen I, Islam M, Shekhar H. Oxidative stress and human health. Advances in Bioscience and Biotechnology. 2012;03:997–1019.
73. Madkour LH. The Roles and Mechanisms of ROS, Oxidative Stress, and Oxidative Damage. In: Madkour LH, editor. Nanoparticles Induce Oxidative and Endoplasmic Reticulum Stresses: Antioxidant Therapeutic Defenses. Cham: Springer International Publishing; 2020. p. 139–91.
74. De Palma C, Clementi E. Reactive Species and Mechanisms of Cell Injury. In: McManus LM, Mitchell RN, editors. Pathobiology of Human Disease. San Diego: Academic Press; 2014. p. 88–96.
75. Graves DB, Bauer G. Key Roles of Reactive Oxygen and Nitrogen Species. In: Metelmann H-R, von Woedtke T, Weltmann K-D, editors. Comprehensive Clinical Plasma Medicine: Cold Physical Plasma for Medical Application. Cham: Springer International Publishing; 2018. p. 71–82.
76. Arnhold J. Chapter 2 – Role of Reactive Species in Destructions. In: Arnhold J, editor. Cell and Tissue Destruction: Academic Press; 2020. p. 23–54.
77. Bhattacharjee S. ROS and Oxidative Stress: Origin and Implication. 2019. p. 1–31.
78. Villamena F. Chemistry of Reactive Species. 2013. p. 1–48.
79. Mittler R. ROS Are Good. Trends in plant science. 2017;22(1):11–9.
80. Dalal V, Sharma DN, Biswas S. Oxidative Stress: Diagnostic Methods and Application in Medical Science. 2017. p. 23–45.
81. Biological Chemistry of Reactive Oxygen Species. Encyclopedia of Radicals in Chemistry, Biology and Materials.
82. Free Radicals, Signal Transduction, and Human Disease. Oxidative Stress in Vertebrates and Invertebrates. p. 17–32.

Part I
Classifications of Oxidative Stress Biomarkers

Chapter 2
Reactive Species

A. Khelfi

Abstract The metabolism of oxygen by cells generates potentially deleterious and highly reactive species (RS). Due to their high reactive nature, these species can be difficult to observe. Under physiological conditions, the rate of oxidant generation is balanced by the amount of oxidant elimination. However, an imbalance between pro-oxidant and antioxidant compounds results in oxidative/reductive stress. Oxidative stress is a process characterized by the reduced capacity of a biological system to overcome an overproduction of RS and free radicals (FR). These species are involved in cell regulation in which the redox status plays an essential role in a plethora of signaling processes that act to control cell physiology. They are increasingly understood to play a key role in the regulation of vital cellular processes such as cell proliferation and differentiation, immune responses, cell motility, and programmed cell death. It is now clear that cells need to achieve a fine-tuned balance between oxidants and antioxidants in order to function properly. Disruptions in cell redox homeostasis are associated with tumor development and are co`nsidered to be a critical component in several chronic diseases. Besides, FR, ignored as biochemical entities not worth of close scrutiny, are now recognized as contributing factors in many pathological conditions.

In this chapter, we start with a description of the most typical RS, which are reactive oxygen species (ROS), reactive nitrogen species (RNS), carbon reactive species, reactive halide species, reactive pseudohalide species, and reactive sulfur species (RSS). Comprehensive discussion of their fundamental chemistry, generation, and roles is provided. Besides, the chapter describes their interactions with each other, which may occur under certain conditions (e.g., pH and the flux of O_2) to generate sometimes more reactive RS and FR.

A. Khelfi (✉)
Faculty of Pharmacy, University of Algiers 1, Algiers, Algeria

National Centre of Toxicology, Algiers, Algeria

Department of Chemistry and Biomolecular Science, Clarkson University, Potsdam, NY, USA
e-mail: ab.khelfi@univ-alger.dz; akhelfi@clarkson.edu

S. Andreescu et al. (eds.), *Biomarkers of Oxidative Stress*,
https://doi.org/10.1007/978-3-031-60738-7_2

Keywords Oxidative stress · Reactive oxygen species (ROS) · Reactive nitrogen species (RNS) · Carbon reactive species · Reactive halide species · Reactive pseudohalide species · Reactive sulfur species (RSS) · Free radicals (FR)

In this chapter, oxidative stress (OS) markers have been divided into six categories: reactive species (RS), markers of oxidative damage, pro-oxidants, antioxidants, autoantibodies (aAbs), transcription factors, and repair system. Nonspecific markers are mentioned in this classification but are not considered as biomarkers of OS.

2.1 Reactive Oxygen Species (ROS)

The well known and most referred to ROS are highly reactive oxygen derivatives that are produced as a result of normal cellular metabolism and can be divided into two groups: radicals and non-radicals. They result from consecutive one-electron reductions of molecular oxygen that take place in restricted cellular systems, leading to the formation of several oxygen derivatives. Figure 1.1 shows the various oxidation states of ROS resulting from one- and two-electron reductions of molecular oxygen to form $O_2^{\bullet-}$ with the concomitant release of hydroperoxide $OH^{\bullet}$ and water as the final two-electron reduction product. Other possibilities for oxygen activation in biological systems include interactions between molecular oxygen and transition metal ions (see Sect. 2.1.2.3) and spin inversion, a reaction that requires energy and that leads to the formation of $^1O_2$. Enzyme-catalyzed reactions leading to ROS generation will be extensively discussed in this section and are shown in Fig. 1.1 as well.

Due to their high chemical reactivity, ROS react at their site of production and can therefore be highly damaging to cells as they can oxidize all kinds of biomolecules. Therefore, an excessive production of or exposure to ROS has to be prevented by various cellular defense mechanisms. On the other hand, at low concentrations, ROS are crucial for normal cellular functions as they trigger essential physiological functions such as cellular activities, immune reactions, and/or gene regulation.

2.1.1 Non-Radicals

2.1.1.1 Singlet Oxygen ($^1O_2$)

Singlet oxygen is produced through spin conversion of one of the two unpaired electrons of the triplet ground-state oxygen molecule ($^3O_2$) by means of energy transfer [1]. Therefore, $^1O_2$ is molecular oxygen in an excited energy state, which is more reactive than normal ground-state molecular oxygen. In a biological environment, photosensitization is the major source of $^1O_2$, for example, in the skin during sunbathing [2]. Indeed, the photosensitization process can import sufficient energy (23 kcal mol^{-1}) to raise one of the unpaired electrons in a ground-state oxygen molecule to higher energy levels [3]. Besides, a series of reactions involving phagocyte enzymes may lead to the formation of $^1O_2$. It is possibly generated by myeloperoxidase (MPO) in phagocytes during respiratory burst and during the conversion of prostaglandin G2 (PGG_2) to prostaglandin H2 (PGH_2) by prostaglandin (PG) peroxidase [4].

Other proposed reactions that can possibly form $^1O_2$ are as follows:

- Spontaneous dismutation of $O_2^{\bullet-}$ (Eqs. 2.1 and 2.2) [5].
- Energy transfer in the mitochondrial respiratory chain [6].
- Disproportionation of hydrogen peroxide (H_2O_2) with peroxynitrite ($ONOO^-$) or hypohalites (Eqs. 2.3, 2.4 and 2.5) and possibly the reaction of hydroperoxides with $ONOO^-$ [7]

$$2O_2 + NADPH \overset{\text{NADPH Oxidase}}{\rightarrow} 2O_2^{\bullet-} + NADP^+ + H^+ \tag{2.1}$$

$$2O_2^{\bullet-} + 2H^+ \overset{\text{Spontaneous}}{\rightarrow} H_2O_2 + {}^1O_2 \tag{2.2}$$

$$2O_2^{\bullet-} + 2H^+ \overset{\text{SOD}}{\rightarrow} H_2O_2 + O_2 \tag{2.3}$$

$$H_2O_2 + Cl^- \overset{\text{MPO}}{\rightarrow} H_2O + OCl^- \tag{2.4}$$

$$H_2O_2 + OCl^- \rightarrow {}^1O_2 + H_2O + Cl^- \tag{2.5}$$

Singlet oxygen is a highly unstable form of molecular oxygen and is extremely reactive toward molecules rich in electrons (half-life of around 10^{-6} s at 37 °C) [8]. Evidence of the actions of $^1O_2$ in biological systems has been reported. For example, this reactive oxygen has the capacity to cause DNA damage, particularly the oxidation of guanine residues leading to G:C to T:A transversions and then to mutations [9]. Besides, addition of $^1O_2$ to diene structures may result in toxic endoperoxides.

2.1.1.2 Ozone (O_3)

Ozone, which is not produced in vivo, is formed in the lower atmosphere by ultraviolet (UV)-mediated homolytic cleavage of dioxygen (Eq. 2.6). In the presence of nitrogen oxidants and some other air pollutants, the formation of O_3 takes place. For example, the cleavage of nitrogen dioxide (NO_2) from car exhausts catalyzes the generation of O_3 by reaction of the oxygen radical and molecular oxygen [10].

$$O + O_2 + M \rightarrow O_3 + M \tag{2.6}$$

M (e.g., volatile hydrocarbons).

Close to the earth's surface, O_3 is a toxic and unwanted oxidant. It is often regarded as one of the most toxic air pollutants [11]. In its ground state, O_3 is relatively stable (half-life from several minutes to several days according to air conditions) [12], but, in a biological environment, its reactivity is extremely high (comparable to that of $OH^{\bullet}$; see Table 1.1). As a powerful oxidizing agent, O_3 is cytotoxic against the respiratory system and will react with cellular macromolecules in its immediate vicinity (proteins, carbohydrates, DNA, and lipids) [13]. Inhalation of O_3 primarily affects the terminal bronchioles and the alveoli (bronchiolar ciliated cells and epithelial alveolar type I cells are severely damaged) [14].

It is believed that O_3 operates via two main mechanisms of action. First, by oxidation of sulfhydryl groups on the amino acids of proteins and, second, by the oxidation of polyunsaturated fatty acids (PUFAs), leading to the formation of peroxides and aldehydes. For instance, the toxic effects of O_3 are often attributed to its ability to cause such disruption in lipid integrity. Ozone reacts with the double bonds in PUFAs, causing them to split through a chemical process known as "ozonolysis," thus leading to the formation of unstable ozonides and subsequently cytotoxic aldehydes (Eq. 2.7) [15].

Regarding its short lifetime in biological systems, the effect of O_3 on DNA was mainly reported in in vitro studies [16, 17].

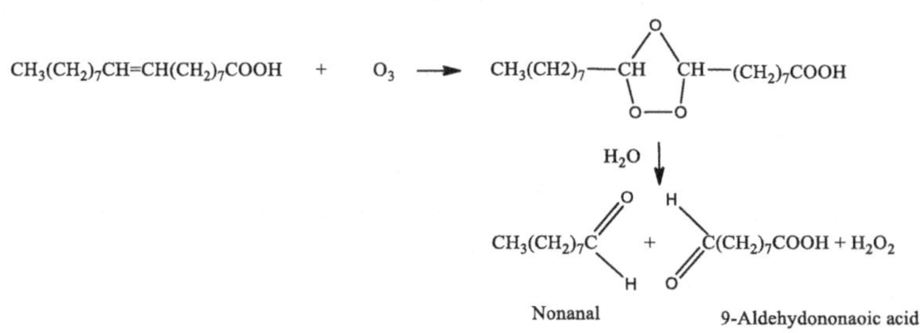

(2.7)

2.1.1.3 Hydrogen Peroxide (H_2O_2)

Hydrogen peroxide can be directly generated via a two-electron reduction of molecular oxygen or via a dismutation reaction of two superoxide radicals either spontaneously or by the accelerated reaction mediated by superoxide dismutase (SOD) (Eqs. 2.8, 2.9 and 2.10) [18]. Ozone is also a possible source of H_2O_2. Under physiological conditions, peroxisomes are the main consumers and producers of H_2O_2.

$$2O_2^{\bullet}- + 2H^+ \overset{\text{Spontaneous}}{\rightarrow} H_2O_2 + {}^1O_2 \tag{2.8}$$

$$O_2^{\bullet}- + O_2^{-} + 2H^+ \overset{\text{Spontaneous}}{\rightarrow} H_2O_2 + O_2 \tag{2.9}$$

$$2O_2^{\bullet}- + 2H^+ \overset{\text{SOD}}{\rightarrow} H_2O_2 + O_2 \tag{2.10}$$

Hydrogen peroxide has the ability to amplify its own production by stimulating ROS precursor generation via enhanced intracellular iron uptake, mitochondrial damage, and a variety of cellular oxidases [19].

In mammalian cells, H_2O_2 is considered an attractive signaling molecule in various physiological processes, particularly autophagy [20]. Moreover, it can act both as a vasodilator and a vasoconstrictor. In high concentrations, it induces vasoconstriction followed by vasodilatation [21].

In contrast to the charged radical $O_2^{\bullet-}$, H_2O_2 is a chemically more stable, relatively poor oxidant, is much less reactive, and, because it is electronically neutral, diffuses more easily across hydrophobic cell membranes [22, 23]. This property enables H_2O_2 to exit the mitochondria and other subcellular organelles freely and makes this species highly deleterious.

Hydrogen peroxide has a longer half-life (measured in seconds in the absence of transition metals) than do other ROS and is poorly reactive with most biomolecules [24]. However, it can break down in the presence of redox-active transition metals, principally Cu^+ and Fe^{2+}, to generate more ROS that are often the highly reactive $OH^{\bullet}$ produced in the Haber–Weiss and Fenton reactions (Eqs. 2.25, 2.26, 2.27 and 2.28) [25]. In the context of inflammation, H_2O_2 also serves as a source of more toxic species such as hypochlorous acid (HOCl) (H_2O_2 has a higher reduction potential than does HOCl ($E°\ H_2O_2/H_2O = 1.77$ and $E°\ HOCl/Cl^- = 1.50$) [26].

In addition to the direct biological reactivity of H_2O_2 being relatively modest, concentrations in vivo are kept extremely low. In an extreme situation of OS, H_2O_2 can directly cause the inactivation of enzymes as well as the oxidation of DNA and keto acids.

Hydrogen peroxide reacts fairly with [Fe–S] clusters and loosely bound metals, releasing iron from heme proteins. It can react slowly with typical thiol groups such as methionine and the cysteine side chains of proteins [27]. Reaction intermediates are sulfoxide and sulfenic acids, which can bind internal or external thiols or cyclize into sulfonamide. The oxidation of protein-containing thiols by H_2O_2 modulates their enzymatic activity.

Other oxidative reactions were determined. For example, H_2O_2 promotes lipid peroxidation and reacts with bicarbonate to possibly form peroxymonocarbonate [28].

Overall, radicals and other secondary ROS released from metalloproteins, along with the products of Fenton chemistry, are the major contributors to the biologically damaging effects of H_2O_2.

Hydrogen peroxide is also produced by xanthine oxidase (XO) (Eq. 2.11), mono-amine, and polyamine oxidases and is generated by peroxisomes by consumption of molecular oxygen in metabolic reactions.

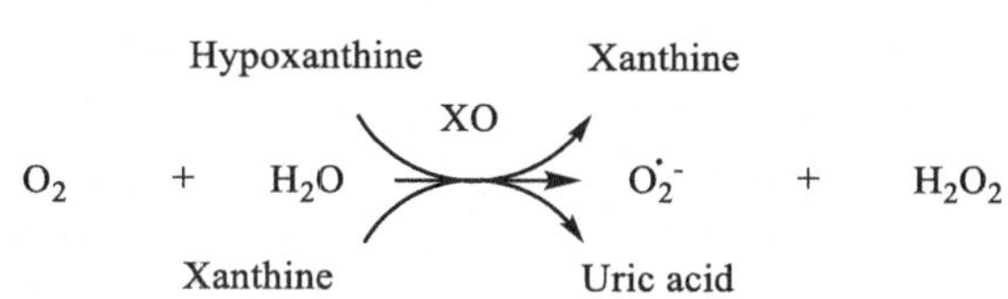

(2.11)

2.1.1.4 Organic Hydroperoxides (ROOH)

Organic hydroperoxides are not radicals but are considered as unstable and reactive products of peroxyl radicals. They are usually formed through a chemical chain reaction known as "lipid peroxidation" (see Sect. 3.4). This process starts with hydrogen abstraction from organic molecules (e.g., lipids) by an oxidant such as $OH^{\bullet}$. The resulting radicals ($R^{\bullet}$) react with molecular oxygen to produce peroxyl radicals ($ROO^{\bullet}$), which, in turn, abstract a hydrogen atom from another molecule, thus propagating the damage by producing a new radical and restarting the cycle. This type of process is also called a radical chain reaction. It will be terminated when two radicals react with one another.

Organic peroxides are chemically related to H_2O_2. They are generally similar to this species in their substrate preferences. There are four principal reactions in which these RS are involved:

- ROOH can react with a free radical (FR) from some source to yield $ROO^{\bullet}$ (Eq. 2.12).
- ROOH can be reduced by transition metal centers to yield either alkoxyl radicals ($RO^{\bullet}$) or $ROO^{\bullet}$ (Eq. 2.13).
- Less importantly, ROOH can be subjected to O–O bond breakdown to directly yield $RO^{\bullet}$ and $OH^{\bullet}$ (Eq. 2.14).
- ROOH can readily react with thiols.

$$ROOH + R^{\bullet\prime} \rightarrow ROO^{\bullet} + R'H \tag{2.12}$$

$$ROOH \rightarrow RO^{\bullet} + OH^{-} \quad (\text{or } RO^{-} + OH^{\bullet}) \tag{2.13}$$

$$ROOH \rightarrow RO^{\bullet} + OH^{\bullet} \tag{2.14}$$

Lipid hydroperoxides, a subgroup of ROOH, are highly unstable and can easily break down into secondary products such as aldehydes (see Sect. 3.4) and isoprostanes. The formation of lipid hydroperoxides derived from fatty acids and their degradation products in the phospholipid bilayer disrupts the membrane and causes leakiness and cell death [29].

2.1.2 *Radicals*

2.1.2.1 The Superoxide Anion ($O_2^{\bullet-}$)

A superoxide is an anion radical formed by the one-electron reduction of molecular oxygen. In an acidic environment, protonation of this species can form the uncharged hydroperoxyl radical $HO_2^{\bullet}$ [30].

The major site for producing $O_2^{\bullet-}$ is the mitochondrial electron transport chain (ETC), upon adenosine triphosphate (ATP) generation. In the physiological context and during energy transduction, electrons are transferred through the ETC for reduction of oxygen to water. However, approximately 1–5% of all electrons leak prematurely from the system (particularly complexes I and III) and produce $O_2^{\bullet-}$ (Eq. 2.15) [24, 31, 32].

$$O_2 + e^- \rightarrow O_2^{\bullet-} \tag{2.15}$$

In certain pathophysiological situations, $O_2^{\bullet-}$ can inactivate the mitochondrial respiratory chain, particularly complexes I and II of the ETC, therefore inducing its own production to extremely high levels.

The superoxide anion can be generated from redox cycling of semiquinones (see Sect. 2.3.2). It can also be generated by "professional" plasma membrane-associated oxidases, of which NOX is the best-known example (Eq. 2.16) [33].

$$2O_2 + NADPH \xrightarrow{\text{NADPH Oxidase}} 2O_2^{\bullet-} + NADP^+ + H^+ \tag{2.16}$$

This enzyme is considered an early biomarker of OS and is found in polynuclear leukocytes (PMNs), monocytes, and macrophages. Upon phagocytosis, these cells produce a burst of $O_2^{\bullet-}$ that leads to the production of several ROS known for their antimicrobial activity. Besides, XO (vessel wall) catalyzes the oxidation of hypoxanthine and xanthine to uric acid (UA) in the terminal steps of purine nucleotide metabolism, which simultaneously leads to the production of $O_2^{\bullet-}$ (Eq. 2.11). Given that XO produces $O_2^{\bullet-}$ in stoichiometric quantities along with UA, it represents one of the major sources of ROS in human physiology. Other possible sources of $O_2^{\bullet-}$ through enzymatic catalysis include cyclooxygenase (COX) and lipoxygenase (LOX) in immune cells.

Certainly, it is notable that excessive $O_2^{\bullet-}$ can generate other ROS, leading to OS in organisms and causing cell injury. As mentioned before, $O_2^{\bullet-}$ can undergo a spontaneous or, more rapidly, an enzymatically catalyzed dismutation, thus generating H_2O_2. It can also give rise to other types of ROS regarding its ambivalent properties.

Interestingly, $O_2^{\bullet-}$ is both a moderate oxidizing agent and, at the same time, a strong reductant that is able to reduce transition metal ions such as Fe^{3+} and Cu^{2+} into their reduced forms (Fe^{2+} and Cu^+, respectively) in a series of reactions, thus resulting in the formation of $OH^\bullet$. It is an unusual radical because despite its reduction potential of 0.94 V, it oxidizes relatively few biological compounds. Its oxidizing capability is highly proton-dependent (see Sect. 2.1.2.3) [34]. As $O_2^{\bullet-}$ is a potent reducing agent, it can reversibly reduce organic redox species, such as quinones to semiquinones, as well [35].

Being a strong base, $O_2^{\bullet-}$ can react with compounds capable of donating H^+ and therefore abstract protons from a variety of compounds to form the uncharged hydroperoxyl radical ($HO_2^\bullet$) in aqueous media (pKa 4.8) [36]. This protonated form is a stronger oxidant. In addition, $O_2^{\bullet-}$ serves as a critical, proximal step in the generation of most forms of reactive nitrogen species (RNS), which play a role in the propagation of OS reactions. For example, one of the potentially detrimental reactions of $O_2^{\bullet-}$ is with the potent vasodilator and cell signaling molecule nitric oxide ($NO^{\bullet}$) to form $ONOO^-$ in the process of endothelial nitric oxide synthase (NOS) uncoupling (see Sect. 2.2.2.1) [37]. It can therefore inactivate $NO^{\bullet}$, causing impaired vasodilation. Besides, $O_2^{\bullet-}$ has the ability to modulate endothelium-dependent relaxation by opening calcium-dependent potassium channels. It can also lower the threshold of platelet aggregation in the presence of thrombin, collagen, and adenosine diphosphate (ADP). This property seems to be uniquely linked to $O_2^{\bullet-}$ as induction and promotion of platelet aggregability are not observed with other related RS such as H_2O_2 [38, 39].

In terms of physicochemical properties, $O_2^{\bullet-}$ has an extremely short life of 10^{-6} s, as a consequence of its rapid conversion to other RS (e.g., superoxide undergoes spontaneous dismutation to H_2O_2, which can be accelerated by 10^4-fold by the enzymatic catalyzed reaction by SOD) [40].

The superoxide anion acts as a powerful nucleophile and is capable of attacking positively charged centers. A superoxide is a moderate oxidant and may initiate hydrogen abstraction from certain molecules such as ascorbic acid and epinephrine.

Being charged, $O_2^{\bullet-}$ has an extremely limited diffusion ability. Therefore, it cannot readily cross lipid membranes and acts mostly close to its site of production. The instability of $O_2^{\bullet-}$ and its inability to readily diffuse through cell membranes make it a weak signaling molecule [41].

The only way that $O_2^{\bullet-}$ can cross lipidic membranes is through anion exchange proteins that are present only in specific cytomembranes and subcellular compartments, such as erythrocytes and endosomes. This feature accounts for the specificity of redox signal transduction and the ability of $O_2^{\bullet-}$ to act on particular targets [42]. Indeed, one of the most important functions of $O_2^{\bullet-}$ is the capacitation of sperm [43].

Evidence supports the expectation that $O_2^{\bullet-}$ is relatively unreactive toward most biomolecules of the cell, although it can be oxidized or reduced in different situations. The only exception is its known interaction with transition metals/sulfur clusters. This did not prevent from considering $O_2^{\bullet-}$ as an RS and from proposing its harmful effects. Several mechanisms have been proposed to explain its toxicity. The best-established one is inactivation of iron/sulfur-containing proteins such as cytochrome *c*, iron regulatory protein-1, aconitase, and the dehydratases required for the synthesis of some amino acids [44].

The superoxide anion has high affinity for iron–sulfur clusters ([Fe–S]), and one-electron oxidation by $O_2^{\bullet-}$ inactivates proteins/enzymes and releases iron to form $OH^{\bullet}$, which can cause further injury [45]. Addition reactions with other RS, such as the tyrosyl radicals ($Tyr^{\bullet}$) at the active site of ribonucleotide reductase, also contribute to $O_2^{\bullet-}$ toxicity. Finally, $O_2^{\bullet-}$ activates at least two major pro-inflammatory transcription factors, nuclear factor kappa B (NF-κB) and activator protein (AP)-1 [46, 47].

A superoxide probably reacts with fatty acid ester links by substitution or lysis, as depicted in Eqs. 2.17 and 2.18:

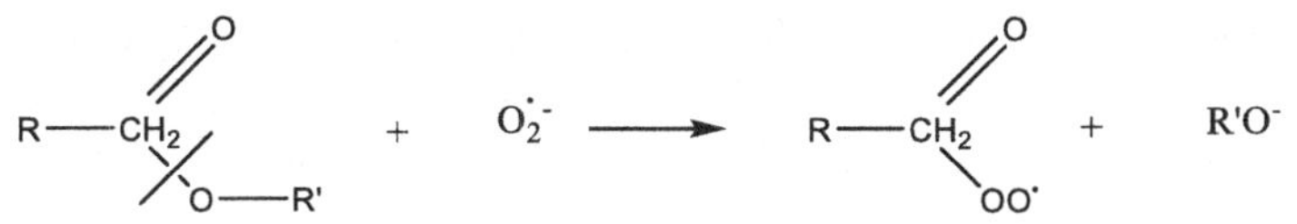

(2.17)

Peroxyl radical

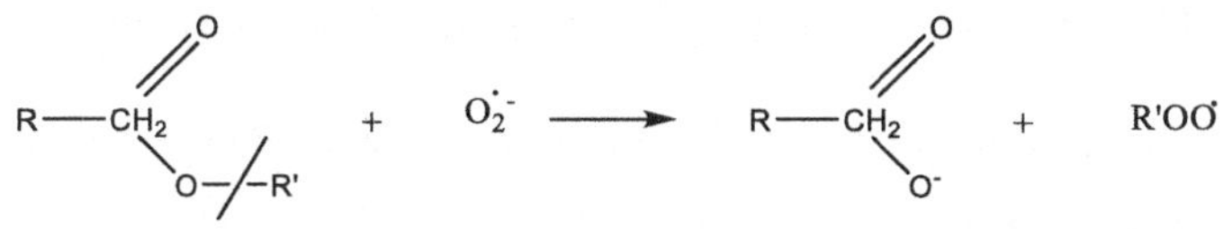

(2.18)

Overall, $O_2^{\bullet-}$ is an unstable chemical entity, which is easily converted into other ROS forms. The reactivity of $O_2^{\bullet-}$ is dependent on the microenvironment. Its mechanism of toxicity mainly involves disassembly of iron–sulfur ([Fe–S]) clusters in proteins, causing release of iron and alterations of –SH residues.

2.1.2.2 The Hydroperoxyl Radical ($HO_2^{\bullet}$)

The hydroperoxyl radical is the conjugate acid or protonated form of $O_2^{\bullet-}$. It is a weak acid, with a pKa of 4.8 [36]. It is generated from $O_2^{\bullet-}$ as the latter can exist in the form of either a superoxide anion or, at low pH, a hydroperoxyl radical (Eq. 2.19).

$$O_2^{\bullet -} + H^+ \overset{pH4.88}{\rightleftharpoons} HO_2^{\bullet} \tag{2.19}$$

At a physiological pH, $O_2^{\bullet-}$ is mostly present in the unprotonated form where only <0.3% of $HO_2^{\bullet}$ is supposed to be present in the cytosol [48]. An increase in proton concentration further enhances the rate of conversion of $O_2^{\bullet-}$ to $HO_2^{\bullet}$, making the reactivity of the latter mostly relevant to slightly acidic pHs, for example, during conditions of cellular acidosis.

Although $HO_2^{\bullet}$ is mainly formed from the conversion of $O_2^{\bullet-}$, other mechanisms of $HO_2^{\bullet}$ formation in biological systems are likely possible such as the dehydrogenation of H_2O_2 by $OH^{\bullet}$ or the two-step oxidation of H_2O_2 by Fe(III) (Eqs. 2.20, 2.21 and 2.22).

$$H_2O_2 + OH^{\bullet} \rightarrow HO_2^{\bullet} + H_2O \tag{2.20}$$

$$Fe^{3+} + H_2O_2 \rightarrow FeOOH^{2+} + H^+ \tag{2.21}$$

$$FeOOH^{2+} \rightarrow Fe^{2+} + HO_2^{\bullet} \tag{2.22}$$

A hydroperoxyl radical can more easily penetrate biological membranes than the charged form. Because of protonation, $HO_2^{\bullet}$ has lower symmetry than does $O_2^{\bullet-}$, making the electronic distribution in $HO_2^{\bullet}$ unequal between the two oxygen atoms.

The implication of this electronic distribution on $HO_2^{\bullet}$ results in a significant difference in the reduction potentials compared to $O_2^{\bullet-}$ where $HO_2^{\bullet}$ has been shown to be a stronger oxidizer than $O_2^{\bullet-}$ with E° 1.06 V and 0.94 V, respectively [49].

As mentioned before, the predominant species is the anion radical at a physiological pH, but a local more acidic environment, for example, in the vicinity of cell membranes, will shift the equilibrium toward the formation of $HO_2^{\bullet}$, which is more reactive than $O_2^{\bullet-}$.

Like $O_2^{\bullet-}$, the most important reaction of $HO_2^{\bullet}$ is dismutation. In this reaction, the hydroperoxyl radical reacts with another hydroperoxyl radical and generates molecular oxygen and H_2O_2 (Eqs. 2.23 and 2.24).

$$O_2^{\bullet-} + HO_2^{\bullet} + H^+ \rightarrow H_2O_2 + O_2 \quad (2.23)$$

$$HO_2^{\bullet} + HO_2^{\bullet} \rightarrow H_2O_2 + O_2 \quad (2.24)$$

In a hydrophilic environment, $HO_2^{\bullet}$ can act as reducing agents capable of reducing ferric ions to ferrous ions; however, the reducing capacity of $HO_2^{\bullet}$ is lower than that of $O_2^{\bullet-}$.

The reactions of $HO_2^{\bullet}$ with organic compounds are generally slow. The slowness of these reaction suggests that $HO_2^{\bullet}$ acts as a sink for intracellularly generated radicals, with $O_2^{\bullet-}$ ultimately converting into more RS by SOD [50]. Therefore, $HO_2^{\bullet}$ can be considered an important species, although under physiological pH, most of the superoxide is in the charged form.

A hydroperoxyl radical can react with fatty acids in living systems. Due to the less polar nature of $HO_2^{\bullet}$ and its capacity to penetrate through the lipid bilayer, this radical has the ability to abstract hydrogen atoms from methylene carbon in PUFAs, thus highlighting its key role in the initiation of lipid peroxidation [30]. This reaction with $O_2^{\bullet-}$ is about five orders of magnitude slower compared to $HO_2^{\bullet}$.

2.1.2.3 The Hydroxyl Radical ($OH^{\bullet}$)

The hydroxyl radical is a neutral form of the hydroxide ion having extremely high reactivity, therefore making it a dangerous radical, the most toxic RS by far [51]. It is the three-electron reduction product of molecular oxygen that is produced by two major biochemical reactions called the Fenton reaction and the Haber–Weiss reaction. In these reactions, H_2O_2 is the source of $OH^{\bullet}$, following interaction with transition metal ions (e.g., iron and copper). The free iron in the Fenton reaction is in its reduced form (ferrous state) and catalyzes electron transfer from donors to H_2O_2 (Eqs. 2.25, 2.26, 2.27 and 2.28). It has been demonstrated that copper can also transfer electrons to H_2O_2, at least in vitro; its impact in vivo seems to be limited to physiological concentrations [52].

$$Me^{n+} + O_2^{\bullet-} \rightarrow Me^{(n-1)+} + O_2 \tag{2.25}$$

$$Me^{(n-1)+} + H_2O_2 \overset{Fenton\_reaction}{\rightarrow} Me^{n+} + OH^{-} + OH^{\bullet} \tag{2.26}$$

$$Fe^{2+} + H_2O_2 \rightarrow Fe^{3+} + OH^{-} + OH^{\bullet} \quad K = 10^2 M^{-1}\ sec^{-1} \tag{2.27}$$

$$Cu^{+} + H_2O_2 \rightarrow Cu^{2+} + OH^{-} + OH^{\bullet} \quad K = 5\ 10^3 M^{-1}\ sec^{-1} \tag{2.28}$$

In 1934, Haber and Weiss illustrated that $O_2^{\bullet-}$ and H_2O_2 can form the highly destructive $OH^{\bullet}$ and initiate the oxidation of organic substrates by the so-called Haber–Weiss reaction [53]. This reaction combines the Fenton reaction with the oxidative action of Fe^{2+} (Eq. 2.29). In the first step of the Haber–Weiss reaction, $O_2^{\bullet-}$ reacts with Fe^{3+} to form Fe^{2+} and O_2. The second step is also known as the Fenton reaction, in which Fe^{2+} reacts with H_2O_2 to form both $OH^{\bullet}$ and OH^{-}.

$$O_2^{\bullet-} + H_2O_2 \underset{Haber\text{-}Weissreaction}{\overset{Me^{n+}/Me^{(n-1)+}}{\rightarrow}} O_2 + OH^{-} + OH^{\bullet} \tag{2.29}$$

In the presence of trace amounts of iron, other transition metals (e.g., copper) also participate in these electron transfer reactions by cycling between oxidized and reduced states. Therefore, the bioavailability of iron is the rate-limiting step in both the Fenton and Haber–Weiss reactions. Even at extremely low concentrations, the recycling of iron from the ferric to the ferrous form by reducing agents such $O_2^{\bullet-}$ can maintain ongoing cycle reactions, leading to the generation of high amounts of $OH^{\bullet}$. Moreover, under OS conditions, $O_2^{\bullet-}$ can accelerate the release of iron from metalloproteins, which reacts with H_2O_2 by the Fenton reaction to form $OH^{\bullet}$.

Another potential source for $OH^{\bullet}$ generation is through the reaction of H_2O_2 with RNS (Eqs. 2.30 and 2.31).

$$NO^{\bullet} + H_2O_2 \rightarrow OH^{\bullet} + HONO \tag{2.30}$$

$$NO_2^{\bullet} + H_2O_2 \rightarrow OH^{\bullet} + HONO_2 \tag{2.31}$$

Other minor sources of $OH^{\bullet}$ deriving from HOCl and $ONOO^{-}$ are illustrated in Fig. 1.1.

The hydroxyl radical is particularly unstable. Owing to its high reactivity, $OH^{\bullet}$ is an extremely short-lived species with a half-life of 10^{-9} s in biological systems. It has an extremely limited diffusion distance and reacts with a large spectrum of biomolecules at close proximity with high affinity [54, 55]. Under physiological conditions, $OH^{\bullet}$ acts as a messenger at extremely low concentrations by quickly reacting with signaling molecules [24].

The hydroxyl radical is defined as one of the most reactive molecules known in biology. This high reactivity makes this FR highly unselective in preferences for co-reactants, leading to the oxidation of a wide variety of organic molecules in the cell including proteins, lipids, carbohydrates, and nucleic acids [56]. Because of these properties, this powerful oxidizing agent is considered highly toxic (E° $OH^{\bullet}$,

H^+/H_2O 2.31 V) and can cause severe damage to cells and intracellular structures even if it has an extremely short half-life [57]. It can also react with chemical groups (nitrite (NO_2^-) and bicarbonate ion (HCO_3^-)) or ions (halide ions). This species is also known for its oxidative damage to DNA, which usually causes mutations and possibly leads to carcinogenesis, if not reversed by DNA repair mechanisms [58]. Damage to proteins leads to enzyme inhibition, denaturation, and protein degradation [59].

The hydroxyl radical can oxidize almost all biological molecules through three main reaction types: hydrogen abstraction, addition, and electron transfer:

- The hydroxyl radical can abstract hydrogen atoms from essentially any hydrogen–carbon bond and undergo addition reactions with aromatic systems, amino acids, guanine, and PUFAs, forming other radicals at a reaction rate close to the diffusion limit (Eqs. 2.32, 2.33 and 2.34). It initiates lipid peroxidation by taking an electron from PUFAs [60].

$$^{+}NH_3\text{-}CH_2\text{-}COO^- + OH^\bullet \rightarrow {}^{\bullet +}NH_3\text{-}CH_2\text{-}COO^- + H_2O \tag{2.32}$$

$$^{+}NH_3\text{-}CH_2\text{-}CH_2OO^- + OH^\bullet \rightarrow {}^{+}NH_3\text{-}\dot{C}H_2\text{-}COO^- + H_2O \tag{2.33}$$

$$RH + OH^\bullet \rightarrow R^\bullet + H_2O \tag{2.34}$$

- The electron transfer mediated by $OH^\bullet$ is represented by Eq. 2.35:

$$Fe^{2+} + OH^\bullet \rightarrow Fe^{3+} + OH^- \tag{2.35}$$

- Addition of $OH^\bullet$ to methylene functions is represented by Eq. 2.36:

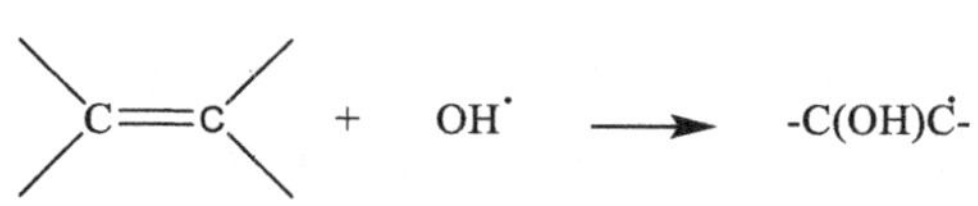

(2.36)

2.1.2.4 Peroxyl Radicals ($ROO^\bullet$)

Peroxyl radicals are intermediate species generated by the addition of molecular oxygen to carbon-centered radicals during lipid oxidation chain reactions. They can also be produced by metal-induced conversion of hydroperoxides. These species are far more stable than either the carbon-centered radical or the related $RO^\bullet$. They typically have half-life times of 7 s at 37 °C and can penetrate biological membranes at moderate distances [3].

These radicals are easily converted into more reactive $RO^\bullet$, and both are important chain-carrying species in auto-oxidation. Peroxyl radicals are considered moderate oxidizing agents (E° 0.77–1.44 V) [61]. They react with numerous biological substrates (lipids, DNA, and proteins) via different mechanisms.

Although peroxyl radicals are much less reactive than $OH^\bullet$, they share some properties in terms of their mode of reaction as they can engage in electron abstraction, hydrogen abstraction, or addition to double bonds [62]. In the latter case, intra- or intermolecular reactions lead to the formation of unstable endoperoxide radical species ($ROOR^\bullet$). Moreover, a peroxyl radical probably reacts with itself to generate $^1O_2$.

Peroxyl radicals are involved in the propagation of a chain reaction during lipid peroxidation. With aliphatic compounds such as fatty acids, $ROO^\bullet$ may transform into ROOH by hydrogen abstraction (Eq. 2.37).

$$ROO^\bullet + R'H \rightarrow ROOH + R^{\bullet\prime} \tag{2.37}$$

This reaction plays a major role when the co-reactant contains weak C–H bonds or other weakly bonded hydrogens, such as thiols (Eq. 2.38):

$$ROO^\bullet + RSH \rightarrow ROOH + RS^\bullet \tag{2.38}$$

Peroxyl radicals with α-hydroxyl or α-amino groups can also undergo rapid unimolecular elimination of $HO_2^\bullet/O_2^{\bullet-}$, leading to carbonyl or imine group formation [63].

Organic peroxyl radicals can undergo combination with other peroxyl radicals such as $HO_2^\bullet$ or $R'OO^\bullet$ to form the tetroxide species ROO-OOR' [64]. Tetroxides are unstable compounds, and their decomposition yields carbonyl and alcohol groups, accompanied by the loss of molecular oxygen. Among the diverse reactions that $ROO^\bullet$ can undergo, some are of particular interest because they contribute to the deterioration of cell membranes during a process called lipid peroxidation.

Peroxyl radicals can react with unsaturated compounds via possible addition to the unsaturated bond or abstraction from allylic or bis-allylic C–H bonds (Eq. 2.39):

$$ROO^\bullet \quad + \quad >C{=}C< \;\rightarrow\; ROO\text{-}CH_2\text{-}C^\bullet < \tag{2.39}$$

Organic peroxyl radicals play an important role in some pathophysiological situations, particularly in the damage induced by ionizing radiation in ischemia and in aging [65].

From the above discussion, it is clear that various peroxyl radicals exhibit moderate or low reactivity toward a number of biological components and may have the capacity to diffuse and attack some targets more selectively compared to other organic radicals.

The reactions of $ROO^\bullet$ with antioxidants are relatively slow, but their reactions with other biomolecules are even slower, and, thus, they may decay by radical–radical reactions in a cytosolic environment. For such radicals, however, reactivity may not be the most important parameter that determines their toxicity and fate, rather it is their diffusion that may be restricted by their hydrophobicity or their position within membrane structures. Thus, they likely react locally even if the rate of reaction is extremely low as it is the case in the propagation step of lipid peroxidation (see Sect. 3.4).

After the reaction of $ROO^{\bullet}$ via any of the above mechanisms, the possibility still exists to produce more RS such as alkoxyl radicals, either by radical–radical decay in the presence of redox metals or following reduction of the hydroperoxide formed.

2.1.2.5 Alkoxyl Radicals ($RO^{\bullet}$)

Alkoxyl radicals are formed as a result of the reduction of peroxyl radicals or alkyl hydroperoxides following a Fenton-type reaction. They undergo a rapid hydrogen shift, resulting in the generation of hydroxyalkyl radicals and, ultimately, in the formation of alcohol by intermolecular hydrogen abstraction. In some cases, when the hydrogen shift is not possible in tertiary alkoxyl radicals, fragmentation reactions occur and yield aldehydes and ketones. These various alkyl radicals, in turn, will react with molecular oxygen to yield $ROO^{\bullet}$. Alkoxyl radicals may react with other organic molecules via hydrogen abstraction, addition, or oxidation. Most of these reactions are more rapid than those of organic $ROO^{\bullet}$. In particular, hydrogen abstraction from PUFAs is about three orders of magnitude more rapid with $RO^{\bullet}$ than with $ROO^{\bullet}$. This is due to the fact that the RO–H bond is considerably stronger than the ROO–H bond [66]. However, in some conditions, the rapid monomolecular rearrangements in $RO^{\bullet}$ often compete with or practically prevent their reactions with other molecules.

Alkoxyl radicals have a half-life of approximatively 10^{-6} s at 37 °C and are thus intermediate in terms of reactivity between hydroxyl and peroxyl radicals [67]. Aliphatic alkoxyl radicals are one of the most unstable organic species, and their reactivity is higher than that of aromatic alkoxyl and peroxyl radicals. This is due to the electron-donating group of aliphatic alkoxyl radicals that enhances the reactivity of alkoxyl radicals [68].

Peroxyl and alkoxyl radicals are strong oxidant species, with relatively high redox potentials of $E°$ ($ROO^{\bullet}/ROOH$) = ~1.00 V and $E°(RO^{\bullet}/ROH)$ = 1.60 V, respectively [69]. The high redox potentials of peroxyl and alkoxyl radicals enable them to react with biomolecules. There are only minute quantities of RO· and ROO· in living systems because of their high reactivity. For the same reason, they are essential for sustaining normal physiological processes and lead to the formation of numerous other important ROS to sustain physiological functions [70, 71]. At high concentrations, they can abstract hydrogen from membrane lipids, ascorbate, and nicotinamide adenine dinucleotide + hydrogen (NADH) [68, 72] and can exacerbate OS through their involvement in lipid peroxidation as reactants or products, under pathological conditions.

2.2 Reactive Nitrogen Species

RNS are nitrogen-containing compounds, which usually possess a high oxidizing potential. Being involved in OS, they are sometimes classified as part of ROS in the literature. They mainly cause nitrosylation of biomolecules such as nucleic acids and proteins, leading to alterations in their structure and function. Nitrosants, as they are also called, can interact to form other oxidizing/nitrosating species, which, in turn, can potentiate their toxicity.

2.2.1 Radicals

2.2.1.1 Nitric Oxide ($NO/NO^{\bullet}$)

Nitric oxide is a highly reactive, small, uncharged molecule containing one unpaired electron. It is one of the most discussed and studied FRs in the literature. This diatomic molecule is formed through the enzymatic conversion of L-arginine to L-citrulline by NO synthase (NOS) (Eq. 2.40). Synthesis of $NO^{\bullet}$ can take place in almost all mammalian cells, including glial cells, endothelium, macrophages, leukocytes, and neurons. Three distinct isoforms of NOS have been identified, two of which, the endothelial NOS (eNOS) and neuronal NOS (nNOS) isoforms, are constitutive, calcium dependent, and regulated by second messengers, whereas one can be upregulated under certain conditions (usually by cytokines) to induce OS (calcium-independent inducible NOS, iNOS).

$$\text{L-Arginine} + 2O_2 + H^+ + 3/2\text{NADPH} \overset{\text{NOS}}{\rightarrow} NO^{\bullet} + \text{Citruline} + 3/2\text{NADP}^+ \quad (2.40)$$

NO has a short half-life of about 10^{-3}–1 s at 37 ° C s and can freely cross biological membranes without channels or receptors. It is considered an intra- and intercellular messenger, soluble in the cytosol with fast diffusion into different cell compartments [73, 74].

Several studies have shown both the cytotoxic and cytoprotective roles of $NO^{\bullet}$ [75, 76]. This depends on the source of $NO^{\bullet}$, its redox state, and the local tissue concentration. The importance of $NO^{\bullet}$ in normal physiology is well understood. At low concentrations, this molecule plays a critical role in a variety of physiological processes such as cell signaling, blood pressure regulation, control vascular tone, relaxation of smooth muscle cells, platelet aggregation, neurotransmission, synaptic plasticity, oxygen consumption, endocrine gland regulation, immune regulation, and other defense strategies of the cell. Therefore, it is considered a vital signaling molecule [77–79].

It is striking to clarify how $NO^{\bullet}$ communicates information to achieve signal transduction. Under physiological conditions, $NO^{\bullet}$ can diffuse from where it is synthesized into surrounding cells, where it will bind and activate guanylate cyclase to produce cyclic guanosine monophosphate (cGMP) [80]. The latter activates cGMP-dependent kinases that regulate diverse cell activities by modulating intracellular calcium levels [81].

The roles of $NO^{\bullet}$ as a neurotransmitter and a vital part of host defense mechanisms are now well established. Nitric oxide performs a number of neurological functions such as the processes of learning and memory.

NO release from eNOS results in vasodilation and inhibits microvascular adhesion, which is a protective action against leukocyte influx.

At physiological concentrations, $NO^{\bullet}$ controls oxygen consumption and mitochondrial respiration, causing reversible inhibition of respiration by modulating cytochrome *c* oxidase activity in complex IV, which is the terminal center of the mitochondrial respiratory chain [82].

Under OS conditions, inducible NOS (iNOS) is upregulated, producing a burst of $NO^{\bullet}$ that far exceeds basal levels, which can cause significant cellular injury via different mechanisms:

- High levels of $NO^{\bullet}$ can directly promote excessive peripheral vasodilation. As $NO^{\bullet}$ is a strong vasodilator, this effect results in vascular decomposition [81].
- Under OS conditions, $NO^{\bullet}$ may upregulate the transcription of NF-κB, leading to inflammatory signaling pathways involved in the synthesis of numerous inflammatory cytokines [83].

Similar to the extremely reactive $OH^{\bullet}$, $NO^{\bullet}$ is highly reactive and will readily react with other molecules in its near environment.

The chemistry of $NO^{\bullet}$ with molecular oxygen is highly complex but ultimately leads to the formation of several species via hydrolysis of various nitrogen oxides such as dinitrogen trioxide (N_2O_3) and dinitrogen tetraoxide (N_2O_4). At a high concentration, when iNOS is stimulated, $NO^{\bullet}$ can readily react with other radicals and exert pathophysiological actions, leading to tissue injury. Of particular interest is the reaction of $NO^{\bullet}$ with $O_2^{\bullet-}$ to produce a higher reactive compound, $ONOO^-$ (Eq. 2.51), which can easily damage cellular components. Nitric oxide can also react with $OH^{\bullet}$ to form the highly reactive nitrous acid (HNO_2). In addition, excessive $NO^{\bullet}$ will disturb electron transfer reactions in cytochrome *c* oxidase and promote $O_2^{\bullet-}$ generation and further production of $ONOO^-$, resulting in oxidative injury of the mitochondria [84].

At extremely high concentrations, $NO^{\bullet}$ may exert genotoxic effects that result in DNA double-strand breaks [85]. This effect is also enhanced in the presence of $O_2^{\bullet-}$, which may extend the adverse effects of $NO^{\bullet}$ to initiate the process of lipid peroxidation and once again enhance the generation of $ONOO^-$. The latter is highly reactive and can damage proteins by reacting with tyrosine and thiol moieties. Other major reactions of $NO^{\bullet}$ are addition reactions with other radicals, including oxygen, thiyl radicals, and phenoxyl radicals, to yield nitrosophenols and lipid peroxyl radicals. Essential biochemical reactions involving $NO^{\bullet}$ are achieved through S-nitrosation of thiols and nitrosylation of transition metal ions.

Nitrosothiol formation on protein thiyl radicals, as shown in Eq. 2.41, is emerging as an important cell regulatory mechanism.

$$RS^{\bullet} + NO^{\bullet} \rightarrow RSNO \tag{2.41}$$

The relevant targets of this action are cysteine residues in proteins in a process called S-nitrosylation, which typically occurs under oxidizing conditions.

Several target macromolecules have been shown to be S-nitrosylated, including enzymes in the regulation of calcium homeostasis (the ryanodine receptor), in the mitochondrial respiratory chain (complex I), and in the glycolytic cycle (glucose-6-phosphate dehydrogenase (G6PD) [86].

NO may alter the hemoglobin structure by direct attachment to heme during the nitrosylation reaction, or by S-nitrosation of the thiol moieties, yielding S-nitrosothiols [87].

With one unpaired electron, $NO^{\bullet}$ is prone to directly react with iron in heme groups and iron centers like in its physiological role of activating guanylate cyclase. It forms nitroso complexes and can possibly be oxidized by oxyhemoglobin to NO_2^- or nitrate (NO_3^-) (Eq. 2.42) [88].

$$Hb^{2+}O_2 + NO^{\bullet} \rightarrow Hb^{3+} + NO_3^{-} \tag{2.42}$$

One typical case occurs in neurons, in what is known as glutamate excitotoxicity. NO-mediated generation of $ONOO^-$ inhibits the activity of the NMDA (*N*-methyl-D-aspartate receptor) receptor through S-nitrosylation, leading to a physiological protective effect on cell life. However, if $ONOO^-$ is persistently overgenerated, it can initiate S-nitrosylation of a variety of cellular targets involved in neuronal injury, which ultimately leads to excitotoxicity [89].

Myeloperoxidase has also been shown to catalyze the removal of $NO^{\bullet}$, although this reaction may be mediated by other radical products of the enzyme [90].

NO^+, NO^-, and HNO are different forms of $NO^{\bullet}$. The latter can be oxidized by oxygen or $O_2^{\bullet-}$ to from nitrosonium cation (NO^+) by removal of a single electron. Under acidic conditions, NO_2^- may be a source of NO^+ as well. Binding of $NO^{\bullet}$ to transition metals can result in either its oxidation to NO^+ or its one-electron reduction to nitroxyl anion (NO^-), both of which depend on the type of metal ion, oxidation state, coordination number, and type of ligand. NO^- can undergo further reactions by interacting with $NO^{\bullet}$ to yield N_2O and $OH^{\bullet}$.

The potential sources of HNO are the breakdown of S-nitrosothiols and metal nitrosyl enzyme complexes as well as the oxidation of *N*-hydroxyarginine (NOHA) during NOS catalytic cycle [91].

In aqueous solutions, NO^+ has an extremely short lifetime estimated at 3×10^{-10} s and is believed to be present only at a highly acidic pH. The half-life of NO^- has not been reported in physiological systems. In biological systems, NO^+ is quickly hydrolyzed to nitrous acid (Eq. 2.43).

$$NO^+ + H_2O \rightarrow H^+ + HNO_2 \tag{2.43}$$

A neuroprotective effect of NO^+ has been identified in animals [92]. NO^+ may nitrosylate *N*-methyl-D-aspartate (NMDA) receptors, which can lead to modulation of associated ion channels. Due to the repulsive effect, NO^- is a poor nitrosylating agent, whereas NO^+ may be transferred to accessible nucleophilic centers like thiols in cysteine residues or GSH and form S-nitrosothiol (R-SNO) by S-nitrosylation (also referred to as nitrosation) [93]. NO^+ transfer to thiol groups modulates the activity of a wide range of proteins involved in a variety of metabolic and signaling pathways. Under certain conditions, S-nitrosothiols are unstable in solutions and preferably form disulfide bonds and release $NO^•$ by hemolysis. NO^+ can react with other nucleophilic centers such as R-OH or RR'NH to produce R-ONO and RR'N-NO, respectively. Finally, low-mass SNOs such as S-nitrosoglutathione (GSNO) and S-nitrosocysteine (CysNO) play a potential role in nitric oxide storage and as carriers in living organisms.

2.2.1.2 Nitrogen Dioxide ($NO_2^•$)

Nitrogen dioxide was previously considered an exogenous FR, but the discovery of the importance of $NO^•$ as a key RS in many physiological processes has increased the attention on the possible production of $NO_2^•$ endogenously as an important biological oxidant.

Nitrogen dioxide is a radical-based RNS that is mainly derived from the interaction of $NO^•$ with molecular oxygen at a relatively fast rate, according to Eq. 2.44:

$$2NO^{\bullet} + O_2 \rightarrow 2NO_2^{\bullet} \tag{2.44}$$

This reaction is termed "auto-oxidation of $NO^•$," which involves an initial, reversible reaction between two nitric oxide molecules and then a reaction of the product with oxygen in the hydrophobic layers of cellular membranes. Nitrogen dioxide is also formed during one-electron reduction of $ONOO^-$ by ferric heme peroxidases and during the decomposition of nitrosoperoxycarbonate (Eq. 2.53). Other relevant sources of $NO_2^•$ are the decomposition of either organic peroxynitrites (RONOO) or peroxynitrous acid (ONOOH) through decomposition and hemolytic cleavage of the N–O bond independently of the presence of transition metals according to Eqs. 2.45 and 2.46:

$$RONOO \rightarrow RO^{\bullet} + NO_2^{\bullet} \tag{2.45}$$

$$HONOO \rightarrow OH^{\bullet} + NO_2^{\bullet} \tag{2.46}$$

In the presence of MPO, it has been shown that H_2O_2 leads to the formation of higher oxygen states of heme and, in the presence of $NO^•$, produces $NO_2^•$ (Eq. 2.47) [94]. Nitrite can also be oxidized by activated MPO to yield $NO_2^•$ [95].

The nitrogen dioxide radical rapidly reacts with most other FRs and reasonably fast with one-electron reductants. As it is formed from $NO^•$, $NO_2^•$ can further react with an additional nitric oxide molecule to form N_2O_3 (Eq. 2.48) [96]. However, reactions between $NO_2^{\bullet}$ and thiols/urate severely limit the possibility of interactions between $NO_2^{\bullet}$ and $NO^{\bullet}$ to take place in the cytoplasm.

$$2NO^{\bullet} + O_2 \rightarrow 2NO_2^{\bullet} \tag{2.47}$$

$$NO_2^{\bullet} + NO^{\bullet} \rightarrow N_2O_3 \tag{2.48}$$

The nitrogen dioxide radical promptly reacts with carbon-centered radicals, in which the radical center is located on the same carbon atom as the hydroxyl group, to yield possibly NO_2^- and the corresponding carbonyls (Eq. 2.49):

$$NO_2^{\bullet} + C^{\bullet}H_2OH \rightarrow NO_2^- + CH_2O + H^+ \tag{2.49}$$

The nitrogen dioxide radical is a moderately strong one-electron oxidant. The reduction potential of the couple $NO_2^{\bullet}/NO_2^-$ is 1.04 V [97]. It is believed that the lifetime of $NO_2^{\bullet}$ in the cytosol is <10 μs [98]. As a strong oxidant, $NO_2^{\bullet}$ can mediate lipid peroxidation reactions. As a FR, $NO_2^{\bullet}$ is capable of reacting via hydrogen abstraction, addition to unsaturated bonds and other FRs, and mediating oxygen atom and electron transfers.

Similar to lipid peroxidation, a radical–radical reaction was proposed for the nitration of PUFAs, which involves initial hydrogen abstraction and subsequent $NO_2^{\bullet}$ addition to the PUFA radical [99]. The dissociation of the nitro-alkyl link can lead to isomerization of the alkene, which could have potential consequences for membrane structure and function.

Radicals with a low reduction potential like $NO_2^{\bullet}$ are more restricted, but they can react with the more oxidizable amino acids such as cysteine, methionine, tyrosine, and tryptophan. Indeed, one of the most relevant reactions of $NO_2^{\bullet}$ in a biological system is nitration of amino acids, which also occurs via a radical–radical reaction, leading, for example, to the formation of nitrotyrosine in peptides and proteins [100].

The reaction of $Tyr^{\bullet}$ with $NO_2^{\bullet}$ (Eq. 2.50) seems to be one of the major routes yielding nitrotyrosine in biological systems [101].

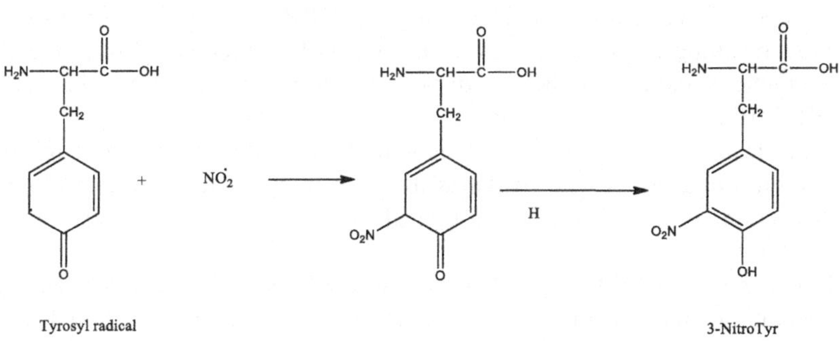

(2.50)

2.2.2 *Non-Radicals*

2.2.2.1 Peroxynitrite ($ONOO^-$)/ONOOH

Peroxynitrite is a highly potent oxidizing and nitrating agent. The standard reduction potentials of the couples $ONOO^-/NO_2^{\bullet}$ (one-electron reduction) and $ONOO^-/NO_2^-$ (two-electron reduction) are 1.6 V and 1.3 V, respectively [102].

Peroxynitrite formation is an inevitable consequence of the simultaneous stimulation of $O_2^{\bullet-}$ synthesis and nitric oxide production during inflammatory processes (Eq. 2.51) [103].

$$NO^{\bullet} + O_2^{\bullet-} \rightarrow ONOO^- \quad (2.51)$$

Peroxynitrite is more reactive than either of its congenitors, but it may not be particularly toxic in normal situations. It has a shorter half-life (10–20 ms at 37 °C) than does $NO^{\bullet}$ but has a sufficient lifetime to be of significance in vivo [104]. It can traverse cell membranes through anion channels [105].

This agent can modulate cellular functions through oxidative protein modifications such as the nitration of amino acid residues and may function as an intracellular signal for the production of interleukin (IL)-8 [106]. Experimental studies have revealed that low concentrations of $ONOO^-$ can cause vasoconstriction, but higher concentrations cause vasodilatation [39]. When overgenerated, however, $ONOO^-$ has numerous deleterious effects, ranging from considerable oxidation and nitration of biomolecules to the disturbance of signaling processes and the induction of cell death [107].

The specificity and selectivity of $ONOO^-$ are attributed to its limited reactivity with some biomolecules. Indeed, $ONOO^-$ specifically targets electron-rich moieties like thiols, iron–sulfur centers, and zinc thiolates, thus affecting cell cycle and signal transduction proteins. Peroxynitrite is known to directly target cysteine and tyrosine residues and initiate one- or two-electron transfers that lead to biological signals [108].

Several reports have demonstrated the ability of $ONOO^-$ to modify proteins with thiol groups, resulting in the generation of nitrosothiols, which can disrupt metal–protein interactions and lead to the loss of protein activity and the release of metal-derived FRs [107].

Peroxynitrite also mediates the formation of thiyl radicals from sulfhydryl groups, particularly on amino acids, and mediates nitration of tyrosine residues. Another possible reaction of $ONOO^-$ in biological systems is the degradation of hyaluronan [109].

Higher generation of $ONOO^-$ further leads to lipid peroxidation, DNA damage and fragmentation, and activation of poly(ADP-ribose) polymerase (PARP)-1, ultimately leading to caspase-independent apoptosis [110–113]. Peroxynitrite is implicated in a range of pathologies such as atherosclerosis, ischemic stroke, asthma, and inflammatory bowel disease (IBD).

As mentioned before, $ONOO^-$ is sufficiently reactive to directly oxidize biomolecules, but, at pathophysiological concentrations, the reaction is a combination of direct and radical-mediated oxidation. In this context, one of most favored reactions of $ONOO^-$ is with carbon dioxide (CO_2) to form the transient intermediate nitrosoperoxycarbonate and, consequently, carbonate and nitrogen dioxide radicals [114, 115].

The high bioavailability of the co-reactant CO_2 increases the contribution of this pathway, as outlined by reactions 2.52 and 2.53, and thereby of the reactions of $NO_2^{\bullet}$ and $CO_3^{\bullet -}$, in mediating several biological effects of $ONOO^-$ and $NO^{\bullet}$. It is noteworthy to know that $NO_2^{\bullet}$ is a radical of less activity than is $ONOO^-$ but of longer diffusion distance. Based on these reactions, it seems that the increased generation of $ONOO^-$ and $NO^{\bullet}$ that accompanies many types of injuries can be mediated by other RS.

$$ONOO^- + CO_2 \rightarrow ONOOCOO^- \tag{2.52}$$

$$ONOOCOO^- \rightarrow NO_2^{\bullet} + CO_3^{\bullet -} \tag{2.53}$$

However, if CO_2 is limiting, then the detrimental effects of $ONOO^-$ via other RS are still possible by forming a protonated form, named peroxynitrous acid (ONOOH, pKa 6.8), which undergoes homolysis and produces other damaging radicals such as $OH^{\bullet}$ and $NO_2^{\bullet}$ (Eq. 2.54), or it can rearrange to NO_3^- [116].

$$HONOO \rightarrow OH^{\bullet} + NO_2^{\bullet} \tag{2.54}$$

The production of $OH^{\bullet}$ without the involvement of transition metal ions is unique (Fig. 1.1). The protonated form of $ONOO^-$ is a strong oxidizing agent that induces depletion of sulfhydryl groups and oxidative damage on most biomolecules. It is responsible for DNA damage, causing strand breaks, protein oxidation, and nitration of amino acid residues in the protein structure (e.g., 3-nitrotyrosine) [117].

Under pathophysiological conditions, $ONOO^-$ can react with other components present in high concentrations, such as H_2O_2, to form adducts that might be responsible for the many deleterious effects seen in biological entities.

2.2.2.2 Nitrite (NO_2^-)/HNO_2 and Nitrate (NO_3^-)

Nitrite is an RS that is introduced into the human body from polluted water or is formed inside the body via the oxidation of $NO^{\bullet}$ (Eqs. 2.55 and 2.56), the reduction of NO_3^-, or the hydrolysis of N_2O_3/N_2O_4 (Eqs. 2.57 and 2.58).

$$NO_2^{\bullet} + NO^{\bullet} \rightarrow N_2O_3 \tag{2.55}$$

$$N_2O_3 + H_2O \rightarrow 2NO_2^- + 2H^+ \tag{2.56}$$

$$2NO_2^{\bullet} \rightarrow N_2O_4 \tag{2.57}$$

$$N_2O_4 + H_2O \rightarrow HNO_3 + HNO_2 \tag{2.58}$$

The nonenzymatic reaction of $NO^{\bullet}$ with O_2 in tissues is quite slow, whereas the oxidation of $NO^{\bullet}$ by the multi-copper oxidase ceruloplasmin in plasma is rapid.

Nitrates are produced by the conversion of $ONOO^-$, the oxidation of NO_2^- (Eq. 2.59), or the hydrolysis of $NO_2^{\bullet}/N_2O_4$ (Eqs. 2.60 and 2.61).

$$3\ HNO_2 \rightarrow HNO_3 + 2\ NO + H_2O \tag{2.59}$$

$$3\ NO_2^{\bullet} + H_2O \rightarrow 2NO_3^- + 2H^+ + NO^{\bullet} \tag{2.60}$$

$$N_2O_4 + H_2O \rightarrow HNO_3 + \ HNO_2 \tag{2.61}$$

Otherwise, NO_3^- is taken up with food and water and absorbed in the gut, part of which intestinal bacteria are able to reduce to NO_2^- [118].

Nitrite is possibly reduced back to $NO^{\bullet}$ under the activity of several enzymatic and nonenzymatic pathways in blood and tissues. The acidic reduction of NO_2^- produces not only $NO^{\bullet}$ but also other nitrogen oxides, with nitrosating (N_2O_3) and nitrating ($NO_2^{\bullet}$) abilities.

Another possibility is the oxidation of NO_2^- that occurs in red blood cell and results in the formation of NO_3^- and methemoglobin (Met-Hb) [119]. In typical cases of poisoning and inflammatory conditions, higher values of NO_2^- can be found in respiratory tract fluids, saliva, gastric juice, and serum.

As mentioned before, NO_2^- is taken up by red blood cells presumably via ionic channels and oxidized by oxyhemoglobin to NO_3^- along with methemoglobin formation. Water poisoning by NO_2^- or nitrite-related compounds may lead to enhanced methemoglobin levels [120]. In parallel, the deoxygenated form of hemoglobin reduces NO_2^- to $NO^{\bullet}$, thus causing vasodilation when the O_2 level is falling [121].

The formation of toxic nitrosamines results from the reaction of NO_2^- with secondary amines. This reaction is promoted under strong acidic conditions in the stomach where nitrous acid (HNO_2) is the main reactant [122].

Nitrosamines can be formed by a reaction between nitrates or nitrites and certain amines. They are considered to be strong carcinogens that may produce cancer in the brain, liver, lungs, kidneys, bladder, stomach, esophagus, and nasal sinus [123].

2.2.2.3 Dinitrogen Trioxide (N_2O_3)

Dinitrogen trioxide results from several sources: acidic nitrite, the reaction of $NO^{\bullet}$ with O_2, or from the $NO^{\bullet}/NO_2^{\bullet}$ reaction (Eq. 2.62).

$$NO_2^{\bullet} + NO^{\bullet} \rightarrow N_2O_3 \tag{2.62}$$

This RNS is a relatively mild oxidant with an oxidation potential of 0.7 V.

The primary mode of chemistry for N_2O_3 is to nitrosate amines to form nitrosamines [124]. It was proposed that N_2O_3 undergoes nitrosation reaction with exocyclic amine groups according to the following reaction (Eqs. 2.63 and 2.64):

$$NH_2R + N_2O_3 \rightarrow RNHNO + HNO_2 \tag{2.63}$$

This reaction seems to be followed by rapid deamination, resulting in a hydroxy group:

$$RNHNO \rightarrow RNNOH \rightarrow ROH + N_2 \tag{2.64}$$

Of particular interest is the N_2O_3-mediated deamination of nucleobases [125]. This effect would result in the conversion of guanine to xanthine, cytosine to uracil, methylcytosine to thymine, and adenine to hypoxanthine.

The ability of N_2O_3 to generate endogenous nitrosamines, which are powerful DNA-alkylating agents, accounts for its potential mutagenicity. Aside from the radical–radical reaction of thiyl radical with $NO^{\bullet}$, another pathway for S-nitrosylation and formation of S-nitrosothiols is the oxidative reaction of thiols that involves N_2O_3 (Eq. 2.65) [96]. This pathway may be important for the regulation of some metabolic processes and oxidative phosphorylation in the mitochondria. In blood circulation, low levels of S-nitrosothiols found in healthy individuals may act as a more stable pool of $NO^{\bullet}$ and represent a good indicator of balanced OS [126].

$$\underset{\text{Protein thiols}}{\text{Protein-SH}} + N_2O_3 \rightarrow \underset{\text{S-nitrothiols}}{\text{Protein-S-NO}} + NO_2^{-} + H^{+} \tag{2.65}$$

The N_2O_3 produced from activated macrophages was observed to undergo S-nitrosylation with GSH (Eq. 2.66).

$$N_2O_3 + GSH \rightarrow NO_2^{-} + GSNO \tag{2.66}$$

The hydrolysis of N_2O_3 (Eq. 2.67) is relatively slower than in S-nitrosylation, so the formation of S-nitrosothiol from N_2O_3 is a relevant mechanism under nitrosative stress conditions.

$$N_2O_3 + H_2O \rightarrow 2NO_2^{-} + 2H^{+} \tag{2.67}$$

Overall, it was estimated that 10% of N_2O_3 produced by activated cells are involved in extracellular nitrosylation [98]. Finally, other RNS can be mentioned such as nitryl chloride (NO_2Cl) and peroxynitrate (O_2NOO^{-}). However, their physiological relevance remains unclear.

2.3 Carbon Reactive Species (Carbon-Centered Radicals)

2.3.1 *Carbonate Radicals*

Carbonate radicals are underrated players in OS biochemistry. These radicals are produced during the reaction of $ONOO^-$, or peroxynitrous acid with CO_2 that is prevalent in physiological fluids, and possibly from the peroxidase activity of metalloenzymes like Cu/Zn SOD [127].

Nitrosoperoxycarbonate ($ONOOCO_2^-$; also referred to as nitrosoperoxocarboxylate) is readily formed as a reactive intermediate from the reaction of $ONOO^-$ with CO_2, which then rapidly decomposes to form a pair of caged radicals [$CO_3^{\bullet -}$/$NO_2^{\bullet}$] [107]. The abundance of the co-reactant CO_2 increases the contributions of this pathway, as outlined by Eqs. 2.68 and 2.69, and thereby of the effects of $NO_2^{\bullet}$ and $CO_3^{\bullet -}$, in mediating any deleterious action following nitric oxide overgeneration [128].

$$ONOO^- + CO_2 \rightarrow ONOOCOO^- \tag{2.68}$$

$$ONOOCOO^- \rightarrow NO_2^{\bullet} + CO_3^{\bullet -} \tag{2.69}$$

A similar reaction series for $ONOO^-$ occurs with CO_2, where a nitrosoperoxocarbonate intermediate can further react with $NO^{\bullet}$ to ultimately produce N_2O_3 and other RNS (Eqs. 2.70 and 2.71) [116].

$$ONOO^- + CO_2 \rightarrow ONOOCOO^- \tag{2.70}$$

$$ONOOCOO^- + NO^{\bullet} \rightarrow NO_2^{\bullet} + NO_2^- + CO_2 \tag{2.71}$$

Another possible source of the carbonate radical is the extremely slow oxidation of bicarbonate by $OH^{\bullet}$ (Eq. 2.72).

$$OH^{\bullet} + HCO_3^- \rightarrow H_2O + CO_3^{\bullet -} \tag{2.72}$$

The physiological importance of the carbonate radical is uncertain. However, this species diffuses much easier than does the much more reactive $OH^{\bullet}$.

The reduction potential of the couple $CO_3^{\bullet -}/CO_3^{-2}$ has been determined to be 1.78 V and seems to be negatively correlated with the pH [129]. Nevertheless, the carbonate radical presents low reactivity toward most extracellular components. Of particular importance is the reactivity exhibited by $CO_3^{\bullet -}$ toward small- and medium-sized biomolecules. The reactivity is the highest for indole heterocyclic compounds, moderate for sulfur-containing compounds, and low for aliphatic amino acids [65].

Hydrogen peroxide also reacts with bicarbonate to form the peroxymonocarbonate ion (HCO_4^-) (Eq. 2.73) [130], which is several times more reactive than free H_2O_2 [131]. This reaction could be of high importance as it reveals a biological mechanism for enhancing H_2O_2 reactivity.

$$H_2O_2 + HCO_3^{\cdot} \rightarrow HCO_4^{\cdot} + H_2O \quad (2.73)$$

2.3.2 *Semiquinone Radicals*

Semiquinone radicals can have extremely long half-lives (up to days at 37 °C) and tend to be neither reactive nor toxic [132]. Problems can arise because of their ability to act as reductants and donate their excess electron to molecular oxygen, thereby generating $O_2^{\bullet-}$ and creating redox cycling. Indeed, if the parent quinone is again reduced by an endogenous enzyme, then a redox-cycling process is accomplished and generates substantial quantities of radicals while depleting cellular supplies of reducing equivalents such as pyridine nucleotides that prevent cytotoxicity (Fig. 2.1) [133].

2.3.3 *Organic Radicals (R*$^{\bullet}$*)*

In biological systems, carbon-centered radicals usually arise following hydrogen abstraction from unsaturated bonds in fatty acids. Most of them are short-lived because they react extremely quickly with molecular oxygen to generate $ROO^{\bullet}$, which enhances chain reactions during the lipid peroxidation process (see Sect. 3.4) [134].

Organic radicals also react rapidly with $O_2^{\bullet-}$. This reaction can occur either by electron transfer to repair the parent compound or by addition (Eq. 2.74) [26].

$$O_2^{\bullet} + R^{\bullet} + H^+ \rightarrow O_2 + RH \quad (2.74)$$

One of the most important organic radicals is the tyrosine-oxidized product, which is produced from the reaction of tyrosine with $NO_2^{\bullet}$ (see Sect. 3.3).

2.4 Reactive Halide Species (RHS)

Reactive halide species (RHS) are a group of small molecules containing a halide atom (Cl, Br, or I), which is not necessarily reactive. In this group of compounds, oxygen is most often reactive, but, for a better understanding, they are

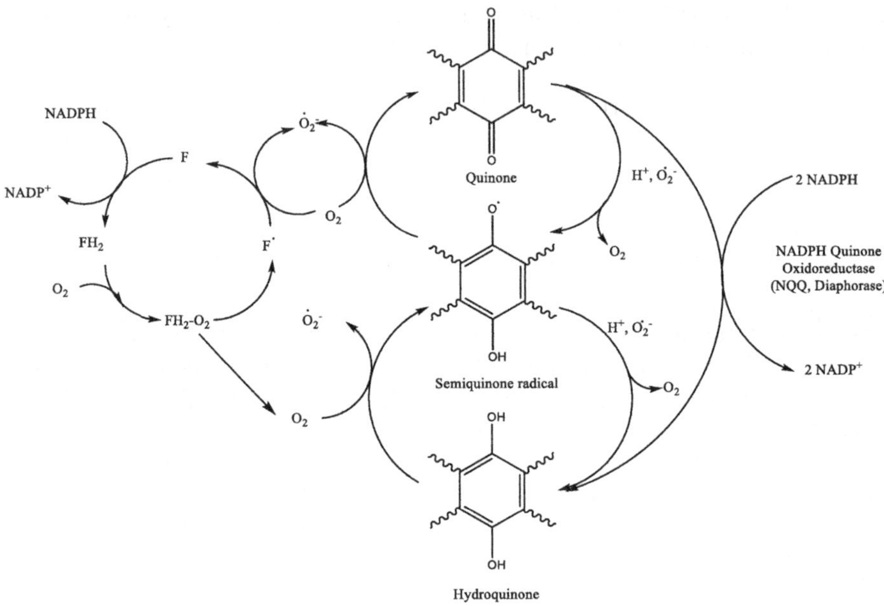

Fig. 2.1 Redox cycling of quinone compounds
Quinone moieties can accept a single electron from endogenous flavin nucleotides, producing semiquinone. The semiquinone can either be further reduced to generate hydroquinone or be reoxidized by molecular oxygen, which is thereby reduced to $O_2^{\bullet-}$. Cycling between semiquinone and hydroquinone is also possible but is usually not energetically favorable. Nicotinamide adenine dinucleotide phosphate (NADPH) quinone oxidoreductase can perform a two-electron reduction of quinones to hydroquinone and therefore prevent redox cycling by quinones

conventionally set as an independent family. The most studied forms in this family are hypochlorous acid (HOCl), hypobromous acid (HOBr), and hypoiodous acid (HOI).

As a general rule, hypohalous acids are generated from their corresponding halides by several peroxidases, particularly MPO, eosinophil peroxidase (EPO), and lactoperoxidase (LPO). They vary greatly in their oxidizing ability according to the following order: hypochlorous > hypobromous > hypoiodous (E° HOCl/H_2O = 1.28 V, E° HOBr/H_2O = 1.13 V, and E° HOI/H_2O = 0.78 V, respectively) [135–137].

2.4.1 *Hypochlorous Acid (HOCl)*

Hypochlorous acid is a major active component of chlorine-based bleaches that are often used to disinfect equipment and domestic areas. In biological systems, HOCl is highly oxidative and plays an important role in killing pathogens in the airways [138].

At inflammatory sites where PMNs are abundant and activated, H_2O_2 is converted in the presence of chloride ions into HOCl. This reaction takes place in PMNs, where it is catalyzed by the well-known enzyme MPO, generally considered a PMN-specific marker (Eq. 2.75) [139]:

$$H_2O_2 + Cl^- \overset{MPO}{\rightarrow} OH^- + HOCl \tag{2.75}$$

Hypochlorous acid is produced in lysosomes. However, there is a possibility that MPO is released by activated neutrophils and catalyzes the same reaction of H_2O_2 with Cl^-. These ions play an important role in inflammation, carcinogenesis, and allergic reactions.

Hypochlorous acid (or its conjugated base hypochlorite, depending on the pH) is endowed with high oxidative abilities and can generate further oxidative products such as chlorinating agents, ROS, RNS, and $Tyr^•$. Numerous reactions of HOCl with different substrates are reported at neutral pH values [140]. Hypochlorous acid is a powerful oxidant and a potent chlorinating agent toward amino acids. The preferred substrates of HOCl are thiols and primary amines on amino acid residues, which become what is called chloramines [141]. Chloramines can further react with another HOCl molecule to form dichloramines ($RNCl_2$). These species are more reactive but less stable than the monochlorinated species [142]. Furthermore, HOCl reacts efficiently with methionine and cysteine residues of proteins. Other targets on proteins are tyrosine, histidine, tryptophan, and lysine.

Of particular interest is the chloractive attack on tyrosine, forming 3-chlorotyrosine, one of the promising markers of chloractive stress [143].

All the abovementioned chloractive attacks can damage the key proteins involved in the inflammatory reaction. For example, the inactivation of the proteolytic inhibitor of elastase, α-1-antiproteinase, by HOCl might greatly potentiate tissue damage during the activation of neutrophils [144].

Hypochlorous acid oxidizes numerous other substrates and undergoes chlorination reactions, for example, with unsaturated lipids to form chlorohydrins and with nucleotides to form chlorinated bases [145, 146]. It also induces oxidative chlorination of proteoglycans and other membrane or intracellular compounds [147].

Hypochlorous acid can damage DNA units in the following ways:

- By induction of DNA–protein interactions.
- By production of pyrimidine oxidation products.
- By addition of chloride to DNA bases.

Finally, HOCl can attack highly hydrophobic macromolecules by chlorinating actions such as cholesterol. It has been suggested that the formation of cholesterol chlorohydrins, possibly during inflammation, could disrupt cell membranes and lead to cell lysis and death [148]. Based on this observation, cholesterol chlorohydrins could represent relevant biomarkers of chloractive stress associated with neutrophil activation.

2.4.2 Hypobromous Acid (HOBr)

Granulocytic enzymes in eosinophils further expand the reactivity of H_2O_2 via EPO by generating another RHS: hypobromous acid (HOBr). This RHS is solely produced in lysosomes in the presence of bromide ion (Br^-) [136] and plays an important role in inflammation and allergic reactions (Eq. 2.76).

$$H_2O_2 + Br^- \xrightarrow{EPO} OH^- + HOBr \tag{2.76}$$

Similar to HOCl, HOBr also reacts with a wide variety of compounds. In addition, HOBr undergoes comparable reactions to HOCl, but the difference in reaction rate and efficiency between chlorination and bromination is not always as great as for HOCl [149].

HOBr is a two-electron oxidant that is less reactive than its parents and more selective for sulfur centers. It reacts with amine groups to form bromamines [150].

For most protein residues, HOBr reacts faster than does HOCl. Cysteine and methionine residues are also preferably oxidized by HOBr but with less efficiency than by HOCl. Ring bromination of tyrosine occurs in a similar way than described with HOCl [151].

In bronchoalveolar lavage fluid from patients with asthma, the level of brominated tyrosine residues increases 25-fold after allergen challenge as compared to the controls. Additionally, the eosinophil count raises by more than 50-fold in these patients after 48 h of allergen challenge [152].

2.4.3 Hypoiodous Acid (HOI)

Iodide (I^-) oxidation by heme peroxidases (e.g., LPO) using probably H_2O_2 is slower compared to chloride and leads to hypoiodous acid (HOI), iodine I_2, I_3^-, and some other reactive iodine species [153, 154]. Hypoiodous acid (HOI) has a much lower redox potential than does HOCl (0.78 V vs. 1.28 V) [155]. Some studies have suggested that iodide, through two-electron oxidation, may result in the formation of iodinium (I^+) [156].

There are numerous interconversions between reactive iodine species depending on the pH and iodide concentration. Because of the low abundance of iodide in biological fluids (about 0.1 mM), iodide oxidation by enzymes slowly happens under in vivo conditions. Most of iodide accumulates in thyroid glands where, as part of thyroid hormone synthesis, it is oxidized and incorporated into tyrosine residues of thyroglobulin by thyroid peroxidase. Hydrogen peroxide acts as an electron acceptor at each step of thyroid hormone synthesis, namely, at iodide oxidation and its organification as well as at the coupling reaction of iodotyrosines [157].

Apart from H_2O_2 and the mentioned iodide compounds, numerous other RS are formed in the thyroid. Of particular interest are Tyr$^•$, diiodotyrosyl residue radical (DITyr$^•$), diiodotyrosyl residue radical in thyroglobulin (Tg-DITyr$^•$), and iodine radical (I$^•$) [158].

2.5 Reactive Pseudohalide Species: Hypothiocyanous Acid (HOSCN)

Hypothiocyanous acid (HOSCN) is produced in biological systems from the pseudohalide thiocyanate (SCN^-) via the action of peroxidase enzymes. Thiocyanate is the main substrate for LPO and the salivary peroxidase (SPO) as well as one of the preferred substrates for MPO and EPO. It is also readily oxidized by gastric peroxidase (GPO) and thyroid peroxidase (TPO) [159–161].

The formation of HOSCN occurs via the stepwise reduction of H_2O_2 by peroxidases in the presence of SCN^-. The reaction mechanism involves initial formation of thiocyanogen ($(SCN)_2$) (Eq. 2.77), which is rapidly hydrolyzed to HOSCN (Eq. 2.78):

$$2\ SCN^- + H_2O_2 \rightarrow (SCN)_2 + 2H_2O \tag{2.77}$$

$$(SCN)_2 + H_2O \rightarrow HOSCN + SCN^- + H^+ \tag{2.78}$$

Further oxidation products of SCN^- are debated, and among them is thiocarbamate-(S)-oxide [162, 163].

In general, SCN^- is the most readily oxidized (pseudo)halide by peroxidases, owing to the lower standard reduction potential for the two-electron $HOSCN/SCN^-$ redox couple of 0.56 V, compared to 0.78 V, 1.13 V, and 1.28 V for I^-, Br^- and Cl^-, respectively, at pH 7 [164].

As a weak oxidant, $OSCN^-$/HOSCN preferentially attacks intracellularly free and protein-bound sulfhydryls [165, 166]. This generates protein–RS-SCN derivatives (Eq. 2.79), which likely decompose to form sulfenic acid intermediates (Eq. 2.80). With small proteins, RS-SCN compounds typically react with a further thiol molecule to form a disulfide bond, resulting in the consumption of two thiol groups per molecule of HOSCN (Eq. 2.81).

$$RSH + HOSCN \rightarrow RSSCN + H_2O \tag{2.79}$$

$$RSSCN + H_2O \rightarrow RSOH + SCN^- + H^+ \tag{2.80}$$

$$RSSCN + R'SH \rightarrow RSSR' + SCN^- + H^+ \tag{2.81}$$

At high concentrations, these properties allow the penetration of HOSCN and specifically disturb the redox status in cells [167, 168].

Hypothiocyanous acid can undergo few other biological reactions. It can react with selenium-containing amino acid residues and Trp residues in various proteins, including albumin, trypsin inhibitor, lysozyme, and myoglobin (Eq. 2.82) [169]:

$$RNH_2 + HOSCN \rightarrow RNH\text{-}SCN + H_2O \quad (2.82)$$

Rapid hydrolysis of RN-SCN intermediates may occur at physiological pH.

Protein carbamylation possibly occurs in the reaction of OCN^- with amino, thiol, carboxyl, imidazole, and phenolic hydroxyl functional groups (Eq. 2.83):

$$RNH_2 + H^+ + OCN^- \rightarrow RNHC(=O)NH_2 \quad (2.83)$$

The mechanism of action of HOSCN in disease is not well understood, although a role of SCN-derived oxidants in cardiovascular disease (CVD) has been suggested by the detection of elevated levels of carbamylated proteins, probably by reactions mediated by cyanate (OCN^-) in atherosclerotic plaques. According to Wang et al., elevated levels of carbamylated proteins in plasma independently predict increased adverse cardiovascular events, including the development of coronary artery disease, potential myocardial infarction, stroke, and death [170].

2.6 Reactive Sulfur Species (RSS)

Sulfur-centered entities (e.g., R-SHs) are an integral part of biological systems and play significant roles in protein structure and function (e.g., cell signaling and enzymatic activities). For this reason, the oxidative modification of such entities has profound implications in cellular function and viability.

Sulfur-centered radicals can be formed from endogenous thiol compounds, such as GSH, and amino acids (methionine, cysteine, etc.). These thiols are easily oxidized to sulfur species with high oxidation states such as thiyl radicals ($RS^•$; one-electron oxidized state), sulfenic (RSOH; two-electron state), sulfinic (RSO_2H; three-electron state), and sulfonic acids (RSO_3H; four-electron state).

2.6.1 Sulfur-Centered Radicals

2.6.1.1 Thiyl Radicals ($RS^•$)

Thiols are favored targets for a wide range of radical species. The initial product resulting from thiol oxidation is the thiyl radical, which is formed from one-electron oxidation of thiols (RSH; with GSH being the simplest form (Eq. 2.84)), which can be catalyzed by a myriad of endogenous compounds such as high-valence transition metal ions, $OH^•$, $NO_2^•$, $ONOO^-$, and oxyhemoglobin.

$$R^{\bullet} + GSH \rightarrow RH + GS^{\bullet} \quad (2.84)$$

It can also be formed after a hemolytic cleavage of disulfide bonds in proteins.

Thiyl radicals are highly reactive with short half-lives in the microsecond range. For this reason, their direct detection in biological systems is challenging [171]. Their oxidizing power has been studied. The value of E°(GS$^{\bullet}$/GSH) was estimated to be 0.92 V [172].

Thiyl radicals have a more electrophilic character than their chemically analogous RO$^{\bullet}$ due to the lower electronegativity of the S-atom compared to oxygen. They are strong oxidants and can undergo a range of reactions, including abstraction of hydrogen from proteins and initiation of lipid peroxidation [173, 174].

Thiyl radicals have the ability to either combine with themselves or with other surrounding molecules to produce other radicals. Some reported reactions include one glutathionyl radical (GS$^{\bullet}$) combination to one glutathionyl anion (GS^{-}) to form the oxidized glutathione disulfide radical anion GSSG$^{\bullet-}$ (Eq. 2.85) [175].

$$GS^{\bullet} + GS^{-} \rightarrow GSSG^{\bullet-} \quad (2.85)$$

Another pathway is possible where the glutathionyl radical reacts with another glutathionyl radical, thus directly yielding glutathione disulfide (GSSG) (Eq. 2.86).

$$GS^{\bullet} + GS^{\bullet} \rightarrow GSSG \quad (2.86)$$

Although a thiyl radical may combine with another thiyl radical to generate a disulfide bond, hydrogen abstraction by a thiyl radical is more likely to happen unless two thiyl radicals are generated in close proximity and oxygen is limiting. Alternatively, if RS$^{\bullet}$ has limited access to other thiols, they can react with molecular oxygen to form peroxysulfenyl radicals (RSOO$^{\bullet}$) and a range of secondary products (Eq. 2.87) [176].

$$GS^{\bullet} + O_2 \rightarrow GSOO^{\bullet} \quad (2.87)$$

Other relevant RS$^{\bullet}$ interactions include that with NO$^{\bullet}$ generating the S-nitrosothiol form (Eq. 2.88).

$$NO^{\bullet} + RS^{\bullet} \rightarrow RS\text{-}NO \quad (2.88)$$

Overall, most glutathionyl radicals produced in biological systems appear to react with GSH, ultimately yielding GSSG. However, the vast majority of GSSG produced is under enzymatic catalysis via the reduction of hydroperoxides by GSH peroxidases (GPx) [177].

As mentioned before, one important property of $RS^{\bullet}$ is their ability to abstract hydrogen atoms, which is relevant for initiating peroxidation and isomerization of fatty acids, degradation of carbohydrates, catalysis of protein backbone fragmentation or epimerization, and catalytic formation of DNA building blocks.

Having this ability, cysteine thiyl radicals were shown to be involved in reversible intramolecular hydrogen transfer reactions with amino acid residues in proteins at a high rate [178]. Similar to thiyl radical-mediated oxidation of DNA sugar moieties, hydrogen abstraction by these species in selected carbohydrates such as 2-deoxy-D-ribose, 2-deoxy-D-glucose, α-D-glucose, and inositol yielded high reaction rates ($\approx 10^4\ M^{-1}\ s^{-1}$) [173].

2.6.1.2 Peroxylsulfenyl Radicals ($RSOO^{\bullet}$)

Peroxysulfenyl radicals, also termed "thiyl peroxyl radicals," are quickly formed upon the interaction of the quite reactive sulfur-centered thiyl radicals with molecular oxygen (Eq. 2.89) [179]. Peroxysulfenyl radicals can react with another thiol molecule, transferring an oxygen atom to form the sulfenic acid, sulfinyl radicals, or further undergo slow rearrangement to form sulfonyl radicals, known as nonreducible thiol species [180].

$$RS^{\bullet} + O_2 \rightarrow RSOO^{\bullet} \tag{2.89}$$

2.6.1.3 Disulfide Anion Radicals ($RSSR^{\bullet -}$)

The best example to illustrate the generation of disulfide anion radicals is the reaction of thiyl radicals of glutathione with the surrounding molecules. For instance, glutathione disulfide anion radical ($GSSG^{\bullet -}$) generation is more likely to occur by the reaction of $GS^{\bullet}$ with GS^{-} (Eq. 2.85).

$GSSG^{\bullet -}$ is a strong reducing radical and can react with molecular oxygen to yield $O_2^{\bullet -}$ and GSSG as the final products according to Eq. 2.90 [181]. Physiologically, this is a major decay route for thiyl radicals.

$$GSSG^{\bullet -} + O_2 \rightarrow GSSG + O_2^{\bullet -} \tag{2.90}$$

In theory, although GSSG is chemically generated from two $GS^{\bullet}$ radicals, it is more likely that is generated in two steps from $GS^{\bullet}$ and GS^{-} to form $GSSG^{\bullet -}$, which may further react to reduce oxygen to generate $O_2^{\bullet -}$ and GSSG. However, the vast majority of GSSG seems to be endogenously produced from the enzymatic pathways involving the GPx.

2.6.2 *Non-Radicals*

2.6.2.1 Sulfenic Acid (RSOH)

Sulfenic acids are basically a heteroatomic version of the analogous ROOH and hydropersulfide (RSSH) compounds. The production of RSOH arises from the two-electron oxidation of thiol moieties by ROOH, H_2O_2, $ONOO^-$, HOCl, or chloramines (Fig. 2.2).

Two-electron oxidation of GSH by H_2O_2 can generate GSOH, which readily transforms to GSSG in the presence of GSH (Eqs. 2.91 and 2.92)

$$GSH + H_2O_2 \rightarrow GSOH + H_2O \tag{2.91}$$

$$GSOH + GSH \rightarrow GSSG + H_2O \tag{2.92}$$

Sulfenic acids are short-lived species due to their ability to undergo combination with another thiol to form disulfide compounds and their quick transformation to sulfenyl radicals. The reduction potential for the $RSO^{\bullet}/RSO^-$ couple was estimated to be 0.74 V, which suggests that $RSO^{\bullet}$ analogs have less oxidative ability than do the $ROO^{\bullet}$ ones (reduction potential of $ROO^{\bullet}$ around 1.05 V versus normal hydrogen electrode (NHE)).

In the presence of RSOH, disulfide formation is likely to occur through unusual non-radical reactions such as the nucleophilic attack of a thiolate at a sulfenic acid moiety [182].

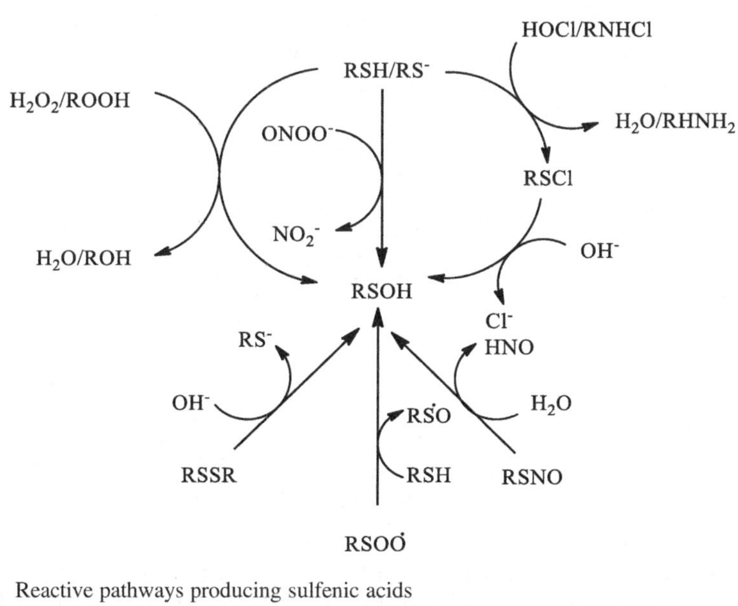

Fig. 2.2 Reactive pathways producing sulfenic acids
Several compounds are the precursors of sulfenic acids, including thiols, nitrothiols, peroxylsulfenyl radicals, and disulfides in the presence of $ONOO^-$, HOCl, and hydroperoxides

2.6.2.2 Hydropersulfides (RSSH) and Polysulfides (RSSnH)

Hydropersulfides and polysulfides have recently received considerable attention [183, 184]. Persulfides (RSSH/RSS$^-$) are sulfane sulfur compounds that are generated from the corresponding thiol through an oxidation reaction (the sulfur in RSSH is of the same oxidation state as that in a disulfide, RSSR). This process is mediated by several oxidants such as H_2O_2 [185].

Several reactions can give rise to persulfides in biological systems (Eqs. 2.93, 2.94, 2.95, 2.96, 2.97 and 2.98). Some are dependent on H_2S, whereas some are not [186]:

$$SH^{\bullet} + RS^{\bullet} \rightarrow RSSH \tag{2.93}$$

$$SH^{\bullet} + RS^{-} \rightarrow RSSH^{\bullet -} \tag{2.94}$$

$$SH^{-} + RS^{\bullet} \rightarrow RSSH^{\bullet -} \tag{2.95}$$

$$RSSH^{\bullet -} + O_2 \rightarrow RSSH + O_2^{\bullet -} \tag{2.96}$$

$$SH^{-} + RSSR' \rightarrow RSSH + R'S^{-} \tag{2.97}$$

$$SH^{-} + RSOH \rightarrow RSSH + OH^{-} \tag{2.98}$$

Interestingly, RSSH is a superior reductant compared with the corresponding thiol. Persulfides are highly reactive and unstable due to their nucleophilic and electrophilic character that facilitates the reactions between them, resulting in a spontaneous decomposition through a disproportionation process. Different products can be formed depending on which sulfur is affected. The interaction with the outer sulfur generates thiol and hydropolysulfides that evolve to elemental sulfur (Eqs. 2.99 and 2.100), whereas the interaction with the inner sulfur yields polysulfides and H_2S (Eq. 2.101):

$$RSSH + R'SS^{-} \rightarrow RSSS^{-} + R'SH \tag{2.99}$$

$$RSSSH \rightarrow RSH + S_n \tag{2.100}$$

$$RSSH + R'SS^{-} \rightarrow RSSSR' + SH^{-} \tag{2.101}$$

References

1. Pospíšil P, Prasad A. Formation of singlet oxygen and protection against its oxidative damage in Photosystem II under abiotic stress. Journal of Photochemistry and Photobiology B: Biology. 2014;137:39–48.
2. Baier J, Maisch T, Maier M, Landthaler M, Bäumler W. Direct Detection of Singlet Oxygen Generated by UVA Irradiation in Human Cells and Skin. Journal of Investigative Dermatology. 2007;127(6):1498–506.

3. Kehrer JP, Klotz LO. Free radicals and related reactive species as mediators of tissue injury and disease: implications for Health. Critical reviews in toxicology. 2015;45(9):765–98.
4. Boelsterli UJMTTmbohcdbt. Xenobiotic-induced oxidative stress: cell injury, signaling and gene regulation. 2007;117:175.
5. Steinbeck M, Khan AU, Karnovsky M. Extracellular production of singlet oxygen by stimulated macrophages quantified using 9,10-diphenylanthracene and perylene in a polystyrene film. The Journal of biological chemistry. 1993;268:15649–54.
6. Bansal M, Kaushal N. Introduction to Oxidative Stress. In: Bansal M, Kaushal N, editors. Oxidative Stress Mechanisms and their Modulation. New Delhi: Springer India; 2014. p. 1–18.
7. Di Mascio P, Briviba K, Sasaki ST, Catalani LH, Medeiros MH, Bechara EJ, et al. The reaction of peroxynitrite with tert-butyl hydroperoxide produces singlet molecular oxygen. Biological chemistry. 1997;378(9):1071–4.
8. Glaeser SP, Berghoff BA, Stratmann V, Grossart H-P, Glaeser J. Contrasting Effects of Singlet Oxygen and Hydrogen Peroxide on Bacterial Community Composition in a Humic Lake. PLOS ONE. 2014;9(3):e92518.
9. Agnez-Lima LF, Melo JTA, Silva AE, Oliveira AHS, Timoteo ARS, Lima-Bessa KM, et al. DNA damage by singlet oxygen and cellular protective mechanisms. Mutation Research/Reviews in Mutation Research. 2012;751(1):15–28.
10. Allamsetty S, Mohapatro S. Prediction of NO and NO2 Concentrations in Ozone Injected Diesel Exhaust after NTP Treatment Using Dimensional Analysis. Energy Procedia. 2019;158:4579–85.
11. Thomas J, Guénette J, Thomson EM. Stress axis variability is associated with differential ozone-induced lung inflammatory signaling and injury biomarker response. Environmental Research. 2018;167:751–8.
12. Moore J, Maier D, Ileleji K. Half-life time of ozone as a function of air movement and conditions in a sealed container. Journal of Stored Products Research. 2013;55:41–7.
13. Cataldo F. Ozone Degradation of Biological Macromolecules: Proteins, Hemoglobin, RNA, and DNA. Ozone: Science & Engineering. 2006;28(5):317–28.
14. Kosmider B, Loader J, Robert M, Mason R. Apoptosis induced by ozone and oxysterols in human alveolar epithelial cells. Free radical biology & medicine. 2010;48:1513–24.
15. Wisthaler A, Weschler CJ. Reactions of ozone with human skin lipids: Sources of carbonyls, dicarbonyls, and hydroxycarbonyls in indoor air. 2010;107(15):6568–75.
16. Cheng T-J, Kao H-P, Chan C-C, Chang WP. Effects of ozone on DNA single-strand breaks and 8-oxoguanine formation in A549 cells. Environmental Research. 2003;93(3):279–84.
17. Díaz-Llera S, González-Hernández Y, Prieto-González EA, Azoy A. Genotoxic effect of ozone in human peripheral blood leukocytes. Mutation Research/Genetic Toxicology and Environmental Mutagenesis. 2002;517(1):13–20.
18. Sullivan-Gunn MJ, Lewandowski PA. Elevated hydrogen peroxide and decreased catalase and glutathione peroxidase protection are associated with aging sarcopenia. BMC Geriatrics. 2013;13(1):104.
19. Cai H. NAD(P)H oxidase-dependent self-propagation of hydrogen peroxide and vascular disease. Circulation research. 2005;96(8):818–22.
20. Chen JW, Ni BB, Li B, Yang YH, Jiang SD, Jiang LS. The Responses of Autophagy and Apoptosis to Oxidative Stress in Nucleus Pulposus Cells: Implications for Disc Degeneration. Cellular Physiology and Biochemistry. 2014;34(4):1175–89.
21. Santiago E, Contreras C, Sacristán A-G, Sanchez A, Rivera L, Climent B, et al. Signaling pathways involved in the H2O2-induced vasoconstriction of rat coronary arteries. Free radical biology & medicine. 2013;60.
22. Bienert G, Schjoerring J, Jahn T. Membrane transport of hydrogen peroxide. Biochimica et biophysica acta. 2006;1758:994–1003.
23. Ouchi Y, Unoura K, Nabika H. Role of Oxidized Lipids in Permeation of H2O2 Through a Lipid Membrane: Molecular Mechanism of an Inhibitor to Promoter Switch. Scientific Reports. 2019;9(1):12497.

24. Huang MZ, Li JY. Physiological regulation of reactive oxygen species in organisms based on their physicochemical properties. 2020;228(1):e13351.
25. Walter M, Schenkeveld WDC, Geroldinger G, Gille L, Reissner M, Kraemer SM. Identifying the reactive sites of hydrogen peroxide decomposition and hydroxyl radical formation on chrysotile asbestos surfaces. Particle and Fibre Toxicology. 2020;17(1):3.
26. Biological Chemistry of Reactive Oxygen Species. Encyclopedia of Radicals in Chemistry, Biology and Materials2012.
27. Winterbourn CC, Metodiewa D. Reactivity of biologically important thiol compounds with superoxide and hydrogen peroxide. Free Radic Biol Med. 1999;27(3-4):322–8.
28. Imlay JA. Cellular defenses against superoxide and hydrogen peroxide. Annual review of biochemistry. 2008;77:755–76.
29. Steinbrenner H, Sies H. Protection against reactive oxygen species by selenoproteins. Biochimica et Biophysica Acta (BBA) - General Subjects. 2009;1790(11):1478–85.
30. de Grey A. HO2·: The forgotten radical. DNA and cell biology. 2002;21:251–7.
31. Boveris A, Chance B. The mitochondrial generation of hydrogen peroxide. General properties and effect of hyperbaric oxygen. The Biochemical journal. 1973;134(3):707–16.
32. Chance B, Sies H, Boveris A. Hydroperoxide metabolism in mammalian organs. Physiological reviews. 1979;59(3):527–605.
33. Rastogi R, Geng X, Li F, Ding Y. NOX Activation by Subunit Interaction and Underlying Mechanisms in Disease. 2017;10(301).
34. Jalilov AS, Zhang C, Samuel ELG, Sikkema WKA, Wu G, Berka V, et al. Mechanistic Study of the Conversion of Superoxide to Oxygen and Hydrogen Peroxide in Carbon Nanoparticles. ACS Appl Mater Interfaces. 2016;8(24):15086–92.
35. Samoilova RI, Crofts AR, Dikanov SA. Reaction of superoxide radical with quinone molecules. J Phys Chem A. 2011;115(42):11589–93.
36. Durand G, Choteau F, Pucci B, Villamena FA. Reactivity of Superoxide Radical Anion and Hydroperoxyl Radical with α-Phenyl-N-tert-butylnitrone (PBN) Derivatives. The Journal of Physical Chemistry A. 2008;112(48):12498–509.
37. Münzel T, Daiber A, Ullrich V, Mülsch A. Vascular Consequences of Endothelial Nitric Oxide Synthase Uncoupling for the Activity and Expression of the Soluble Guanylyl Cyclase and the cGMP-Dependent Protein Kinase. 2005;25(8):1551–7.
38. Allen CL, Bayraktutan U. Oxidative stress and its role in the pathogenesis of ischaemic stroke. International journal of stroke: official journal of the International Stroke Society. 2009;4(6): 461–70.
39. Modak JM, McCullough LD. Chapter 49 - Oxidative and Nitrosative Stress. In: Caplan LR, Biller J, Leary MC, Lo EH, Thomas AJ, Yenari M, et al., editors. Primer on Cerebrovascular Diseases (Second Edition). San Diego: Academic Press; 2017. p. 240–6.
40. Madkour LH. Sources of Oxidants and Function of ROS Inside the Macromolecule Cells. In: Madkour LH, editor. Nanoparticles Induce Oxidative and Endoplasmic Reticulum Stresses: Antioxidant Therapeutic Defenses. Cham: Springer International Publishing; 2020. p. 99–138.
41. D'Autréaux B, Toledano MB. ROS as signalling molecules: mechanisms that generate specificity in ROS homeostasis. Nature reviews Molecular cell biology. 2007;8(10):813–24.
42. Oakley FD, Abbott D, Li Q, Engelhardt JF. Signaling components of redox active endosomes: the redoxosomes. Antioxid Redox Signal. 2009;11(6):1313–33.
43. de Lamirande E, Gagnon C. A positive role for the superoxide anion in triggering hyperactivation and capacitation of human spermatozoa. Int J Androl. 1993;16(1):21–5.
44. Hanson ES, Leibold EA. Regulation of the iron regulatory proteins by reactive nitrogen and oxygen species. Gene Expr. 1999;7(4-6):367–76.
45. Popović-Bijelić A, Mojović M, Stamenković S, Jovanović M, Selaković V, Andjus P, et al. Iron-sulfur cluster damage by the superoxide radical in neural tissues of the SOD1G93A ALS rat model. Free Radical Biology and Medicine. 2016;96:313–22.

46. Ndengele MM, Muscoli C, Wang ZQ, Doyle TM, Matuschak GM, Salvemini D. Superoxide potentiates NF-kappaB activation and modulates endotoxin-induced cytokine production in alveolar macrophages. Shock (Augusta, Ga). 2005;23(2):186–93.
47. Cuzzocrea S, Pisano B, Dugo L, Ianaro A, Ndengele M, Salvemini D. Superoxide-related signaling cascade mediates nuclear factor-kappaB activation in acute inflammation. Antioxidants & redox signaling. 2004;6(4):699–704.
48. Zhou D, Shao L, Spitz DR. Reactive oxygen species in normal and tumor stem cells. Adv Cancer Res. 2014;122:1–67.
49. Healy EF, Roth-Rodriguez A, Toledo S. A model for gain of function in superoxide dismutase. Biochemistry and Biophysics Reports. 2020;21:100728.
50. Winterbourn CC. Superoxide as an intracellular radical sink. Free Radic Biol Med. 1993;14 (1):85–90.
51. Lipinski B. Hydroxyl Radical and Its Scavengers in Health and Disease. Oxidative Medicine and Cellular Longevity. 2011;2011:809696.
52. De Palma C, Clementi E. Reactive Species and Mechanisms of Cell Injury. In: McManus LM, Mitchell RN, editors. Pathobiology of Human Disease. San Diego: Academic Press; 2014. p. 88–96.
53. Haber F, Weiss J, Pope WJ. The catalytic decomposition of hydrogen peroxide by iron salts. 1934;147(861):332–51.
54. Qu X, Song W, Shen Z. A Highly Selective NIR Fluorescent Turn-on Probe for Hydroxyl Radical and Its Application in Living Cell Images. Front Chem. 2019;7:598.
55. Ly A, Aguilera JA, Milligan JR. Kinetic behavior of the reaction between hydroxyl radical and the SV40 minichromosome. Radiat Phys Chem Oxf Engl 1993. 2007;76(6):982–7.
56. Hrycay EG, Bandiera SM. Chapter Two - Involvement of Cytochrome P450 in Reactive Oxygen Species Formation and Cancer. In: Hardwick JP, editor. Advances in Pharmacology. 74: Academic Press; 2015. p. 35–84.
57. Valachová K, Mach M, Šoltés L. Oxidative Degradation of High-Molar-Mass Hyaluronan: Effects of Some Indole Derivatives to Hyaluronan Decay. 2020;21(16):5609.
58. Wu Y, Mundy CJ, Colvin ME, Car R. On the Mechanisms of OH Radical Induced DNA-Base Damage: A Comparative Quantum Chemical and Car−Parrinello Molecular Dynamics Study. The Journal of Physical Chemistry A. 2004;108(15):2922–9.
59. Stadtman E, Moskovitz J, Levine R. Oxidation of Methionine Residues of Proteins: Biological Consequences. Antioxid Redox Signal. 2003;5:577–82.
60. Ferreira CA, Ni D, Rosenkrans ZT, Cai W. Scavenging of reactive oxygen and nitrogen species with nanomaterials. Nano Research. 2018;11(10):4955–84.
61. Masood N, Fatima K, Luqman S. A Modified Method for Studying Behavioral Paradox of Antioxidants and Their Disproportionate Competitive Kinetic Effect to Scavenge the Peroxyl Radical Formation. The Scientific World Journal. 2014;2014:931581.
62. Galano A, Muñoz-Rugeles L, Alvarez-Idaboy JR, Bao JL, Truhlar DG. Hydrogen Abstraction Reactions from Phenolic Compounds by Peroxyl Radicals: Multireference Character and Density Functional Theory Rate Constants. The Journal of Physical Chemistry A. 2016;120 (27):4634–42.
63. Sonntag Cv, Dowideit P, Fang X, Mertens R, Pan X, Schuchmann MN, et al. The fate of peroxyl radicals in aqueous solution. Water Science and Technology. 1997;35(4):9–15.
64. Khursan SL. Organic tetroxides and mechanism of peroxy radical recombination. PATAI'S Chemistry of Functional Groups2014. p. 1–34.
65. Huie RE, Neta P. Chemistry of Reactive Oxygen Species. In: Gilbert DL, Colton CA, editors. Reactive Oxygen Species in Biological Systems: An Interdisciplinary Approach. Boston, MA: Springer US; 2002. p. 33–73.
66. Tsang W. Heats of Formation of Organic Free Radicals by Kinetic Methods. In: Martinho Simões JA, Greenberg A, Liebman JF, editors. Energetics of Organic Free Radicals. Dordrecht: Springer Netherlands; 1996. p. 22–58.

67. Kehrer JP, Robertson JD, Smith CV. 1.14 - Free Radicals and Reactive Oxygen Species. In: McQueen CA, editor. Comprehensive Toxicology (Second Edition). Oxford: Elsevier; 2010. p. 277–307.
68. Bietti M, DiLabio GA, Lanzalunga O, Salamone M. Electron Transfer Properties of Alkoxyl Radicals. A Time-Resolved Kinetic Study of the Reactions of the tert-Butoxyl, Cumyloxyl, and Benzyloxyl Radicals with Alkyl Ferrocenes. The Journal of Organic Chemistry. 2010;75 (17):5875–81.
69. Apak R, Özyürek M, Güçlü K, Çapanoğlu E. Antioxidant Activity/Capacity Measurement. 1. Classification, Physicochemical Principles, Mechanisms, and Electron Transfer (ET)-Based Assays. Journal of Agricultural and Food Chemistry. 2016;64(5):997–1027.
70. Greene LE, Lincoln R, Cosa G. Rate of Lipid Peroxyl Radical Production during Cellular Homeostasis Unraveled via Fluorescence Imaging. Journal of the American Chemical Society. 2017;139(44):15801–11.
71. Garrec J, Monari A, Assfeld X, Mir LM, Tarek M. Lipid Peroxidation in Membranes: The Peroxyl Radical Does Not "Float". The Journal of Physical Chemistry Letters. 2014;5(10): 1653–8.
72. Antosiewicz J, Spodnik J, Teranishi M, Herman-Antosiewicz A, Kurono C, Soji T, et al. NADH-generating substrates reduce peroxyl radical toxicity in RL34 cells. Folia morphologica. 2009;68.
73. Thomas DD, Liu X, Kantrow SP, Lancaster JR, Jr. The biological lifetime of nitric oxide: implications for the perivascular dynamics of NO and O2. Proc Natl Acad Sci U S A. 2001;98 (1):355–60.
74. Hughes MN. Chemistry of nitric oxide and related species. Methods in enzymology. 2008;436:3–19.
75. Wiley J. The many faces of nitric oxide: Cytotoxic, cytoprotective or both. Neurogastroenterology and motility: the official journal of the European Gastrointestinal Motility Society. 2007;19:541–4.
76. Rapozzi V, Della Pietra E, Bonavida B. Dual roles of nitric oxide in the regulation of tumor cell response and resistance to photodynamic therapy. Redox Biol. 2015;6:311–7.
77. Costa D, Benincasa G, Lucchese R, Infante T, Nicoletti GF, Napoli C. Effect of nitric oxide reduction on arterial thrombosis. Scandinavian Cardiovascular Journal. 2019;53(1):1–8.
78. Tejero J, Shiva S, Gladwin MT. Sources of Vascular Nitric Oxide and Reactive Oxygen Species and Their Regulation. 2019;99(1):311–79.
79. Ahmad A, Dempsey SK, Daneva Z, Azam M, Li N, Li PL, et al. Role of Nitric Oxide in the Cardiovascular and Renal Systems. International journal of molecular sciences. 2018;19(9).
80. Mónica FZ, Bian K, Murad F. The Endothelium-Dependent Nitric Oxide-cGMP Pathway. Advances in pharmacology (San Diego, Calif). 2016;77:1–27.
81. Sandoo A, van Zanten JJ, Metsios GS, Carroll D, Kitas GD. The endothelium and its role in regulating vascular tone. The open cardiovascular medicine journal. 2010;4:302–12.
82. Zhang J, Jin B, Li L, Block ER, Patel JM. Nitric oxide-induced persistent inhibition and nitrosylation of active site cysteine residues of mitochondrial cytochrome-c oxidase in lung endothelial cells. American journal of physiology Cell physiology. 2005;288(4):C840–9.
83. Clancy RM, Gomez PF, Abramson SB. Nitric oxide sustains nuclear factor kappaB activation in cytokine-stimulated chondrocytes. Osteoarthritis and Cartilage. 2004;12(7):552–8.
84. Poderoso JJ, Helfenberger K, Poderoso C. The effect of nitric oxide on mitochondrial respiration. Nitric oxide: biology and chemistry. 2019;88:61–72.
85. Baydoun HH, Cherian MA, Green P, Ratner L. Inducible nitric oxide synthase mediates DNA double strand breaks in Human T-Cell Leukemia Virus Type 1-induced leukemia/lymphoma. Retrovirology. 2015;12(1):71.
86. Mikami Y, Kanemaru K, Okubo Y, Nakaune T, Suzuki J, Shibata K, et al. Nitric Oxide-induced Activation of the Type 1 Ryanodine Receptor Is Critical for Epileptic Seizure-induced Neuronal Cell Death. EBioMedicine. 2016;11:253–61.

87. Tsikas D, Dehnert S, Urban K, Surdacki A, Meyer HH. GC–MS analysis of S-nitrosothiols after conversion to S-nitroso-N-acetyl cysteine ethyl ester and in-injector nitrosation of ethyl acetate. Journal of Chromatography B. 2009;877(28):3442–55.
88. Pannu AK, Bhalla A, Vishnu RI, Dhibar DP, Sharma N, Vijayvergiya R. Organophosphate induced delayed neuropathy after an acute cholinergic crisis in self-poisoning. Clinical toxicology (Philadelphia, Pa). 2021;59(6):488–92.
89. Luo P, Li X, Wu X, Dai S, Yang Y, Xu H, et al. Preso regulates NMDA receptor-mediated excitotoxicity via modulating nitric oxide and calcium responses after traumatic brain injury. Cell Death & Disease. 2019;10(7):496.
90. Eiserich JP, Baldus S, Brennan ML, Ma W, Zhang C, Tousson A, et al. Myeloperoxidase, a leukocyte-derived vascular NO oxidase. Science. 2002;296(5577):2391–4.
91. Cline MR, Chavez TA, Toscano JP. Oxidation of N-hydroxy-l-arginine by hypochlorous acid to form nitroxyl (HNO). Journal of inorganic biochemistry. 2013;118:148–54.
92. Vidwans AS, Kim S, Coffin DO, Wink DA, Hewett SJ. Analysis of the neuroprotective effects of various nitric oxide donor compounds in murine mixed cortical cell culture. Journal of neurochemistry. 1999;72(5):1843–52.
93. Gaston BM, Carver J, Doctor A, Palmer LA. S-nitrosylation signaling in cell biology. Molecular interventions. 2003;3(5):253–63.
94. Abu-Soud HM, Hazen SL. Nitric oxide is a physiological substrate for mammalian peroxidases. J Biol Chem. 2000;275(48):37524–32.
95. Byun J, Mueller DM, Fabjan JS, Heinecke JW. Nitrogen dioxide radical generated by the myeloperoxidase-hydrogen peroxide-nitrite system promotes lipid peroxidation of low density lipoprotein. FEBS Letters. 1999;455(3):243–6.
96. Ali AA, Coulter JA, Ogle CH, Migaud MM, Hirst DG, Robson T, et al. The contribution of N_2O_3 to the cytotoxicity of the nitric oxide donor DETA/NO: an emerging role for S-nitrosylation. Biosci Rep. 2013;33(2):e00031.
97. Shafirovich V, Cadet J, Gasparutto D, Dourandin A, Geacintov NE. Nitrogen Dioxide as an Oxidizing Agent of 8-Oxo-7,8-dihydro-2'-deoxyguanosine but Not of 2'-Deoxyguanosine. Chemical Research in Toxicology. 2001;14(2):233–41.
98. Schiller E. Reactive Oxygen and Nitrogen Species. In: Schiller E, editor. Free Radicals and Inhalation Pathology: Respiratory System Mononuclear Phagocyte System Hypoxia and Reoxygenation Pneumoconioses and other Granulomatoses Cancer. Berlin, Heidelberg: Springer Berlin Heidelberg; 2004. p. 75–141.
99. Buchan GJ, Bonacci G, Fazzari M, Salvatore SR, Gelhaus Wendell S. Nitro-fatty acid formation and metabolism. Nitric oxide: biology and chemistry. 2018;79:38–44.
100. Radi R. Protein tyrosine nitration: biochemical mechanisms and structural basis of functional effects. Acc Chem Res. 2013;46(2):550–9.
101. Lepoivre M, Houée-Levin C, Coeytaux K, Decottignies P, Auger G, Lemaire G. Nitration of the tyrosyl radical in ribonucleotide reductase by nitrogen dioxide: A gamma radiolysis study. Free Radical Biology and Medicine. 2005;38(11):1511–7.
102. Möller MN, Rios N, Trujillo M, Radi R, Denicola A, Alvarez B. Detection and quantification of nitric oxide-derived oxidants in biological systems. The Journal of biological chemistry. 2019;294(40):14776–802.
103. Radi R. Oxygen radicals, nitric oxide, and peroxynitrite: Redox pathways in molecular medicine. 2018;115(23):5839–48.
104. Radi R. Peroxynitrite, a stealthy biological oxidant. The Journal of biological chemistry. 2013;288(37):26464–72.
105. Denicola A, Souza JM, Radi R. Diffusion of peroxynitrite across erythrocyte membranes. Proc Natl Acad Sci U S A. 1998;95(7):3566–71.

106. Khreiss T, József L, Potempa LA, Filep JG. Loss of Pentameric Symmetry in C-Reactive Protein Induces Interleukin-8 Secretion Through Peroxynitrite Signaling in Human Neutrophils. 2005;97(7):690–7.
107. Szabó C, Ischiropoulos H, Radi R. Peroxynitrite: biochemistry, pathophysiology and development of therapeutics. Nature Reviews Drug Discovery. 2007;6(8):662–80.
108. Chen W, Druhan LJ, Chen C-A, Hemann C, Chen Y-R, Berka V, et al. Peroxynitrite induces destruction of the tetrahydrobiopterin and heme in endothelial nitric oxide synthase: transition from reversible to irreversible enzyme inhibition. Biochemistry. 2010;49(14):3129–37.
109. Corsaro MM, Pietraforte D, Di Lorenzo AS, Minetti M, Marino G. Reaction of Peroxynitrite with Hyaluronan and Related Saccharides. Free radical research. 2004;38(4):343–53.
110. Radi R, Beckman JS, Bush KM, Freeman BA. Peroxynitrite-induced membrane lipid peroxidation: The cytotoxic potential of superoxide and nitric oxide. Archives of Biochemistry and Biophysics. 1991;288(2):481–7.
111. Islam BU, Habib S, Ahmad P, Allarakha S, Moinuddin, Ali A. Pathophysiological Role of Peroxynitrite Induced DNA Damage in Human Diseases: A Special Focus on Poly (ADP-ribose) Polymerase (PARP). Indian J Clin Biochem. 2015;30(4):368–85.
112. Szabó É, Virág L, Bakondi E, Gyüre L, Haskó G, Bai P, et al. Peroxynitrite Production, DNA Breakage, and Poly(ADP-ribose) Polymerase Activation in a Mouse Model of Oxazolone-Induced Contact Hypersensitivity. Journal of Investigative Dermatology. 2001;117(1):74–80.
113. Levrand S, Vannay-Bouchiche C, Pesse B, Pacher P, Feihl F, Waeber B, et al. Peroxynitrite is a major trigger of cardiomyocyte apoptosis in vitro and in vivo. Free Radic Biol Med. 2006;41 (6):886–95.
114. Pavlovic R, Santaniello E. Peroxynitrite and nitrosoperoxycarbonate, a tightly connected oxidizing-nitrating couple in the reactive nitrogen-oxygen species family: New perspectives for protection from radical-promoted injury by flavonoids. The Journal of pharmacy and pharmacology. 2008;59:1687–95.
115. Gunaydin H, Houk KN. Mechanisms of peroxynitrite-mediated nitration of tyrosine. Chemical research in toxicology. 2009;22(5):894–8.
116. Espey MG, Miranda KM, Thomas DD, Xavier S, Citrin D, Vitek MP, et al. A chemical perspective on the interplay between NO, reactive oxygen species, and reactive nitrogen oxide species. Annals of the New York Academy of Sciences. 2002;962:195–206.
117. Bartesaghi S, Radi R. Fundamentals on the biochemistry of peroxynitrite and protein tyrosine nitration. Redox Biol. 2018;14:618–25.
118. Tiso M, Schechter AN. Nitrate Reduction to Nitrite, Nitric Oxide and Ammonia by Gut Bacteria under Physiological Conditions. PLOS ONE. 2015;10(3):e0119712.
119. Das D, Sen S, Sen K. Caffeine and catechin towards prevention of drug induced oxidation of hemoglobin: A spectroscopic study. Spectrochimica Acta Part A: Molecular and Biomolecular Spectroscopy. 2020;232:118167.
120. Richard AM, Diaz JH, Kaye AD. Reexamining the risks of drinking-water nitrates on public health. Ochsner J. 2014;14(3):392–8.
121. Minneci PC, Deans KJ, Shiva S, Zhi H, Banks SM, Kern S, et al. Nitrite reductase activity of hemoglobin as a systemic nitric oxide generator mechanism to detoxify plasma hemoglobin produced during hemolysis. Am J Physiol Heart Circ Physiol. 2008;295(2):H743–H54.
122. Song P, Wu L, Guan W. Dietary Nitrates, Nitrites, and Nitrosamines Intake and the Risk of Gastric Cancer: A Meta-Analysis. Nutrients. 2015;7(12):9872–95.
123. Robles H. Nitrosamines. In: Wexler P, editor. Encyclopedia of Toxicology (Third Edition). Oxford: Academic Press; 2014. p. 584–5.
124. Breider F, Salihu I, Gunten U. Formation of N-nitrosamines by micelle-catalysed nitrosation of aliphatic secondary amines. Environmental Science: Processes and Impacts. 2018.
125. Dong M, Vongchampa V, Gingipalli L, Cloutier JF, Kow YW, O'Connor T, et al. Development of enzymatic probes of oxidative and nitrosative DNA damage caused by reactive nitrogen species. Mutation research. 2006;594(1-2):120–34.

126. Maron BA, Tang S-S, Loscalzo J. S-nitrosothiols and the S-nitrosoproteome of the cardiovascular system. Antioxid Redox Signal. 2013;18(3):270–87.
127. Goss SP, Singh RJ, Kalyanaraman B. Bicarbonate enhances the peroxidase activity of Cu,Zn-superoxide dismutase. Role of carbonate anion radical. J Biol Chem. 1999;274(40):28233–9.
128. Pryor WA, Houk KN, Foote CS, Fukuto JM, Ignarro LJ, Squadrito GL, et al. Free radical biology and medicine: it's a gas, man! 2006;291(3):R491–R511.
129. Ocampo R, Rivera-Utrilla J, Sánchez-Polo M, López-Peñalver J, Ramos R. Degradation of antineoplastic cytarabine in aqueous solution by gamma radiation. Chemical Engineering Journal. 2011;174:1–8.
130. Richardson DE, Yao H, Frank KM, Bennett DA. Equilibria, Kinetics, and Mechanism in the Bicarbonate Activation of Hydrogen Peroxide: Oxidation of Sulfides by Peroxymonocarbonate. Journal of the American Chemical Society. 2000;122(8):1729–39.
131. Medinas DB, Cerchiaro G, Trindade DF, Augusto O. The carbonate radical and related oxidants derived from bicarbonate buffer. 2007;59(4-5):255–62.
132. Vejerano E, Lomnicki SM, Dellinger B. Formation and stabilization of combustion-generated, environmentally persistent radicals on Ni(II)O supported on a silica surface. Environ Sci Technol. 2012;46(17):9406–11.
133. Circu ML, Maloney RE, Aw TY. Disruption of pyridine nucleotide redox status during oxidative challenge at normal and low-glucose states: implications for cellular adenosine triphosphate, mitochondrial respiratory activity, and reducing capacity in colon epithelial cells. Antioxid Redox Signal. 2011;14(11):2151–62.
134. Pratt DA, Tallman KA, Porter NA. Free radical oxidation of polyunsaturated lipids: New mechanistic insights and the development of peroxyl radical clocks. Acc Chem Res. 2011;44(6):458–67.
135. Squadrito GL, Postlethwait EM, Matalon S. Elucidating mechanisms of chlorine toxicity: reaction kinetics, thermodynamics, and physiological implications. Am J Physiol Lung Cell Mol Physiol. 2010;299(3):L289–L300.
136. Flemmig J, Zschaler J, Remmler J, Arnhold J. The fluorescein-derived dye aminophenyl fluorescein is a suitable tool to detect hypobromous acid (HOBr)-producing activity in eosinophils. J Biol Chem. 2012;287(33):27913–23.
137. Nicolussi A, Dunn JD, Mlynek G, Bellei M, Zamocky M, Battistuzzi G, et al. Secreted heme peroxidase from Dictyostelium discoideum: Insights into catalysis, structure, and biological role. J Biol Chem. 2018;293(4):1330–45.
138. Hakim H, Thammakarn C, Suguro A, Ishida Y, Kawamura A, Tamura M, et al. Evaluation of sprayed hypochlorous acid solutions for their virucidal activity against avian influenza virus through in vitro experiments. J Vet Med Sci. 2015;77(2):211–5.
139. Schürmann N, Forrer P, Casse O, Li J, Felmy B, Burgener A-V, et al. Myeloperoxidase targets oxidative host attacks to Salmonella and prevents collateral tissue damage. Nature Microbiology. 2017;2(4):16268.
140. Arnhold J. Chapter 2—Role of Reactive Species in Destructions. In: Arnhold J, editor. Cell and Tissue Destruction: Academic Press; 2020. p. 23–54.
141. Szabó M, Simon F, Fábián I. The formation of N-chloramines with proteinogenic amino acids. Water Research. 2019;165:114994.
142. Coker MS, Hu WP, Senthilmohan ST, Kettle AJ. Pathways for the decay of organic dichloramines and liberation of antimicrobial chloramine gases. Chem Res Toxicol. 2008;21(12):2334–43.
143. Kang JJI, Neidigh JW. Hypochlorous Acid Damages Histone Proteins Forming 3-Chlorotyrosine and 3,5-Dichlorotyrosine. Chemical Research in Toxicology. 2008;21(5):1028–38.

144. Whiteman M, Szabó C, Halliwell B. Modulation of peroxynitrite- and hypochlorous acid-induced inactivation of alpha1-antiproteinase by mercaptoethylguanidine. Br J Pharmacol. 1999;126(7):1646–52.
145. Schröter J, Schiller J. Chlorinated Phospholipids and Fatty Acids: (Patho)physiological Relevance, Potential Toxicity, and Analysis of Lipid Chlorohydrins. Oxidative medicine and cellular longevity. 2016;2016:8386362-.
146. Yudhistira T, Mulay SV, Kim Y, Halle MB, Churchill DG. Imaging of Hypochlorous Acid by Fluorescence and Applications in Biological Systems. 2019;14(18):3048–84.
147. Akeel A, Sibanda S, Martin SW, Paterson AWJ, Parsons BJ. Chlorination and oxidation of heparin and hyaluronan by hypochlorous acid and hypochlorite anions: effect of sulfate groups on reaction pathways and kinetics. Free Radical Biology and Medicine. 2013;56:72–88.
148. Carr AC, van den Berg JJM, Winterbourn CC. Chlorination of Cholesterol in Cell Membranes by Hypochlorous Acid. Archives of Biochemistry and Biophysics. 1996;332(1):63–9.
149. Pattison D, Davies M. Reactions of Myeloperoxidase-Derived Oxidants with Biological Substrates:Gaining Chemical Insight into Human Inflammatory Diseases. Current medicinal chemistry. 2006;13:3271–90.
150. Chapman AL, Skaff O, Senthilmohan R, Kettle AJ, Davies MJ. Hypobromous acid and bromamine production by neutrophils and modulation by superoxide. The Biochemical journal. 2009;417(3):773–81.
151. Bathish B, Paumann-Page M, Paton LN, Kettle AJ, Winterbourn CC. Peroxidasin mediates bromination of tyrosine residues in the extracellular matrix. 2020.
152. Wu W, Samoszuk MK, Comhair SA, Thomassen MJ, Farver CF, Dweik RA, et al. Eosinophils generate brominating oxidants in allergen-induced asthma. J Clin Invest. 2000;105(10):1455–63.
153. Davies MJ. Myeloperoxidase: Mechanisms, reactions and inhibition as a therapeutic strategy in inflammatory diseases. Pharmacology & Therapeutics. 2021;218:107685.
154. Bafort F, Parisi O, Perraudin JP, Jijakli MH. Mode of Action of Lactoperoxidase as Related to Its Antimicrobial Activity: A Review. Enzyme Research. 2014;2014:517164.
155. Gamon LF, Dieterich S, Ignasiak MT, Schrameyer V, Davies MJ. Iodide modulates protein damage induced by the inflammation-associated heme enzyme myeloperoxidase. Redox Biol. 2020;28:101331.
156. Kopp P, Solis-S JC. Chapter 3 - Thyroid Hormone Synthesis. In: Wondisford FE, Radovick S, editors. Clinical Management of Thyroid Disease. Philadelphia: W.B. Saunders; 2009. p. 19–41.
157. Karbownik M, Lewinski A. The role of oxidative stress in physiological and pathological processes in the thyroid gland; possible involvement in pineal-thyroid interactions. Neuro endocrinology letters. 2003;24(5):293–303.
158. Karbownik-Lewińska M, Kokoszko-Bilska A. Oxidative damage to macromolecules in the thyroid - experimental evidence. Thyroid Research. 2012;5(1):25.
159. van Dalen CJ, Whitehouse MW, Winterbourn CC, Kettle AJ. Thiocyanate and chloride as competing substrates for myeloperoxidase. The Biochemical journal. 1997;327(Pt 2):487–92.
160. Chandler JD, Day BJ. Biochemical mechanisms and therapeutic potential of pseudohalide thiocyanate in human health. Free radical research. 2015;49(6):695–710.
161. Flemmig J, Gau J, Schlorke D, Arnhold J. Lactoperoxidase as a potential drug target. Expert opinion on therapeutic targets. 2016;20(4):447–61.
162. Ashby M. Hypothiocyanite. Advances in Inorganic Chemistry. 2012;64:263–303.
163. Arlandson M, Decker T, Roongta VA, Bonilla L, Mayo KH, MacPherson JC, et al. Eosinophil peroxidase oxidation of thiocyanate. Characterization of major reaction products and a potential sulfhydryl-targeted cytotoxicity system. J Biol Chem. 2001;276(1):215–24.

164. Arnhold J, Monzani E, Furtmüller PG, Zederbauer M, Casella L, Obinger C. Kinetics and Thermodynamics of Halide and Nitrite Oxidation by Mammalian Heme Peroxidases. 2006;2006(19):3801–11.
165. Skaff O, Pattison DI, Davies MJ. Hypothiocyanous acid reactivity with low-molecular-mass and protein thiols: absolute rate constants and assessment of biological relevance. The Biochemical journal. 2009;422(1):111–7.
166. Skaff O, Pattison David I, Morgan Philip E, Bachana R, Jain Vimal K, Priyadarsini KI, et al. Selenium-containing amino acids are targets for myeloperoxidase-derived hypothiocyanous acid: determination of absolute rate constants and implications for biological damage. Biochemical Journal. 2011;441(1):305–16.
167. Hawkins CL. The role of hypothiocyanous acid (HOSCN) in biological systems. Free radical research. 2009;43(12):1147–58.
168. Barrett TJ, Hawkins CL. Hypothiocyanous Acid: Benign or Deadly? Chemical Research in Toxicology. 2012;25(2):263–73.
169. Hawkins CL, Pattison DI, Stanley NR, Davies MJ. Tryptophan residues are targets in hypothiocyanous acid-mediated protein oxidation. The Biochemical journal. 2008;416(3): 441–52.
170. Wang Z, Nicholls SJ, Rodriguez ER, Kummu O, Hörkkö S, Barnard J, et al. Protein carbamylation links inflammation, smoking, uremia and atherogenesis. Nature medicine. 2007;13(10):1176–84.
171. Stoyanovsky DA, Maeda A, Kagan VE. Assessments of Thiyl Radicals in Biosystems: Difficulties and New Applications. Analytical Chemistry. 2011;83(17):6432–8.
172. Madej E, Wardman P. The oxidizing power of the glutathione thiyl radical as measured by its electrode potential at physiological pH. Arch Biochem Biophys. 2007;462(1):94–102.
173. Schöneich C. Thiyl radicals and induction of protein degradation. Free radical research. 2016;50(2):143–9.
174. Tartaro Bujak I, Mihaljević B, Ferreri C, Chatgilialoglu C. The influence of antioxidants in the thiyl radical induced lipid peroxidation and geometrical isomerization in micelles of linoleic acid. Free radical research. 2016;50(sup1):S18–S23.
175. Hofstetter D, Nauser T, Koppenol WH. Hydrogen exchange equilibria in glutathione radicals: rate constants. Chemical research in toxicology. 2010;23(10):1596–600.
176. Pou S, M. Rosen G. Generation of thiyl radical by nitric oxide: a spin trapping study. Journal of the Chemical Society, Perkin Transactions 2. 1998(6):1507–12.
177. Durmaz A, Dikmen N. Homocysteine effects on cellular glutathione peroxidase (GPx-1) activity under in vitro conditions. Journal of Enzyme Inhibition and Medicinal Chemistry. 2007;22(6):733–8.
178. Mozziconacci O, Williams TD, Kerwin BA, Schöneich C. Reversible Intramolecular Hydrogen Transfer between Protein Cysteine Thiyl Radicals and αC−H Bonds in Insulin: Control of Selectivity by Secondary Structure. The Journal of Physical Chemistry B. 2008;112(49): 15921–32.
179. Lafleur MV, Retèl J. Contrasting effects of SH-compounds on oxidative DNA damage: repair and increase of damage. Mutation research. 1993;295(1):1–10.
180. Radical-Based Damage of Sulfur-Containing Amino Acid Residues. Encyclopedia of Radicals in Chemistry, Biology and Materials 2012.
181. Winterbourn CC, Hampton MB. Thiol chemistry and specificity in redox signaling. Free Radic Biol Med. 2008;45(5):549–61.
182. Jacob C, Giles GI, Giles NM, Sies H. Sulfur and Selenium: The Role of Oxidation State in Protein Structure and Function. 2003;42(39):4742–58.
183. Fukuto JM, Ignarro LJ, Nagy P, Wink DA, Kevil CG, Feelisch M, et al. Biological hydropersulfides and related polysulfides—a new concept and perspective in redox biology. 2018;592(12):2140–52.

184. Steudel R, Chivers TJCSR. The role of polysulfide dianions and radical anions in the chemical, physical and biological sciences, including sulfur-based batteries. 2019;48(12):3279–319.
185. Ono K, Akaike T, Sawa T, Kumagai Y, Wink DA, Tantillo DJ, et al. Redox chemistry and chemical biology of H2S, hydropersulfides, and derived species: Implications of their possible biological activity and utility. Free Radical Biology and Medicine. 2014;77:82–94.
186. Benchoam D, Cuevasanta E, Möller MN, Alvarez B. Hydrogen Sulfide and Persulfides Oxidation by Biologically Relevant Oxidizing Species. Antioxidants (Basel). 2019;8(2):48.

Chapter 3
Biomarkers of Oxidative Damage

A. Khelfi

Abstract Overwhelming evidence indicates that oxidative stress, via the overproduction of reactive species (RS), can impact on all major classes of biochemical entities including nucleic acids, proteins, amino acids, lipids, and carbohydrates. Oxidation of these entities can result in tissue injury. Many of the resulting breakdown products (8-OHdG, protein carbonyls, 4-hydroxy-2-nonenal (4-HNE), malondialdehyde (MDA), isoprostanes (IsoPs), OxLDL, etc.) are themselves highly reactive and threaten cellular integrity and function. More importantly, some of them decompose to free radicals (FR) that can continue to propagate the vicious lipid peroxidation chain reaction. Since the oxidized end-products from oxidative stress are generally more stable than the oxidants themselves, measurement of these biomarkers often showed consistent results in determining the level of the redox status in biological samples.

This chapter is a comprehensive illustration of the damage that RS are capable of doing. It addresses the pathological consequences of oxidation of most important entities in cells (nucleic acids, carbohydrates, proteins, and lipids). It presents new mechanistic insights into how oxidative damage of biomolecules occurs. In addition to the extensive examination of the specific mechanisms hypothesized for the diverse pathologies, this chapter also includes a basic review of the chemistry accounting for the generation of the oxidative end-products.

Keywords Oxidative damage · Nucleic acids · 8-OHdG · Carbohydrates · Protein carbonyls · Malondialdehyde (MDA) · 4-hydroxy-2-nonenal (4-HNE) · Isoprostanes (IsoPs) · OxLDL

A. Khelfi (✉)
Faculty of Pharmacy, University of Algiers1, Algiers, Algeria

National Centre of Toxicology, Algiers, Algeria

Department of Chemistry and Biomolecular Science, Clarkson University, Potsdam, NY, USA
e-mail: ab.khelfi@univ-alger.dz; akhelfi@clarkson.edu

S. Andreescu et al. (eds.), *Biomarkers of Oxidative Stress*,
https://doi.org/10.1007/978-3-031-60738-7_3

3.1 Oxidative Damage to Nucleic Acids

The interaction of reactive species (RS) with DNA can affect the integrity and regulation of certain genes, which may result in detrimental changes to cells. These interactions may involve direct modifications of DNA by oxidation of bases, resulting in the generation, for example, of 8-oxo-2'-deoxyguanosine (8-oxo-dG) moieties or adduct formation between lipid/protein-derived products and bases [1]. Oxidative attacks on DNA also occur at sugar moieties, leading to single- or double-stranded DNA breaks, removal of nucleotides and generation of base-free moieties, deletions, DNA–DNA cross-links, and a variety of modifications in the organic bases of the nucleotides [2, 3]. Overall, RS can directly modify both DNA bases and sugars.

DNA damage may also emerge from changes in transcription factors or enzymes involved in regulating gene expression or repair. Additionally, specific regions of the DNA can be targeted during oxidative stress (OS). For example, RS may exert deleterious effects on microsatellite (short tandem repeated sequences) regions as redox-active metal ions and $OH^{\bullet}$ can increase microsatellite instability [4].

Oxidative damage to DNA can result in cell death through apoptosis, as it is the case with radiation damage [5]. Even though single-stranded DNA breaks caused by oxidant injury can be easily tolerated by cells, double-stranded DNA breaks induced, for example, by ionizing radiation pose a significant threat to cell survival.

Oxidative modifications of DNA can also lead to the process of tumor initiation. Changes in the nucleotides of one strand can result in mutations (deletions or translocations) and translation errors with frameshifts as well as mismatches with the nucleotides of another strand, thus yielding subsequent mutations. Thus, DNA damage causes alterations in the encoded proteins, resulting in their dysfunction or complete inactivation and potentially in the initiation of the tumor process. Besides their possible role in tumor initiation, oxygen free radicals (FRs) and related reactive oxygen species (ROS) have been implicated in the process of tumor promotion. Indeed, DNA damage can result in the arrest, induction or errors of transcription, induction or inhibition of signal transduction pathways, replication errors, or genome instability.

As mentioned before, the interactions of RS with DNA may lead to mixed forms of DNA damage such as DNA–protein cross-links, DNA–lipid cross-links, and chromosomal arrangements. DNA–protein cross-links are formed when $OH^{\bullet}$ exerts deleterious actions on either DNA or the proteins that can be associated with it. If replication or transcription precedes repair, then DNA–protein cross-links may be lethal due to subsequent mutations.

With regard to DNA–lipid cross-links, the endoperoxides produced during lipid peroxidation (see Sect. 3.4) can decompose to form a range of secondary products like β-hydroxyacrolein, MDA, and 4-HNE, which are strong binding compounds that react with high efficiency with DNA [6]. For example, reactions between β-hydroxyacrolein/MDA and DNA bases produce adducts such as M1C, M1A, and M1G that are formed from the DNA bases cytosine, adenosine, and guanine, respectively (see Sect. 3.4) [7]. It has been reported that M1G, M1A, and M1C are highly mutagenic compounds in mammalian cells and are carcinogenic in rats [8]. In addition, 4-HNE is also considered as a mutation promoter [9]. The reasonable explanation for the mutagenic effect of 4-HNE lies in the fact that it forms the 4-HNE-dG adduct following interaction with DNA [10]. The resulting adduct induces G:C to T:A mutations in human cells. These mutations are preferentially formed at codon 249 of p53, a mutational hot spot in human cancers [11].

Lipid peroxidation also results in the generation of exocyclic DNA adducts such as etheno-adducts (etheno-dA, etheno-dC, and etheno-dG) and propane adducts [12]. Etheno-dA results in the transition to G, and etheno-dC induces transversion to A and transition to T. Propano-adducts can be formed following the reaction of DNA with acrolein, which is generated by lipid peroxidation. Propano-adducts usually induce base pair substitution mutations [13].

All of these adducts reveal unusual reactive groups of DNA that could participate in the formation of DNA–DNA inter-strand cross-links and DNA–protein cross-links and are thus a potential initiator of human carcinogenesis, particularly hepatocellular carcinoma and lung cancer. Besides, elevated etheno-DNA adducts were found in tissues of patients suffering from chronic inflammatory processes, whereas increased levels of urinary εdA were observed in subjects and workers exposed to diesel engine exhaust [14, 15].

Mitochondrial DNA (mtDNA) damage has also been observed during OS [16] and has attracted substantial interest for the following reasons:

- Mitochondria are the major source of oxygen-derived FRs.
- DNA is exposed to high levels of FRs.
- Mitochondria are poorly endowed with DNA repair processes.

Mitochondrial DNA is about 100 times more susceptible to oxidative damage compared with nuclear DNA (nDNA) because of the lack of protective histones, lack of repair mechanisms, high information density, and close proximity to ROS generation by the respiratory chain [17]. For example, the incidence of 8-hydroxyguanine (8-OHG) is 10- to 20-fold higher in mitochondrial DNA (mtDNA) than in nDNA [18]. Therefore, ROS are potentially harmful to mtDNA integrity. This feature can deeply affect the transcription and translation of important sequences in coded mtDNA, which underlie the synthesis of several subunits belonging to the mitochondrial respiratory chain (except complex II). A vicious cycle is then established in which oxidized mtDNA becomes dysfunctional and produces a high rate of ROS in mitochondria, thus leading to further mitochondrial impairment especially when ROS attack mtDNA in other mitochondria. This condition can ultimately result in severe nuclear DNA damage and cell death [19]. Finally, oxidative damage to mtDNA can lead to deletions and sequence rearrangements. For example, 4977 base pair deletions have been recorded in many tissues and found to increase with age [20].

Excess ROS production, resulting in oxidative damages to DNA and a variety of modifications in the organic bases of the nucleotides, was reported in various pathological contexts. Most of these DNA modifications are highly relevant to carcinogenesis, aging, and neurodegenerative and autoimmune diseases. This is also the case for ionizing radiations, tobacco smoke, and heavy metals such as iron, cadmium, chrome, and arsenic, which seem to be involved in carcinogenesis by generating FR-mediated damage.

Such damage has also been observed with a number of therapeutic drugs, many of which are cytostatic agents. A prominent example is Adriamycin (doxorubicin), which, like other anthracycline derivatives, is metabolized to a semiquinone compound that binds to DNA [21]. Another mechanism has been proposed to explain DNA damage by such a compound, which is based on their capability to activate molecular oxygen when undergoing redox cycling [22]. Nucleic acid damage through redox cycling and generation of RS is not restricted to this compound metabolized to semiquinones, as bleomycin, for example, forms a complex with iron ions, which, in turn, is subjected to redox cycling [23].

Having developed the possible targets for RS, the following are the main inducers of DNA damage and their mechanisms of action. The main cause of oxidative damage to DNA is through the generation of $OH^{\bullet}$ from hydrogen peroxide (H_2O_2) because the latter is more diffusible to the nucleus and can easily form the former under OS. The hydroxyl radical is known to directly react with all components of DNA (purine, pyrimidine bases, and deoxyribose backbone), which generates various products from DNA. The hydroxyl radical attacks the DNA backbone by causing eight classes of oxidative damage: oxidized and modified bases, abasic sites, DNA–DNA intra-strand links, single-strand breaks (SSBs), double-strand breaks (DSBs), deletions and rearrangements, DNA–protein cross-links, and DNA–lipid cross-links.

The mechanisms of DNA oxidative damage by $OH^{\bullet}$ could proceed along two different ways. The first way is through abstraction of hydrogen from the sugar moiety (C4–C5) or the methyl group of thymine to subsequently form the hydroxymethylated nucleobase [24]. This mode of reaction between $OH^{\bullet}$ and the sugar-phosphate backbone of nucleic acids results in the generation of a deoxyribose radical that further reacts to produce strand scissions and possibly chromosome breakage.

The second mode of action involves $OH^{\bullet}$ addition to the electron-rich moieties (e.g., double bonds) in purine and pyrimidine bases [25, 26]. Reactions of $OH^{\bullet}$ with purines and pyrimidines result in multiple adduct products in DNA. Of particular interest is the frequent oxidative damage to guanine and thymine due to the high oxidation potential of these nucleobases relative to other bases. Examples of oxidatively modified nucleobases are 8-hydroxyguanine, thymine glycol, and 5-hydroxymethyl uracil [27]. Other strong ROS can mediate deleterious effects on DNA such as organic hydroperoxides ($ROO^{\bullet}$), which contribute to DNA oxidation. Under in vitro conditions, it has been demonstrated that $O_2^{\bullet-}$ can interact with iron–sulfur clusters and release a ferrous ion that binds to DNA, causing potential mutagenesis [28].

Reactive nitrogen species (RNS) (e.g., nitric oxide ($NO^•$), nitrogen dioxide ($NO_2^•$), nitrous acid (HNO_2), peroxynitrite ($ONOO^-$), and dinitrogen trioxide (N_2O_3)) can mediate oxidative damage to DNA as well. They are considered potentially mutagenic as they induce nitration, nitrosylation, and deamination of DNA bases [29]. The main product of such an oxidative attack has been identified as 8-nitro-2'-deoxyguanosine (8-NO_2dG), and several of these species result in subsequent strand breaks of DNA [30]. Reactive halide species (RHS), particularly hypochlorous acid (HOCl), can interact with DNA as well and result in deleterious effects and dysfunctions in different types of cells.

Overall, both nitrogenous bases and sugars are highly susceptible to attacks by electrophiles, particularly $OH^•$, $ONOO^-$, and HOCl. Another possibility resulting from oxidative attacks is hydrogen abstraction, which leads to the generation of an oxidized abasic site where the nucleotide is still connected but without the base and with the sugar ring open.

3.1.1 Biomarkers of Oxidative Damage to Nucleic Acids

Overall, bases of DNA are particularly sensitive to FRs, leading to a great number of derivatives by addition, oxidation, or fragmentation. ROS-mediated base modifications lead to the formation of several common DNA adducts that begin with the generation of purine and pyrimidine radicals. Based on their redox properties, reaction partners, and the presence or absence of oxygen, these radicals undergo oxidation or reduction reactions to yield numerous products. Among the four nucleobases in DNA, guanine is the most susceptible to oxidative damage because of its low redox potential. Various chemical reactions have been reported, which occur in a cascade manner and mainly produce guanine lesions through oxidation, nitration, halogenation, and alkylation [31]. The main products of its oxidation are 8-hydroxyguanine (8-OHG) and 8-hydroxy-2'-deoxyguanosine (8-OHdG) and can result in the modification of base pairs.

3.1.1.1 8-Hydroxy-2'-Deoxyguanosine (8-OHdG)

8-Hydroxy-2'-deoxyguanosine is the most commonly observed and well-known oxidative DNA by-product. It is the most prominent example among more than 20 base modifications and represents a potential biomarker of carcinogenesis [32]. It is also a good indicator of OS in a number of diseases along with other metabolic markers. This compound is mostly formed through the addition of $OH^•$ to position 8 of guanine in the DNA.

8-OHG and 8-OHdG are the most commonly used markers of DNA damage as they easily accumulate. This modification occurs in 1 in 10^5 G residues in physiological conditions but increases considerably under OS and with age [33].

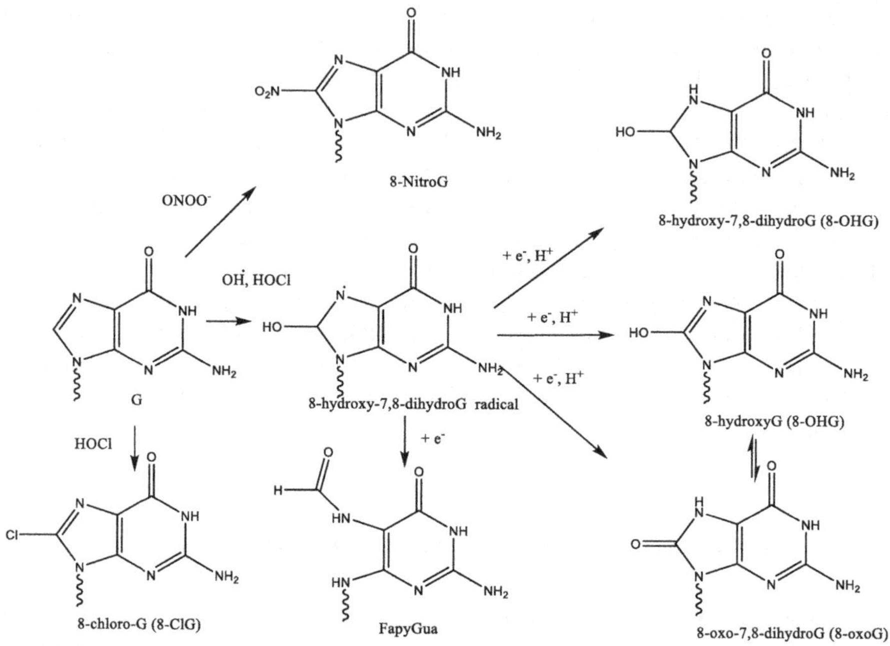

Fig. 3.1 Modification of guanine nucleobases by reactive species

High levels of 8-OHdG have been detected in cells associated with tumorigenesis, indicating that OS may play an important role in tumor development [34]. 8-OHdG is generated following an $OH^{\bullet}$ attack on guanine at its C-8 position (Fig. 3.1). It has been reported that RNS (e.g., $ONOO^{-}$) also react with DNA, forming the adduct 8-oxo-dG [35].

8-OHdG is highly mutagenic and carcinogenic. Hydroxylation changes its structure, so it can no longer form three hydrogen bonds with its usual partner cytosine but instead forms two hydrogen bonds with adenine. If not repaired, then this mismatch will promote base pair transversion from G-C to T-A in the next cycle of DNA synthesis, resulting in mutations [36]. These mutations may still be repaired, otherwise the cell may undergo apoptosis or necrosis pathway, or in the worst scenario where the mutations occur in a proto-oncogene or a tumor-suppressor gene (such as *p53*), could start along with this event the first step of tumor formation. Besides the mutagenic effect of 8-OHdG, its formation at transcription factor-binding sites can modify the binding of transcription factors, which leads to changes in the expression of related genes as has been shown for activator protein (AP)-1 and Sp-1 target sequences [37]. This effect is due to the promoter regions of genes, which contain guanine–cytosine (GC)-rich sequences for transcription factors and is known to be susceptible to oxidant attacks. Reduction could occur for 8-OHdG and result in ring opening and subsequent formation of 2,6-diamino-4-hydroxy-5-formamidopyrimidine (FapyGua). The latter could lead to base pairing mismatch and adduct generation with a protein amino side chain [38]. Overall, 8-OHdG represents one of the most widely used markers of oxidative damage to DNA, which is mainly formed by the reaction of $OH^{\bullet}$ with a guanine base in DNA. It is the most easily formed base alteration with mutagenic and carcinogenic properties. It is a good biomarker of carcinogenesis as well.

3.1.1.2 8-Oxo-2'-Deoxyguanosine (8-Oxo-dG)

Under severe OS, guanine bases can be converted to higher toxic adducts, resulting in a deleterious effect on the nucleus machinery. Further oxidation of dG and 8-OHdG in the presence of $OH^{\bullet}$ has the tendency to yield 8-oxo-2'-deoxyguanosine (8-oxo-dG), which has been a widely studied DNA lesion in recent years. Similarly, $ONOO^{-}$ and HOCl can interact with guanine to yield 8-oxo-dG. Due to its mutagenic ability, 8-oxo-dG is known for its contribution to human cancers. In fact, the accumulation of 8-oxo-dG is oncogenic and causes mitochondrial dysfunction [39]. Moreover, it has been shown that the enzyme human mutT homologue (MTH1) that detoxifies oxidized nucleotides is a potential target in cancer therapy [40]. The major mutagenic lesion of 8-oxo-dG is its ability to base pair with adenine rather than with cytosine (GC → GA) and thus generate transversion mutations during DNA replication [41, 42].

3.1.1.3 8-Nitro-2'-Deoxyguanosine (8-NO_2dG)

8-Nitro-2'-deoxyguanosine (8-NO_2dG)/8-nitroguanosine (8-NO_2G) is generated by nitration of guanine in the free forms or in DNA/RNA strands by RNS such as oxides of nitrogen and $ONOO^{-}$ [43–45]. This adduct is one of the most widely used approaches to quantify OS in animals and humans and forms the basis for many assays aimed at measuring oxidative DNA damage. Unlike normal tissues, enhanced levels of 8-NO_2dG have been detected in the nuclei of inflammatory and epithelial cells in inflamed tissues [46].

8-NO_2dG has a tendency to depurinate and form an apurinic site, a location in DNA that has neither a purine nor a pyrimidine base [44]. This leads to G:T transversions and carcinogenesis with a poor prognosis, as shown in the lungs of humans and mice [44]. Besides this effect, 8-NO_2dG has also been found to inhibit enzymes involved in DNA repair such as formamidopyrimidine (Fapy) DNA glycosylase, which catalyzes the removal of 8-oxo-G [47].

3.1.1.4 8,5'-Cyclo-2'-Deoxyadenosine (Cyclo-dA)

Besides deoxyguanosine, $OH^{\bullet}$ can attack other bases like adenine to yield 8 (or 2, 4-, 5-)-hydroxy-deoxyadenine and 8,5'-cyclo-2'-deoxyadenosine (cyclo-dA) [48] (Fig. 3.2). As with 8-OHdG, cyclo-dA has also been shown to inhibit transcription

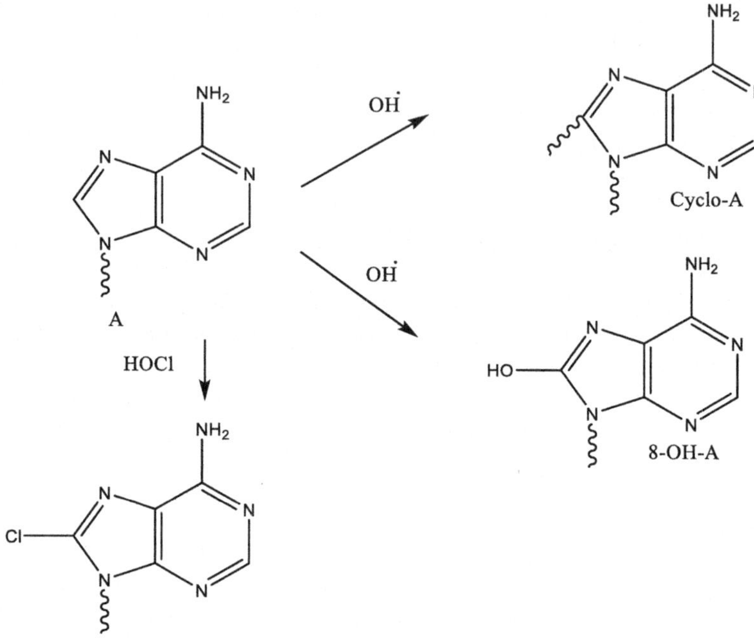

8-chloroA (8-ClA)

Fig. 3.2 Modification of adenine nucleobases by reactive species

from a reporter gene in a cell system by impairing the binding of the TATA-binding protein [49]. Both 2-hydroxy-deoxyadenine (2-OHdA) and 8-hydroxy-deoxyadenine (8-OHdA) were found to be mutagenic due to their miscoding potential [50].

3.1.1.5 Chlorinated Bases

It has been shown that the reaction of dG with HOCl results in the generation of 8-oxo-dG and 8-chloro-2'-deoxyguanosine (8-CldG) (Fig. 3.1) [51, 52]. The reaction of nucleic acids with HOCl, which originated from human neutrophils, gives rise to various chlorinated products, including 5-chlorodeoxycytidine (5-CldC), 8-chloro-2'-deoxyguanosine (8-CldG), and 8-chlorodeoxyadenosine (8-CldA) (Fig. 3.2) [24]. In another study undertaken by Whiteman et al., exposure of calf thymus DNA to HOCl yielded 8-CldA and, to a lesser extent, 5-CldC. The latter is a more reliable biomarker than is 8-CldA due to its susceptibility to degradation by other RS such as RNS or ROS [53].

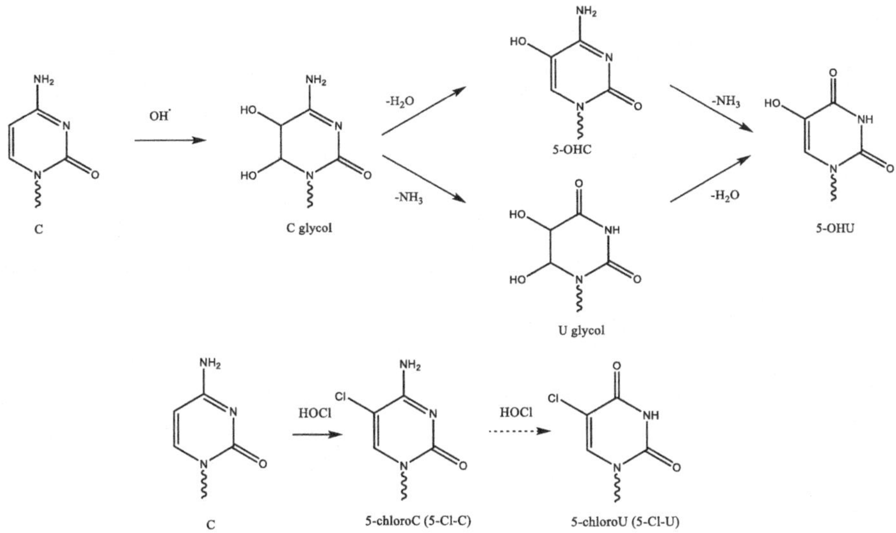

Fig. 3.3 Modification of cytosine nucleobases by reactive species

3.1.1.6 Other Oxidized Bases

ROS, particularly OH•, are also involved in the modification of other bases besides guanine. The hydroxyl radical can lead to the production of adenine N(1) oxide, 5-hydroxyuracil (5-OHU), and 5-hydroxycytosine (5-OHC) (Fig. 3.3), which are associated with certain disorders such as atherosclerosis and carcinogenesis [54]. These adducts are found to be pro-mutagenic due to their miscoding potential [50].

The OH• radical is also able to undergo addition reactions with the double bonds of DNA bases to abstract hydrogen from the methyl group of thymine and each of the C atoms of deoxyribose. The addition of the hydroxyl radical to thymidine at the 5,6-double bond yields dihydroxythymine (Fig. 3.4), whereas addition to the C-8 position of the purine ring of adenosine (dA) forms 8-hydroxy-deoxyadenosine (8-OHdA) as mentioned previously (Fig. 3.2) [55]. On top of producing 8-NO_2dG, it has been shown that $ONOO^-$ is able to make several other modifications. Increased levels of both oxidized and deaminated base products, including 5-hydroxyhydantoin, thymine glycol, 5-(hydroxymethyl)-uracil (5-OHMeU), 4,6-diamino-5-formamide-pyrimidine, 2,6-diamino-5-formamidopyrimidine, 8-oxo-G, 8-oxo-A, hypoxanthine, and xanthine, were found following the treatment of calf thymus DNA with $ONOO^-$ [27].

DNA methylation, which is an epigenetic process modulating gene expression, also seems to be affected by OS. Abnormal DNA methylation patterns induced by RS may lead to aberrant DNA repair activity. Methylation at CpG islands in DNA is an important epigenetic process that may result in gene silencing. Conversion of

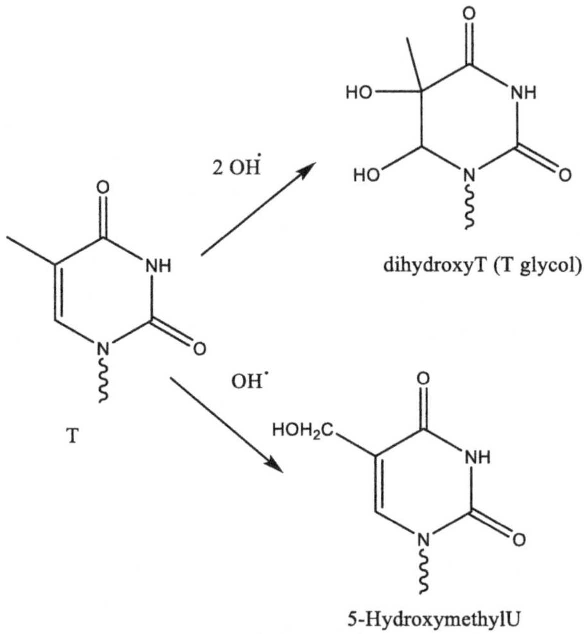

Fig. 3.4 Modification of thymine nucleobases by reactive species

5-methylcytosine (5-MeC) to 5-OHMeU may occur via deamination/oxidation reactions and therefore affects gene expression [56]. In addition, DNA methylation seems to affect chromatin organization and DNA repair activity [4].

3.1.2 Effects of Oxidative Stress on RNA Molecules and Chromatin

RNA is subject to oxidation as well, with implications in the functional processes of cells. Evidence suggests that RNA oxidation is involved in several diseases, including diabetes mellitus (DM), cardiovascular disease (CVD), and Alzheimer's diseases (AD) [57, 58].

Oxidation of RNA bases primarily leads to strand breaks and base modifications. Of particular importance is the oxidation of messenger RNA (mRNA), which results in decreased and abnormal protein production. Cells can recognize this problem and degrade the affected proteins but if extensive and/or prolonged, then the apoptotic pathway may be triggered.

Micro-RNAs were also found to be the target of OS actions. For example, oxidative modification of micro-RNA-184 enables it to target the mRNA of the B-cell lymphoma proteins Bcl-xL and Bcl-w, thereby blocking their translation, which leads to the initiation of apoptosis [59]. On the other hand, it has been shown that a subset of micro-RNAs, termed "redoximiRs," can regulate redox pathways

[60]. Another illustrative example is micro-RNA-15b, which regulates mitochondrial $O_2^{•-}$ production through sirtuin-4 translation, as demonstrated in photoaged skin [61].

The oxidative effects exerted on the cytoskeleton and the nuclear matrix affect chromatin configuration and structure, two important factors that strongly affect gene expression. Indeed, the cytoskeleton guarantees a connection of chromatin with the nuclear membrane, thus ensuring genome expression in eukaryotic cells. Modifications in membrane fluidity induced by RS can therefore bring about important changes in the structure of chromatin, which then affects gene expression. Additionally, oxidative damage to the cytoskeleton can favor the release of immature RNA from the nucleus, as these entities provide anchoring points for pre-mRNA processing [62].

3.2 Oxidative Damage to Carbohydrates

Advances in FR chemistry indicate that no biological compound is safe from FR attacks and the relations of carbohydrates with OS are manifold. Indeed, cells contain several types of carbohydrates that are susceptible to attacks by FRs. They include sugar moieties in DNA, RNA and proteins, and mono- and polysaccharides. It is not surprising that glucose and other related sugars undergo the OS process when conditions are appropriate.

Some carbohydrates can be part of a multitude of oxidation reactions, resulting in the formation of advanced glycation end products (AGEs) [63]. AGEs are a heterogeneous group of proteins, lipids, and nucleic acids that are nonenzymatically and irreversibly glycosylated. They are constantly produced in the body but are markedly increased in conditions of hyperglycemia and OS [64]. For the process of AGE formation, glucose is the predominant substrate, but other sugars, such as fructose, can react with the protein side chain and generate distinct forms of AGEs through the modification of fructoselysine and creation of *Nε*-(carboxymethyl)lysine (CML) and pentosidine [63]. AGE-modified proteins are initially formed via the sugar addition reaction, followed by further reactions, rearrangements, dehydrations, and cleavages, which is known as nonenzymatic glycation of proteins by a reducing sugar. Otherwise, sugars like glucose in the presence of oxygen and some ROS ($O_2^{•-}$, H_2O_2, and $OH^•$) with its α-hydroxyaldehyde structure can enolize and undergo metal-catalyzed auto-oxidation to produce reactive monocarbonyls/hydroxyaldehydes (e.g., glycolaldehyde) and dicarbonyls/ketones (e.g., glyoxal, methylglyoxal, glucosone, 3-deoxyglucosone, and MDA). These products are nothing more than sugar fragments of glycolaldehyde and short-chain species whose chains are too short to cyclize but can auto-oxidate. These mono- and dicarbonyls, formed from the oxidation of glucose followed by a series of intramolecular rearrangements, dehydration, and condensation reactions, along with those formed following the so-called Maillard reaction resulting in an Amadori product, react with oxidized proteins, DNA, and lipids to yield AGEs. The glycation process of proteins requires the interaction of basic amino acid side chains, usually the amino moieties of lysine and arginine, with mono- or dicarbonyl moieties and compounds. This results in cross-linking of carbonyls with amino groups to form AGEs. Therefore, AGEs are the end products of the Maillard reaction between reducing sugars and proteins.

The Maillard reaction is simply a condensation of the carbonyl moiety in a modified sugar with the amino groups of protein amino acids to form Schiff-base adducts. The modified proteins undergo Schiff-base rearrangement to form more stable glycated products, called Amadori products. Its subsequent rearrangement leads to formation of stable and irreversible AGE compounds.

Additional pathways that form AGEs include methylglyoxal production through the glycolysis pathway, 3-deoxyglucosone production by reaction with fructosamine-3-kinase, glucosone, glyoxal, and MDA generation by auto-oxidation of glucose, and glycolaldehyde formation by auto-oxidation and from HOCl via myeloperoxidase (MPO) produced by activated neutrophils. These products may react with other kinds of macromolecules, leading to structural and functional modifications. Several investigations support the notion that the glycation process may in fact be mediated by FR reactions. For instance, glucose is prone to auto-oxidation by transition metal catalysis to produce RS such as H_2O_2 and $OH^{\bullet}$, which, in turn, promote glucose auto-oxidation [65–67]. Glycation of proteins is an irreversible posttranslational modification, leading to decomposition of proteins in biological systems. For example, the reaction of glucose oxidation end products with proteins results in the formation of one of the well-known AGEs called pentosidine (Pent), which is associated with a number of inflammatory pathologies [68, 69].

"AGEs" do not denote a single compound, but the term is rather a general term for the end products of glycation and glyco-oxidation. Many compounds have so far been identified as AGEs, including, but not limited, to *Nε*-(carboxymethyl)lysine (CML), crossline, pyrraline, GA-pyridine, Arg-pyrimidine, *Nω*-(carboxymethyl)-arginine (CMA), *Nε*-(carboxyethyl)lysine (CEL), glyoxal-lysine dimer (GOLD), methylglyoxal-lysine dimer (MOLD), deoxyglucosone lysine dimer (DOLD), and 2-(2-furoyl)-4(5)-(2-furanyl)-1H-imidazole (Fig. 3.5) [70]. It is noteworthy to know that CML and pentosidine have always been regarded as representative biomarkers of AGEs.

Sugar oxidation not only gives rise to these short-chain intermediates and AGEs but also generates $O_2^{\bullet-}$ and H_2O_2, which could exacerbate ROS production and further form the well-known mutagens α- and β-dicarbonyls [71, 72]. Indeed, AGEs may stimulate the generation of ROS and promote further OS in cells by activating a nicotinamide adenine dinucleotide phosphate (NADPH) oxidase-coupled receptor called RAGE [73]. In turn, binding and activation of RAGE can also promote the production of AGEs via the NADPH oxidase-mediated production of ROS. It is believed that RAGEs are responsible for the cellular effects of AGEs and for mediating their deleterious actions [74, 75]. Growing evidence shows that AGEs and their receptor (RAGE) are relevant biomarkers of glyco-oxidative stress [76].

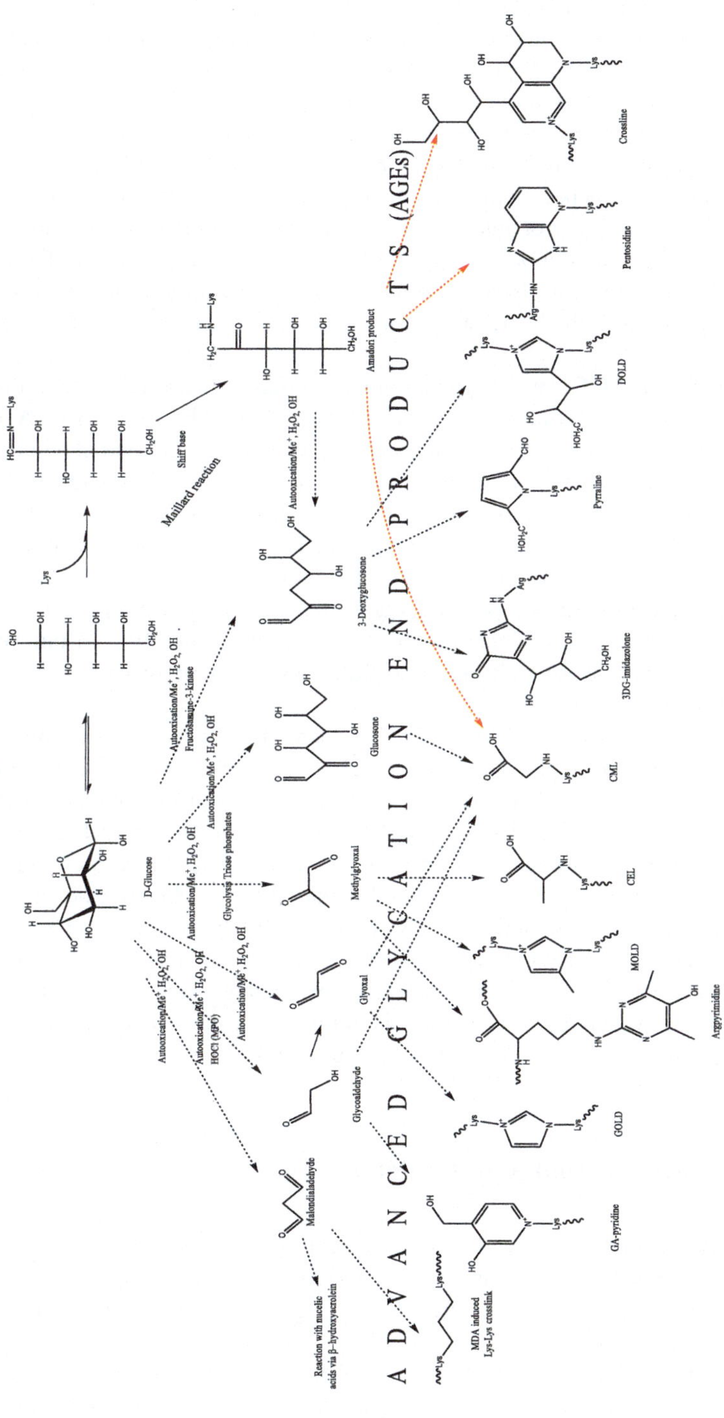

Fig. 3.5 An overview of the generation of advanced glycation end products (AGEs) and the Maillard reaction

AGEs have been implicated in several diseases. It has been shown that they accumulate almost constantly in the plasma and tissues in accelerated aging, DM, and renal failure [77, 78]. Glycation in skin collagen and the corneal layer of the epidermis accelerates formation of protein cross-linkages, which reduce elasticity, and results in advancing senility. The quantity of AGEs in skin collagen increases with age and diabetic complications [79, 80]. The development of diabetic complications (retinopathy, nephropathy, and neuropathy) can be attributed to the significant glycation of many proteins and the production of AGEs under diabetogenic conditions [81, 82]. It occurs in many tissues and organs, especially in the kidney and liver. The presence of glycated proteins was reported in sclerotic tissues of the renal glomerulus and in the renal arteries of patients with diabetic nephropathy [83, 84].

Proteins rich in AGEs and D-type amino acids play an important role in diabetic complications such as retinopathy and keratopathy [85]. The amounts of AGEs in the skin, nails, and hair are greater in diabetic patients than in nondiabetic patients [86]. Under hyperglycemic conditions, these products are formed extensively as a result of the reaction between the amino residues of proteins with the aldehydes of the reducing glucose in a nonenzymatic manner. They are considered as biomarkers of glycemic control and risk factors associated with diabetes. The suggested routes by which AGEs promote the progression of diabetic complications include functional disorders as a result of the modification of proteins involved in intracellular signalization, organ disorders from intracellular accumulation of AGEs, and induction of RAGE [87].

Formation of AGEs in the body is a risk factor for other types of pathologies, including coronary heart diseases [88–90]. The oxidative reaction of glyoxal and methylglyoxal with proteins is associated with a number of inflammatory pathologies and neurodegenerative diseases [68, 91]. In the ophthalmological field as well, proteins rich in AGEs are clustered in cataract, spheroid degeneration, pinguecula, and macular degeneration as in age-related ocular changes [85]. Besides the generation of AGEs, the interaction of ROS with structural polysaccharides like hyaluronic acid, a major component of extracellular fluids, results in their depolymerization and degradation. Such a process seems to play a key role in inflammatory diseases of the joints. For example, mucopolysaccharides of synovial fluid are degraded through the action of RS produced by immune cells.

3.3 Oxidative Damage to Proteins

RS have a high affinity to molecules with an electrophilic character, including amino acids. A range of reactive oxygen, nitrogen, chlorine, and carbon species are capable of modifying proteins, such as $OH^{\bullet}$, $CO_3^{\bullet-}$, OCl^-, and products formed from reactions involving $ONOO^-$.

The amino acid residues that are the most vulnerable to an RS attack are unsaturated, cyclic, and sulfur-containing amino acids such as arginine, glutamate, histidine, phenylalanine, tryptophan, tyrosine, valine, cysteine, proline, threonine, and methionine. Protein oxidation could produce RS that cause damage to other biomolecules. For example, oxidation can inactivate metalloproteins and convert the metal into a secondary RS, generating further RS when conditions are appropriate. High levels of RS can cause FR-mediated peptide cleavage of the peptide chain, alteration of the electrical charge of proteins, cross-linked protein aggregates, and direct oxidation of amino acids, leading to increased susceptibility to proteolysis by degradation by specific proteases [4, 92]. Enzymes that have metals on or close to their active sites are likely more susceptible to metal-catalyzed oxidation and loss of their activities.

Xenobiotic-induced RS formation can lead to oxidation of the side chains of amino acid residues in proteins, resulting in the formation of adducts, protein–protein cross-links, or protein fragmentation when amide links are targeted. Protein–protein cross-links result from different kinds of inter- and intra-protein reactions, such as the addition of a lysine amino group to a carbonyl group of the oxidized protein, the interaction of two carbon-centered radicals of protein backbones, the oxidation of cysteine residues to form reversible disulfide (S–S) cross-links, or the oxidation of tyrosine residues to form Tyr–Tyr cross-links. Oxidation of cysteine residues under OS conditions can also lead to reversible mixed disulfides, particularly with glutathione (GSH) (S-glutathionylation), resulting in the accumulation of what is known as S-glutathionylated proteins [93]. Similarly, cross-links may be formed between proteins and other entities like lipid and carbohydrate derivatives. Oxidized lipids and carbohydrates that react with oxidized proteins promote protein aggregation and inhibit proteasome degradation. A number of xenobiotics and related electrophilic metabolites can covalently bind to cellular proteins. These compounds are metabolized to FRs that undergo covalent binding with proteins. Examples include carbon tetrachloride (CCl_4), halothane, paracetamol, and Adriamycin. Covalent binding was also evident for reactive metabolites that are electrophilic epoxides rather than FRs (e.g., vinyl chloride, trichloroethylene, and other halogenated alkanes).

High levels of ROS may lead to direct and indirect modifications of proteins. These modifications can compromise both structural and functional properties of proteins and pose potentially serious problems for the cell. The biological consequences of oxidative damage to proteins include the following [94]:

- Loss of enzymatic activity and thermolability
- Loss of ligand–receptor specificity
- Intra- and extracellular aggregation of proteins
- Enhanced or decreased susceptibility to proteolysis
- Abnormal cellular uptake (e.g., low-density lipoprotein (LDL))
- Impaired signaling pathways
- Metabolic disturbances
- Modification of gene transcription (e.g., SoxR and IκB)
- Denaturation of epitopes and increased immunogenicity

FRs are capable of initiating oxidative damage to soluble and/or membrane-bound proteins, yielding dimers or larger aggregates seen in neurodegenerative diseases. The direct modification of the protein backbone under an RS attack through different forms of oxidation/nitration influences the protein activity and, subsequently, the metabolism and results in functional disruption. Modification of proteins can also lead to protein unfolding and altered conformation, altered interactions with biological partners, and modified turnover [95]. As a result of oxidative damage, increased susceptibility of proteins to proteolysis may occur. The consequence of peptide bond oxidation of structural proteins, in general, is a loss of physiological function. Other possible effects of RS on structural proteins are the initiation of depolymerization reactions as is the case with collagen. Cross-linking and covalent binding to proteins may lead to a loss of protein biological function, for example, enzymes may be unable to carry out their catalytic function and specific receptors may be inactivated. Enzymes containing metals may undergo metal-catalyzed oxidation that could lead to enzyme inactivation. For example, oxidation of iron–sulfur centers by $O_2^{\bullet-}$ in some metalloenzymes is irreversible and leads to enzyme inactivation [96, 97]. A covalent binding of a reactive metabolite to proteins may also trigger allergic reactions, since a protein that has been structurally modified following adduction might exhibit antigenic characteristics. The main consequence of cross-links is a change in the higher-order structure of the protein, which results in a loss of biological function of globular proteins. Modifications of amino acids leading to the formation of aldehydes or ketones (e.g., carbonyls), hydroxy species, or ring cleavage in histidine or tryptophan residues usually impair signaling, and both the structural and functional activities of proteins, thereby having significant impacts on a myriad of disorders [98].

Common ROS that pose an oxidative threat to proteins include $^1O_2$, $O_2^{\bullet-}$, $HO_2^{\bullet}$, H_2O_2, $OH^{\bullet}$, $ROO^{\bullet}$, and $RO^{\bullet}$. Out of all these ROS, $OH^{\bullet}$ is the most powerful oxidant causing nonspecific modifications of proteins, which can result from the direct action of ROS on proteins through, for example, carbonylation, disulfide bond formation, and glutathionylation or through indirect actions on other biochemical entities by conjugation with breakdown products of fatty acid peroxidation. The well-known ROS-induced oxidative modifications of proteins fall into the following categories:

- Modification or cleavage of the peptide bond
- Direct oxidation of amino acids, mainly thiol-containing amino acids, leading to disulfide bond formation and glutathionylation
- Oxidation of thiol-containing amino acid residues and formation of amino acid sulfoxide
- Ring cleavage of aromatic amino acids
- Hydroxylation of amino acids
- Oxidation of amino acids and function of carbonyl groups in the carbon skeleton
- ROS interaction with metallocofactors

Apart from these conventional protein oxidations, there are other processes of protein modification involving other types of RS:

- Conjugation with breakdown products of fatty acid peroxidation
- Adduct formation by glycoxidation of proteins
- Nitrosylation: Protein reaction with RNS
- Chlorination: Protein reaction with RHS

Most of these oxidative damages to proteins are chemically stable and widely used biomarkers of protein oxidation [99, 100].

3.3.1 Biomarkers of Oxidative Damage to Proteins

3.3.1.1 Protein Carbonyls

The most common and widely used biomarkers of oxidative damage to proteins are aldehydes and ketones, usually termed "protein carbonyls" (CO). They can be easily tracked experimentally both in vitro and in vivo and are considered one of the best markers of OS but represent low specific oxidation products of proteins.

Carbonyl groups are generated by many different mechanisms either directly by irreversible oxidation of certain amino acid side chains, mainly lysine, arginine, threonine, and proline, in the presence of ROS (usually $OH^{\bullet}$), or through the glycation process, or indirectly during oxidation of lipids. The second pathway relies on lipid peroxidation where protein carbonyls can be produced by the cleavage of the peptide backbone and Michael addition reactions of nucleophilic side chains of Lys, His, and Cys residues with lipid-derived aldehydes (e.g., 4-HNE, MDA, or acrolein) or through glycoxidation reactions of glucose-derived products with the Lys amino group, forming AGEs (see Sect. 3.2) [101].

Carbonyl groups are also formed through oxidative cleavage of protein backbones at the level of glutamyl side chains and the oxidative deamination of lysine [102]. Since protein carbonyls can arise from different mechanisms and include other indicators of OS (e.g., AGEs), their concentration is commonly higher than that of other biomarkers. High levels of carbonylated proteins have been detected in several human pathologies, including rheumatoid arthritis (RA), neurodegenerative diseases (Alzheimer's disease (AD)), hypercholesterolemia, DM, sepsis, chronic hepatitis, acute pancreatitis, thalassemia major, and chronic renal failure [102, 103]. They are believed to be the early biomarkers of the diagnosis of these diseases [103]. Protein carbonyl levels also increase with age and in some health conditions such as obesity [104, 105]. Some chemical and physical toxicants can generate carbonyl groups such as γ-rays and ozone (O_3) [56]. Overall, carbonyl groups are mainly produced by reactions between amino acid side groups (usually Lys, Arg, Pro, or Thr) and $OH^{\bullet}$. They can be generated during oxidative insult, resulting in the cleavage of the polypeptide backbones. Because carbonyls are not a specific biomarker of OS, their use as a marker of oxidative protein damage remains somewhat controversial (Fig. 3.6).

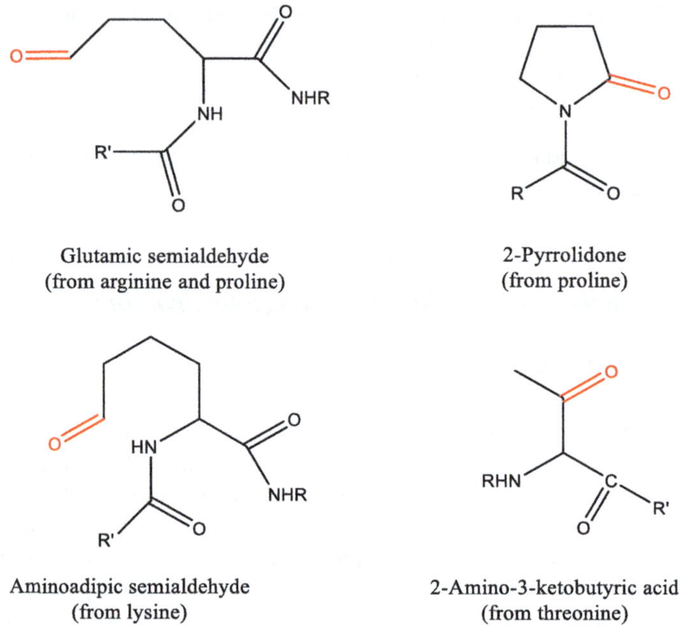

Fig. 3.6 Carbonyl derivatives generated from the oxidation of specific protein residues

3.3.1.2 Modified Amino Acids

Individual amino acids can undergo oxidative damage with varied susceptibilities if they interact with FRs and non-radical RS. Furthermore, most amino acid residues within proteins can be the targets of ROS and RNS, among which, unsaturated, aromatic, and sulfur-containing amino acids are the most susceptible to oxidation [106]. This is also true for those that are relatively "resistant" like proline and lysine, which are subject to hydroxylation by $OH^{•}$ in an extreme condition of OS. In general, oxidative modifications of amino acids include oxidation of sulfur-containing residues, hydroxylation of aromatic and aliphatic groups, ring cleavage of aromatic groups, nitration of tyrosine residues, nitrosylation and glutathionylation of cysteine residues, chlorination of aromatic groups and primary amino groups, and conversion of some amino acid residues to carbonyl derivatives.

3.3.1.3 Oxidized Tyrosine

Nitration of proteins in inflamed tissues has been extensively reported in the literature. This modification can be generated through several pathways that include reactions with RNS and ROS. Under conditions of OS, chemical species such as $ONOO^-$ and $NO_2^{•}$ can be generated and result in nitration of susceptible tyrosine residues to form 3-nitrotyrosine (3-NO_2Tyr), which is a footprint of in vivo nitration of proteins and one of the best markers of oxidative protein damage [107–109]. The generation of 3-NO_2Tyr, the main product of tyrosine oxidation, occurs within a polypeptide sequence or in free tyrosine amino acids. A second pathway of 3-NO_2Tyr formation involves MPO [110]. In the presence of H_2O_2 and NO_2^-, MPO catalyzes the production of $NO_2^•$. In this process, nitration is usually a two-step reaction, in which tyrosine is first oxidized, resulting in a tyrosine radical, and, second, the latter undergoes a radical–radical reaction with $NO_2^•$ [111, 112].

Nitration of tyrosine modulates the structure and function of proteins in many disease states [113, 114]. It decreases the activity of key enzymes and may have detrimental effects on various cellular functions [101]. Nitration of tyrosine residues in cells and inflamed tissues has been extensively reported in the literature [115, 116]. Its role in inflammation may be explained by the relatively stable nature of the modification, probably representing a good candidate for neoantigen formation (Fig. 3.7) [117].

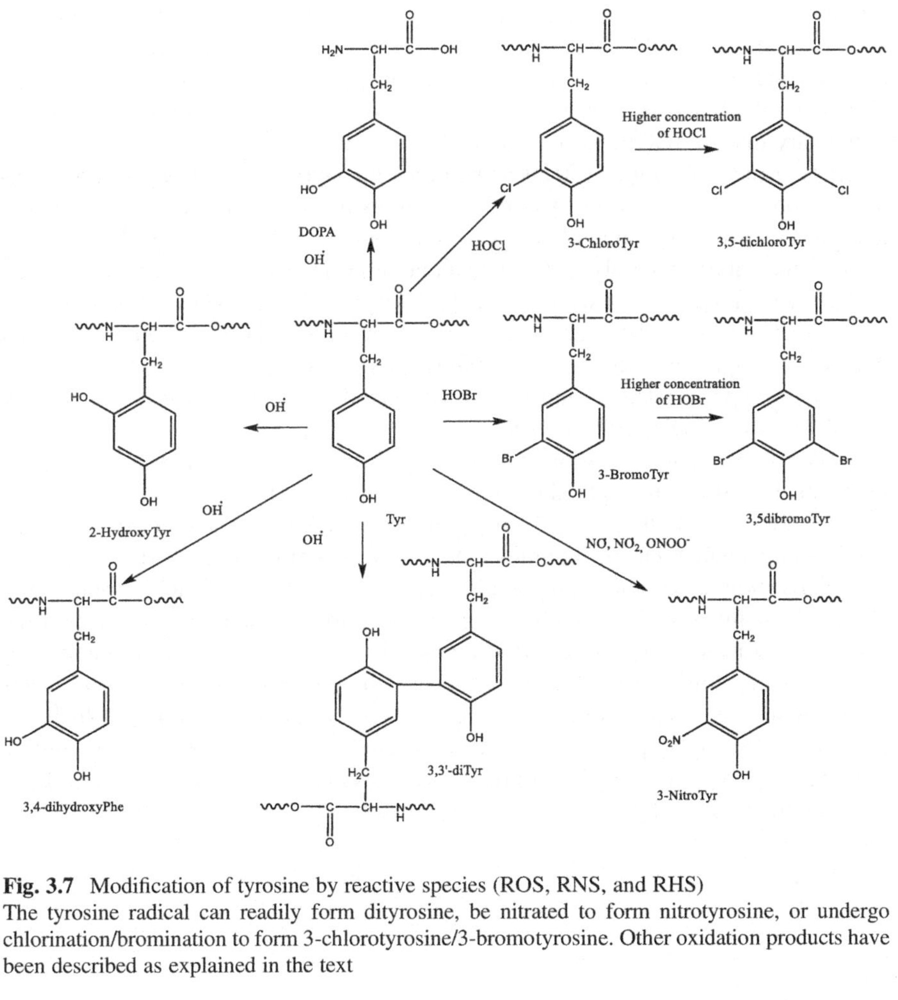

Fig. 3.7 Modification of tyrosine by reactive species (ROS, RNS, and RHS)
The tyrosine radical can readily form dityrosine, be nitrated to form nitrotyrosine, or undergo chlorination/bromination to form 3-chlorotyrosine/3-bromotyrosine. Other oxidation products have been described as explained in the text

3-NO_2Tyr has been described as a stable marker in several diseases [118], but its utility in clinical practice is still questioned. Some studies showed elevated plasma levels of 3-NO_2Tyr in diseases associated with OS, such as atherosclerosis, myocardial ischemia, asthma, DM, amyotrophic lateral sclerosis (ALS), and inflammatory bowel disease (IBD), as well as in toxic and carcinogenic models [119, 120]. Increases in NO_2Tyr in patients with celiac disease, children with chronic renal failure, patients with acute respiratory distress, and smokers have been observed [121–124]. Some of these studies found reduced levels of this marker following therapeutic treatments. Moreover, an involvement of 3-NO_2Tyr in age-related neurodegenerative diseases has been suggested, whereas other studies have reported modified tyrosine to be highly excreted in the urine of patients with DM [125–127]. Protein carbonylation and nitration (an increase in 3-NO_2Tyr levels) were also observed in Huntington's disease (HD) [128, 129].

Nitrotyrosine has been detected at elevated levels in serum and synovial fluid from patients with RA [130]. Smith et al. have reported that protein-bound nitrotyrosine and dityrosine increase by eight- and fivefold, respectively, in the affected regions of the brain in Alzheimer's disease patients relative to the age-matched normal human brain [131]. Finally, NO_2Tyr could be a disease-specific biomarker of OS in osteoarthritis, systemic lupus erythematosus, chronic obstructive pulmonary disease (COPD), and septic shock. The successful in vivo detection of 3-nitrotyrosine-containing proteins using proteomic approaches offers an early diagnostic biomarker for diseases by defining patterns of aberrant proteins and will help address the specificity value of NO_2Tyr as a biomarker of diseases [132].

ROS may attack the polypeptide backbone of proteins to form a carbon-centered radical, which may react with molecular oxygen to form $ROO^{\bullet}$. For example, $OH^{\bullet}$, along with some RS, can induce the formation of the carbon-centered tyrosyl radical through hydrogen abstraction and tyrosine oxidation to yield 3,4-dihydroxyphenylalanine and levodopa (L-DOPA) [133]. Tyrosyl radicals can also derive from $RO^{\bullet}$ and $ROO^{\bullet}$ RS.

Under physiological conditions, $Tyr^{\bullet}$ is involved in a wide variety of biological reactions, including prostaglandin (PG) biosynthesis [134], and can readily undergo bimolecular coupling to form 3,3'-dityrosine by reacting with another tyrosyl radical, forming stable inter- or intramolecular cross-links (Fig. 3.7) [135].

Chlorination is another modification that can occur on proteins, potentially altering their function. This modification has been most widely reported in tyrosine residues. It is mostly caused by MPO and usually reveals activation of immune cells such as neutrophils [136]. Hypochlorous acid, the product of MPO activity, reacts with tyrosine residues, yielding 3-Cl-Tyr, 5-Cl-Tyr, and 3,5-Cl-Tyr, which are involved in impaired protein function associated with atherosclerosis as well as reduced airway function in children with cystic fibrosis (CF) [137, 138]. Measurement of Cl-Tyr can provide a valuable marker to trace halogenations during OS.

3.3.1.4 Oxidized Thiol-Containing Amino Acids

Among the most sensitive biomolecules for radical attack are sulfur centers, and oxidation of free thiols is an early event in protein oxidation. Sulfur-containing amino acids such as cysteine and methionine are particularly susceptible to ROS oxidation. Oxidation of these residues through simple thiol oxidation or the formation of mixed disulfides bonds can lead to a loss of critical sulfhydryl groups, which may induce conformational changes, protein unfolding, and degradation [4].

The oxidation of methionine residues by ROS forms methionine sulfoxide (Met-SO) and, subsequently, methionine sulfone (Eq. 3.1). Methionine could also be irreversibly oxidized by other types of RS such as $ONOO^-$ and HOCl to methionine sulfoxide.

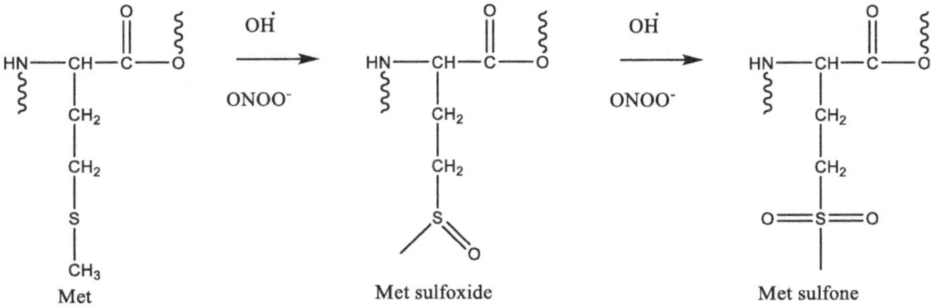

(3.1)

Cysteine residues on proteins can be readily modified due to the free thiol (–SH) group, which can be reversibly oxidized to form disulfide bonds. Activated oxygen can abstract hydrogen atoms from cysteine residues to form a thiyl radical that will cross-link to the second free thiyl radical to form a disulfide bridge. This radical coupling reaction forms either inter- (P1-S-S-P2) or intramolecular (P1-S-S-P1) disulfide cross-linkages. Besides this cascade of reaction, cysteine can be oxidized to cysteine sulfenic acid derivatives (Fig. 3.8). This oxidation occurs in the presence of strong oxidants such as HOCl, H_2O_2, and $OH^{\bullet}$ and represents another route of disulfide production.

The change of the sulfhydryl side chains of cysteine, by an initial RS-mediated two-electron oxidation, is the most commonly recognized redox posttranslational modification of proteins [139]. Generally, such reactions use the electrophilic nature of a particular RS, such as a nucleophilic attack by deprotonated thiols releases either H_2O_2 or Cl^- to form sulfenic acid, which can further form internal or mixed disulfides with other products. Indeed, sulfenic acids are reactive intermediates that can be trapped by internal thiols to form intra-protein disulfides. They may react with GSH or other external thiols to form intermolecular disulfide species or cyclize into sulfenamide derivatives [140, 141].

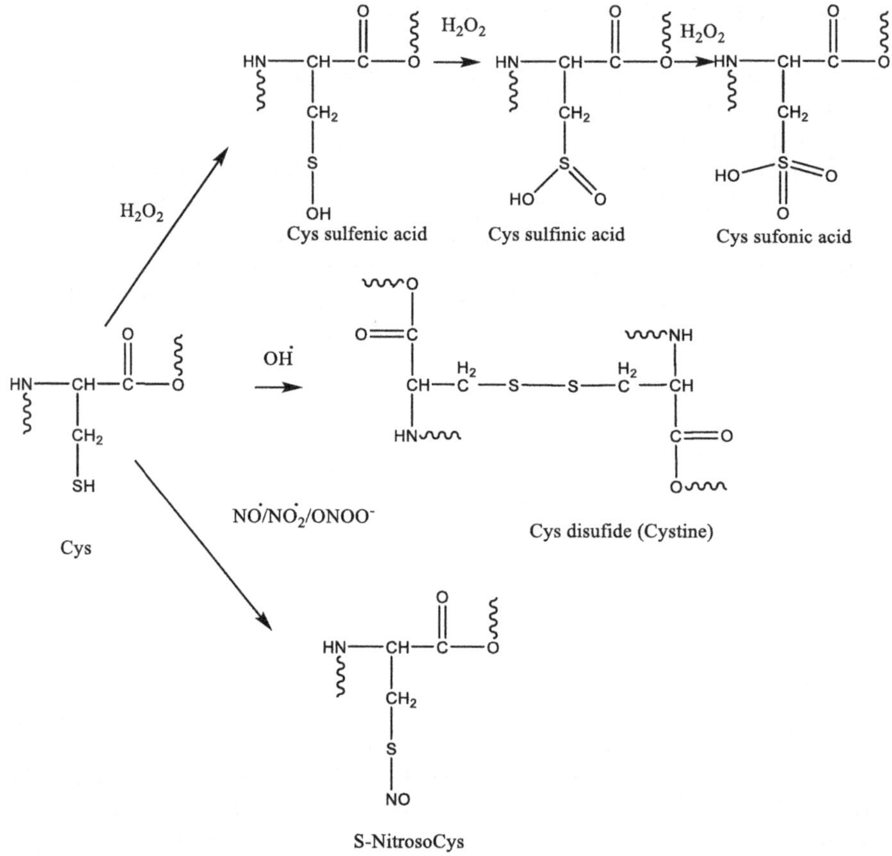

Fig. 3.8 Modification of cysteine by reactive species (ROS and RNS)

Cysteine sulfinic and sulfonic acids (cysteic acid (RSO_3H)) can be formed by subsequent two- and four-electron oxidations of sulfenic acid congeners in the presence of a strong RS as well. For example, cysteine oxidation by HOCl leading to the formation of cysteine sulfonic acid is mediated by a sulfenyl chloride intermediate (RSCl) reaction with water. The discovery of sulfiredoxin proteins, which can convert sulfinic acids back to thiols, reveals an additional mechanism of redox regulation at this higher oxidation state [142, 143]. Cysteine oxidation by HOCl usually leads to the formation of cystine (RSSR) via additional cysteine residues.

ROS production can result in the oxidation of cystine residues within proteins, resulting in sulfenic, sulfinic, and sulfonic acid derivatives as well as Cys disulfide. RNS usually results in the formation of S-nitrosocysteine. Disulfide bonds interfere with the folding and refolding processes of native proteins and disturb protein homeostasis by causing their aggregation in living cells [144]. By enhancing the aggregation tendency of proteins, disulfide bonds are believed to participate in the progression of some neurodegenerative diseases [145].

Disulfide bond formation in proteins also includes glutathionylation by disulfide exchange catalyzed by thiol transferase (Eq. 3.2) [146].

$$\text{Protein-SH} + \text{GSSG} \rightarrow \text{Protein-SSG} + \text{GSH} \tag{3.2}$$

A nonenzymatic pathway leading to glutathionylation is also possible when $OH^{\bullet}$ reacts with the cysteine residues in proteins to form thiyl radicals, which may react with GSH to form an intermediate that can undergo a redox reaction with molecular oxygen to form a glutathionylated protein and $O_2^{\bullet-}$ (Eqs. 3.3, 3.4 and 3.5).

$$\text{Protein-SH} + OH^{\bullet} \rightarrow \underset{\text{Thiyl radical}}{\text{Protein-S}^{\bullet}} + H_2O \tag{3.3}$$

$$\text{Protein-S}^{\bullet} + \text{GSH} \rightarrow \underset{\text{Dithiyl radical}}{\text{Protein-SS}^{\bullet-}} + \text{G} + H^+ \tag{3.4}$$

$$\text{Protein-SS}^{\bullet-} + \text{G} + O_2 \rightarrow \underset{\text{Glutathionylated protein}}{\text{Protein-SS-G}} + O_2^{\bullet-} \tag{3.5}$$

Overall, modifications of sulfur-containing amino acids include reversible processes such as the oxidation of critical sulfhydryl groups on cysteine residues through formation of homogeneous or mixed disulfides, simple thiol oxidation to higher oxidation states, or irreversible oxidation of methionine residues to the corresponding sulfoxides.

Under nitrosative stress, cysteine could be one of the important targets undergoing S-nitrosation. Peroxynitrite can react with thiols and undergo irreversible oxidation of cysteine, resulting in the nitrosylation of sulfhydryl groups [147]. Although this modification is required for cellular signaling, S-nitrosothiols could represent neo-epitopes capable of eliciting an immune response, given the unstable nature of the S–N bond (Fig. 3.8) [117].

3.3.1.5 Oxidized Aromatic Amino Acids

The oxidation of aromatic amino acids is often the more common event after that of thiols. It includes the irreversible modifications of phenylalanine, histidine, and tryptophan rings. Electrophilic attacks of $OH^{\bullet}$ on the aromatic ring of Phe, His, and Trp are carried out through a variety of pathways, including hydroxylation (hydroxylation of phenylalanine with the formation of *o*-, *m*-, *p*-tyrosine, and

3,4-dihydroxyphenylalanine) or formation of urea functionality in the form of 2-oxohistidine (Eqs. 3.6 and 3.7).

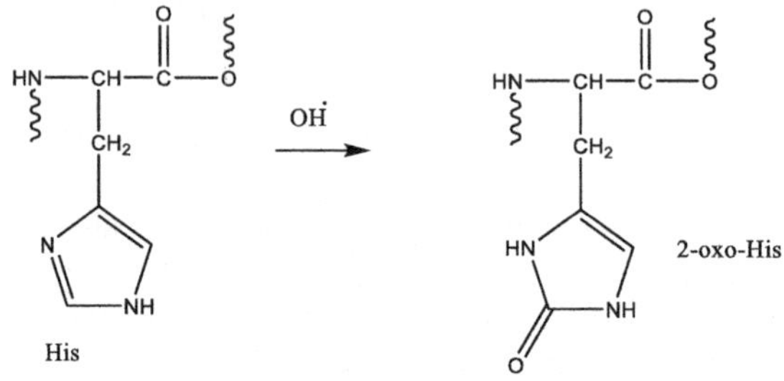

(3.6)

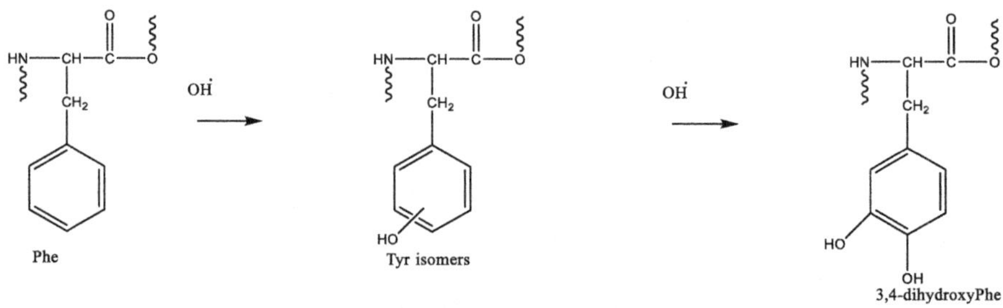

(3.7)

The interaction of ROS with tryptophan results in tryptophan oxidation to hydroxytryptophan, giving rise to *N*-formylkynurenine or kynurenine (Fig. 3.9).

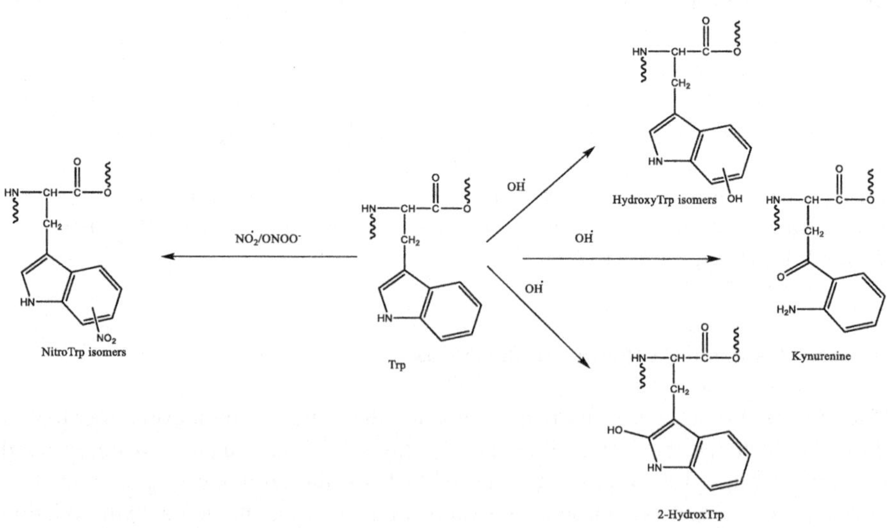

Fig. 3.9 Modification of tryptophane by reactive species (ROS and RNS)
RNS mediate oxidative modification of Trp, resulting in the formation of nitrated isomers, and ROS result in the formation of different oxidized derivatives

N-formylkynurenine can easily interact with protein amino groups in a Schiff-base reaction between proteins.

Peroxynitrite is also reactive against aromatic amino acids such as Trp and Phe, whose reaction typically involves prior formation of the aromatic radical species. Nitration of tryptophan residues in inflamed tissues has been extensively reported in the literature. This modification modulates the structure and function of proteins in many disease states, particularly atopic dermatitis [113, 114, 148, 149].

Chloramine generation with histidine residues has also been reported. Chloramines constitute important intermediates in stable end products of chlorination [150]. Under extreme conditions, oxidations of aromatic amino acids may result in ring cleavage in histidine or tryptophan [151].

3.3.1.6 Other Modified Amino Acids

The aberrant oxidation of lysine, arginine, proline, and threonine residues, usually catalyzed by redox-cycling metal ions (e.g., Fe^{2+} and Cu^{2+}), can result in the conversion of the side chain amines to carbonyls (aldehydes from lysine, arginine, and proline; ketone from threonine), potentially promoting protein dysfunction. As mentioned before, increasing protein carbonyl content in several biological fluids serves as a marker of the OS status in the organism. Irreversible oxidation of some amino acids leads to hydrolysis of the peptide bond and protein backbone cleavage like in the case of proline oxidation and acrylamide formation from asparagine. Proline oxidation mainly occurs in collagen, which is particularly rich in proline and hydroxyproline and seems to be one of the most affected proteins during OS [152]. Lysine residues can be subject to oxidative attack or modification to Schiff's bases in the presence of glucose, which is considered a relevant player in OS (see Sect. 3.2). Finally, chloramines on lysine residues constitute important intermediates in stable end products of chlorination [150, 153].

3.3.1.7 Mixed Forms of Oxidized Proteins

Biomarkers of Carbohydrate-Modified Proteins

Several families of stable oxidized protein moieties have been characterized, including AGEs. Functional groups of proteins can react with several products resulting from ROS-induced glycoxidation, generating inactive adduct derivatives [154]. The reaction between reducing sugars, their oxidative products (carbonyls), and proteins is called glycation, and AGEs are formed as late-stage products of this reaction [155].

The reactive electrophiles, mono- and dicarbonyls, are usually formed during glucose oxidation, a process that is enhanced under OS. ROS undergo many complicated reactions with glucose, including dehydration, condensation, oxidation, and rearrangement, to produce highly electrophile reactive dicarbonyl compounds such as 3-deoxyglucosone (3DG), glyoxal, and methylglyoxal [156]. This process is increased under OS. Therefore, AGEs are a heterogeneous group of molecules, with CML and pentosidine as the main products [157].

AGEs increase with age, and their formation has been related to the level of carbohydrates; so, they have been linked to DM and obesity as well as other diseases, including atherosclerosis, Alzheimer's disease, and renal insufficiency [158–160]. The role of the receptor of AGEs (RAGE) and the RAGE–nuclear factor kappa B (NF-κB) axis in neuroinflammation is in line with the nonenzymatic glycosylation theory of aging, suggesting a central role of the AGEs in age-related cognitive decline [161]. Additionally, the soluble receptors of AGEs (sRAGEs) have been found to play an important role in the pathogenesis of acute respiratory distress syndrome (ARDS) [162].

Autoantibodies (aAbs) to immunoglobulin (Ig)G-AGE were detected in patients with rheumatoid arthritis, suggesting that glycation of IgG results in the generation of immunogenic epitopes, potentially inducing an immune response with circulating aAbs [163]. Turk et al. also suggested a role for AGE immune complexes in the atherosclerotic process [164].

The same group demonstrated that both diabetic and nondiabetic patients with coronary artery disease had higher concentrations of circulating immune complexes containing the AGE moiety compared to healthy subjects, whereas only diabetics had higher anti-AGE antibodies [164]. Overall, modifications of the protein backbone under an ROS attack can occur through the glycoxidation process. It should be pointed out that carbonyl group formation usually occurs during the glycation and glycoxidation reactions of lysine residues. Lysine ε-amino groups react with reducing sugars and their oxidative products to generate several carbonyl derivatives.

Biomarkers of Lipid-Modified Proteins

Besides being altered by direct glycol-oxidation, proteins may be modified as a consequence of lipid peroxidation. This structural change can be considered as a covalent modification of proteins, generating inactive adduct derivatives classified as advanced lipoxidation end products (ALEs). Several secondary lipid peroxidation products can initiate oxidative damage to proteins. In the presence of redox-cycling metal ions, fatty acid hydro- or endoperoxides ultimately decompose to reactive by-products such as MDA, acrolein, and the α,β-unsaturated aldehydes, 4-hydroxy-2-hexenal (4-HHE), 4-hydroxy-2-nonenal (4-HNE), and 4-oxononenal (4-ONE) (Fig. 3.10) [165]. In particular, the aldehydic electrophiles, 4-HHE, 4-HNE, and MDA react with nucleophiles such as the cysteine thiol or, to a lesser extent, lysine and histidine residues to form a protein carbonyl adduct through a Michael addition reaction (see Sect. 3.4) [166]. 4-HNE or MDA can bind in a Michael-type reaction to thiol groups as well.

Hexanoyl lysine and propanoyl lysine have attracted attention because they are derived from the reaction of lysine with the lipid hydroperoxides of ω-6 and ω-3 polyunsaturated fatty acids (PUFAs), respectively [167–169]. Hexanoyl lysine is also one of the parameters measured for the OS profile in the "anti-aging dock" [70].

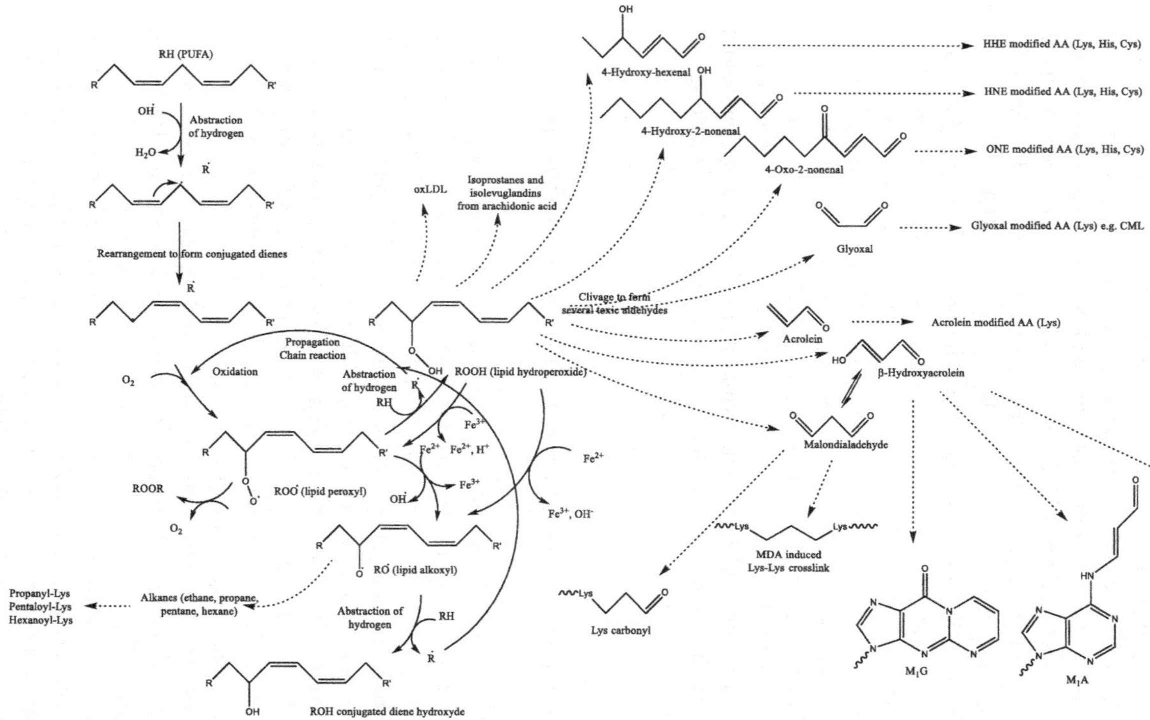

Fig. 3.10 An overview of the RS-induced lipid peroxidation process with cascade radical chain reactions involving initiation, propagation, and termination steps
Following hydrogen abstraction by RS (particularly OH˙) and diene conjugation of the resulting alkyl radical, the addition of molecular oxygen produces a peroxyl radical, which may promote a radical chain reaction. Indeed, the peroxyl radical can further propagate the peroxidation reaction by attacking a new lipid molecule. It can also be reduced to a more stable hydroperoxide. The latter may react with Fe^{2+}, generating the alkoxyl radical, which may establish a further chain reaction or be degraded to hydrocarbons (pentane). A series of further rearrangements of hydroperoxide results in degradation of the oxidized fatty acid, concomitantly releasing reactive aldehydes such as MDA, 4-HNE, acrolein, and glyoxal. This process can result in several dozen propagation steps before it is stopped by a termination reaction. Examples of such termination reactions are the recombination of peroxyl radicals with other FRs. Generation of adducts between lipid-derived products and protein may occur, especially for membranous proteins

In general, peroxidized lipid-modified lysine is excreted in greater quantities in the urine of patients with diabetes mellitus compared to healthy subjects [167]. Overall, protein may be modified as a consequence of lipid peroxidation and the formation of Schiff's bases with the resulting aldehydes.

Advanced Oxidation Protein Products (AOPPs)
Advanced oxidation protein products (AOPPs), which share several homologies with AGEs, are rich in tyrosine moieties and have a high molecular mass, limiting their biological clearance. These oxidation products are generally considered as protein aggregates formed by disulfide bridges and/or dityrosine cross-linking [170]. The reaction of proteins with chlorinated oxidants such as HOCl results in chlorination of amino acid residues and formation of 3-Cl-Tyr and 3,5- Cl-Tyr as the main products. These chlorinated residues are also classified as AOPPs.

AOPPs are markers of OS, which mainly reflect chronic renal complications and participate in the progression of chronic renal failure [171, 172]. They have been identified as markers of inflammation in many diseases as well [173, 174]. Chlorotyrosine is generally formed during inflammation following the action of MPO and contributes to the generation of an AOPP, which may act as a mediator of the inflammation process and monocyte activation [175]. In correlation with AOPP levels, 3-Cl-Tyr has been detected in the airways of preterm infants, in children with cystic fibrosis, and in patients with atherosclerosis and rheumatoid arthritis [176–179]. High levels of AOPPs have also been detected in other diseases such as diabetes, uremia, systemic sclerosis, atherosclerosis, and CVD [152].

3.3.2 Reversible Versus Irreversible Modification of Proteins

As mentioned, RS may cause modification of proteins in a variety of ways, where some are direct and others indirect. Direct modification involves modulation of a protein's activity through carbonylation, nitrosylation, disulfide bond formation, and glutathionylation. Indirect modifications include conjugation with breakdown products of carbohydrate and fatty acid peroxidation. Modification of proteins can be classified as a result of reversible processes, such as nitrosylation, glutathionylation, oxidation of critical sulfhydryl groups on cysteine residues through formation of mixed disulfides, simple thiol oxidation to higher oxidation states, or oxidation of methionine residues to the corresponding sulfoxides.

Irreversible oxidations of amino acids may occur and include the formation of carbonyls, hydroperoxides and their reduced hydroxy species, or ring cleavage in histidine or tryptophan. Carboxylation of lysine and arginine, nitration of tyrosine and tryptophan, and protein–protein cross-linking are irreversible in nature as well.

Finally, carbonylation of proteins and the loss of the membrane protein thiol level (MPTL) are two important signs of irreversible oxidation processes that lead to loss of protein function and associated metabolic impairment.

3.4 Oxidative Damage to Lipids

Lipids are a major class of biological molecules and play many key roles in cells. An imbalance in the oxidant and antioxidant ratio can lead to oxidative damage to many lipids. When ROS levels are high, for example, beyond the cellular homeostatic threshold, they can induce lipid peroxidation in both cellular and organellar membranes. This process is a cyclical chain reaction and involves a myriad of RS (Fig. 3.10).

The usual targets are membrane constituents (e.g., phospholipids and cholesterol) and circulating lipids (e.g., lipoproteins and unesterified cholesterol). The unsaturated bond on phospholipids and the ester linkage between glycerol and the fatty acid are the common sites of RS-induced oxidation. FRs initiate lipid peroxidation by abstracting a hydrogen atom from methylene carbon in the allylic position in the side chain of lipid structures. The homo-conjugated double bonds found in the fatty acid of a lipid molecule make it easier for FRs, called initiators, to remove the hydrogen atom and to generate lipid-derived radicals. For this reason, PUFAs (labeled LH) are more sensitive to FR oxidation than are monounsaturated fatty acids (MUFAs) or saturated fatty acids (SFAs). Other lipids containing several double bonds like cholesterol are also susceptible to FR-mediated reactions.

Compounds that can initiate lipid peroxidation include $OH^{\bullet}$, $RO^{\bullet}$, and $ROO^{\bullet}$. When the hydrogen atom is removed from the methylene group ($-CH_2-$), an unpaired electron is left behind on the carbon atom ($-^{\bullet}CH-$). In some cases, hydrogen abstraction leads to a homolytic cleavage of the covalent bond. Singlet oxygen can directly react with fatty acids rather than extracting a hydrogen atom to initiate a chain reaction, but this route represents a minor lipid oxidation pathway.

Overall, the C–H bond in the allylic position is particularly vulnerable to oxidation reactions. Hydrogen atoms in such positions can react with FRs, called initiators, and are therefore easily abstracted from the C–H bond. All these reactions during the lipid peroxidation process are conventionally grouped under the name of "initiation step." This step results in the formation of a lipid alkyl radical ($L^{\bullet}$) (Eqs. 3.8 and 3.9).

$$\mathrm{A{-}B} \overset{\text{Initiation}}{\rightarrow} \mathrm{A}^{\bullet} + \mathrm{B}^{\bullet} \tag{3.8}$$

$$\mathrm{A}^{\bullet} + \mathrm{LH} \overset{\text{Initiation}}{\rightarrow} \mathrm{AH} + \mathrm{L}^{\bullet} \tag{3.9}$$

The initiation step is followed by conformational changes and an isomerization reaction, leading to the transformation of an isolated double bond into a conjugated one called conjugated dienes. The lipid alkyl radical then reacts with molecular oxygen to produce a key and powerful oxidant called peroxyl radical ($LOO^{\bullet}$). This newly formed reactive radical can then, in turn, extract hydrogen atoms from the surrounding molecules or neighboring unsaturated fatty acid and initiates another chain reaction of lipid peroxidation. The peroxyl radical is, therefore, converted to lipid hydroperoxide (LOOH) and yields a new lipid alkyl radical. Secondary alkoxyl radicals can also be generated from $LOO^{\bullet}$ and accentuate hydrogen abstraction from PUFAs, leading to the "reinitiation" of the chain reaction (Eqs. 3.10 and 3.11). Once a species with unpaired electrons is created, radical reactions are able to propagate themselves. Consequently, the process of lipid peroxidation may proceed in an autocatalytical manner. The vast majority of FRs generated during lipid peroxidation is the result of these reactions known as the propagation step.

$$L^{\bullet} + O_2 \overset{\text{Propagation}}{\rightarrow} LOO^{\bullet} \tag{3.10}$$

$$LOO^{\bullet} + LH \overset{\text{Propagation}}{\rightarrow} LOOH + L^{\bullet} \tag{3.11}$$

After hydrogen abstraction, lipid alkoxyl is reduced to the chemically more stable corresponding alcohol form (LOH) or can fragment, as through metal-catalyzed β-scission reactions. Lipid hydroperoxides are semi-stable compounds, except in the presence of transitional metals or metal complexes. Indeed, the role of metal ions is not restricted to the initiation reaction where they contribute to the overproduction of $OH^{\bullet}$; they can also interact with lipid hydroperoxides and accelerate their decomposition or promote the generation of secondary radicals such as alkoxyl and peroxyl radicals (Eqs. 3.12 and 3.13).

$$LOOH + Fe^{2+} \overset{\text{Propagation}}{\rightarrow} LO^{\bullet} + OH^{-} + Fe^{3+} \tag{3.12}$$

$$LOOH + Fe^{3+} \overset{\text{Propagation}}{\rightarrow} LOO^{\bullet} + H^{+} + Fe^{2+} \tag{3.13}$$

Radical chain reactions are halted when termination reactions take place. These are reactions between two radical intermediates, resulting in the generation of non-radical products (Eqs. 3.14, 3.15 and 3.16).

$$LOO^{\bullet} + LOO^{\bullet} \overset{\text{Termination}}{\rightarrow} LOOL + O_2 \tag{3.14}$$

$$L^{\bullet} + L^{\bullet} \overset{\text{Termination}}{\rightarrow} L\text{—}L \tag{3.15}$$

$$LOO^{\bullet} + B^{\bullet} \overset{\text{Termination}}{\rightarrow} LOOB \quad \text{B can be R or L} \tag{3.16}$$

Through a termination process, alkyl, alkoxyl, and peroxyl radicals in neighboring molecules may combine, leading to the formation of lipid–lipid cross-linkages.

The peroxidation reaction is also terminated when it reaches other biological entities like proteins and nucleic acids. These damaged biomolecules are usually detected and removed from cells. Lipid hydroperoxides are semi-stable compounds and ultimately decompose to several aldehyde and non-aldehyde products.

Usually in the presence of reducing metal ions, $LOO^{\bullet}$ can react with molecular oxygen by cyclization to produce a cyclic peroxide, which can undergo a second cyclization to form a bicyclic peroxide and then endoperoxides. Endoperoxides, via a series of reactions and fragmentation of the acyl chain, produce highly reactive and toxic aldehydes, with the major ones being MDA and 4-HNE, but also including 4-oxononenal (4-ONE), alkadienals, crotonaldehyde, acrolein, α,β-unsaturated aldehydes, glyoxal, hexanal (derived from ω-6 fatty acids), propanal, 4-HHE (derived from ω-3 fatty acids), and isolevuglandins (IsoLGs) (Fig. 3.10) [180]. Lipid hydroperoxide can easily decompose into non-aldehyde species, including alkanes, lipid epoxides, and alcohols. Besides, lipid hydroperoxides may undergo a series of rearrangement and cyclization processes to generate a well-known species called IsoPs. Cholesterol peroxyl has also been shown to undergo oxidation, to give rise to a variety of epoxides and alcohols.

The most important products from the analytical point of view are as follows: MDA and other thiobarbituric acid reactive substances (TBARS); 4-HNE, hexanal, propanal, and 4-HHE, alkanes (ethane, pentane, isoprene, etc.); and IsoPs. Several of these degradation products, including IsoPs, 4-HNE, and MDA, have been shown to be present in the exhaled breath and breath condensates of humans.

Lipids play structural and functional roles in membranes. Their peroxidation causes impairment of biological membrane functioning by decreasing fluidity and inactivating membrane-bound proteins (enzymes, ionic canals, and receptors). Disruption of membrane lipid bilayer arrangement and permeability, through the peroxidation process and covalent binding, can lead to necrosis or the induction of apoptosis of cells [181]. Peroxidation of lipids may also affect subcellular membranes and lead to an impairment of their function and biological properties. As peroxidation of lipids progresses and products accumulate, ion channels may be affected, membrane-bound receptors, membrane transport proteins, or enzymes may be inactivated, or the lipid bilayer may itself become more permeable.

The side products of lipid peroxidation can also exert toxic effects, even in distant locations from the primary peroxidation event. Malondialdehyde and a large group of hydroxyalkenals, primarily 4-HNE and 4-HHE, are best known. Such aldehydes can interact with protein residues and DNA. All modified proteins are cumulatively designated as advanced lipid peroxidation end products ("ALEs"). They result from unstable lipid peroxides that can easily break down in the presence of RS and transition metals. Isoprostanes are another group of lipid peroxidation products that are generated via the peroxidation of arachidonic acid.

Lipid peroxidation is considered a major mechanism in the generation of diseases and disorders such as cancer, neurodegenerative diseases, and CVD, which function through alteration of the structural integrity, fluidity, and permeability of cell membranes [182]. Under pathological conditions (e.g., atherosclerosis and ischemia), oxidation of PUFAs by means of ROS and FRs is believed to be exacerbated by the presence of divalent metal ions. Upon interaction with iron, copper, chromium, cadmium, and nickel, RS are believed to change to highly toxic compounds and play an active role in lipid peroxidation and in the promotion of carcinogenesis [183].

3.4.1 Biomarkers of Oxidative Damage to Lipids

3.4.1.1 Malondialdehyde (MDA)

Malondialdehyde is a three-carbon compound and one of the most relevant end products of lipid peroxidation. It is formed from peroxidized PUFAs, mainly arachidonic acid in cell membranes and low-density lipoproteins (LDLs). It is a known mutagen due to its ability to react with nucleic acids (see Sect. 3.1) [184, 185]. MDA is a volatile molecule that reacts via Schiff-base formation with a number of nucleophilic amino groups of protein and DNA, resulting in intermolecular cross-links. MDA can react with a number of amino acid residues in proteins, including lysine, arginine, methionine, tyrosine, and histidine, primarily through 1,4-Michael addition reactions and is, therefore, capable of inactivating many essential macromolecules that lead to cell membrane damage [186].

MDA is easily detected in several biological samples (blood, urine, cerebrospinal fluid, etc.) and has been widely used as a measure of OS. The popularity of MDA as a primary biomarker of OS is evident from the number of diseases in which MDA levels have been reported to be increased. It is commonly found in high levels in atherosclerosis, DM, liver disease, Alzheimer's and Huntington's diseases, amyotrophic lateral sclerosis (ALS), preeclampsia, and cancer [129, 187–190]. High levels of MDA and cytoplasmic lipofuscin were also observed in HD [128, 129].

3.4.1.2 4-Hydroxy-2-Nonenal and 4-Hydroxy-2-Hexenal

4-Hydroxy-2-nonenal (4-HNE) and 4-hydroxy-2-hexenal (4-HHE) were first reviewed by Esterbauer et al. as aldehyde end products of lipid peroxidation [191]. 4-HNE is generated by the nonenzymatic oxidation of ω-6 PUFAs, whereas 4-HHE is formed from ω-3 PUFAs. More precisely, 4-HNE is generated by the degradation of lipid hydroperoxides—hydroperoxyoctadecadienoate (HPODE) from linoleate or hydroperoxyeicosatetraenoate (HPETE) from arachidonate—after cyclization and β-scission.

4-HNE is the major electrophilic product of lipid peroxidation and is a highly reactive diffusible compound, known to induce/regulate various cellular events [9]. Numerous studies have shown that HNE at physiological levels stimulates cell proliferation and differentiation as well as induces a cytoprotective response via multiple signaling pathways: activation of protein kinase C (PKC), of the family of mitogen-activated protein kinase (MAPK), and of tyrosine kinase receptors as well as inhibition of Ca-ATPases with subsequent accumulation of Ca^{2+} in the cytoplasm [192, 193].

4-HNE and 4-HHE are more reactive than MDA with proteins, and the resulting proteins are not degraded efficiently by the proteasome. Of particular interest is 4-HNE as it can chemically target cysteine residues and nucleophilic amino moieties (e.g., lysine, histidine, less probably arginine, methionine, and tyrosine) in proteins and generate alkylated and cross-linked proteins through 1,4-Michael-type adducts or Schiff bases. In a way, HNE could display second messenger behavior as it may activate kinases and stimulate cell proliferation (e.g., vascular endothelial cells (ECs) and smooth muscle cells) [194]. This reactive aldehyde can form adducts with both intracellular proteins, such as histone deacetylase (HDAC) 2 and extracellular proteins, such as collagen and fibronectin, thus altering their function, which can impact cell function and stimulate apoptotic pathways [195–197].

Since 4-HNE along with 4-HHE forms stable adducts with side chains of the proteins, they also have several cytotoxic, genotoxic, and chemotactic effects [198]. 4-HNE also causes depletion of intracellular GSH, enhanced production of peroxide and fibronectin, and activation of the epidermal growth factor receptor [199–201]. Overproduction of 4-HNE may have a detrimental effect on a variety of signaling molecules, such as Nrf-2, HDAC2, and IκB kinase (NF-κB activation).

4-HNE is a major reactive compound produced by lipid peroxidation at sites of inflammation along with MDA. At high concentrations, 4-HNE has been proposed to play a key role in DM, cardiovascular disease, and some neurodegenerative diseases [194, 202]. 4-HNE is commonly found in high concentrations in the brains of patients with Alzheimer's disease. It is a neurotoxic agent causing apoptosis or altering the microtubule structure [203]. It is also prone to react with lipoid acid and to form adducts with proteins (target amino acids are cysteine, histidine, and lysine) in the brains of patients with Alzheimer's disease [204–206]. In particular, 4-HNE induces modifications in the conformation of Tau proteins following adduction reactions, which favor neurofibrillary tangle formation [207]. As in Alzheimer's disease, the presence of 4-HNE-modified proteins has also been confirmed in Parkinson's disease (PD) [208]. In ALS, OS has also been evidenced by elevated levels of 4-HNE [209, 210].

Many of the effects of ROS in airways are mediated by the secondary release of 4-HNE. It has been demonstrated that 4-HNE-modified proteins (protein carbonyls) are present in high levels in the airways (i.e., alveolar epithelial cells, airway endothelial cells (ECs), and neutrophils) in smokers with airway obstruction compared to control subjects [211]. This reveals not only the presence of 4-HNE but also that 4-HNE modifies proteins in lung cells to a great extent in patients with COPD. The increased levels of 4-HNE adducts are inversely correlated with the forced expiratory volume in the first second, suggesting a role of 4-HNE in the pathogenesis of COPD.

3.4.1.3 Isoprostanes (IsoPs)

Isoprostanes are cyclic peroxides formed from PUFAs with at least three double bonds, including linolenic acid, arachidonic acid, eicosapentaenoic acid (F_3-IsoPs), and docosahexaenoic acid (DHA) (F_4-IsoPs, sometimes called neuroprostanes). F_2-IsoPs are a group of lipid degradation products that are generated during the OS process and originate from the peroxidation of arachidonic acid (an ω-6 PUFA). They comprise a series of bioactive prostaglandin-like compounds formed during auto-oxidation and nonenzymatic FR-initiated peroxidation of arachidonic acid (normally enzymatically metabolized by cyclooxygenase (COX) to form a wide variety of prostanoids (prostaglandins and thromboxanes)). Indeed, it had been demonstrated that potent RS, such as $ROO^•$, $OH^•$, $NO_2^•$, $ONOO^−$, and higher oxidation states of heme and hemoproteins (ferryl heme) are capable of initiating peroxidation of arachidonic acid [212]. The initial $H^•$ abstraction from arachidonic acid can occur at different points in the carbon chain (Fig. 3.11). Roberts and Morrow found that these types of RS also induced IsoP formation in biological systems [213]. As depicted in Fig. 3.11, transformation of arachidonic acid yields racemic mixtures of IsoPs that are structurally similar to endoperoxides called PGs [214]. The mechanism of formation of IsoPs is illustrated in Fig. 3.11, which shows the pathways that lead to the formation of four regioisomers (I–IV) of IsoPs according to the position of the hydroxyl function. The endoperoxides can undergo reduction to form F_2-IsoPs or rearrangements to produce E_2-IsoPs, D_2-IsoPs, and/or isothromboxanes. F_2 stands for the final number of insaturations that arachidonic acid is capable of giving. They are reliable markers of lipid peroxidation and possess a 1,3-dihydroxycyclopentane ring with the OH groups in the syn position. Overall, there are 64 possible isomers of IsoPs, which can be generated from the oxidation of the 4 regioisomers, with each containing 16 diastereomers. It is believed that the structure of each IsoP isomer depends upon the initial site to attack by the RS (Fig. 3.11) [214, 215]. One representative pathway is the formation of G_2-IsoPs, which are similar to prostaglandin G2 (PGG_2) produced from the COX pathway. G_2-IsoPs are reduced to the hydroxyl analogue (H_2-IsoPs), and their complete reduction generates the 64 regioisomers and stereoisomers of the F_2-IsoPs mentioned earlier (Fig. 3.11).

In 1990, Morrow and coworkers first discovered these nonenzymatically produced F_2-like prostanoid derivatives of arachidonic acid, and, because these oxidized lipids are isomers of COX-generated prostaglandin F_2, they were collectively referred as F_2-IsoPs [216]. 15-F_{2t}-IsoP or 8-*iso*-$PGF_2\alpha$ is one abundant endogenous F_2-IsoP, and, at present, represents one of the most thoroughly investigated markers of OS. The letter "t" stands for the trans-position of the oriented side chain to the prostane ring [217, 218].

Analogous products resulting from the oxidation of DHA (called neuroprostanes) have also been identified, but, in this case, 8 regiomers and a total of 128 possible isomers can be formed [219]. Subsequent studies demonstrated that IsoPs are formed in situ within phospholipids and are then cleaved and released in their free form into

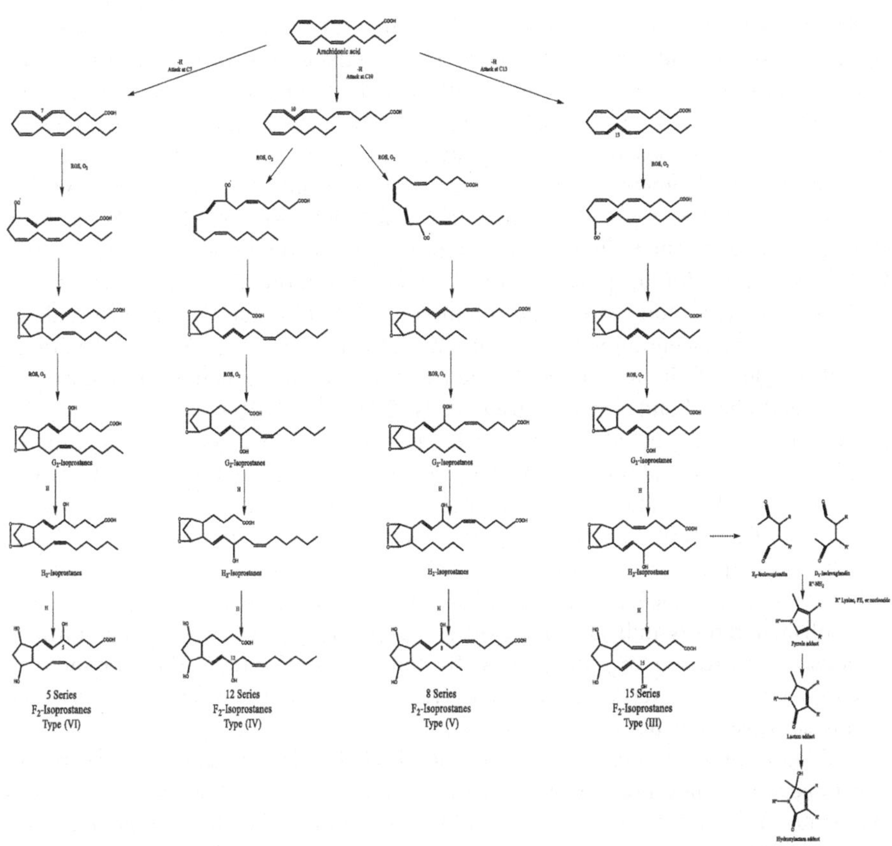

Fig. 3.11 Different types and series of isoprostanes formed from the oxidation of arachidonic acid A variety of potent oxidants such as hydroxyl and peroxyl radicals initiate lipid peroxidation by abstracting allylic hydrogen atoms from arachidonic acid. The resulting addition of oxygen to lipid radicals, followed by endocyclization and reduction, results in the production of isoprostanes. Four major types of isoprostanes are formed by the FR attack on arachidonic acid. The most commonly measured isoprostanes in clinical laboratory are those belonging to type III (F_2 isoprostane 15 series). Type III isoprostanes are formed when FR attacks occur at carbon 13 of arachidonic acid Specific IsoLG regioisomers differ by the position of hydroxyl groups on the side chains (5-, 8-, 12-, and 15-IsoLGs) and the relative orientation of their keto and aldehyde moieties (D_2-IsoLGs and E_2-IsoLGs)

circulation by the action of phospholipase A_2 [220]. Generally, IsoPs are stable in the blood environment and are also excreted in the urine. Overall, F_2-IsoPs are generated by the peroxidation of arachidonic acid via the generation of several ROO$^•$ isomers, which undergo endocyclization to prostaglandin-like compounds. In theory, up to 64 isomeric F_2-IsoPs can be formed. Part of the formed IsoPs is released from phospholipids by the action of phospholipases. However, the bulk of IsoPs of plasmatic lipoproteins are in the esterified form.

The isoprostanes that were originally identified as biomarkers of OS have been shown to exert potent biological effects. They mediate most of the RS effects and have the ability to affect signaling pathways, including those that regulate the apoptotic form of cell death [221]. There has been considerable literature showing the biological activities of IsoPs in platelet activation [222–224]. They are also vasoconstrictors in many species and in various vascular beds [225–227]. Of the 64 isomers of IsoPs, 8-*iso*-PGF_2 (also known as 15-F_2IsoP or $iPF_{2\alpha}$–III) has been shown to have potent pulmonary and renal vasoconstrictor activity [155]. 8-*iso*-$PGF_{2\alpha}$ is known for its powerful renal vasoconstriction effect. For example, it decreases kidney blood flow and glomerular filtration rate by almost half at extremely low concentrations. The biological effects of IsoPs also include modulation of platelet activity and monocyte adhesion as well as induction of the proliferation of endothelial and smooth muscle cells [223, 228–230].

In 2014, Jamil et al. investigated the role of one of the oxygenated eicosapentaenoic metabolites formed by nonenzymatic oxidation (5-F_{3t}-IsoPs) in the regulation of excitatory neurotransmission. In this study, the group revealed the beneficial role of this compound by reducing aspartate neurotransmitter release, thereby slowing the progression of ocular neuropathic disease by modulating K^+-induced glutamate release by 5-epi-5-F_{3t}-IsoP in isolated bovine retina [231].

Published reports indicate that IsoP levels are increased shortly after birth in response to increased oxygen tension by the activation of the thromboxane A_2 receptor, which may represent a novel physiological signal to stimulate postnatal ductus arteriosus closure [232–234].

Oxidation of arachidonate-based phospholipids by RS can produce other bioactive molecules. They present pro-inflammatory abilities affecting both monocytes and neutrophils, which causes increased EC interaction and high cytokine release. It is suggested that they may play an important role in a number of chronic inflammatory diseases [235]. As mentioned before, IsoPs are similar to prostaglandins and carry out numerous biological functions. However, they may form protein adducts with lysine side chains. These compounds have also been shown to be biologically active as mediators of oxidant injury [236].

Since IsoP formation is mediated by RS attacks on PUFAs and they possess potent biological activity, their measurements in plasma and urine represent the gold standard for monitoring OS in vivo. There are several other criteria explaining why among all the available compounds, IsoPs are one of the most reliable attractive markers for assessing OS. The measurement of F_2-IsoPs is the closest to be considered a "gold standard" for the following reasons:

- These compounds are specific products of arachidonic acid as they are not produced by any known enzymatic pathways, including the COX or lipoxygenase (LOX) pathways.
- They are quite stable in vivo and ex vivo and hence are more reproducibly measured.
- F_2-IsoP levels can be quantified in extracellular fluids such as plasma, urine, and exhaled breath condensate (EBC), making the measurement of these compounds

a relatively noninvasive approach, particularly useful for human and whole animal studies. The measurement of F_2-IsoPs and their metabolites in these matrices may prove to be a valuable assay of whole-body lipid peroxidation.
- Many of the available methods are sensitive, robust, and reproducible, and more development appear in the literature continuously.
- Positive correlations have been reported between IsoP concentrations and the incidence of several diseases known to be affected by OS [237–241].

Even if the validity of F_2-IsoPs as biomarkers of OS is well-established, it is technically quite complicated to measure F_2-IsoPs and their metabolites in biological fluids, and some limitations with respect to their measurement have to be taken into account [242]. Although F_2-IsoPs are chemically stable, once generated and released into the circulation, they are rapidly metabolized and excreted in the urine. Their rapid disappearance from plasma may hamper their practical application and usually requires highly sensitive methods.

Out of the numerous nonenzymatic oxygenated metabolites investigated, the 15-F_{2t}-IsoP (8-*iso*-$PGF_{2\alpha}$) has been suggested as a specific, reliable, and noninvasive marker of lipid peroxidation in vivo [243, 244]. These IsoPs can easily be measured in most biological fluids such as plasma, urine, EBC, and induced sputum. Their metabolite in the urine, 2,3,dinor-4,5-dihydro-8-*iso*-$PGF_{2\alpha}$, has also been shown to be a reliable noninvasive marker of OS in vivo [245].

Besides being OS biomarkers, IsoPs are considered to be indicators of many disease states. The increased concentrations of 8-*iso*-$PGF_{2\alpha}$ are suggested to play a promoting role in OS-associated pathologies such as CVD, hepatic cirrhosis, Alzheimer's disease, and inflammatory conditions [246–248]. F_2- and F_4-IsoPs are other markers of OS in Alzheimer's disease and are found in elevated levels in the brain of Alzheimer's disease patients [249–251]. Isoprostanes have also been found to be elevated in plasma, exhaled breath condensate (EBC), and bronchoalveolar lavage fluid in asthma, cystic fibrosis, and COPD patients [252–255]. Urinary excretion of IsoPs has been found to be elevated in smokers and in patients with scleroderma, hypercholesterolemia, hypertension, and diabetes mellitus [256–261].

It has been proposed that increased concentrations of F_2-IsoPs in blood circulation may contribute to fatal hepatorenal syndrome, a disorder characterized by the development of kidney failure and severe liver disease [262]. Another important consideration of IsoP detection is that these oxidized lipids increase in response to the known FR generators diquat and CCl_4, as shown in an experimental study conducted by Roberts and Morrow as well as by Awad and Morrow [213, 263]. The multi-laboratory validation study (Biomarkers of Oxidative Stress Study (BOSS)) showed that IsoP measurement was the most accurate approach to assess the OS induced by CCl_4 poisoning [264].

3.4.1.4 Isolevuglandins (IsoLGs)

Isolevuglandins (Fig. 3.11) are highly reactive compounds, which interact readily and irreversibly with primary amines (e.g., protein lysyl residues and phosphatidylethanolamine) in cells to form pyrrole (lactam) and oxidized pyrrole (hydroxylactam) adducts [265]. They have been found to induce inflammatory pathways, immune responses, and cell death as well as inhibit ion channel function in cultured cells [266–269]. These biological effects suggest that IsoLGs could contribute to disease processes, including inflammation, hypertension, arrhythmia, atherosclerosis, and neurodegeneration [265, 269].

3.4.1.5 Alkanes

Several alkanes resulting from lipid peroxidation (e.g., ethane and pentane) were found to be present in the exhaled breath and breath condensates of humans [270]. Changes in their concentrations in some pathological conditions were also demonstrated, as exhaled ethane concentrations have been shown to be increased in interstitial lung disease and scleroderma [271, 272]. Additionally, the volatile organic compound index, which includes all volatile alkanes, has been shown to be significantly elevated in diabetic patients and individuals with unstable angina [273, 274]. The levels of 8-IsoPs and alkanes, such as ethane and pentane, are increased in exhaled air condensates in smokers and in patients with COPD [275–277].

3.4.1.6 Acrolein

Acrolein is another example of a reactive aldehyde. It is not only produced as a product of lipid peroxidation but is also a well-known environmental toxin, which is present in cigarette smoke, diesel exhaust, charcoal smoke, etc. This compound forms covalent adducts with both proteins and DNA, the reason behind being a suspected carcinogen [278].

Acrolein has a high affinity toward cysteine, histidine, and lysine residues and may form adducts with both intracellular (histone deacetylase) and extracellular proteins (collagen and fibronectin), altering their function and disturbing cell function [279]. Acrolein reacts with Lys residues of apolipoprotein A-I (apoA-I), the major protein of HDL, which plays a relevant role in mobilizing cholesterol from artery wall macrophages. Additionally, it is a new marker of lipid peroxidation and is commonly measured for the monitoring of environmental pollutions.

3.4.1.7 Oxidized LDLs (OxLDLs)

Lipoproteins can also be oxidized. The process of LDL oxidation possesses the general characteristics of lipid peroxidation reactions and free radical reactions and leads to the production of what is known as oxidized LDLs (oxLDLs).

The peroxidation of LDL is believed to be a critical step in the development of atherosclerosis. Indeed, there is strong evidence that modification of an LDL is necessary for its unregulated uptake by the macrophages [280]. This process establishes the initiation of atheroma formation. Once oxidized, LDLs are internalized by macrophages through the "scavenger receptors" producing foam cells, which can release molecules that stimulate the development of atherosclerotic lesions. OxLDLs can also facilitate per se atherogenic progression by multiple mechanisms, ultimately leading to the increase of the arterial content of inflammatory cells.

Besides several pro-inflammatory and pro-atherosclerotic effects, oxLDLs are highly immunogenic and can produce several autoantibodies (aAbs) in some patients [281]. Although the oxidation of LDLs mainly occurs at the subendothelial sites of the arterial wall, small but detectable amounts of oxLDLs are present in blood circulation. These levels have been shown to correlate positively with subclinical atherosclerosis as well as with clinical coronary heart disease [282, 283]. Importantly, the presence of high concentrations of oxLDLs has been shown to be a significant predictor of future myocardial infarction [284, 285]. In a study conducted by Itabe et al., the level of oxLDLs was shown to be eightfold higher in hemodialysis patients compared to normal subjects [286].

3.4.1.8 Oxidized Cholesterol

Cholesterol can be peroxidized at different positions (5α, 6α, 6β, 7α, and 7β) by an FR or $^1O_2$ attack and yield cytotoxic products. In a non-radical pathway, $^1O_2$ undergoes an ene-type reaction with cholesterol, leading to the production of cholesterol hydroperoxides (ChOOH), with 5α-OOH-Chol as the major product and 6α-/6β-/7α-/7β-OOH-Chol as the minor products (Fig. 3.12) [287].

In the presence of FRs like LOO$^•$, formation of ChOOH occurs through hydrogen abstraction at the C-7 position to form the cholesteryl radical (Ch$^•$). The latter readily reacts with molecular oxygen to form 7α- or 7β-cholesteryl peroxyl radicals and ultimately the cholesteryl hydroperoxides 7α- or 7β-OOH-chol after subsequent hydrogen abstraction from the surrounding PUFAs (Fig. 3.12). Of particular interest are the cholesterol-derived compounds bearing epoxide function, resulting mainly from auto-oxidation and giving rise to the epimeric 5,6-epoxides (also called also 5,6-epoxysterols, of interest oxysterols 5,6a-EC). These compounds are unstable because of the high reactivity of the epoxide ring toward nucleophiles, including amines, thiols, and the hydroxyl group, thus exhibiting antioxidant properties. On the other hand, lipid peroxyl radicals (LOO$^•$) can also attack the cholesterol double bond to yield the 7α-/7β-oxyl radical, which can undergo β-hydrogen scission to yield 7-ketone [287].

Under extreme OS conditions, subsequent breakage of the B-ring will produce aldehydes called secosterols, which induce protein misfolding [166]. A common product in atherosclerotic plaques is 27-hydroxy-cholesterol, which promotes atherosclerosis via pro-inflammatory processes [288].

3.4.1.9 Hydroxyeicosatetraenoic Acid (HETE)

Fatty acid hydroperoxides are not only generated from lipid peroxidation but they can also be produced in great abundance by various LOX/COX, which represent one of the major pathways for the production of PUFA hydroperoxides and hydroxides

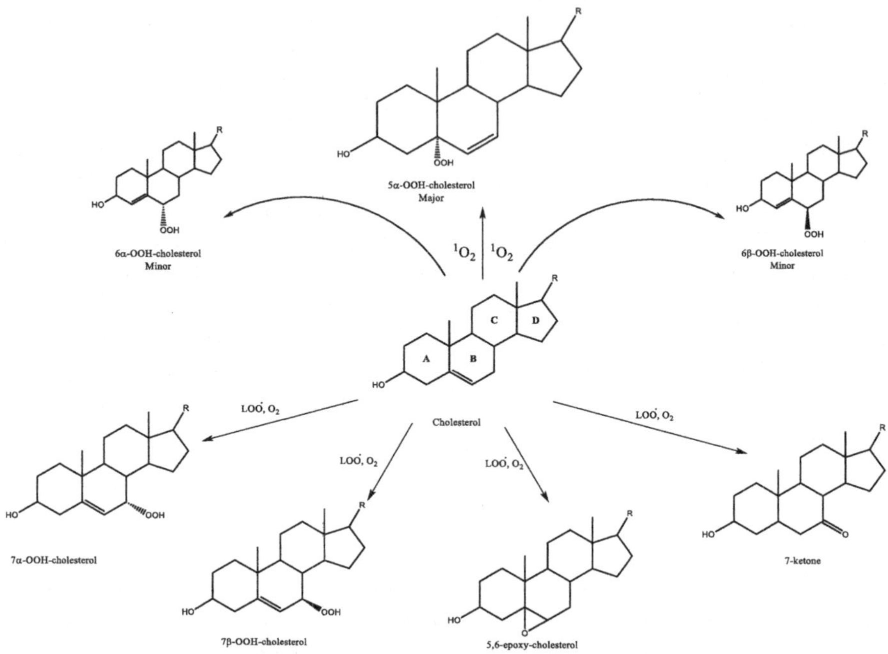

Fig. 3.12 Oxidation of cholesterol with major and minor products

from arachidonic acid [289]. Examples of these PUFA hydroperoxides and hydroxides are 15-hydroperoxyeicosatetraenoic acid (15-HpETE) and the subsequent 15-hydroxyeicosatetraenoic acid (15-HETE), as depicted in Fig. 3.13.

Further transformation of the hydroxyl fatty acid, which is biologically active, could form isoleukotrines, which are potent inflammatory mediators. An internal rearrangement in the hydroperoxyl intermediate could form epoxy alcohol (Fig. 3.13). HpETE hydroperoxides are unstable and can yield $OH^{\bullet}$ and $RO^{\bullet}$ via O–O bond cleavage.

The ability of HETE to measure levels represents an important development in analytical chemistry for the purpose of monitoring clinically relevant oxidative lipid damage. HETE is mostly a chemotactic agent for neutrophils and facilitates calcium uptake and protein kinase C mobilization [290]. 12-HETE participates in adrenocorticotropin and parathyroid secretion, modulation of mitogenic processes, and lymphocyte function, whereas 15-HETE is specialized in inhibition of neutrophil migration across the cytokine-activated endothelium [291, 292].

Linoleic acid is another target of oxidant compounds. It is the most abundant PUFA found in mammals, and the derived oxidative products likely account for the majority of lipid–protein carbonyls formed in the organism [293]. Upon lipid peroxidation, the membrane-bound linoleic acid is oxidized to yield 9- and 13-hydroperoxy-octadecadienoic acids (HPODE). These hydroperoxides undergo

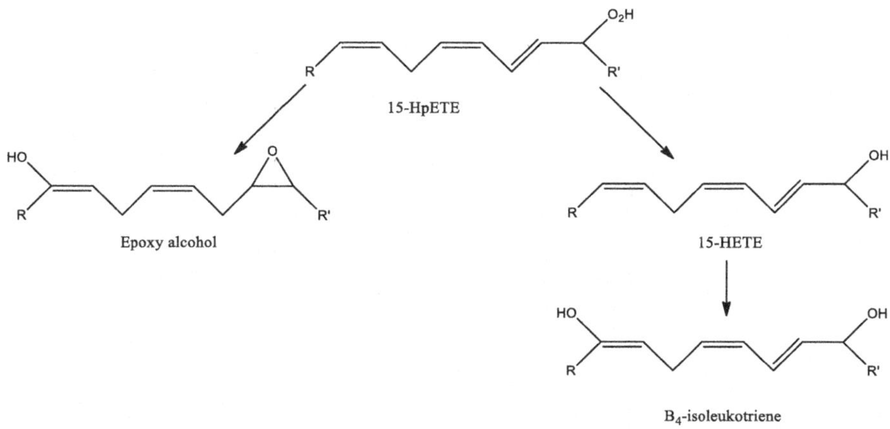

Fig. 3.13 Generation of HETE from HpETE

a fragmentation process to form α,β-unsaturated aldehydes that readily modify the surrounding proteins.

Finally, after cyclization and β-scission, lipid hydroperoxides derived from arachidonic acid and linoleic acid (HpETE and HPODE, respectively) form some of the most deleterious compounds toward cells known as carbonyl and bi-carbonyl lipid end products, one of which is HNE. The levels of plasma lipid peroxides and TBARS have been shown to be elevated in asthma and COPD patients and negatively correlated with the forced expiratory volume in patients, suggesting a role of increased lipoxidative stress in the pathogenesis of airflow obstruction in these diseases [294, 295]. Further studies are needed to define the source, significance, and specificity of these peroxidation products.

3.4.1.10 Mixed Forms of Oxidized Lipids

Being chemically reactive and potentially toxic, secondary products of lipid peroxidation (e.g., MDA, 4-HNE, and acrolein) aggravate OS through production of lipid-derived radicals that themselves can react with and damage essential molecules such as proteins and DNA [198].

MDA and other unsaturated aldehydes are capable of altering many cellular proteins by forming protein cross-linkages. Another possibility leading to protein cross-linkages is that radical intermediates formed within proteins following direct interaction with RS can interact with lipid-derived radicals, yielding protein–lipid bridges. The resulting lipid–protein adducts are called ALEs. The consequence of ALE generation is a usual functional impairment of the respective enzymes, receptors, and structural proteins. In addition, aldehydes generated from lipid peroxidation were shown to decrease protein biosynthesis and increase the osmotic fragility of lysosomes and other subcellular organelles [22].

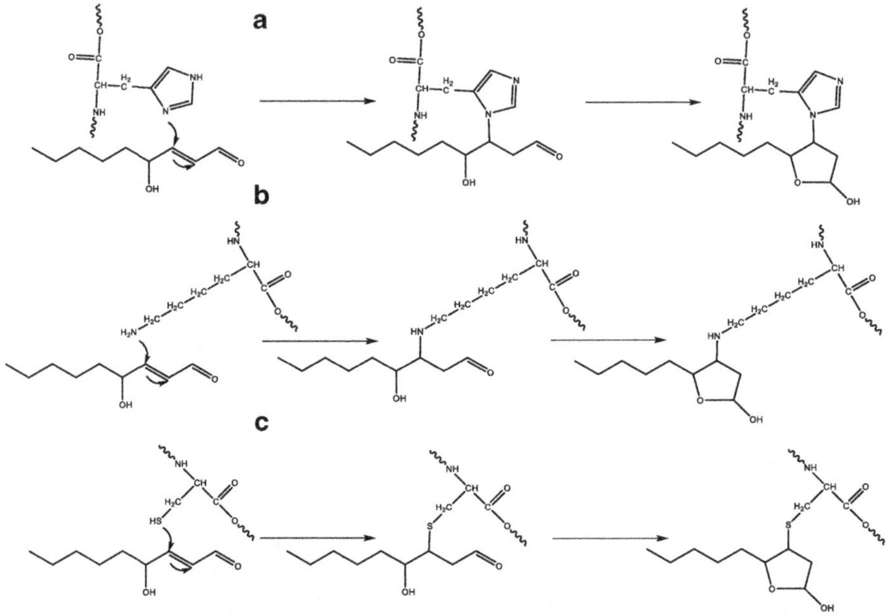

Fig. 3.14 Covalent binding of 4-HNE on protein residues via Michael addition. (**a**) formation of the 4-HNE-His adduct; (**b**) formation of the 4-HNE-Lys adduct; and (**c**) formation of the 4-HNE-Cys adduct. The hemi-acetal structure is common for these adducts

4-Hydroxy-2-nonenal is a highly reactive substance and easily forms adducts with proteins by covalent binding to cysteine, lysine, and histidine residues. This leads to the formation of stable Michael adducts with a hemi-acetal structure (Fig. 3.14).

Lipid peroxidation intermediates and products can also covalently modify nucleic acids, leading to cellular damage and inhibition of many cellular functions. The link between DNA oxidative damage, faulty repair of DNA, proto-oncogene activation, and the ability of some by-products of lipid peroxidation to act as promoters of carcinogenesis has been widely discussed in recent reports [296, 297].

Another class of reactive ALE precursors is represented by α-oxoaldehydes (e.g., glyoxal). They are basically derived from sugars in hyperglycemic conditions; however, they are also generated from lipid peroxidation and can react with Lys residues in proteins [180].

Overall, the oxidation of PUFAs through lipid peroxidation involves a chain reaction that progresses through three stages: initiation, propagation, and termination. Lipid peroxidation as a process is characterized by two features: first, the original radical-induced damage is amplified and propagated in a chain reaction-like manner, spreading the damage across the membrane. Second, the products of lipid peroxidation (toxic aldehydes, MDA, 4-HNE, etc.) can themselves be highly reactive and may cause further damage to other entities such as proteins and DNA (Fig. 3.15).

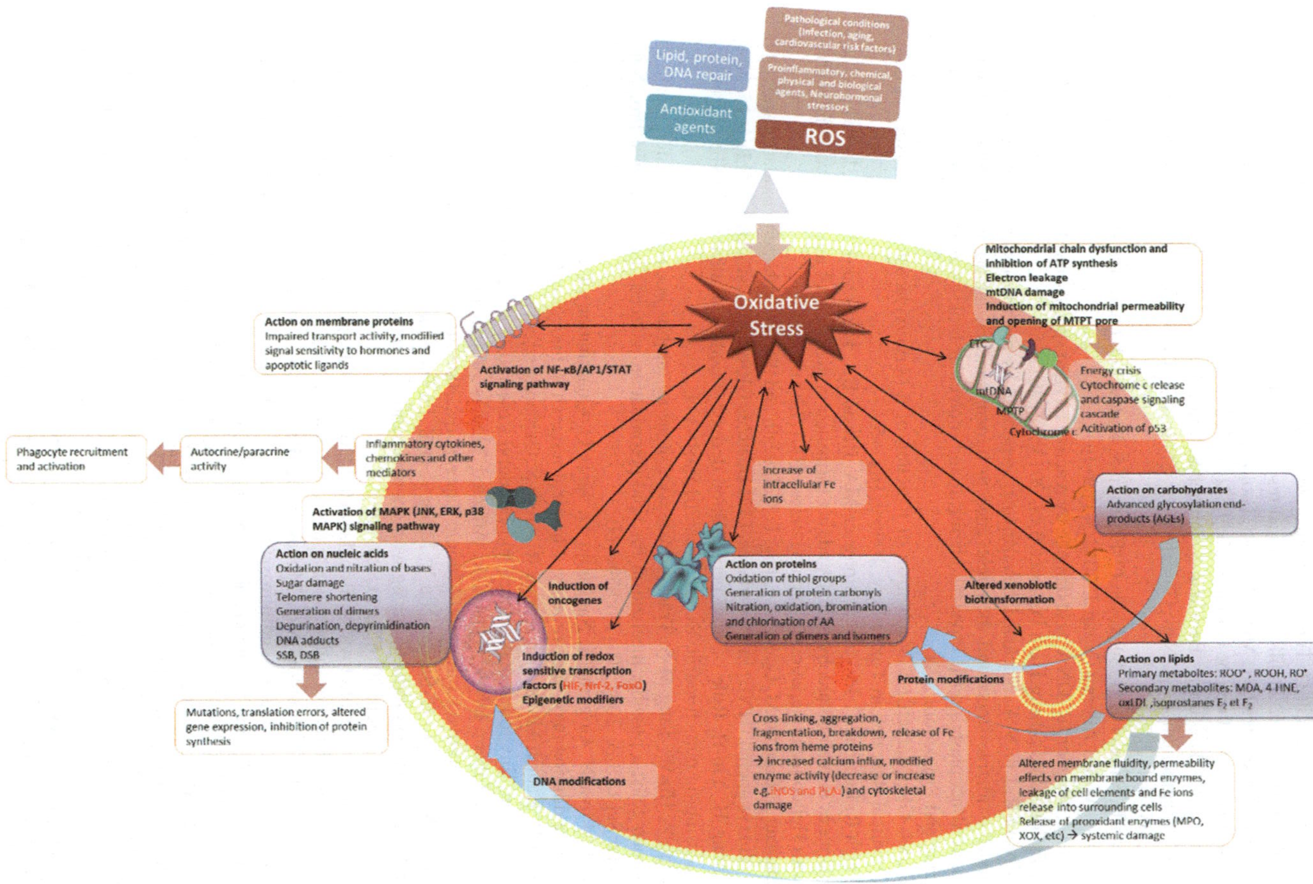

Fig. 3.15 ROS-induced oxidative damage to cells (lipids, proteins, carbohydrates, and DNA)
When RS and FRs interact with lipids, proteins, carbohydrates, and DNA, there is the potential for numerous changes that could affect cell function and lead to injury

References

1. Winczura A, Zdżalik D, Tudek B. Damage of DNA and proteins by major lipid peroxidation products in genome stability. Free radical research. 2012;46:442–59.
2. Aitken RJ, De Iuliis GN, McLachlan RI. Biological and clinical significance of DNA damage in the male germ line. Int J Androl. 2009;32(1):46–56.
3. Agarwal A, Majzoub A, Baskaran S, Panner Selvam MK, Cho CL, Henkel R, et al. Sperm DNA Fragmentation: A New Guideline for Clinicians. The world journal of men's health. 2020;38(4):412–71.
4. Birben E, Sahiner UM, Sackesen C, Erzurum S, Kalayci O. Oxidative stress and antioxidant defense. World Allergy Organ J. 2012;5(1):9–19.
5. Cadet J, Wagner JR. DNA base damage by reactive oxygen species, oxidizing agents, and UV radiation. Cold Spring Harb Perspect Biol. 2013;5(2):a012559.
6. Nilsson R, Liu N-A. Nuclear DNA damages generated by reactive oxygen molecules (ROS) under oxidative stress and their relevance to human cancers, including ionizing radiation-induced neoplasia part I: Physical, chemical and molecular biology aspects. Radiation Medicine and Protection. 2020.
7. Reddy GR, Marnett LJ. Mechanism of Reaction of β-(Aryloxy)acroleins with Nucleosides. Chemical Research in Toxicology. 1996;9(1):12–5.
8. Singh R, Leuratti C, Josyula S, Sipowicz MA, Diwan BA, Kasprzak KS, et al. Lobe-specific increases in malondialdehyde DNA adduct formation in the livers of mice following infection with Helicobacter hepaticus. Carcinogenesis. 2001;22(8):1281–7.
9. Dalleau S, Baradat M, Guéraud F, Huc L. Cell death and diseases related to oxidative stress:4-hydroxynonenal (HNE) in the balance. Cell Death & Differentiation. 2013;20(12):1615–30.
10. Zhong H, Yin H. Role of lipid peroxidation derived 4-hydroxynonenal (4-HNE) in cancer: Focusing on mitochondria. Redox Biol. 2015;4:193–9.
11. Feng Z, Hu W, Tang M-S. Trans-4-hydroxy-2-nonenal inhibits nucleotide excision repair in human cells: a possible mechanism for lipid peroxidation-induced carcinogenesis. Proc Natl Acad Sci U S A. 2004;101(23):8598–602.
12. Tretyakova N, Goggin M, Sangaraju D, Janis G. Quantitation of DNA adducts by stable isotope dilution mass spectrometry. Chemical research in toxicology. 2012;25(10):2007–35.
13. Kaur R, Kaur J, Mahajan J, Kumar R, Arora S. Oxidative stress—implications, source and its prevention. Environmental Science and Pollution Research. 2014;21(3):1599–613.
14. Increased Etheno-DNA Adducts in Affected Tissues of Patients Suffering from Crohn's Disease, Ulcerative Colitis, and Chronic Pancreatitis. 2006;8(5–6):1003–10.
15. Bin P, Shen M, Li H, Sun X, Niu Y, Meng T, et al. Increased levels of urinary biomarkers of lipid peroxidation products among workers occupationally exposed to diesel engine exhaust. Free radical research. 2016;50(8):820–30.
16. Shokolenko I, Venediktova N, Bochkareva A, Wilson GL, Alexeyev MF. Oxidative stress induces degradation of mitochondrial DNA. Nucleic Acids Res. 2009;37(8):2539–48.
17. Mangialasche F, Polidori MC, Monastero R, Ercolani S, Camarda C, Cecchetti R, et al. Biomarkers of oxidative and nitrosative damage in Alzheimer's disease and mild cognitive impairment. Ageing research reviews. 2009;8(4):285–305.
18. Gamelin L, Harry P. Rodenticides. EMC – Toxicologie-Pathologie. 2005;2(3):89–97.
19. Lenaz G. Mitochondria and reactive oxygen species. Which role in physiology and pathology? Advances in experimental medicine and biology. 2012;942:93–136.
20. Lezza AM, Mecocci P, Cormio A, Beal MF, Cherubini A, Cantatore P, et al. Mitochondrial DNA 4977 bp deletion and OH8dG levels correlate in the brain of aged subjects but not Alzheimer's disease patients. FASEB journal : official publication of the Federation of American Societies for Experimental Biology. 1999;13(9):1083–8.
21. Jawad B, Poudel L, Podgornik R, Steinmetz NF, Ching W-Y. Molecular mechanism and binding free energy of doxorubicin intercalation in DNA. Physical Chemistry Chemical Physics. 2019;21(7):3877–93.

22. Younes M. Chapter 5 – Free Radicals and Reactive Oxygen Species. In: Marquardt H, Schäfer SG, McClellan R, Welsch F, editors. Toxicology. San Diego: Academic Press; 1999. p. 111–25.
23. Chevion M, Berenshtein E, Zhu B-Z, Gilbert D, Colton C. The Role of Transition Metal Ions in Free Radical-Mediated Damage. 1999. p. 103–31.
24. Villamena F. Chemistry of Reactive Species. 2017. p. 13–64.
25. Boiteux S, Gajewski E, Laval J, Dizdaroglu M. Substrate specificity of the Escherichia coli Fpg protein formamidopyrimidine-DNA glycosylase: excision of purine lesions in DNA produced by ionizing radiation or photosensitization. Biochemistry. 1992;31(1):106–10.
26. Kanvah S, Joseph J, Schuster GB, Barnett RN, Cleveland CL, Landman U. Oxidation of DNA: Damage to Nucleobases. Acc Chem Res. 2010;43(2):280–7.
27. Whiteman M, Jenner A, Halliwell B. Hypochlorous Acid-Induced Base Modifications in Isolated Calf Thymus DNA. Chemical Research in Toxicology. 1997;10(11):1240–6.
28. Keyer K, Imlay JA. Superoxide accelerates DNA damage by elevating free-iron levels. Proc Natl Acad Sci U S A. 1996;93(24):13635–40.
29. Jones LE, Jr., Ying L, Hofseth AB, Jelezcova E, Sobol RW, Ambs S, et al. Differential effects of reactive nitrogen species on DNA base excision repair initiated by the alkyladenine DNA glycosylase. Carcinogenesis. 2009;30(12):2123–9.
30. Islam BU, Habib S, Ahmad P, Allarakha S, Moinuddin, Ali A. Pathophysiological Role of Peroxynitrite Induced DNA Damage in Human Diseases: A Special Focus on Poly (ADP-ribose) Polymerase (PARP). Indian J Clin Biochem. 2015;30(4):368–85.
31. Jena NR. DNA damage by reactive species: Mechanisms, mutation and repair. Journal of biosciences. 2012;37(3):503–17.
32. Fenga C, Gangemi S, Teodoro M, Rapisarda V, Golokhvast K, Docea AO, et al. 8-Hydroxydeoxyguanosine as a biomarker of oxidative DNA damage in workers exposed to low-dose benzene. Toxicology Reports. 2017;4:291–5.
33. Gutteridge BHaJM. Oxidative stress: adaptation, damage, repair and death. 3 ed. New York, NY: Oxford Science Publications; 1999. p. 36–104.
34. Iida T, Furuta A, Kawashima M, Nishida J, Nakabeppu Y, Iwaki T. Accumulation of 8-oxo-2'-deoxyguanosine and increased expression of hMTH1 protein in brain tumors. Neuro-oncology. 2001;3(2):73–81.
35. Yu H, Venkatarangan L, Wishnok JS, Tannenbaum SR. Quantitation of four guanine oxidation products from reaction of DNA with varying doses of peroxynitrite. Chemical research in toxicology. 2005;18(12):1849–57.
36. Neeley WL, Essigmann JM. Mechanisms of formation, genotoxicity, and mutation of guanine oxidation products. Chem Res Toxicol. 2006;19(4):491–505.
37. Ghosh R, Mitchell DL. Effect of oxidative DNA damage in promoter elements on transcription factor binding. Nucleic Acids Res. 1999;27(15):3213–8.
38. Cadet J, Douki T, Ravanat J-L, Di Mascio P. Sensitized formation of oxidatively generated damage to cellular DNA by UVA radiation. Photochemical & Photobiological Sciences. 2009;8(7):903–11.
39. Leon J, Sakumi K, Castillo E, Sheng Z, Oka S, Nakabeppu Y. 8-Oxoguanine accumulation in mitochondrial DNA causes mitochondrial dysfunction and impairs neuritogenesis in cultured adult mouse cortical neurons under oxidative conditions. Scientific Reports. 2016;6:22086.
40. Bräutigam L, Pudelko L, Jemth A-S, Gad H, Narwal M, Gustafsson R, et al. Hypoxic Signaling and the Cellular Redox Tumor Environment Determine Sensitivity to MTH1 Inhibition. Cancer research. 2016;76(8):2366–75.
41. Hepel M, Andreescu S. Oxidative Stress and Human Health. Oxidative Stress: Diagnostics, Prevention, and Therapy Volume 2. ACS Symposium Series. 1200: American Chemical Society; 2015. p. 1–33.
42. Freudenthal BD, Beard WA, Perera L, Shock DD, Kim T, Schlick T, et al. Uncovering the polymerase-induced cytotoxicity of an oxidized nucleotide. Nature. 2015;517(7536):635–9.

43. Jena N, Mishra PC. Formation of 8-nitroguanine and 8-oxoguanine due to reactions of peroxynitrite with guanine. Journal of computational chemistry. 2007;28:1321–35.
44. Hiraku Y. Formation of 8-nitroguanine, a nitrative DNA lesion, in inflammation-related carcinogenesis and its significance. Environmental health and preventive medicine. 2010;15 (2):63–72.
45. Hiraku Y, Kawanishi S, Ichinose T, Murata M. The role of iNOS-mediated DNA damage in infection-and asbestos-induced carcinogenesis. Annals of the New York Academy of Sciences. 2010;1203:15–22.
46. Hiraku Y, Nishikawa Y, Ma N, Afroz T, Mizobuchi K, Ishiyama R, et al. Nitrative DNA damage induced by carbon-black nanoparticles in macrophages and lung epithelial cells. Mutation Research/Genetic Toxicology and Environmental Mutagenesis. 2017;818:7–16.
47. Laval F, Wink DA, Laval J. A discussion of mechanisms of NO genotoxicty: Implication of inhibition of DNA repair proteins. Reviews of Physiology Biochemistry and Pharmacology, Volume 131: Special Issue on Membrane-Mediated Cellular Responses: The Roles of Reactive Oxygens, NO, CO, II. Berlin, Heidelberg: Springer Berlin Heidelberg; 1997. p. 175–91.
48. Brooks PJ, Wise DS, Berry DA, Kosmoski JV, Smerdon MJ, Somers RL, et al. The oxidative DNA lesion 8,5'-(S)-cyclo-2'-deoxyadenosine is repaired by the nucleotide excision repair pathway and blocks gene expression in mammalian cells. J Biol Chem. 2000;275(29): 22355–62.
49. Marietta C, Gulam H, Brooks P. A single 8,5′-cyclo-2′-deoxyadenosine lesion in a TATA box prevents binding of the TATA binding protein and strongly reduces transcription in vivo. DNA repair. 2002;1:967–75.
50. Olinski R, Gackowski D, Foksinski M, Rozalski R, Roszkowski K, Jaruga P. Oxidative DNA damage: assessment of the role in carcinogenesis, atherosclerosis, and acquired immunodeficiency syndrome1 1This article is part of a series of reviews on "Oxidative DNA Damage and Repair." The full list of papers may be found on the homepage of the journal. Free Radical Biology and Medicine. 2002;33(2):192–200.
51. Sassa A, Kamoshita N, Matsuda T, Ishii Y, Kuraoka I, Nohmi T, et al. Miscoding properties of 8-chloro-2'-deoxyguanosine, a hypochlorous acid-induced DNA adduct, catalysed by human DNA polymerases. Mutagenesis. 2013;28(1):81–8.
52. Suzuki T, Ohshima H. Nicotine-modulated formation of spiroiminodihydantoin nucleoside via 8-oxo-7,8-dihydro-2'-deoxyguanosine in 2'-deoxyguanosine-hypochlorous acid reaction. FEBS Lett. 2002;516(1–3):67–70.
53. Matthew Whiteman Andrew Jenner Barry H. 8-Chloroadenine: a novel product formed from hypochlorous acid-induced damage to calf thymus DNA. Biomarkers. 1999;4(4):303–10.
54. Zuo L, Zhou T, Pannell BK, Ziegler AC, Best TM. Biological and physiological role of reactive oxygen species–the good, the bad and the ugly. Acta physiologica (Oxford, England). 2015;214(3):329–48.
55. Joseph J, Schuster GB. One-electron oxidation of DNA: reaction at thymine. Chemical Communications. 2010;46(42):7872–8.
56. Madkour LH. The Roles and Mechanisms of ROS, Oxidative Stress, and Oxidative Damage. In: Madkour LH, editor. Nanoparticles Induce Oxidative and Endoplasmic Reticulum Stresses: Antioxidant Therapeutic Defenses. Cham: Springer International Publishing; 2020. p. 139–91.
57. Poulsen HE, Specht E, Broedbaek K, Henriksen T, Ellervik C, Mandrup-Poulsen T, et al. RNA modifications by oxidation: A novel disease mechanism? Free Radical Biology and Medicine. 2012;52(8):1353–61.
58. Nunomura A, Tamaoki T, Motohashi N, Nakamura M, McKeel DW, Jr., Tabaton M, et al. The earliest stage of cognitive impairment in transition from normal aging to Alzheimer disease is marked by prominent RNA oxidation in vulnerable neurons. J Neuropathol Exp Neurol. 2012;71(3):233–41.
59. Wang JX, Gao J, Ding SL, Wang K, Jiao JQ, Wang Y, et al. Oxidative Modification of miR-184 Enables It to Target Bcl-xL and Bcl-w. Mol Cell. 2015;59(1):50–61.

60. Cheng X, Ku CH, Siow RC. Regulation of the Nrf2 antioxidant pathway by microRNAs: New players in micromanaging redox homeostasis. Free Radic Biol Med. 2013;64:4–11.
61. Lang A, Grether-Beck S, Singh M, Kuck F, Jakob S, Kefalas A, et al. MicroRNA-15b regulates mitochondrial ROS production and the senescence-associated secretory phenotype through sirtuin 4/SIRT4. Aging (Albany NY). 2016;8(3):484–505.
62. Gille G, Sigler K. Oxidative stress and living cells. Folia Microbiologica. 1995;40(2):131–52.
63. Rai AK, Jaiswal N, Maurya CK, Sharma A, Ahmad I, Ahmad S, et al. Fructose-induced AGEs-RAGE signaling in skeletal muscle contributes to impairment of glucose homeostasis. The Journal of Nutritional Biochemistry. 2019;71:35–44.
64. Yamazaki S, Masaki C, Nodai T, Tsuka S, Tamura A, Mukaibo T, et al. The effects of hyperglycaemia on peri-implant tissues after osseointegration. Journal of Prosthodontic Research. 2020;64(2):217–23.
65. Tao Z, Raffel RA, Souid A-K, Goodisman J. Kinetic studies on enzyme-catalyzed reactions: oxidation of glucose, decomposition of hydrogen peroxide and their combination. Biophys J. 2009;96(7):2977–88.
66. Ali ME, Rahman MM, Sarkar SM, Hamid SBA. Heterogeneous Metal Catalysts for Oxidation Reactions. Journal of Nanomaterials. 2014;2014:192038.
67. Chace KV, Carubelli R, Nordquist RE. The role of nonenzymatic glycosylation, transition metals, and free radicals in the formation of collagen aggregates. Archives of Biochemistry and Biophysics. 1991;288(2):473–80.
68. Miyata T, Ishiguro N, Yasuda Y, Ito T, Nangaku M, Iwata H, et al. Increased Pentosidine, an Advanced Glycation End Product, in Plasma and Synovial Fluid from Patients with Rheumatoid Arthritis and Its Relation with Inflammatory Markers. Biochemical and Biophysical Research Communications. 1998;244(1):45–9.
69. Arnold P, Njemini R, Vantieghem S, Gorus E, Pool-Goudzwaard A, Buyl R, et al. Reaction time in healthy elderly is associated with chronic low-grade inflammation and advanced glycation end product. Experimental Gerontology. 2018;108:118–24.
70. Naito Y, Lee M, Kato Y, Nagai R, Yonei Y. Oxidative Stress Markers. Anti-aging Medicine. 2010;7:36–44.
71. Benov L, Beema A. Superoxide-dependence of the short chain sugars-induced mutagenesis. Free radical biology & medicine. 2003;34:429–33.
72. Bierhaus A, Hofmann MA, Ziegler R, Nawroth PP. AGEs and their interaction with AGE-receptors in vascular disease and diabetes mellitus. I. The AGE concept. Cardiovascular Research. 1998;37(3):586–600.
73. Ott C, Jacobs K, Haucke E, Navarrete Santos A, Grune T, Simm A. Role of advanced glycation end products in cellular signaling. Redox Biol. 2014;2:411–29.
74. Wautier MP, Chappey O, Corda S, Stern DM, Schmidt AM, Wautier JL. Activation of NADPH oxidase by AGE links oxidant stress to altered gene expression via RAGE. American journal of physiology Endocrinology and metabolism. 2001;280(5):E685–94.
75. Wendt TM, Tanji N, Guo J, Kislinger TR, Qu W, Lu Y, et al. RAGE drives the development of glomerulosclerosis and implicates podocyte activation in the pathogenesis of diabetic nephropathy. The American journal of pathology. 2003;162(4):1123–37.
76. Palma-Duran SA, Kontogianni MD, Vlassopoulos A, Zhao S, Margariti A, Georgoulis M, et al. Serum levels of advanced glycation end-products (AGEs) and the decoy soluble receptor for AGEs (sRAGE) can identify non-alcoholic fatty liver disease in age-, sex- and BMI-matched normo-glycemic adults. Metabolism. 2018;83:120–7.
77. Saulnier P-J, Wheelock KM, Howell S, Weil EJ, Tanamas SK, Knowler WC, et al. Advanced Glycation End Products Predict Loss of Renal Function and Correlate With Lesions of Diabetic Kidney Disease in American Indians With Type 2 Diabetes. 2016;65(12):3744–53.
78. Akhter F, Chen D, Akhter A, Yan SF, ShiDu Yan S. Age-dependent accumulation of dicarbonyls and advanced glycation endproducts (AGEs) associates with mitochondrial stress. Free Radical Biology and Medicine. 2021.

79. Nowotny K, Castro JP, Hugo M, Braune S, Weber D, Pignitter M, et al. Oxidants produced by methylglyoxal-modified collagen trigger ER stress and apoptosis in skin fibroblasts. Free Radical Biology and Medicine. 2018;120:102–13.
80. Monnier VM, Bautista O, Kenny D, Sell DR, Fogarty J, Dahms W, et al. Skin collagen glycation, glycoxidation, and crosslinking are lower in subjects with long-term intensive versus conventional therapy of type 1 diabetes: relevance of glycated collagen products versus HbA1c as markers of diabetic complications. DCCT Skin Collagen Ancillary Study Group. Diabetes Control and Complications Trial. Diabetes. 1999;48(4):870–80.
81. Ducos C, Rigo M, Larroumet A, Delyfer MN, Korobelnik JF, Monlun M, et al. Diabetic retinopathy in well-controlled type 2 diabetes: Role of glycaemic memory. Diabetes & Metabolism. 2020.
82. Zabad IEM, Amin MN, El-Shishtawy MM. Protective effect of vanillin on diabetic nephropathy by decreasing advanced glycation end products in rats. Life Sciences. 2019;239:117088.
83. Miyata S, Monnier V. Immunohistochemical detection of advanced glycosylation end products in diabetic tissues using monoclonal antibody to pyrraline. J Clin Invest. 1992;89(4): 1102–12.
84. Niwa T, Katsuzaki T, Miyazaki S, Miyazaki T, Ishizaki Y, Hayase F, et al. Immunohistochemical detection of imidazolone, a novel advanced glycation end product, in kidneys and aortas of diabetic patients. J Clin Invest. 1997;99(6):1272–80.
85. Kaji Y. [Ocular diseases caused by accumulation of proteins with post-translational modifications]. Nippon Ganka Gakkai Zasshi. 2009;113(3):424–42; discussion 43.
86. Sueki H, Nozaki S, Fujisawa R, Aoki K, Kuroiwa Y. Glycosylated Proteins of Skin, Nail and Hair: Application as an Index for Long-term Control of Diabetes Mellitus. The Journal of Dermatology. 1989;16(2):103–10.
87. Yonekura H, Yamamoto Y, Sakurai S, Watanabe T, Yamamoto H. Roles of the Receptor for Advanced Glycation Endproducts in Diabetes-Induced Vascular Injury. Journal of Pharmacological Sciences. 2005;97(3):305–11.
88. Nagai R, Fujiwara Y, Mera K, Motomura K, Iwao Y, Tsurushima K, et al. Usefulness of antibodies for evaluating the biological significance of AGEs. Annals of the New York Academy of Sciences. 2008;1126:38–41.
89. Yonei Y, Yagi M, Hibino S, Matsuura N. Herbal extracts inhibit Maillard reaction, and reduce chronic diabetic complications risk in streptozotocin-induced diabetic rats. ANTI-AGING MEDICINE. 2008;5(10):93–8.
90. Kilhovd BK, Juutilainen A, Lehto S, Rönnemaa T, Torjesen PA, Hanssen KF, et al. Increased serum levels of advanced glycation endproducts predict total, cardiovascular and coronary mortality in women with type 2 diabetes: a population-based 18 year follow-up study. Diabetologia. 2007;50(7):1409–17.
91. Mohamed H, Perrotte M, Le Page A, Camponova P, Fulop T, Ramassamy C. P 141 – Peripheral oxidative profile and specific advanced glycation end products can be a signature of cognitive decline in Alzheimer's disease. Free Radical Biology and Medicine. 2017;108: S66.
92. Biomarkers of oxidative stress and cellular-based assays of indirect antioxidant measurement. Measurement of Antioxidant Activity & Capacity 2018. p. 165–86.
93. Dalle-Donne I, Rossi R, Colombo G, Giustarini D, Milzani A. Protein S-glutathionylation: a regulatory device from bacteria to humans. Trends in biochemical sciences. 2009;34(2):85–96.
94. Gu M, Imlay JA. The SoxRS response of Escherichia coli is directly activated by redox-cycling drugs rather than by superoxide. Mol Microbiol. 2011;79(5):1136–50.
95. Davies MJ. Protein oxidation and peroxidation. The Biochemical journal. 2016;473(7): 805–25.
96. Imlay JA. Iron-sulphur clusters and the problem with oxygen. 2006;59(4):1073–82.
97. Gu M, Imlay J. Superoxide poisons mononuclear iron enzymes by causing mismetallation. Mol Microbiol. 2013;89.

98. Spickett CM, Pitt AR. Protein oxidation: role in signalling and detection by mass spectrometry. Amino Acids. 2012;42(1):5–21.
99. Chang YT, Chang WN, Tsai NW, Huang CC, Kung CT, Su YJ, et al. The roles of biomarkers of oxidative stress and antioxidant in Alzheimer's disease: a systematic review. BioMed research international. 2014;2014:182303.
100. Shah D, Mahajan N, Sah S, Nath S, Paudyal B. Oxidative stress and its biomarkers in systemic lupus erythematosus. Journal of biomedical science. 2014;21:23.
101. Biomarkers of Oxidative Stress in Neurodegenerative Diseases. Molecular Basis of Oxidative Stress 2013. p. 359–76.
102. Dalle-Donne I, Rossi R, Giustarini D, Milzani A, Colombo R. Protein carbonyl groups as biomarkers of oxidative stress. Clinica chimica acta; international journal of clinical chemistry. 2003;329:23–38.
103. Bansal M, Kaushal N. Introduction to Oxidative Stress. In: Bansal M, Kaushal N, editors. Oxidative Stress Mechanisms and their Modulation. New Delhi: Springer India; 2014. p. 1–18.
104. Gil L, Siems W, Mazurek B, Gross J, Schroeder P, Voss P, et al. Age-associated analysis of oxidative stress parameters in human plasma and erythrocytes. Free radical research. 2006;40 (5):495–505.
105. Sultana R, Perluigi M, Butterfield DA. Protein oxidation and lipid peroxidation in brain of subjects with Alzheimer's disease: insights into mechanism of neurodegeneration from redox proteomics. Antioxid Redox Signal. 2006;8(11–12):2021–37.
106. Stadtman ER, Levine R. Free radical-mediated oxidation of free amino acids and amino acid residues in proteins. Amino acids. 2004;25:207–18.
107. Zou MH, Cohen R, Ullrich V. Peroxynitrite and vascular endothelial dysfunction in diabetes mellitus. Endothelium : journal of endothelial cell research. 2004;11(2):89–97.
108. Tsikas D. Analytical methods for 3-nitrotyrosine quantification in biological samples: the unique role of tandem mass spectrometry. Amino Acids. 2012;42(1):45–63.
109. Ghesquière B, Colaert N, Helsens K, Dejager L, Vanhaute C, Verleysen K, et al. In vitro and in vivo protein-bound tyrosine nitration characterized by diagonal chromatography. Mol Cell Proteomics. 2009;8(12):2642–52.
110. Gaut JP, Byun J, Tran HD, Lauber WM, Carroll JA, Hotchkiss RS, et al. Myeloperoxidase produces nitrating oxidants in vivo. J Clin Invest. 2002;109(10):1311–9.
111. Bartesaghi S, Ferrer-Sueta G, Peluffo G, Valez V, Zhang H, Kalyanaraman B, et al. Protein tyrosine nitration in hydrophilic and hydrophobic environments. Amino Acids. 2007;32(4): 501–15.
112. Souza JM, Peluffo G, Radi R. Protein tyrosine nitration–functional alteration or just a biomarker? Free Radic Biol Med. 2008;45(4):357–66.
113. Frijhoff J, Winyard PG, Zarkovic N, Davies SS, Stocker R, Cheng D, et al. Clinical Relevance of Biomarkers of Oxidative Stress. Antioxid Redox Signal. 2015;23(14):1144–70.
114. Kawasaki H, Tominaga M, Shigenaga A, Kamo A, Kamata Y, Iizumi K, et al. Importance of tryptophan nitration of carbonic anhydrase III for the morbidity of atopic dermatitis. Free Radical Biology and Medicine. 2014;73:75–83.
115. Kooy NW, Lewis SJ, Royall JA, Ye YZ, Kelly DR, Beckman JS. Extensive tyrosine nitration in human myocardial inflammation: evidence for the presence of peroxynitrite. Critical care medicine. 1997;25(5):812–9.
116. Chazotte-Aubert L, Hainaut P, Ohshima H. Nitric Oxide Nitrates Tyrosine Residues of Tumor-Suppressor p53 Protein in MCF-7 Cells. Biochemical and Biophysical Research Communications. 2000;267(2):609–13.
117. Madkour LH. Sources of Oxidants and Function of ROS Inside the Macromolecule Cells. In: Madkour LH, editor. Nanoparticles Induce Oxidative and Endoplasmic Reticulum Stresses: Antioxidant Therapeutic Defenses. Cham: Springer International Publishing; 2020. p. 99–138.
118. Meredith S, Parekh G, Towler J, Schouten J, Davis P, Griffiths H, et al. P87 – Mapping nitrotyrosine modifications in fibrinogen by mass spectrometry as a biomarker for inflammatory disease. Free Radical Biology and Medicine. 2014;75:S50.

119. Parastatidis I, Thomson L, Fries DM, Moore RE, Tohyama J, Fu X, et al. Increased Protein Nitration Burden in the Atherosclerotic Lesions and Plasma of Apolipoprotein A–I Deficient Mice. 2007;101(4):368–76.
120. Ceriello A, Mercuri F, Quagliaro L, Assaloni R, Motz E, Tonutti L, et al. Detection of nitrotyrosine in the diabetic plasma: Evidence of oxidative stress. Diabetologia. 2001;44: 834–8.
121. ter Steege JC, Koster-Kamphuis L, van Straaten EA, Forget PP, Buurman WA. Nitrotyrosine in plasma of celiac disease patients as detected by a new sandwich ELISA. Free Radic Biol Med. 1998;25(8):953–63.
122. Kari JA, Donald AE, Vallance DT, Bruckdorfer KR, Leone A, Mullen MJ, et al. Physiology and biochemistry of endothelial function in children with chronic renal failure. Kidney Int. 1997;52(2):468–72.
123. Gole MD, Souza JM, Choi I, Hertkorn C, Malcolm S, Foust RF, 3rd, et al. Plasma proteins modified by tyrosine nitration in acute respiratory distress syndrome. Am J Physiol Lung Cell Mol Physiol. 2000;278(5):L961–7.
124. Pignatelli B, Li CQ, Boffetta P, Chen Q, Ahrens W, Nyberg F, et al. Nitrated and oxidized plasma proteins in smokers and lung cancer patients. Cancer research. 2001;61(2):778–84.
125. Yeo WS, Kim YJ, Kabir MH, Kang JW, Ahsan-Ul-Bari M, Kim KP. Mass spectrometric analysis of protein tyrosine nitration in aging and neurodegenerative diseases. Mass spectrometry reviews. 2015;34(2):166–83.
126. Sultana R, Poon HF, Cai J, Pierce WM, Merchant M, Klein JB, et al. Identification of nitrated proteins in Alzheimer's disease brain using a redox proteomics approach. Neurobiology of disease. 2006;22(1):76–87.
127. Kato Y, Dozaki N, Nakamura T, Kitamoto N, Yoshida A, Naito M, et al. Quantification of Modified Tyrosines in Healthy and Diabetic Human Urine using Liquid Chromatography/ Tandem Mass Spectrometry. Journal of clinical biochemistry and nutrition. 2009;44(1):67–78.
128. Browne SE, Ferrante RJ, Beal MF. Oxidative stress in Huntington's disease. Brain pathology (Zurich, Switzerland). 1999;9(1):147–63.
129. Túnez I, Sánchez-López F, Agüera E, Fernández-Bolaños R, Sánchez FM, Tasset-Cuevas I. Important role of oxidative stress biomarkers in Huntington's disease. Journal of medicinal chemistry. 2011;54(15):5602–6.
130. Misko TP, Radabaugh MR, Highkin M, Abrams M, Friese O, Gallavan R, et al. Characterization of nitrotyrosine as a biomarker for arthritis and joint injury. Osteoarthritis and Cartilage. 2013;21(1):151–6.
131. Smith MA, Harris PLR, Sayre LM, Beckman JS, Perry GJJoN. Widespread peroxynitrite-mediated damage in Alzheimer's disease. 1997;17(8):2653–7.
132. Zhao Y, Zhang Y, Sun H, Maroto R, Brasier AR. Selective Affinity Enrichment of Nitrotyrosine-Containing Peptides for Quantitative Analysis in Complex Samples. J Proteome Res. 2017;16(8):2983–92.
133. Bartesaghi S, Wenzel J, Trujillo M, López M, Joseph J, Kalyanaraman B, et al. Lipid peroxyl radicals mediate tyrosine dimerization and nitration in membranes. Chemical research in toxicology. 2010;23(4):821–35.
134. Huang MZ, Li JY. Physiological regulation of reactive oxygen species in organisms based on their physicochemical properties. 2020;228(1):e13351.
135. Mukherjee S, Kapp EA, Lothian A, Roberts AM, Vasil'ev YV, Boughton BA, et al. Characterization and Identification of Dityrosine Cross-Linked Peptides Using Tandem Mass Spectrometry. Analytical Chemistry. 2017;89(11):6136–45.
136. Odobasic D, Kitching AR, Holdsworth SR. Neutrophil-Mediated Regulation of Innate and Adaptive Immunity: The Role of Myeloperoxidase. Journal of Immunology Research. 2016;2016:2349817.
137. Shao B, Oda MN, Oram JF, Heinecke JW. Myeloperoxidase: an oxidative pathway for generating dysfunctional high-density lipoprotein. Chem Res Toxicol. 2010;23(3):447–54.

138. Thomson E, Brennan S, Senthilmohan R, Gangell CL, Chapman AL, Sly PD, et al. Identifying peroxidases and their oxidants in the early pathology of cystic fibrosis. Free Radic Biol Med. 2010;49(9):1354–60.
139. Dickinson BC, Chang CJ. Chemistry and biology of reactive oxygen species in signaling or stress responses. Nature chemical biology. 2011;7(8):504–11.
140. Salmeen A, Andersen JN, Myers MP, Meng T-C, Hinks JA, Tonks NK, et al. Redox regulation of protein tyrosine phosphatase 1B involves a sulphenyl-amide intermediate. Nature. 2003;423 (6941):769–73.
141. van Montfort RLM, Congreve M, Tisi D, Carr R, Jhoti H. Oxidation state of the active-site cysteine in protein tyrosine phosphatase 1B. Nature. 2003;423(6941):773–7.
142. Chang TS, Jeong W, Woo HA, Lee SM, Park S, Rhee SG. Characterization of mammalian sulfiredoxin and its reactivation of hyperoxidized peroxiredoxin through reduction of cysteine sulfinic acid in the active site to cysteine. J Biol Chem. 2004;279(49):50994–1001.
143. Biteau B, Labarre J, Toledano MB. ATP-dependent reduction of cysteine–sulphinic acid by S. cerevisiae sulphiredoxin. Nature. 2003;425(6961):980–4.
144. Jacobson T, Navarrete C, Sharma SK, Sideri TC, Ibstedt S, Priya S, et al. Arsenite interferes with protein folding and triggers formation of protein aggregates in yeast. 2012;125(21): 5073–83.
145. Mossuto MF. Disulfide bonding in neurodegenerative misfolding diseases. Int J Cell Biol. 2013;2013:318319-.
146. Gallogly MM, Starke DW, Mieyal JJ. Mechanistic and kinetic details of catalysis of thiol-disulfide exchange by glutaredoxins and potential mechanisms of regulation. Antioxid Redox Signal. 2009;11(5):1059–81.
147. Ghezzi P. Oxidoreduction of protein thiols in redox regulation. Biochemical Society Transactions. 2005;33(6):1378–81.
148. Iizumi K, Kawasaki H, Shigenaga A, Tominaga M, Otsu A, Kamo A, et al. Tryptophan nitration of immunoglobulin light chain as a new possible biomarker for atopic dermatitis. Journal of clinical biochemistry and nutrition. 2018;63(3):197–204.
149. Nuriel T, Hansler A, Gross SS. Protein nitrotryptophan: formation, significance and identification. Journal of proteomics. 2011;74(11):2300–12.
150. Pattison DI, Davies MJ. Kinetic Analysis of the Role of Histidine Chloramines in Hypochlorous Acid Mediated Protein Oxidation. Biochemistry. 2005;44(19):7378–87.
151. Kehrer JP, Klotz LO. Free radicals and related reactive species as mediators of tissue injury and disease: implications for Health. Critical reviews in toxicology. 2015;45(9):765–98.
152. Marrocco I, Altieri F, Peluso I. Measurement and Clinical Significance of Biomarkers of Oxidative Stress in Humans. Oxidative Medicine and Cellular Longevity. 2017;2017: 6501046.
153. Pattison DI, Hawkins CL, Davies MJ. Hypochlorous Acid-Mediated Protein Oxidation: How Important Are Chloramine Transfer Reactions and Protein Tertiary Structure? Biochemistry. 2007;46(34):9853–64.
154. Marrocco I, Altieri F. Measurement and Clinical Significance of Biomarkers of Oxidative Stress in Humans. 2017;2017:6501046.
155. Biomarkers of Oxidative Stress in Plasma and Urine. Analytical Techniques for Clinical Chemistry. p. 555–94.
156. Nevin C, McNeil L, Ahmed N, Murgatroyd C, Brison D, Carroll M. Investigating the Glycating Effects of Glucose, Glyoxal and Methylglyoxal on Human Sperm. Scientific Reports. 2018;8(1):9002.
157. Monnier V, Sell D. Prevention and Repair of Protein Damage by the Maillard Reaction In Vivo. Rejuvenation research. 2006;9:264–73.
158. Loomis SJ, Chen Y, Sacks DB, Christenson ES, Christenson RH, Rebholz CM, et al. Cross-sectional Analysis of AGE-CML, sRAGE, and esRAGE with Diabetes and Cardiometabolic Risk Factors in a Community-Based Cohort. Clinical Chemistry. 2017;63(5):980–9.

159. Sánchez E, Betriu À, Arroyo D, López C, Hernández M, Rius F, et al. Skin Autofluorescence and Subclinical Atherosclerosis in Mild to Moderate Chronic Kidney Disease: A Case-Control Study. PLOS ONE. 2017;12(1):e0170778.
160. Lohou E, Sasaki NA, Boullier A, Sonnet P. Multifunctional diamine AGE/ALE inhibitors with potential therapeutical properties against Alzheimer's disease. European Journal of Medicinal Chemistry. 2016;122:702–22.
161. Sawikr Y, Yarla NS, Peluso I, Kamal MA, Aliev G, Bishayee A. Neuroinflammation in Alzheimer's Disease: The Preventive and Therapeutic Potential of Polyphenolic Nutraceuticals. Advances in protein chemistry and structural biology. 2017;108:33–57.
162. Gu W, Xu Z, Qi F, Sang Z, Wang C, Li F. Plasma levels of soluble receptor for advanced glycation end products in patients with acute respiratory distress syndrome. Int J Clin Exp Med [Internet]. 2014 2014; 7(12):[5558–62 pp.]. Available from: http://europepmc.org/abstract/MED/25664071, https://europepmc.org/articles/PMC4307518, https://europepmc.org/articles/PMC4307518?pdf=render.
163. Ahmad S, Habib S, Moinuddin, Ali A. Preferential recognition of epitopes on AGE–IgG by the autoantibodies in rheumatoid arthritis patients. Human Immunology. 2013;74(1):23–7.
164. Products of advanced glycation in patients with type 2 diabetes and vascular disease. Annals of clinical biochemistry. 2003;40(5):552–9.
165. Doorn JA, Petersen DR. Covalent modification of amino acid nucleophiles by the lipid peroxidation products 4-hydroxy-2-nonenal and 4-oxo-2-nonenal. Chem Res Toxicol. 2002;15(11):1445–50.
166. Bieschke J, Zhang Q, Bosco DA, Lerner RA, Powers ET, Wentworth P, et al. Small Molecule Oxidation Products Trigger Disease-Associated Protein Misfolding. Acc Chem Res. 2006;39 (9):611–9.
167. Kato Y, Yoshida A, Naito M, Kawai Y, Tsuji K, Kitamura M, et al. Identification and quantification of N(epsilon)-(Hexanoyl)lysine in human urine by liquid chromatography/tandem mass spectrometry. Free Radic Biol Med. 2004;37(11):1864–74.
168. Arai H, Kato Y, Fukunaga K, Mohri S, Nakamura K. Formation of N.EPSILON.-(Hexanonyl)lysine in Oxidized Human Very-low-density Lipoprotein. Seibutsu Butsuri Kagaku. 2004;48:37–40.
169. Hisaka S, Kato Y, Kitamoto N, Yoshida A, Kubushiro Y, Naito M, et al. Chemical and immunochemical identification of propanoyllysine derived from oxidized n-3 polyunsaturated fatty acid. Free Radic Biol Med. 2009;46(11):1463–71.
170. Güntaş G, Engin B, Ekmekçi ÖB, Kutlubay Z, Ekmekci H, Songür A, et al. Evaluation of advanced oxidation protein products, prooxidant-antioxidant balance, and total antioxidant capacity in untreated vitiligo patients. Ann Dermatol. 2015;27(2):178–83.
171. Cao W, Hou FF, Nie J. AOPPs and the progression of kidney disease. Kidney International Supplements. 2014;4(1):102–6.
172. Du S-l, Zeng X-z, Tian J-w, Ai J, Wan J, He J-x. Advanced oxidation protein products in predicting acute kidney injury following cardiac surgery. Biomarkers. 2015;20(3):206–11.
173. Cristani M, Speciale A, Saija A, Gangemi S, Minciullo PL, Cimino F. Circulating Advanced Oxidation Protein Products as Oxidative Stress Biomarkers and Progression Mediators in Pathological Conditions Related to Inflammation and Immune Dysregulation. Curr Med Chem. 2016;23(34):3862–82.
174. Taylor EL, Armstrong KR, Perrett D, Hattersley AT, Winyard PG. Optimisation of an Advanced Oxidation Protein Products Assay: Its Application to Studies of Oxidative Stress in Diabetes Mellitus. Oxid Med Cell Longev. 2015;2015:496271.
175. Witko-Sarsat V, Friedlander M, Nguyen Khoa T, Capeillère-Blandin C, Nguyen AT, Canteloup S, et al. Advanced oxidation protein products as novel mediators of inflammation and monocyte activation in chronic renal failure. Journal of immunology (Baltimore, Md : 1950). 1998;161(5):2524–32.

176. Hazen SL, Heinecke JW. 3-Chlorotyrosine, a specific marker of myeloperoxidase-catalyzed oxidation, is markedly elevated in low density lipoprotein isolated from human atherosclerotic intima. J Clin Invest. 1997;99(9):2075–81.
177. Wu SM, Pizzo SV. α2-Macroglobulin from Rheumatoid Arthritis Synovial Fluid: Functional Analysis Defines a Role for Oxidation in Inflammation. Archives of Biochemistry and Biophysics. 2001;391(1):119–26.
178. Kettle AJ, Chan T, Osberg I, Senthilmohan R, Chapman AL, Mocatta TJ, et al. Myeloperoxidase and protein oxidation in the airways of young children with cystic fibrosis. American journal of respiratory and critical care medicine. 2004;170(12):1317–23.
179. Buss IH, Senthilmohan R, Darlow BA, Mogridge N, Kettle AJ, Winterbourn CC. 3-Chlorotyrosine as a Marker of Protein Damage by Myeloperoxidase in Tracheal Aspirates From Preterm Infants: Association With Adverse Respiratory Outcome. Pediatric Research. 2003;53(3):455–62.
180. Negre-Salvayre A, Coatrieux C, Ingueneau C, Salvayre R. Advanced lipid peroxidation end products in oxidative damage to proteins. Potential role in diseases and therapeutic prospects for the inhibitors. Br J Pharmacol. 2008;153(1):6–20.
181. Wong-Ekkabut J, Xu Z, Triampo W, Tang IM, Tieleman DP, Monticelli L. Effect of lipid peroxidation on the properties of lipid bilayers: a molecular dynamics study. Biophys J. 2007;93(12):4225–36.
182. Niki E. Lipid peroxidation: physiological levels and dual biological effects. Free Radic Biol Med. 2009;47(5):469–84.
183. Kasprzak KSJCi. Possible role of oxidative damage in metal-induced carcinogenesis. 1995;13 (4):411–30.
184. Feng Z, Hu W, Marnett LJ, Tang MS. Malondialdehyde, a major endogenous lipid peroxidation product, sensitizes human cells to UV- and BPDE-induced killing and mutagenesis through inhibition of nucleotide excision repair. Mutation research. 2006;601(1–2):125–36.
185. He L, Yang X, Xu K, Lin W. A mitochondria-targeted fluorescent probe for imaging endogenous malondialdehyde in HeLa cells and onion tissues. Chemical Communications. 2017;53(29):4080–3.
186. Ozougwu J. The Role of Reactive Oxygen Species and Antioxidants in Oxidative Stress. International Journal of Pharmacy and Biosciences. 2016;3:1–8.
187. Collin F. Chemical Basis of Reactive Oxygen Species Reactivity and Involvement in Neurodegenerative Diseases. International journal of molecular sciences. 2019;20(10).
188. Mahreen R, Mohsin M, Nasreen Z, Siraj M, Ishaq M. Significantly increased levels of serum malonaldehyde in type 2 diabetics with myocardial infarction. Int J Diabetes Dev Ctries. 2010;30(1):49–51.
189. Guzik B, Sagan A, Ludew D, Mrowiecki W, Chwała M, Bujak-Gizycka B, et al. Mechanisms of oxidative stress in human aortic aneurysms — Association with clinical risk factors for atherosclerosis and disease severity. International Journal of Cardiology. 2013;168(3): 2389–96.
190. Rašić I, Rašić A, Akšamija G, Radović S. The relationship between serum level of malondialdehyde and progression of colorectal cancer. Acta Clin Croat. 2018;57(3):411–6.
191. Esterbauer H, Schaur RJ, Zollner H. Chemistry and biochemistry of 4-hydroxynonenal, malonaldehyde and related aldehydes. Free Radical Biology and Medicine. 1991;11(1): 81–128.
192. Forman HJ, Fukuto JM, Miller T, Zhang H, Rinna A, Levy S. The chemistry of cell signaling by reactive oxygen and nitrogen species and 4-hydroxynonenal. Archives of biochemistry and biophysics. 2008;477(2):183–95.
193. Jang EJ, Jeong HO, Park D, Kim DH, Choi YJ, Chung KW, et al. Src Tyrosine Kinase Activation by 4-Hydroxynonenal Upregulates p38, ERK/AP-1 Signaling and COX-2 Expression in YPEN-1 Cells. PLOS ONE. 2015;10(10):e0129244.

194. Chapple SJ, Cheng X, Mann GE. Effects of 4-hydroxynonenal on vascular endothelial and smooth muscle cell redox signaling and function in health and disease. Redox Biol. 2013;1(1): 319–31.
195. Marwick JA, Kirkham PA, Stevenson CS, Danahay H, Giddings J, Butler K, et al. Cigarette smoke alters chromatin remodeling and induces proinflammatory genes in rat lungs. American journal of respiratory cell and molecular biology. 2004;31(6):633–42.
196. Kirkham PA, Spooner G, Rahman I, Rossi AG. Macrophage phagocytosis of apoptotic neutrophils is compromised by matrix proteins modified by cigarette smoke and lipid peroxidation products. Biochemical and Biophysical Research Communications. 2004;318(1):32–7.
197. Kirkham P, Spooner G, Ffoulkes-Jones C, Calvez R. Cigarette smoke triggers macrophage adhesion and activation: Role of lipid peroxidation products and scavenger receptor. Free radical biology & medicine. 2003;35:697–710.
198. Poli G, Schaur RJ, Siems WG, Leonarduzzi G. 4-hydroxynonenal: a membrane lipid oxidation product of medicinal interest. Medicinal research reviews. 2008;28(4):569–631.
199. Ji Y, Dai Z, Wu G, Wu Z. 4-Hydroxy-2-nonenal induces apoptosis by activating ERK1/2 signaling and depleting intracellular glutathione in intestinal epithelial cells. Scientific Reports. 2016;6(1):32929.
200. Tsukagoshi H, Kawata T, Shimizu Y, Ishizuka T, Dobashi K, Mori M. 4-Hydroxy-2-Nonenal Enhances Fibronectin Production by IMR-90 Human Lung Fibroblasts Partly via Activation of Epidermal Growth Factor Receptor-Linked Extracellular Signal-Regulated Kinase p44/42 Pathway. Toxicology and Applied Pharmacology. 2002;184(3):127–35.
201. Vatsyayan R, Chaudhary P, Sharma A, Sharma R, Rao Lelsani PC, Awasthi S, et al. Role of 4-hydroxynonenal in epidermal growth factor receptor-mediated signaling in retinal pigment epithelial cells. Experimental eye research. 2011;92(2):147–54.
202. Pradeep AR, Agarwal E, Bajaj P, Rao NS. 4-Hydroxy-2-nonenal, an oxidative stress marker in crevicular fluid and serum in type 2 diabetes with chronic periodontitis. Contemp Clin Dent. 2013;4(3):281–5.
203. Mihalas BP, De Iuliis GN, Redgrove KA, McLaughlin EA, Nixon B. The lipid peroxidation product 4-hydroxynonenal contributes to oxidative stress-mediated deterioration of the ageing oocyte. Scientific Reports. 2017;7(1):6247.
204. Hardas SS, Sultana R, Clark AM, Beckett TL, Szweda LI, Murphy MP, et al. Oxidative modification of lipoic acid by HNE in Alzheimer disease brain. Redox Biol. 2013;1(1):80–5.
205. Gęgotek A, Skrzydlewska E. Biological effect of protein modifications by lipid peroxidation products. Chemistry and Physics of Lipids. 2019;221:46–52.
206. Sayre LM, Zelasko DA, Harris PL, Perry G, Salomon RG, Smith MA. 4-Hydroxynonenal-derived advanced lipid peroxidation end products are increased in Alzheimer's disease. Journal of neurochemistry. 1997;68(5):2092–7.
207. Liu Q, Smith MA, Avilá J, DeBernardis J, Kansal M, Takeda A, et al. Alzheimer-specific epitopes of tau represent lipid peroxidation-induced conformations. Free Radical Biology and Medicine. 2005;38(6):746–54.
208. Selley ML. (E)-4-hydroxy-2-nonenal may be involved in the pathogenesis of Parkinson's disease. Free Radic Biol Med. 1998;25(2):169–74.
209. Smith RG, Henry YK, Mattson MP, Appel SH. Presence of 4-hydroxynonenal in cerebrospinal fluid of patients with sporadic amyotrophic lateral sclerosis. Annals of neurology. 1998;44 (4):696–9.
210. Devos D, Moreau C, Kyheng M, Garçon G, Rolland AS, Blasco H, et al. A ferroptosis–based panel of prognostic biomarkers for Amyotrophic Lateral Sclerosis. Scientific Reports. 2019;9 (1):2918.
211. Rahman I, van Schadewijk AA, Crowther AJ, Hiemstra PS, Stolk J, MacNee W, et al. 4-Hydroxy-2-nonenal, a specific lipid peroxidation product, is elevated in lungs of patients with chronic obstructive pulmonary disease. American journal of respiratory and critical care medicine. 2002;166(4):490–5.

212. Tarpey MM, Wink DA, Grisham MB. Methods for detection of reactive metabolites of oxygen and nitrogen: in vitro and in vivo considerations. Am J Physiol Regul Integr Comp Physiol. 2004;286(3):R431–44.
213. Roberts LJ, Morrow JD. Measurement of F(2)-isoprostanes as an index of oxidative stress in vivo. Free Radic Biol Med. 2000;28(4):505–13.
214. Milne GL, Yin H, Hardy KD, Davies SS, Roberts LJ, 2nd. Isoprostane generation and function. Chemical reviews. 2011;111(10):5973–96.
215. Langhorst ML, Hastings MJ, Yokoyama WH, Hung S-C, Cellar N, Kuppannan K, et al. Determination of F2-Isoprostanes in Urine by Online Solid Phase Extraction Coupled to Liquid Chromatography with Tandem Mass Spectrometry. Journal of Agricultural and Food Chemistry. 2010;58(11):6614–20.
216. Morrow JD, Hill KE, Burk RF, Nammour TM, Badr KF, Roberts LJ, 2nd. A series of prostaglandin F2-like compounds are produced in vivo in humans by a non-cyclooxygenase, free radical-catalyzed mechanism. Proc Natl Acad Sci U S A. 1990;87(23):9383–7.
217. Bellisario V, Mengozzi G, Grignani E, Bugiani M, Sapino A, Bussolati G, et al. Towards a formalin-free hospital. Levels of 15-F2t-isoprostane and malondialdehyde to monitor exposure to formaldehyde in nurses from operating theatres. Toxicology Research. 2016;5(4):1122–9.
218. Roberts LJ, 2nd, Milne GL. Isoprostanes. J Lipid Res. 2009;50 Suppl(Suppl):S219-S23.
219. Signorini C, De Felice C, Durand T, Galano JM, Oger C, Leoncini S, et al. Relevance of 4-F (4t)-neuroprostane and 10-F(4t)-neuroprostane to neurological diseases. Free Radic Biol Med. 2018;115:278–87.
220. Dennis EA, Cao J, Hsu Y-H, Magrioti V, Kokotos G. Phospholipase A2 enzymes: physical structure, biological function, disease implication, chemical inhibition, and therapeutic intervention. Chemical reviews. 2011;111(10):6130–85.
221. Wójcik P, Žarković N, Gęgotek A, Skrzydlewska E. Involvement of Metabolic Lipid Mediators in the Regulation of Apoptosis. Biomolecules. 2020;10(3):402.
222. Leitinger N, Blazek I, Sinzinger H. The influence of isoprostanes on ADP-induced platelet aggregation and cyclic AMP-generation in human platelets. Thrombosis Research. 1997;86 (4):337–42.
223. Cranshaw JH, Evans TW, Mitchell JA. Characterization of the effects of isoprostanes on platelet aggregation in human whole blood. Br J Pharmacol. 2001;132(8):1699–706.
224. Khasawneh FT, Huang J-S, Mir F, Srinivasan S, Tiruppathi C, Le Breton GC. Characterization of isoprostane signaling: Evidence for a unique coordination profile of 8-iso-PGF2α with the thromboxane A2 receptor, and activation of a separate cAMP-dependent inhibitory pathway in human platelets. Biochemical pharmacology. 2008;75(12):2301–15.
225. Montero A, Munger KA, Khan RZ, Valdivielso JM, Morrow JD, Guasch A, et al. F_2-isoprostanes mediate high glucose-induced TGF-β synthesis and glomerular proteinuria in experimental type I diabetes. Kidney International. 2000;58(5):1963–72.
226. Pfister SL, Nithipatikom K, Campbell WB. Role of superoxide and thromboxane receptors in acute angiotensin II-induced vasoconstriction of rabbit vessels. Am J Physiol Heart Circ Physiol. 2011;300(6):H2064-H71.
227. Chen SP, Chung YT, Liu TY, Wang YF, Fuh JL, Wang SJ. Oxidative stress and increased formation of vasoconstricting F2-isoprostanes in patients with reversible cerebral vasoconstriction syndrome. Free Radic Biol Med. 2013;61:243–8.
228. Kumar A, Kingdon E, Norman J. The isoprostane 8-iso-PGF2α suppresses monocyte adhesion to human microvascular endothelial cells via two independent mechanisms. 2005;19(3):1–24.
229. Miggin SM, Kinsella BT. Thromboxane A2 receptor mediated activation of the mitogen activated protein kinase cascades in human uterine smooth muscle cells. Biochimica et Biophysica Acta (BBA) – Molecular Cell Research. 2001;1539(1):147–62.
230. Giustarini D, Dalle-Donne I, Tsikas D, Rossi R. Oxidative stress and human diseases: Origin, link, measurement, mechanisms, and biomarkers. Critical reviews in clinical laboratory sciences. 2009;46(5–6):241–81.

231. Jamil J, Bankhele P, Salvi A, Mannix JE, Oger C, Guy A, et al. Role of the non-enzymatic metabolite of eicosapentaenoic acid, 5-epi-5-F3t-isoprostane in the regulation of [(3)H]D-aspartate release in isolated bovine retina. Neurochemical research. 2014;39(12):2360–9.
232. Weinberger B, Nisar S, Anwar M, Ostfeld B, Hegyi T. Lipid peroxidation in cord blood and neonatal outcome. Pediatrics International. 2006;48(5):479–83.
233. Rogers MS, Wang CC, Lau TK, Xiao X, Zhou XG, Fok TF, et al. Relationship between Isoprostane Concentrations, Metabolic Acidosis, and Morbid Neonatal Outcome. Clinical Chemistry. 2005;51(7):1271–4.
234. Comporti M, Signorini C, Leoncini S, Buonocore G, Rossi V, Ciccoli L. Plasma F2-isoprostanes are elevated in newborns and inversely correlated to gestational age. Free Radic Biol Med. 2004;37(5):724–32.
235. Rahman I. Reactive Oxygen Species and Antioxidant Therapeutic Approaches. 2009. p. 293–312.
236. Roy J, Galano J-M, Durand T, Le Guennec J-Y, Chung-Yung Lee J. Physiological role of reactive oxygen species as promoters of natural defenses. 2017;31(9):3729–45.
237. Sagun G, Oguz A, Mesci B, Isbilen B, Kavala M, Keskin H, et al. Levels of F2 isoprostane in Behcet's disease: Correlation with cardiometabolic risk factors. Redox Report. 2015;20(5): 223–7.
238. Makila-Mabe BG, Kikandau KJ, Sombo TM, Okitundu DL, Mwanza J-C, Boivin MJ, et al. Serum 8,12-iso-iPF2α-VI Isoprostane Marker of Oxidative Damage and Cognition Deficits in Children with Konzo. PLOS ONE. 2014;9(9):e107191.
239. Mir F, Lee D, Ray H, Sadiq SA. CSF isoprostane levels are a biomarker of oxidative stress in multiple sclerosis. Neurology – Neuroimmunology Neuroinflammation. 2014;1(2):e21.
240. Stathopoulos D, Loukides S, Syrigos K. 8-Isoprostane in Exhaled Breath Condensate of Patients with Non-small Cell Lung Cancer: The Effect of Chemotherapy. Anticancer Research. 2014;34(9):5143–5.
241. Inonu H, Doruk S, Sahin S, Erkorkmaz U, Celik D, Celikel S, et al. Oxidative Stress Levels in Exhaled Breath Condensate Associated With COPD and Smoking. Respiratory Care. 2012;57 (3):413–9.
242. Edzard S, Ralf AB, Rainer HB, Dimitrios T. Mass Spectrometric Analysis of F2-Isoprostanes: Markers and Mediators in Human Disease. Current Pharmaceutical Analysis. 2007;3(1): 39–51.
243. Karamouzi A, Kovachev D, Karamouzis I, Antoniadou-Hitoglou M, Tsikoulas I, Aggelopoulou-Sakadami N. Saliva Levels of 15-F2t-Isoprostane as Biomarker of Lipid Peroxidation in Autistic Children. 2007;5(3):141–4.
244. Freitas BdJeSdA, Lloret GR, Visacri MB, Tuan BT, Amaral LS, Baldini D, et al. High 15-F2t-Isoprostane Levels in Patients with a Previous History of Nonmelanoma Skin Cancer: The Effects of Supplementary Antioxidant Therapy. BioMed research international. 2015;2015:963569.
245. Roberts LJ, Moore KP, Zackert WE, Oates JA, Morrow JD. Identification of the Major Urinary Metabolite of the F2-isoprostane 8-Iso-prostaglandin F2α in Humans*. Journal of Biological Chemistry. 1996;271(34):20617–20.
246. Roest M, Voorbij HAM, Van der Schouw YT, Peeters PHM, Teerlink T, Scheffer PG. High levels of urinary F2-isoprostanes predict cardiovascular mortality in postmenopausal women. Journal of Clinical Lipidology. 2008;2(4):298–303.
247. de Faria AP, Modolo R, Moreno H. Plasma 8-Isoprostane as a Biomarker and Applications to Cardiovascular Disease. In: Patel VB, Preedy VR, editors. Biomarkers in Cardiovascular Disease. Dordrecht: Springer Netherlands; 2016. p. 467–88.
248. Tsikas D, Rode I, Becker T, Nashan B, Klempnauer J, Frölich J. Elevated plasma and urine levels of ADMA and 15(S)-8-iso-PGF 2α in end-stage liver disease [5]. Hepatology (Baltimore, Md). 2003;38:1063–4.

249. Montine TJ, Markesbery WR, Morrow JD, Roberts LJ, 2nd. Cerebrospinal fluid F2-isoprostane levels are increased in Alzheimer's disease. Annals of neurology. 1998;44 (3):410–3.
250. Nourooz-Zadeh J, Liu EH, Yhlen B, Anggård EE, Halliwell B. F4-isoprostanes as specific marker of docosahexaenoic acid peroxidation in Alzheimer's disease. Journal of neurochemistry. 1999;72(2):734–40.
251. Reich EE, Markesbery WR, Roberts LJ, 2nd, Swift LL, Morrow JD, Montine TJ. Brain regional quantification of F-ring and D-/E-ring isoprostanes and neuroprostanes in Alzheimer's disease. The American journal of pathology. 2001;158(1):293–7.
252. Borrill ZL, Roy K, Singh D. Exhaled breath condensate biomarkers in COPD. The European respiratory journal. 2008;32(2):472–86.
253. Shahid SK, Kharitonov SA, Wilson NM, Bush A, Barnes PJ. Exhaled 8-isoprostane in childhood asthma. Respiratory Research. 2005;6(1):79.
254. Wan G-H, Yan D-C, Tseng H-Y, Tung T-H, Lin S-J, Lin Y-W. Cysteinyl leukotriene levels correlate with 8-isoprostane levels in exhaled breath condensates of atopic and healthy children. Pediatric Research. 2013;74(5):584–91.
255. Montuschi P, Kharitonov SA, Ciabattoni G, Corradi M, van Rensen L, Geddes DM, et al. Exhaled 8-isoprostane as a new non-invasive biomarker of oxidative stress in cystic fibrosis. Thorax. 2000;55(3):205–9.
256. Cracowski J-L. Isoprostanes as a tool to investigate oxidative stress in scleroderma spectrum disorders—advantages and limitations. Rheumatology. 2006;45(7):922–3.
257. van der Plas A, Pouly S, Bourdonnaye G, Baker G, Lüdicke F. Influence of smoking on levels of urinary 8-iso Prostaglandin F2α. Toxicology Reports. 2018;6.
258. Taylor AW, Bruno RS, Traber MG. Women and smokers have elevated urinary F(2)-isoprostane metabolites: a novel extraction and LC-MS methodology. Lipids. 2008;43(10): 925–36.
259. Carmella SG, Heskin AK, Tang MK, Jensen J, Luo X, Le CT, et al. Longitudinal stability in cigarette smokers of urinary eicosanoid biomarkers of oxidative damage and inflammation. PLOS ONE. 2019;14(4):e0215853.
260. Schwedhelm E, Bartling A, Lenzen H, Tsikas D, Maas R, Brümmer J, et al. Urinary 8-iso-Prostaglandin F2α as a Risk Marker in Patients With Coronary Heart Disease. 2004;109(7):843–8.
261. Reilly MP, Praticò D, Delanty N, DiMinno G, Tremoli E, Rader D, et al. Increased Formation of Distinct F2 Isoprostanes in Hypercholesterolemia. 1998;98(25):2822–8.
262. Aruoma OI. Free radicals, oxidative stress, and antioxidants in human health and disease. Journal of the American Oil Chemists' Society. 1998;75(2):199–212.
263. Awad JA, Morrow JD. Excretion of f2-isoprostanes in bile: A novel index of hepatic lipid peroxidation. Hepatology. 1995;22(3):962–8.
264. Kadiiska MB, Gladen BC, Baird DD, Graham LB, Parker CE, Ames BN, et al. Biomarkers of oxidative stress study: III. Effects of the nonsteroidal anti-inflammatory agents indomethacin and meclofenamic acid on measurements of oxidative products of lipids in CCl4 poisoning. Free Radical Biology and Medicine. 2005;38(6):711–8.
265. Salomon RG, Bi W. Isolevuglandin adducts in disease. Antioxid Redox Signal. 2015;22(18): 1703–18.
266. Brame CJ, Boutaud O, Davies SS, Yang T, Oates JA, Roden D, et al. Modification of proteins by isoketal-containing oxidized phospholipids. J Biol Chem. 2004;279(14):13447–51.
267. Fukuda K, Davies SS, Nakajima T, Ong BH, Kupershmidt S, Fessel J, et al. Oxidative mediated lipid peroxidation recapitulates proarrhythmic effects on cardiac sodium channels. Circulation research. 2005;97(12):1262–9.
268. Guo L, Chen Z, Cox BE, Amarnath V, Epand RF, Epand RM, et al. Phosphatidylethanolamines modified by γ-ketoaldehyde (γKA) induce endoplasmic reticulum stress and endothelial activation. J Biol Chem. 2011;286(20):18170–80.

269. Kirabo A, Fontana V, de Faria AP, Loperena R, Galindo CL, Wu J, et al. DC isoketal-modified proteins activate T cells and promote hypertension. J Clin Invest. 2014;124(10):4642–56.
270. Ross BM, Glen I. Breath ethane concentrations in healthy volunteers correlate with a systemic marker of lipid peroxidation but not with omega-3 Fatty Acid availability. Metabolites. 2014;4 (3):572–9.
271. Kanoh S, Kobayashi H, Motoyoshi K. Exhaled ethane: an in vivo biomarker of lipid peroxidation in interstitial lung diseases. Chest. 2005;128(4):2387–92.
272. Cope KA, Solga SF, Hummers LK, Wigley FM, Diehl AM, Risby TH. Abnormal exhaled ethane concentrations in scleroderma. Biomarkers. 2006;11(1):70–84.
273. Phillips M, Cataneo RN, Cheema T, Greenberg J. Increased breath biomarkers of oxidative stress in diabetes mellitus. Clinica Chimica Acta. 2004;344(1):189–94.
274. Phillips M, Cataneo RN, Greenberg J, Grodman R, Salazar M. Breath markers of oxidative stress in patients with unstable angina. Heart disease (Hagerstown, Md). 2003;5(2):95–9.
275. Paredi P, Kharitonov SA, Leak D, Ward S, Cramer D, Barnes PJ. Exhaled Ethane, a Marker of Lipid Peroxidation, Is Elevated in Chronic Obstructive Pulmonary Disease. American journal of respiratory and critical care medicine. 2000;162(2):369–73.
276. Euler DE, Davé SJ, Guo H. Effect of cigarette smoking on pentane excretion in alveolar breath. Clinical Chemistry. 1996;42(2):303–8.
277. Montuschi P, Collins JV, Ciabattoni G, Lazzeri N, Corradi M, Kharitonov SA, et al. Exhaled 8-isoprostane as an in vivo biomarker of lung oxidative stress in patients with COPD and healthy smokers. American journal of respiratory and critical care medicine. 2000;162(3 Pt 1): 1175–7.
278. Wang H-T, Hu Y, Tong D, Huang J, Gu L, Wu X-R, et al. Effect of carcinogenic acrolein on DNA repair and mutagenic susceptibility. The Journal of biological chemistry. 2012;287(15): 12379–86.
279. Kirkham PA, Spooner G, Ffoulkes-Jones C, Calvez R. Cigarette smoke triggers macrophage adhesion and activation: role of lipid peroxidation products and scavenger receptor. Free Radic Biol Med. 2003;35(7):697–710.
280. Steinberg D. A critical look at the evidence for the oxidation of LDL in atherogenesis. Atherosclerosis. 1997;131 Suppl:S5–7.
281. Pawlak K, Mysliwiec M, Pawlak D. Oxidized LDL to autoantibodies against oxLDL ratio – The new biomarker associated with carotid atherosclerosis and cardiovascular complications in dialyzed patients. Atherosclerosis. 2012;224(1):252–7.
282. Hulthe J, Fagerberg B. Circulating Oxidized LDL Is Associated With Subclinical Atherosclerosis Development and Inflammatory Cytokines (AIR Study). 2002;22(7):1162–7.
283. Holvoet P, Mertens A, Verhamme P, Bogaerts K, Beyens G, Verhaeghe R, et al. Circulating Oxidized LDL Is a Useful Marker for Identifying Patients With Coronary Artery Disease. 2001;21(5):844–8.
284. Holvoet P, Kritchevsky SB, Tracy RP, Mertens A, Rubin SM, Butler J, et al. The metabolic syndrome, circulating oxidized LDL, and risk of myocardial infarction in well-functioning elderly people in the health, aging, and body composition cohort. Diabetes. 2004;53(4): 1068–73.
285. van den Berg VJ, Vroegindewey MM, Kardys I, Boersma E, Haskard D, Hartley A, et al. Anti-Oxidized LDL Antibodies and Coronary Artery Disease: A Systematic Review. Antioxidants (Basel). 2019;8(10):484.
286. Itabe H, Yamamoto H, Imanaka T, Shimamura K, Uchiyama H, Kimura J, et al. Sensitive detection of oxidatively modified low density lipoprotein using a monoclonal antibody. J Lipid Res. 1996;37(1):45–53.
287. Xu L, Porter NA. Free radical oxidation of cholesterol and its precursors: Implications in cholesterol biosynthesis disorders. Free radical research. 2015;49(7):835–49.
288. Umetani M, Ghosh P, Ishikawa T, Umetani J, Ahmed M, Mineo C, et al. The cholesterol metabolite 27-hydroxycholesterol promotes atherosclerosis via proinflammatory processes mediated by estrogen receptor alpha. Cell Metab. 2014;20(1):172–82.

289. Hoobler EK, Holz C, Holman TR. Pseudoperoxidase investigations of hydroperoxides and inhibitors with human lipoxygenases. Bioorganic & Medicinal Chemistry. 2013;21(13): 3894–9.
290. Grant GE, Rubino S, Gravel S, Wang X, Patel P, Rokach J, et al. Enhanced formation of 5-oxo-6,8,11,14-eicosatetraenoic acid by cancer cells in response to oxidative stress, docosahexaenoic acid and neutrophil-derived 5-hydroxy-6,8,11,14-eicosatetraenoic acid. Carcinogenesis. 2011;32(6):822–8.
291. Joulain C, Meskini N, Anker G, Lagarde M, Prigent A-F. Esterification of 12(s)-hydroxy-5,8,10,14-eicosatetraenoic acid into the phospholipids of human peripheral blood mononuclear cells: Inhibition of the proliferative response. 1995;164(1):154–63.
292. Takata S, Papayianni A, Matsubara M, Jimenez W, Pronovost PH, Brady HR. 15-Hydroxyeicosatetraenoic acid inhibits neutrophil migration across cytokine-activated endothelium. The American journal of pathology. 1994;145(3):541–9.
293. Slade PG, Williams MV, Brahmbhatt V, Dash A, Wishnok JS, Tannenbaum SR. Proteins modified by the lipid peroxidation aldehyde 9,12-dioxo-10(E)-dodecenoic acid in MCF7 breast cancer cells. Chemical research in toxicology. 2010;23(3):557–67.
294. Tsukagoshi H, Shimizu Y, Iwamae S, Hisada T, Ishizuka T, Iizuka K, et al. Evidence of oxidative stress in asthma and COPD: potential inhibitory effect of theophylline. Respiratory medicine. 2000;94(6):584–8.
295. Rahman I, Morrison D, Donaldson K, MacNee W. Systemic oxidative stress in asthma, COPD, and smokers. American journal of respiratory and critical care medicine. 1996;154 (4 Pt 1):1055–60.
296. Long J, Zhang C-J, Zhu N, Du K, Yin Y-F, Tan X, et al. Lipid metabolism and carcinogenesis, cancer development. American journal of cancer research. 2018;8(5):778–91.
297. Barrera G. Oxidative Stress and Lipid Peroxidation Products in Cancer Progression and Therapy. ISRN Oncology. 2012;2012:137289.

Chapter 4
Pro-oxidants

A. Khelfi

Abstract Reactive species (RS) are important factors involved in the vast majority of cell disorders. The generated oxidative damages are underlying factors implicated in many diseases including cancer, diabetes, aging. It is without question that pro-oxidants (pro-oxidant enzymes and non-enzymatic pro-oxidants) mediate oxidative stress and are at the epicenter of these pathophysiological events.

Pro-oxidant enzymes (e.g., NADPH oxidase (NOX), myeloperoxidase (MPO), nitric oxide synthase (NOS), xanthine oxidase (XO), and cyclooxygenase (COX)) are of central interest in a variety of biological phenomena. A disturbance in the activity of these enzymes is in most cases in favor of the initiation of oxidative stress. Besides, several pro-oxidant components with no enzymatic activities were identified (e.g., hemoproteins, homocysteine, and transition metal ions) and can have harmful effects on cells.

This chapter is an overview of the basic functions of pro-oxidant agents in biological systems. It provides a valuable summary of their physiological roles and discusses the biological consequences of their accumulation during oxidative stress.

Keywords NADPH oxidase (NOX) · Myeloperoxidase (MPO) · Nitric oxide synthase (NOS) · Xanthine oxidase (XO) · Cyclooxygenase (COX) · Hemoproteins · Homocysteine · Transition metal ions

4.1 Pro-oxidant Enzymes

The important biological sources responsible for reactive species (RS) production include several enzymes produced in different organelles of the cell (Fig. 4.1).

A. Khelfi (✉)
Faculty of Pharmacy, University of Algiers, Algiers, Algeria

National Centre of Toxicology, Algiers, Algeria

Department of Chemistry and Biomolecular Science, Clarkson University, Potsdam, NY, USA
e-mail: ab.khelfi@univ-alger.dz; akhelfi@clarkson.edu

S. Andreescu et al. (eds.), *Biomarkers of Oxidative Stress*,
https://doi.org/10.1007/978-3-031-60738-7_4

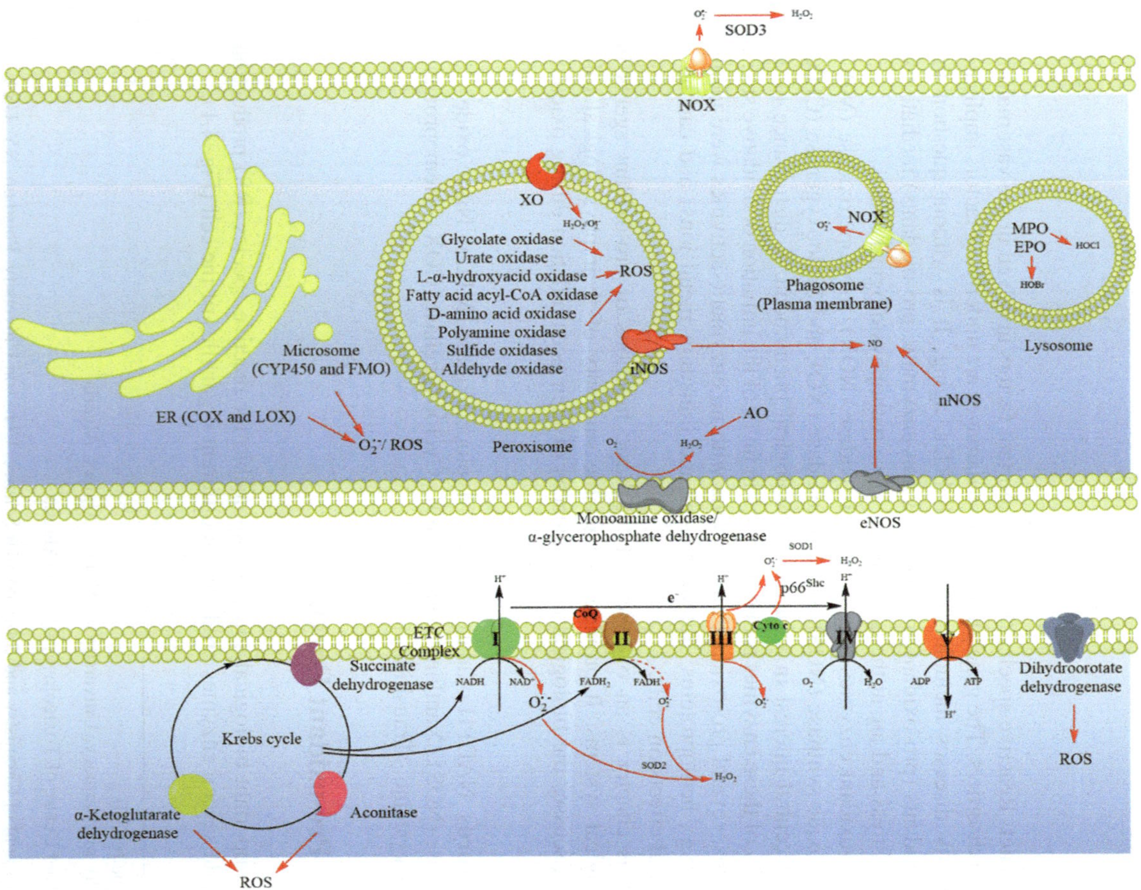

Fig. 4.1 Major sources of RS generation in cells and pro-oxidant enzymes
Intracellular ROS are mainly generated in mitochondria, ER, peroxisomes, lysosomes, phagosomes, the cytosol, and the plasma membrane.

Reactive species are produced by cells through the action of various soluble and membrane-bound enzymes.

Mitochondrion Electrons donated by nicotinamide adenine dinucleotide + hydrogen (NADH) and flavin adenine dinucleotide (FADH) pass through the ETC and ultimately reduce O_2 to form H_2O at complex IV. ROS are produced from the leakage of electrons from the ETC that react with molecular oxygen to form $O_2^{\bullet-}$ at complexes I, II, and III. The superoxide anion is produced within a matrix at complexes I and II, whereas at complex III, $O_2^{\bullet-}$ is released toward both the matrix and the intermembrane space. Once generated, $O_2^{\bullet-}$ is dismutated to H_2O_2 by sodium dismutase (SOD)1 in the intermembrane space and by SOD2 in the matrix. Then, H_2O_2 is converted into various ROS forms.

The superoxide anion is generated in low levels under physiological states, but its production is greatly enhanced under pathological situations via a dysfunctional mitochondrial respiratory chain.

The mitochondrial electron transport chain is the main producer of $O_2^{\bullet-}$ in the mammalian cell. Other sources include aconitase, α-ketoglutarate dehydrogenase, succinate dehydrogenase, monoamine oxidases, dihydroorotate dehydrogenase, eNOS, and α-glycerophosphate dehydrogenase.

Endoplasmic reticulum Subcellular fractions containing the ER are known to produce $O_2^{\bullet-}$ and H_2O_2, which largely arise from the cytochrome P450 system.

Plasma membrane and phagosomes NOX generates $O_2^{\bullet-}$ by transferring electrons from the intracellular electron donor NADPH across the membrane and by coupling a single electron to molecular oxygen to produce $O_2^{\bullet-}$.

Endoplasmic reticulum and microsomes Cytochrome P450 reductase transfers an electron from NADPH to molecular oxygen, forming ROS in the presence of CYP450 oxidoreductase.

Peroxisomes Peroxisomes contain many of the cellular enzymes that generate ROS.

Molecular oxygen as an electron accepter is converted into $O_2^{\bullet-}$ through a process in which XO catalyzes the oxidation of hypoxanthine to generate xanthine. Other peroxisomal ROS-generating enzymes include urate oxidase, l-α-hydroxyacid oxidase, fatty acid acyl-CoA oxidase, D-amino acid oxidase, polyamine oxidase, sulfide oxidases, AO, and iNOS.

Lysosomes Lysosomes produce many different kinds of reactive oxygen and nitrogen species (RONS) for oxygen-dependent killing. Indeed, hypochlorous acid (HOCl) is generated by MPO

Mitochondria are the main intracellular source of RS in mammalian cells due to the presence of pro-oxidant enzymes in their electron transport chain (ETC). Besides the ETC, there are several mitochondrial enzymes responsible for reactive oxygen species (ROS) formation, including Krebs cycle enzymes (aconitase, α-ketoglutarate dehydrogenase, and succinate dehydrogenase), several oxidoreductases located in the intermediate space (p66Shc), and enzymes of the mitochondrial membrane (monoamine oxidases, dihydroorotate dehydrogenase, endothelial nitric oxide synthase (eNOS), and α-glycerophosphate dehydrogenase) [1, 2].

Peroxisomes are major sources of total hydrogen peroxide (H_2O_2) produced by cells, as they are enriched by a variety of enzymes that produce RS:

- Xanthine oxidase (XO)
- Urate oxidase
- L-α-hydroxyacid oxidase
- Fatty acid acyl-coenzyme A (CoA) oxidase
- D-amino acid oxidase
- Polyamine oxidase
- Sulfide oxidase
- Aldehyde oxidase (AO)
- Inducible NOS (iNOS)

Lysosomes also contain enzymes that are implicated in the production of oxidants:

- Myeloperoxidase (MPO)
- Eosinoperoxidase (EPO)

The phagosome enzyme responsible for ROS generation and increased oxidative stress (OS) is nicotinamide adenine dinucleotide phosphate (NADPH) oxidase (NOX). The endoplasmic reticulum (ER), and particularly microsomes, is enriched by a variety of pro-oxidant enzymes, including cytochrome P450 (CYP450)-dependent monooxygenase, NADPH-dependent cytochrome reductases, and flavin-dependent monooxygenases. The membrane-bound oxidative enzymes responsible for ROS generation and increased OS include:

- NOX
- XO
- iNOS
- eNOS
- ETC enzymes
- Monoamine oxidase
- α-Glycerophosphate dehydrogenase
- Dihydroorotate dehydrogenase
- Succinate dehydrogenase

Soluble oxidative enzymes implicated in the production of oxidants are as follows:

- MPO
- EPO
- Lipoxygenase (LOX)
- Cyclooxygenase (COX)
- CYP450
- Flavin monooxygenase (FMO)
- Neuronal NOS (nNOS)
- AO
- Aconitase
- α-Ketoglutarate dehydrogenase

4.1.1 Enzymes of Mitochondrial Chain Complexes

Mitochondrial respiration is a major biological source of OS [3]. It contains multiple potential sites for ROS generation. The vast majority of ROS from mitochondria is continually produced by the complexes of the respiratory chain that transfers electrons to oxygen. Mechanistically, ROS production occurs during the physiological process of adenosine triphosphate (ATP) generation by the respiratory chain enzymes.

Mitochondrial electron transfer is normally a tightly coupled process. This tight coupling is reflected by the minimal leakage of electrons seen under conditions of maximal electron transfer. Nevertheless, some leakage occurs and seems to be increased under extreme physiological conditions.

The main ETC sources of ROS from mitochondria are complexes I and III (Fig. 4.1), particularly under conditions of high membrane potential [4]. These redox centers leak electrons to oxygen, thereby partially reducing the oxygen molecule to $O_2^{\bullet -}$. During this process, the molecular ground-state oxygen is activated to form $^1O_2$, by means of energy transfer or by electron transfer, leading to superoxide formation.

Complex I (NADH–ubiquinone oxidoreductase) is considered to be the major source of $O_2^{\bullet -}$ in which this species has been proposed to be formed from the fully reduced flavin to oxygen [5]. Complex III (ubiquinone–cytochrome *c* reductase) catalyzes the electron transfer from ubiquinol to cytochrome *c* through the Q-cycle. The semiquinone formed, which reacts with oxygen, is another source of superoxide production. However, coenzyme Q is unlikely to be a major source of free radicals (FRs) and may function as an antioxidant, protecting cells from free radical damage (see Sect. 5.2) [6].

Recently, it has been reported that complex II might be involved in ROS generation through auto-oxidation of coenzyme Q from the succinate–coenzyme Q complex [7]. During normal oxidative phosphorylation, 1–5% of all molecular oxygen consumed cannot be completely reduced and is converted to the free radical $O_2^{\bullet -}$ in the intact mitochondria [8]. In the physiological context, mitochondrial formation of ROS participates in signal transduction pathways that control cell hypertrophy, hypoxia, proliferation, and apoptosis. Maximal respiration and mitochondrial uncoupling due to xenobiotic exposure results in a considerable amount of protons, which bypass the ATP synthase pathway and leak back to the mitochondrial matrix, leading to high production of RS. Aged and xenobiotic-exposed mitochondria can produce higher quantities of RS, including $O_2^{\bullet-}$ and H_2O_2. This effect has been observed with doxorubicin, rotenone, mercury, cigarette smoke, asbestos, and silver nanoparticle (AgNP) exposures. The RS initially damage the mitochondria. Over time, cumulative damage to these organelles impairs mitochondrial function, eventuating in the decline of cellular function and cell death.

4.1.2 NADPH Oxidase (NOX)

NADPH oxidase (EC 1.6.3.1) is a transmembrane enzyme complex, which represents an important physiological source of RS in the body. It is located in various cells, including lymphocytes, neutrophils, monocytes, macrophages, myocytes, and chondrocytes, as well as in cells of the vascular wall, particularly endothelial cells (ECs), smooth muscle cells, and fibroblasts [9].

The NADPH oxidase family includes seven isoforms (NOX1–5 and DUOX1 and 2 (dual oxidase)), with each having a distinct catalytic subunit. Four NOX members (NOX1, NOX2, NOX4, and NOX5) are expressed in cardiovascular cells:

- NOX1 is expressed in vascular smooth muscle cells (VSMCs).
- NOX2 is expressed in ECs, cardiomyocytes, fibroblasts, VSMCs, and phagocytic cells (neutrophils and monocytes).
- NOX4 is expressed in all cardiovascular cell types.
- NOX5 is expressed in ECs and VSMCs.

The best-characterized NOX is the phagocytic one (NOX2), which has an important role in host defense against microorganisms. The importance of NOX-generated ROS in the normal functioning of the immune system is exemplified by the enhanced susceptibility to infections of patients suffering from chronic granulomatous disease. This disorder is caused by a defect in the NOX activity in phagocytes, which impairs elimination of the invading organisms [10].

Recently, NOX-generated RS has been found to play different roles in host defense; in smooth muscle cells, NOX generation of RS contributes to hypertension or to cell growth responses [11]. Several other functions are related to NOX4. Expression of this isoform is known to be enhanced by several stimuli, including mitochondrial dysfunction, ER stress, shear stress, arterial injury, and tissue hypoxia [12, 13]. NOX4 has been found to be related to the cellular senescence of ECs and age-related cardiac remodeling. Moreover, NOX4 promotes angiogenesis and flow recovery from limb ischemia in an eNOS-dependent manner, augmenting vasodilatation [13–15].

Overall, depending on the isoform, organism, and cell type, activation of these enzymes can be accomplished by hormones, cytokines, shear stress, or growth factors [16, 17] and have diverse effects, including production of inflammatory mediators as well as cell growth and differentiation.

As mentioned earlier, NOX is one of the major sources of superoxide formation in cells. This function is accomplished by NADPH, which serves as an electron donor to an activated NOX to produce $O_2^{\bullet-}$ from the oxygen molecule (Eq. 2.16). NOX is involved in the development of numerous pathologies, including Alzheimer's disease (AD), atherosclerosis, ischemia–reperfusion injury, ischemic stroke, hypertension, sepsis, inflammatory bowel disease (IBD), asthma, and diabetes mellitus (DM) [18–26]. Finally, activation of NOX has been noted in asbestos and ultraviolet (UV) light exposures.

4.1.3 Myeloperoxidase (MPO)

Myeloperoxidase (EC 1.11.2.2) is a hemoprotein that is mainly present in neutrophils and at lower levels in monocytes. Upon infections and inflammatory conditions, MPO is released from leukocytes and results in an oxidative insult due to its high production of RS. Due to the high toxicity of MPO-produced RS, neutrophils are the main cells responsible for killing pathogens. The myeloperoxidase system is involved in other physiological functions including the following:

- Reducing the bioavailability of endothelium-derived $NO^{\bullet}$ and its vasodilatory functions
- Stimulating neutrophil adherence to ECs, interleukin (IL)-8 formation by monocytes, and platelet aggregation
- Activating matrix metalloproteinases (MMPs) and other protease cascades
- Promoting advanced glycation end product (AGE) formation

Myeloperoxidase is a considerable source of RS. It catalyzes the halogenation of H_2O_2 with Cl^-, resulting in the formation of HOCl (Eq. 2.75). Besides, it is involved in nitrite conversion to nitryl chloride (NO_2Cl) and $NO_2^{\bullet}$, as evidenced by tyrosine nitration. Moreover, MPO can potentiate nitrosation reactions in the presence of $NO^{\bullet}$ and H_2O_2 via NO^+ and nitrogen oxides like species formation [27].

The MPO–HOCl system is intimately implicated in the atherosclerotic process. Indeed, enzymatically active MPO and hypochlorite-modified lipoproteins are present in human atherosclerotic lesions, where they co-localize with macrophages and ECs [28].

Myeloperoxidase can be used as a prognostic marker of cardiovascular disease (CVD). For instance, the initial measurement of plasma MPO independently predicts the early risk of myocardial infarction in patients with chest pain [29]. Besides, elevated activity of MPO has been noted in different situations, including Alzheimer's disease, atherosclerosis, chronic obstructive pulmonary disease (COPD), inflammatory bowel disease (IBD), sepsis, and rheumatoid arthritis, as well as benzene exposure based on genetic polymorphism analysis (Table 22.1) [30–36].

4.1.4 Eosinoperoxidase (EPO)

Eosinoperoxidase (EC 1.11.1.7) is an enzyme found within eosinophil granulocytes. The biological activities of EPO are partly related to its peroxidase activity. This peroxidase catalyzes halogenation reactions, leading to the formation of the long-acting reactive halogen species reactive oxygen and nitrogen species (RHS) hypobromous acid (Eq. 2.76), which plays an important role in inflammation and allergic reactions. Eosinoperoxidase also contributes to OS by modification of proteins by halogenations, nitration, and protein cross-links via tyrosyl radicals.

4.1.5 Nitric Oxide Synthase (NOS)

Nitric oxide synthase (EC 1.14.13.39) produces $NO^{\bullet}$ from O_2 and L-arginine, in the presence of NADPH, calcium, and/or biopterin as cofactors (Eq. 2.40). This enzyme exists in three forms, namely, endothelial NOS (eNOS), neuronal NOS (nNOS), and inducible NOS (iNOS). eNOS and nNOS are constitutive enzymes, which maintain the physiological levels of $NO^{\bullet}$ needed for vascular tone and neurotransmission. These constitutively expressed NOS isoforms are regulated by the intracellular levels of calcium. iNOS induction is calcium-independent, leading to the high-output production of $NO^{\bullet}$.

iNOS exists at extremely low tissue levels under basal conditions. However, it can be massively induced and activated during inflammatory conditions in several cells, including macrophages, neutrophils, ECs, adipocytes, and hepatocytes, as well as in smooth, skeletal, and cardiac myocytes. iNOS can release much higher amounts of $NO^{\bullet}$ after stimulation by inflammatory cytokines (IL-6, tumor necrosis factor-α (TNF-α), and interferon-γ (IFN-γ)), endotoxins, lipopolysaccharides, and other relevant inflammatory mediators, thus playing a crucial role in immune response.

Convergent signals from cytokine-activated pathways, such as Janus kinase (JAK), mitogen-activated protein kinase (MAPK), and IkappaB kinase (IKK)/nuclear factor kappa B (NF-κB), induce iNOS expression through the binding of transcription factors to the iNOS gene promoter [37].

eNOS activity is generally considered to be beneficial and produces low levels of $NO^{\bullet}$. This physiological production of $NO^{\bullet}$ is responsible for transmission of signals and is sufficient to maintain a basal vasodilator tone. Nitric oxide as a vasodilator has been shown to inhibit leukocyte interaction with the endothelium, control cell proliferation, and inhibit platelet aggregation and cell adhesion [38].

nNOS produces $NO^{\bullet}$ in the nervous system and acts as a signaling molecule in the central nervous system (CNS) where it influences neurotransmitter release, neurodevelopment, and controls learning and memory processes. In contrast to other signaling molecules, which act through receptors, $NO^{\bullet}$ diffuses out of the cell where it is fabricated and diffuses in target cells to interact with its molecular target, for example, nucleic acids, proteins, and other RS like $O_2^{\bullet-}$. The inducible form of NOS is considered a marker of OS. Once activated, it generates large amounts of $NO^{\bullet}$, which is consumed and thereby may cause various problems.

Increased and extensive levels of iNOS have been associated with human tumorigenesis and infiltrating macrophages at the sites of chronic active inflammation, leading to massive production of $NO^{\bullet}$ [39]. Clinical triggers for iNOS include acute ischemic stroke, sepsis, IBD, Alzheimer's and Parkinson's diseases, COPD, and cardiac allograft rejection [40–46]. High exposure to bleomycin, lead, and metal nanoparticles (NPs) may also increase iNOS activity.

4.1.6 Xanthine Oxidase (XO)

Xanthine oxidase (EC 1.17.3.2) belongs to the molybdenum hydroxylase family within the oxidoreductase class of enzymes, which are cytosolic and peroxisomal proteins present in capillary ECs as well as epithelial cells in lactating mammary glands [47]. The enzyme exists in two interconvertible forms: xanthine oxidase (XO) that can reduce molecular oxygen directly to $O_2^{\bullet-}$ and H_2O_2 and xanthine dehydrogenase (XDH) that uses NAD^+ as an electron acceptor.

Xanthine oxidase is a prominent source of OS, especially under hypoxic conditions. Under normal physiological conditions, the enzyme primarily exists in the dehydrogenase form. However, during hypoxia, it is converted to the oxidase form that can produce RS [48]. Briefly, XO catalyzes the oxidation of hypoxanthine to xanthine, where O_2 is required as an electron acceptor, and that of xanthine to uric acid, thus producing H_2O_2, and transferring electrons to O_2, producing $O_2^{\bullet-}$ (Eq. 2.11), which is proposed as a central mechanism of oxidative injury. Some studies suggest that XO can catalyze the reaction of bicarbonate and O_2 to produce $CO_3^{\bullet-}$ anions and H_2O_2 in special oxidation conditions [49].

XO-derived ROS play a physiological role in innate immunity, specifically in the inflammatory response and in antimicrobial defense of the gastrointestinal tract. Additionally, XO is likely to play a part in the metabolism of xenobiotics and has been shown to mediate the secretion of milk fat globules [50]. Finally, XO activity was reported to play an important role in some pathological conditions such as ischemic stroke, IBD, and sepsis (Table 22.1) [51–54].

4.1.7 Cyclooxygenase (COX) and Lipoxygenase (LOX)

Arachidonic acid released from membrane phospholipids is metabolized by membrane-associated enzymes called "cyclooxygenase and lipoxygenase." There are at least three isoenzymes of COX: COX-1, COX-2, and COX-3. COX-1 is a constitutively expressed enzyme important in the physiological functions of the gastrointestinal and vascular systems. It is highly expressed in interstitial cells, blood vessels, smooth muscles cells, platelets, and mesothelial cells, whereas COX-2 is an inducible enzyme that is upregulated during inflammatory conditions. It is ubiquitous in parenchymal cells with the exception of the heart [55]. COX-3 enzymes have been described as being fully expressed in the spinal cord, brain, and heart. The primary function of this isoform is to regulate pain responses and fever.

Both COX-1 and COX-2 function as prostaglandin H synthase catalyzing the bis-dioxygenation of arachidonic acid to form hydroxyperoxy endoperoxide (PGG_2). Its two-electron reduction to hydroxyperoxy endoperoxide (PGH_2) occurs through the COX peroxidase activity. Overall, COX enzymes are involved in the synthesis of bioactive prostanoids, including prostaglandins, prostacyclin, and thromboxane, which modulate platelet aggregation as well as inflammatory and chemotactic responses.

Lipoxygenase catalyzes the insertion of molecular oxygen into arachidonic acid as the first step in the formation of leukotrienes and hydroxyeicosatetraenoic acids. LOX products are involved in mediating vasoconstriction, vascular permeability, and bronchoconstriction. In addition to lipid oxidation, the metabolism of arachidonic acid also induces ROS production [56]. The series of reactions include many oxidation steps, with generation of FR intermediates. For instance, oxygenation of arachidonic acid by COX-1 via the FR mechanism has been described to produce peroxyl radicals [57]. Even with the formation of hydroperoxides, one can envision their homolytic cleavage to form alkoxyl and hydroxyl radicals. In the pathological context, activation and polymorphism of COX were reported to play a critical role in ischemic stroke [58–60].

4.1.8 Cytochrome P450 (CYP450)

The endoplasmic reticulum (microsomes) of cells, particularly hepatocytes, contains a family of mixed-function enzymes called "cytochrome P450 or CYP450" that can oxidize a wide range of endogenous substances. These enzymes are also involved in the detoxification of several drugs and toxic agents. This process is of great physiological relevance and represents the most important process for the elimination of xenobiotics.

The reaction mechanisms have been extensively studied and involve the reduction of the heme moiety by the transfer of an electron from an NADPH-specific flavoprotein enzyme. Cytochrome P450 is a large family consisting of multiple subfamilies. There are 18 mammalian CYP450 families, which encode 57 genes in the human genome [61]. Cytochrome P450 families are classified by similar gene sequences; they are assigned a family number (e.g., CYP1, CYP2) and a subfamily letter (e.g., CYP1A, CYP2D) and are differentiated by a number for the isoform (e.g., CYP1A1, CYP2D6). Several studies have suggested that CYP450 plays an important role in ROS formation [62, 63].

These enzymes are capable of activating molecular oxygen. This involves the transfer of electrons from NADPH via cytochrome P450 reductase to heme at the CYP-active site. Once reduced, the heme binds and splits oxygen, releasing one atom of oxygen as water and leaving the other one bound to heme. The latter is a powerful oxidant that readily reacts with the bound xenobiotic substrate. Although, under basal conditions, this process is tightly coupled, the disruption of the mixed-function oxidase metabolism through injury or the presence of high levels of xenobiotics may cause leakage of electrons and high generation of ROS. Among a variety of CYP450 proteins, CYP3A4 was the most active one in the production of $O_2^{\bullet -}$ and H_2O_2. CYP1A1 was the next most reactive CYP, whereas CYP1A2 and 2B6 display a comparable, lower activity [64]. In addition, some xenobiotic products formed by CYP450 metabolism may yield FRs in subsequent reactions. As an example, we can cite the case of ethyl alcohol, carbon tetrachloride (CCl_4), and acetaminophen. Moreover, many xenobiotics are capable of modulating OS as a consequence of biotransformation to quinones (e.g., benzene and benzo(a)pyrene). CYP450 are important actors for such activity since they initiate and participate in the redox cycling of these compounds.

4.1.9 *Flavin Monooxygenases (FMOs)*

Flavin monooxygenases are a distinct from of CYP, in that they contain flavin (FAD) but no heme, and in that they take NADPH as a direct co-substrate (rather than receive its electrons via a reductase, as in the case of CYP). Flavin monooxygenases, such as Ero1, present in the ER are significant sources of RS [65]. Like CYP, FMO is also linked to the generation of ROS by the leakage of electrons. Following reduction by NADPH and interaction with molecular oxygen, FAD is present as an FAD hydroperoxide (FAD-OOH) at the active site of FMO, oxygenating substrates that usually harbor soft oxygen acceptors, such as nitrogen or sulfur [66]. Otherwise, FAD-OOH may decompose to release $O_2^{\bullet -}$ and H_2O_2 that, interestingly, appears to be more pronounced in the presence of a substrate [67]. For instance, FMO can directly reduce O_2 to $O_2^{\bullet -}$ and H_2O_2 while oxidizing purines.

Another mechanism of ROS generation has been described, where Ero1 oxidizes thiol substrates to disulfides during protein folding. Reduced $FADH_2$ can transfer two electrons to molecular oxygen to form equimolar amounts of H_2O_2, according to Eqs. 4.1 and 4.2 [65]:

$$\text{Ero-FAD} + 2\,\text{RSH} \rightarrow \text{Ero-FADH}_2 + \text{RSSR} \qquad (4.1)$$

$$\text{Ero-FADH}_2 + O_2 \rightarrow \text{Ero-FAD} + H_2O_2 \qquad (4.2)$$

4.1.10 Aldehyde Oxidase (AO)

Aldehyde oxidase belongs to the same family of xanthine oxidoreductases. Although AO has structural similarities to XO, they differ in substrate specificity and inhibitors. Being a metabolizing enzyme, AO is highly ubiquitous in the liver but can also be found in other tissues such as the spleen, lungs, and heart.

Aldehyde oxidase can effectively hydroxylate most aldehydes to carboxylic acids with the production of $O_2^{\bullet-}$ and H_2O_2 (Eq. 4.3).

$$2\mathrm{RCOH} + 2\mathrm{O_2} \xrightarrow{\mathrm{AO}} 2\mathrm{RCOOH} + \mathrm{O_2^{\bullet-}} \tag{4.3}$$

Additionally, AO also uses NADH as a substrate to form $O_2^{\bullet-}$. Nitrite reduction by AO has also been described to form $NO^{\bullet}$ and has been shown to function as superoxide-generating NOX in which O_2 reduction to $O_2^{\bullet-}$ occurs at the FAD site [68–70].

There is evidence for an increasing relevance of AO in the metabolism and clearance of numerous important drugs, including the antimalarial agent quinine, and some cancer chemotherapeutic agents such as methotrexate [71]. Besides, AO was reported to play a critical role in alcohol-mediated liver injury due to its ability to generate ROS during the oxidation of acetaldehyde or NADH. AO is also implicated in the development of some OS-mediated diseases such as an increased risk of breast cancer [72]. AO-derived ROS have been shown to play a critical role in IBD and diabetic cardiovascular complications [73, 74]. However, the pathophysiological importance of this pathway for diabetic vascular disease warrants further study.

4.1.11 Other Pro-oxidant Enzymes

A variety of enzymes, including dopamine β-hydroxylase, urate oxidase, fatty acyl-CoA oxidase, D-amino acid oxidase, polyamine oxidase, α-hydroxyacid oxidase (glycolate oxidase), tryptophan dioxygenase, lactoperoxidase (LPO), glycerol-3-phosphate oxidase (GPO), thyroid peroxidase (TPO), and sulfide oxidases, have been shown to produce $O_2^{\bullet-}$ and H_2O_2 that are essential to metabolic processes. They have not been well-studied in terms of FR toxicology, and little is known of their contributions to tissue injury.

The same is true for enzymes involved in the mitochondrial machinery, including Krebs cycle enzymes (α-ketoglutarate dehydrogenase, aconitase, and succinate dehydrogenase), and glycerol-3-phosphate dehydrogenase, monoamine oxidase (MAO), and dihydroorotate dehydrogenase, located in the mitochondrial membrane, as well as p66Shc, an adaptor protein found in the intermembrane space (Fig. 4.1). For instance, monoamine oxidase generates H_2O_2 according to the following reaction:

$$RCH_2NH_2 + H_2O + O_2 \xrightarrow{MAO} RCHO + NH_3 + H_2O_2 \tag{4.4}$$

4.2 Nonenzymatic Pro-oxidants

The simplest biological systems can activate molecular oxygen during auto-oxidation reactions with a group of proteins and compounds, including hemoproteins (e.g., hemoglobin), thiols (e.g., homocysteine), catecholamines (e.g., dopamine and epinephrine), transition-metal ions, glucose, flavins, tetrahydropterins, and hydroquinones. Superoxide anions and other ROS are therefore generated as by-products and can mediate cell injury in stress conditions.

4.2.1 Hemoproteins

Auto-oxidation of heme proteins such as hemoglobin (Eq. 4.5), myoglobin, or eNOS results in the release of bound oxygen as $O_2^{\bullet-}$ [75, 76].

Hemoglobin in erythrocytes is estimated to oxidize at a rate of a few percentages per day, so it represents an appreciable source of $O_2^{\bullet-}$ in blood circulation.

$$\text{Hb-Fe}^{2+}O_2 \rightarrow \text{Hb-Fe}^{3+} + O_2^{\bullet\,-} \tag{4.5}$$

4.2.2 Homocysteine

Homocysteine is a sulfur-containing, nonprotein amino acid, which is highly sensitive to oxidation. In biological systems, it can be found in both bound and free forms. Protein disulfide-bound homocysteine accounts for >70% of total homocysteine, and the remaining is found as low-molecular-weight disulfide forms, including homocysteine (homocysteine homodimer) and homocysteine–cysteine mixed disulfide [77].

Most thiols auto-oxidize in the presence of transition-metal catalysts and molecular oxygen, leading to the formation of ROS. Homocysteine, as a thiol molecule, can undergo similar auto-oxidative chemistry. During this process, homocysteine exerts pro-oxidant actions through the production of $O_2^{\bullet-}$ and H_2O_2. In the presence of $NO^{\bullet}$, $O_2^{\bullet-}$ can form the powerful oxidant $ONOO^-$ [78].

During oxidation, homocysteine generates ROS, which can oxidize low-density lipoproteins (LDLs), especially in the presence of metals such as iron or copper. Indeed, oxidation of homocysteine reduced by iron and the subsequent formation of

ROS and oxidized LDLs (oxLDLs) result in a cytotoxic effect on ECs [79]. This initiates inflammatory responses and is involved in the establishment of foam cells, a key atherosclerotic lesion component. Alternative hypotheses with respect to oxidation, regarding homocysteine toxicity, include protein homocysteinylation, induction of pro-oxidant enzymes, and substitution effects. Protein homocysteinylation occurs through various mechanisms [77]:

- Oxidation of thiol groups (in particular cysteine), mediated by homocysteine in its free form, which occurs through the formation of a thiolate anion
- The misincorporation of S-nitroso-homocysteine (formed from $NO^{\bullet}$ and homocysteine) in place of methionine in the backbone of proteins during protein biosynthesis
- Post-biosynthetic acylation of free amino groups, in particular the ε amino group of lysine residues and the terminal amino group of proteins (mediated by homocysteine thiolactone, a dehydrated homocysteine derivative)

Regarding induction of pro-oxidant enzymes, homocysteine enhances superoxide production and NOX assembly by human neutrophils. In addition, it enhances production of H_2O_2 and stimulates migration of neutrophils [80].

With regard to substitution effects, cyclization of homocysteine and formation of the thiolactone derivative displace thioretinamide from binding to cobalamin, thus forming thioco, a complex that inhibits oxidative phosphorylation and leads to accumulation of oxygen radicals and enhanced OS [81].

Overall, hyperhomocysteinemia is considered an independent risk factor for CVD (e.g., atherosclerosis, stroke, myocardial infarction, and arterial and venous thrombotic events) [77]. In fact, elevated levels of homocysteine were found to be associated with atherosclerosis, stroke, thrombosis, and occlusive vascular disease [82–84]. Plasma homocysteine levels are also elevated in several chronic neuropsychiatric disorders, including Alzheimer's disease [85].

4.2.3 Other Pro-oxidant Compounds

In stress conditions, ROS are produced by auto-oxidation of some oxygen-utilizing compounds. However, these compounds are not considered as biomarkers of OS because they are involved in other important metabolic, physiological, and pathophysiological pathways. The following are some examples with the corresponding RS production.

Auto-oxidation of catecholamines (dopamine and epinephrine) results in the generation of $O_2^{\bullet-}$. This process can lead to tissue injury in an isolated organ or cell systems and has been implicated in neurotoxicity in vivo [86]. The possibility that the oxidation of dopamine is involved in Parkinson's disease received extensive attention a few decades ago.

Transition-metal ions (e.g., copper and iron) also promote RS formation by catalyzing auto-oxidation reactions as well as Fenton and Haber–Weiss reactions (see Sect. 2.1.2.3). Their involvement in AD has been largely discussed earlier (see Sect. 2.1.2.3).

Auto-oxidation of glucose may occur in the presence of metallic ions, releasing several types of aldehydes (e.g., ketoaldehydes and glyoxal) and possibly some RS (H_2O_2 and $OH^{\bullet}$). This oxidation leads to the cleavage of proteins and their glycation by attachment of ketoaldehydes (see Sect. 3.2). Glyoxal binds to the amino group of proteins with the consequence of the appearance of carboxymethyl-lysine residues. These compounds have the capacity to bind copper and induce a process of lipid peroxidation, which can, in turn, increase the production of glyoxal.

References

1. Sharma P, Jha A, Dubey R, Pessarakli M. Reactive Oxygen Species, Oxidative Damage, and Antioxidative Defense Mechanism in Plants under Stressful Conditions. Journal of Botany. 2012;2012.
2. Quinlan CL, Perevoshchikova IV, Hey-Mogensen M, Orr AL, Brand MD. Sites of reactive oxygen species generation by mitochondria oxidizing different substrates. Redox Biol. 2013;1 (1):304–12.
3. Kausar S, Wang F, Cui H. The Role of Mitochondria in Reactive Oxygen Species Generation and Its Implications for Neurodegenerative Diseases. Cells. 2018;7(12).
4. Zhao R-Z, Jiang S, Zhang L, Yu Z-B. Mitochondrial electron transport chain, ROS generation and uncoupling (Review). Int J Mol Med. 2019;44(1):3–15.
5. Dröse S, Brandt U. The mechanism of mitochondrial superoxide production by the cytochrome bc1 complex. J Biol Chem. 2008;283(31):21649–54.
6. Bentinger M, Brismar K, Dallner G. The antioxidant role of coenzyme Q. Mitochondrion. 2007;7:S41–S50.
7. Grivennikova VG, Kozlovsky VS, Vinogradov AD. Respiratory complex II: ROS production and the kinetics of ubiquinone reduction. Biochimica et Biophysica Acta (BBA) – Bioenergetics. 2017;1858(2):109–17.
8. Huang MZ, Li JY. Physiological regulation of reactive oxygen species in organisms based on their physicochemical properties. 2020;228(1):e13351.
9. Bansal M, Kaushal N. Introduction to Oxidative Stress. In: Bansal M, Kaushal N, editors. Oxidative Stress Mechanisms and their Modulation. New Delhi: Springer India; 2014. p. 1–18.
10. Giardino G, Cicalese MP, Delmonte O, Migliavacca M, Palterer B, Loffredo L, et al. NADPH Oxidase Deficiency: A Multisystem Approach. Oxidative Medicine and Cellular Longevity. 2017;2017:4590127.
11. Santillo M, Colantuoni A, Mondola P, Guida B, Damiano S. NOX signaling in molecular cardiovascular mechanisms involved in the blood pressure homeostasis. Front Physiol. 2015;6: 194.
12. Chen K, Kirber MT, Xiao H, Yang Y, Keaney JF, Jr. Regulation of ROS signal transduction by NADPH oxidase 4 localization. The Journal of cell biology. 2008;181(7):1129–39.
13. Craige SM, Chen K, Pei Y, Li C, Huang X, Chen C, et al. NADPH Oxidase 4 Promotes Endothelial Angiogenesis Through Endothelial Nitric Oxide Synthase Activation. 2011;124(6): 731–40.
14. Lener B, Kozieł R, Pircher H, Hütter E, Greussing R, Herndler-Brandstetter D, et al. The NADPH oxidase Nox4 restricts the replicative lifespan of human endothelial cells. The Biochemical journal. 2009;423(3):363–74.

15. Wang M, Zhang J, Walker SJ, Dworakowski R, Lakatta EG, Shah AM. Involvement of NADPH oxidase in age-associated cardiac remodeling. Journal of molecular and cellular cardiology. 2010;48(4):765–72.
16. Ritsick DR, Edens WA, Finnerty V, Lambeth JD. Nox regulation of smooth muscle contraction. Free Radic Biol Med. 2007;43(1):31–8.
17. Clempus RE, Sorescu D, Dikalova AE, Pounkova L, Jo P, Sorescu GP, et al. Nox4 is required for maintenance of the differentiated vascular smooth muscle cell phenotype. Arteriosclerosis, thrombosis, and vascular biology. 2007;27(1):42–8.
18. Madrigal-Matute J, Fernandez-Garcia C-E, Blanco-Colio LM, Burillo E, Fortuño A, Martinez-Pinna R, et al. Thioredoxin-1/peroxiredoxin-1 as sensors of oxidative stress mediated by NADPH oxidase activity in atherosclerosis. Free Radical Biology and Medicine. 2015;86: 352–61.
19. Karim AS, Reese SR, Wilson NA, Jacobson LM, Zhong W, Djamali A. Nox2 is a mediator of ischemia reperfusion injury. Am J Transplant. 2015;15(11):2888–99.
20. Ciancarelli I, De Amicis D, Di Massimo C, Carolei A, Ciancarelli MG. Oxidative stress in post-acute ischemic stroke patients after intensive neurorehabilitation. Current neurovascular research. 2012;9(4):266–73.
21. Gumanova NG, Teplova NV, Ryabchenko AU, Denisov EN. Serum nitrate and nitrite levels in patients with hypertension and ischemic stroke depend on diet: A multicenter study. Clinical Biochemistry. 2015;48(1):29–32.
22. Huang X, Sun M, Li D, Liu J, Guo H, Dong Y, et al. Augmented NADPH oxidase activity and p22phox expression in monocytes underlie oxidative stress of patients with type 2 diabetes mellitus. Diabetes Research and Clinical Practice. 2011;91(3):371–80.
23. Park MW, Cha HW, Kim J, Kim JH, Yang H, Yoon S, et al. NOX4 promotes ferroptosis of astrocytes by oxidative stress-induced lipid peroxidation via the impairment of mitochondrial metabolism in Alzheimer's diseases. Redox Biol. 2021;41:101947.
24. Tiseo G, Cavarretta E, Forniti A, Nocella C, Sciarretta S, Spagnolello O, et al. Interplay between Nox2 Activity and Platelet Activation in Patients with Sepsis and Septic Shock: A Prospective Study. Oxid Med Cell Longev. 2020;2020:4165358.
25. Grasberger H, Magis AT, Sheng E, Conomos MP, Zhang M, Garzotto LS, et al. DUOX2 variants associate with preclinical disturbances in microbiota-immune homeostasis and increased inflammatory bowel disease risk. J Clin Invest. 2021.
26. Wan WY, Hollins F, Haste L, Woodman L, Hirst RA, Bolton S, et al. NADPH Oxidase-4 Overexpression Is Associated With Epithelial Ciliary Dysfunction in Neutrophilic Asthma. Chest. 2016;149(6):1445–59.
27. Lakshmi VM, Nauseef WM, Zenser TV. Myeloperoxidase potentiates nitric oxide-mediated nitrosation. J Biol Chem. 2005;280(3):1746–53.
28. Nuszkowski A, Gräbner R, Marsche G, Unbehaun A, Malle E, Heller R. Hypochlorite-modified low density lipoprotein inhibits nitric oxide synthesis in endothelial cells via an intracellular dislocalization of endothelial nitric-oxide synthase. J Biol Chem. 2001;276(17):14212–21.
29. Mocatta TJ, Pilbrow AP, Cameron VA, Senthilmohan R, Frampton CM, Richards AM, et al. Plasma Concentrations of Myeloperoxidase Predict Mortality After Myocardial Infarction. Journal of the American College of Cardiology. 2007;49(20):1993–2000.
30. Wu CY, Bawa KK, Ouk M, Leung N, Yu D, Lanctôt KL, et al. Neutrophil activation in Alzheimer's disease and mild cognitive impairment: A systematic review and meta-analysis of protein markers in blood and cerebrospinal fluid. Ageing research reviews. 2020;62:101130.
31. Iana A, Sirbu E. Linking myeloperoxidase with subclinical atherosclerosis in adults with metabolic syndrome. Wiener klinische Wochenschrift. 2020;132(5–6):150–4.
32. Guo XX, Deng NS, Chen QH, Yu HY, Ding XH, Hu SP, et al. Application of Inflammatory Markers in Induced Sputum in Stable Chronic Obstructive Pulmonary Disease Patients with Positive Bronchodilation Tests. Current medical science. 2019;39(4):560–7.

33. Chami B, Martin NJJ, Dennis JM, Witting PK. Myeloperoxidase in the inflamed colon: A novel target for treating inflammatory bowel disease. Archives of Biochemistry and Biophysics. 2018;645:61–71.
34. Törnblom S, Nisula S, Vaara ST, Poukkanen M, Andersson S, Pettilä V, et al. Early prolonged neutrophil activation in critically ill patients with sepsis. Innate immunity. 2021;27(2):192–200.
35. Alisik T, Alisik M, Nacir B, Ayhan FF, Genc H, Erel O. Evaluation of dysfunctional high-density lipoprotein levels with myeloperoxidase/paraoxonase-1 ratio in rheumatoid arthritis. International journal of clinical practice. 2021:e14172.
36. Sun P, Qian J, Zhang ZB, Wan JX, Wu F, Jin XP, et al. Polymorphisms in phase I and phase II metabolism genes and risk of chronic benzene poisoning in a Chinese occupational population. Carcinogenesis. 2008;29(12):2325–9.
37. Salim T, Sershen CL, May EE. Investigating the Role of TNF-α and IFN-γ Activation on the Dynamics of iNOS Gene Expression in LPS Stimulated Macrophages. PloS one. 2016;11(6): e0153289-e.
38. Emami Z, Mesbah Namin A, Kojuri J, Mashayekhi Jalali F, Rasti M. Expression and Activity of Platelet Endothelial Nitric Oxide Synthase Are Decreased in Patients with Coronary Thrombosis and Stenosis. Avicenna J Med Biotechnol. 2019;11(1):88–93.
39. Perrotta C, Cervia D, Di Renzo I, Moscheni C, Bassi MT, Campana L, et al. Nitric Oxide Generated by Tumor-Associated Macrophages Is Responsible for Cancer Resistance to Cisplatin and Correlated With Syntaxin 4 and Acid Sphingomyelinase Inhibition. Front Immunol. 2018;9:1186.
40. Yu X, Ge L, Niu L, Lian X, Ma H, Pang L. The Dual Role of Inducible Nitric Oxide Synthase in Myocardial Ischemia/Reperfusion Injury: Friend or Foe? Oxidative medicine and cellular longevity. 2018;2018:8364848.
41. Kumar S, Gupta E, Gupta N, Kaushik S, Srivastava VK, Kumar S, et al. Functional role of iNOS-Rac2 interaction in neutrophil extracellular traps (NETs) induced cytotoxicity in sepsis. Clinica Chimica Acta. 2021;513:43–9.
42. Pieper GM, Roza AM. The complex role of iNOS in acutely rejecting cardiac transplants. Free radical biology & medicine. 2008;44(8):1536–52.
43. Baranipour S, Amini Kadijani A, Qujeq D, Shahrokh S, Haghazali M, Mirzaei A, et al. Inducible nitric oxide synthase as a potential blood-based biomarker in inflammatory bowel diseases. Gastroenterology and hepatology from bed to bench. 2018;11(Suppl 1):S124–s8.
44. Jiang WT, Liu XS, Xu YJ, Ni W, Chen SX. Expression of Nitric Oxide Synthase Isoenzyme in Lung Tissue of Smokers with and without Chronic Obstructive Pulmonary Disease. Chinese medical journal. 2015;128(12):1584–9.
45. Behairi N, Belkhelfa M, Mesbah-Amroun H, Rafa H, Belarbi S, Tazir M, et al. All-trans-retinoic acid modulates nitric oxide and interleukin-17A production by peripheral blood mononuclear cells from patients with Alzheimer's disease. Neuroimmunomodulation. 2015;22(6):385–93.
46. Wu D, Tseng IJ, Yuan RY, Hsieh CY, Hu CJ. Memory consolidation and inducible nitric oxide synthase expression during different sleep stages in Parkinson disease. Sleep medicine. 2014;15 (1):116–20.
47. Bachmann K. Chapter 8 – Drug Metabolism. In: Hacker M, Messer W, Bachmann K, editors. Pharmacology. San Diego: Academic Press; 2009. p. 131–73.
48. Poss WB, Huecksteadt TP, Panus PC, Freeman BA, Hoidal JR. Regulation of xanthine dehydrogenase and xanthine oxidase activity by hypoxia. The American journal of physiology. 1996;270(6 Pt 1):L941–6.
49. Bonini MG, Miyamoto S, Di Mascio P, Augusto O. Production of the carbonate radical anion during xanthine oxidase turnover in the presence of bicarbonate. J Biol Chem. 2004;279(50): 51836–43.
50. Harrison R. Physiological roles of xanthine oxidoreductase. Drug metabolism reviews. 2004;36 (2):363–75.

51. Yen FS, Hsu CC, Li HL, Wei JC, Hwu CM. Urate-lowering therapy may mitigate the risks of hospitalized stroke and mortality in patients with gout. PLoS One. 2020;15(6):e0234909.
52. Dawson J, Broomfield N, Dani K, Dickie DA, Doney A, Forbes K, et al. Xanthine oxidase inhibition for the improvement of long-term outcomes following ischaemic stroke and transient ischaemic attack (XILO-FIST) – Protocol for a randomised double blind placebo-controlled clinical trial. European stroke journal. 2018;3(3):281–90.
53. Meijer B, Seinen ML, Hosman T, Linskens RK, Kneppelhout JK, Peters GJ, et al. High inter-individual variability of serum xanthine oxidoreductase activity in IBD patients. Nucleosides, nucleotides & nucleic acids. 2018;37(6):317–23.
54. Ibrahim YF, Fadl RR, Ibrahim S, Gayyed MF, Bayoumi A, Refaie M. Protective effect of febuxostat in sepsis-induced liver and kidney injuries after cecal ligation and puncture with the impact of xanthine oxidase, interleukin 1β, and c-Jun N-terminal kinases. Human & experimental toxicology. 2020;39(7):906–19.
55. Ferrer MD, Busquets-Cortés C, Capó X, Tejada S, Tur JA, Pons A, et al. Cyclooxygenase-2 Inhibitors as a Therapeutic Target in Inflammatory Diseases. Curr Med Chem. 2019;26(18): 3225–41.
56. Cho KJ, Seo JM, Kim JH. Bioactive lipoxygenase metabolites stimulation of NADPH oxidases and reactive oxygen species. Molecules and cells. 2011;32(1):1–5.
57. Mukherjee A, Brinkley DW, Chang K-M, Roth JP. Molecular Oxygen Dependent Steps in Fatty Acid Oxidation by Cyclooxygenase-1. Biochemistry. 2007;46(13):3975–89.
58. Oliveira-Filho J, Ornellas AC, Zhang CR, Oliveira LM, Araújo-Santos T, Borges VM, et al. COX-2 rs20417 Polymorphism Is Associated with Stroke and White Matter Disease. Journal of stroke and cerebrovascular diseases: the official journal of National Stroke Association. 2015;24 (8):1817–22.
59. Arsene D, Vasilescu F, Toader C, Bălan A, Popa C, Ardeleanu C. Clinico-pathological correlations in fatal ischemic stroke. An immunohistochemical study of human brain penumbra. Romanian journal of morphology and embryology = Revue roumaine de morphologie et embryologie. 2011;52(1):29–38.
60. Chen GZ, Shan XY, Cheng GP, Tao HM. Cyclooxygenase-2 genetic polymorphism and stroke subtypes in Chinese. Journal of molecular neuroscience: MN. 2013;51(2):467–73.
61. Nebert DW, Wikvall K, Miller WL. Human cytochromes P450 in health and disease. Philos Trans R Soc Lond B Biol Sci. 2013;368(1612):20120431.
62. Zangar RC, Davydov DR, Verma S. Mechanisms that regulate production of reactive oxygen species by cytochrome P450. Toxicol Appl Pharmacol. 2004;199(3):316–31.
63. Veith A, Moorthy B. ROLE OF CYTOCHROME P450S IN THE GENERATION AND METABOLISM OF REACTIVE OXYGEN SPECIES. Curr Opin Toxicol. 2018;7:44–51.
64. Puntarulo S, Cederbaum AI. Production of reactive oxygen species by microsomes enriched in specific human cytochrome P450 enzymes. Free Radic Biol Med. 1998;24(7–8):1324–30.
65. Gross E, Sevier CS, Heldman N, Vitu E, Bentzur M, Kaiser CA, et al. Generating disulfides enzymatically: Reaction products and electron acceptors of the endoplasmic reticulum thiol oxidase Ero1p. Proc Natl Acad Sci U S A. 2006;103(2):299.
66. Krueger SK, Williams DE. Mammalian flavin-containing monooxygenases: structure/function, genetic polymorphisms and role in drug metabolism. Pharmacology & therapeutics. 2005;106 (3):357–87.
67. Siddens LK, Krueger SK, Henderson MC, Williams DE. Mammalian flavin-containing monooxygenase (FMO) as a source of hydrogen peroxide. Biochemical pharmacology. 2014;89(1):141–7.
68. Li H, Cui H, Kundu TK, Alzawahra W, Zweier JL. Nitric oxide production from nitrite occurs primarily in tissues not in the blood: critical role of xanthine oxidase and aldehyde oxidase. The Journal of biological chemistry. 2008;283(26):17855–63.
69. Li H, Kundu TK, Zweier JL. Characterization of the magnitude and mechanism of aldehyde oxidase-mediated nitric oxide production from nitrite. J Biol Chem. 2009;284(49):33850–8.

70. Kundu TK, Velayutham M, Zweier JL. Aldehyde oxidase functions as a superoxide generating NADH oxidase: an important redox regulated pathway of cellular oxygen radical formation. Biochemistry. 2012;51(13):2930–9.
71. Garattini E, Fratelli M, Terao M. The mammalian aldehyde oxidase gene family. Hum Genomics. 2009;4(2):119–30.
72. Wright RM, McManaman JL, Repine JE. Alcohol-induced breast cancer: a proposed mechanism. Free Radic Biol Med. 1999;26(3–4):348–54.
73. Idewaki Y, Iwase M, Fujii H, Ohkuma T, Ide H, Kaizu S, et al. Association of Genetically Determined Aldehyde Dehydrogenase 2 Activity with Diabetic Complications in Relation to Alcohol Consumption in Japanese Patients with Type 2 Diabetes Mellitus: The Fukuoka Diabetes Registry. PloS one. 2015;10(11):e0143288-e.
74. Liu Y, Hou Y, Wang G, Zheng X, Hao H. Gut Microbial Metabolites of Aromatic Amino Acids as Signals in Host–Microbe Interplay. Trends in Endocrinology & Metabolism. 2020;31(11): 818–34.
75. Vásquez-Vivar J, Kalyanaraman B, Martásek P, Hogg N, Masters BS, Karoui H, et al. Superoxide generation by endothelial nitric oxide synthase: the influence of cofactors. Proc Natl Acad Sci U S A. 1998;95(16):9220–5.
76. Rifkind JM, Mohanty JG, Nagababu E. The pathophysiology of extracellular hemoglobin associated with enhanced oxidative reactions. 2015;5(500).
77. Perna AF, Ingrosso D, De Santo NG. Homocysteine and oxidative stress. Amino Acids. 2003;25(3–4):409–17.
78. Loscalzo J. The oxidant stress of hyperhomocyst(e)inemia. J Clin Invest. 1996;98(1):5–7.
79. McCully KS. Chemical pathology of homocysteine. IV. Excitotoxicity, oxidative stress, endothelial dysfunction, and inflammation. Annals of clinical and laboratory science. 2009;39(3): 219–32.
80. Alvarez-Maqueda M, El Bekay R, Monteseirín J, Alba G, Chacón P, Vega A, et al. Homocysteine enhances superoxide anion release and NADPH oxidase assembly by human neutrophils. Effects on MAPK activation and neutrophil migration. Atherosclerosis. 2004;172(2):229–38.
81. McCully KS. Chemical pathology of homocysteine. II. Carcinogenesis and homocysteine thiolactone metabolism. Annals of clinical and laboratory science. 1994;24(1):27–59.
82. Sule AA, Chin TJ, Khien LH. Recurrent unprovoked venous thromboembolism in a young female patient with high levels of homocysteine. Int J Angiol. 2012;21(2):95–8.
83. Brattström L, Wilcken DE. Homocysteine and cardiovascular disease: cause or effect? The American journal of clinical nutrition. 2000;72(2):315–23.
84. Kim J-M, Park K-Y, Shin D-W, Park M-S, Kwon O-S. Relation of serum homocysteine levels to cerebral artery calcification and atherosclerosis. Atherosclerosis. 2016;254:200–4.
85. Chen C-S, Chou M-C, Yeh Y-C, Yang Y-H, Lai C-L, Yen C-F, et al. Plasma Homocysteine Levels and Major Depressive Disorders in Alzheimer Disease. The American Journal of Geriatric Psychiatry. 2010;18(11):1045–8.
86. Napolitano A, Manini P, d'Ischia M. Oxidation chemistry of catecholamines and neuronal degeneration: an update. Curr Med Chem. 2011;18(12):1832–45.

Chapter 5
Antioxidants

A. Khelfi

Abstract A new picture is emerging based on extensive research that cells harbor a highly complex network of redox-sensitive processes, which are constantly adjusting the metabolism and physiology of cells to the prevailing conditions of the environment. Antioxidants, whether synthesized endogenously or exogenously administered, are reducing agents that neutralize oxidants before they can cause damage to biomolecules. These substances act through powerful antioxidant defense systems within the cell to minimize the damaging effects caused by reactive species (RS). The first line of this defense system is the deployment of small antioxidant substances such as vitamin C (ascorbic acid), vitamin E (tocopherols), glutathione (GSH), coenzyme Q10 (CoQ10), and carotenoids. Thereafter, a battery of antioxidant enzymatic defenses (superoxide dismutase (SOD), glutathione peroxidase (GPx), glutathione reductase (GR), thioredoxin (Trx), and catalase (CAT)) prevents the free radical (FR) cascade and terminates the lipid peroxidation cycle. The enzymatic defenses are regarded as crucial in that only they can prevent FR toxicity when the antioxidant molecule supply is exhausted. Unfortunately, the imbalance between pro-oxidant and antioxidant systems in cells in favor of the former usually results in oxidative stress.

In this chapter, we review the antioxidant defenses of cells. Extensive coverage is given to enzymatic antioxidant that acts in a concerted and sequential manner to terminate lipid peroxidation chain reaction and to prevent oxidative damage to proteins. We provided the reaction schemes and kinetics of these enzymes, the preference and range of substrates catalyzed, the structure of the native protein and subunits, and the diversity of isoenzymes. The chapter also gives an excellent account of antioxidant compounds involved in the protection of the human body. It serves as an overview of antioxidant molecules that are components of natural products.

A. Khelfi (✉)
Faculty of Pharmacy, University of Algiers 1, Algiers, Algeria

National Centre of Toxicology, Algiers, Algeria

Department of Chemistry and Biomolecular Science, Clarkson University, Potsdam, NY, USA
e-mail: ab.khelfi@univ-alger.dz; akhelfi@clarkson.edu

S. Andreescu et al. (eds.), *Biomarkers of Oxidative Stress*,
https://doi.org/10.1007/978-3-031-60738-7_5

Keywords Superoxide dismutase (SOD) · Glutathione peroxidase (GPx) · Glutathione reductase (GR) · Thioredoxin (Trx) · Catalase (CAT) · Glutathione (GSH) · Coenzyme Q10 (CoQ10) · Vitamin E (tocopherol)

The human body is equipped with a variety of antioxidants serving to counterbalance the effect of oxidants. They either neutralize or scavenge reactive species (RS), and some of them are able to regenerate other antioxidants, thus establishing an antioxidant network.

For practical purposes, antioxidants can be divided into two categories: enzymatic and nonenzymatic (small molecules). In living cells, both enzymatic and nonenzymatic antioxidants have evolved and, together, provide sufficient protection against damage induced by free radicals (FRs) under normal circumstances.

An increasing body of evidence from experimental studies as well as clinical practice concerning their effect suggests that antioxidants need not always play a positive role. This fact must be kept in mind, especially during therapeutic administration of antioxidants as too high doses of antioxidants can push the bodily equilibrium into a state of so-called reductive stress, which has been reported as being as harmful as oxidative stress (OS) [1] and can even cause disease [2–5]. On the other hand, in disease, protective antioxidant mechanisms are reduced, leading to an imbalance between RS production and removal, resulting in accumulation of end products of FR action and slow progressive tissue damage [6].

5.1 Antioxidant Enzymes

The major enzymatic antioxidants are superoxide dismutase (SOD), catalase (CAT), glutathione (GSH) peroxidase (GPx), glutathione reductase (GR), and glutathione S-transferase (GST). In addition to these major enzymes, other antioxidants, including heme oxygenase-1 (HO-1), and redox proteins, such as Trx and peroxiredoxins, have also been found to play crucial roles in antioxidant defenses.

5.1.1 Superoxide Dismutase (SOD)

Superoxide dismutases (EC 1.15.1.1) are a family of metalloenzymes that catalyze the dismutation of two superoxide anion radicals to non-radical molecules: triplet oxygen and hydrogen peroxide (H_2O_2) (Eq. 5.1). The rate of this enzymatic dismutation is roughly 10,000 times greater than the spontaneous rate [7]. Since $O_2^{\bullet-}$ is the primary reactive oxygen species (ROS) produced from a variety of sources, its dismutation by SOD is of primary importance for each cell.

$$2\dot{O}_2^{\cdot} + 2H^+ \overset{SOD}{\rightarrow} H_2O_2 + O_2 \tag{5.1}$$

Three isoforms of SOD can be found in mammalian tissues. They mainly differ in terms of metal content: cytosolic Cu/Zn-SOD (SOD1), mitochondrial Mn-SOD (SOD2), and extracellular Cu/Zn-SOD (SOD3) [8]. All three are highly expressed in all tissues, mainly in the renal tubules of healthy kidneys [9]. The fact that these enzymes are present in high concentrations, together with their high catalytic efficiency, provides a high degree of cellular protection against $O_2^{\bullet-}$ under normal conditions.

Interestingly, SOD2 is an inducible enzyme that is probably critical for survival under stress conditions, including tumor necrosis factor stimulation, ultraviolet (UV) irradiation, and oxidative stress (OS) [10]. SOD3 is most active in the vessel wall. It enhances the availability of $NO^{\bullet}$ by scavenging $O_2^{\bullet-}$ [11]. It may play a crucial role in defense from $O_2^{\bullet-}$ generated from sources such as phagocytic cells.

5.1.2 Glutathione Peroxidase (GPx)

The enzyme GPx (EC 1.11.1.9) occurs in two forms: selenium-dependent and selenium-independent GPx, differing in terms of the number of subunits, the selenium bond at the active center, and the catalytic mechanism.

Selenium-independent GPx (glutathione S-transferase, GST) catalyzes detoxification of various xenobiotics. On the other hand, selenium-dependent GPx are a family of tetrameric enzymes that contain the unique amino acid selenocysteine within the active sites and use low-molecular-weight thiols, such as GSH. Four selenium-dependent GPx have been identified in humans, out of which two occur in the blood: the cytosolic form (GPx-1) in erythrocytes and the extracellular form (GPx-3) in plasma [12]. Of these, GPx-1 (cellular GPx; in the cytosol and mitochondria) is the ubiquitous and key antioxidant enzyme within most cells, including the endothelium, critical for preserving endothelial function in arteries exposed to OS [13]. GPx-1 reduces H_2O_2 and fatty acid hydroperoxides but not esterified peroxyl lipids. Esterified lipids are reduced by membrane-bound GPx-4 (phospholipid hydroperoxide GPx). This GPx is of particular biological relevance, as it specifically catalyzes the glutathione-mediated reduction of hydroperoxides in membrane phospholipids (phospholipid hydroperoxide GPx).

GPx-2 (gastrointestinal GPx) is localized in gastrointestinal epithelial cells where it serves to reduce dietary peroxides.

GPx-3 (extracellular GPx) is the only member of the GPx family that resides in the extracellular compartment and is believed to be one of the most important extracellular antioxidant enzymes in mammals. It is predominantly synthesized in the renal proximal tubules, and, therefore, its activity is significantly reduced in patients with chronic kidney disease [14].

Overall, GPx catalyzes both the reduction of H_2O_2 and ROOH to water and the corresponding alcohols, respectively, using a selenium atom at the active site and tripeptide glutathione (Eqs. 5.2 and 5.3). In tissues, this enzyme works in concert with SOD to detoxify $O_2^{\bullet -}$.

$$H_2O_2 + 2GSH \xrightarrow{\text{GPx}} 2H_2O + GSSG \tag{5.2}$$

$$ROOH + 2GSH \xrightarrow{\text{GPx}} ROH + GSSG + H_2O \tag{5.3}$$

Glutathione peroxidase effectively functions at low concentrations of H_2O_2 and reduces lipid hydroperoxides in biological membranes, thereby inhibiting the accumulation of a potential source of initiation of radical chain lipid peroxidation. Although acting on membrane-bound peroxides, the enzyme exists in cytosolic, mitochondrial, and membrane-bound forms (see above) [15]. High GPx activity was detected in the liver, medium activity in the heart, lungs, and brain, and low activity in muscles. Glutathione peroxidases also use GSH to reduce $ONOO^-$ to $NO^{\bullet}$, acting as a peroxynitrite reductase [16]. They also detoxify homocysteine-induced OS. According to Schnabel et al., the effect of homocysteine, a well-known risk factor for atherosclerosis, may be mediated through a reduction in GPx activity [17].

5.1.3 Glutathione Reductase (GR)

Under physiological conditions, glutathione in cells mostly exists in its reduced form (GSH). The ratio glutathione:glutathione disulfide (GSH:GSSG) is 100:1 [18]. This is due to the action of GR (EC 1.6.4.2) in cells, which catalyzes the nicotinamide adenine dinucleotide phosphate (NADPH)-driven reduction of GSSG, mainly produced by GPx (Eq. 5.4).

$$GSSG + 2NADPH^{+} \xrightarrow{\text{GR}} 2GSH + 2NADP^{+} \tag{5.4}$$

Therefore, GSH is continually regenerated by GR. The latter can reduce not only GSSG but also "mixed" disulfides (GSSR). The cofactor of GR is NADPH, which is mainly produced in the pentose cycle by glucose-6-phosphate dehydrogenase (G6PD).

5.1.4 Thioredoxin (Trx) and Trx Reductase (TrxR)

Besides the GPx system, disposal of H_2O_2 is closely associated with several thiol-containing enzymes, namely, thioredoxin (Trx-1 and Trx-2) (EC 1.8.4.10), thioredoxin reductases (TrxR) (EC 1.8.1.9), and peroxiredoxins (Prx, which are thioredoxin peroxidases).

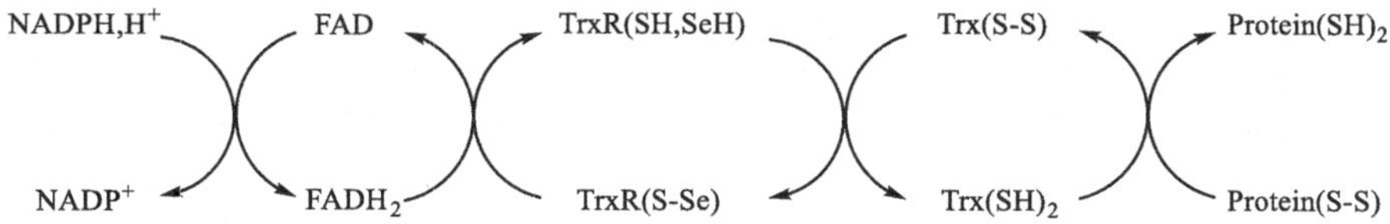

Fig. 5.1 Overview of the Trx/TrxR system

Thioredoxin is a small protein with a redox-active dithiol at its active site that can be oxidized to the corresponding disulfide. It is critical for the redox regulation of protein function. The main function of Trx is to keep protein thiols in the reduced state. For this reason, Trx is often regarded as an antioxidant enzyme (Eq. 5.5).

$$\mathrm{Trx(SH)_2 + H_2O_2 \rightarrow Trx(S\text{-}S) + H_2O} \tag{5.5}$$

Thioredoxin reductase is a selenium-dependent enzyme, endogenously found as a homodimer, which regenerates reduced Trx from the dithiol form with the assistance of flavin adenine dinucleotide (FADH2) and NADPH (Fig. 5.1).

Trx and TrxR have been characterized in human cells. They are either predominantly cytosolic (Trx-1 or TrxR-1) or mitochondrial (Trx-2 and TrxR-2) [19].

These Trx catalyze thiol–disulfide exchange reactions required for many functions, including H_2O_2 disposal and general regulation of protein function by thiol redox control. Trx-1 can act as an indirect antioxidant via peroxiredoxin-mediated reduction of H_2O_2. Trx-2 can regulate protein activity via thiol–disulfide exchange reactions with protein mixed disulfides (PSSX), which, coupled with reduction by TrxR, result in the net reduction of PSSX by NADPH.

Plasma/serum concentrations of Trx are elevated in OS-associated disorders, and, therefore, their levels can serve as markers of OS [20]. The Trx system can be induced by a variety of stress conditions, including viral infection, ischemic insult, exposure to UV light, and X-ray irradiation [21].

5.1.5 Peroxiredoxins (Prx)

Peroxiredoxins (EC 1.11.1.15) are antioxidant proteins containing essential catalytic cysteine residues that use Trx to scavenge lipid hydroperoxides and $ONOO^-$. The couple Prx/Trx is called protein thiol–disulfide oxidoreductases. Peroxiredoxins act as a direct antioxidant by reducing H_2O_2 (Eqs. 5.6 and 5.7).

$$\mathrm{Prx(SH)_2 + H_2O_2 \rightarrow Prx(S\text{-}S) + H_2O} \tag{5.6}$$

$$\mathrm{Prx(SH)_2 + ROOH \rightarrow Prx(S\text{-}S) + H_2O + ROH} \tag{5.7}$$

Six different Prx (Prx-1–6) have been found in human cells, differing in their ultrastructural compartmentalization. They are distributed at sites of ROS formation, including the cytosol, mitochondria, and peroxisomes [22]. Prx-1 is expressed in the cytosol of many types of cells and tissues and has been found to be induced in macrophages exposed to OS. Prx-5 has been shown to function as a peroxynitrite reductase [23].

5.1.6 *Catalase (CAT)*

Catalase (EC 1.11.1.6) is expressed in all important organs, especially in the liver and erythrocytes. In cells, CAT is a heme-containing enzyme located in the peroxisomes, whereas mitochondria and the endoplasmic reticulum (ER) only contain a negligible amount of CAT. This enzyme is composed of four subunits containing heme (tetrameric hemoprotein) with a coordinated Fe(III) atom at their active site.

Catalase is also considered a dismutase enzyme. It catalyzes the decomposition of H_2O_2 to water and oxygen and hence finishes the detoxification reaction started by SOD. It is most effective at degrading high concentrations of H_2O_2, such as might be found in peroxisomes [24]. It has one of the highest turnover numbers known for an enzyme. The mechanism of the catalytic reaction is expressed by Eqs. 5.8, 5.9 and 5.10.

$$2H_2O_2 \xrightarrow{\text{CAT}} O_2 + 2H_2O \tag{5.8}$$

$$\text{CAT-FeIII} + H_2O_2 \rightarrow \text{Compound I} \tag{5.9}$$

$$\text{Compound I} + H_2O_2 \rightarrow \text{CAT-FeIII} + H_2O + O_2 \tag{5.10}$$

The composition of the intermediate is not exactly known, but it is supposed to contain Fe(III) − HOOH or Fe(III) = O complex [25].

Decomposition of H_2O_2 is accomplished via the conversion between two conformations of the enzyme: catalase–ferricatalase (iron coordinated with water) and compound I (iron complexed with an oxygen atom). Catalase also uses NADPH as a reducing equivalent to prevent its oxidative inactivation (formation of compound II) by H_2O_2 as it is reduced to water [75].

5.1.7 *Glutathione S-Transferase (GST)*

Organic free radicals (FRs) and electrophilic metabolites of organic molecules (unsaturated aldehydes and epoxides) can directly react with GSH. Such reactions can be catalyzed by glutathione S-transferases (EC 2.5.1.18), a group of cytosolic enzymes with overlapping substrate specificity. These enzymes can also serve as selenium-independent GPx capable of reducing the ROOH already formed (Eq. 5.11).

$$ROOH + 2\ GSH \xrightarrow{GST} ROH + GSSG + H_2O \quad (5.11)$$

Three major families of GSTs have been described: cytosolic GST, mitochondrial GST, and membrane-associated microsomal GST that has a role in eicosanoid and GSH metabolisms [26, 27]. Seven classes of cytosolic GST are identified in mammals, designated Alpha, Mu, Pi, Sigma, Theta, Omega, and Zeta [28]. One interesting isoform is GSTP1, which physically interacts with Prx-6 and leads to the recovery of Prx enzyme activity via glutathionylation of the oxidized protein [29].

5.1.8 Heme Oxygenase-1 (HO-1)

Heme oxygenase (EC 1.14.15.20), a heat-shock protein, is the inducible isoform of the rate-limiting enzyme of heme degradation. Heme oxygenase converts heme into carbon monoxide, biliverdin, and iron. The protective effect of HO against OS is indirect since it is related to the fact that, once formed, biliverdin turns into bilirubin, which has powerful antioxidant activities. In addition, the iron produced by the activity of HO stimulates the synthesis of ferritin, which is also involved in the antioxidant response. However, the activity of HO can have detrimental effects in the short term since the same iron can act as a pro-oxidant through its catalytic action in the Fenton reaction.

The stress-responsive and antioxidant enzyme HO-1 is known to be induced by various stimuli, including heat shock, hyperoxia, and oxidative agents. It represents a powerful endogenous protective mechanism against FRs in a variety of pathological conditions. For example, HO-1 mRNA showed the most marked response to organic extracts of diesel exhaust particles that induce high levels of OS in rat heart microvascular endothelial cells (ECs) [30].

5.2 Nonenzymatic Antioxidants

Nonenzymatic antioxidants include low-molecular-weight agents that scavenge FRs. Several compounds, such as vitamins (vitamins A, C, and E), coenzyme Q_{10} (CoQ_{10}), and GSH, a tripeptide that comprises a thiol (sulfhydryl) group, have been reported. Some of these antioxidants are water-soluble and act in the cell's soluble fraction or in the plasma, whereas others, being lipid-soluble, act within membranes, while amphipathic molecules can act in both environments. These antioxidants are present in extra- and intracellular fluids, and some of them come from the diet (vitamins A, E, and C). A decrease in small-molecular-weight antioxidants is a common occurrence after oxidative insult, and their measurement is frequently recommended and used as a complimentary estimation of OS.

5.2.1 Glutathione (GSH)

Glutathione is an important water-soluble antioxidant, both as a cofactor and as a direct-acting scavenger. This thiol-containing tripeptide (γ-glutamyl-cysteinyl-glycine) is present in virtually all living cells in high concentrations (millimolar concentrations within cells). The intracellular concentration of GSH is 500 times higher than its extracellular concentration [25]. It is primarily localized in the cytosol, nuclei, and mitochondria. The most important source of GSH is the liver, where 90% of the circulating GSH is synthesized de novo [25]. Glutathione homeostasis is regulated by several enzymes such as GR, GPx, and GST.

Glutathione is the major soluble antioxidant and one of the most important redox buffers of several eukaryotic cell types [31]. It represents the major free thiol found in cells and participates in a number of cell functions, including amino acid transport, detoxification of xenobiotics, and protection against FRs and reactive species (RS). This is due to its reactive cysteine-thiol, providing an abundant target for radical attacks. The free thiol group in this molecule is nucleophilic and will rapidly react with many electrophiles. Glutathione S-transferases significantly accelerate these reactions in some cases, but they can also occur without enzymatic intervention.

The protective mechanism against radicals constantly results in an increased formation of intracellular glutathione disulfide (oxidized form; GSSG). GSSG is again reduced, and GSH is regenerated in a redox cycle involving GR using the electron donor NADPH, which is produced in the pentose cycle of glucose degradation. Besides being the oxidized form of GSH, GSSG reaction with thiol groups of proteins can form mixed disulfides, thus changing their physiological function. In this respect, GSSG appears to act as a nonspecific signaling molecule [32].

Glutathione shows its antioxidant effects in several ways. It is a coenzyme for several detoxifying enzymes such as glutathione peroxidases and transferases, which play an essential protective role against H_2O_2 and lipid peroxides' oxidative actions (see Sect. 5.1). Glutathione, besides scavenging RS enzymatically, can also directly scavenge radicals like $^1O_2$, H_2O_2, $OH^{\bullet}$, $RO^{\bullet}$, and $ROO^{\bullet}$. In such reactions, the thiol group easily loses a single electron and hydrogen atom, resulting in the generation of a thiyl radical (Eq. 5.12).

$$GSH + OH^{\cdot} \rightarrow GS^{\cdot} + H_2O \tag{5.12}$$

Glutathione donates its electron to H_2O_2 to reduce it into H_2O and O_2 and donates protons to membrane lipids and protects them from oxidant attacks (Eq. 5.13).

$$R^{\cdot} + GSH \rightarrow RH + GS^{\cdot} \tag{5.13}$$

Thiyl radicals can, in turn, combine to form oxidized glutathione disulfide (Eq. 5.14).

$$GS^{\cdot} + GS^{\cdot} \rightarrow GSSG \tag{5.14}$$

Considering its own antioxidant capacity, GSH is also involved in the recycling of other antioxidants. Indeed, GSH is able to regenerate vitamins C and E to their active forms and therefore participate in neutralization of hydrophobic radicals.

Overall, GSH plays a significant role in the protection of an organism against oxidative damage for the following reasons:

- GSH is a cofactor of some enzymes participating in the detoxification mechanisms of OS, e.g., GPx and GST. It detoxicates H_2O_2 and lipoperoxides during the catalytical action of GPx.
- GSH participates in amino acid transfer through the cell membrane.
- GSH is a direct trapper of $OH^{\bullet}$ and $^1O_2$.
- GSH can reduce the tocopheroxyl radical either directly or indirectly during the reduction of semidehydroascorbate to ascorbate via dehydroascorbate reductase, regenerating these important antioxidants back to their active form. GSH plays an important role as an antioxidant in the intracellular space and also indirectly affects the antioxidative status in the extracellular space and membranes, where, in cooperation with α-tocopherol, it can inhibit lipid peroxidation.
- GSH efficiently reduces oxidized proteins and toxic aldehydes (e.g., 4-hydroxy-2-nonenal (4-HNE)).

The correct ratio of GSH/GSSG is a suitable marker of OS and significantly contributes to the total redox state of the cell [18, 33] and the downregulation of the GPx/GR ratio with increased OS [34, 35]. Many pathologies, including Parkinson's disease (PD), acquired immunodeficiency syndrome (AIDS), and cerebral ischemia, are associated with a decrease in intracellular GSH levels [36–38].

5.2.2 *Coenzyme Q_{10} (CoQ_{10})*

Coenzyme Q_{10} is a redox-cycling and lipid-soluble molecule that is essential for the mitochondrial machinery as it works as an electron carrier in the respiratory chain. It is also present in serum lipoproteins, endomembranes, and the plasma membrane. Chemically, coenzyme Q_{10} is constituted by a benzoquinone ring and a lipid side chain with several isoprenoid units (in microorganisms 6, in rats 9, in humans 10) [39]. Its redox-cycling reactions can be driven either by the simultaneous transfer of two electrons in one stage or by two sequential steps of one-electron transfer through a partially reduced semiquinone intermediate with the potential danger of $O_2^{\bullet-}$ formation (Eqs. 5.15 and 5.16).

$$CoQH_2(\text{ubiquinol}) + O_2^{\cdot -} \rightarrow CoQ^{\cdot -}(\text{semiquinone}) + H_2O_2 \tag{5.15}$$

$$CoQ^{\cdot -}(\text{semiquinone}) + O_2 \rightarrow CoQ\ (\text{ubiquinone}) + O_2^{\cdot -} \tag{5.16}$$

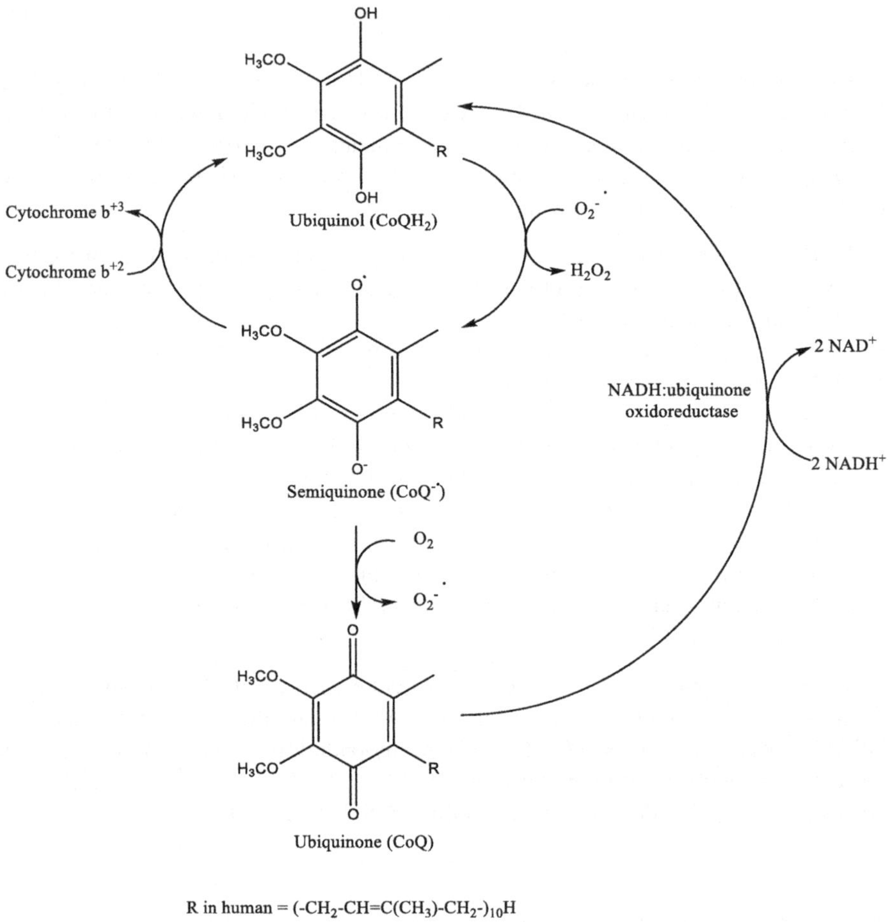

Fig. 5.2 Redox cycling of coenzyme Q_{10}

The formed semiquinone can be reduced back to ubiquinol ($CoQH_2$) in the respiratory chain in mitochondria (by cytochrome b). NADPH quinone oxidoreductase-1 can perform a two-electron reduction of quinones to the hydroquinone form (Fig. 5.2).

Fortunately, the regenerated ubiquinol can trap the $O_2^{\bullet-}$ radical (Eq. 5.17) and perferryl radicals (Fe^{3+}-$O_2^{\bullet-}$) as follows [76]:

$$CoQH_2 + Fe^{+3} - O_2^{\cdot-} \rightarrow CoQ^{\cdot} + Fe^{+3} + H_2O_2 \tag{5.17}$$

Hydrogen peroxide is subsequently depleted by CAT and GPx. Besides, ubiquinol is believed to be an important antioxidant for protecting biological membranes. Alongside vitamins C and E, ubiquinol is protective against radical damage toward mitochondrial and plasma membranes as well as plasma lipoproteins [40, 41]. It can prevent lipid peroxidation (membranous lipids and circulating low-density lipoproteins (LDLs)) by directly acting as a lipid radical scavenger or by regenerating vitamin E:

- Ubiquinol can directly react with carbon-centered lipid radicals, as well as lipid peroxyl radicals, generating its semiquinone form (Eqs. 5.18 and 5.19):

$$CoQH_2 + L^{\cdot} \rightarrow CoQ^{\cdot\cdot} + LH + H^+ \tag{5.18}$$

$$CoQH_2 + LOO^{\cdot} \rightarrow CoQ^{-\cdot} + LOOH + H^+ \tag{5.19}$$

- The ubiquinol molecule has been demonstrated to be an efficient regenerator of α-tocopherol from the α-tocopheroxyl radical (Eq. 5.20) [42]. This is due to the neighboring location of both compounds in the membrane lipid bilayer.

$$CoQH_2 + \alpha - TO^{\cdot} \rightarrow CoQ^{-\cdot} + \alpha - TOH + H^+ \tag{5.20}$$

Coenzyme Q_{10} is supposed to act as an antioxidant by other mechanisms. It was found that ubiquinol reduces the highly toxic ferryl myoglobin (Mb – Fe(IV) = O) to either the less toxic metmyoglobin (Mb – Fe(III)) or the physiological oxymyoglobin (MbO_2) [41]. It is assumed that by this mechanism, ubiquinol could protect cardiomyocytes against the harmful effect of oxidants.

Besides the mitochondrial respiratory chain, several enzymes catalyze coenzyme Q_{10} reduction to achieve its antioxidant activity in eukaryotic cells. NADH-cytochrome b5 reductase can reduce coenzyme Q_{10} through a one-electron reaction mechanism [43], and the soluble DT-diaphorase can reduce quinones by a two-electron reaction and maintain the reduced state of coenzyme Q_{10} [44]. A stipulated mechanism, on which the antioxidant activity of coenzyme Q_{10} relies, has been described in earlier investigations, reporting the interaction of SOD with hydroquinones. In conjunction with DT-diaphorase, the former inhibits the auto-oxidation of hydroquinones and possibly that of ubiquinone [45].

Like many antioxidants, the redox status of coenzyme Q_{10} is often used as a marker of OS. This compound has been shown to decrease with aging and in many pathologies, including human immunodeficiency virus (HIV) infection and myocardial disease [46, 47]. Besides, coenzyme Q_{10} has been found to be able to contain ROS generation and DNA damage promoted by UVA in human keratinocytes [48].

5.2.3 Vitamin A (Carotenoids)

Carotenoids are pigments of plant origin. At present, around 600 derivatives of carotenoids are known (e.g., lutein, zeaxanthin, canthaxanthin, β-cryptoxanthin, lycopene, α- and β-carotene) [49]. Only a small number of carotenoids are present in the plasma and tissue in sufficient amounts. β-Carotene, sometimes also called pro-vitamin A as it can be converted into vitamin A, is found in the membranes of certain tissues (e.g., the retina) in high concentrations [50].

β-Carotene is a highly effective $^1O_2$ quencher, with the latter being responsible for UV light damage of the skin, cataract, and macular degeneration. It can return the excited oxygen to its basic energetic state. The role of β-carotene in systemic photoprotection, originating from its antioxidant properties, is opposed by some reported pro-oxidant effects [51].

Of the biologically important natural carotenoids, lycopene, the open-chain analog of β-carotene, was shown to be the most efficient quencher of $^1O_2$. Indeed, it has been found that the antioxidant activity of lycopene is the best of all 600 natural carotenoids [52, 53]. Besides, carotenoids can directly react with FRs (e.g., with $ROO^•$, $OH^•$, and $O_2^{•-}$) following three possible mechanisms: electron transfer, hydrogen abstraction, and radical addition. In this manner, carotenoids are capable of interrupting peroxidative chain reactions like vitamin E.

It has been reported that lycopene possesses better reactivity than does β-carotene against $ROO^•$ [54]. The scavenging potential of lycopene toward $NO_2^•$, as well as thiyl and sulfonyl radicals, has been confirmed as well [55]. Moreover, lycopene has been confirmed to scavenge $ONOO^-$, which is known for its damaging oxidizing potential in vitro and in vivo [56]. Furthermore, it was postulated that lycopene might improve the antioxidant protection of cells by regenerating both vitamins E and C from their correspondent radical forms, on the basis of its capacity to reduce tocopheroxyl radicals [57].

β-Carotene deficiency is found in some groups of people, including smokers or drinkers, and is also associated with some conditions and pathologies, including malabsorption, long-term enteral nutrition, and liver diseases. A number of studies have also focused on the involvement of lycopene in several diseases, including cancer, as it was found that its plasma and serum concentration is inversely associated with cancer risk [58]. Comparable correlations have been established for markers of cardiovascular disease (CVD), osteoporosis, cognitive function, and body weight [59–63].

5.2.4 Vitamin C (Ascorbic Acid (AH_2))

Ascorbic acid is a water-soluble vitamin of high biological relevance. It is present in different forms and carries out important redox functions in cells (Fig. 5.3).

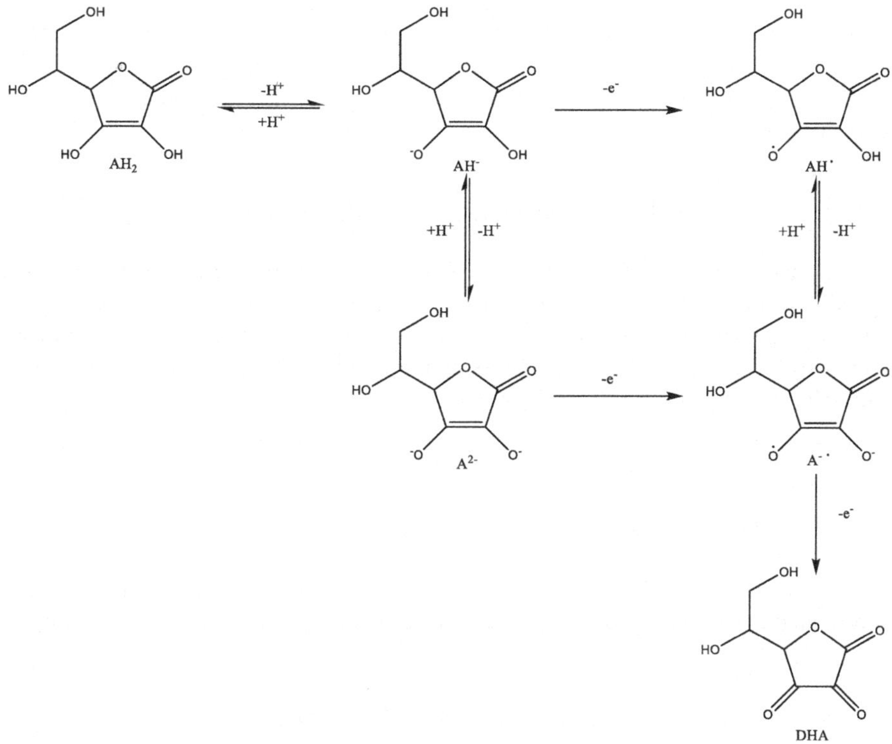

Fig. 5.3 Different forms of vitamin C in cells

Vitamin C provides an aqueous-phase antioxidant capability to complement the lipophilic antioxidant capacities of α-tocopherol (see below). Chemically, this vitamin efficiently scavenges RS such as $O_2^{\bullet -}$, $HO_2^{\bullet}$, $OH^{\bullet}$, H_2O_2, $^1O_2$, HOCl, and $GS^{\bullet}$. It is also reactive with nitrogenated species. It prevents lipid peroxidation by scavenging peroxyl and alkoxyl radicals in the aqueous phase.

When ascorbic acid (AH_2) encounters FR, the result is the formation of a stable semidehydroascorbate radical ($AH^{\bullet}$) (Eqs. 5.21 and 5.22).

$$AH_2 + OH^{\cdot} \rightarrow AH^{\cdot} + H_2O \tag{5.21}$$

$$AH_2 + O_2^{\cdot -} + H^+ \rightarrow AH^{\cdot} + H_2O_2 \tag{5.22}$$

The semidehydroascorbate radical possibly disproportionates, thus generating ascorbate (AH_2) and dehydroascorbate (DHA) (Eq. 5.23).

$$2AH^{\cdot} \rightarrow DHA + AH_2 \tag{5.23}$$

In the presence of AH_2, ascorbate peroxidase (APX) catalyzes the reduction of H_2O_2 to H_2O (Eq. 5.24):

$$AH_2 + H_2O_2 \xrightarrow{APX} DHA + 2H_2O \tag{5.24}$$

Dehydroascorbate can be enzymatically recycled to the reduced state by dehydroascorbate reductase in the presence of GSH (Eq. 5.25).

$$DHA + 2GSH \xrightarrow{DHAR} AH_2 + GSSG \tag{5.25}$$

Although ascorbic acid is not a strong scavenger of lipophilic radicals, it has a synergistic effect with lipid-soluble tocopherol, in $ROO^{\bullet}$ removal. At the lipid–aqueous interphase, ascorbic acid reduces the membrane-bound oxidized tocopheroxyl radical to form active tocopherol, able to accomplish its antioxidant roles (Eq. 5.26) [64]:

$$AH^{-} + TO^{\cdot} \rightarrow A^{\cdot -} + TOH \tag{5.26}$$

AH^{-} is the ascorbate mono-anion, and $A^{\bullet -}$ is a highly stable ascorbate anion radical. Ascorbic acid is also able to recycle other important antioxidants from their radical forms, including GSH. However, vitamin C can also function as a pro-oxidant, particularly in the presence of transition-metal ions (ion and copper) (Eq. 5.27) [65].

$$2Fe^{2+} + A^{2-} \rightarrow 2Fe^{3+} + DHA \tag{5.27}$$

Since ascorbate can reduce these ions, thus producing ionic transition-metal species, it is possibly capable of interacting with oxygen in a redox cycle, generating $O_2^{\bullet -}$ and H_2O_2 (Eqs. 5.28 and 5.29). The latter enters into a Fenton-type reaction to generate the highly oxidizing $OH^{\bullet}$ and OH^{-} (Eqs. 2.25, 2.26, 2.27 and 2.28).

$$Fe^{2+} + O_2 \rightarrow Fe^{3+} + O_2^{\cdot -} \tag{5.28}$$

$$2O_2^{\cdot -} + 2H^{+} \rightarrow O_2 + H_2O_2 \tag{5.29}$$

Although excessive doses of ascorbate may be pro-oxidant, physiological amounts have always been proven to be antioxidant even in the presence of metal ions.

The ascorbate/dehydroascorbate redox radio is commonly used as a marker of OS. Plasma levels of ascorbic acid have been shown to decrease with oxidative insults, such as cigarette smoke, and in pathological conditions, such as cystic fibrosis (CF) [66, 67].

5.2.5 Vitamin E (Tocopherol)

"Vitamin E" is a term used for a group of compounds composed of a mixture of eight different derivatives (α-, β-, γ-, δ-tocopherols and α-, β-, γ-, δ-tocotrienols) [68]. Their main structural characteristics are the 6-chromanol ring and the lipophilic isoprene side chains. α-Tocopherol is the most active form of vitamin E, with α- and γ-tocopherols being the most abundant forms found in mammalian cells [69, 70].

Vitamin E is the most important lipid-soluble antioxidant in humans and the major membrane-bound antioxidant in cells. This vitamin resides almost exclusively on or at the surface of cell membranes and plays a vital role in protecting them from oxidative damage. In addition, it is one of the major antioxidants involved in protecting lipoproteins against lipid peroxidation. It mainly prevents lipid peroxidation of polyunsaturated fatty acids (PUFAs) and acts by breaking the propagation of radical chain reactions [71].

Its primary activity is to scavenge lipid radicals (mainly $ROO^{\bullet}$ and $RO^{\bullet}$) produced during lipid peroxidation in cellular membranes (Eq. 5.30). It is also able to directly trap oxygen radicals (e.g., $OH^{\bullet}$).

$$TOH + ROO^{\cdot} \rightarrow TO^{\cdot} + ROOH \tag{5.30}$$

During these reactions, the chromanol group of vitamin E, which is necessary for the antioxidant activity of the molecule, donates an electron and a proton, terminating the propagation of peroxidation while converting tocopherol to its radical form (the tocopheroxyl radical $TO^{\bullet}$). The tocopheroxyl radical can be reduced back to active vitamin E by endogenous compounds such as ascorbate (Eq. 5.26), ubiquinol (Eq. 5.20), and GSH (Eq. 5.31).

$$TO^{\cdot} + GSH \rightarrow TOH + GS^{\cdot} \tag{5.31}$$

Because vitamin E is readily recycled by other antioxidants, it may be one of the last antioxidants to be depleted during times of OS.

The tocopheroxyl radical can be further oxidized to other products by the transfer of a second electron to lipid peroxyl radicals, generating the corresponding 8-α-(alkyldioxy)tocopherones [72]. In addition, the tocopheroxyl radical can be eliminated in a true termination reaction by radical coupling with peroxyl radicals.

Vitamin E has been shown to be depleted after exposure to many different types of OS, including exercise and UV light [73]. The contribution of vitamin E to stabilizing atherosclerotic plaque is a well-known function of this compound [74].

5.3 Other Antioxidants

The abovementioned antioxidants constitute an important barrier against oxidative effects and represent reliable biomarkers of OS (Fig. 5.4). Several other elements and molecules express antioxidant properties, but they are not considered as biomarkers

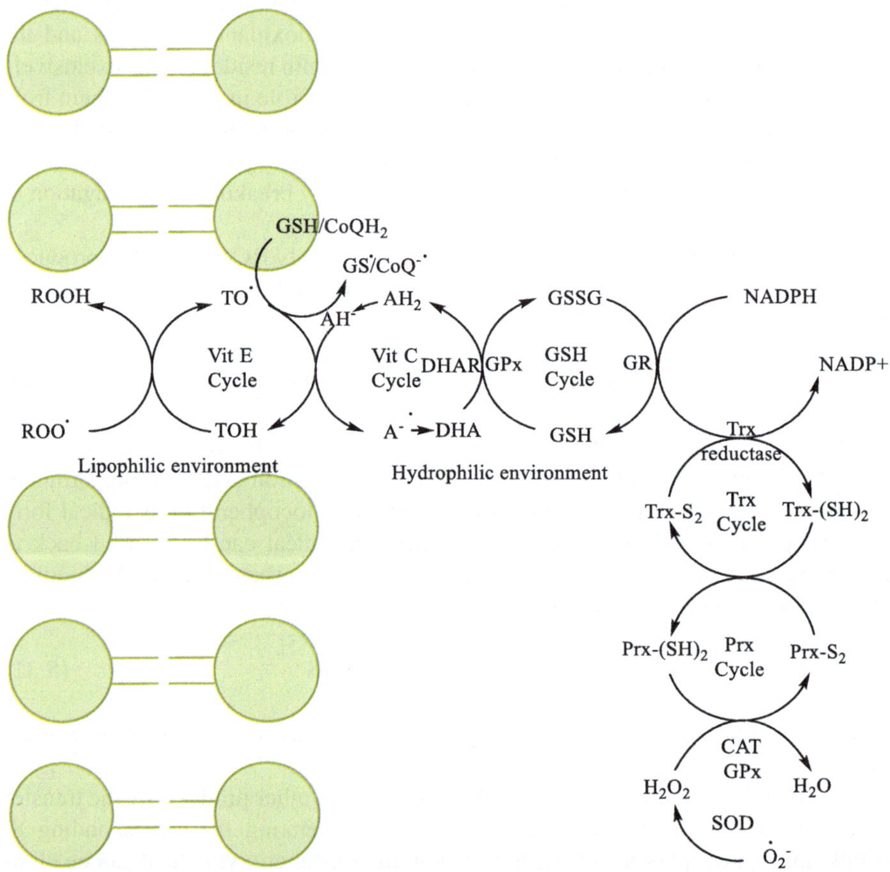

Fig. 5.4 Tightly connected ROS scavenging network (antioxidant system)
Lipid peroxyl radicals ($ROO^{\bullet}$) are scavenged by α-tocopherol to form the vitamin E radical $TO^{\bullet}$. The latter can be regenerated back to its reduced state by vitamin C (AH_2), leaving a vitamin C radical ($A^{\bullet-}$). $A^{\bullet-}$ is transformed to a more stable dehydroascorbate (DHA). DHA is reduced back to AH_2 by the action of dehydroascorbate reductase (DHAR) using reduced glutathione as a substrate, simultaneously forming oxidized glutathione. The enzyme GR reduces GSSG to GSH using NADPH. Trx–Prx requires NADPH as a source of reducing equivalents to reduce H_2O_2 to water. Prx are the key players in H_2O_2 removal and redox regulation. Redox-regulated proteins can function as additional redox relay bases between Prx and Trx.
Catalase, when acting in the catalytic mode, requires no reducing equivalents to remove H_2O_2. The physiological concentration of H_2O_2 is maintained by CAT.
The superoxide radical can be transformed to H_2O_2 by SOD.
Different thiol compounds can be present under different redox states. The significance of a specific antioxidant biomarker will depend on its intracellular localization, tissue expression, and redox potential

of OS as they ensure other activities. The following are examples of such antioxidants:

- Proteins and enzymes: Paraoxonase, arylesterase, glutaredoxin, carrier proteins (metallothionein, albumin, ferritin, transferrin, lactoferrin, ceruloplasmin, haptoglobin, and hemopexine).
- Low-weight molecules: Vitamin P, bilirubin, uric acid, lipoic acid, melatonin, folate, pyruvate, taurine, hypotaurine, and lactate.
- Metals and oligo-elements: Selenium (cofactor of GPx, Trx, etc.), copper (cofactor of SOD1 and SOD3), zinc (cofactor of SOD1 and SOD3, protector of thiol groups, and inducer of metallothionein expression), and manganese (cofactor of SOD2).

References

1. Castagné V, Lefèvre K, Natero R, Becker DA, Clarke PGH. An optimal redox status for the survival of axotomized ganglion cells in the developing retina. Neuroscience. 1999;93(1): 313–20.
2. Lloret A, Fuchsberger T, Giraldo E, Vina J. Reductive Stress: A New Concept in Alzheimer's Disease. Curr Alzheimer Res. 2016;13(2):206–11.
3. Brewer AC, Mustafi SB, Murray TV, Rajasekaran NS, Benjamin IJ. Reductive stress linked to small HSPs, G6PD, and Nrf2 pathways in heart disease. Antioxid Redox Signal. 2013;18(9): 1114–27.
4. Dimitriadis F, Tsounapi P, Zachariou A, Kaltsas A, Sokolakis I, Hatzichristodoulou G, et al. Therapeutic Effects of Micronutrient Supplements on Sperm Parameters: Fact or Fiction? Current pharmaceutical design. 2021;27(24):2757–69.
5. Ma WX, Li CY, Tao R, Wang XP, Yan LJ. Reductive Stress-Induced Mitochondrial Dysfunction and Cardiomyopathy. Oxid Med Cell Longev. 2020;2020:5136957.
6. Bisht S, Dada R. Oxidative stress: Major executioner in disease pathology, role in sperm DNA damage and preventive strategies. Front Biosci (Schol Ed). 2017;9(3):420–47.
7. Afzal F, Khurshid R, Ashraf M, Gul Kazi A. Chapter 13—Reactive Oxygen Species and Antioxidants in Response to Pathogens and Wounding. In: Ahmad P, editor. Oxidative Damage to Plants. San Diego: Academic Press; 2014. p. 397–424.
8. Tamayo D, Muñoz JF, Lopez Á, Urán M, Herrera J, Borges CL, et al. Identification and Analysis of the Role of Superoxide Dismutases Isoforms in the Pathogenesis of Paracoccidioides spp. PLOS Neglected Tropical Diseases. 2016;10(3):e0004481.
9. Fujita H, Fujishima H, Chida S, Takahashi K, Qi Z, Kanetsuna Y, et al. Reduction of renal superoxide dismutase in progressive diabetic nephropathy. J Am Soc Nephrol. 2009;20(6): 1303–13.
10. Pudlarz AM, Czechowska E, M SK, Ranoszek-Soliwoda K, Tomaszewska E, Celichowski G, et al. The effect of immobilized antioxidant enzymes on the oxidative stress in UV-irradiated rat skin. Nanomedicine (London, England). 2020;15(1):23–39.
11. Yan Z, Spaulding HR. Extracellular superoxide dismutase, a molecular transducer of health benefits of exercise. Redox Biol. 2020;32:101508.
12. Burk RF, Hill KE. 4.13—Glutathione Peroxidases. In: McQueen CA, editor. Comprehensive Toxicology (Second Edition). Oxford: Elsevier; 2010. p. 229–42.
13. Sultan CS, Saackel A, Stank A, Fleming T, Fedorova M, Hoffmann R, et al. Impact of carbonylation on glutathione peroxidase-1 activity in human hyperglycemic endothelial cells. Redox Biol. 2018;16:113–22.

14. Pang P, Abbott M, Abdi M, Fucci Q-A, Chauhan N, Mistri M, et al. Pre-clinical model of severe glutathione peroxidase-3 deficiency and chronic kidney disease results in coronary artery thrombosis and depressed left ventricular function. Nephrol Dial Transplant. 2018;33(6): 923–34.
15. Conrad M, Schneider M, Seiler A, Bornkamm G. Physiological role of phospholipid hydroperoxide glutathione peroxidase in mammals. Biological chemistry. 2007;388:1019–25.
16. Selles B, Hugo M, Trujillo M, Srivastava V, Wingsle G, Jacquot JP, et al. Hydroperoxide and peroxynitrite reductase activity of poplar thioredoxin-dependent glutathione peroxidase 5: kinetics, catalytic mechanism and oxidative inactivation. The Biochemical journal. 2012;442 (2):369–80.
17. Schnabel R, Lackner KJ, Rupprecht HJ, Espinola-Klein C, Torzewski M, Lubos E, et al. Glutathione Peroxidase-1 and Homocysteine for Cardiovascular Risk Prediction: Results from the AtheroGeneStudy. Journal of the American College of Cardiology. 2005;45(10): 1631–7.
18. Zitka O, Skalickova S, Gumulec J, Masarik M, Adam V, Hubalek J, et al. Redox status expressed as GSH:GSSG ratio as a marker for oxidative stress in paediatric tumour patients. Oncol Lett. 2012;4(6):1247–53.
19. Zhang H, Du Y, Zhang X, Lu J, Holmgren A. Glutaredoxin 2 reduces both thioredoxin 2 and thioredoxin 1 and protects cells from apoptosis induced by auranofin and 4-hydroxynonenal. Antioxid Redox Signal. 2014;21(5):669–81.
20. Soyama T, Masutani H, Hirata C, Iwai-Kanai E, Inamoto T. Thioredoxin as a novel sensitive marker of biological stress response in smoking. Journal of clinical biochemistry and nutrition. 2020.
21. Okuyama H, Shimahara Y, Nakamura H, Araya S, Kawada N, Yamaoka Y, et al. Thioredoxin prevents thioacetamide-induced acute hepatitis. Comparative hepatology. 2004;3 Suppl 1:S6.
22. Pirson M, Clippe A, Knoops B. The curious case of peroxiredoxin-5: what its absence in aves can tell us and how it can be used. BMC Evolutionary Biology. 2018;18(1):18.
23. Dubuisson M, Stricht D, Clippe A, Etienne F, Nauser T, Kissner R, et al. Human peroxiredoxin 5 is a peroxynitrite reductase. FEBS letters. 2004;571:161–5.
24. Heck DE, Shakarjian M, Kim HD, Laskin JD, Vetrano AM. Mechanisms of oxidant generation by catalase. Annals of the New York Academy of Sciences. 2010;1203:120–5.
25. Durackova Z, Gvozdjakova A. Oxidants, Antioxidants and Oxidative Stress. 2008. p. 19–54.
26. Raza H. Dual localization of glutathione S-transferase in the cytosol and mitochondria: implications in oxidative stress, toxicity and disease. FEBS J. 2011;278(22):4243–51.
27. Allocati N, Masulli M, Di Ilio C, Federici L. Glutathione transferases: substrates, inihibitors and pro-drugs in cancer and neurodegenerative diseases. Oncogenesis. 2018;7(1):8.
28. da Fonseca RR, Johnson WE, O'Brien SJ, Vasconcelos V, Antunes A. Molecular evolution and the role of oxidative stress in the expansion and functional diversification of cytosolic glutathione transferases. BMC evolutionary biology. 2010;10:281.
29. Zhang J, Ye Z-W, Singh S, Townsend DM, Tew KD. An evolving understanding of the S-glutathionylation cycle in pathways of redox regulation. Free Radic Biol Med. 2018;120: 204–16.
30. Hirano S, Furuyama A, Koike E, Kobayashi T. Oxidative-stress potency of organic extracts of diesel exhaust and urban fine particles in rat heart microvessel endothelial cells. Toxicology. 2003;187(2):161–70.
31. Aoyama K, Nakaki T. Glutathione in Cellular Redox Homeostasis: Association with the Excitatory Amino Acid Carrier 1 (EAAC1). Molecules. 2015;20(5):8742–58.
32. Ghanta S, Chattopadhyay S. Glutathione as a signaling molecule: another challenge to pathogens. Plant Signal Behav. 2011;6(6):783–8.
33. Rose S, Melnyk S, Pavliv O, Bai S, Nick TG, Frye RE, et al. Evidence of oxidative damage and inflammation associated with low glutathione redox status in the autism brain. Translational Psychiatry. 2012;2(7):e134.

34. Xin L, Mekle R, Fournier M, Baumann PS, Ferrari C, Alameda L, et al. Genetic Polymorphism Associated Prefrontal Glutathione and Its Coupling With Brain Glutamate and Peripheral Redox Status in Early Psychosis. Schizophr Bull. 2016;42(5):1185–96.
35. Torres L, Quaglio N, Tavares de Souza G, Garcia R, Dati L, Moreira W, et al. Peripheral Oxidative Stress Biomarkers in Mild Cognitive Impairment and Alzheimer's Disease. Journal of Alzheimer's disease: JAD. 2011;26:59–68.
36. Morris D, Ly J, Chi P-T, Daliva J, Nguyen T, Soofer C, et al. Glutathione synthesis is compromised in erythrocytes from individuals with HIV. Front Pharmacol. 2014;5:73.
37. Smeyne M, Smeyne RJ. Glutathione metabolism and Parkinson's disease. Free radical biology & medicine. 2013;62:13–25.
38. Yabuki Y, Fukunaga K. Oral administration of glutathione improves memory deficits following transient brain ischemia by reducing brain oxidative stress. Neuroscience. 2013;250:394–407.
39. Kawamukai M. Biosynthesis and bioproduction of coenzyme Q(10) by yeasts and other organisms. Biotechnology and applied biochemistry. 2009;53:217–26.
40. Takahashi T, Mine Y, Okamoto T. Extracellular coenzyme Q(10) (CoQ(10)) is reduced to ubiquinol-10 by intact Hep G2 cells independent of intracellular CoQ(10) reduction. Arch Biochem Biophys. 2019;672:108067.
41. Littarru GP, Tiano L. Bioenergetic and Antioxidant Properties of Coenzyme Q10: Recent Developments. Molecular biotechnology. 2007;37:31–7.
42. Freisleben HJ, Packer L. Free-radical scavenging activities, interactions and recycling of antioxidants. Biochem Soc Trans. 1993;21(2):325–30.
43. Hyun D-H, Lee G-H. Cytochrome b5 reductase, a plasma membrane redox enzyme, protects neuronal cells against metabolic and oxidative stress through maintaining redox state and bioenergetics. Age (Dordr). 2015;37(6):122.
44. Ross D, Siegel D. Functions of NQO1 in Cellular Protection and CoQ(10) Metabolism and its Potential Role as a Redox Sensitive Molecular Switch. Front Physiol. 2017;8:595.
45. Pisoschi AM, Pop A. The role of antioxidants in the chemistry of oxidative stress: A review. Eur J Med Chem. 2015;97:55–74.
46. Marcheggiani F, Cirilli I, Orlando P, Silvestri S, Vogelsang A, Knott A, et al. Modulation of Coenzyme Q10 content and oxidative status in human dermal fibroblasts using HMG-CoA reductase inhibitor over a broad range of concentrations. From mitohormesis to mitochondrial dysfunction and accelerated aging. Aging (Albany NY). 2019;11(9):2565–82.
47. Zozina VI, Covantev S, Goroshko OA, Krasnykh LM, Kukes VG. Coenzyme Q10 in Cardiovascular and Metabolic Diseases: Current State of the Problem. Curr Cardiol Rev. 2018;14(3): 164–74.
48. Inui M, Ooe M, Fujii K, Matsunaka H, Yoshida M, Ichihashi M. Mechanisms of inhibitory effects of CoQ10 on UVB-induced wrinkle formation in vitro and in vivo. BioFactors (Oxford, England). 2008;32:237–43.
49. Tapiero H, Townsend DM, Tew KD. The role of carotenoids in the prevention of human pathologies. Biomed Pharmacother. 2004;58(2):100–10.
50. Grune T, Lietz G, Palou A, Ross AC, Stahl W, Tang G, et al. Beta-carotene is an important vitamin A source for humans. J Nutr. 2010;140(12):2268S–85S.
51. Ha D-O, Park CU, Kim M-J, Lee J. Antioxidant and prooxidant activities of β-carotene in accelerated autoxidation and photosensitized model systems. Food Science and Biotechnology. 2012;21(2):607–11.
52. Zhang Y, Liu Y, Lv Q. DFT study on the quenching mechanism of singlet oxygen by lycopene. RSC Advances. 2016;6(100):98498–505.
53. Amarowicz R. Lycopene as a natural antioxidant. 2011;113(6):675–7.
54. Galano A, Francisco-Marquez M. Reactions of OOH Radical with β-Carotene, Lycopene, and Torulene: Hydrogen Atom Transfer and Adduct Formation Mechanisms. The Journal of Physical Chemistry B. 2009;113(32):11338–45.

55. Mortensen A, Skibsted LH, Sampson J, Rice-Evans C, Everett SA. Comparative mechanisms and rates of free radical scavenging by carotenoid antioxidants. FEBS Letters. 1997;418(1): 91–7.
56. Muzandu K, Ishizuka M, Sakamoto K, Ibrahim Z, Bohi K, Kazusaka A, et al. Effect of lycopene and β-carotene on peroxynitrite-mediated cellular modifications. Toxicology and applied pharmacology. 2006;215:330–40.
57. Upaganlawar A, Gandhi H, Balaraman R. Effect of vitamin E alone and in combination with lycopene on biochemical and histopathological alterations in isoproterenol-induced myocardial infarction in rats. J Pharmacol Pharmacother. 2010;1(1):24–31.
58. Chen P, Zhang W, Wang X, Zhao K, Negi DS, Zhuo L, et al. Lycopene and Risk of Prostate Cancer: A Systematic Review and Meta-Analysis. Medicine (Baltimore). 2015;94(33):e1260-e.
59. Foy C, Passmore AP, Vahidassr M, Young I, Lawson J. Plasma chain-breaking antioxidants in Alzheimer's disease, vascular dementia and Parkinson's disease. QJM: monthly journal of the Association of Physicians. 1999;92:39–45.
60. Mecocci P, Polidori MC, Cherubini A, Ingegni T, Mattioli P, Catani M, et al. Lymphocyte oxidative DNA damage and plasma antioxidants in Alzheimer disease. Archives of neurology. 2002;59(5):794–8.
61. Polidori MC, Mattioli P, Aldred S, Cecchetti R, Stahl W, Griffiths H, et al. Plasma antioxidant status, immunoglobulin g oxidation and lipid peroxidation in demented patients: relevance to Alzheimer disease and vascular dementia. Dementia and geriatric cognitive disorders. 2004;18 (3-4):265–70.
62. Maggio D, Polidori MC, Barabani M, Tufi A, Ruggiero C, Cecchetti R, et al. Low levels of carotenoids and retinol in involutional osteoporosis. Bone. 2006;38(2):244–8.
63. Andersen LF, Jacobs DR, Jr., Gross MD, Schreiner PJ, Dale Williams O, Lee DH. Longitudinal associations between body mass index and serum carotenoids: the CARDIA study. The British journal of nutrition. 2006;95(2):358–65.
64. Bursać-Mitrović M, Milovanović DR, Mitić R, Jovanović D, Sovrlić M, Vasiljević P, et al. EFFECTS OF L-ASCORBIC ACID AND ALPHA-TOCOPHEROL ON BIOCHEMICAL PARAMETERS OF SWIMMING-INDUCED OXIDATIVE STRESS IN SERUM OF Guinea PIGS. Afr J Tradit Complement Altern Med. 2016;13(4):29–33.
65. Poljšak B, Ionescu J. Pro-oxidant vs. Antioxidant effects of Vitamin C. 2009. p. 153–83.
66. Prasad S, Kaisar MA, Cucullo L. Unhealthy smokers: scopes for prophylactic intervention and clinical treatment. BMC Neuroscience. 2017;18(1):70.
67. Brown R, Wyatt H, Price J, Kelly FJERJ. Pulmonary dysfunction in cystic fibrosis is associated with oxidative stress. 1996;9(2):334–9.
68. Pein H, Ville A, Pace S, Temml V, Garscha U, Raasch M, et al. Endogenous metabolites of vitamin E limit inflammation by targeting 5-lipoxygenase. Nature Communications. 2018;9(1): 3834.
69. Atkinson J, Manor D, Parker R. Vitamin E. In: Lennarz WJ, Lane MD, editors. Encyclopedia of Biological Chemistry (Second Edition). Waltham: Academic Press; 2013. p. 545–50.
70. Bruno RS, Mah E. Vitamin E. Reference Module in Biomedical Sciences: Elsevier; 2014.
71. Catalgol B, Ozer NK. Protective effects of vitamin E against hypercholesterolemia-induced age-related diseases. Genes & Nutrition. 2012;7(1):91–8.
72. Kumar A, Prasad A, Pospíšil P. Formation of α-tocopherol hydroperoxide and α-tocopheroxyl radical: relevance for photooxidative stress in Arabidopsis. Scientific Reports. 2020;10(1): 19646.
73. Han D, Loukianoff S, McLaughlin L. Part VII • Chapter 17—Oxidative stress indices: analytical aspects and significance. In: Sen CK, Packer L, Hänninen OOP, editors. Handbook of Oxidants and Antioxidants in Exercise. Amsterdam: Elsevier Science B.V.; 2000. p. 433–83.

74. Sozen E, Demirel T, Ozer NK. Vitamin E: Regulatory role in the cardiovascular system. 2019;71(4):507–15.
75. Birben E, Sahiner UM, Sackesen C, Erzurum S, Kalayci O. Oxidative stress and antioxidant defense. World Allergy Organ J. 2012;5(1):9–19.
76. Bentinger M, Brismar K, Dallner G. The antioxidant role of coenzyme Q. Mitochondrion. 2007;7:S41–S50.

Chapter 6
Other Biomarkers

A. Khelfi

Abstract Oxidative stress is a state where a biological system has a diminished capacity to counteract an overproduction or invasion of reactive species (RS) and free radicals (FR). This process can deeply affect cell physiology and lead to multiple dysfunctions and serious injuries to tissues. Oxidative events are usually characterized using conventional biomarkers, which have been proven to have great benefit in multiple clinical settings. Meanwhile, several new biomarkers of oxidative stress have been identified. These include auto-antibodies, transcription factors, Nrf-2, NF-κB, repair system. Several reports emphasize the involvement of transcription factors in pathological processes mediated by oxidative stress which include the leakage of cell membranes, dysfunction of mitochondria, depletion of glutathione (GSH), disturbed redox states of cells, and the depletion of ATP. RS are now known to induce the production of auto-antibodies, which can be determined in blood samples. In response to these deleterious effects, repair systems are triggered and can also serve as remarkable indicators of oxidative deployment in cells.

This chapter provides a good account of the new identified biomarkers of oxidative stress and the situations when these markers can have great value in clinical settings.

Keywords Auto-antibodies · Transcription factors · Nrf-2 · NF-κB · Repair system

A. Khelfi (✉)
Faculty of Pharmacy, University of Algiers 1, Algiers, Algeria

National Centre of Toxicology, Algiers, Algeria

Department of Chemistry and Biomolecular Science, Clarkson University, Potsdam, NY, USA
e-mail: ab.khelfi@univ-alger.dz; akhelfi@clarkson.edu

S. Andreescu et al. (eds.), *Biomarkers of Oxidative Stress*,
https://doi.org/10.1007/978-3-031-60738-7_6

6.1 Autoantibodies

During OS, antibodies can be generated against RS-oxidized molecules and therefore initiate and amplify an inflammatory reaction. From an analytical point of view, these antibodies can be used as highly specific biomarkers of OS. This is the case in atherosclerotic and DNA oxidation processes.

Autoantibodies are involved at different stages of atherosclerosis development. One major antigen for these antibodies is oxLDL, which is taken up by macrophages in situ, leading to the formation of foam cells. It is noteworthy that antibodies against oxLDL (aoxLDL) may be found in healthy individuals and a growing body of evidence indicates that aoxLDL could function as a protective mediator against atherosclerosis development, especially at early stages of disease [1]. In late-stage diseases, such antibodies could be pathogenic. Antibodies against phosphatidylcholine (aPC), especially IgM, appear to protect against the formation of atherosclerotic plaque, possibly by preventing the uptake of oxLDL in macrophages. Antiphospholipid (aPL) antibodies also play a complex role in CVD. Their role in atherosclerosis is not clear, but they possibly promote atherothrombosis and thus CVD [2]. Taken together, aoxLDL, aPL, and aPC are examples of aAb that could play a critical role in atherogenesis and could be used as high indicators of OS.

Few studies have reported the existence of aAb against oxidized bases. They have been used as indicators of oxidative damage in certain conditions. This is the case of plasma anti-5-hydroxymethyl-2′-deoxyuridine (HMdU) aAb which have been found at different levels in post-menopausal women [3], women who had a family history of breast cancer [4], workers exposed to heavy metals [5], and cigarette smokers [6].

Finally, the presence of anti-3-nitrotyrosine antibodies has been reported by Khan and Siddiqui in the joint synovial fluid of patients with RA and osteoarthritis [7]. Moreover, the authors found higher levels of anti-3-nitrotyrosine antibodies in the synovial fluid in the later stages of systemic lupus erythematosus when compared to the early stages.

6.2 Transcription Factors

6.2.1 Nrf-2

Nuclear factor erythroid 2-related factor-2 (Nrf-2) is a conserved transcription factor and a master regulator of the antioxidant system controlling the expression of more than 250 genes [8]. Some of these genes encode antioxidant enzymes including those involved in the synthesis and use of GSH and Trx, or participate in metabolic pathways that generate cell reductive power in the form of NADPH (i.e., SOD, CAT, GPx, GST, glutathione synthetase, HO-1, and NOX). Thus, Nrf-2 is related to the cellular defense against RS.

6.2.2 NF-κB

Nuclear factor-κB (NF-κB) is a transcriptional factor implicated in inflammatory reactions and immune activation. Diverse stimuli, including cytokines, oxidants, microbial infections, and mitogens, lead to the activation of IκB causing the nuclear translocation of NF-κB, where it stimulates the expression of target genes including genes coding for ROS-generating enzymes (COX-2 and iNOS) and proinflammatory cytokines. This signaling pathway can occur in different types of cells, like macrophages, which can lead to OS [9].

6.3 Biomarkers of Repair System

The 8-oxoguanine DNA glycosylase 1 (OGG-1) is a base-excision DNA-repair enzyme. It is the primary enzyme responsible for the recognition and excision of 8-oxo-G/8-oxo-dG, mutagenic base by-products that occur as a result of DNA exposure to RS. Measure of OGG-1 would be a good indicator of oxidative damage to DNA.

OGG-1 localizes to both the nucleus and mitochondria and has been investigated for its role in many diseases, including various cancers, and neurodegenerative diseases [10]. Indeed, polymorphism of OGG-1 plays a critical role in several cancers including breast [11, 12], lung [13, 14], cervical [15], bladder [16], gastric [17], colorectal [18], and thyroid [19] cancers. Besides, OGG-1 seems to be involved in divers pathological situations such as Alzheimer's disease [20, 21], Parkinson's disease [22, 23], Huntington's disease [24, 25], RA [26], asbestos exposure [27], and cataract [28].

Another important actor in cellular repair system is heat shock protein (HSP), which act in the following way:

- Direct action: HSP prevents tissue aggregation of proteins.
- Indirect action: HSP27 promotes the activity of GPx, GR, and GST, thus increasing the availability of GSH.

References

1. Fiskesund R, Su J, Bulatovic I, Vikström M, de Faire U, Frostegård J. IgM phosphorylcholine antibodies inhibit cell death and constitute a strong protection marker for atherosclerosis development, particularly in combination with other auto-antibodies against modified LDL. Results in Immunology. 2012;2:13–8.
2. Sherer Y, Shoenfeld Y. Antiphospholipid antibodies: are they pro-atherogenic or an epiphenomenon of atherosclerosis? Immunobiology. 2003;207(1):13–6.

3. Cooney RV, Maskarinec G, Franke AA, Okinaka L, Karkoszka J, Wilkens LR, et al. Association of tocopherols with circulating autoantibody levels against an oxidized DNA nucleoside in humans. Free Radical Biology and Medicine. 2001;31(4):460–8.
4. Frenkel K, Karkoszka J, Glassman T, Dubin N, Toniolo P, Taioli E, et al. Serum autoantibodies recognizing 5-hydroxymethyl-2'-deoxyuridine, an oxidized DNA base, as biomarkers of cancer risk in women. Cancer epidemiology, biomarkers & prevention: a publication of the American Association for Cancer Research, cosponsored by the American Society of Preventive Oncology. 1998;7(1):49–57.
5. Frenkel K, Karkoszka J, Cohen B, Barański B, Jakubowski M, Cosma G, et al. Occupational exposures to Cd, Ni, and Cr modulate titers of antioxidized DNA base autoantibodies. Environ Health Perspect. 1994;102 Suppl 3(Suppl 3):221–5.
6. Mooney LA, Perera FP, Van Bennekum AM, Blaner WS, Karkoszka J, Covey L, et al. Gender differences in autoantibodies to oxidative DNA base damage in cigarette smokers. Cancer epidemiology, biomarkers & prevention: a publication of the American Association for Cancer Research, cosponsored by the American Society of Preventive Oncology. 2001;10(6):641–8.
7. Khan F, Siddiqui AA. Prevalence of anti-3-nitrotyrosine antibodies in the joint synovial fluid of patients with rheumatoid arthritis, osteoarthritis and systemic lupus erythematosus. Clin Chim Acta. 2006;370(1-2):100–7.
8. Cuadrado A, Manda G, Hassan A, Alcaraz MJ, Barbas C, Daiber A, et al. Transcription Factor NRF2 as a Therapeutic Target for Chronic Diseases: A Systems Medicine Approach. 2018;70 (2):348–83.
9. Kwon DJ, Ju SM, Youn GS, Choi SY, Park J. Suppression of iNOS and COX-2 expression by flavokawain A via blockade of NF-κB and AP-1 activation in RAW 264.7 macrophages. Food and chemical toxicology: an international journal published for the British Industrial Biological Research Association. 2013;58:479–86.
10. Komakula SSB, Tumova J, Kumaraswamy D, Burchat N, Vartanian V, Ye H, et al. The DNA Repair Protein OGG1 Protects Against Obesity by Altering Mitochondrial Energetics in White Adipose Tissue. Scientific Reports. 2018;8(1):14886.
11. Sterpone S, Cornetta T, Padua L, Mastellone V, Giammarino D, Testa A, et al. DNA repair capacity and acute radiotherapy adverse effects in Italian breast cancer patients. Mutation Research/Fundamental and Molecular Mechanisms of Mutagenesis. 2010;684(1):43–8.
12. Kim K-Y, Han W, Noh D-Y, Kang D, Kwack K. Impact of genetic polymorphisms in base excision repair genes on the risk of breast cancer in a Korean population. Gene. 2013;532(2): 192–6.
13. De Ruyck K, Szaumkessel M, De Rudder I, Dehoorne A, Vral A, Claes K, et al. Polymorphisms in base-excision repair and nucleotide-excision repair genes in relation to lung cancer risk. Mutation Research/Genetic Toxicology and Environmental Mutagenesis. 2007;631(2):101–10.
14. Li Z, Guan W, Li M-x, Zhong Z-y, Qian C-y, Yang X-q, et al. Genetic Polymorphism of DNA Base-excision Repair Genes (APE1, OGG1 and XRCC1) and Their Correlation with Risk of Lung Cancer in a Chinese Population. Archives of Medical Research. 2011;42(3):226–34.
15. Niwa Y, Matsuo K, Ito H, Hirose K, Tajima K, Nakanishi T, et al. Association of XRCC1 Arg399Gln and OGG1 Ser326Cys polymorphisms with the risk of cervical cancer in Japanese subjects. Gynecologic Oncology. 2005;99(1):43–9.
16. Gangwar R, Ahirwar D, Mandhani A, Mittal RD. Do DNA repair genes OGG1, XRCC3 and XRCC7 have an impact on susceptibility to bladder cancer in the North Indian population? Mutation Research/Genetic Toxicology and Environmental Mutagenesis. 2009;680(1):56–63.
17. Yousaf S, Khan AU, Akram Z, Kayani MA, Nadeem I, Begum B, et al. Expression deregulation of DNA repair pathway genes in gastric cancer. Cancer Genetics. 2019;237:39–50.
18. Przybylowska K, Kabzinski J, Sygut A, Dziki L, Dziki A, Majsterek I. An association selected polymorphisms of XRCC1, OGG1 and MUTYH gene and the level of efficiency oxidative DNA damage repair with a risk of colorectal cancer. Mutation Research/Fundamental and Molecular Mechanisms of Mutagenesis. 2013;745–746:6–15.

19. García-Quispes WA, Pastor S, Galofré P, Biarnés F, Castell J, Velázquez A, et al. Influence of DNA-repair gene variants on the micronucleus frequency in thyroid cancer patients. Mutation Research/Genetic Toxicology and Environmental Mutagenesis. 2013;750(1):34–9.
20. Dinçer Y, Akkaya Ç, Mutlu T, Yavuzer S, Erkol G, Bozluolcay M, et al. DNA repair gene OGG1 polymorphism and its relation with oxidative DNA damage in patients with Alzheimer's disease. Neuroscience Letters. 2019;709:134362.
21. Sliwinska A, Kwiatkowski D, Czarny P, Toma M, Wigner P, Drzewoski J, et al. The levels of 7,8-dihydrodeoxyguanosine (8-oxoG) and 8-oxoguanine DNA glycosylase 1 (OGG1)—A potential diagnostic biomarkers of Alzheimer's disease. Journal of the Neurological Sciences. 2016;368:155-9.
22. Aslan NS, Orhan G, Karahalil B. The impacts of prominent gene polymorphisms in DNA repair enzymes on Parkinson's disease. Neuroscience Letters. 2020;735:135203.
23. Coppedè F, Ceravolo R, Migheli F, Fanucchi F, Frosini D, Siciliano G, et al. The hOGG1 Ser326Cys polymorphism is not associated with sporadic Parkinson's disease. Neuroscience Letters. 2010;473(3):248–51.
24. Berger F, Vaslin L, Belin L, Asselain B, Forlani S, Humbert S, et al. The impact of single-nucleotide polymorphisms (SNPs) in OGG1 and XPC on the age at onset of Huntington disease. Mutation Research/Genetic Toxicology and Environmental Mutagenesis. 2013;755(2):115–9.
25. Coppedè F, Migheli F, Ceravolo R, Bregant E, Rocchi A, Petrozzi L, et al. The hOGG1 Ser326Cys polymorphism and Huntington's disease. Toxicology. 2010;278(2):199–203.
26. Mohamed RH, El-Shal AS, El-Shahawy EE, Abdel Galil SM. Association of XRCC1 and OGG1 DNA repair gene polymorphisms with rheumatoid arthritis in Egyptian patients. Gene. 2016;578(1):112–6.
27. Dianzani I, Gibello L, Biava A, Giordano M, Bertolotti M, Betti M, et al. Polymorphisms in DNA repair genes as risk factors for asbestos-related malignant mesothelioma in a general population study. Mutation Research/Fundamental and Molecular Mechanisms of Mutagenesis. 2006;599(1):124–34.
28. Zhang Y, Zhang L, Song Z, Sun DL, Liu HR, Fu SB, et al. Genetic Polymorphisms in DNA Repair Genes OGG1, APE1, XRCC1, and XPD and the Risk of Age-Related Cataract. Ophthalmology. 2012;119(5):900–6.

Part II
Conventional and Advanced Measurement Techniques of Oxidative Stress

Chapter 7
Oxidative Stress Assessment

A. Khelfi, M. K. Panner Selvam, and R. Henkel

Abstract It is without question that pro-oxidant agents mediate oxidative stress (OS) and are at the epicenter of biological functions and pathophysiological events. Because of the importance of oxidative events to normal and abnormal cell functions, it has become imperative to develop appropriate analytical techniques to facilitate the estimation and quantitation of the redox state in biological systems (e.g., cells, tissues, the whole organism). Thus, reactive species (RS), free radicals (FR), and breakdown end-products should be measured precisely by analytical methods that offer greater sensitivities and suitability for each and every biomarker. The biomedical application of these methods is extensive, and future influence will be matchless. However, the determination and assessment of OS biomarkers is usually complex and cumbersome. An interdisciplinary approach is therefore needed and can allow for the monitoring of OS in various models for real clinical and biomedical applications.

Monitoring OS can be done by assaying the end-product of oxidative damage in clinical samples or by investigating the antioxidant potential of an organism, tissue, or body fluids.

This chapter covers a wide range of techniques ranging from portable hand-held detectors to sophisticated technology such as fluorescent imaging, proteomics, and

A. Khelfi (✉)
Faculty of Pharmacy, University of Algiers, Algiers, Algeria

National Centre of Toxicology, Algiers, Algeria

Department of Chemistry and Biomolecular Science, Clarkson University, Potsdam, NY, USA
e-mail: ab.khelfi@univ-alger.dz; akhelfi@clarkson.edu

M. K. Panner Selvam
Department of Urology, Tulane University School of Medicine, New Orleans, LA, USA
e-mail: mpannerselvam@tulane.edu

R. Henkel
LogixX Pharma Ltd., Berkshire, UK

Department of Medical Bioscience, University of the Western Cape, Bellville, South Africa

Department of Metabolism, Digestion & Reproduction, Imperial College, London, UK
e-mail: Ralf.henkel@logixxpharma.com; rhenkel@uwc.ac.za; r.henkel@imperial.ac.uk

S. Andreescu et al. (eds.), *Biomarkers of Oxidative Stress*,
https://doi.org/10.1007/978-3-031-60738-7_7

chromatographic fingerprints in order to evaluate perturbations in cell function resulting from increased OS. This chapter also discusses the advantages and limitations of the state-of-the-art methodologies that are currently available in redox assessment.

Keywords Oxidative stress (OS) · Techniques · Assessment · Methods · Biomarkers · Fluorescent imaging · Proteomics · Chromatographic fingerprints

The presence of OS in biological systems can be determined by markers, both enzymatic and nonenzymatic, in different samples (blood, urine, EBC, semen, and tissues). It is possible to measure/assess all OS components (RS, pro-oxidants, antioxidants, markers of oxidative damage, aAb, transcription factors, and repair enzymes), but not equally simple.

In practice, the analytical measurement of OS markers is difficult due to the short half-life (usually in seconds) of some compounds. Because of their instability, reagents are usually required for their detection. Reactive species and secondary products (e.g., protein, lipid, carbohydrates, or nucleotide radicals) in in vitro and in vivo systems can be directly and indirectly detected. A global marker of oxidative damage is not yet available, and the assessment of oxidative status may require a combination of several methods to measure several OS biomarkers.

Many approaches allow the evaluation and demonstration of the participation of OS in biochemical events with distinct advantages and disadvantages (Table 7.1). The battery of methods available to assess any of the components of OS is enormous, and a variety of instruments have been employed for their detection. It is important to know exactly what each method is measuring, as the most appropriate method will depend on the question being investigated, the time of investigation, and the type of sample available.

In this section, we do not intend to outline in detail the analytical methods but describe their classification and general utility in the evaluation of OS. We will present different approaches of investigation, which are important to scientists in toxicological, chemical, biochemical, and pharmacological fields. The methods currently employed can be broadly divided into 13 chapters.

In another angle of view, Table 7.2 illustrates a variety of compounds that are important in OS assessment (e.g., RS, markers of oxidative damage, pro-oxidants, antioxidants, transcription factors, markers of the repair system) with different methods of detection in most relevant biological systems.

Overall, evaluation of the OS holds great significance and has been a major challenge due to lack of sensitive and robust methods to accurately measure the levels of RS and their targets. The transient and reactive nature of RS makes it even harder to monitor them in biological matrices. The main emphasis of this section is to develop efficient and sensitive methods to analyze the OS and also to make the measurement more reliable, stable, and convenient. These methods not only can delineate mechanisms by which OS works but also help draw the correlations between the impact of OS and its respective clinical outcomes.

For FC, see Table 21.3.

Table 7.1 Advantages and disadvantages of common techniques in oxidative stress assessment

Techniques	Analysis spectrum	Advantages	Disadvantages
Spectrophotometry	LMA, lipid hydroperoxides, conjugated dienes, protein carbonyls, GSH/GSSG, protein thiols, NO_2^-, NO_3^-, S-nitrosothiols, $O_2^{\bullet-}$, HOCl, H_2O_2, AOPP, vitamin C, and carotenoid	Simple, fast, easy to perform, and inexpensive	Low specificity and sensitivity Limited to compounds that fall under UV–visible absorption
Chemiluminescence	$O_2^{\bullet-}$, $NO^{\bullet}$, $ONOO^-$, HOCl, H_2O_2, lipid peroxides, and $^1O_2$	Simple, fast, easy to perform High sensitivity Real time monitoring and imaging	Low specificity Test system can be expensive Quantifying the results can be challenging The stability of the substrates affects assay performance Time-consuming
Fluorimetry and fluorescent imaging	H_2O_2, $NO^{\bullet}$, $O_2^{\bullet-}$, S-nitrosothiols, Protein carbonyls, MDA, AGEs, and GSH	Simple, fast, easy to perform High sensitivity Non-invasive Real time monitoring and imaging	Low specificity Limited spatial resolution High background interference
Electron paramagnetic resonance/electron spin resonance	ROS and RNS	Simple, fast, easy to perform High sensitivity Real time monitoring and imaging	Low specificity Since the specificity depends on unpaired electrons, it may not apply to all systems
Electrochemical sensing	ROS and RNS	Fast analysis High sensitivity Covers oxidative and reductive stress Low cost Ease of manufacture Small size Ability to detect multiple elements simultaneously	Temperature and pH-dependent technique

(continued)

Table 7.1 (continued)

Techniques	Analysis spectrum	Advantages	Disadvantages
Chromatography	Theorically all low weight compounds	High specificity High sensitivity	High cost Sample derivatization often required Frequent artificial oxidation during sample pretreatment Time-consuming procedures Qualified manipulators required
Immunoassays (ELISA)	Protein, carbohydrate, lipid, and DNA end products Pro-oxidant and antioxidant enzymes	Simple and fast High sensitivity Fast and high-throughput analysis Cover a wide variety of compounds Economical assays Qualified manipulators not required	Low specificity due to cross-reactivity of antibodies Possible heterogeneity of detected compounds and specificity of antibodies
Western blot	Protein, lipid, and DNA end products Pro-oxidant and antioxidant enzymes Transcription factors	High sensitivity Determination of OS mechanisms	Low specificity due to cross-reactivity of antibodies Possible heterogeneity of detected compounds and specificity of antibodies Time-consuming procedures
Immunohistochemistry	Protein, carbohydrate, lipid, and DNA end products Pro-oxidant enzymes Transcription factors and repair system	High sensitivity High stability of biomarkers in sections Determination of OS mechanisms	Only qualitative/semi-quantitative analysis Time-consuming procedures
Enzyme kinetics	Pro-oxidant and antioxidant enzymes	Fast analysis High sensitivity	Interferences with RS
Enzyme-based assays for RS determination	H_2O_2, $O_2^{\bullet -}$, and ROOH Ascorbic acid	Fast analysis High sensitivity	Interferences Limited application
Proteomics	Enzymes and proteins	High specificity High sensitivity	High cost Qualified manipulators required

MiOXSYS	Measurement of electron transfer from antioxidants to oxidants Measures seminal OS	Quick snapshot of the redox status of a liquid Only a small (30 μL) sample volume is necessary Fast analysis High sensitivity Covers oxidative and reductive stress	Temperature and pH dependent Oxidation–reduction potential values can be affected by viscosity of the semen sample
TUNEL assays	DNA damage	Sensitive, reliable with minimal inter-observer variability Low volume requirement Effective even in low concentration samples Reference sample not required. Easy to perform and inexpensive using optical microscopy or fluorescent microscopy	Need for standardization between laboratories Subjected to inter-observer, intra-observer, inter-assay, and intra-assay variability Different protocols available which makes the TUNEL assay less comparable
SCSA	DNA damage	Reliable estimate of the percentage of DNA-damaged Statistically robust Highly standardized methodology Robust and sensitive method	Expensive instrumentation Highly skilled technicians required
COMET assay	Oxidized bases and repair system	High specificity High sensitivity	Specialized instrumentation Limited application Time-consuming procedures Qualified manipulators required
PCR	Pro-oxidant and antioxidant enzymes Transcription factors	High specificity High sensitivity	Specialized instrumentation Limited application Time-consuming procedures High cost Qualified manipulators required

(continued)

Table 7.1 (continued)

Techniques	Analysis spectrum	Advantages	Disadvantages
Capillary electrophoresis	Theorically all charged low weight compounds	High specificity High sensitivity	Specialized instrumentation Limited application Qualified manipulators required
Flow cytometry	RS and thiol groups	High sensitivity Real time assessment	Limited application Interferences
NMR/Metabolomics	Theorically all low weight compounds	High specificity Simultaneous analysis of a wide range of metabolic end products Identification of novel biomarkers	Low sensitivity
Magnetic resonance imaging (MRI)	ROS	Imaging of OS	Low specificity
Raman spectroscopy	ROS	High sensitivity	Low specificity
Phosphorescence	$^1O_2$	High sensitivity Real time monitoring	Low specificity
Photoluminescence	NO and H_2O_2	High sensitivity	Low specificity
Ultrasound contrast agents	ROS	High sensitivity Imaging of OS	Low specificity
Radio assays	Protein thiols and S-thiolation	High sensitivity	Low specificity

Table 7.2 Major oxidative stress biomarkers and methods of detection

Type of biomarker		Technique	Most appropriate samples for analysis	Stability	Limit of quantification	Values in healthy individuals	References
RS	Hydrogen peroxide	Colorimetric assay	–	–	–	–	–
		Fluorimetric assay	–	–	–	–	
		Enzymatic method-CAT	–	–	–	–	–
		Enzymatic method-HRP	–	–	–	–	–
		Enzymatic method-GPx/GR	–	–	–	–	–
		Enzymatic method-Cytochrome c peroxidase	–	–	–	–	–
		Electrochemical assay	–	–	–	–	–
		Enzymatic method-HRP	Exhaled breath condensate	4 weeks (–80 °C)	130 nM	510 ± 260	[1]
	Superoxide anion	Luminol-chemiluminescence	Blood	–	–	50 kilo-counts/min	[2]
		Enzymatic method-Aconitase	–	–	–	–	–
		Electrochemical assay	–	–	–	–	–
	Hydroxyl radical	HPLC-UV- Salicylic acid	Plasma	–	19.45 μmol/L	1114.9 ± 1114.9 μmol/L	[3]
		HPLC-EC-Salicylic acid	Plasma	–	–	2,3-DHBA 2 h: 13.6–113 nmol/L	[4]
		GC-MS-Salicylic acid	Plasma	–	–	–	[5]
		HPLC-FL-Phenylalanine	Serum	–	–	18.42 ± 1.89 mM	[6]

(continued)

Table 7.2 (continued)

Type of biomarker		Technique	Most appropriate samples for analysis	Stability	Limit of quantification	Values in healthy individuals	References
	Nitric oxide	Electrochemical assay	–	–	–	–	–
	Nitrite/Nitrate	Colorimetric (Griess)	Plasma	–	–	2.0 μmol/L	[7]
		Colorimetric (Griess)	Plasma	–	–	378 nmol/L	[8]
		Colorimetric (Griess)	Erythrocytes	–	–	65.0 nmol/10^9 cells	[7]
		Fluorimetric assay	–	–	–	–	–
		Capillary electrophoresis	Plasma	–	Nitrite 1.4 μmol/L Nitrate 1.9 μmol/L	Nitrite 6.1 ± 2.3 μmol/L Nitrate 40.2 ± 10.3 μmol/L	[9]
		Capillary electrophoresis	Urine	–	–	Nitrate 611 ± 389 μmol/g creatinine	[10]
		HPLC-UV	Plasma	–	NO_X 1 μM	NO_X 10.0–80.6 μM	[11]
		HPLC-FL	Serum	–	Nitrite 0.98 pg/mL Nitrate 0.99 pg/mL	Nitrite 11–223 ng/mL Nitrate 96–1084 ng/mL	[12]
		GC-MS	Plasma	–	–	NO_X 19.9–85.0 μM	[11]
	Other markers	ROOH (e.g., 7-Hydroperoxylcholestrol): Colorimetric assay/iodometric assay, Fluorimetric assay, Electrochemical assay, Chemiluminescence assay, Enzymatic assay (COX and GPx), HPLC-Chemiluminescence, HPLC-UV, HPLC-FL, HPLC-EC, LC-MS, and GC-MS					

Products of DNA oxidation	8-OHdG	Electrochemistry	–	–	–	–	–
		HPLC-FL	Urine	–	–	9.8 ± 5.8 ng/mL	[13]
		HPLC-EC	Serum	–	4.3 pg/mL	25.5 ± 13.8 pg/mL	[14]
		LC-MS/MS	Plasma	–	0.02 ng/mL	14.3 ± 5.9 pg/mL	[15]
		GC-MS	Urine	–	0.072 mg/L	1.6–4.5 µmol/mmol creatinine	[16]
		LC-MS	Urine	–	0.113 ng/mL	2.9 ng/mL	[17]
		LC-MS/MS	Urine	–	1.0 ng/mL	1.9–28.6 ng/mg creatinine	[18]
		UPLC-MS/MS	Urine	–	1.0 nM	11.17 ± 0.45 ng/mg creatinine	[19]
		UPLC-MS/MS	Leucocyte DNA extract	–	1.0 nM	10.50 ± 5.87 per 10^5 dG	[19]
		ELISA	–	–	–	–	–
		Immunohistochemistry	–	Several years	–	–	[20]
		COMET-FPG	Lymphocyte DNA	–	–	–	[21]
		Capillary electrophoresis	Urine	–	8.63 nmol/L	10.15 ± 3.85 nmol/L	[22]
	8-oxo-dG	Electrochemistry	–	–	–	–	–
		HPLC-EC	Urine	–	–	281.7 ± 179.1 pmol 8-oxo-dG/day/kg weight	[23]
		GC-MS	Urine	–	–	2.10 ± 0.44 nmol/mol creatinine	[24]
		UPLC-MS/MS	Plasma	–	–	0.14 ± 0.05 nM	[25]
		UPLC-MS/MS	Serum	–	–	0.08 ± 0.02 nM	[25]
		UPLC-MS/MS	Urine	–	0.9 pmol/mL	16.11 ± 8.84 nM	[25]

(continued)

Table 7.2 (continued)

Type of biomarker		Technique	Most appropriate samples for analysis	Stability	Limit of quantification	Values in healthy individuals	References
		LC-MS/MS	Urine	−80 °C more many years	3.30 nmol/L	1.18 ± 0.57 nmol/mmol creatinine	[26]
		ELISA	Plasma	–	–	21.7 ± 9.2 pg/mL	[27]
		ELISA	Urine	–	–	6.2 ± 4.8 ng/mL	[27]
		ELISA	Urine	–	–	1.73 ± 0.49 nmol/mmol creatinine	[28]
		LC-MS/MS	DNA extract from lymphocyte	–	0.2 fmol/g DNA, or ~0.4 adducts 8-oxo-dG/106 dG	1.57 ± 0.88 adducts per 10^6 dG	[29]
		COMET-OGG-1	Lymphocyte	–	–	–	[30]
	Others	8-OHG, 8-oxo-G, 8-NO_2G, 8-NO_2dG, 8-OHdA, 2-OHdA, 5-hydroxymethylU, 5-ClU, 5-CldC, 8-CldG, 8-CldA, thymidine glycol, cyclo-dA, 8 (or 2, 4-, 5-)-OHdA, 5-OHU, 5-OHC, εdA, εdC, M1dG, 5-OHMeU, 4,6-diamino-5-formamide-pyrimidine, and 2,6-diamino-5-formamidepyrimidine. These compounds can be detected in urine by GC-MS and LC-MS/MS Auto-antibodies against oxidized DNA (ELISA)					
Products of carbohydrate oxidation	Pentosidine	HPLC-FL	Serum	–	0.005 μmol/L	1.5 mmol/mol	[31]
		ELISA	Serum	–	–	246.03 ± 55.05 pg/mL	[32]
	CML	LC-MS/MS	Plasma	15 days (−20 °C)	10 ng/mL	10.1–32.6 ng/mL	[33]
		LC-MS	Serum	–	0.01 μmol/L	9.9 mmol/mol	[31]
		LC-MS	Serum, plasma	–	0.16 μM	2.543 μM	[34]
		GC-MS	Plasma	–	–	0.090 ± 0.016 nmol/molLys	[35]
		ELISA	Serum	–	4 ng/L	–	[36]

		Western blot	Tissues	–	–	–	[37]
		Immunohistochemistry	Tissues	Several years	–	–	[38]
	Glyoxal	UPLC-FL	Serum	2 months (–80 °C)	2.5 nM	0.4–1 μM	[39]
		GC-MS	Plasma	–	15.3 ng/mL	17.17 ng/mL	[40]
		UPLC-MS/MS	Plasma	120 min in ambient temperature	200 fmol in column	406 ± 26 nmol/L	[41]
	Methylglyoxal	UHPLC-FL	Serum	2 months (–80 °C)	10 nM	1.1–2.7 μM	[39]
		GC-MS	Plasma	–	31.2 ng/mL	<LOQ	[40]
		UPLC-MS/MS	Plasma	120 min in ambient temperature	17 fmol in column	212 ± 8 nmol/L	[41]
	CEL	LC-MS/MS	Serum	–	0.015 μmol/L	93.9 mmol/mol	[31]
	Other markers	Clossline, pyrraline, GA-pyridine, Arg-pyrimidine, CMA, GOLD, MOLD, DOLD, and 2-(2-furoyl)-4(5)-(2-furanyl)-1H-imidazole. These compounds can be detected in plasma/serum by LCMS/MS					
Products of protein oxidation	Protein carbonyl	Colorimetric assay	Plasma	Months to years (−80 °C)	–	0.52 nmol/mg protein	[42]
		Colorimetric assay	Plasma	–	–	1.0 nmol/mg protein	[43]
		Colorimetric assay	Plasma	–	–	0.76 nmol/mg protein	[44]
		Colorimetric assay	Erythrocytes	–	–	86.8 nmol/10^9 cells	[7]
		ELISA	Plasma	–	–	0.86 nmol/mg protein	[45]
		ELISA	Plasma	–	–	Around 0.4 nmol/mg protein	[46]
		Western blot	Plasma	–	–	–	[47]
		Immunohistochemistry	Tissues	Several years	–	–	[48]

(continued)

Table 7.2 (continued)

Type of biomarker		Technique	Most appropriate samples for analysis	Stability	Limit of quantification	Values in healthy individuals	References
	$3\text{-}NO_2Tyr$	HPLC-FL	Serum	–	0.1 μM	–	[49]
		HPLC-EC	Plasma	–	–	–	[50]
		GC-MS	Plasma	–	125 pM	1.4–4.2 nM	[51]
		LC-MS/MS	Plasma	–	5 pg/mL	224–962 Total 3-NT (pg/mL)	[52]
		LC-MS/MS	Plasma	–	0.6 μg/mL	–	[53]
		LC-MS/MS	Plasma	3 months (–20 °C)	1.0 ng/mL	2.0–40.1 ng/mL	[54]
		ELISA	Serum	–	0.04 nM nitrated BSA	2.4 nM nitrated BSA	[55]
		HPLC-EC	Urine	–	4 nM	0.517 ± 0.0372 μM	[56]
		Western blotting LIE	Tissue	–	–	–	[57]
		Immunohistochemistry	Tissues	Several years	–	–	[58]
		LC-MS/MS	Exhaled breath condensate	–	3.9 pg/mL	11.7–172 pg/mL	[59]
		Immunoassay	Urine	–	20 ng/mL	1.20 μg/mL	[60]
		GC-MS	Urine	–	125 pM	8.4 ± 10.4 nM	[61]
		LC-MS/MS	Urine	–	5 pg/mL	Free: 48.9–587 pg/mL Total: 63.5–751 pg/mL	[52]

	3-Cl-Tyr	GC-MS	Plasma	–	–	8 ± 1 nM	[62]
		LC-MS/MS	Plasma	–	0.3 μg/mL	–	[53]
		LC-MS/MS	Serum	2 weeks (–20 °C)	2.50 ng/mL	<LOQ	[63]
		GC-MS	Urine	–	–	1.5 ± 1.0 ng/mg-creatinine	[64]
	Dityrosine	HPLC-FL	Plasma	–	–	–	[65]
		GC-MS	LDL	–	–	–	[66]
		LC-MS/MS	Urine	–	0.03 μM	10.1 ± 0.4 μmol/mol creatinine (smokers)	[67]
		LC-MS/MS	Urine	–	0.10 ng/mL	< 2.4 μmol mol^{-1} Cr	[68]
	AOPP	Spectrophotometry	Plasma	–	–	52 μmol/L	[69]
	Other markers	2-oxoHis (ion exchange HPLC, HPLC-EC, and LC-MS), p, m, o Tyr (HPLC-FL and GC-MS), Leu hydroxides and hydroperoxides (HPLC), Met sulfoxide (HPLC), 5-hydroxy-2-aminovaleric acid (GC-MS), Val hydroxides and hydro-peroxides (HPLC), DOPA (HPLC, GC-MS), DOPA-protein (HPLC-FL and HPLC-EC), protein thiols (colorimetric assay and LC-MS), oxidized Trp (HPLC-UV and autofluorescence), anti-3-NO_2Tyr (ELISA), and asymmetric dimethylarginine (ADMA) (LC-MS/MS)					
Products of lipid oxidation	MDA	Colorimetric-TBARS	Plasma	–	–	1.7–2.7 μmol/L	[70]
		Colorimetric-TBARS	Plasma	–	–	2.30 μmol/L	[71]
		Colorimetric-TBARS	Plasma	–	–	1.33 nmol/L	[72]
		Colorimetric-TBARS	Plasma	–	–	6.20 nmol/L	[73]
		Colorimetric-TBARS	Plasma	–	–	5.80 μmol/L	[74]
		Colorimetric-TBARS	Erythrocytes	–	–	2.43 pmol/mgHb	[75]
		Fluorimetric assay	Plasma	–	–	2.80 μmol/L	[76]
		Fluorimetric assay	Plasma	–	–	0.66 μmol/L	[77]
		Chemiluminescence assay	–	–	–	–	–
		HPLC-UV	Plasma	–	2.86 μM	18.42 μM	[78]
		HPLC-FL	Plasma	–	0.049 μM	0.97 μM	[79]

(continued)

Table 7.2 (continued)

Type of biomarker		Technique	Most appropriate samples for analysis	Stability	Limit of quantification	Values in healthy individuals	References
		LC-MS/MS	Plasma	–	39 pg/mL	352 ng/mL	[80]
		GC-MS	Plasma	–	0.02 μM	–	[81]
		GC-FID	Plasma	–	2.48 ng/mL	–	[82]
		Immunohistochemistry	Tissue	Years	–	–	[83]
		HPLC-UV	Urine	–	4.7 ng/mL	84.05 ng/mL	[84]
		HPLC-FL	Urine	–	0.80 nM	24.62–72.80 nM	[85]
		LC-MS/MS	Urine	–	105 pg/mL	6.8 μg/mmol creat	[80]
		HPLC-FL	Exhaled breath condensate	–	3 nM	8.475.1 nmol	[86]
		LC-MS/MS	Exhaled breath condensate	–	32 ng/L	45 ng/L	[80]
		Capillary electrophoresis	Exhaled breath condensate	–	3.8 nM	13.0–21.6 nM	[87]
	4-HNE	HPLC-FL	Serum	–	–	0.780–1.14 nmol/mL	[88]
		HPLC-EC	Cultured cells	–	–	–	[89]
		GC-MS	Plasma	–	2.0 nmol/L	0.119 ± 0.052 μmol/L	[90]
		LC-MS/MS	Plasma	–	45 pg/mL	1736 ng/mL	[80]
		ELISA	Plasma and serum	–	24.6 ± 7.4 pmol/mg protein	Around 20 pmol/mg protein	[91]
		HPLC-UV	Urine	–	9.0 nmol/mL	30 nmol/mL	[92]
		Immunohistochemistry	Tissues	Several years	–	–	[93]
		LC-MS/MS	Exhaled breath condensate	–	39 ng/L	165 ng/L	[80]

		GC-MS	Urine	–	75 pg/mL	0.39 ± 0.11 ng mg – 1 creatinine	[94]
		LC-MS/MS	Urine	–	112 pg/mL	211 µg/mmol creat	[80]
	15-F_{2t}-Isoprostane	GC-MS	Plasma	–	–	58–166 pg/mL	[95]
		LC-MS/MS	Plasma	–	8 pg/mL	46 pg/mL	[80]
		ELISA	Plasma	–	4 pg/mL	31.8 ± 5.5 pg/mL	[96]
		RIA	Plasma	–	–	79 ± 9.6 pmol/L	[97]
		LC-MS/MS	Exhaled breath condensate	–	6 ng/L	52 ng/L	[80]
		LC-MS/MS	Exhaled breath condensate	–	10 ng/L	4.44 ± 2.01 ng/L	[98]
		LC-MS/MS	Exhaled breath condensate	8 months (– 80 °C)	4 ng/L	<LOQ	[99]
		Competitive immunoassay	Exhaled breath condensate	2 months (– 70 °C)	6.1 ng/L	81.9 ng/L	[100]
		Sandwich ELISA	Exhaled breath condensate	–	6.1 pg/mL	14.8 pg/mL	[101]
		GC-MS	Urine	–	50 pg/mL	0.811 ± 0.359 ng/mg creatinine	[102]
		LC-MS/MS	Urine	–	21 pg/mL	25 ng/mmol creat	[80]
		RIA	Urine	–	–	3.4 ± 0.3 nmol/L	[97]
		ELISA	Urine	–	4 pg/mL	2.9 ± 2.0 ng/mg creatinine	[96]

(continued)

Table 7.2 (continued)

Type of biomarker		Technique	Most appropriate samples for analysis	Stability	Limit of quantification	Values in healthy individuals	References
	Acrolein	HPLC-FL	Plasma	–	16.8 ng/mL	–	[103]
		GC-MS	Urine	–	–	–	[104]
		LC-MS/MS	Urine	2 weeks	3-hydroxypropyl mercapturic acid 0.163 ng/mL	3-hydroxypropyl mercapturic acid 3.47 ng/mL	[105]
		ELISA	Urine	–	–	3-hydroxypropyl mercapturic acid (median) 2.11 μM	[106]
		Immunohistochemistry	Tissue	Years	–	–	[107]
	Ethane	GC-MS	Exhaled breath condensate	–	–	–	[108]
		GC-FID	Exhaled breath condensate	72 h at 4 °C	3.84 ng/mL	<10 ng/mL	[109]
	Pentane	GC-FID	Exhaled breath condensate	–	–	–	[110]
		GC-MS	Exhaled breath condensate	–	0.21 pg/mL	0.25–48.89 pg/mL	[111]
	Lipid hydroperoxides	Spectrophotometry	Plasma	–	–	0.82 μmol/L	[112]
	Other markers	Conjugated-diene (second derivative spectrometry, HPLC-UV, and GC-MS), isolevuglandin (LC-MS and LC-MS/MS), oxLDL (ELISA and capillary electrophoresis), oxLDLab (ELISA), MDAdG, M1G, M1C, M1A (ELISA and GC-MS), 4-HNE-protein adducts (ELISA, Western blot), and etheno-DNA adducts (LC-MS/MS)					
Pro-oxidants	MPO	Enzyme kinetics-colorimetric assay	Serum	–	–	50.90–88.50 IU/L	[113]
		Enzyme kinetics-fluorimetric assay	–	–	–	–	–

		ELISA	Plasma	–	15 ng/mL	Male: 42.9 ± 11.9 ng/mL Female: 40.7 ± 9.8 ng/mL	[114]
	EPO	Enzyme kinetics-colorimetric assay	–	–	–	–	–
	XO	Enzyme kinetics	Serum	–	–	207.0 ± 86.1 IU/mL	[115]
	NOS	Enzyme kinetics	–	–	–	–	–
		Immunohistochemistry	Tissues	Several years	–	–	[58]
	NOX	Western blot	Tissues	–	–	–	[116]
		RT-PCR	Tissues	–	–	–	[117]
		Immunohistochemistry	Tissues	Several years	–	–	[118]
	Homocysteine	HPLC-UV	Plasma	–	0.05 μ mol/L	5.16–13.23 μ mol/L	[119]
		HPLC-FL	Plasma	–	0.05 μM	8.85 μM	[120]
		HPLC-EC	Plasma	One week (−20 °C)	0.8 ng/mL	1.31 ± 0.62 μg/mL	[121]
		UPLC-MS/MS	Plasma	–	4 nM Homocystine 7 nM	Free 0.15 ± 0.01 μM Total 6.71 ± 0.53 μM Bound 6.56 ± 0.53 μM Homocystine 15.2 ± 2.12 nM	[122]
		Capillary electrophoresis	Plasma	–	2.5 μM	9.2 μM	[123]
		HPLC-UV	Urine	–	0.05 μ mol/L	m 3.14 to 13.64 μ mol/L	[119]
	Other markers	Iron (serum/plasma by SAA) COX, LOX, and AO (serum/plasma by immunoassays)					

(continued)

Table 7.2 (continued)

Type of biomarker		Technique	Most appropriate samples for analysis	Stability	Limit of quantification	Values in healthy individuals	References
Antioxidants	SOD	Enzyme kinetics	Serum	–	–	67.65 ± 7.26 IU/mL	[124]
		ELISA	Serum	–	–	Males: 52.7 ± 18.4 ng/mL Females: 58.1 ± 18.9 ng/mL	[125]
	GPx	Enzyme kinetics	Erythrocytes	–	–	36.1 ± 6.5 IU/g of Hb	[126]
		ELISA	Plasma	–	–	3.8–43.8 μg/mL	[127]
		RIA	Serum	10 months (−20 °C)	–	0.3–6.3 mg/L	[128]
	GR	Enzyme kinetics	Blood	–	–	8.91 ± 1.04 IU/g of Hb	[129]
	CAT	Enzyme kinetics	Blood	–	–	1642.07 IU/mL	[130]
	Trx	ELISA	Serum	–	–	43.8 ± 1.4 ng/mL	[131]
	Prx1	ELISA	Serum	–	–	13.2 11.9 ng/mL	[132]
		ELISA	Plasma	–	–	16.77 ± 3.899 ng/mL	[133]
	Prx2	ELISA	Serum	–	–	≈ 2 ng/mL	[134]
		ELISA	Plasma	–	–	20.37 ± 8.61 ng/mL	[133]
	Prx3	ELISA	Serum	–	–	52.3 ± 26.8 ng/mL	[135]
		RT-PCR	Tissues	–	–	–	[136]
	Prx4	ELISA	Serum	–	–	1088·0 ± 753·3 pg/mL	[137]
		ELISA	Plasma	–	–	2696 ± 1972 pg/mL	[133]
	Prx6	ELISA	Serum	–	–	8.43 ± 5.03 ng/mL	[138]
		ELISA	Plasma	–	–	238 ± 111 pg/mL	[133]

	Prx 3,5, and 6	Western blot	Tissues	–	–	–	[139]
	TrxR	Enzyme kinetics	–	–	–	–	–
	GSH/GSSG	Colorimetric assay	Plasma	Short	–	8.1 μmol/L	[140]
		Colorimetric assay	Plasma	Short	–	50. 6 mg/dL	[75]
		Colorimetric assay	Blood	Short	–	257.4 mmol/L	[71]
		Colorimetric assay	Erythrocytes	Short	–	70. 1 mg/100 mL RBC	[141]
		Fluorimetric assay	Erythrocytes	Short	–	44.9 μmol/L	[142]
		Electrochemical assay	–	–	–	–	–
		HPLC-UV	Plasma	–	0.05 μ mol/L	2.89 to 28.9 μ mol/L	[119]
		HPLC-FL	Plasma	–	0.05 μM	10.9 μM	[120]
		HPLC-EC	Plasma	One week (−20 °C)	GSH 30 ng/mL GSSG 96 ng/mL	GSH 2.07 ± 0.87 μg/mL GSSG 0.88 ± 0.05 μg/mL	[121]
		HPLC-EC	Erythrocytes	One week (−20 °C)	GSH 30 ng/mL GSSG 96 ng/mL	GSH 9.23 ± 1.93 μg/mL GSSG 9.06 ± 1.72 μg/mL	[121]
		GC-MS	Erythrocytes	30 days (+4 °C)	8.58 μM	1682.66 μM/g HB	
		UPLC-MS/MS	Plasma	–	GSH 0.5 nM GSSG 2 nM	Free 3.89 ± 0.65 μM Total 7.61 ± 0.51 μM Bound 3.72 ± 0.36 μM GSSG 7.09 ± 0.51 μM	[122]
		LC-MS/MS	Blood	7 days −80 °C	GSH 0.5 μM GSSG 0.0625 μM	GSH 865.5 ± 266.5 μM GSSG 80.7 ± 39.3 μM	[143]

(continued)

Table 7.2 (continued)

Type of biomarker		Technique	Most appropriate samples for analysis	Stability	Limit of quantification	Values in healthy individuals	References
		LC-MS/MS	Blood	3 years – 80 °C	GSH 1.5 μM GSSG 0.1 μM	GSH 900 ± 140 μM GSSG 1.17 ± 0.43 μM	[144]
		ELISA	–	–	–	–	–
		Western blotting LIE	Isolated cells	–	–	–	[145]
		LC-MS	Blood	–	–	3.15 ± 1.99%	[146]
		Capillary electrophoresis	Blood	–	GSH 16.6 μmol/L GSSG 3 μmol/L	GSH 1078.1 ± 17.0 μmol/L GSSG 110.6 ± 7.3 μmol/L	[147]
		Capillary electrophoresis	Exhaled breath condensate	–	0.59 nM	0.88 nM	[148]
	β-Carotene	Spectrophotometry	Plasma	–	–	1.71 μmol/L	[8]
		Electrochemistry	–	–	–	–	–
		HPLC-UV	Plasma	3 months (–80 °C)	0.02 μM	0.92 μmol/L	[149]
		LC-MS/MS	Plasma	–	0.84 fmol Iin column	699 ± 550 nmol/L	[150]
	Ascorbic acid	Spectrophotometry	Plasma	–	–	56.9 μmol/L	[141]
		Spectrophotometry	Plasma	–	–	0.86 μg/dL	[151]
		Fluorimetric assay	–	–	–	–	–
		Chemiluminescence	–	–	–	–	–
		Electrochemical assay	–	–	–	–	–
		Enzymatic assay	–	–	–	–	–
		HPLC-UV	Plasma	–	1.02 nmol/mL	53 ± 23 μmol/L	[152]
		HPLC-FL	Serum	–	–	6.4 ng per 2 μL	[153]
		HPLC-EC	Plasma	One week (–20 °C)	2 ng/mL	4.29 ± 0.36 μg/mL	[121]

		LC-MS	Plasma	–	–	–	[154]
		GC-MS	Plasma	–	–	7–12 μg/mL (40–70 μM)	[155]
		Capillary electrophoresis	Serum	–	–	32.15 μmol/L	[156]
		Capillary electrophoresis	Plasma	–	–	4.748 mg/L	[157]
		Capillary electrophoresis	Urine	–	0.020 μg/mL	0.50 μg/mL	[158]
	α-Tocopherol	Spectrophotometry	Plasma	–	–	25.7 μmol/L	[8]
		HPLC-UV	Plasma	3 months (−80 °C)	2.1 μM	27.2 μmol/L	[149]
		HPLC-FL	Serum	–	0.562 μmol/L	8–42 μmol/L	[159]
		HPLC-EC	Serum	–	–	8.9 ± 2.3 μg/mL	[160]
		GC-MS	Plasma	–	0.417 μM	31.175–48.217 μM	[161]
		GC-MS	Serum	–	0.15 μg/mL	3.2 μg/mL	[162]
		LC-MS/MS	Serum	–	9.20 μg/L	12.038 ± 2.501 μg/L	[163]
	Ubiquinone/ubiquinol	HPLC-UV	Plasma	3 months (−80 °C)	0.15 μM	1.21 μmol/L	[149]
		HPLC-UV-EC	Plasma	–	–	Ubiquinone <0.1 nmol/mL Ubiquinol <0.1 nmol/mL	[164]
		HPLC-EC	Plasma	–	–	Ubiquinone 0.117 ± 0.239 mg/L Ubiquinol 0.712 ± 0.239 mg/L	[165]
		LC-MS/MS	Serum	–	Ubiquinone 5.49 ng/mL Ubiquinol 15.80 ng/mL	Ubiquinone 65.75 ng/mL Ubiquinol 835 ng/mL	[166]

(continued)

Table 7.2 (continued)

Type of biomarker		Technique	Most appropriate samples for analysis	Stability	Limit of quantification	Values in healthy individuals	References
	Other markers	GST (serum/plasma by immunoassays) HO-1 (in situ by immunohistochemistry) Selenium, zinc, and manganese, and copper (serum/plasma by SAA) Paraoxonase, arylesterase, glutaredoxin, carrier proteins (metallothionein, albumin, ferritin, transferrin, lactoferrin, ceruloplasmin, haptoglobin, and hemopexine) (ELISA) Albumin reduced (HMA), mixed disulfide between albumin and low molecular mass thiols (HNAl,), SH groups in albumin oxidized to sulfenic, sulfinic, and sulfonic acid (HNA2) (HPLC-FL) Lycopene, bilirubin, uric acid, lipoic acid, melatonin, folate, pyruvate, taurine, hypotaurine, and lactate (spectrophotometric assay, HPLC-UV, HPLC-EC, GC-MS, and LC-MS)					
Transcription factors	Nrf-2 Months (−80 °C)	Immunohistochemistry	Tissues	Several years	–	–	[167]
		RT-PCR	Peripheral blood mononuclear cell (PBMC)	–	–	–	[168]
	NF-kB	RT-PCR	Tissue	–	–	–	[169]
Repair system	OGG-1	COMET	Mammalian cells	–	–	–	[170]
		Immunohistochemistry	Tissue	Several years	–	–	[20]

References

1. Svensson S, Olin A-C, Lärstad M, Ljungkvist G, Torén K. Determination of hydrogen peroxide in exhaled breath condensate by flow injection analysis with fluorescence detection. Journal of Chromatography B. 2004;809(2):199–203.
2. Aizawa H, Makita Y, Sumitomo K, Aburakawa Y, Katayama T, Nakatani-Enomoto S, et al. Edaravone diminishes free radicals from circulating neutrophils in patients with ischemic brain attack. Intern Med. 2006;45(1):1–4.
3. C. C, C. M, S. B, Faure H, Favier A. A simple HPLC-spectrophotomeric assay of salicylic acid in biological fluids. J Chromatogr Sci. 1996;34:166–73.
4. Coudray C, Talla M, Martin S, Fatome M, Favier A. High-Performance Liquid Chromatography-Electrochemical Determination of Salicylate Hydroxylation Products as an in Vivo Marker of Oxidative Stress. Analytical Biochemistry. 1995;227(1):101–11.
5. Luo X, Lehotay DC. Determination of hydroxyl radicals using salicylate as a trapping agent by gas chromatography-mass spectrometry. Clinical Biochemistry. 1997;30(1):41–6.
6. Biondi R, Brancorsini S, Poli G, Egidi MG, Capodicasa E, Bottiglieri L, et al. Detection and scavenging of hydroxyl radical via D-phenylalanine hydroxylation in human fluids. Talanta. 2018;181:172–81.
7. Sureda A, Batle JM, Tauler P, Ferrer MD, Tur JA, Pons A. Vitamin C supplementation influences the antioxidant response and nitric oxide handling of erythrocytes and lymphocytes to diving apnea. Eur J Clin Nutr. 2006;60(7):838–46.
8. Zhou J-F, Xiao W-Q, Zheng Y-C, Dong J, Zhang S-M. Increased oxidative stress and oxidative damage associated with chronic bacterial prostatitis. Asian J Androl. 2006;8(3): 317–23.
9. Žunić G, Spasić S, Jelić-Ivanović Z. Simple and rapid method for the measurement of nitrite and nitrate in human plasma and cerebrospinal fluid by capillary electrophoresis. Journal of Chromatography B: Biomedical Sciences and Applications. 1999;727(1):73–9.
10. Friedberg MA, Hinsdale ME, Shihabi ZK. Analysis of nitrate in biological fluids by capillary electrophoresis. Journal of chromatography A. 1997;781(1–2):491–6.
11. Romitelli F, Santini SA, Chierici E, Pitocco D, Tavazzi B, Amorini AM, et al. Comparison of nitrite/nitrate concentration in human plasma and serum samples measured by the enzymatic batch Griess assay, ion-pairing HPLC and ion-trap GC–MS: The importance of a correct removal of proteins in the Griess assay. Journal of Chromatography B. 2007;851(1):257–67.
12. Akyüz M, Ata Ş. Determination of low level nitrite and nitrate in biological, food and environmental samples by gas chromatography–mass spectrometry and liquid chromatography with fluorescence detection. Talanta. 2009;79(3):900–4.
13. Katayama M, Matsuda Y, Kobayashi K, Kaneko S, Ishikawa H. Monitoring of 8-oxo-7,8-dihydro-2′-deoxyguanosine in urine by high-performance liquid chromatography after pre-column derivatization with glyoxal reagents. Biomedical chromatography: BMC. 2006;20(8):800–5.
14. Koide S, Kinoshita Y, Ito N, Kimura J, Yokoyama K, Karube I. Determination of human serum 8-hydroxy-2′-deoxyguanosine (8-OHdG) by HPLC-ECD combined with solid phase extraction (SPE). Journal of Chromatography B. 2010;878(23):2163–7.
15. Wang C-J, Yang N-H, Chang C-C, Liou S-H, Lee H-L. Rapid and simple one-step membrane extraction for the determination of 8-hydroxy-2′-deoxyguanosine in human plasma by a combination of on-line solid phase extraction and LC–MS/MS. Journal of Chromatography B. 2011;879(30):3538–43.
16. Omran MM, Rashed RE, Darwish H, Belal AA, Mohamed FZ. Development of a gas chromatography–mass spectrometry method for breast cancer diagnosis based on nucleoside metabolomes 1-methyl adenosine, 1-methylguanosine and 8-hydroxy-2′-deoxyguanosine. 2020;34(1):e4713.

17. Gan H, Xu H. A novel aptamer-based online magnetic solid phase extraction method for simultaneous determination of urinary 8-hydroxy-2′-deoxyguanosine and monohydroxylated polycyclic aromatic hydrocarbons. Talanta. 2019;201:271–9.
18. He L, Liu X, Zhang J. Simultaneous quantification of urinary 6-sulfatoxymelatonin and 8-hydroxy-2′-deoxyguanosine using liquid chromatography-tandem mass spectrometry. Journal of Chromatography B. 2018;1095:119–26.
19. Wu D, Liu B, Yin J, Xu T, Zhao S, Xu Q, et al. Detection of 8-hydroxydeoxyguanosine (8-OHdG) as a biomarker of oxidative damage in peripheral leukocyte DNA by UHPLC–MS/MS. Journal of Chromatography B. 2017;1064:1–6.
20. Inokuchi S, Itoh S, Yoshizumi T, Yugawa K, Yoshiya S, Toshima T, et al. Mitochondrial expression of the DNA repair enzyme OGG1 improves the prognosis of pancreatic ductal adenocarcinoma. Pancreatology. 2020;20(6):1175–82.
21. Gelaleti RB, Damasceno DC, Lima PH, Salvadori DM, Calderon Ide M, Peraçoli JC, et al. Oxidative DNA damage in diabetic and mild gestational hyperglycemic pregnant women. Diabetology & metabolic syndrome. 2015;7:1.
22. Zhang S, Song X, Zhang W, Luo N, Cai L. Determination of low urinary 8-hydroxy-2`-deoxyguanosine excretion with capillary electrophoresis and molecularly imprinted monolith solid phase microextraction. Science of The Total Environment. 2013;450–451:266–70.
23. Lengger C, Schöch G, Topp H. A High-Performance Liquid Chromatographic Method for the Determination of 8-Oxo-7,8-dihydro-2′-deoxyguanosine in Urine from Man and Rat. Analytical Biochemistry. 2000;287(1):65–72.
24. Foksinski M, Rozalski R, Guz J, Ruszkowska B, Sztukowska P, Piwowarski M, et al. Urinary excretion of dna repair products correlates with metabolic rates as well as with maximum life spans of different mammalian species. Free Radical Biology and Medicine. 2004;37(9): 1449–54.
25. Lam PMW, Mistry V, Marczylo TH, Konje JC, Evans MD, Cooke MS. Rapid measurement of 8-oxo-7,8-dihydro-2′-deoxyguanosine in human biological matrices using ultra-high-performance liquid chromatography–tandem mass spectrometry. Free Radical Biology and Medicine. 2012;52(10):2057–63.
26. Topic A, Francuski D, Markovic B, Stankovic M, Dobrivojevic S, Drca S, et al. Gender-related reference intervals of urinary 8-oxo-7,8-dihydro-2′-deoxyguanosine determined by liquid chromatography-tandem mass spectrometry in Serbian population. Clinical Biochemistry. 2013;46(4):321–6.
27. Hu C-W, Huang Y-J, Li Y-J, Chao M-R. Correlation between concentrations of 8-oxo-7,8-dihydro-2′-deoxyguanosine in urine, plasma and saliva measured by on-line solid-phase extraction LC-MS/MS. Clinica Chimica Acta. 2010;411(17):1218–22.
28. Rossner P, Orhan H, Koppen G, Sakai K, Santella RM, Ambroz A, et al. Urinary 8-oxo-7,8-dihydro-2′-deoxyguanosine analysis by an improved ELISA: An inter-laboratory comparison study. Free Radical Biology and Medicine. 2016;95:169–79.
29. Boysen G, Collins LB, Liao S, Luke AM, Pachkowski BF, Watters JL, et al. Analysis of 8-oxo-7,8-dihydro-2′-deoxyguanosine by ultra high pressure liquid chromatography–heat assisted electrospray ionization–tandem mass spectrometry. Journal of Chromatography B. 2010;878(3):375–80.
30. Dinçer Y, Akkaya Ç, Mutlu T, Yavuzer S, Erkol G, Bozluolcay M, et al. DNA repair gene OGG1 polymorphism and its relation with oxidative DNA damage in patients with Alzheimer's disease. Neuroscience Letters. 2019;709:134362.
31. Palma-Duran SA, Kontogianni MD, Vlassopoulos A, Zhao S, Margariti A, Georgoulis M, et al. Serum levels of advanced glycation end-products (AGEs) and the decoy soluble receptor for AGEs (sRAGE) can identify non-alcoholic fatty liver disease in age-, sex- and BMI-matched normo-glycemic adults. Metabolism. 2018;83:120–7.
32. Kerkeni M, Saïdi A, Bouzidi H, Yahya SB, Hammami M. Elevated serum levels of AGEs, sRAGE, and pentosidine in Tunisian patients with severity of diabetic retinopathy. Microvascular Research. 2012;84(3):378–83.

33. Kuang L, Jing Z, Wang J, Ma L, Liu X, Yang J. Quantitative determination of ε-N-carboxymethyl-l-lysine in human plasma by liquid chromatography–tandem mass spectrometry. Journal of Pharmaceutical and Biomedical Analysis. 2014;90:1–6.
34. Rankin NJ, Burgess K, Weidt S, Wannamethee G, Sattar N, Welsh P. High-throughput quantification of carboxymethyl lysine in serum and plasma using high-resolution accurate mass Orbitrap mass spectrometry. Annals of clinical biochemistry. 2019;56(3):397–407.
35. Miyata T, Fu M-X, Kurokawa K, van Ypersele De Strihou C, Thorpe SR, Baynes JW. Autoxidation products of both carbohydrates and lipids are increased in uremic plasma: Is there oxidative stress in uremia? Kidney International. 1998;54(4):1290–5.
36. Finco AB, Machado-de-Ávila RA, Maciel R, De Moura J, Billiald P, Stinghen AEM, et al. Generation and characterization of monoclonal antibody against Advanced Glycation End Products in chronic kidney disease. Biochemistry and Biophysics Reports. 2016;6:142–8.
37. Inoue R, Sakata N, Nakai K, Aikawa H, Tsutsumi M, Nii K, et al. Nε-(Carboxymethyl)lysine in Debris from Carotid Artery Stenting: Multiple versus Nonmultiple Postoperative Lesions. Journal of Stroke and Cerebrovascular Diseases. 2014;23(10):2827–33.
38. Sakata N, Imanaga Y, Meng J, Tachikawa Y, Takebayashi S, Nagai R, et al. Immunohistochemical localization of different epitopes of advanced glycation end products in human atherosclerotic lesions. Atherosclerosis. 1998;141(1):61–75.
39. El-Maghrabey MH, Nakatani T, Kishikawa N, Kuroda N. Aromatic aldehydes as selective fluorogenic derivatizing agents for α-dicarbonyl compounds. Application to HPLC analysis of some advanced glycation end products and oxidative stress biomarkers in human serum. Journal of Pharmaceutical and Biomedical Analysis. 2018;158:38–46.
40. Wu M-Y, Chen B-G, Chang CD, Huang M-H, Wu T-G, Chang D-M, et al. A novel derivatization approach for simultaneous determination of glyoxal, methylglyoxal, and 3-deoxyglucosone in plasma by gas chromatography–mass spectrometry. Journal of Chromatography A. 2008;1204(1):81–6.
41. Scheijen JL, Schalkwijk CG. Quantification of glyoxal, methylglyoxal and 3-deoxyglucosone in blood and plasma by ultra performance liquid chromatography tandem mass spectrometry: evaluation of blood specimen. Clinical chemistry and laboratory medicine. 2014;52(1):85–91.
42. Lemarechal H, Allanore Y, Chenevier-Gobeaux C, Ekindjian OG, Kahan A, Borderie D. High redox thioredoxin but low thioredoxin reductase activities in the serum of patients with rheumatoid arthritis. Clin Chim Acta. 2006;367(1–2):156–61.
43. Morillas-Ruiz JM, Villegas García JA, López FJ, Vidal-Guevara ML, Zafrilla P. Effects of polyphenolic antioxidants on exercise-induced oxidative stress. Clinical nutrition (Edinburgh, Scotland). 2006;25(3):444–53.
44. Zurawa-Janicka D, Renke J, Popadiuk S, Skorko-Glonek J, Szumera M, Plata-Nazar K, et al. Preferential immunoglobulin oxidation in children with juvenile idiopathic arthritis. Scand J Rheumatol. 2006;35(3):193–200.
45. Alamdari DH, Kostidou E, Paletas K, Sarigianni M, Konstas AGP, Karapiperidou A, et al. High sensitivity enzyme-linked immunosorbent assay (ELISA) method for measuring protein carbonyl in samples with low amounts of protein. Free Radical Biology and Medicine. 2005;39(10):1362–7.
46. Mohanty JG, Bhamidipaty S, Evans MK, Rifkind JM. A fluorimetric semi-microplate format assay of protein carbonyls in blood plasma. Analytical Biochemistry. 2010;400(2):289–94.
47. Shacter E, Williams JA, Lim M, Levine RL. Differential susceptibility of plasma proteins to oxidative modification: Examination by western blot immunoassay. Free Radical Biology and Medicine. 1994;17(5):429–37.
48. Baraibar MA, Barbeito AG, Muhoberac BB, Vidal R. A mutant light-chain ferritin that causes neurodegeneration has enhanced propensity toward oxidative damage. Free Radic Biol Med. 2012;52(9):1692–7.
49. Iwasaki Y, Mochizuki K, Nakano Y, Maruya N, Goto M, Maruyama Y, et al. Comparison of fluorescence reagents for simultaneous determination of hydroxylated phenylalanine and nitrated tyrosine by high-performance liquid chromatography with fluorescence detection. 2012;26(1):41–50.

50. Ohshima H, Celan I, Chazotte L, Pignatelli B, Mower HF. Analysis of 3-Nitrotyrosine in Biological Fluids and Protein Hydrolyzates by High-Performance Liquid Chromatography Using a Postseparation, On-line Reduction Column and Electrochemical Detection: Results with Various Nitrating Agents. Nitric oxide: biology and chemistry. 1999;3(2):132–41.
51. Schwedhelm E, Tsikas D, Gutzki F-M, Frölich JC. Gas Chromatographic–Tandem Mass Spectrometric Quantification of Free 3-Nitrotyrosine in Human Plasma at the Basal State. Analytical Biochemistry. 1999;276(2):195–203.
52. Radabaugh MR, Nemirovskiy OV, Misko TP, Aggarwal P, Mathews WR. Immunoaffinity liquid chromatography–tandem mass spectrometry detection of nitrotyrosine in biological fluids: Development of a clinically translatable biomarker. Analytical Biochemistry. 2008;380(1):68–76.
53. Song Y, Liao J, Zha C, Wang B, Liu CC. Simultaneous determination of 3-chlorotyrosine and 3-nitrotyrosine in human plasma by direct analysis in real time–tandem mass spectrometry. Acta Pharmaceutica Sinica B. 2015;5(5):482–6.
54. Li X, Li S, Kellermann G. A novel mixed-mode solid phase extraction coupled with LC–MS/MS for the re-evaluation of free 3-nitrotyrosine in human plasma as an oxidative stress biomarker. Talanta. 2015;140:45–51.
55. Knight AR, Taylor EL, Lukaszewski R, Jensen KT, Jones HE, Carré JE, et al. A high-sensitivity electrochemiluminescence-based ELISA for the measurement of the oxidative stress biomarker, 3-nitrotyrosine, in human blood serum and cells. Free Radical Biology and Medicine. 2018;120:246–54.
56. Saravanabhavan G, Blais E, Vincent R, Kumarathasan P. A high performance liquid chromatography-electrochemical array method for the measurement of oxidative/nitrative changes in human urine. Journal of chromatography A. 2010;1217(19):3269–74.
57. Sultana R, Reed T, Butterfield DA. Detection of 4-hydroxy-2-nonenal- and 3-nitrotyrosine-modified proteins using a proteomics approach. Methods Mol Biol. 2009;519:351–61.
58. Chimenti C, Scopelliti F, Vulpis E, Tafani M, Villanova L, Verardo R, et al. Increased oxidative stress contributes to cardiomyocyte dysfunction and death in patients with Fabry disease cardiomyopathy. Human pathology. 2015;46(11):1760–8.
59. Göen T, Müller-Lux A, Dewes P, Musiol A, Kraus T. Sensitive and accurate analyses of free 3-nitrotyrosine in exhaled breath condensate by LC–MS/MS. Journal of Chromatography B. 2005;826(1):261–6.
60. Jin J, Wang C, Tao Y, Tan Y, Yang D, Gu Y, et al. Determination of 3-nitrotyrosine in human urine samples by surface plasmon resonance immunoassay. Sensors and Actuators B: Chemical. 2011;153(1):164–9.
61. Tsikas D, Mitschke A, Suchy M-T, Gutzki F-M, Stichtenoth DO. Determination of 3-nitrotyrosine in human urine at the basal state by gas chromatography–tandem mass spectrometry and evaluation of the excretion after oral intake. Journal of Chromatography B. 2005;827(1):146–56.
62. Gaut JP, Byun J, Tran HD, Heinecke JW. Artifact-Free Quantification of Free 3-Chlorotyrosine, 3-Bromotyrosine, and 3-Nitrotyrosine in Human Plasma by Electron Capture–Negative Chemical Ionization Gas Chromatography Mass Spectrometry and Liquid Chromatography–Electrospray Ionization Tandem Mass Spectrometry. Analytical Biochemistry. 2002;300(2):252–9.
63. Crow BS, Quiñones-González J, Pantazides BG, Perez JW, Winkeljohn WR, Garton JW, et al. Simultaneous Measurement of 3-Chlorotyrosine and 3,5-Dichlorotyrosine in Whole Blood, Serum and Plasma by Isotope Dilution HPLC-MS-MS. Journal of analytical toxicology. 2016;40(4):264–71.
64. Mita H, Higashi N, Taniguchi M, Higashi A, Kawagishi Y, Akiyama K. Urinary 3-bromotyrosine and 3-chlorotyrosine concentrations in asthmatic patients: lack of increase in 3-bromotyrosine concentration in urine and plasma proteins in aspirin-induced asthma after intravenous aspirin challenge. Clinical and experimental allergy: journal of the British Society for Allergy and Clinical Immunology. 2004;34(6):931–8.

65. Colombo G, Clerici M, Giustarini D, Portinaro N, Badalamenti S, Rossi R, et al. A central role for intermolecular dityrosine cross-linking of fibrinogen in high molecular weight advanced oxidation protein product (AOPP) formation. Biochimica et Biophysica Acta (BBA) – General Subjects. 2015;1850(1):1–12.
66. Leeuwenburgh C, Rasmussen JE, Hsu FF, Mueller DM, Pennathur S, Heinecke JW. Mass Spectrometric Quantification of Markers for Protein Oxidation by Tyrosyl Radical, Copper, and Hydroxyl Radical in Low Density Lipoprotein Isolated from Human Atherosclerotic Plaques*. Journal of Biological Chemistry. 1997;272(6):3520–6.
67. Orhan H, Coolen S, Meerman J. Quantification of urinary o,o′-dityrosine, a biomarker for oxidative damage to proteins, by high performance liquid chromatography with triple quadrupole tandem mass spectrometry: A comparison with ion-trap tandem mass spectrometry. Journal of chromatography B, Analytical technologies in the biomedical and life sciences. 2005;827:104–8.
68. Martinez-Moral M-P, Kannan K. How stable is oxidative stress level? An observational study of intra- and inter-individual variability in urinary oxidative stress biomarkers of DNA, proteins, and lipids in healthy individuals. Environment International. 2019;123:382–9.
69. Fialová L, Malbohan I, Kalousová M, Soukupová J, Krofta L, Stípek S, et al. Oxidative stress and inflammation in pregnancy. Scandinavian journal of clinical and laboratory investigation. 2006;66(2):121–7.
70. Adiga IK, Nair RR. A positive association between cardiomyocyte volume and serum malondialdehyde levels. Int J Cardiol. 2007;115(2):246–8.
71. Ahmed RS, Suke SG, Seth V, Jain A, Bhattacharya SN, Banerjee BD. Impact of oral vitamin E supplementation on oxidative stress & lipid peroxidation in patients with polymorphous light eruption. Indian J Med Res. 2006;123(6):781–7.
72. Bednarek-Tupikowska G, Tworowska U, Jedrychowska I, Radomska B, Tupikowski K, Bidzinska-Speichert B, et al. Effects of oestradiol and oestroprogestin on erythrocyte antioxidative enzyme system activity in postmenopausal women. Clinical endocrinology. 2006;64(4):463–8.
73. Ece A, Gürkan F, Kervancioğlu M, Kocamaz H, Güneş A, Atamer Y, et al. Oxidative stress, inflammation and early cardiovascular damage in children with chronic renal failure. Pediatric nephrology (Berlin, Germany). 2006;21(4):545–52.
74. Hanta I, Kocabas A, Canacankatan N, Kuleci S, Seydaoglu G. Oxidant-antioxidant balance in patients with COPD. Lung. 2006;184(2):51–5.
75. Manoharan S, Kolanjiappan K, Suresh K, Panjamurthy K. Lipid peroxidation & antioxidants status in patients with oral squamous cell carcinoma. Indian J Med Res. 2005;122(6):529–34.
76. Agil A, Durán R, Barrero F, Morales B, Araúzo M, Alba F, et al. Plasma lipid peroxidation in sporadic Parkinson's disease. Role of the L-dopa. J Neurol Sci. 2006;240(1–2):31–6.
77. Schepens MA, Roelofs HM, Peters WH, Wanten GJ. No evidence for oxidative stress in patients on home parenteral nutrition. Clinical nutrition (Edinburgh, Scotland). 2006;25(6): 939–48.
78. Moselhy HF, Reid RG, Yousef S, Boyle SP. A specific, accurate, and sensitive measure of total plasma malondialdehyde by HPLC. J Lipid Res. 2013;54(3):852–8.
79. Cheng G-W, Wu H-L, Huang Y-L. Simultaneous determination of malondialdehyde and ofloxacin in plasma using an isocratic high-performance liquid chromatography/fluorescence detection system. Analytica chimica acta. 2008;616(2):230–4.
80. Syslová K, Kačer P, Kuzma M, Najmanová V, Fenclová Z, Vlčková Š, et al. Rapid and easy method for monitoring oxidative stress markers in body fluids of patients with asbestos or silica-induced lung diseases. Journal of Chromatography B. 2009;877(24):2477–86.
81. Sobsey CA, Han J, Lin K, Swardfager W, Levitt A, Borchers CH. Development and evaluation of a liquid chromatography–mass spectrometry method for rapid, accurate quantitation of malondialdehyde in human plasma. Journal of Chromatography B. 2016;1029–1030:205–12.
82. Malaei R, Ramezani AM, Absalan G. Analysis of malondialdehyde in human plasma samples through derivatization with 2,4-dinitrophenylhydrazine by ultrasound-assisted dispersive

liquid–liquid microextraction-GC-FID approach. Journal of Chromatography B. 2018;1089: 60–9.
83. García-Valero J, Olloquequi J, Rodríguez E, Martín-Satué M, Texidó L, Ferrer J. Decreased Expression of EC-SOD and Fibulin-5 in Alveolar Walls of Lungs From COPD Patients. Archivos de Bronconeumología. 2021.
84. Fashi A, Cheraghi M, Badiee H, Zamani A. An analytical strategy based on the combination of ultrasound assisted flat membrane liquid phase microextraction and a smartphone reader for trace determination of malondialdehyde. Talanta. 2020;209:120618.
85. Li P, Ding G, Deng Y, Punyapitak D, Li D, Cao Y. Determination of malondialdehyde in biological fluids by high-performance liquid chromatography using rhodamine B hydrazide as the derivatization reagent. Free Radical Biology and Medicine. 2013;65:224–31.
86. Zhang X, Staimer N, Gillen DL, Tjoa T, Schauer JJ, Shafer MM, et al. Associations of oxidative stress and inflammatory biomarkers with chemically-characterized air pollutant exposures in an elderly cohort. Environmental Research. 2016;150:306–19.
87. Lačná J, Foret F, Kubáň P. Sensitive determination of malondialdehyde in exhaled breath condensate and biological fluids by capillary electrophoresis with laser induced fluorescence detection. Talanta. 2017;169:85–90.
88. El-Maghrabey MH, Kishikawa N, Ohyama K, Kuroda N. Analytical method for lipoperoxidation relevant reactive aldehydes in human sera by high-performance liquid chromatography-fluorescence detection. Anal Biochem. 2014;464:36–42.
89. Goldring C, Casini AF, Maellaro E, Del Bello B, Comporti M. Determination of 4-hydroxynonenal by high-performance liquid chromatography with electrochemical detection. Lipids. 1993;28(2):141–5.
90. Zelzer S, Mangge H, Oberreither R, Bernecker C, Gruber HJ, Prüller F, et al. Oxidative stress: Determination of 4-hydroxy-2-nonenal by gas chromatography/mass spectrometry in human and rat plasma. Free radical research. 2015;49(10):1233–8.
91. Weber D, Milkovic L, Bennett SJ, Griffiths HR, Zarkovic N, Grune T. Measurement of HNE-protein adducts in human plasma and serum by ELISA—Comparison of two primary antibodies. Redox Biol. 2013;1(1):226–33.
92. Ligor M, Ligor T, Gadzała-Kopciuch R, Buszewski B. The chromatographic assay of 4-hydroxynonenal as a biomarker of diseases by means of MEPS and HPLC technique. Biomedical chromatography: BMC. 2015;29(4):584–9.
93. Ozeki M, Miyagawa-Hayashino A, Akatsuka S, Shirase T, Lee WH, Uchida K, et al. Susceptibility of actin to modification by 4-hydroxy-2-nonenal. J Chromatogr B Analyt Technol Biomed Life Sci. 2005;827(1):119–26.
94. Stopforth A, Burger BV, Crouch AM, Sandra P. Urinalysis of 4-hydroxynonenal, a marker of oxidative stress, using stir bar sorptive extraction–thermal desorption–gas chromatography/mass spectrometry. Journal of Chromatography B. 2006;834(1):134–40.
95. Nourooz-Zadeh J, Gopaul NK, Barrow S, Mallet AI, Anggård EE. Analysis of F2-isoprostanes as indicators of non-enzymatic lipid peroxidation in vivo by gas chromatography-mass spectrometry: development of a solid-phase extraction procedure. Journal of chromatography B, Biomedical applications. 1995;667(2):199–208.
96. Bielecki A, Saravanabhavan G, Blais E, Vincent R, Kumarathasan P. An efficient sample preparation method for high-throughput analysis of 15(S)-8-iso-PGF2α in plasma and urine by enzyme immunoassay. Journal of analytical toxicology. 2012;36(9):595–600.
97. Basu S. Radioimmunosassay of 8-iso-prostaglandin F2α: an index for oxidative injury via free radical catalysed lipid peroxidation. Prostaglandins, Leukotrienes and Essential Fatty Acids. 1998;58(4):319–25.
98. Wang C-J, Yang N-H, Liou S-H, Lee H-L. Fast quantification of the exhaled breath condensate of oxidative stress 8-iso-prostaglandin F2α using on-line solid-phase extraction coupled with liquid chromatography/electrospray ionization mass spectrometry. Talanta. 2010;82(4): 1434–8.

99. Janicka M, Kubica P, Kot-Wasik A, Kot J, Namieśnik J. Sensitive determination of isoprostanes in exhaled breath condensate samples with use of liquid chromatography–tandem mass spectrometry. Journal of Chromatography B. 2012;893–894:144–9.
100. Hoffmeyer F, Raulf-Heimsoth M, Harth V, Bünger J, Brüning T. Comparative analysis of selected exhaled breath biomarkers obtained with two different temperature-controlled devices. BMC pulmonary medicine. 2009;9(1):48.
101. Monsé C, Hagemeyer O, van Kampen V, Raulf M, Weiss T, Menne E, et al. Human Inhalation Study with Zinc Oxide: Analysis of Zinc Levels and Biomarkers in Exhaled Breath Condensate. Advances in experimental medicine and biology. 2020.
102. Mas E, Michel F, Guy A, Bultel V, Falquet Y, Chardon P, et al. Quantification of urinary F2-isoprostanes with 4(RS)-F4t-neuroprostane as an internal standard using gas chromatography–mass spectrometry: Application to polytraumatized patients. Journal of Chromatography B. 2008;872(1):133–40.
103. Paci A, Rieutord A, Guillaume D, Traoré F, Ropenga J, Husson HP, et al. Quantitative high-performance liquid chromatographic determination of acrolein in plasma after derivatization with Luminarin® 3. Journal of Chromatography B: Biomedical Sciences and Applications. 2000;739(2):239–46.
104. Takamoto S, Sakura N, Yashiki M, Kojima T. Determination of acrolein by headspace solid-phase microextraction gas chromatography and mass spectrometry. Journal of Chromatography B: Biomedical Sciences and Applications. 2001;758(1):123–8.
105. Zhang X, Xiong W, Shi L, Hou H, Hu Q. Simultaneous determination of five mercapturic acid derived from volatile organic compounds in human urine by LC–MS/MS and its application to relationship study. Journal of Chromatography B. 2014;967:102–9.
106. Sakamoto A, Uemura T, Terui Y, Yoshida M, Fukuda K, Nakamura T, et al. Development of an ELISA for Measurement of Urinary 3-Hydroxypropyl Mercapturic Acid (3-HPMA), the Marker of Stroke. Medical sciences (Basel, Switzerland). 2020;8(3).
107. Shao B, O'Brien KD, McDonald TO, Fu X, Oram JF, Uchida K, et al. Acrolein Modifies Apolipoprotein A-I in the Human Artery Wall. 2005;1043(1):396–403.
108. Silva LIB, Freitas AC, Rocha-Santos TAP, Pereira ME, Duarte AC. Breath analysis by optical fiber sensor for the determination of exhaled organic compounds with a view to diagnostics. Talanta. 2011;83(5):1586–94.
109. Dale O, Bergum H, Lund T, Nilsen T, Aadahl P, Stenseth R. A validated method for rapid analysis of ethane in breath and its application in kinetic studies in human volunteers. Free radical research. 2003;37(8):815–21.
110. Drury JA, Nycyk JA, Cooke RWI. Pentane Measurement in Ventilated Infants Using a Commercially Available System. Free Radical Biology and Medicine. 1997;22(5):895–900.
111. Grabowska-Polanowska B, Faber J, Skowron M, Miarka P, Pietrzycka A, Śliwka I, et al. Detection of potential chronic kidney disease markers in breath using gas chromatography with mass-spectral detection coupled with thermal desorption method. Journal of Chromatography A. 2013;1301:179–89.
112. Argüelles S, Machado MJ, Ayala A, Machado A, Hervías B. Correlation between circulating biomarkers of oxidative stress of maternal and umbilical cord blood at birth. Free radical research. 2006;40(6):565–70.
113. Tejaswi G, Suchitra MM, Rajasekhar D, Kiranmayi VS, Srinivasa Rao PVLN. Myeloperoxidase, protein carbonyls and oxidative stress in coronary artery disease. Journal of Indian College of Cardiology. 2017;7(4):149–52.
114. Chang P-Y, Wu T-L, Hung C-C, Tsao K-C, Sun C-F, Wu LL, et al. Development of an ELISA for myeloperoxidase on microplate: Normal reference values and effect of temperature on specimen preparation. Clinica Chimica Acta. 2006;373(1):158–63.
115. Simavli H, Tosun M, Bucak YY, Erdurmus M, Ocak Z, Onder HI, et al. Serum and aqueous xanthine oxidase levels, and mRNA expression in anterior lens epithelial cells in pseudoexfoliation. Graefe's archive for clinical and experimental ophthalmology = Albrecht von Graefes Archiv fur klinische und experimentelle Ophthalmologie. 2015;253(7):1161–7.

116. Muñoz M, Martínez MP, López-Oliva ME, Rodríguez C, Corbacho C, Carballido J, et al. Hydrogen peroxide derived from NADPH oxidase 4- and 2 contributes to the endothelium-dependent vasodilatation of intrarenal arteries. Redox Biol. 2018;19:92–104.
117. Salles N, Szanto I, Herrmann F, Armenian B, Stumm M, Stauffer E, et al. Expression of mRNA for ROS-generating NADPH oxidases in the aging stomach. Experimental Gerontology. 2005;40(4):353–7.
118. Eun HS, Chun K, Song I-S, Oh C-H, Seong I-O, Yeo M-K, et al. High nuclear NADPH oxidase 4 expression levels are correlated with cancer development and poor prognosis in hepatocellular carcinoma. Pathology. 2019;51(6):579–85.
119. Stachniuk J, Kubalczyk P, Furmaniak P, Głowacki R. A versatile method for analysis of saliva, plasma and urine for total thiols using HPLC with UV detection. Talanta. 2016;155:70–7.
120. Cevasco G, Piątek AM, Scapolla C, Thea S. An improved method for simultaneous analysis of aminothiols in human plasma by high-performance liquid chromatography with fluorescence detection. Journal of Chromatography A. 2010;1217(14):2158–62.
121. Khan A, Khan MI, Iqbal Z, Shah Y, Ahmad L, Nazir S, et al. A new HPLC method for the simultaneous determination of ascorbic acid and aminothiols in human plasma and erythrocytes using electrochemical detection. Talanta. 2011;84(3):789–801.
122. Sutton TR, Minnion M, Barbarino F, Koster G, Fernandez BO, Cumpstey AF, et al. A robust and versatile mass spectrometry platform for comprehensive assessment of the thiol redox metabolome. Redox Biol. 2018;16:359–80.
123. Ivanov AV, Virus ED, Luzyanin BP, Kubatiev AA. Capillary electrophoresis coupled with 1,1′-thiocarbonyldiimidazole derivatization for the rapid detection of total homocysteine and cysteine in human plasma. Journal of Chromatography B. 2015;1004:30–6.
124. Li J-M, Zhang C. Value of serum OPN levels in patients with acute cerebral hemorrhage for assessment of nerve function impairment. Journal of Acute Disease. 2016;5(3):222–6.
125. Adachi T, Ohta H, Yamada H, Futenma A, Kato K, Hirano K. Quantitative analysis of extracellular-superoxide dismutase in serum and urine by ELISA with monoclonal antibody. Clinica Chimica Acta. 1992;212(3):89–102.
126. Lafont A, Marwick TH, Chisolm GM, Van Lente F, Vaska KJ, Whitlow PL. Decreased free radical scavengers with reperfusion after coronary angioplasty in patients with acute myocardial infarction. American heart journal. 1996;131(2):219–23.
127. Jacobson GA, Narkowicz C, Tong YC, Peterson GM. Plasma glutathione peroxidase by ELISA and relationship to selenium level. Clinica Chimica Acta. 2006;369(1):100–3.
128. Huang W, Åkesson B. Radioimmunoassay of glutathione peroxidase in human serum. Clinica Chimica Acta. 1993;219(1):139–48.
129. Desai PB, Manjunath S, Kadi S, Chetana K, Vanishree J. Oxidative stress and enzymatic antioxidant status in rheumatoid arthritis: a case control study. European review for medical and pharmacological sciences. 2010;14(11):959–67.
130. Fekecs T, Kádár Z, Battyáni Z, Kalmár-Nagy K, Szakály P, Horváth ÖP, et al. Changes in Oxidative Stress in Patients Screened for Skin Cancer After Solid-Organ Transplantation. Transplantation Proceedings. 2010;42(6):2336–8.
131. Das KC, White CW. Detection of thioredoxin in human serum and biological samples using a sensitive sandwich ELISA with digoxigenin-labeled antibody. Journal of Immunological Methods. 1998;211(1):9–20.
132. Cai CY, Zhai LL, Wu Y, Tang ZG. Expression and clinical value of peroxiredoxin-1 in patients with pancreatic cancer. European Journal of Surgical Oncology (EJSO). 2015;41(2):228–35.
133. Al-Masri AA, El Eter E, Tayel S, Zamil H. Differential associations of circulating peroxiredoxins levels with indicators of glycemic control in type 2 diabetes mellitus. European review for medical and pharmacological sciences. 2014;18(5):710–6.
134. Zhang Y, Ying X, Zhao Q, Ma J, Zhang D, He C, et al. Identification of Protein Expression Changes in Hepatocellular Carcinoma through iTRAQ. Dis Markers. 2020;2020:2632716.

135. Ismail S, Mayah W, Battia HE, Gaballah H, Jiman-Fatani A, Hamouda H, et al. Plasma nuclear factor kappa B and serum peroxiredoxin 3 in early diagnosis of hepatocellular carcinoma. Asian Pacific journal of cancer prevention: APJCP. 2015;16(4):1657–63.
136. Qian C, Shen Z, Zhang J, Ji F, Liu H. Identification of cholesterol gallstone with two-dimensional gel electrophoresis and mass spectrometry. Annals of Hepatology. 2015;14 (2):251–8.
137. Gateva A, Assyov Y, Velikova T, Kamenov Z. Increased peroxiredoxin 4 levels in patients with prediabetes compared to normal glucose tolerance subjects. Clinical endocrinology. 2016;85(4):551–5.
138. Fraisier C, Papa A, Almeras L. High-mobility group box-1, promising serological biomarker for the distinction of human WNV disease severity. Virus Research. 2015;195:9–12.
139. Petersen DR, Orlicky DJ, Roede JR, Shearn CT. Aberrant expression of redox regulatory proteins in patients with concomitant primary Sclerosing cholangitis/inflammatory bowel disease. Experimental and Molecular Pathology. 2018;105(1):32–6.
140. Misra R, Mangi S, Joshi S, Mittal S, Gupta SK, Pandey RM. LycoRed as an alternative to hormone replacement therapy in lowering serum lipids and oxidative stress markers: a randomized controlled clinical trial. J Obstet Gynaecol Res. 2006;32(3):299–304.
141. Attri S, Sharma N, Jahagirdar S, Thapa BR, Prasad R. Erythrocyte metabolism and antioxidant status of patients with Wilson disease with hemolytic anemia. Pediatr Res. 2006;59(4 Pt 1): 593–7.
142. Gil L, Siems W, Mazurek B, Gross J, Schroeder P, Voss P, et al. Age-associated analysis of oxidative stress parameters in human plasma and erythrocytes. Free radical research. 2006;40 (5):495–505.
143. Squellerio I, Caruso D, Porro B, Veglia F, Tremoli E, Cavalca V. Direct glutathione quantification in human blood by LC–MS/MS: comparison with HPLC with electrochemical detection. Journal of Pharmaceutical and Biomedical Analysis. 2012;71:111–8.
144. Moore T, Le A, Niemi A-K, Kwan T, Cusmano-Ozog K, Enns GM, et al. A new LC–MS/MS method for the clinical determination of reduced and oxidized glutathione from whole blood. Journal of Chromatography B. 2013;929:51–5.
145. Han J, Weisbrod RM, Shao D, Watanabe Y, Yin X, Bachschmid MM, et al. The redox mechanism for vascular barrier dysfunction associated with metabolic disorders: Glutathionylation of Rac1 in endothelial cells. Redox Biol. 2016;9:306–19.
146. Mandal A, Woodi M, Sood V, Krishnaswamy P, Rao A, Ballal S, et al. Quantitation and characterization of glutathionyl haemoglobin as an oxidative stress marker in chronic renal failure by mass spectrometry. Clinical biochemistry. 2007;40:986–94.
147. Žunić G, Spasić S. Capillary electrophoresis method optimized with a factorial design for the determination of glutathione and amino acid status using human capillary blood. Journal of Chromatography B. 2008;873(1):70–6.
148. Hodáková J, Preisler J, Foret F, Kubáň P. Sensitive determination of glutathione in biological samples by capillary electrophoresis with green (515nm) laser-induced fluorescence detection. Journal of Chromatography A. 2015;1391:102–8.
149. Boulet L, Alex B, Clavey N, Martinez J, Ducros V. Simultaneous analysis of retinol, six carotenoids, two tocopherols, and coenzyme Q10 from human plasma by HPLC. Journal of Chromatography B. 2020;1151:122158.
150. Bukowski MR, Voeller K, Jahns L. Simple and sensitive dilute-and-shoot analysis of carotenoids in human plasma. Journal of Chromatography B. 2018;1095:32–8.
151. Block G, Dietrich M, Norkus E, Jensen C, Benowitz NL, Morrow JD, et al. Intraindividual variability of plasma antioxidants, markers of oxidative stress, C-reactive protein, cotinine, and other biomarkers. Epidemiology (Cambridge, Mass). 2006;17(4):404–12.
152. Ferin R, Pavão ML, Baptista J. Rapid, sensitive and simultaneous determination of ascorbic and uric acids in human plasma by ion-exclusion HPLC-UV. Clinical Biochemistry. 2013;46 (7):665–9.

153. Iwata T, Yamaguchi M, Hara S, Nakamura M, Ohkura Y. Determination of total ascorbic acid in human serum by high-performance liquid chromatography with fluorescence detection. Journal of Chromatography B: Biomedical Sciences and Applications. 1985;344:351–5.
154. Borras E, Schrumpf L, Stephens N, Weimer BC, Davis CE, Schelegle ES. Novel LC-MS-TOF method to detect and quantify ascorbic and uric acid simultaneously in different biological matrices. Journal of Chromatography B. 2021;1168:122588.
155. Deutsch JC, Kolhouse JF. Ascorbate and dehydroascorbate measurements in aqueous solutions and plasma determined by gas chromatography-mass spectrometry. Analytical Chemistry. 1993;65(4):321–6.
156. Tao Y, Zhang X, Wang J, Wang X, Yang N. Simultaneous determination of cysteine, ascorbic acid and uric acid by capillary electrophoresis with electrochemiluminescence. Journal of Electroanalytical Chemistry. 2012;674:65–70.
157. Zinellu A, Carru C, Sotgia S, Deiana L. Optimization of ascorbic and uric acid separation in human plasma by free zone capillary electrophoresis ultraviolet detection. Analytical Biochemistry. 2004;330(2):298–305.
158. el-Hady DA, Gotti R, el-Maali NA. Selective determination of catechin, epicatechin and ascorbic acid in human urine using chiral capillary electrophoresis. Journal of separation science. 2008;31(12):2252–9.
159. Thibeault D, Su H, MacNamara E, Schipper HM. Isocratic rapid liquid chromatographic method for simultaneous determination of carotenoids, retinol, and tocopherols in human serum. Journal of Chromatography B. 2009;877(11):1077–83.
160. Wakabayashi H, Yamato S, Nakajima M, Shimada K. Simultaneous determination of oxidized and reduced coenzyme Q and alpha-tocopherol in biological samples by high performance liquid chromatography with platinum catalyst reduction and electrochemical detection. Biol Pharm Bull. 1994;17(8):997–1002.
161. Zerbinati C, Galli F, Regolanti R, Poli G, Iuliano L. Gas chromatography–mass spectrometry microanalysis of alpha- and gamma-tocopherol in plasma and whole blood. Clinica Chimica Acta. 2015;446:156–62.
162. Řimnáčová L, Hušek P, Šimek P. A new method for immediate derivatization of hydroxyl groups by fluoroalkyl chloroformates and its application for the determination of sterols and tocopherols in human serum and amniotic fluid by gas chromatography–mass spectrometry. Journal of Chromatography A. 2014;1339:154–67.
163. Yu S, Yin Y, Wang D, Xie S, Cheng Q, Cheng X, et al. Development and validation of a simple isotope dilution-liquid chromatography-tandem mass spectrometry method for detecting vitamins A and E in serum and amniotic fluid. International Journal of Mass Spectrometry. 2019;435:118–23.
164. Lang JK, Packer L. Quantitative determination of vitamin E and oxidized and reduced coenzyme Q by high-performance liquid chromatography with in-line ultraviolet and electrochemical detection. Journal of Chromatography A. 1987;385:109–17.
165. Lee BL, Ong CN. Comprehensive high-performance liquid chromatographic method for the measurements of lipophilic antioxidants in human plasma. Journal of Chromatography A. 2009;1216(15):3131–7.
166. Ruiz-Jiménez J, Priego-Capote F, Mata-Granados JM, Quesada JM, Luque de Castro MD. Determination of the ubiquinol-10 and ubiquinone-10 (coenzyme Q10) in human serum by liquid chromatography tandem mass spectrometry to evaluate the oxidative stress. Journal of Chromatography A. 2007;1175(2):242–8.
167. Sun M-J, Cheng Y-S, Liu C-S, Sun R. Changes in the PGC-1α and mtDNA copy number may play a role in the development of pelvic organ prolapse in pre-menopausal patients. Taiwanese Journal of Obstetrics and Gynecology. 2019;58(4):526–30.
168. Fattah A, Amiri F, Mohammadian M, Alipourfard I, Valilo M, Taheraghdam A, et al. Dysregulation of body antioxidant content is related to initiation and progression of Parkinson's disease. Neuroscience Letters. 2020;736:135297.

169. Marta Ż-N, Agnieszka W, Jacek P, Jeleń A, Adrian K, Dagmara S-K, et al. NFKB2 gene expression in patients with peptic ulcer diseases and gastric cancer. Molecular Biology Reports. 2020;47(3):2015–21.
170. Collins AR. Investigating oxidative DNA damage and its repair using the comet assay. Mutation Research/Reviews in Mutation Research. 2009;681(1):24–32.

Chapter 8
Spectrophotometric Methods

A. Khelfi

Abstract Spectrophotometry is one of the most used analytical techniques in clinical practice due to its accessibility, accuracy, versatility, and low cost. This technique has been widely used for the measurement of multiple oxidative biomarkers such as malondialdehyde (MDA), conjugated dienes, protein carbonyls, glutathione (GSH), protein thiols, advanced oxidation protein products (AOPP), and vitamin C. Several spectrophotometric methods based on the reaction of reactive species (RS) (e.g., nitrites, superoxide, and hydrogen peroxide) with a variety of sensors have been developed and are now available on the market. However, production of RS by the sensors themselves and perturbation of the studied systems by the probes represent a major limitation for these methods.

Here, we present multiple spectrophotometric methods that can be applied in diverse clinical settings for use in diagnosis and prognosis. We specifically consider their advantages and limitations and offer solutions in unpredictable situations.

Keywords Malondialdehyde (MDA) · Conjugated dienes · Protein carbonyls · Glutathione (GSH) · Protein thiols · Nitrites · Superoxide · Hydrogen peroxide · Advanced oxidation protein products (AOPP) · Vitamin C

8.1 Introduction to Spectroscopic Methods

The direct analysis of RS in biological samples is difficult due to their high reactivity and short half-lives.

Several spectroscopic methods based on the reaction of RS with a variety of sensors (i.e., compounds that change their light absorption and/or emission or magnetic characteristics upon reaction with RS) have been developed and are now

A. Khelfi (✉)
Faculty of Pharmacy, University of Algiers, Algiers, Algeria

National Centre of Toxicology, Algiers, Algeria

Department of Chemistry and Biomolecular Science, Clarkson University, Potsdam, NY, USA
e-mail: ab.khelfi@univ-alger.dz; akhelfi@clarkson.edu

S. Andreescu et al. (eds.), *Biomarkers of Oxidative Stress*,
https://doi.org/10.1007/978-3-031-60738-7_8

available. The use of these sensors for the detection of RS has some limitations such as low stability due to undesired reactions or formation of products that may be metabolized in cellular systems. Production of RS by the sensors themselves and perturbation of the studied systems by the probes represent additional limitations [1]. The simultaneous formation of several RS complicates the monitoring especially the discrimination between reactive oxygen species (ROS). Because of these artifacts, methods currently available for the direct assessment of RS are of limited applicability in analytical biochemistry. Therefore, it is common to measure not the RS themselves, but biomarkers of cell damage that they may emerge from the interaction between RS and the biological system. There is some justification for this approach as ultimately oxidative damage caused by RS matters rather than the total amount of RS.

Another solution to this is to perform affordable spectroscopic methods that measure redox states in cells and the whole body. For example, the redox state of heme proteins can be monitored continuously by following specific absorbance wavelengths. The redox state of proteins and other redox changes are now used in real-time redox imaging. Spectroscopic analysis of the redox state of cytosolic and mitochondrial cytochromes has become a field of interest as well.

Due to the fact that there is a rising need to specifically monitor RS production, assess redox states, and discriminate between different RS molecules, in the following, spectroscopic methodologies will be briefly summarized, with especial emphasis on limitations and advantages.

8.2 Spectrophotometric Methods

8.2.1 Malondialdehyde

Malondialdehyde is one of the most widely used and popular biomarkers to assess lipid peroxidation in the whole organism. In the last stage of the peroxidation process, lipid peroxides are decomposed to aldehydes like MDA, which can be detected by spectrophotometry. Other end products, for example, hexanal, can also be measured. All of these aldehydes are termed "thiobarbituric acid reactive substances" (TBARS) [2].

One of the most common measurements of MDA is the TBARS test. This test was introduced already several decades ago by Kohn and Liversedge [3]. Briefly, the assay is based on MDA derivatization with TBA (thiobarbituric acid) forming a TBA-MDA adduct (Fig. 8.1). The latter is a stable chromophore that can be quantified easily using spectrophotometry [4].

In order to form TBARS, a biological specimen must be heated at 70–100 °C with TBA under acidic conditions (HCl). Heating and acidic conditions are required to break the covalent bonds between proteins and MDA and therefore release polymerized forms of MDA. This assay can be performed after a previous precipitation of protein using trichloroacetic acid in total plasma [5]. After cooling, the specimen is centrifuged and absorbance of the resulting pink TBA-MDA adduct (Fig. 8.1) is

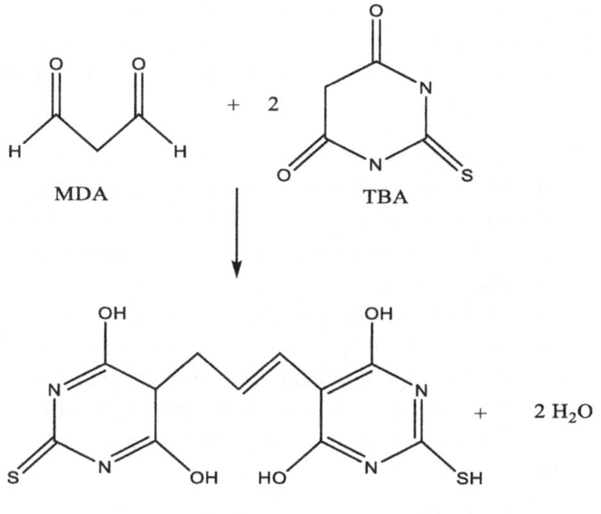

Fig. 8.1 Reaction of MDA with TBA forming a TBARS

measured at 532–535 nm, with absorbance being proportional to the concentration of MDA. The MDA-TBA adduct has a high molar absorption ($\lambda_{absorption} = 532$ nm; $\varepsilon = 153{,}000$), allowing for its measurement with conventional spectrophotometers [6].

It is noteworthy to know that MDA represents only a minor product of lipid peroxidation (less than 1%). Moreover, MDA is not exclusively formed from lipid peroxidation. It is formed only from a particular process out of a myriad of several lipid decomposition reactions [7]. Therefore, caution is required in interpreting TBARS data.

Although TBA tests are easy, rapid, and have been an important tool for assessment in in vitro systems, their use in biological systems has been heavily criticized. They are not all specific for oxidative stress (OS) and are at risk of artifacts. Indeed, it can be difficult to distinguish between TBARS formed as a result of lipid peroxidation and those formed during degradation of sugars (e.g., sucrose, deoxyglucose, and 2-desoxyribose) or free radical attack on deoxyribose and sialic acid making experimental results difficult to interpret [8]. Additionally, there are too many interfering factors to make these reliable tests, for example, the chemical reaction to detect TBARS is influenced by hemoglobin, biliverdin, sialic acid, bilirubin, ascorbate, acetaldehyde, amino acids (glutamic acid, proline, arginine, and methionine), and urea present in the sample, causing overestimation of MDA concentration [9]. Moreover, cells contain many endogenous pigments that absorb at 532 nm, resulting in artificially high values of molecular absorbance. An additional pitfall with TBA test is the interference of hemolysis that falsely increases the results of MDA content. Other interfering factors include the presence of metal ions that can increase the rate of MDA decomposition, whereas metal-chelating molecules can decrease the rate of this decomposition, making reliability a problem. Moreover, common anticoagulants used to prevent blood clotting in clinical practice (e.g., heparin and EDTA) may yield different values even on the same sample.

Another significant issue with TBA assays is that they are highly influenced by reaction conditions:

The first operational factor that can greatly affect the apparent concentration of TBARS is the iron content of reagents used for sample analysis, although the addition of EDTA for iron chelation may reduce variability. It will be more appropriate to limit this pitfall, using reagents which do not contain measurable quantities of iron.

Second, heating conditions during the assay which may further provoke lipid peroxidation leading to MDA generation (up to 98% of detected MDA can be formed by high-temperature conditions) and rapid metabolization of MDA (by aldehyde dehydrogenase to form acetyl-CoA). Heating conditions render these techniques totally artifactual and leads to the overestimation of the results [10]. Further MDA generation may be prevented by reducing the heating time. Adding an antioxidant such as butyl hydroxytoluene (BHT) is preferable to limit lipid peroxidation during sample heating. Interestingly, a derivative has been developed by R and D System Company, which reacts specifically with MDA at 40 °C, resulting in a stable chromophore ($\lambda_{absorption} = 586$ nm) and preventing the decomposition of peroxides by high temperature.

Besides reaction conditions, aldehydes can be absorbed from the diet resulting in the overestimation of MDA coming from lipid peroxidation. Indeed, some foods contain MDA–amino acid adducts that can be absorbed through the gut [11, 12]. It follows that measurements of MDA can be influenced by diet; therefore, MDA levels should not be used as an index of whole-body lipid peroxidation unless diet is also controlled.

Overall, it is scientifically unsound to equate increased levels of TBARS alone with the occurrence of FR generation in biological samples. Due to many different protocols for TBA assay, MDA levels are markedly variable in the literature. The lack of standardization of TBA assays has made comparison of TBA measurements almost impossible. Due to its low specificity, the colorimetric TBA assay is regarded as unacceptable in modern research and clinical practice, simply because several TBA-reactive compounds found in human samples are not related to lipid peroxidation. It appears prudent to consider a positive TBA test response per se as an empirical indicator of the potential occurrence of peroxidative lipid injury rather than as a measure of lipid peroxidation.

Despite the issues that may occur with TBA assay, it usually works adequately when applied to measurements of peroxidation in isolated membrane fractions such as liposomes and microsomes, or in precipitated lipoprotein fractions [13, 14]. A number of commercial assay kits are now available for these applications. It is noteworthy to mention that the colorimetric MDA assay is also used in the food industry to monitor food spoilage [15].

Finally, many of the mentioned problems associated with MDA quantitation in biological samples have been addressed by using more specific techniques like HPLC and GC-MS (see Chap. 15).

8.2.2 Lipid (hydro)peroxides

Lipid peroxidation is a complex process whereby polyunsaturated fatty acids (PUFAs) in cellular membranes undergo reactions with RS and oxygen to yield lipid hydroperoxides (LOOH). LOOH decompose subsequently to form several end products including alkenals, hydroxyalkenals, alkanals, MDA, and volatile hydrocarbons. LOOH are, therefore, considered the major initial products of lipid peroxidation. They can be measured in blood/plasma samples by numerous photometric methods. These methods are based on the oxidation of Fe^{2+} [16], iodide [17], or 10-(N-methylcarbamoyl)-3,7-(dimethylamino)-phenothiazine (MCDP) by total lipid (hydro)peroxides present in the plasma [18], resulting in oxidation products that can be quantified spectrophotometrically.

A frequently used method is based on the ability of LOOH to convert ferrous ions to ferric ions, which form with thiocyanate a red complex ($FeSCN^{2+}$) absorbing at λ max = 500 nm and depending on ligand concentration [19]. Lipid extraction is highly recommended, since amino acid hydroperoxides, endogenous ferric ions, and H_2O_2 may be a source of error. This assay is also influenced by iron-complexing agents (e.g., phenolic compounds), which may drastically change the Fe(II)-Fe(III) oxidation potential.

Monitoring of LOOH has also been achieved by the so-called ferrous oxidation-xylenol orange (FOX) assay [16]. FOX assay uses Fenton chemistry to generate $OH^{\bullet}$ from the O–O homolytic cleavage in LOOH, which decolorizes the xylenol orange dye. Unfortunately, H_2O_2 is capable of responding to FOX assay [20]. Vassalle et al. performed an adaptation of the colorimetric assay for LOOH on a Synchron CX9 Pro automated analyzer (Beckman Coulter) [21], and a colorimetric assay (d-ROM) was obtained from Diacron (Grosseto). This assay is based on the capacity of transition metals to catalyze the conversion of hydroperoxides into alkoxy and peroxy radicals, which react with an amine compound (N, N-diethyl-para-phenylenediamine) and form a colored complex that can be measured at 520 nm.

A similar method, called iodometric assay, utilizes the ability of hydroperoxides to convert iodide (e.g., potassium iodide) to triiodide (I_3^-) in acidic solutions. Triiodide is a strong chromophore and can easily be measured spectrophotometrically [17]:

$$LOOH + 2H^+ + 3I^- \rightarrow LOH + H_2O + I_3^- \quad (8.1)$$

Triodide can also be titrated with standard thiosulfate solution using starch as the end-point indicator. Unfortunately, many other substances including ascorbate, acetone, and oxygen can interfere with the iodine assay. To avoid non-specific reactions, lipids are first extracted before the addition of iodide solution. Enzymatic hydrolysis of triglycerides, cholesterol esters, and phospholipids is often required before extraction with ethyl acetate solvent. Another variant has been developed where the specimen of interest is extracted with methanol:chloroform. After incubation with potassium iodide, chloroform:acetic acid mixture (2:3 v/v) is used, followed by the addition of 0.25% cadmium acetate. After centrifugation, the upper layer can be analyzed at 373 nm. The iodometric assay needs careful precaution to prevent auto-oxidation by air of the excess unreacted iodide [22].

Iodometric assay like the FOX assay gives rise to high levels of peroxides in plasma and can be useful for measurements in isolated LDL [23]. The results can be influenced by lipid peroxides coming from dietary fat, especially those coming from PUFAs used for cooking.

Some commercially available assay kits rely on the reaction of LOOH with a heme compound, concomitantly oxidizing a precursor to produce methylene blue which can be measured spectrophotometrically. Although simple, quick, and sensitive (picomolar range), results do not correlate well with other measures of lipid peroxidation in human plasma. Treatment with reducing agents like sodium borohydride eliminates the chemiluminescent signal.

8.2.3 Conjugated Dienes

Lipid peroxidation of unsaturated lipids results during its early steps in the formation of conjugated dienes. These compounds are characterized by a double–single–double bond and are represented by endoperoxides and hydroperoxides, which possess this conjugated diene structure. As primary products of lipid oxidation (see Chap. 3), conjugated dienes would be important substances to measure. Since the diene remains after the associated peroxide has degraded, quantitation of conjugated diene negates one of the criticisms of peroxide quantitation.

The conjugated diene moiety of LOOH is a strong chromophore that can be detected spectrophotometrically as it absorbs readily in the ultraviolet (UV) range (215–250 nm). Measurement of this absorbance is extremely useful as it represents an index of peroxidation in pure lipid systems and in tissue preparations. Moreover, signal intensities are proportional to conjugated diene concentrations.

Conjugated dienes are easy to measure in body fluids after solvent extraction. Briefly, membrane lipids or human plasma is extracted with a mixture of methanol: chloroform. The lower layer (organic layer) is usually dried, and the residue remaining after reconstructing with cyclohexane can be used for absorbance measurement at 234 nm against cyclohexane blank [24]. This assay has also been applied on tissue extracts, making it a potentially important tool in OS research [25]. A noteworthy application is the quantitation of baseline diene conjugation in circulating LDLs [26].

There are, however, limitations in measuring conjugated dienes in biological samples because many of other endogen compounds (e.g., PUFAs themselves, heme proteins, carbonyl compounds, purines, and pyrimidines) absorb strongly in the same wavelengths and create a high background, even if this is partly eliminated by extraction using chloroform:methanol solvents. Another issue is that the absorption maximum is not intense and superimpose on the end absorption of lipid extract. The conjugated diene absorbance appears as a shoulder on the PUFA absorbance spectrum, which makes the height of the absorption maximum often difficult to measure [27]. Because of the non-specificity and non-sensitivity of the assay, several methods have been proposed to circumvent these drawbacks, for example, second derivative spectrophotometry, based on the differential of the first derivative spectrum, which gives greater resolution power [28]. The second derivative spectrophotometry allows for higher sensitivity, since the conjugated diene shoulder appearing in ordinary spectra translates into a sharp minimum peak that is more easy to measure. The increased resolution of spectra may allow for discrimination between different conjugated diene structures present. However, even under these conditions, interpretation of high absorbance at 234 nm is difficult [29], compared to the common application of the test to monitor peroxidation of lipoproteins and pure lipids [30]. It seems that most of the absorbing materials consists of non-oxygen-containing isomers of linoleic acid (octadeca-9(cis), 11-dienoic acid) that may not arise from lipid peroxidation [31]. Conjugated dienes from dietary sources like milk products can also obfuscate measurements. In human fluids, molecules taken from the diet and acquired from bacteria such as octadeca-9(cis), 11-dienoic acid interfere with conjugated diene values [9]. Thus, the application of simple diene conjugation methods to plasma or extracts is a questionable measurement of lipid peroxidation and usually leads to an overestimation of conjugated diene levels. On the other hand, conjugated diene compounds readily break down to other products which may underestimates the total amount in the analyzed samples, which requires a short sample analysis time.

Overall, since many molecules absorb at 234 nm, a technical problem emerges with the spectrophotometric measurement of conjugated dienes and results in a lack of specificity with tendency to give artificially high values. The presence of conjugated dienes does not necessarily imply that hydroperoxides exist in samples. Thus, as a general suggestion, it is recommended that attention should be paid to interpreting their presence as a marker of ongoing lipid peroxidation in biological materials, unless additional and more specific analyses are performed. Consequently, conjugated dienes are frequently measured by HPLC-UV as a biomarker of lipid peroxidation (see Chap. 15).

8.2.4 Protein Carbonyls

Generation of carbonyl molecules is one of the most widely used markers of protein oxidation in both in vitro and in vivo studies. Protein carbonyls result from the reaction of RS with amino groups that leads to the appearance of aldehyde and ketone moieties on proteins (see Chap. 3). The chemical stability of protein carbonyls makes them suitable for laboratory measurements. Their storage for almost 10 years at −80 °C allows for several sample analyses.

Protein carbonylation is often quantified in plasma samples by detection of carbonyl moieties after derivatization with diphenylhydrazine. Quantification can be achieved by direct spectrophotometric determination at 360 nm [32, 33]. However, simple detection techniques measuring optical absorption is subject to significant interference and are, therefore, not suitable for applications on biological fluids and tissues. Another problem with spectrophotometric determination of protein carbonyls is that it does not distinguish well between oxidation and protein glycation from other processes.

Other popular reagents can be used such as 2,4-dinitrophenylhydrazine (DNPH) and N′-aminooxymethylcarbonylhydrazine which have a hydrazine-like moiety and react specifically with carbonyl groups, producing stable hydrazones (N′--aminooxymethylcarbonylhydrazone and 2,4,-dinitrophenyl(DNP)hydrazone, respectively) that can be measured spectrophotometrically (characteristic absorption at 365–375 nm). During DNPH test, plasma is first treated with DNPH and then deproteinized using trichloroacetic acid.

The pellet is washed with ethyl acetate:ethanol and finally solubilized in guanidine hydrochloride. The resulting yellow product, DNPhydrazone, is measured at 370 nm. The results are expressed in nmol/mg of protein by referring to BSA (bovine serum albumin) measured at 280 nm [34]. Other chemicals have been used for carbonyl detection such as tritiated sodium borohydride and biotin-containing probes.

By using these assays, researchers have demonstrated that the body content of protein carbonyls increases with age [35] and in pathological conditions like Alzheimer's disease [36].

Carbonyl assays, like the TBA assay, have also been heavily criticized for its lack of reproducibility and specificity:

- First, protein carbonyls themselves have been questioned, whether or not they are specific markers of OS. Carbonyl moieties are formed not only as a consequence of OS but also as a result of enzymatic and non-enzymatic protein glycosylation [37]. Carbonyl moieties are also found in lipids and nucleic acids; therefore, protein carbonyls must first be extracted out using complex extraction procedures. Indeed, DNA materials that are not completely extracted from analyzed samples can cause erroneously high carbonyl measurements [38].
- Second, carbonyl assays are greatly influenced by reaction conditions. The drawbacks are being work-intensive and time-consuming. Carbonyl values can be influenced by the timing of DNPH addition to the reactive mixture, as well as by the time separating the reaction to the measurements [39]. Other experimental pitfalls are loss of acid soluble proteins, which often occur during the washing of unbound DNPH from the protein pellet. Indeed, most of the assays require washing steps to eliminate free DNPH and 10 mg protein weight for optimal

results [40]. In some circumstances, an additional wash using acidified acetone avoids the interference due to high amounts of chromophore (myoglobin, hemoglobin, and retinoids) and lipids when applied to tissue extracts [41]. Thus, due to complex procedures with numerous washing steps, these assays are not convenient for high throughput measurements.

- The most troubling aspect of the carbonyl assay is its lack of reproducibility that has been reported in the literature [38]. Studies that reported increases in protein carbonyls with age and in neuron diseases have not been reproduced [38, 39]. Cao et al. stated that carbonyl assays are unreliable and irreproducible due to the inability to completely rid samples of interfering nucleic acids and free DNPH. The problems of reproducibility have led some researchers to seriously question the validity of the carbonyl assay [38].

Finally, the difficulties encountered with the spectrophotometric methods can be assessed by the reaction with DNPH and application of fluorescence or HPLC procedures (see Chap. 15). Mono- and polyclonal antibodies against carbonyl-modified proteins are actually available and can be used for ELISA and Western blot analysis (see Chaps. 16 and 21).

8.2.5 GSH/GSSH

GSH is the major low molecular weight thiol found in mammalian cells. The evaluation of OS is better done by measuring the ratio between its reduced and oxidized forms. This ratio provides a reliable indicator of the redox potential in cells. It can be determined spectrophotometrically in plasma and tissue extracts. Numerous methods are available. The choice depends on sample nature, the equipment available, and whether determination of oxidized glutathione (GSSG) is required.

Sample Preparation

Unfortunately, reduced glutathione is very unstable; therefore, samples must be prepared few minutes after blood collection. They should be properly treated to prevent auto-oxidation after collection. Because the GSH/GSSG ratio is greater than 100:1, even a tiny amount of GSH oxidation to GSSG can underestimate its redox state. Moreover, GSH is reactive at neutral and basic pH resulting in crosslinks with other thiols. To prevent GSH oxidation, samples should be acidified immediately after sample collection. Acids like m-phosphoric acid, perchloric acid, 5-sulfosalicylic acid, and picric acid are commonly used for this purpose [42]. Chelating agents like ethylenediamine tetra-acetic acid can also be used to limit GSH oxidation. Acidification is also important to minimize reduction of GSSG by glutathione reductase (GR) and reduce transpeptidase activity, which cleaves the γ-glutamyl peptide bond of GSH. Blood samples also have to be stabilized quickly by deproteinization and freezing at low temperatures, or directly by adding the thiol-reagent in the sampling tube. Samples that are treated with acids and properly stored (−80 °C) could remain stable for months [42].

Spectrophotometric Assay

The glutathione content can be determined photometrically by reacting GSH with Ellman's reagent, (5,5′-dithiobis-(2-nitrobenzoic acid); DTNB), which results in its reduction to a yellow colored 5-thio-2-nitrobenzoic acid (TNB). The reaction of DTNB with GSH is shown in Eqs. 8.2 and 8.3:

$$\text{GSH} + \text{DTNB} \rightarrow \text{GSTNB} + \text{TNB} \quad (8.2)$$

$$\text{GSH} + \text{GSTNB} \rightarrow \text{GSSG} + \text{TNB} \quad (8.3)$$

Total glutathione can be determined in the presence of GR. The oxidized glutathione (GSSG) is recycled by GR and NADPH, which further reacts with DTNB to produce TNB, which can be analyzed spectrophotometrically at 412 nm. TNB has an extremely high extinction coefficient ($\varepsilon = 13{,}600$ at 412 nm) [43]. NADPH is used as the reducing equivalent. The probable reaction mechanism is as follows:

$$\text{GSSG} + \text{NADPH} + \text{H}^+ \xrightarrow{\text{GSSG reductase}} 2\text{GSH} + \text{NADP}^+ \quad (8.4)$$

$$\text{GSTNB} + \text{NADPH} + \text{H}^+ \xrightarrow{\text{GSSG reductase}} \text{GSH} + \text{TNBH} + \text{NADP}^+ \quad (8.5)$$

The rate of color formation provides a quantitative estimation of total glutathione content (GSH and GSSG) in samples. High specificity of this assay is primarily due to the specificity of GR for GSH [44].

Two sets of samples must be analyzed to obtain both GSH and GSSG levels. After total glutathione estimation, GSH is chemically neutralized using derivatization with either N-ethylmaleimide (NEM) or 2-vinylpyridine (2-VP) [45], leaving only GSSG to be determined. GSH values are obtained by subtraction of GSSG levels from total glutathione values. The sensitivity of the assay is around 100 picomoles. Some questions remain on whether NEM or 2-VP is better to neutralize GSH. NEM reacts quickly with GSH (1–2 min), thereby making it unlikely for GSH to auto-oxidize to GSSG during derivatization. However, excess NEM inhibits GR; therefore, it must be removed by either column separation (Sep-Pak C18 cartridge) or solvent extraction, both of which are somewhat laborious procedures. On the other hand, 2-VP does not need to be removed and does not interfere with GR activity [45]. However, the derivatization process of 2-VP takes a couple of hours, thus providing a greater likelihood of GSH oxidation to GSSG. This has led some authors to suggest that 2-VP gives artificially high GSSG values and recommend NEM instead [9].

DTNB reacts with all thiol groups, including protein thiols and small molecular weight thiols; therefore, protein content of samples must first be separated before GSH levels can be measured. Thiol measurements of non-protein fractions are often used as an estimation of GSH levels in biological samples. However, these measurements always overestimate GSH levels.

Finally, many other techniques can measure glutathione with high specificity including ion exchange chromatography, HPLC (see Chap. 15), and enzymatic methods (see Chap. 14).

8.2.6 *Protein Thiols*

Most intracellular proteins are found in their reduced state. During OS, protein sulfhydryl groups (protein thiols) are oxidized to their disulfide forms. A common approach used to assess protein oxidation would be the monitoring of the disappearance of protein thiols or the determination of the thiol/disulfide protein redox status. This status is difficult to directly estimate and, in most cases, investigators only measure protein thiol levels. Protein thiols in biological samples can be determined with the general thiol reagent, DTNB, as previously discussed in Sect. 8.2.5 (Eq. 8.6) [46].

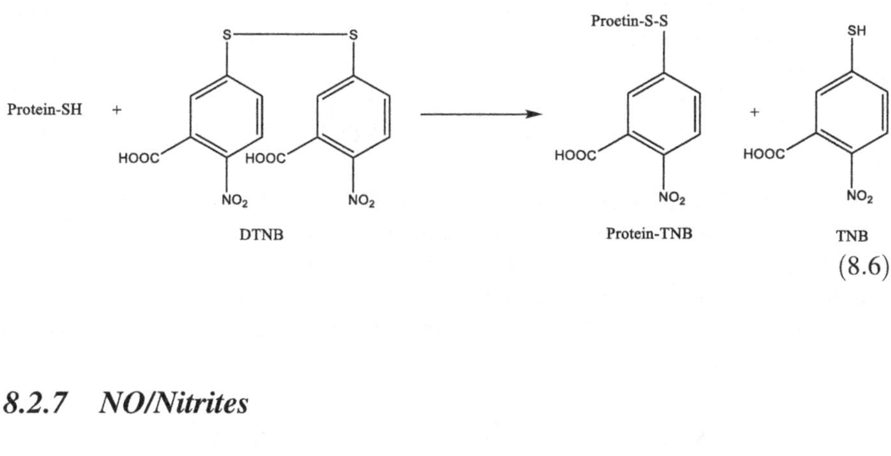

(8.6)

8.2.7 *NO/Nitrites*

8.2.7.1 Griess Method

$NO^{\bullet}$ is an extremely difficult RS to measure due to its short half-life and a very low concentration in biological samples. It is readily metabolized into the stable end products NO_2^- and NO_3^- in most body fluids including plasma.

In routine practice, a simple spectrophotometric method is used to determine its oxidation end products, NO_2^- and NO_3^-, as an indirect measurement of nitric oxide production. It is a two-step diazotization reaction known as "Griess assay," which first has been used for the assessment of NO_2^- and NO_3^- in freshwater or saltwater and later has been adapted of the determination of these derivatives in physiological fluids.

In the first step, NO_2^- turns into nitrous acid in the presence of hydrogen (Eq. 8.7) and reacts with sulfanilic acid/sulfanilamide to produce a diazonium ion. During the second step, the resulting product is coupled with α-naphthylamine/N-(1-naphthlyl)-

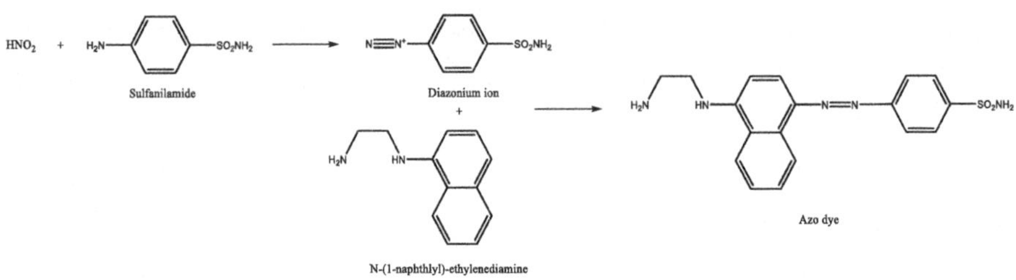

Fig. 8.2 The Griess reaction. The nitrosating agent N_2O_3 generated from acidified HNO_2 (or from the auto-oxidation of NO) reacts with sulfanilamide to form a diazonium derivative. This intermediate interacts with N-(1-naphthlyl)-ethylenediamine to yield a colored diazo product

ethylenediamine in the presence of phosphoric acid to form a red-pink azo dye (Fig. 8.2). The reaction can be monitored at 548 nm in conventional spectrophotometers and can be adapted for a microplate reader using a 550-nm filter. The value of absorbance is proportional to the amount of NO_2^- in the sample. Standard curves are generated using $NaNO_2^-$. Deproteination of samples improves the measurement by removing the turbidity caused by precipitations observed in protein-rich specimens.

$$NO_2^- + H^+ \rightarrow HNO_2 \tag{8.7}$$

Griess diazotization reaction is specific for NO_2^- but does not detect NO_3^-. In systems where NO_3^- is present, prior reduction of NO_3^- to NO_2^- is required to measure the total $NO^•$ content by the treatment of samples with sodium formate, cadmium, or NADH dependent-nitrate reductase (Eq. 8.8) [47]. Nitrate reductase from *Aspergillus* is highly efficient at reducing very small amounts of NO_3^- to NO_2^-.

$$NO_3^- + NADPH \xrightarrow{\text{Nitrate reductase}} NO_2^- + NADP^+ \tag{8.8}$$

Because NADH/NADPH interferes with the subsequent reactions, excess of NADH is removed by lactate dehydrogenase (LDH)-catalyzed reaction with pyruvate [48]. An alternative way for removing any unreacted NADPH is to replace the LDH/pyruvate system with 1 mM potassium ferricyanide [49]. Nitrate levels are calculated by the subtraction of nitrite levels from the total reactive content.

Griess assay is highly sensitive and provide a large linear range for routine analysis of $NO^•$ [50]. The detection limit is between 100 nM and 1 μM [51]. This assay is inexpensive and easy to perform. The reagents are accessible, and numerous commercial kits are available for this assay.

Other methods have been developed for the specific evaluation of nitric oxide levels in biological samples including HPLC (see Chap. 15), capillary electrophoresis (see Chap. 21), and anion-exchange chromatography (see Chap. 15).

8.2.7.2 Hemoglobin Oxidation Method

$NO^{\bullet}$ can be measured by using oxyhemoglobin as illustrated in reaction 8.9. Oxidation of hemoglobin to methemoglobin can be detected spectrophotometrically with a detection threshold of 1 nmol.

$$NO + Hb(Fe^{2+})O_2 \rightarrow Hb(Fe^{3+}) + NO_3^- \tag{8.9}$$

When there is concern that $O_2^{\bullet-}$ may compete with $NO^{\bullet}$, addition of superoxide dismutase (SOD) will increase the specificity of the assay and limits $O_2^{\bullet-}$dependent redox cycling of hemoglobin [52]. An alternative method is to simultaneously assess $NO^{\bullet}$ and $O_2^{\bullet-}$ using both metHb formation and cytochrome c reduction [53].

Several methods have been proposed to determine the rate of NO-dependent formation of metHb. Among the most sensitive ones is dual-wavelength spectroscopy, yielding a detection limit of 1 nmol/L. The change in concentration of metHb is represented by the following formula:

$$\Delta CmetHb = \frac{\Delta(\Delta A401 - 410)}{\Delta\varepsilon 401 - 410(metHb - oxyHb)}$$

Appropriate controls should be used to ensure that metHb formation is only attributable to $NO^{\bullet}$. In cases of tissue extracts, NOS inhibitors should prevent metHb production. The pH of incubation should also remain constant during measurements for two main reasons. First, the absorption spectrum of metHb is pH-dependent. Second, the experimental conditions under which $NO^{\bullet}$ is formed may be altered by pH fluctuations.

8.2.7.3 S-Nitrosothiols

Determination of S-nitrothiol (RSNO) levels often employs Saville reaction, which involves the displacement of the nitrosonium ion (NO^+) by mercury salts (Fig. 8.3)

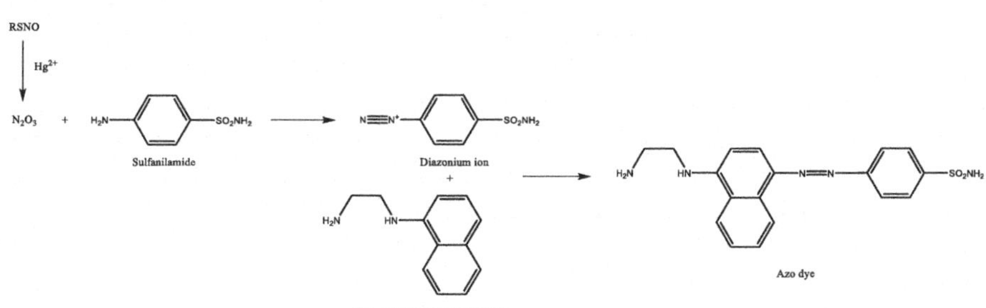

Fig. 8.3 Detection of S-nitrosothiols by Saville reaction
Formation of N_2O_3 from the interaction between RSNO and $HgCl_2$. The presence of the Griess reagents results in the formation of the same diazonium as described earlier

[54]. This reaction is followed by Griess reaction to quantify the NO_2^- formed from the hydrolysis of NO^+. The colorimetric reaction uses the same chemistry as described previously, in which NO^+-derived N_2O_3 generated by the interaction between RSNO and mercuric chloride reacts with sulfanilamide to form a diazonium ion (Fig. 8.3). The resulting diazonium then reacts with N-(1-naphthlyl)-ethylenediamine to form the colored azo complex. The reported detection range was 0.5–100 μM [55].

The reactions may be conducted at neutral rather than acidic pH, which removes the interference of contaminating NO_2^- and allows for the specific measurement of the RSNO-derived nitrosating agents.

Variations of this assay have been developed to quantify low and high molecular weight RSNOs in human plasma as well as S-nitrosated derivatives of human hemoglobin [56].

8.2.8 $O_2^{\bullet -}$

8.2.8.1 Cytochrome c Reduction Method

Several UV-Vis spectrophotometric methods were developed for $O_2^{\bullet -}$ detection based on one-electron transfer reactions due to the high rate constants observed for this type of reaction.

Reduction of ferricytochrome c^{3+} to ferrocytochrome c^{2+} has been widely used to measure $O_2^{\bullet -}$ generated in isolated enzymes, tissue extracts, and living cells (e.g., neutrophils) [57]. The method is based on the following reaction:

$$Fe^{3+}\text{-cyt c} + O_2^{\bullet -} \rightarrow Fe^{2+}\text{-cyt c} + O_2 \quad (8.10)$$

The reaction is followed kinetically by continuously recording the increase in absorbance at 550 nm. The extinction coefficient of cytochrome c is estimated to be 19.1 10^3 M^{-1} cm^{-1}. Because of the inability of cytochrome c to penetrate cells due to its charged character and large size, the assay can be used only to measure extracellular $O_2^{\bullet -}$. Since the intracellular access of cytochrome c is restricted, it is incapable of reporting low concentrations of $O_2^{\bullet -}$ generated in the cytoplasm by organelles in intact cells [58]. Therefore, cytochrome c reduction is a time-honored and accurate method for monitoring large amounts of $O_2^{\bullet -}$ released by cells into the extracellular space. In many studies, this assay was applied to detect $O_2^{\bullet -}$ levels. For instance, a temporal relationship between $O_2^{\bullet -}$ production and the contractile activity of muscles was found in perfused rat diaphragm. Moreover, the rate of cytochrome c reduction was subject to fatigue conditions, suggesting the link between $O_2^{\bullet -}$ and muscle activity [59].

Nevertheless, there are several precautions when using this reaction:

- First, cytochrome c assay is not absolutely specific for $O_2^{\bullet-}$. Tissue extracts contain compounds such as GSH and ascorbate that can reduce cytochrome c as well as cellular reductases that catalyze cytochrome c reduction. In addition, enzymes such as xanthine oxidase (XO) are capable of reducing quinones, which reduced forms are able to directly reduce cytochrome *c*. For this reason, measurements have to be performed in the presence and absence of exogenous SOD to allow for the estimation of the SOD-sensitive part of cytochrome c reduction, which is specifically mediated by $O_2^{\bullet-}$ [58]. As an application, $O_2^{\bullet-}$ generated from NAD(P)H oxidase (NOX) has been detected through the SOD-inhibiting rate of cytochrome c reduction [60]. The use of SOD-cytochrome c assay in vascular tissues would be helpful for evaluating effective strategies to prevent vascular complications in hypertension [61].
- Second, reduced cytochrome c can be oxidized again by cytochrome oxidases, peroxidases, and a variety of oxidants (e.g., H_2O_2 and $ONOO^-$) leading to an underestimation of superoxide formation [62]. This diminishes apparent rates of cytochrome c reduction since the rate constant of this reaction ($\sim 10^5\ M^{-1}\ s^{-1}$) is low. Inhibitors of cytochrome oxidase (10 μmol/L CN-) or scavengers of RS (100 IU/mL CAT for H_2O_2 and 10 mmol/L urate for $ONOO^-$) can be added to limit this pitfall.
- Third, to enhance specificity of the assay for $O_2^{\bullet-}$, cytochrome c can be acetylated. Acetylation of lysine residues within cytochrome c decreases electron transfer by microsomal and mitochondrial reductases to cytochrome c and from cytochrome c to cytochrome oxidase while maintaining the ability of $O_2^{\bullet-}$ to reduce cytochrome c. Succinoylation of cytochrome c is more effective in diminishing reduction of cytochrome c by CYP450 reductase or cytochrome b5 as well as oxidation by cytochrome c oxidase. However, the rate constant for reduction of succinoylated cytochrome c by $O_2^{\bullet-}$ is only 10% of native cytochrome c, requiring, therefore, higher concentrations of succinoylated cytochrome c in the system [63]. As mentioned before, improved specificity of the cytochrome c assay for $O_2^{\bullet-}$ can be achieved by measuring SOD-inhibitable cytochrome c reduction. The level of $O_2^{\bullet-}$ is reflected indirectly by the extent of inhibition of cytochrome c reduction by exogenous SOD.
- Fourth, the restricted intracellular access of cytochrome c due to charge and size considerations results in the underestimation of the rate of reaction when attempting to measure rates of $O_2^{\bullet-}$ formation within intact cells, as opposed to cellular or tissue extracts. Therefore, this assay cannot be used to detect $O_2^{\bullet-}$ released by the mitochondria in the cytoplasm of intact cells.
- Fifth, cytochrome c reduction is the "gold standard" for quantifying $O_2^{\bullet-}$ released by isolated enzymes or during the respiratory burst of neutrophils; however, its relative insensitivity to detect low rates of $O_2^{\bullet-}$ production limits its applicability for in vivo detection of this RS. Moreover, in in vitro experiments, it is difficult to detect the small quantities of $O_2^{\bullet-}$ generated by non-phagocytic cells such as endothelial cells and vascular smooth muscle cells [61].

- Sixth, reactions of cytochrome c with H_2O_2 and various commonly used pharmacological agents such as apocynin, oxypurinol, and L-NAME can attenuate its reduction and cause artifacts in $O_2^{\bullet-}$ detection.

8.2.8.2 Nitro-blue Tetrazolium Solution

Tetrazolium salts, especially p-nitrotetrazolium blue (NBT), are widely used for the spectrophotometric detection of $O_2^{\bullet-}$ in biological samples.

NBT-based methods use the reduction of NBT by $O_2^{\bullet-}$ to form a colored monoformazan, which formation can be measured spectrophotometrically at 550–560 nm.

These assays are inexpensive and readily available and consist of two-step process. The first step occurs through a one-electron process to form an intermediate NBT radical ($NBT^{\bullet+}$):

$$NBT^{2+} + O_2^{\bullet -} \rightarrow NB\dot{T}^{+} + O_2 \quad (8.11)$$

NBT radical undergoes a dismutation reaction with another NBT^{+} or accepts a second electron yielding the blue–black monoformazan ($NBTH^{+}$):

$$NB\dot{T}^{+} + NB\dot{T}^{+} + H^{+} \rightarrow NBT^{2+} + NBTH^{+} \quad (8.12)$$

Or

$$NB\dot{T}^{+} + O_2^{\bullet -} + H^{+} \rightarrow NBTH^{+} + O_2 \quad (8.13)$$

The NBT assay reflects the superoxide generating activity in the cytoplasm of cells. It serves as a good preliminary assessment for infertile men with suspected high ROS as the incriminated etiology. This assay has been used to detect $O_2^{\bullet-}$ produced by macrophages and monocytes [64], although low levels of $O_2^{\bullet-}$ were out of its detection limit. Moreover, it has been applied for the diagnosis of CGD (chronic granulomatous disease), a defect of peripheral granulocytes in killing certain bacteria. The diagnosis is based on the diminished ability of the reduction of NBT to blue formazan among patients with CGD [65, 66]. Nevertheless, NBT assay has been criticized for its capacity to generate artifactually $O_2^{\bullet-}$ under aerobic conditions [67]. Indeed, numerous other substances including several tissue reductases can donate an electron to NBT^{2+}, forming the NBT radical, which can react with O_2 to generate $O_2^{\bullet-}$:

$$NBT^{2+} + e^{-} \rightarrow NB\dot{T}^{+} \quad (8.14)$$

$$NB\dot{T}^{+} + O_2 \rightarrow NBT^{2+} + O_2^{\bullet -} \tag{8.15}$$

Thus, the artifactually formed $O_2^{\bullet-}$ can then generate the NBT monoformazan. Since also NBT reduction occurs independently from $O_2^{\bullet-}$, only the SOD-sensitive part can be used to measure specifically $O_2^{\bullet-}$.

Besides, formazans are hardly water soluble and precipitate, resulting in some limited identification of radical-generating cells in tissues. This drawback causes irreproducibility in absorbance readings as well. Consequently, NBT should not be considered as a first choice to measure $O_2^{\bullet-}$. One solution that could be used is the solubilization of formazans with aqueous DMSO + NaOH to enable colorimetric determination at 620 nm using a microplate reader [68].

To conclude, despite the fact that the classic NBT assay is commonly used in many diseases, its sensitivity is challenged by new types of dyes. Considering the susceptibility of NBT to intracellular reductases, the assay is not specific enough for a single radical detection. For these reasons, measurements of $O_2^{\bullet-}$ in biological samples should not rely exclusively on NBT assay but should be supplemented with an independent measure of $O_2^{\bullet-}$.

8.2.8.3 Adrenochrome Formation

Linear rates of oxidation of epinephrine to adrenochrome can be mediated by $O_2^{\bullet-}$ and, therefore, can be used to monitor this radical spectrophotometrically at 485 to 575 nm. This can be achieved using a dual-wavelength spectrophotometer or a single wavelength spectrophotometer at 480 nm. Because of a limited specificity for $O_2^{\bullet-}$ due to epinephrine auto-oxidation, it is imperative to measure SOD-inhibitable rates of adrenochrome formation [63].

8.2.9 Hypochlorous Acid

Hypochlorous acid can be detected by trapping this RS using taurine compound (Eq. 8.16). Taurine chloramine can then be spectrophotometrically measured using 5-thio-2-nitrobenzoic acid (TNB) [7].

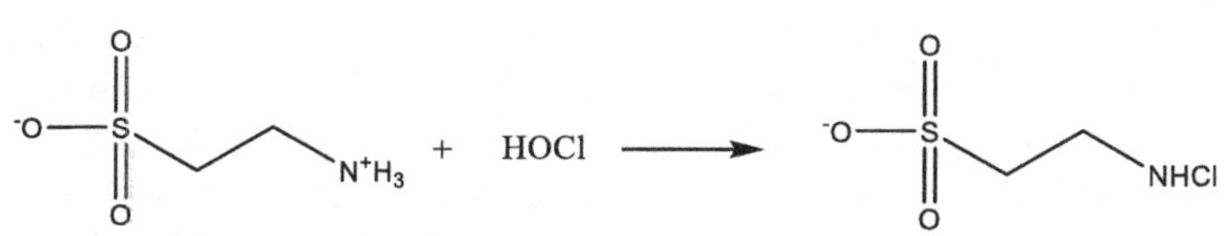

(8.16)

Taurine — Taurine chloramine

This method has some limitations such as the need to make predetermination of chloramine concentration for accurate measurements, light sensitivity of TNB, and poor selectivity since other oxidants can react with TNB. Iodide was proposed as an alternative to TNB. In this reaction, TMB (3,3′,5,5′-tetramethylbenzidine) is used as chromophore due to its ability to be oxidized by hypoiodous acid (HIO) with a sensitivity limit of 1 μM of taurine chloramine. Dihydrorhodamine was also used as chromophore giving ten-fold higher sensitivity than TMB [69].

8.2.10 H_2O_2

3,3′Diaminobenzidine (DAB) is a colorimetric reagent designed for in situ imaging of H_2O_2. DAB reacts with this species in the presence of peroxidase to form an insoluble brown precipitate, which can be microscopically imaged at tissue and cellular levels [70]. The main advantage of this sensor over others is the stability of the resulting polymeric products even under temperature and illumination changes [71]. The resulting product can be accurately located in cells, facilitating the in situ detection of H_2O_2 due to its insolubility and stability. A group of researchers have used DAB sensor during a photodynamic therapy in cells in order to locate ROS production sites with a cellular resolution at submicrometric level [71].

The DAB assay is inexpensive and easy to perform with no requirements on lab working conditions. Thus, it is reasonably proposed as a great complement for other techniques aiming at sub-cellular localization of ROS production.

Another reagent has been proposed to detect H_2O_2 spectrophotometrically. The method is based on the oxidation capability of H_2O_2 on ferrous iron to form its ferric state. Addition of thiocyanate (SCN^-) to the ferric ions yields a red complex absorbing in the 450–480 nm range. This assays has been used to estimate H_2O_2 formation in hepatic microsomes, endothelial cells, and fibroblasts [63].

8.2.11 Advanced Oxidation Protein Products (AOPPs)

The reaction of proteins with chlorinated oxidants like HOCl results in the chlorination of amino acid residues and the formation of 3-Cl-Tyr and 3,5- Cl-Tyr as main products. These oxidation products are generally considered as advanced oxidation protein products (AOPPs) and represent the protein aggregation levels generated by tyrosine cross-linking with di-tyrosine moieties and/or disulfide bridges. AOPPs have high molecular mass, limiting their clearance and seem to be a marker of highly oxidized proteins. Colorimetric tests were proposed for their quantification in plasma using a chloramine standard or human serum albumin derivatives [72].

AOPPs can easily be applied in human diseases or experimental models in order to highlight protein oxidation due to neutrophil-derived oxidant reaction (HOCl) or advanced Maillard reaction.

8.2.12 Vitamin C

The colorimetric assay for ascorbic acid determination is based on the oxidation of ascorbic acid by copper to form dehydroascorbic and 2,3-diketogulonic acids. These products react in sulfuric medium with DNPH to form DNPhydrazones, which absorb at 520 nm [73].

Another colorimetric assay was developed based on the reaction of ascorbate with 2,6-dichlorophenolindophenol. Although the assay is fairly sensitive, there is a good chance of non-specific reactions in biological systems. This assay should be limited to simple systems since Baker et al. have found a 12% higher value in spiked plasma due to non-specific reactions [9].

8.2.13 Other Compounds

Total plasma carotenoid concentration can be determined, after extraction, by measuring absorbance at 453 nm. This assay lacks specificity.

References

1. Monaghan P, Metcalfe NB, Torres R. Oxidative stress as a mediator of life history trade-offs: mechanisms, measurements and interpretation. Ecol Lett. 2009;12(1):75–92.
2. Aguilar Diaz De Leon J, Borges CR. Evaluation of Oxidative Stress in Biological Samples Using the Thiobarbituric Acid Reactive Substances Assay. J Vis Exp. 2020(159).
3. Henry Irving K, Margaret L. ON A NEW AEROBIC METABOLITE WHOSE PRODUCTION BY BRAIN IS INHIBITED BY APOMORPHINE, EMETINE, ERGOTAMINE, EPINEPHRINE, AND MENADIONE. Journal of Pharmacology and Experimental Therapeutics. 1944;82(3):292.
4. Muñoz AHS, Puga MP, Wrobel K, Sevilla MEG, Wrobel K. Micro Assay for Malondialdehyde in Human Serum by Extraction-Spectrophotometry Using an Internal Standard. Microchimica Acta. 2004;148(3):285–91.
5. Olszewska-Słonina DM, Mątewski D, Czajkowski R, Olszewski KJ, Woźniak A, Odrowąż-Sypniewska G, et al. The concentration of thiobarbituric acid reactive substances (TBARS) and paraoxonase activity in blood of patients with osteoarthrosis after endoprosthesis implantation. Med Sci Monit. 2011;17(9):Cr498–504.
6. Tsikas D. Assessment of lipid peroxidation by measuring malondialdehyde (MDA) and relatives in biological samples: Analytical and biological challenges. Anal Biochem. 2017;524:13–30.
7. Villamena F. Chemistry of Reactive Species. 2013. p. 1–48.
8. Draper HH, Squires EJ, Mahmoodi H, Wu J, Agarwal S, Hadley M. A comparative evaluation of thiobarbituric acid methods for the determination of malondialdehyde in biological materials. Free Radical Biology and Medicine. 1993;15(4):353–63.
9. Han D, Loukianoff S, McLaughlin L. Part VII • Chapter 17 – Oxidative stress indices: analytical aspects and significance. In: Sen CK, Packer L, Hänninen OOP, editors. Handbook of Oxidants and Antioxidants in Exercise. Amsterdam: Elsevier Science B.V.; 2000. p. 433–83.

10. Halliwell B, Gutteridge JM. Free radicals in biology and medicine: Oxford university press, USA; 2015.
11. Wilson R, Lyall K, Smyth L, Fernie CE, Riemersma RA. Dietary hydroxy fatty acids are absorbed in humans: implications for the measurement of 'oxidative stress' in vivo. Free Radic Biol Med. 2002;32(2):162–8.
12. Cohn JS. Oxidized fat in the diet, postprandial lipaemia and cardiovascular disease. Curr Opin Lipidol. 2002;13(1):19–24.
13. Amioka N, Miyoshi T, Otsuka H, Yamada D, Takaishi A, Ueeda M, et al. Serum malondialdehyde-modified low-density lipoprotein levels on admission predict prognosis in patients with acute coronary syndrome undergoing percutaneous coronary intervention. J Cardiol. 2019;74(3):258–66.
14. Gutteridge JMC. The measurement of malondialdehyde in peroxidised ox-brain phospholipid liposomes. Analytical Biochemistry. 1977;82(1):76–82.
15. Papastergiadis A, Mubiru E, Van Langenhove H, De Meulenaer B. Malondialdehyde Measurement in Oxidized Foods: Evaluation of the Spectrophotometric Thiobarbituric Acid Reactive Substances (TBARS) Test in Various Foods. Journal of Agricultural and Food Chemistry. 2012;60(38):9589–94.
16. Jiang ZY, Hunt JV, Wolff SP. Ferrous ion oxidation in the presence of xylenol orange for detection of lipid hydroperoxide in low density lipoprotein. Anal Biochem. 1992;202(2):384–9.
17. Cramer GL, Miller JF, Jr., Pendleton RB, Lands WE. Iodometric measurement of lipid hydroperoxides in human plasma. Anal Biochem. 1991;193(2):204–11.
18. Yagi K. Simple procedure for specific assay of lipid hydroperoxides in serum or plasma. Methods Mol Biol. 1998;108:107–10.
19. Mihaljević B, Katusin-Razem B, Razem D. The reevaluation of the ferric thiocyanate assay for lipid hydroperoxides with special considerations of the mechanistic aspects of the response. Free Radic Biol Med. 1996;21(1):53–63.
20. Li D, Wang M, Cheng N, Xue X, Wu L, Cao W. A modified FOX-1 method for Micro-determination of hydrogen peroxide in honey samples. Food Chemistry. 2017;237:225–31.
21. Vassalle C. An easy and reliable automated method to estimate oxidative stress in the clinical setting. Methods Mol Biol. 2008;477:31–9.
22. Favier AE. How to demonstrate the occurrence of an oxidative stress in human? In: Favier AE, Cadet J, Kalyanaraman B, Fontecave M, Pierre JL, editors. Analysis of Free Radicals in Biological Systems. Basel: Birkhäuser Basel; 1995. p. 99–117.
23. Kodali ST, Kauffman P, Kotha SR, Yenigalla A, Veeraraghavan R, Pannu SR, et al. Oxidative Lipidomics: Analysis of Oxidized Lipids and Lipid Peroxidation in Biological Systems with Relevance to Health and Disease. In: Berliner LJ, Parinandi NL, editors. Measuring Oxidants and Oxidative Stress in Biological Systems. Cham: Springer International Publishing; 2020. p. 61–92.
24. Devasagayam TP, Boloor KK, Ramasarma T. Methods for estimating lipid peroxidation: an analysis of merits and demerits. Indian J Biochem Biophys. 2003;40(5):300–8.
25. Situnayake RD, Crump BJ, Zezulka AV, Davis M, McConkey B, Thurnham DI. Measurement of conjugated diene lipids by derivative spectroscopy in heptane extracts of plasma. Annals of clinical biochemistry. 1990;27 (Pt 3):258–66.
26. Ahotupa M, Asankari TJ. Baseline diene conjugation in LDL lipids: an indicator of circulating oxidized LDL. Free Radic Biol Med. 1999;27(11–12):1141–50.
27. Palmieri B, Sblendorio V. Oxidative stress tests: overview on reliability and use. Part I. European review for medical and pharmacological sciences. 2007;11(5):309–42.
28. Corongiu FP, Banni S. [30] Detection of conjugated dienes by second derivative ultraviolet spectrophotometry. Methods in enzymology. 233: Academic Press; 1994. p. 303–10.
29. Banni S, Lucchi L, Baraldi A, Botti B, Cappelli G, Corongiu F, et al. No direct evidence of increased lipid peroxidation in hemodialysis patients. Nephron. 1996;72(2):177–83.
30. Esterbauer H, Striegl G, Puhl H, Rotheneder M. Continuous monitoring of in vitro oxidation of human low density lipoprotein. Free radical research communications. 1989;6(1):67–75.

31. Thérond P, Bonnefont-Rousselot D, Davit-Spraul A, Conti M, Legrand A. Biomarkers of oxidative stress: an analytical approach. Current opinion in clinical nutrition and metabolic care. 2000;3(5):373–84.
32. Yan LJ, Forster MJ. Chemical probes for analysis of carbonylated proteins: a review. J Chromatogr B Analyt Technol Biomed Life Sci. 2011;879(17–18):1308–15.
33. Soglia F, Baldi G, González-Mohino A, Díaz-Velasco S, Petracci M, Estévez M. Spectrophotometric Analysis of Protein CarbonylsCarbonyls. In: Lorenzo JM, Domínguez R, Pateiro M, Munekata PES, editors. Methods to Assess the Quality of Meat Products. New York, NY: Springer US; 2022. p. 107–14.
34. Reznick AZ, Packer L. Oxidative damage to proteins: spectrophotometric method for carbonyl assay. Methods in enzymology. 1994;233:357–63.
35. Stadtman ER. Protein oxidation and aging. Science. 1992;257(5074):1220–4.
36. Greilberger J, Fuchs D, Leblhuber F, Greilberger M, Wintersteiger R, Tafeit E. Carbonyl proteins as a clinical marker in Alzheimer's disease and its relation to tryptophan degradation and immune activation. Clin Lab. 2010;56(9–10):441–8.
37. Levine RL, Garland D, Oliver CN, Amici A, Climent I, Lenz AG, et al. Determination of carbonyl content in oxidatively modified proteins. Methods in enzymology. 1990;186:464–78.
38. Cao G, Cutler RG. Protein oxidation and aging. I. Difficulties in measuring reactive protein carbonyls in tissues using 2,4-dinitrophenylhydrazine. Arch Biochem Biophys. 1995;320(1): 106–14.
39. Lyras L, Evans PJ, Shaw PJ, Ince PG, Halliwell B. Oxidative damage and motor neurone disease difficulties in the measurement of protein carbonyls in human brain tissue. Free radical research. 1996;24(5):397–406.
40. Quinlan GJ, Evans TW, Gutteridge JM. Oxidative damage to plasma proteins in adult respiratory distress syndrome. Free radical research. 1994;20(5):289–98.
41. Fagan JM, Sleczka BG, Sohar I. Quantitation of oxidative damage to tissue proteins. The international journal of biochemistry & cell biology. 1999;31(7):751–7.
42. Roberts JC, Francetic DJ. The Importance of Sample Preparation and Storage in Glutathione Analysis. Analytical Biochemistry. 1993;211(2):183–7.
43. Araujo ARTS, Saraiva MLMFS, Lima JLFC. Determination of total and oxidized glutathione in human whole blood with a sequential injection analysis system. Talanta. 2008;74(5):1511–9.
44. Rahman I, Kode A, Biswas SK. Assay for quantitative determination of glutathione and glutathione disulfide levels using enzymatic recycling method. Nat Protoc. 2006;1(6):3159–65.
45. McGill MR, Jaeschke H. A direct comparison of methods used to measure oxidized glutathione in biological samples: 2-vinylpyridine and N-ethylmaleimide. Toxicol Mech Methods. 2015;25 (8):589–95.
46. Hansen RE, Østergaard H, Nørgaard P, Winther JR. Quantification of protein thiols and dithiols in the picomolar range using sodium borohydride and 4,4′-dithiodipyridine. Anal Biochem. 2007;363(1):77–82.
47. Guevara I, Iwanejko J, Dembińska-Kieć A, Pankiewicz J, Wanat A, Anna P, et al. Determination of nitrite/nitrate in human biological material by the simple Griess reaction. Clinica Chimica Acta. 1998;274(2):177–88.
48. Bryan NS, Grisham MB. Methods to detect nitric oxide and its metabolites in biological samples. Free Radic Biol Med. 2007;43(5):645–57.
49. Grisham MB, Johnson GG, Lancaster JR, Jr. Quantitation of nitrate and nitrite in extracellular fluids. Methods in enzymology. 1996;268:237–46.
50. Marzinzig M, Nussler AK, Stadler J, Marzinzig E, Barthlen W, Nussler NC, et al. Improved methods to measure end products of nitric oxide in biological fluids: nitrite, nitrate, and S-nitrosothiols. Nitric oxide: biology and chemistry. 1997;1(2):177–89.
51. Tracey WR, Linden J, Peach MJ, Johns RA. Comparison of spectrophotometric and biological assays for nitric oxide (NO) and endothelium-derived relaxing factor (EDRF): nonspecificity of the diazotization reaction for NO and failure to detect EDRF. The Journal of pharmacology and experimental therapeutics. 1990;252(3):922–8.

52. Sutton HC, Roberts PB, Winterbourn CC. The rate of reaction of superoxide radical ion with oxyhaemoglobin and methaemoglobin. Biochemical Journal. 1976;155(3):503–10.
53. Kelm M, Dahmann R, Wink D, Feelisch M. The Nitric Oxide/Superoxide Assay: INSIGHTS INTO THE BIOLOGICAL CHEMISTRY OF THE NO/O 2− 2 INTERACTION. Journal of Biological Chemistry. 1997;272(15):9922–32.
54. Saville B. A scheme for the colorimetric determination of microgram amounts of thiols. Analyst. 1958;83(993):670–2.
55. Wink DA, Kim S, Coffin D, Cook JC, Vodovotz Y, Chistodoulou D, et al. [21] Detection of S-nitrosothiols by fluorometric and colorimetric methods. Methods in enzymology. 301: Academic Press; 1999. p. 201–11.
56. Jourd'heuil D, Gray L, Grisham MB. S-nitrosothiol formation in blood of lipopolysaccharide-treated rats. Biochemical and biophysical research communications. 2000;273(1):22–6.
57. Minor T, Isselhard W, Berghaus K. Parenchymal and vascular endothelial cell injury in the hypoxic and reperfused rat liver. Evidence for superoxide anion generation by perfusion with ferricytochrome c. Biomedicine & Pharmacotherapy. 1993;47(5):213–8.
58. Tarpey MM, Wink DA, Grisham MB. Methods for detection of reactive metabolites of oxygen and nitrogen: in vitro and in vivo considerations. Am J Physiol Regul Integr Comp Physiol. 2004;286(3):R431–44.
59. Kolbeck RC, She ZW, Callahan LA, Nosek TM. Increased superoxide production during fatigue in the perfused rat diaphragm. American journal of respiratory and critical care medicine. 1997;156(1):140–5.
60. Jones OT, Hancock JT. Assays of plasma membrane NADPH oxidase. Methods in enzymology. 1994;233:222–9.
61. Landmesser U, Dikalov S, Price SR, McCann L, Fukai T, Holland SM, et al. Oxidation of tetrahydrobiopterin leads to uncoupling of endothelial cell nitric oxide synthase in hypertension. J Clin Invest. 2003;111(8):1201–9.
62. Tarpey MM, Fridovich I. Methods of detection of vascular reactive species: nitric oxide, superoxide, hydrogen peroxide, and peroxynitrite. Circulation research. 2001;89(3):224–36.
63. Tarpey MM, Fridovich I. Methods of Detection of Vascular Reactive Species. 2001;89(3): 224–36.
64. Choi HS, Kim JW, Cha YN, Kim C. A quantitative nitroblue tetrazolium assay for determining intracellular superoxide anion production in phagocytic cells. J Immunoassay Immunochem. 2006;27(1):31–44.
65. Baehner RL, Nathan DG. Quantitative nitroblue tetrazolium test in chronic granulomatous disease. The New England journal of medicine. 1968;278(18):971–6.
66. Holland SM. Chronic granulomatous disease. Clinical reviews in allergy & immunology. 2010;38(1):3–10.
67. Auclair C, Torres M, Hakim J. Superoxide anion involvement in NBT reduction catalyzed by NADPH-cytochrome P-450 reductase: a pitfall. FEBS Lett. 1978;89(1):26–8.
68. Apak R, Özyürek M, Güçlü K, Çapanoğlu E. Antioxidant Activity/Capacity Measurement. 3. Reactive Oxygen and Nitrogen Species (ROS/RNS) Scavenging Assays, Oxidative Stress Biomarkers, and Chromatographic/Chemometric Assays. Journal of Agricultural and Food Chemistry. 2016;64(5):1046–70.
69. Thomas EL, Grisham MB, Margaret Jefferson M. [40] Preparation and characterization of chloramines. Methods in enzymology. 132: Academic Press; 1986. p. 569–85.
70. Thordal-Christensen H, Zhang Z, Wei Y, Collinge D. Subcellular localization of H2O2 in plants. H2O2 accumulation in papillae and hypersensitive response during the barley-powdery mildew interaction. The Plant Journal 11(6): 1187–1194. 1997;11:1187–94.
71. Stockert JC, Blázquez-Castro A. Establishing the subcellular localization of photodynamically-induced ROS using 3,3′-diaminobenzidine: A methodological proposal, with a proof-of-concept demonstration. Methods. 2016;109:175–9.

72. Taylor EL, Armstrong KR, Perrett D, Hattersley AT, Winyard PG. Optimisation of an Advanced Oxidation Protein Products Assay: Its Application to Studies of Oxidative Stress in Diabetes Mellitus. Oxid Med Cell Longev. 2015;2015:496271.
73. Pachla LA, Reynolds DL, Kissinger PT. Analytical Methods for Determining Ascorbic Acid in Biological Samples, Food Products, and Pharmaceuticals. Journal of Association of Official Analytical Chemists. 1985;68(1):1–12.

Chapter 9
Chemiluminescence Methods

A. Khelfi

Abstract Chemiluminescence corresponds to the emission of photons (electromagnetic radiation as light) when chemically excited molecules decay to ground state following a chemical reaction. Chemiluminescence-based methods are accurate and can be successfully used for oxidative stress (OS) monitoring. Chemiluminescent sensors gain access to the intracellular space, present minimal cellular toxicity, and show sensitivity when compared with other chemical probes. Most of the chemiluminescent-based assays rely on the reaction of luminol and lucigenin with an oxidant to form light that can be monitored using a sensitive luminometer. Unfortunately, in most cases, chemiluminescence usually does not allow a specific identification of the oxidant that has been produced (e.g., superoxide, NO, peroxynitrite, hydrogen peroxide, and singlet oxygen). Nevertheless, the global production of oxidants can be measured and imaged on freshly isolated cells. This chapter, which reviews the various excited species that can lead to light generation, is therefore important for scientists who would like to excel in the field of luminescence assays. Recent advances in nanotechnology, which led to the development of new probes for monitoring OS status in medical science using nanosensors, are also presented.

Keywords Superoxide · NO · Peroxynitrite · Hydrogen peroxide · Singlet oxygen · Luminol · Lucigenin · Nanotechnology · Nanosensors

Chemiluminescence is the emission of light as a result of a chemical reaction. Methods based on chemiluminescence have been used in several applications because some sensors gain access to the intracellular space, present minimal cellular toxicity, and show sensitivity when compared with other chemical probes. Indeed,

A. Khelfi (✉)
Faculty of Pharmacy, University of Algiers, Algiers, Algeria

National Centre of Toxicology, Algiers, Algeria

Department of Chemistry and Biomolecular Science, Clarkson University, Potsdam, NY, USA
e-mail: ab.khelfi@univ-alger.dz; akhelfi@clarkson.edu

S. Andreescu et al. (eds.), *Biomarkers of Oxidative Stress*,
https://doi.org/10.1007/978-3-031-60738-7_9

chemiluminescence is usually more sensitive than fluorescence by two orders of magnitude [1].

9.1 Luminometry

Luminometry consists of introducing an enhancer emitting light during its reaction with the radical of interest. It is considered as a good method and an attractive way for OS assessment. Nevertheless, the obtained light signals are relatively low when biological samples are analyzed, requiring high sensitive luminometers. Most of the available microplate systems, even when equipped with a luminometer mode, are not sensitive enough to detect such signals. Luminometry suffers from another major difficulty of reproducibility since there is no reliable standard to compare the emission of light from day to day. To address these issues, most luminometers contain photodiodes or photomultiplier tubes to detect chemiluminescent signals. These detectors are either placed to the side or below the sample tubes. They are designed to measure the signals as electric current or photons when the photons strike the photomultiplier tube. The results are expressed as millivolts/sec, relative light units (RLUs), or counted photons/min (cpm) [2].

Different sensors have been used in chemiluminescent assays. Each sensor specifically measures a particular type of free radicals (FR) in the sample and whether they are intracellular and/or extracellular species. Luminol and lucigenin are the most widely employed sensors to measure reactive species (RS) using chemiluminescence technology.

Luminol can measure the global level of RS under physiological conditions as it is not charged and permeable through the membrane. Therefore, both extra- and intracellular RS including H_2O_2, $OH^{\bullet}$, $O_2^{\bullet -}$, and $ONOO^-$ are measured. On the other hand, lucigenin measures only extracellular RS as it is positively charged and impermeable. Lucigenin is specific for $O_2^{\bullet -}$ and $OH^{\bullet}$ [3].

Chemiluminescent assays have been widely used in patients with idiopathic infertility (see Chap. 6 of volume 2). In advanced research clinical laboratories, the use of chemiluminescent assays allows the assessment of RS in the fresh semen sample. The results correlate with fertilizing potential of sperm in vitro and in vivo [4]. It is a useful tool in clinical andrology laboratory and allows for a quick identification of infertile men with unexplained infertility.

9.1.1 Superoxide

Lucigenin is currently the most frequently used enhancer for $O_2^{\bullet -}$ detection. Unfortunately, this sensor undergoes redox cycling at high concentrations (more than 5 μmol/L) or in the present of NADPH/NADH [5]. It is crucial to realize that NADPH- or NADH-driven lucigenin chemiluminescence does not reflect only RS generation but rather the reaction of the enhancer with the reductant, which is by some means catalyzed by an unknown mechanism [6].

The obtained light signal with 5 μmol/L of lucigenin is low; therefore, several other enhancers have been developed and evaluated recently. For example, coelenterazine and analogs (e.g., CLA) were advocated being superior to lucigenin since they do not exhibit any redox cycling in biological systems [7, 8]. Nevertheless, auto-oxidation and undesirable signal to background ratio limit the use of these enhancers.

L-012 is an enhancer, which has the capacity to replace lucigenin as a primary system for $O_2^{\bullet-}$ detection [9, 10]. Since L-012 does not undergo redox cycling reactions with its environmental medium, it can be used in concentrations up to 100 μmol/L, yielding a robust signal [11]. Nevertheless, L-012 is not specific for $O_2^{\bullet-}$ and detects $ONOO^-$ as well. Moreover, NADPH and NADH can exert quenching effects on L-012 signals [12].

9.1.2 NO

As mentioned earlier, NO_2^- may be a reliable indicator of nitric oxide production. Many publications have made possible a consensus for physiological concentrations of NO_2^- in human fluids, at least as measured by chemiluminescence [13].

A special chemiluminometer for gas was developed, which allows for the measurement of $NO^\bullet$ after extraction by a stream of inert gas and a reaction with O_3 [14]. In this assay, $NO^\bullet$ reacts with O_3 to form NO_2 in an excited state. When this NO_2 species returns to ground state, light is emitted (Eq. 9.1):

$$O_3 + NO \rightarrow O_2 + NO_2^* \rightarrow NO_2 + h\nu \qquad (9.1)$$

The rapid rates of reaction of $NO^\bullet$ with O_3 and the sensitivity of detection of the emitted light by NO_2* (100 pmol/L) allow for continuous and real-time monitoring of nitric oxide formation. Although very sensitive, this technique requires a sensitive photomultiplier tube and a nitric oxide analyzer equipped with ozone generator. Moreover, purging of $NO^\bullet$ from the analyzed sample is required.

Although gaseous $NO^\bullet$ can be detected, most of the biologically formed $NO^\bullet$ is rapidly oxidized to mainly two metabolites: NO_2^- and NO_3^-. To detect these metabolites in biological samples, they must first undergo reduction reactions. This can be achieved using nitrate reductase (Eq. 8.8) to reduce NO_3^- to NO_2^-, which is then further reduced to $NO^\bullet$ using iodide under acidic conditions. An alternative way to detect nitric oxide metabolites is the use of acidic vanadium (III) at room temperature. After the return of the chemiluminescent tracing to baseline, the temperature is raised to 90 °C, which results in vanadium-dependent reduction of NO_3^- to $NO^\bullet$ [15].

9.1.3 Peroxynitrite

Peroxynitrite reacts with luminol to yield a chemiluminescent signal, which can be enhanced by CO_2 due to the formation of nitrosoperoxocarbonate ($ONOOCO_2^-$) present in bicarbonate-containing systems [16, 17]. Superoxide dismutase (SOD) can inhibit peroxynitrite-dependent luminol chemiluminescence. The mechanism for SOD inhibition is based on the inhibition of superoxide-mediated oxidation of the preformed luminol radical [16, 18]. Excess $NO^•$ can cause inhibition of peroxynitrite-dependent luminol chemiluminescence as well, via its reaction with luminol radicals [18]. Although not well described, $ONOO^-$ also rapidly reacts with another chemiluminescent sensors used for the detection of $O_2^{•-}$, coelenterazine [19]. A good strategy to discriminate between chemiluminescence coming from $O_2^{•-}$ or $ONOO^-$ would be the determination of lucigenin chemiluminescence under identical conditions, since $ONOO^-$ does not result in chemiluminescence with lucigenin [17].

9.1.4 H_2O_2 and HOCl

Monitoring chemiluminescence of luminol in the presence of HOCl is a sensitive, non-enzymatic method to measure H_2O_2 release from cells as described by Mueller and Arnhold [20].

9.1.5 Lipid Peroxides

Peroxy radicals convert to cyclic peroxides (dioxetanes) which decompose to excited carbonyls. The emitted photons can be detected by a photomultiplier [21]. Lipid hydroperoxides can be decomposed in the presence of hemin, giving chemiluminescence in the presence of luminol [22].

9.1.6 Singlet Oxygen and Peroxides

Production of some reactive oxygen species (ROS) is accompanied with the emission of light of wavelengths located in the near-infrared and infrared range (spontaneous chemiluminescence). Among ROS, relaxation of $^1O_2$ to the ground state produces an ultraweak photon emission that can be measured at 1268 nm. These ultralow light emissions can be measured using a chemiluminescence spectrometer equipped with a germanium photodiode detector cooled by liquid nitrogen [23]. This detector offers high sensitivity to near-infrared radiation from 1000 to 1700 nm. This technique requires a singlet oxygen calibration standard obtained by the reaction of H_2O_2 with HOCl. The emission of light can be measured directly in cells, tissues, and whole animals using a special apparatus equipped with a red-sensitive photomultiplier [24]. The emission of light has been observed during stress realized in hepatic microsomes and isolated hepatocytes [25], as well as in whole liver and heart after application of various stress [26], or after ischemic reoxygenation [27]. This technique has also been used to demonstrate the production of FR by human cells when stimulated by TNF [28]. Finally, direct luminescence of fresh human urine has been shown to be related to the amount of peroxides in samples [29].

9.2 Chemiluminescence-Based Imaging Techniques

The global production of ROS can be measured by chemiluminescence technology on freshly isolated cells. To increase the sensitivity, chemiluminescent sensors are used but are highly sensible to artifacts. Similar to fluorescence assays, chemiluminescent sensors are sensitive to radicals and easy to operate. The sensors react with RS to form photons, which are captured by a scintillation counter or a photometer without the need of an excitation light source [7]. Unfortunately, this chemiluminescence usually does not allow a specific identification of the RS that have been produced.

9.2.1 Luminol

Luminol (5-amino-2,3-dihydro-l,4-phtalazinedione) is sensitive to various RS including H_2O_2, $^1O_2$, $OH^•$, HOCl, $CO_3^{•-}$, and $ONOO^-$. Its sensitivity to $O_2^{•-}$ production is low.

As shown in Fig. 9.1, there are multiple steps during the conversion of luminol to the final luminescent product. The first one-electron oxidizing reaction involving the conversion of luminol (LH^-) to the intermediate radical ($LH^•$) is not specific, and oxidants such as $O_2^{•-}$, $OH^•$, and $CO_3^{•-}$ have been proved to be competent in this process [30]. The intermediate radical is an unstable endoperoxide that decomposes to an electronically excited product. This excited aminophthalate releases a photon as it falls to the ground state.

A variant of the luminol-based chemiluminescence assay has been developed. The assay relies on the oxidation of luminol by NaOCl. Hydrogen peroxide reacts with this oxidized product, yielding an excited compound capable of luminescence. This assay is specific for H_2O_2, unlike luminol-based chemiluminescence and can detect nanomolar concentrations of H_2O_2 [20].

Luminol-dependent chemiluminescence has many advantages. It is easy, inexpensive, sensitive, and can measure RS in intact cells over time. Luminol has been widely used to measure FR production by isolated and stimulated neutrophils [31]. However, the separation step of cells is time-consuming and can change the

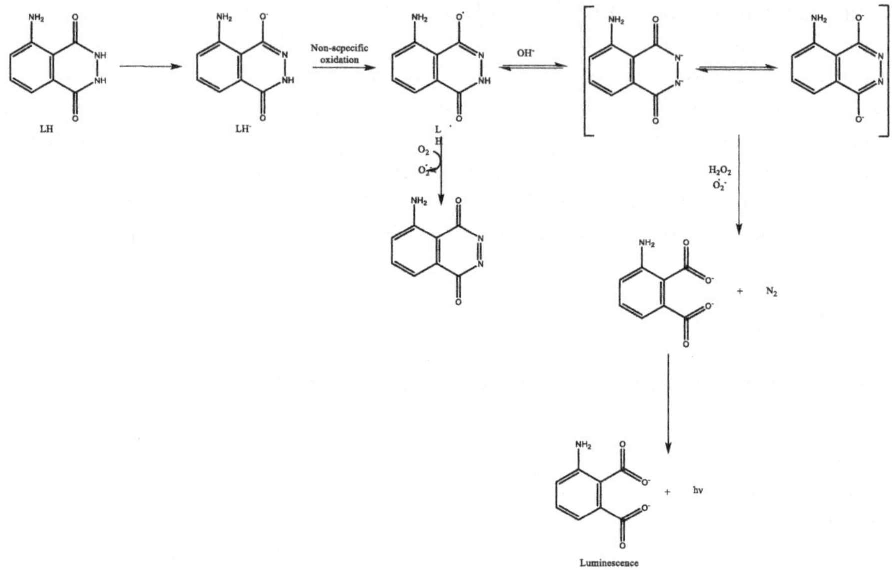

Fig. 9.1 Reaction of luminol with ROS
In this reaction, luminol is converted under basic conditions into a resonance-stabilized dianion, which is oxidized by $H_2O_2/O_2^{\bullet-}$ into the dicarboxylate ion, along with the release of N_2. The formed compound is in an excited electronic state and loses its extra energy by emitting photons, allowing the structure to go to its ground state form

native production of RS and the sensitivity of the cells to stimuli. Certain authors obtained reliable results when assessing the production of total blood after stimulation by phorbol myristate acetate (PMA) [32]. The assay has given interesting results in clinical situations such as hemodialyzed patients after dialysis [33] and has demonstrated an overproduction of RS in AIDS [34], and a priming effect of coronary angioplasty [35]. Another advantage of chemiluminescence is that it can be applied to measure FR production in human biopsies. For instance, catalase-inhibitable chemiluminescence produced by biopsy specimens was obtained during colonoscopy from patients with ulcerative colitis [36].

On the other hand, there are several limitations to luminol-dependent chemiluminescence:

- First, with a lower rate than that with $O_2^{\bullet-}$, LH was reported to spontaneously reduce O_2 to $O_2^{\bullet-}$ in the presence of any univalent oxidant, acting, therefore, as an artificial source of $O_2^{\bullet-}$ [37]. Moreover, luminol can mediate $O_2^{\bullet-}$ formation by a variety of oxidants including OCl^- and ferricyanide [38].
- Second, the anti-oxidative system can interfere with this probe in cells since it could compete with radicals that react with luminol radical or even directly react with luminol radical.

- Third, under neutral pH conditions, luminol cannot react with FR efficiently.
- Fourth, luminol emission at 400 nm is prone to many interferences as auto-absorption of biomolecules in cells or biological systems could happen [39].
- Fifth, the indiscriminate recognition of various FR by luminol is the main factor limiting its broad application in many cases.

Considering these factors influencing the chemiluminescence of luminol, it is unwise to monitor the dynamics of FR generation in cells or systems with this probe alone. Finally, this probe has been used in combination with lucigenin to monitor inflammation and OS in whole animal following PMA administration (Fig. 9.3) [40].

9.2.2 Lucigenin

Lucigenin (10,10′-dimethyl-9,9′-biacridinium dinitrate) is an alternative probe for detecting ROS production. According to some authors, this sensor is unable to enter the cell, but others claim it specifically measures the production of ROS by the mitochondria [41, 42]. Lucigenin is often said to be more specific for the detection of $O_2^{\bullet-}$ than luminol. It is the most commonly used probe for $O_2^{\bullet-}$ production by neutrophils, macrophages as well as by extracellular isolated enzymes [43].

Lucigenin is first reduced to a cation radical ($LC^{\bullet+}$). Then, it reacts with $O_2^{\bullet-}$ to produce an unstable dioxetane, which breaks down to form two molecules of excited state N-methylacridone, which emit light when they return to their ground state (Fig. 9.2). Reductants that can generate $LC^{\bullet+}$ are H_2O_2, flavoproteins, eNOS, NADPH reductases, and CYP450 [44]. Therefore, this probe cannot measure individual FR but measures global ROS levels in the sample.

Controversy has surrounded lucigenin in many literature reports:

- First of all, redox cycling of lucigenin can lead to extra $O_2^{\bullet-}$ production (Fig. 9.2) [45], which has raised doubts about its use in measuring rates of $O_2^{\bullet-}$ production. LC^{2+} is reduced by flavoproteins and NADPH reductase to $LC^{\bullet+}$, which reacts with O_2 to generate artifactually $O_2^{\bullet-}$ [8, 46, 47]. Vasquez-Vivar et al. also demonstrated that lucigenin stimulates $O_2^{\bullet-}$ production by NOS [48]. Barbacanne et al. suggested that the presence of electron donors such as NADH and NADPH influences both lucigenin redox cycling and chemiluminescent intensities resulting from large amounts of $O_2^{\bullet-}$ generation [49]. The use of low concentrations of lucigenin has been suggested as a mean to limit this issue. Indeed, Li et al. have shown that lucigenin produces $O_2^{\bullet-}$ when employed above a threshold concentration [50]. Under a threshold concentration, they demonstrated that lucigenin effectively detects low levels of $O_2^{\bullet-}$ without causing additional $O_2^{\bullet-}$ production. However, since it is not possible to predict the extent of redox cycling in different systems, the threshold may be difficult to determine accurately [38]. Although it has been suggested that 5 μM of lucigenin will eliminate the potential for redox cycling [7, 51], other investigations have shown that redox cycling

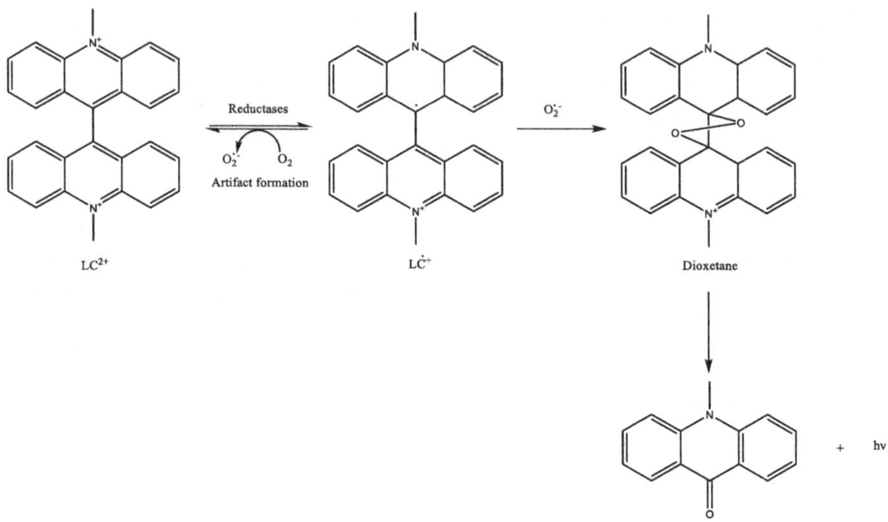

Fig. 9.2 Reaction of lucigenin with $O_2^{\bullet-}$ resulting in chemiluminescence

accompanied with increased oxygen consumption and enhanced formation of $O_2^{\bullet-}$ may occur at this concentration [49, 52].

- Second, $O_2^{\bullet-}$ is not the only radical that can react with lucigenin and other reducing molecules or nucleophiles like H_2O_2 can also induce the luminescence of this sensor, through the generation of the lucigenin radical [30].
- Third, lucigenin assay is sensitive to changes of reductase activity since cellular reductases play an important role in the generation of lucigenin radical.
- Fourth, given the reactivity between lucigenin and $O_2^{\bullet-}$, which is rather low and slow, the sensor is unable to measure very low levels of $O_2^{\bullet-}$. Indeed, the conversion of lucigenin to the corresponding radical by $O_2^{\bullet-}$ requires other cellular-reducing systems (e.g., mitochondrial electron transport chain (ETC), xanthine oxidase (XO), or phagocyte NAD(P)H oxidase (NOX)). This introduces an obvious complexity in interpreting results [53].
- Fifth, another problem with the use of luminol for the detection of $O_2^{\bullet-}$ is that the intermediate radical auto-oxidizes as well [53].

Overall, luminol luminescence is also not a reliable indicator of $O_2^{\bullet-}$, even when $O_2^{\bullet-}$ is involved in the reaction leading to light emission. The level of lucigenin that can be used to measure $O_2^{\bullet-}$ and not undergo redox cycling must be determined for each new set of experimental conditions through alternative techniques like measurement of oxygen consumption (Fig. 9.3).

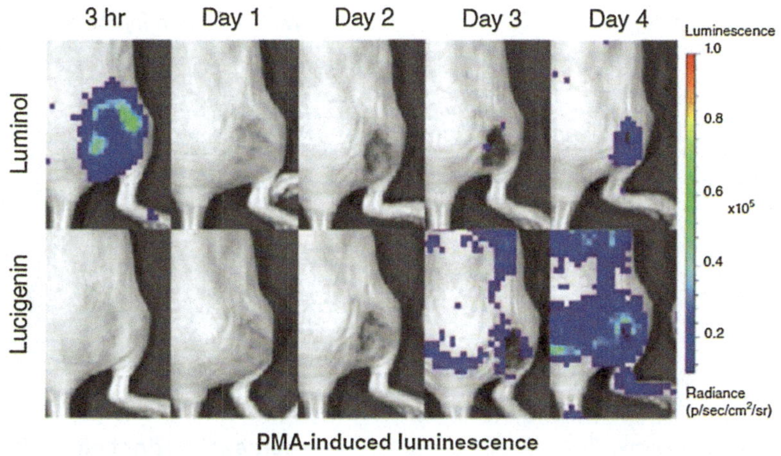

Fig. 9.3 Luminol and lucigenin images of different stages of inflammation
Subcutaneous injection of PMA caused a rapid skin irritation and inflammation. Daily imaging of both lucigenin and luminol luminescence was performed for 4 days. Using an IVIS® Spectrum imaging system, significant luminol chemiluminescence from the injection sites was observed as early as 3 h post-PMA injection. However, from day 3, as wound contraction and scar formation were visible, a steady increase of lucigenin luminescence was observed. Therefore, luminol luminescence is associated with the acute phase of inflammation, while lucigenin luminescence is presumably associated with the late phases of inflammation [40]

9.2.3 L-012

L-012 (8-amino-5-chloro-2,3-dihydro-7-phenylpyrido[3,4-*d*] pyridazine sodium salt) shows higher sensitivity and recognition ability toward $O_2^{\bullet -}$ than its luminol analog. L-012 is also more sensitive than luminol for the detection of $ONOO^-$ [54]. This sensor has been used to analyze $O_2^{\bullet -}$ production in many cells and systems such as whole blood, isolated leukocytes (e.g., neutrophils), vascular cells, mitochondria, as well as chronic inflammation systems [12, 55, 56]. $O_2^{\bullet -}$ intestinal production (Fig. 9.4) and skin (Fig. 9.5) induced inflammation have been assessed using whole animals [54, 57].

L-012 ($LumH_2$) does not react directly with $O_2^{\bullet -}$. The first step consists of the generation of an intermediate radical ($LumH^{\bullet}$) following the oxidation of the sensor by H_2O_2 or $ONOO^-$. In the second step, $LumH^{\bullet}$ reacts with $O_2^{\bullet -}$ and releases N_2 to form a luminescent aminophthalic acid. An alternative pathway may occur where $LumH^{\bullet}$ reacts with O_2 to generate $O_2^{\bullet -}$ and L-012 quinone (q-Lum). The latter would form a small portion of LumH-OOH in the presence of H_2O_2, resulting in an overestimated results of $O_2^{\bullet -}$ (Fig. 9.6) [10].

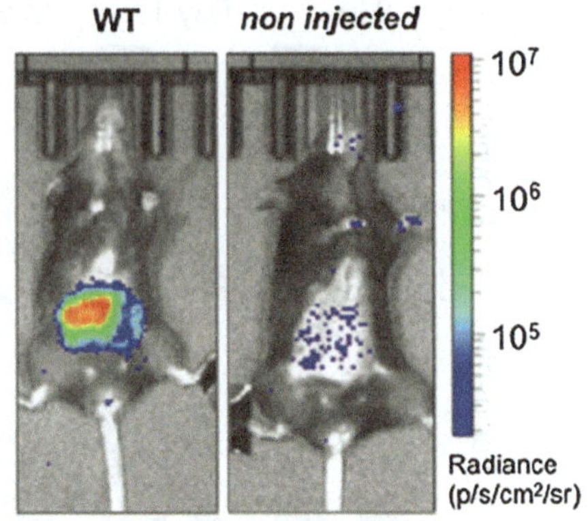

Fig. 9.4 Superoxide detection in wild-type mice 3 min post i.p. administration of L-012 (20 mg/kg), compared with a non-injected mouse. Luminescence is expressed as radiance (p/s/cm^2/sr) [54]

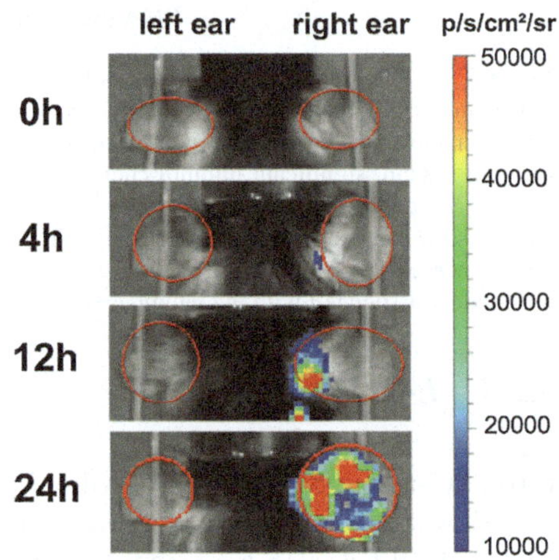

Fig. 9.5 ROS production in C57BL/6 wild-type mice. Temporal assessment of ROS production (L-012 optical imaging; left ear: control, right ear: TNCB-induced cutaneous contact hypersensitivity). The peak of the signal intensity was obtained at 24 h [57]

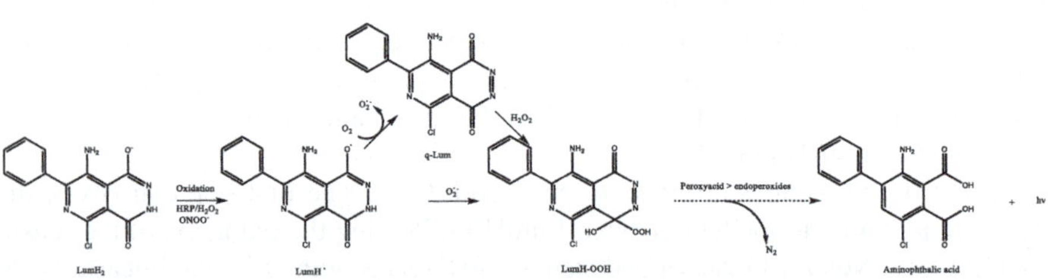

Fig. 9.6 Mechanism of reaction of L-012 (LumH2) with ROS/RNS

9.2.4 *Coelenterazine and Analogs*

In an effort to make use of the advantage of chemiluminescent techniques in terms of sensitivity, other non-redox cycling compounds have been designed and studied for

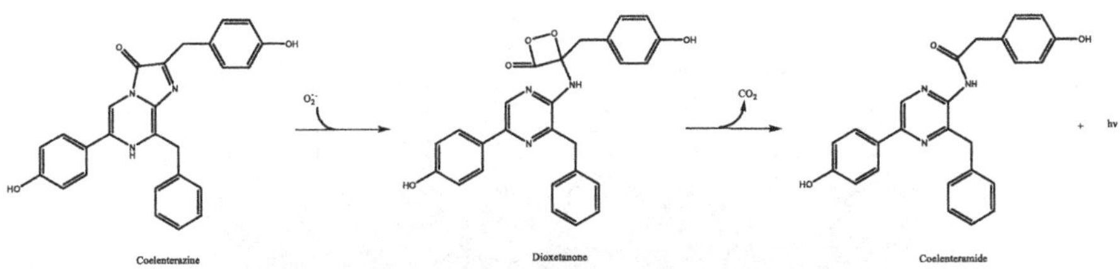

Fig. 9.7 Reaction of coelenterazine with $O_2^{\bullet-}$

their ability to specifically react with $O_2^{\bullet-}$. Most of these sensors are structurally similar to luciferins, which are derived from bioluminescent marine life.

One of the most successful chemiluminescent sensors is coelenterazine (2-(4-hydroxybenzyl)-6-(4-hydroxyphenyl)-8-benzyl-3,7-dihydroimidazo[1,2-α] pyrazin-3-one), which is a lipophilic luminophore originally isolated from the coelenterate *Aequorea* [58]. This compound is brighter than Lucigenin and does not undergo redox cycling reactions. Superoxide-stimulated chemiluminescence occurs after a direct oxidation of the acetamidopyrazine anion, which is protonated to form a neutral light-emitting excited species, thereby avoiding the redox cycling-dependent artifactual generation of $O_2^{\bullet-}$ associated with lucigenin (Fig. 9.7) [45].

Although the intensity of the emitted light from the interaction between coelenterazine and $O_2^{\bullet-}$ is greater compared to that from lucigenin, coelenterazine-dependent chemiluminescence is not entirely specific for $O_2^{\bullet-}$ because $ONOO^-$ also results in luminescence in the presence of coelenterazine [59]. Selective use of NOS inhibitors or peroxynitrite scavengers may help to differentiate between the contributions of peroxynitrite- and superoxide-mediated chemiluminescence. Coelenterazine-dependent chemiluminescence has been used to detect $O_2^{\bullet-}$ in vascular tissue [60], cultured cells [61], neutrophils [62], and isolated mitochondria [63].

Coelenterazine has been used to assess ROS production in several types of cancer cells as well as in implanted animals with tumor cells (Figs. 9.8 and 9.9) [64]. ROS imaging in an animal model of the inflammatory bowel disease has also been achieved using this probe (Fig. 9.10) [59].

Several structural analogs of coelenterazine have been used in several systems for chemiluminescent detection of $O_2^{\bullet-}$. They include CLA (2-methyl-6-phenyl-3,7-dihydroimidazo[1,2-α]pyrazin-3-one) and MCLA (2-methyl-6-(4-methoxyphenyl)-3,7-dihydroimidazo[1,2-α]pyrazin-3-one) (from *Cypridina* luciferin), of which MCLA is the most sensitive sensor for $O_2^{\bullet-}$ (Figs. 9.11 and 9.12) [65].

CLA and MCLA have been used to assess $O_2^{\bullet-}$ production in a variety of conditions including cultured cells [66, 67], macrophage and neutrophil activation [68], and vascular tissue [69]. Of note, MCLA assay have been used to detect $O_2^{\bullet-}$ formation on the surface of liver [70], within the intestine [71], and heart [72] in vivo.

As for coelenterazine, MCLA is not entirely specific toward $O_2^{\bullet-}$, as the presence of $^1O_2$ or peroxyl radicals will result in chemiluminescence [73]. MCLA also has a

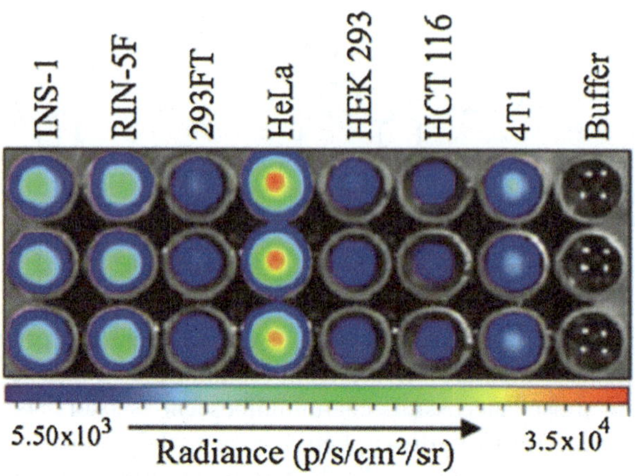

Fig. 9.8 Coelenterazine chemiluminescence in different cancer cell lines
The levels of ROS/RNS vary significantly among cell lines. 5×10^5 cells of each line were imaged with 10 µmol/l coelenterazine. The 4T1 cell line is a mouse breast adenocarcinoma line, HCT 116 is a human colon carcinoma line, HeLa is a human cervical cancer line, HEK 293 and 293FT are human embryonic kidney lines, and RIN-5F and INS-1 are rat β cell lines. The chemiluminescence of 5×10^5 cells of each line was quantified following 60 min of incubation with native coelenterazine [64]

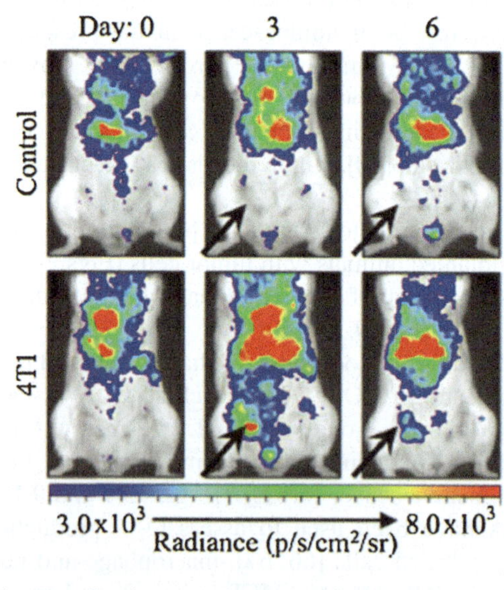

Fig. 9.9 Coelenterazine chemiluminescence in tumor sites
Images of a representative control and 4T1-implanted mouse prior to (day 0) and 3 and 6 days post-implantation. Arrows indicate the site of vehicle control or 4T1 cancer cell implantation [64]

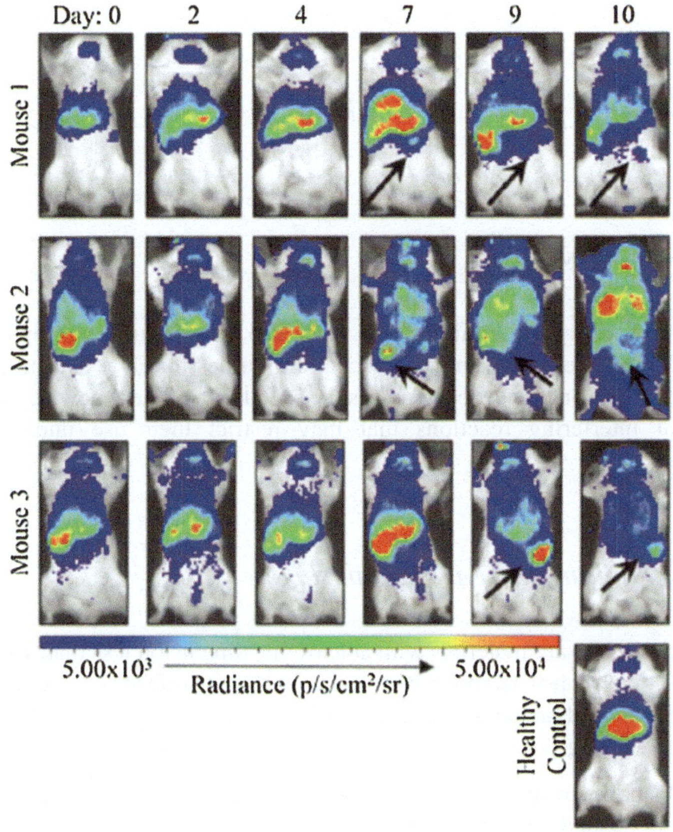

Fig. 9.10 Coelenterazine detects changes in regional ROS/RNS production in response to IBD Chemiluminescent images of three representative mice before treatment (day 0) and every 48 h thereafter. A healthy mouse was imaged on day 10 for comparison. Arrows indicate new regions of chemiluminescent signal generation [59]

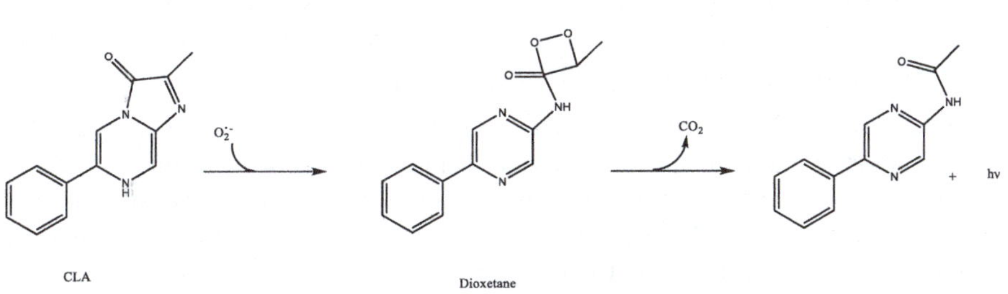

Fig. 9.11 Reaction of CLA with $O_2^{\bullet-}$

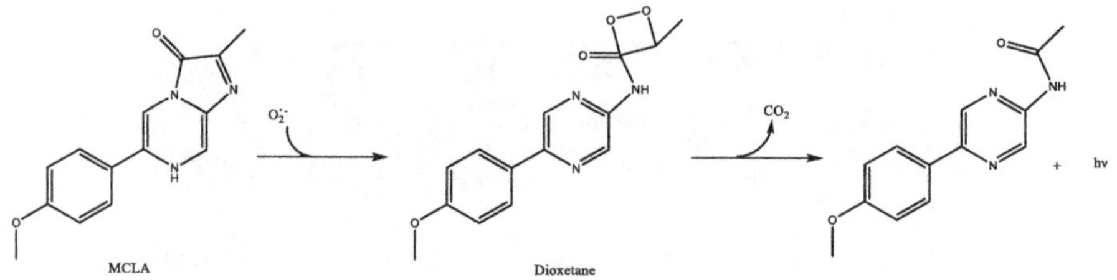

Fig. 9.12 Reaction of MCLA with $O_2^{\bullet-}$

high level of background luminescence. It is also temperature- and light-sensitive and can auto-oxidize, adding more practical difficulties for $O_2^{\bullet-}$ detection.

Overall, more work is needed to assess the specificity of these new sensors as well as potential interfering reactions that may restrict their use under particular conditions.

9.2.5 Chemiluminescent Nanoparticles

Nanoparticles (NPs) are materials with at least one dimension less than 100 nm. NPs come in different shapes and properties and have a high surface area-to-volume ratio. NPs have several advantages over small sensors that are currently used in cellular imaging and sensing [74]:

- First, NPs have stronger luminescent emission due to the large number of sensors that can be loaded into each particle.
- Second, the high surface area-to-volume ratio of NPs offers a higher capacity for RS detection.
- Third, NPs may protect the sensors from external interferences such as undesirable non-specific uptake by proteins and enzymatic reactions.
- Fourth, it is easy to target NPs to cells or their sub-cellular compartments by adding appropriate ligand moieties onto their surface. Conjugation and encapsulation of different molecules such as luminescent probes, DNA, or proteins provide infinite possibilities in NP design for specific functions.

In view of all of the mentioned properties, NPs are becoming increasingly used as sensing tools in the medical and biological field of sensing and imaging [75, 76].

Advances in developing different kinds of luminescence NP probes have greatly enabled monitoring of ROS in both cells and animals. Lee et al. developed peroxalate-based NPs fabricated from fluorescent sensors and peroxalate esters to image H_2O_2 in vivo with high sensitivity and specificity [77]. Peroxalate NPs were capable of imaging H_2O_2 in the peritoneal cavity of animals during a LPS-induced inflammatory response. The same team improved the method by reducing the size of NPs and modifying their content to detect H_2O_2 at biological concentrations [78, 79]. Luminescent NPs have also been used for in vivo targeting and imaging of tumor tissues. In one study, chemiluminescent NPs were successfully developed to detect H_2O_2 as a tumor signal molecule [80]. Such sensors improve the stability of peroxalates in biological systems and are sensitive to physiological concentrations of H_2O_2. This way of monitoring H_2O_2 would be helpful for the clinical diagnosis of several RS-related diseases.

References

1. Ramirez-Rios S, Michallet S, Peris L, Barette C, Rabat C, Feng Y, et al. A New Quantitative Cell-Based Assay Reveals Unexpected Microtubule Stabilizing Activity of Certain Kinase Inhibitors, Clinically Approved or in the Process of Approval. Front Pharmacol. 2020;11:543.
2. Kashou AH, Sharma R, Agarwal A. Assessment of oxidative stress in sperm and semen. Methods Mol Biol. 2013;927:351–61.
3. Firoozbakhtian A, Hosseini M. Chemiluminescence Sensors in Bioanalysis. In: Narayan R, editor. Encyclopedia of Sensors and Biosensors (First Edition). Oxford: Elsevier; 2023. p. 341–56.
4. Agarwal A, Ahmad G, Sharma R. Reference values of reactive oxygen species in seminal ejaculates using chemiluminescence assay. J Assist Reprod Genet. 2015;32(12):1721–9.
5. Brandes RP, Janiszewski M. Direct detection of reactive oxygen species ex vivo. Kidney International. 2005;67(5):1662–4.
6. Baker MA, Krutskikh A, Curry BJ, McLaughlin EA, Aitken RJ. Identification of cytochrome P450-reductase as the enzyme responsible for NADPH-dependent lucigenin and tetrazolium salt reduction in rat epididymal sperm preparations. Biol Reprod. 2004;71(1):307–18.
7. Münzel T, Afanas'ev IB, Kleschyov AL, Harrison DG. Detection of Superoxide in Vascular Tissue. Arteriosclerosis, thrombosis, and vascular biology. 2002;22(11):1761–8.
8. Tarpey MM, White CR, Suarez E, Richardson G, Radi R, Freeman BA. Chemiluminescent detection of oxidants in vascular tissue. Lucigenin but not coelenterazine enhances superoxide formation. Circulation research. 1999;84(10):1203–11.
9. Nishinaka Y, Aramaki Y, Yoshida H, Masuya H, Sugawara T, Ichimori Y. A new sensitive chemiluminescence probe, L-012, for measuring the production of superoxide anion by cells. Biochem Biophys Res Commun. 1993;193(2):554–9.
10. Zielonka J, Lambeth JD, Kalyanaraman B. On the use of L-012, a luminol-based chemiluminescent probe, for detecting superoxide and identifying inhibitors of NADPH oxidase: a reevaluation. Free Radic Biol Med. 2013;65:1310–4.
11. Sohn HY, Gloe T, Keller M, Schoenafinger K, Pohl U. Sensitive superoxide detection in vascular cells by the new chemiluminescence dye L-012. J Vasc Res. 1999;36(6):456–64.
12. Daiber A, Oelze M, August M, Wendt M, Sydow K, Wieboldt H, et al. Detection of superoxide and peroxynitrite in model systems and mitochondria by the luminol analogue L-012. Free radical research. 2004;38(3):259–69.
13. Giustarini D, Dalle-Donne I, Tsikas D, Rossi R. Oxidative stress and human diseases: Origin, link, measurement, mechanisms, and biomarkers. Critical reviews in clinical laboratory sciences. 2009;46(5–6):241–81.
14. Favier AE. How to demonstrate the occurrence of an oxidative stress in human? In: Favier AE, Cadet J, Kalyanaraman B, Fontecave M, Pierre JL, editors. Analysis of Free Radicals in Biological Systems. Basel: Birkhäuser Basel; 1995. p. 99–117.
15. Braman RS, Hendrix SA. Nanogram nitrite and nitrate determination in environmental and biological materials by vanadium(III) reduction with chemiluminescence detection. Analytical Chemistry. 1989;61(24):2715–8.
16. Radi R, Cosgrove TP, Beckman JS, Freeman BA. Peroxynitrite-induced luminol chemiluminescence. The Biochemical journal. 1993;290 (Pt 1)(Pt 1):51–7.

17. Radi R, Denicola A, Freeman BA. [37] Peroxynitrite reactions with carbon dioxide-bicarbonate. Methods in enzymology. 301: Academic Press; 1999. p. 353–67.
18. Castro L, Alvarez MN, Radi R. Modulatory role of nitric oxide on superoxide-dependent luminol chemiluminescence. Arch Biochem Biophys. 1996;333(1):179–88.
19. Valez V, Cassina A, Batinic-Haberle I, Kalyanaraman B, Ferrer-Sueta G, Radi R. Peroxynitrite formation in nitric oxide-exposed submitochondrial particles: detection, oxidative damage and catalytic removal by Mn-porphyrins. Arch Biochem Biophys. 2013;529(1):45–54.
20. Mueller S, Arnhold J. Fast and sensitive chemiluminescence determination of H2O2 concentration in stimulated human neutrophils. J Biolumin Chemilumin. 1995;10(4):229–37.
21. Murphy ME, Sies H. Visible-range low-level chemiluminescence in biological systems. Methods in enzymology. 1990;186:595–610.
22. Pastorino AM, Zamburlini A, Zennaro L, Maiorino M, Ursini F. Measurement of lipid hydroperoxides in human plasma and lipoproteins by kinetic analysis of photon emission. Methods in enzymology. 300: Academic Press; 1999. p. 33–43.
23. Di Mascio P, Kaiser S, Sies H. Lycopene as the most efficient biological carotenoid singlet oxygen quencher. Archives of Biochemistry and Biophysics. 1989;274(2):532–8.
24. Cadenas E, Sies H, Nastainczyk W, Ullrich V. Singlet oxygen formation detected by low-level chemiluminescence during enzymatic reduction of prostaglandin G2 to H2. Hoppe-Seyler's Zeitschrift fur physiologische Chemie. 1983;364(5):519–28.
25. Gonzalez Flecha B, Llesuy S, Boveris A. Hydroperoxide-initiated chemiluminescence: an assay for oxidative stress in biopsies of heart, liver, and muscle. Free Radic Biol Med. 1991;10(2): 93–100.
26. Slawinski J, Ezzahir A, Godlewski M, Kwiecinska T, Rajfur Z, Sitko D, et al. Stress-induced photon emission from perturbed organisms. Experientia. 1992;48(11–12):1041–58.
27. Barsacchi R, Coassin M, Maiorino M, Pelosi G, Simonelli C, Ursini F. Increased ultra weak chemiluminescence emission from rat heart at postischemic reoxygenation: protective role of vitamin E. Free Radic Biol Med. 1989;6(6):573–9.
28. Ferlat S, Favier A. Tumor necrosis factor et radicaux libres: consequences potentielles pour l'immunite cellulaire. CR Soc Biol. 1993;187:296–307.
29. Lissi EA, Salim-Hanna M, Sir T, Videla LA. Is spontaneous urinary visible chemiluminescence a reflection of in vivo oxidative stress? Free Radic Biol Med. 1992;12(4):317–22.
30. Zhang Y, Dai M, Yuan Z. Methods for the detection of reactive oxygen species. Analytical Methods. 2018;10(38):4625–38.
31. Stolarek R, Białasiewicz P, Nowak D. N-Acetylcysteine Effect on the Luminol-dependent Chemiluminescence Pattern of Reactive Oxygen Species Generation by Human Polymorphonuclear Leukocytes. Pulmonary pharmacology & therapeutics. 2002;15(4):385–92.
32. Descamps-Latscha B, Nguyen AT, Golub RM, Feuillet-Fieux MN. Chemiluminescence in microamounts of whole blood for investigation of the human phagocyte oxidative metabolism function. Ann Immunol (Paris). 1982;133c(3):349–64.
33. Nguyen AT, Lethias C, Zingraff J, Herbelin A, Naret C, Descamps-Latscha B. Hemodialysis membrane-induced activation of phagocyte oxidative metabolism detected in vivo and in vitro within microamounts of whole blood. Kidney Int. 1985;28(2):158–67.
34. Guillard O, Sonnet J, Lauwerys R. Reactive oxygen species production in whole blood from AIDS patients. Clin Chim Acta. 1989;185(1):113–4.
35. Steg PG, Pasquier C, Huu TP, Chollet-Martin S, Juliard JM, Himbert D, et al. Evidence for priming and activation of neutrophils early after coronary angioplasty. Eur J Med. 1993;2(1): 6–10.
36. Simmonds NJ, Allen RE, Stevens TR, Van Someren RN, Blake DR, Rampton DS. Chemiluminescence assay of mucosal reactive oxygen metabolites in inflammatory bowel disease. Gastroenterology. 1992;103(1):186–96.
37. Faulkner K, Fridovich I. Luminol and lucigenin as detectors for O2. Free Radic Biol Med. 1993;15(4):447–51.

38. Armstrong JS, Whiteman M. Measurement of Reactive Oxygen Species in Cells and Mitochondria. Methods in Cell Biology. 80: Academic Press; 2007. p. 355–77.
39. Fuloria S, Subramaniyan V, Karupiah S, Kumari U, Sathasivam K, Meenakshi DU, et al. Comprehensive Review of Methodology to Detect Reactive Oxygen Species (ROS) in Mammalian Species and Establish Its Relationship with Antioxidants and Cancer. Antioxidants (Basel). 2021;10(1).
40. Tseng J-C, Peterson JD. Whole-Body Chemiluminescence and Fluorescence Imaging of Inflammation. In: Man F, Cleary SJ, editors. Imaging Inflammation. Cham: Springer International Publishing; 2023. p. 143–89.
41. Kitagawa RR, Raddi MS, Khalil NM, Vilegas W, da Fonseca LM. Effect of the isocoumarin paepalantine on the luminol and lucigenin amplified chemiluminescence of rat neutrophils. Biol Pharm Bull. 2003;26(6):905–8.
42. Rembish SJ, Trush MA. Further evidence that lucigenin-derived chemiluminescence monitors mitochondrial superoxide generation in rat alveolar macrophages. Free Radic Biol Med. 1994;17(2):117–26.
43. Tseng J-C, Kung Andrew L. In Vivo Imaging of Inflammatory Phagocytes. Chemistry & Biology. 2012;19(9):1199–209.
44. Wardman P. Fluorescent and luminescent probes for measurement of oxidative and nitrosative species in cells and tissues: progress, pitfalls, and prospects. Free Radic Biol Med. 2007;43(7): 995–1022.
45. Liochev SI, Fridovich I. Lucigenin luminescence as a measure of intracellular superoxide dismutase activity in Escherichia coli. Proc Natl Acad Sci U S A. 1997;94(7):2891–6.
46. Sohn HY, Keller M, Gloe T, Crause P, Pohl U. Pitfalls of using lucigenin in endothelial cells: implications for NAD(P)H dependent superoxide formation. Free radical research. 2000;32(3): 265–72.
47. Liochev SI, Fridovich I. Lucigenin (bis-N-methylacridinium) as a mediator of superoxide anion production. Arch Biochem Biophys. 1997;337(1):115–20.
48. Vásquez-Vivar J, Hogg N, Pritchard KA, Jr., Martasek P, Kalyanaraman B. Superoxide anion formation from lucigenin: an electron spin resonance spin-trapping study. FEBS Lett. 1997;403 (2):127–30.
49. Barbacanne MA, Souchard JP, Darblade B, Iliou JP, Nepveu F, Pipy B, et al. Detection of superoxide anion released extracellularly by endothelial cells using cytochrome c reduction, ESR, fluorescence and lucigenin-enhanced chemiluminescence techniques. Free Radic Biol Med. 2000;29(5):388–96.
50. Li Y, Zhu H, Kuppusamy P, Roubaud V, Zweier JL, Trush MA. Validation of Lucigenin (Bis-N-methylacridinium) as a Chemilumigenic Probe for Detecting Superoxide Anion Radical Production by Enzymatic and Cellular Systems*. Journal of Biological Chemistry. 1998;273 (4):2015–23.
51. Skatchkov MP, Sperling D, Hink U, Mülsch A, Harrison DG, Sindermann I, et al. Validation of lucigenin as a chemiluminescent probe to monitor vascular superoxide as well as basal vascular nitric oxide production. Biochem Biophys Res Commun. 1999;254(2):319–24.
52. Janiszewski M, Souza HP, Liu X, Pedro MA, Zweier JL, Laurindo FR. Overestimation of NADH-driven vascular oxidase activity due to lucigenin artifacts. Free Radic Biol Med. 2002;32(5):446–53.
53. Tarpey MM, Fridovich I. Methods of detection of vascular reactive species: nitric oxide, superoxide, hydrogen peroxide, and peroxynitrite. Circulation research. 2001;89(3):224–36.
54. Conroy E, Aviello G. Imaging Intestinal ROS in Homeostatic Conditions Using L-012. Methods Mol Biol. 2019;1982:313–27.
55. Daiber A, August M, Baldus S, Wendt M, Oelze M, Sydow K, et al. Measurement of NAD(P)H oxidase-derived superoxide with the luminol analogue L-012. Free Radic Biol Med. 2004;36 (1):101–11.

56. Kielland A, Blom T, Nandakumar KS, Holmdahl R, Blomhoff R, Carlsen H. In vivo imaging of reactive oxygen and nitrogen species in inflammation using the luminescent probe L-012. Free Radic Biol Med. 2009;47(6):760–6.
57. Schwenck J, Mehling R, Thaiss WM, Kramer D, Menendez IG, Öz HH, et al. Temporal Dynamics of Reactive Oxygen and Nitrogen Species and NF-κB Activation During Acute and Chronic T Cell-Driven Inflammation. Mol Imaging Biol. 2020;22(3):504–14.
58. Magalhães CM, Esteves da Silva JCG, Pinto da Silva L. Study of coelenterazine luminescence: Electrostatic interactions as the controlling factor for efficient chemiexcitation. Journal of Luminescence. 2018;199:339–47.
59. Bronsart L, Nguyen L, Habtezion A, Contag C. Reactive Oxygen Species Imaging in a Mouse Model of Inflammatory Bowel Disease. Mol Imaging Biol. 2016;18(4):473–8.
60. Hink U, Li H, Mollnau H, Oelze M, Matheis E, Hartmann M, et al. Mechanisms underlying endothelial dysfunction in diabetes mellitus. Circulation research. 2001;88(2):E14–22.
61. Duerrschmidt N, Wippich N, Goettsch W, Broemme HJ, Morawietz H. Endothelin-1 induces NAD(P)H oxidase in human endothelial cells. Biochem Biophys Res Commun. 2000;269(3): 713–7.
62. Lucas M, Solano F. Coelenterazine is a superoxide anion-sensitive chemiluminescent probe: its usefulness in the assay of respiratory burst in neutrophils. Anal Biochem. 1992;206(2):273–7.
63. Raha S, McEachern GE, Myint AT, Robinson BH. Superoxides from mitochondrial complex III: the role of manganese superoxide dismutase. Free Radic Biol Med. 2000;29(2): 170–80.
64. Bronsart LL, Stokes C, Contag CH. Multimodality Imaging of Cancer Superoxide Anion Using the Small Molecule Coelenterazine. Mol Imaging Biol. 2016;18(2):166–71.
65. Abo M, Weerapana E. Chemical Probes for Redox Signaling and Oxidative Stress. Antioxidants & redox signaling. 2019;30(10):1369–86.
66. Ishii M, Shimizu S, Yamamoto T, Momose K, Kuroiwa Y. Acceleration of oxidative stress-induced endothelial cell death by nitric oxide synthase dysfunction accompanied with decrease in tetrahydrobiopterin content. Life Sci. 1997;61(7):739–47.
67. Tanaka M, Sotomatsu A, Yoshida T, Hirai S, Nishida A. Detection of superoxide production by activated microglia using a sensitive and specific chemiluminescence assay and microglia-mediated PC12h cell death. Journal of neurochemistry. 1994;63(1):266–70.
68. Geuijen C, Tacken P, Wang LC, Klooster R, van Loo PF, Zhou J, et al. A human CD137×PD-L1 bispecific antibody promotes anti-tumor immunity via context-dependent T cell costimulation and checkpoint blockade. Nat Commun. 2021;12(1):4445.
69. Skatchkov MP, Sperling D, Hink U, Anggard E, Münzel T. Quantification of superoxide radical formation in intact vascular tissue using a Cypridina luciferin analog as an alternative to lucigenin. Biochem Biophys Res Commun. 1998;248(2):382–6.
70. Uehara K, Maruyama N, Huang C-K, Nakano M. The first application of a chemiluminescence probe, 2-methyl-6-[P-methoxyphenyl]-3,7-dihydroimidazo[1,2-α]pyrazin-3-one (MCLA), for detecting O2− production, in vitro, from Kupffer cells stimulated by phorbol myristate acetate. FEBS Letters. 1993;335(2):167–70.
71. Saitoh D, Kadota T, Okada Y, Masuda Y, Ohno H, Inoue M. Direct evidence for the occurrence of superoxide radicals in the small intestine of the burned rat. Am J Emerg Med. 1995;13(1): 37–40.
72. Ushiroda S, Maruyama Y, Nakano M. Continuous detection of superoxide in situ during ischemia and reperfusion in the rabbit heart. Japanese heart journal. 1997;38(1):91–105.
73. Tampo Y, Tsukamoto M, Yonaha M. The antioxidant action of 2-methyl-6-(p-methoxyphenyl)-3,7-dihydroimidazo[1,2-alpha]pyra z in-3-one (MCLA), a chemiluminescence probe to detect superoxide anions. FEBS Lett. 1998;430(3):348–52.
74. Uusitalo LM, Hempel N. Recent advances in intracellular and in vivo ROS sensing: focus on nanoparticle and nanotube applications. International journal of molecular sciences. 2012;13 (9):10660–79.

75. Ryvolova M, Chomoucka J, Drbohlavova J, Kopel P, Babula P, Hynek D, et al. Modern micro and nanoparticle-based imaging techniques. Sensors (Basel). 2012;12(11):14792–820.
76. Santra S, Xu J, Wang K, Tan W. Luminescent nanoparticle probes for bioimaging. J Nanosci Nanotechnol. 2004;4(6):590–9.
77. Lee D, Khaja S, Velasquez-Castano JC, Dasari M, Sun C, Petros J, et al. In vivo imaging of hydrogen peroxide with chemiluminescent nanoparticles. Nat Mater. 2007;6(10):765–9.
78. Dasari M, Lee D, Erigala VR, Murthy N. Chemiluminescent PEG-PCL micelles for imaging hydrogen peroxide. J Biomed Mater Res A. 2009;89(3):561–6.
79. Lee I-J, Hwang O, Yoo D-H, Khang G-S, Lee D-W. Detection of hydrogen peroxide in vitro and in vivo using peroxalate chemiluminescent micelles. Bulletin of the Korean Chemical Society. 2011;32(7):2187–92.
80. Chen R, Zhang L, Gao J, Wu W, Hu Y, Jiang X. Chemiluminescent nanomicelles for imaging hydrogen peroxide and self-therapy in photodynamic therapy. J Biomed Biotechnol. 2011;2011:679492.

Chapter 10
Fluorimetric Methods

A. Khelfi

Abstract Fluorimetry is an analytical technique that measures the fluorescence at a specific emission wavelength in compounds after induction with a light pulse of a specific excitation wavelength. Fluorescent probes used in oxidative stress (OS) assessment are based on redox processes between OS biomarker (hydrogen peroxide, $NO^{\bullet}$, superoxide, S-nitrosothiols, protein carbonyls, malondialdehyde (MDA), advanced glycation end-products (AGEs), glutathione (GSH)) and the reduced probe, which fluoresces upon oxidation. Although fluorescence is among the most sensitive techniques for radical detection, it has multiple limitations. Due to the lack of selectivity of most sensors (e.g., dichlorodihydrofluorescein), these chemicals are more appropriately called "redox probes" to indicate their general reactivity to numerous reactive species (RS). The probe should be compatible with multicolor detection to combine RS detection with other parameters such as pH indicators or cellular differentiation markers. Many probes for the detection of RS are commercially available, but most of them do not meet the ideal criteria. In this chapter, we describe the most important fluorimetric sensors for quantifying OS biomarkers. We discuss their advantages and limitations as well as the reaction involved in the generation of the fluorescence.

Keywords Hydrogen peroxide · Dichlorodihydrofluorescein · NO · Superoxide · S-nitrosothiols · Protein carbonyls · Malondialdehyde (MDA) · Advanced glycation end-products (AGEs) · Glutathione (GSH)

A. Khelfi (✉)
Faculty of Pharmacy, University of Algiers, Algiers, Algeria

National Centre of Toxicology, Algiers, Algeria

Department of Chemistry and Biomolecular Science, Clarkson University, Potsdam, NY, USA
e-mail: ab.khelfi@univ-alger.dz; akhelfi@clarkson.edu

S. Andreescu et al. (eds.), *Biomarkers of Oxidative Stress*,
https://doi.org/10.1007/978-3-031-60738-7_10

10.1 Introduction

Fluorescence occurs when photons are absorbed by a fluorophore resulting in its excitation with a subsequent emission of light. Fluorescent probes used in OS assessment are based on redox processes between RS and the reduced probe, which fluoresces upon oxidation. Although fluorescence is among the most sensitive techniques for radical detection (around 1 nM), caution is needed for their application. Fluorescence probes are capable of detecting several RS simultaneously via one- or two-electron redox reactions and suffer from selectivity as they may compete with endogenous antioxidants such as ascorbate, thiols, and urate. Moreover, some probes can generate $O_2^{\bullet-}$ via the generation of an active intermediate after reaction with RS. Due to the lack of selectivity, these chemicals are more appropriately called redox probes to indicate their general reactivity to numerous RS.

At cellular scale, RS signaling is controlled in a spatiotemporal manner, which requires the studying of the involved dynamics at sub-cellular locations. The ideal RS probe should be highly specific for one RS and not interfere with or be affected by environmental factors like salt concentration, pH, or endogenous enzymes. It should have fast reaction kinetics and high sensitivity in a range that allows the detection of the expected RS concentrations. If this condition cannot be achieved, a panel of probes operating over a wide range of sensitivities would be ideal to explore the full range of concentrations that may occur. The probe should be very bright, allowing the use of low concentration to avoid interference with the surrounding environment. The probe should be compatible with multicolor detection to combine RS detection with other parameters such as pH indicators or cellular differentiation markers. The probe should also be targetable to specific organs, cells, or subcellular locations.

Many probes for the detection of several RS are commercially available, but most of them do not meet the ideal criteria (Table 10.1) that would allow for the precise and accurate detection of redox events.

The simplest instrument to measure the output of these sensors is the fluorescence microplate reader, where data are presented as increases or decreases in relative fluorescence. However, plate readers measure total fluorescence and cannot distinguish between intracellular and extracellular fluorescence if the studied system involves chemical reactions taking place in cultured cells. Moreover, the quality and sensitivity of the instruments commercially available vary tremendously, and the requirement for additional excitation/emission filters often makes them expensive. Newer models are now available that do not require expensive filters but work on a dual monochromator principle with good flexibility.

Table 10.1 The ideal fluorescent probe for RS detection

Properties	Explanation
Sensitivity	High sensitivity to detect physiological concentrations of RS/free radicals (FR) (nano-micromolar range)
Selectivity/ Specificity	Selectivity and no cross-reactivity with other RS to avoid disambiguity of the type of RS involved in the reaction Eliminates any signal coming from interfering (endogenous or exogenous) compounds
Signal-to-noise ratio	Low fluorescence of free probes to avoid false signals coming from their accumulation in the cell
Linear response	Large linear relationship between the fluorescent signal to the RS concentration to allow for quantitative analysis of RS generation in physiological and stress conditions
Defined spectra and peaks	Narrow peaks of excitation and emission spectra to allow for simultaneous detection of several probes during the same analysis
Two-photon analysis	Compatibility with two-photon instruments to allow for prolonged observation and deep tissue penetration imaging without tissue damage
Photostability	Low photo-oxidation and photo-bleaching to facilitate imaging on the microscope
In vivo capability	Possibility of the probe usage in in vivo studies to allow redox reaction studies in animal models
Toxicity	Non- to very low toxicity of the probe without interfering with the underlying biological processes
Membrane permeability	High membrane permeability but moderate diffusion of the product to allow for ideal localization of redox reactions
Post-fixation retention	Retention after fixation to allow for simultaneous detection of the probe and the antibody for co-localization studies

10.2 Fluorometry

10.2.1 Hydrogen Peroxide

10.2.1.1 Dichlorodihydrofluorescin

The oxidation of 2′-7′dichlorodihydrofluorescin (DCFH) to 2′-7′dichlorofluorescein (DCF) has been used extensively for the detection and quantification of H_2O_2. The diacetate form of DCFH, dichlorofluorescin diacetate (DCFH-DA), and its acetomethyl ester (DCFH-DA-AM) have also been used to measure H_2O_2 [1].

DCFH-DA and DCFH-DA-AM are taken up by cells where nonspecific cellular esterases act upon them to cleave off the lipophilic groups, resulting in a charged compound that remains inside the cell. Oxidation of DCFH by RS converts the molecule to 2′, 7′dichlorodihydrofluorescein (DCF), which is a highly fluorescent compound (Fig. 11.1). The reported wavelengths for the measurement of DCF fluorescence are 498 nm for excitation and 522 nm for emission.

Originally, DCF was thought to be specific for H_2O_2, but several evidence have shown that other RS such as NO_3^- and HOCl can oxidize DCFH [2]. Most important is the fact that hydrogen peroxide-dependent oxidation of DCFH to DCF requires ferrous iron [3]. In addition, as DCF is no longer a charged compound, it is not precluded from migrating out of the cell and accumulating in the environmental media, where it is free to interact with extracellular oxidants/antioxidants.

10.2.1.2 Amplex Red

In the presence of horseradish peroxidase (HRP), Amplex Red reagent (10-acetyl-3,7-dihydroxyphenoxazine) reacts with H_2O_2 with a 1:1 stoichiometry to produce a highly fluorescent resorufin product ($\lambda_{excitation} = 563$ nm and $\lambda_{emission} = 587$ nm) (Fig. 11.17). The concentration of H_2O_2 is proportional to the intensity of resorufin. This way of employing Amplex Red is sensitive to H_2O_2 with a detection limit of 5 pmol. The Amplex Red/HRP assay has been found to detect levels of H_2O_2 produced by NAD(P)H oxidase (NOX) and other oxidases [4].

When the assay is applied on isolated mitochondria, fluorescence is measured in a microplate reader, prewarmed to 37 °C, at 15 s intervals, for a total time of 30 min. The initial linear phase of increase in absorbance is used to calculate the rates of hydrogen peroxide production per milligram of mitochondrial protein. Results are calculated as relative fluorescence units (RFUs), or they can be expressed as pmol H_2O_2/mg of protein/30 min [5]. The fluorescence microplate reader is set at excitation wavelength 560 nm and emission wavelength 590 nm.

10.2.1.3 Homovanillic Acid

Homovanillic acid dimerizes when oxidized by H_2O_2 through HRP catalysis. As with Amplex Red, homovanillic acid monomer is nonfluorescent, but as a dimmer, it possesses a peak excitation wavelength of 315 nm, with an emission wavelength of 425 nm (Eq. 10.1).

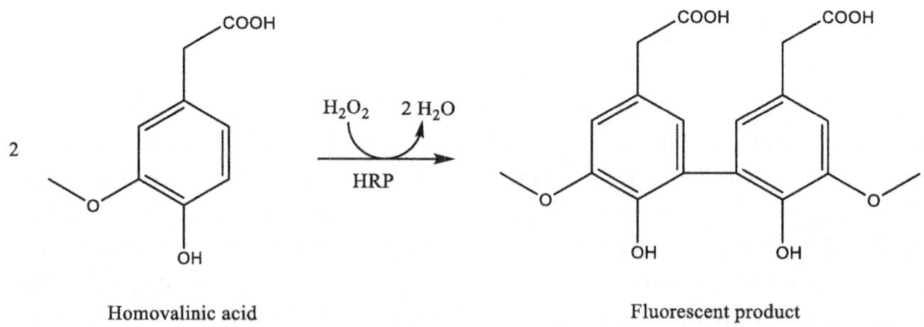

(10.1)

10.2.2 NO

10.2.2.1 Diaminonaphthalene Assay

In attempts to improve the sensitivity of Griess assay for $NO^{\bullet}$, a number of fluorometric methods has been developed. Like the Griess reaction, fluorometric assays are dependent on N_2O_3 or HNO_2, which are formed spontaneously by the acidification of NO_2^-. The most commonly used method employs a nonfluorescent reagent, 2, 3-diaminonaphthalene (DAN), to react with nitrite- or nitric oxide-derived N_2O_3 generated from the acid-catalyzed formation of HNO_2 or the auto-oxidation of $NO^{\bullet}$ to form 2, 3 naphthotriazole (NAT), which is highly fluorescent with an excitation wavelength of 375 nm and an emission wavelength of 415 nm (Fig. 10.1). This method can be applied in polluted waters and physiological fluids [6].

DAN assay offers several advantages including sensitivity, specificity, and versatility. It is capable of detecting as little as 10–30 nM naphthotriazole; therefore, it can be used to quantify $NO^{\bullet}$ generated under physiological conditions at neutral pH with minimal interference resulting from nitrite decomposition [7, 8]. As with Griess reaction, DAN assay can be used to continually quantify nitric oxide production in biological fluids, culture media, and organ culture supernatants.

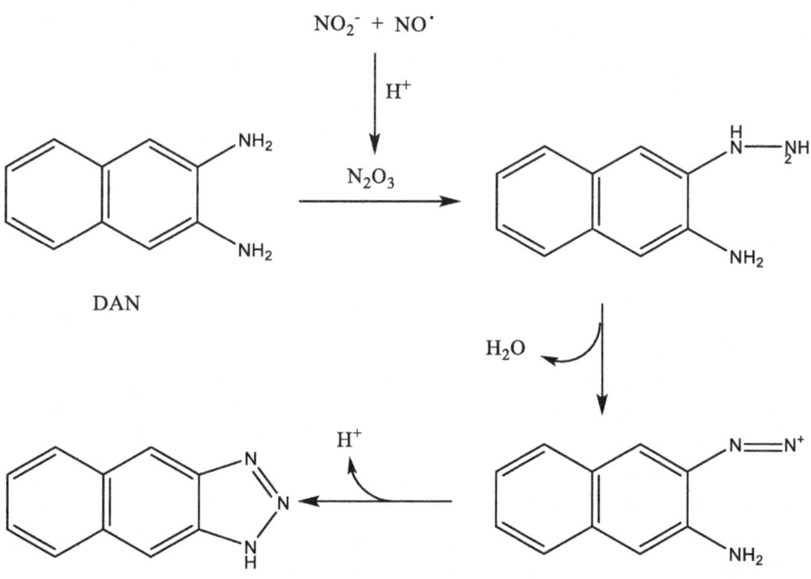

Fig. 10.1 Fluorimetric detection of $NO^{\bullet}$ and NO_2^- using DAN probe
The nitrosating agent N_2O_3 generated from the auto-oxidation of $NO^{\bullet}$ or from the acidification of NO_2^- reacts with DAN to yield a highly fluorescent product, naphthotriazole (NAT)

10.2.2.2 Diaminofluoroscein-2 Assay

Diaminofluoroscein-2 (DAF-2) is a commonly used reagent for the detection of $NO^{\bullet}$ in in vitro, in vivo, and in situ studies [9, 10]. Like DAN, the N-nitrosation of the diamino groups results in a nitrosamine, which through an internal rearrangement forms a highly fluorescent triazole ($\lambda_{excitation} = 495$ nm, $\lambda_{emission} = 515$ nm).

The main advantages of this compound are that wavelength associated with fluorescein can be used, making the instruments currently used for other bioassays as well as cell and tissue imaging easily adapted to nitric oxide detection. Although nitric oxide-derived N_2O_3 is thought to be the primary compound by which cells and tissues generate the triazole derivative, some investigators suggest that oxidative nitrosylation may represent an alternative pathway for triazole generation [11, 12].

Since DAF-2 can be oxidized via a one-electron reaction with $NO_2^{\bullet}$, it may first oxidize the diamino complex to an aromatic radical, which then undergoes radical–radical coupling reaction to form nitrosamine. The latter subsequently rearranges to the fluorescent triazole. This implies that, under biological conditions, the triazole can be formed from both oxidative and nitrosative chemistry. In culture media, these two pathways can be differentiated using nitrosative inhibitors or scavengers like azide, which prevent nitrosative but not oxidative reactions [12]. Urate can be used since it quenches oxidative nitrosylation but not nitrosative stress. Using different scavengers, the chemistry of DAF-2 formation can be identified in different experiments. Although DAF-2 was thought to be a relevant indicator of $NO^{\bullet}$, the compound has shown reactivity with nitroxyl (HNO), giving higher yields of triazole than $NO^{\bullet}$. In light of the potential presence of HNO in biological systems, it would be more appropriate to assume the possibility that some of the $NO^{\bullet}$ detected by DAF-2 is in fact HNO.

10.2.3 $O_2^{\bullet-}$

Dihydroethidium (DHE) can be used to measure mitochondrial $O_2^{\bullet-}$ production in isolated mitochondria. DHE is capable of getting inside the cells and be oxidized by $O_2^{\bullet-}$ to form the specific red fluorescent 2-hydroxyethidium (2-OH-E^+) (Fig. 11.13). This sensor can be oxidized by other oxidants including H_2O_2, $OH^{\bullet}$, and $ONOO^-$ to generate a nonspecific red fluorescent ethidium (E^+). Both 2-OH-E^+ and E^+ can be captured indiscriminately by fluorescence-based instruments. Fluorescence can be measured at 600 nm when excited at 520 nm in monochromatic fluorescence plate reader. Considering the disparities between the fluorescent products (mixture of 2-OH-E^+ and E^+) and the concentration of $O_2^{\bullet-}$, DHE assay alone is not enough for $O_2^{\bullet-}$ quantification. Distinguishing the fluorescence coming from 2-OH-E^+ and E^+ has been achieved by the application of separation techniques like HPLC and LC-MS [13, 14]. Finally, it should be noted that the reaction stoichiometry between

DHE and $O_2^{\bullet -}$ is not all the same among different types of cells and systems and the generation rate of 2-OH-E^+ does not always reflect the release of $O_2^{\bullet -}$ [13, 15].

10.2.4 S-Nitrosothiols

The detection of RSNO often employs Saville reaction [16]. The fluorometric assay is based on the reaction of DAN with NO^+-derived N_2O_3, liberated from RSNO following mercuric chloride addition, which leads to the conversion of DAN to its fluorescent triazole derivative (Fig. 8.3). This assay can be conducted at neutral rather than acidic pH, which eliminates the interference coming from NO_2^- and allows for the specific detection of the RSNO-derived nitrosating agent. The assay has been found to be effective in the range of 50–1000 nM RSNO [17].

Variations of this assay have been used to quantify low and high molecular weight RSNOs in human plasma as well as the S-nitrosated derivatives of the human hemoglobin [18].

10.2.5 Protein Carbonyls

Fluorimetric assays for the quantification of protein carbonyls are based on derivatization reactions using some common fluorophores like 2,4-dinitrophenylhydrazine (DNP) and 7-hydrazino-4-nitrobenzo-2,1,3-oxadiazole (NBDH) [19, 20].

10.2.6 MDA

TBA-MDA adduct (Fig. 8.1) can be measured fluorometrically. For increased sensitivity and specificity, the fluorescent chromogen can be extracted into an organic solvent such as butanol and measured using a 530 nm excitation wavelength and a 550 nm emission wavelength [21].

10.2.7 Advanced Glycation End Products (AGEs)

Noninvasive spectrographic auto-fluorescence readers can be applied in a clinical setting to detect AGEs in serum and tissues. This approach has been used to assess AGE concentrations in human skin. However, it should be standardized in terms of the analyzed body region, the surrounding light, and the possible presence of skin areas with tattoos. Elevated skin auto-fluorescence has been demonstrated in kidney disease, diabetes, and in patients with arterial stiffness [22].

10.2.8 GSH

Detection of GSH by fluorimetry is relatively a straightforward process since a number of methods are now available for GSH measurement using multiple fluorophores (e.g., orthophthalaldehyde, N-chlorodansylamine (NCDA), dansyl chloride, monobromobimanes (MBBs), and monochlorobimanes (MCBs)) [23, 24]. These sensors have all proved useful in determining intracellular concentrations of GSH. Derivatization usually causes some GSH to oxidize to its disulfide form.

MCBs are versatile, membrane permeable, and nonfluorescent probes that bind irreversibly to thiol groups yielding a fluorescent product [25]. Fluorescence can be measured in a fluorimeter with excitation and emission wavelengths set at 395 and 480 nm, respectively. Interestingly, very low reactivity of MCBs with protein sulfhydryl residues has been recorded [26]. Nevertheless, it is advisable to measure GSH levels using a protein-free assay.

Orthophthalaldehyde can accurately assess total glutathione levels in protein precipitated cell lysates [27]. It is, therefore, possible to estimate the ratio of GSH/GSSG from experiments using both MCBs and orthophthalaldehyde.

References

1. Paraidathathu T, de Groot H, Kehrer JP. Production of reactive oxygen by mitochondria from normoxic and hypoxic rat heart tissue. Free Radic Biol Med. 1992;13(4):289–97.
2. Hoffman A, Spetner LM, Burke M. Ramifications of a redox switch within a normal cell: its absence in a cancer cell. Free Radic Biol Med. 2008;45(3):265–8.
3. Rothe G, Valet G. Flow cytometric analysis of respiratory burst activity in phagocytes with hydroethidine and 2′,7′-dichlorofluorescin. J Leukoc Biol. 1990;47(5):440–8.
4. Zhou M, Diwu Z, Panchuk-Voloshina N, Haugland RP. A Stable Nonfluorescent Derivative of Resorufin for the Fluorometric Determination of Trace Hydrogen Peroxide: Applications in Detecting the Activity of Phagocyte NADPH Oxidase and Other Oxidases. Analytical Biochemistry. 1997;253(2):162–8.
5. Armstrong JS, Whiteman M. Measurement of Reactive Oxygen Species in Cells and Mitochondria. Methods in Cell Biology. 80: Academic Press; 2007. p. 355–77.
6. Bryan NS, Grisham MB. Methods to detect nitric oxide and its metabolites in biological samples. Free Radic Biol Med. 2007;43(5):645–57.
7. Miles AM, Wink DA, Cook JC, Grisham MB. Determination of nitric oxide using fluorescence spectroscopy. Methods in enzymology. 268: Academic Press; 1996. p. 105–20.
8. Nims RW, Cook JC, Krishna MC, Christodoulou D, Poore CM, Miles AM, et al. Colorimetric assays for nitric oxide and nitrogen oxide species formed from nitric oxide stock solutions and donor compounds. Methods in enzymology. 1996;268:93–105.
9. Kojima H, Nakatsubo N, Kikuchi K, Kawahara S, Kirino Y, Nagoshi H, et al. Detection and Imaging of Nitric Oxide with Novel Fluorescent Indicators: Diaminofluoresceins. Analytical Chemistry. 1998;70(13):2446–53.
10. Kojima H, Sakurai K, Kikuchi K, Kawahara S, Kirino Y, Nagoshi H, et al. Development of a fluorescent indicator for nitric oxide based on the fluorescein chromophore. Chem Pharm Bull (Tokyo). 1998;46(2):373–5.

11. Espey MG, Miranda KM, Thomas DD, Wink DA. Ingress and reactive chemistry of nitroxyl-derived species within human cells. Free Radic Biol Med. 2002;33(6):827–34.
12. Espey MG, Thomas DD, Miranda KM, Wink DA. Focusing of nitric oxide mediated nitrosation and oxidative nitrosylation as a consequence of reaction with superoxide. Proc Natl Acad Sci U S A. 2002;99(17):11127–32.
13. Fink B, Laude K, McCann L, Doughan A, Harrison DG, Dikalov S. Detection of intracellular superoxide formation in endothelial cells and intact tissues using dihydroethidium and an HPLC-based assay. American journal of physiology Cell physiology. 2004;287(4):C895–902.
14. Zhao H, Joseph J, Fales HM, Sokoloski EA, Levine RL, Vasquez-Vivar J, et al. Detection and characterization of the product of hydroethidine and intracellular superoxide by HPLC and limitations of fluorescence. Proc Natl Acad Sci U S A. 2005;102(16):5727–32.
15. Zielonka J, Zhao H, Xu Y, Kalyanaraman B. Mechanistic similarities between oxidation of hydroethidine by Fremy's salt and superoxide: stopped-flow optical and EPR studies. Free Radic Biol Med. 2005;39(7):853–63.
16. Saville B. A scheme for the colorimetric determination of microgram amounts of thiols. Analyst. 1958;83(993):670–2.
17. Wink DA, Kim S, Coffin D, Cook JC, Vodovotz Y, Chistodoulou D, et al. Detection of S-nitrosothiols by fluorometric and colorimetric methods. Methods in enzymology. 1999;301: 201–11.
18. Jourd'heuil D, Hallén K, Feelisch M, Grisham MB. Dynamic state of S-nitrosothiols in human plasma and whole blood. Free Radic Biol Med. 2000;28(3):409–17.
19. Weber D, Davies MJ, Grune T. Determination of protein carbonyls in plasma, cell extracts, tissue homogenates, isolated proteins: Focus on sample preparation and derivatization conditions. Redox Biol. 2015;5:367–80.
20. Stocker P, Ricquebourg E, Vidal N, Villard C, Lafitte D, Sellami L, et al. Fluorimetric screening assay for protein carbonyl evaluation in biological samples. Anal Biochem. 2015;482:55–61.
21. Lee R, Margaritis M, Channon KM, Antoniades C. Evaluating oxidative stress in human cardiovascular disease: methodological aspects and considerations. Curr Med Chem. 2012;19 (16):2504–20.
22. Frijhoff J, Winyard PG, Zarkovic N, Davies SS, Stocker R, Cheng D, et al. Clinical Relevance of Biomarkers of Oxidative Stress. Antioxid Redox Signal. 2015;23(14):1144–70.
23. Ellman GL. Tissue sulfhydryl groups. Archives of Biochemistry and Biophysics. 1959;82(1): 70–7.
24. Neuschwander-Tetri BA, Roll FJ. Glutathione measurement by high-performance liquid chromatography separation and fluorometric detection of the glutathione-orthophthalaldehyde adduct. Analytical Biochemistry. 1989;179(2):236–41.
25. Kosower EM, Kosower NS. Bromobimane probes for thiols. Methods in enzymology. 1995;251:133–48.
26. Curtin JF, Donovan M, Cotter TG. Regulation and measurement of oxidative stress in apoptosis. Journal of Immunological Methods. 2002;265(1):49–72.
27. Kaul N, Choi J, Forman HJ. Transmembrane redox signaling activates NF-kappaB in macrophages. Free Radic Biol Med. 1998;24(1):202–7.

Chapter 11
Fluorescence-Based Imaging Techniques

A. Khelfi

Abstract Fluorescence imaging of reactive species (RS) using a variety of highly sensitive sensors allows temporal and spatial tracking of oxidants in uni- and multicellular organisms. Fluorescent probes are often oxidant sensitive, and they are non-fluorescent before being oxidized by RS. Upon reaction between probes and RS, the oxidized sensor becomes a highly fluorescent molecule. There is a wide range of chemical probes used in the biomedical field (e.g., dichlorodihydrofluorescein diacetate (DCFH-DA), dihydrorhodamine, dihydroethidium, Amplex Red, BODIPY, boronates, DanePy, DAF-2-DA, etc.) that exploit the existing oxidative chemistry to examine the production of RS in living cells. Genetically encoded biosensors based on fluorescent proteins (e.g. green fluorescent proteins (GFPs)), which provide a mean to real-time and dynamic detection of redox-status change without special requirements for permeability into target cells, are new and useful sensors to monitor changes in redox homeostasis during distinct physiological or pathological states.

This chapter covers the up-to-date fluorescence imaging techniques for basic and clinical applications. It provides a concise, yet thorough, description of the state-of-the-art fluorescent probes that any investigator working in the field needs to access with their advantages and limitations. It also deals with nanodiagnostic methods for measuring oxidative stress (OS) with a special focus on the recent development in this field.

Keywords Dichlorodihydrofluorescein diacetate (DCFH-DA) · Dihydrorhodamine · Dihydroethidium · Amplex Red · BODIPY · Boronates · DanePy · DAF-2-DA · Nanodiagnostic methods · Green fluorescent proteins (GFPs)

A. Khelfi (✉)
Faculty of Pharmacy, University of Algiers1, Algiers, Algeria

National Centre of Toxicology, Algiers, Algeria

Department of Chemistry and Biomolecular Science, Clarkson University, Potsdam, NY, USA
e-mail: ab.khelfi@univ-alger.dz; akhelfi@clarkson.edu

S. Andreescu et al. (eds.), *Biomarkers of Oxidative Stress*,
https://doi.org/10.1007/978-3-031-60738-7_11

11.1 Introduction

Given the short-lived nature of most radicals and oxidants, it is often difficult to measure OS through direct detection methods. As a result, in order to quantify the level of OS in cellular systems, one must rely upon sentinel probes that change fluorescence in response to RS [1, 2]. Indeed, in vitro OS assays based on fluorescent probes are the most commonly used mean of quantifying OS in cell culture systems. Fluorescent probes are often oxidant sensitive, and they are nonfluorescent before being oxidized by RS. Upon reaction between probes and RS, the oxidized sensor becomes a highly fluorescent molecule. Most of these probes are oxidized by a one-electron free radical (FR) mechanism, yielding a probe radical intermediate and resulting in fluorescent products. There is a wide range of chemical probes that exploit the existing oxidative chemistry to examine the production of reactive oxygen species (ROS)/reactive nitrogen species (RNS) in living cells and high throughput fluorescent-based platforms. Among them, cell-permeable compounds are helpful to reflect the oxidative state of cellular compartments, providing information on RS production under stimulus. Cell-permeable probes are necessary for the applicability of exogenous RS-sensing fluorophores. Generally, the water/octanol partition coefficient could be utilized to quantify membrane. For a quantitative analysis, it is necessary to know the reaction mechanism in detail, as well as possible interfering side effects and the cellular localization of these probes. An overview of detailed chemical reaction schemes is given in this chapter.

Signal from the single-wavelength emitting probes can be affected by the concentration of the probe, which questions the quantification analysis of these probes. Ratiometric probes monitor changes in the fluorescence in ratio to the probe itself, which corrects the artifacts caused by the variable concentration of the probe or photo-bleaching.

In fluorescent imaging, two main approaches can be used:

- Addition of exogenous fluorescence probes, which penetrate into the cell and change their fluorescence properties due to reaction with ROS.
- Expression of ROS-sensitive fluorescent proteins, mostly variants of the GFP, which act as real-time redox reporters for use in cells and entire animals.

The use of fluorophores offers a promising tool because it permits the application of recently developed advanced techniques of time- and space-resolved fluorescence microscopy for in vivo studies. Bottom-reading machines have the advantage that the cells can be measured in situ without the need for trypsinization or cell scraping, processes that themselves generate cellular OS and result in artifactual changes in fluorescence. In intact cells, organelle-derived RS could be detected by confocal microscopy using conventional RS-specific probes or protein-based ROS reporter molecules that can be selectively expressed in different organelles (Table 11.1) [3]. Cells can be loaded with fluorescent probes and viewed in real time in culture chambers at 37 °C. The intracellular location of RS can be visualized, and the role of mitochondrial, endoplasmic reticulum (ER), or lysosomal events in OS may be

Table 11.1 Commonly used instruments for the detection of RS using fluorescence technology

Detection technology	Suitability
Wide field fluorescence microscopy with filter changer for ratiometric microscopy	Cells in culture and tissue sections
Confocal microscopy	Cells in culture and tissue sections
2 hν microscopy	Tissue (depth up to 1 mm) with two-photon imaging of fluorescent sensors
Fluorescence lifetime imaging (FLIM)	Cells in culture and tissue sections
Endomicroscopy	Organs in living animals

visualized using counter stains. Besides, flow cytometry can be used for the detection of RS with fluorescent probes (see Chap. 21).

Despite all these advantages, several criticisms have been addressed to fluorescence imaging techniques developed for RS identification in cells. This is because most of them are not specific to a particular RS and are sensible to RS-independent oxidation.

Nevertheless, through variable design of these probes, some of them can be imparted with a certain amount of RS specificity, allowing for an appropriate measurement and visualization of ROS/RNS (e.g., $O_2^{\bullet-}$, $ROO^{\bullet}$, H_2O_2, $OH^{\bullet}$, and $NO^{\bullet}$). Indeed, various commercial assays are available for specific quantification/visualization of RS in vitro. Regardless of the exact target molecule of interest, the assays should meet the following criteria:

- It should be able to detect the corresponding target RS/FR in most physiological conditions.
- It should be highly selective for the target of interest (without producing false signals).
- It should have low detection limit that allows for signal detection under physiological concentrations [2].

It is noteworthy to know that assaying oxidative activity in living cells with fluorogenic probes is complicated by the frequent presence of multiple RS in the same cell. Scavengers and enzymes such as superoxide dismutase (SOD) and catalase (CAT) are useful knockdown reagents for triaging the optical response of RS probes. Fluorescent assays for general OS assessment are available as well. For example, 2′-7′-dichlorofluorescein (DFC) fluorescence assay, and ratio of antioxidant/oxidized antioxidant are commonly used for measuring OS levels in cells [4, 5].

Tables 11.2 and 11.3 list a selection of probes used in OS research for the detection of various RS in cells. Most of the mentioned fluorophores are not truly specific to certain RS. However, as exogenous probes typically respond in a certain oxidative potential range, appropriate mixtures of probes improve the selectivity of the assay. These assays are more selective than those utilizing just a single probe.

Along with an expanding field of redox biology and probes chemistry, a considerable interest has been invested in the improvement of fluorescent RS probes. The following section outlines some of the most common probes used with a description of their use. In the next paragraphs, we will also review the recent advancements in RS fluorescence measurements in cells both in vitro and in vivo.

Table 11.2 RS and their detection using fluorescent imaging

ROS/RNS	Detection reagents
H_2O_2	DCFH-DA, CM-DCFH-DA, Amplex Red, Calcein-AM, DHR, PL1, PCL-1, DHE, RPF1, Boronate probes, FP-H_2O_2-NO, roGFP2-Orp1, HyPer, rxRFP, PF6-AM, OxyFRET, and PerFRET
$OH^{\bullet}$	APF, HPF, AuNC@HPF, DCFH-DA, CM-DCFH-DA, 3-CCA, CONER, AuNC@HPF, DHR, DHE, C-11-BODIPY581/591, Hydro-Cy3, and Hydro-Cy5
$^1O_2$	DHR, APF, HPF, and DanePy
$O_2^{\bullet-}$	DHE, mito-SOX, Amplex Red, Hydro-Cy3, Hydro-Cy5, and cpYFP
HOCl	DHR, APF, DHR123, DCFH-DA, DHE, and Boronate probes
NO	DAF-2-DA, 2,3-Diaminonaphthalene, DCFH-DA, MNIP-Cu, CuFL1, CuFL1E, CuFLDEX, CuSNFL1, CuSNFL1E, CuRBT, NO550, Cou-Rho-NO, FP-H_2O_2-NO, fluoresceinamine-SiNW probe, and FRET metallothionein-Yap1
$ONOO^-$	APF, HPF, DHR, DCFH-DA, DHR, DHE, Amplex Red, C-11-BODIPY581/591, and Boronate probes
$NO_2^{\bullet}$	DHR, DCFH-DA
$CO_3^{\bullet-}$	DCFH-DA
$LO^{\bullet}$	C-11-BODIPY581/591, and DCFH-DA
$LOO^{\bullet}$/LOOH	DPPP, Cis-Parinaric acid, C-11-BODIPY581/591, DCFH-DA, and OHSer
2 GSH/GSSG ratio	roGFP1, roGFP2, Grx1-roGFP2-iL, rxYFP, rxYFP-Grx1P, rxYFP 3R, and cpRFP

11.2 Fluorescent Sensors

11.2.1 Dichlorodihydrofluorescein Diacetate (DCFH-DA) and Derivatives

11.2.1.1 General Aspects

The DCFH-DA assay is the most popular method for OS determination in cells. This assay uses 2′,7′-dichlorodihydrofluorescein diacetate (DCFH-DA), which is commonly used for the direct measurement of intracellular redox states. It is a relatively nonfluorescent molecule, hydrolyzed by esterases to a membrane-permeable probe, 2′,7′-dicholodihydrofluorescin (DCFH). Through the loss of the acetate groups, this probe accumulates mainly in the cytosol. To avoid any cytotoxicity, cells should be loaded with DCFH-DA at low concentrations. With a variety of cell types, it has been found that loading at 1–10 μM for 0.75–1 h is adequate [6]. When necessary, DCFH could be obtained by treating DCFH-DA with either esterase (like acetylesterase) or a strong alkaline solution (like NaOH) [7]. Upon two-electron oxidation, DCFH is then oxidized by nonspecific oxygen radicals to form the highly fluorescent (DCF), after the generation of a nonfluorescent intermediate $DCF^{\bullet}$. Accumulation of DCF can be visualized in cells at 522 nm after excitation at 498 nm (Fig. 11.1). Research studies have proved that the fluorescence intensity of

Table 11.3 Selected fluorescent probes used for RS detection and imaging

Name of the probe	$\lambda_{excitation}$ (nm)	$\lambda_{emission}$ (nm)	Ratiometric	Two-photon	Animal model
PF6-AM	492	517	–	–	–
"Click" boronate probe	400	475	–	–	–
RS-BE	510	580	–	–	–
RPF1	420	517/464	+	–	–
PL1	410	540/475	+	–	–
RF1	495	503	–	–	–
PCL-1	NA	612	–	–	+
DMACA	400	484/566	+	–	–
NPF1	598	660	–	–	–
PN1	750 fs pulses	500–550/396–465	–	+	–
sulfo-QCy7	595	>635	–	–	+
MitoPY1	510	528	–	–	–
SHP-Mito	750	530–600/400–470	+	+	–
Mito-SOX	369	580	–	–	–
NucPE1	514	525–555	–	–	+
DAF-DA	495	515	–	–	–
DCFH-DA	492–495	515–527	–	–	–
Oxyburst probes	492–495	515–527	–	–	–
DHE	369	580	–	–	–
MNIP-Cu	NA	330–385	–	–	–
CuFL1E and CuFLDex	506	526	–	–	–
CuSNFL1 and CuSNFL1E	496	538	–	–	–
CuRBT	488	585	–	–	–
NO550	470	550	–	–	–
Cou-Rho-NO	365	583/473	+	–	–
FP-H_2O_2-NO: NO H_2O_2 NO + H_2O_2	 550 400 400/550	 580 460 580	–	–	–

DCF is correlated with the levels of OS in cells, showing the reliability of this probe in reflecting redox levels [8].

The oxidation of DCFH to the fluorescent compound DCF was initially thought to be a relatively specific indicator of intracellular formation of H_2O_2 and has been used extensively in the detection of H_2O_2 produced during the respiratory burst in inflammatory cells, either in cell lysates or in intact cells using flow cytometry [9]. Although it is often cited as a probe for H_2O_2, DCFH does not react directly

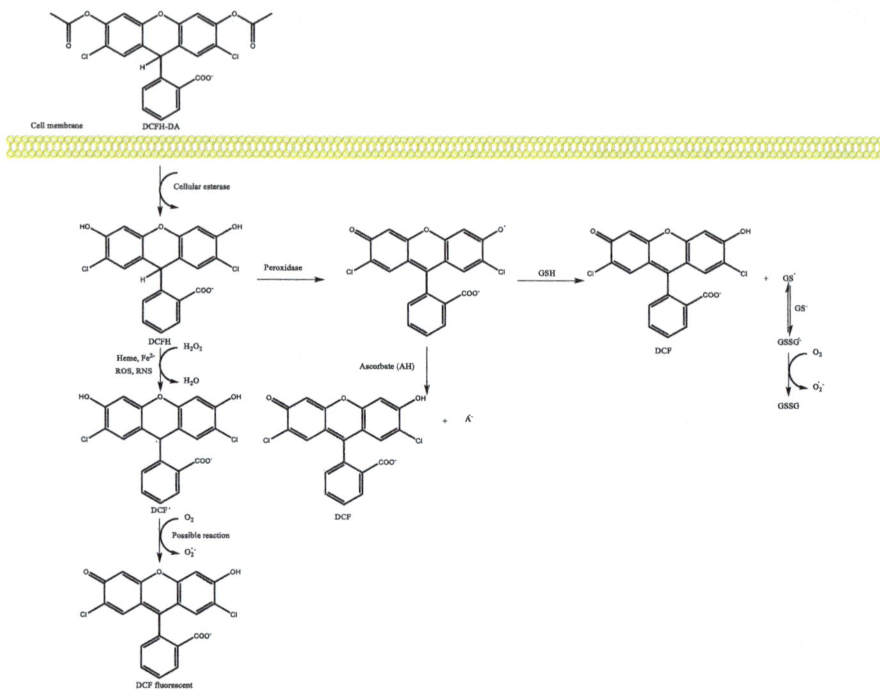

Fig. 11.1 DCFH-DA intracellular reactions
DCFH-DA is hydrolyzed by cellular esterases to DCFH, which oxidation by several RS yields florescent DCF via an intermediate radial, $DCF^{\bullet}$. Peroxidases can also convert DCF into a phenoxyl radical that can interact with antioxidants such as ascorbic acid (AH) or GSH. Thiyl radical resulting from the latter reaction can lead to $O_2^{\bullet-}$ generation

with H_2O_2 to produce the fluorescent product but rather requires a catalyst like cytochrome c or a redox-active metal ion. There are, however, some serious questions about DCFH-DA's use as a specific probe for H_2O_2. Indeed, other substances are capable of directly inducing DCF formation in the absence of H_2O_2, including $NO^{\bullet}$, $ONOO^{-}$, and HOCl [10, 11]. The response sensitivity of DCFH to $ONOO^{-}$ has been proven to be higher than that to other RS.

In the presence of heme-containing compounds, lipid peroxides are also capable of generating DCF fluorescence [12]. $ROO^{\bullet}$, $RO^{\bullet}$, $CO_3^{\bullet-}$, $NO_2^{\bullet}$, and $OH^{\bullet}$ radicals can also oxidize DCFH [6, 13]. Oxidation of DCFH is reportedly not sensitive to $^1O_2$ directly, but $^1O_2$ can indirectly contribute to the formation of DCF through its reaction with cellular substrates that yield peroxyl radicals [14]. In a cell-free system, DCFH has been shown to be oxidized by HRP and iron in the absence of H_2O_2 [13]. The addition of iron accelerates the oxidation of DCFH by H_2O_2, suggesting that $OH^{\bullet}$ may be involved in DCFH oxidation [15]. It has been reported that the hydrogen peroxide-dependent oxidation of DCFH to DCF occurs slowly in the absence of ferrous iron and can be completely inhibited by deferoxamine [16].

Moreover, in bovine aortic endothelial cells, the generation of a signal from DCFH-DA upon addition of H_2O_2 required the uptake of extracellular iron from the medium [17].

DCFH-DA has been widely used for the evaluation of hypoxia-reperfusion in several types of cells including HepG2 cells and H9c2 cardiomyocytes (Figs. 11.2 and 11.3) [18, 19]. Generation of ROS following menadione exposure was evaluated using DCFH-DA, as illustrated in Fig. 11.4 [20].

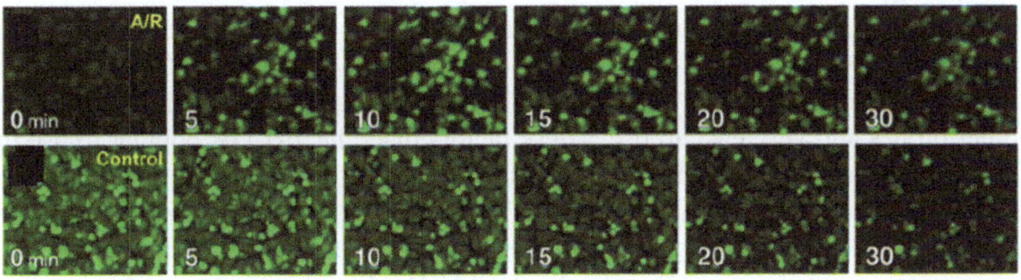

Fig. 11.2 Real-time visualization of OS during in vitro anoxia-reoxygenation
Formation of DCF from DCFH-DA (100 μM) in HepG2 cells during 30 min of reoxygenation following 4 h of anoxia. Representative fluorescence microscopy images are shown for reperfused cells following anoxia and control conditions [18]

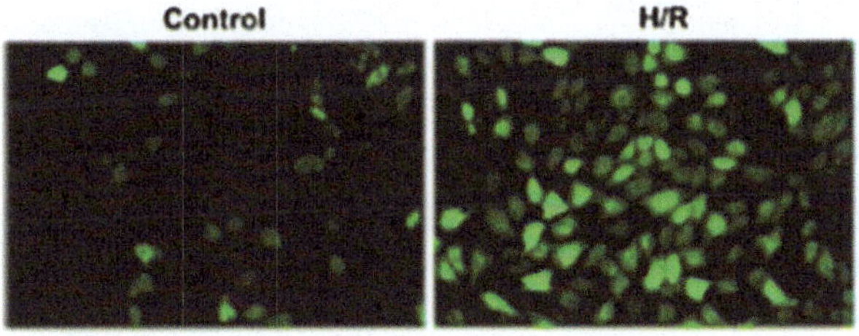

Fig. 11.3 Hypoxia-reperfusion-induced OS in H9c2 cardiomyocytes
Intracellular ROS production was detected using DCFH-DA method [19]

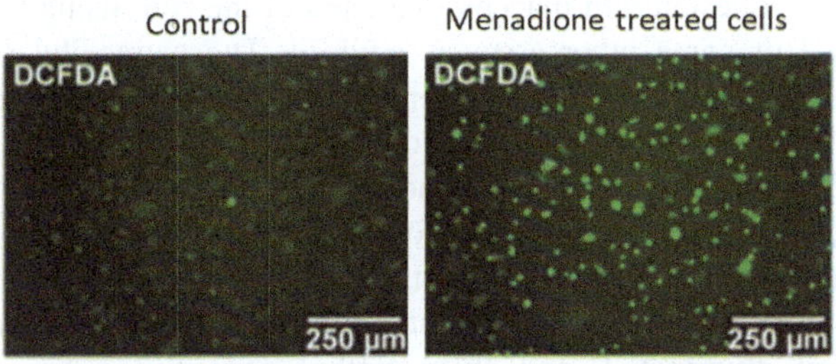

Fig. 11.4 DCF-DA detection of ROS in bovine aortic endothelial cells (control) and menadione-treated cells [20]

Besides, the dehydrogenation probe DCF-DA is also responsive to radicals, while its recognition range of RS is wider than that of DCFH-DA. DCF-DA is also cell permeable, and the enzymatic cleavage of the diacetate by cytoplasmic esterase is similar to that of DCFH-DA. The oxidation reaction to form DCF could be mediated by H_2O_2, $O_2^{\bullet -}$, or $NO^{\bullet}$ [21, 22].

11.2.1.2 Advantages and Inconveniences

Advantages

DCFH-DA has many advantages over other probes. It is inexpensive, very easy to use, extremely sensitive to changes of the redox state of cells, and can be used to follow changes in RS levels over time. DCFH-DA is relatively nontoxic compared to other probes and very easy to use. These factors made DCFH-DA popular to use in methods to measure oxidants in cells. Indeed, DCFH-DA has been used in many clinical and experimental applications. The probe has been used quantitatively to detect ROS produced by PMA (Phorbol-12-Myristate-13-Acetate)-stimulated phagocytes, as well as in cultured cells. DCFH-DA has also been used to visualize oxidative changes in carbon tetrachloride-perfused rat liver and in venular endothelium during neutrophil activation [23], as well as to examine the effect of ischemia and reperfusion in lung and heart tissues [24, 25]. It has been suggested using DCFH that processes such as signal transduction and β-amyloid toxicity involve H_2O_2 [26, 27].

Inconveniences

Despite the popularity of DCFH-DA assay, it cannot be reliably used to measure intracellular H_2O_2 and other RS for the following reasons:

- First of all, different types of cells show different permeability to this probe, resulting in different final equilibrium concentrations inside and outside cells.
- Second, the activity of esterase may vary a lot among cells. Some cell types have low esterase activity. This can have important implications for detecting RS because DCFH-DA needs to be hydrolyzed to DCFH by cellular esterases. Therefore, careful consideration should be given to the levels of esterase activity in cells before using this probe. In cells with low esterase activity or where esterases are sequestered in inaccessible parts of the cell, using the deacetylated probe DCFH or some other assay is advised. The extracellular hydrolysis of DCFH-DA could eliminate the differences in esterase activities among different cells, thus facilitating better detection of RS levels.
- Third, DCF is membrane permeable and can leak out of cells over time. Detecting slow hydrogen peroxide production over time can be difficult [28]. The low intracellular retention of DCF limits its use as a quantitative tool or in systems with slow oxidation. To overcome the cellular retention flaw, various analogs (carboxylated, fluorinated, and chloromethyl derivatives) were designed that retain their fluorescent properties but are charged at physiological pH, which allows them to accumulate in the cell even during slow hydrogen peroxide production [29].

- Fourth, the fluorescence intensity of DCF is not directly correlated with any specific FR. As mentioned earlier, radicals like $NO^•$, HOCl, and $ONOO^-$ can oxidize DCFH to DCF, thus interfering with the determination of the hydrogen peroxide level [30]. To distinguish RNS from ROS production, inhibitors of NOS can be used to prevent $NO^•$ mediated DCFH oxidation [11].
- Fifth, the probe radical intermediate $DCF^•$ reacts easily with oxygen to form $O_2^{•-}$ and then additional H_2O_2, amplifying the results of H_2O_2 [31]. Moreover, the reaction of DCFH with peroxidase forms DCF radicals, with the subsequent generation of $O_2^{•-}$ [6]. The dismutation of $O_2^{•-}$ to H_2O_2 would then yield artifactual DCFH oxidation and amplification of DCF fluorescence. In addition, cellular fluorescence of DCFH itself may also auto-oxidize to form H_2O_2.
- Sixth, as mentioned earlier numerous factors like cellular peroxidases (e.g., myeloperoxidase (MPO)), horseradish peroxidase (HRP), cytochrome c, hematin, heme proteins, as well as transition metals (e.g. ferrous ions) can promote and catalyze the oxidation of DCFH to DCF. This can give false increase in fluorescence values even in the absence of H_2O_2 [32, 33]. Hence, cellular peroxidase level and heme protein content are important variables to consider when interpreting studies with this probe [16]. For example, Ohashi et al. showed that manipulating heme content of cells by adding hemin or d-aminolevulinic acid affected rates of DCFH oxidation [34]. One-electron oxidation of DCFH by various peroxidases is likely to produce intermediate radicals (Fig. 11.1) that can interact with cellular antioxidants such as GSH and ascorbate, and with NADH to create more FR [35].
- Seventh, chromium (V), mitoxantrone, pyocyanin, and ametantrone can directly oxidize DCFH and cause an artifactual signal, and the possibility of such direct oxidations must always be checked before using DCFH-DA to measure OS in cells exposed to various toxic agents [36].
- Eighth, DCFH is easily photo-oxidized and photobleached, which results in auto-oxidation and false turn-on fluorescence response.
- Ninth, DCF itself can act as a photosensitizer for DCFH-DA oxidation, both priming and accelerating the formation of DCF [14].
- Tenth, it has been found that DCFH itself is a substrate for many enzymes including oxidases (e.g., xanthine oxidase (XO) and glucose oxidase) and peroxidases [15, 37].
- Eleventh, DCF can reportedly be further oxidized to both DCF semiquinone FR and DCF phenoxyl radical in a peroxidase (HRP)-catalyzed reaction, and this reaction may complicate the interpretation of results obtained with this probe in cells undergoing OS [35].

Overall, DCFH-DA is not competent enough to specifically detect a particular RS in cells due to its indiscriminate recognition of various oxygen radicals and oxidants. For this reason, this agent should not be used as a reliable measure of H_2O_2. The points mentioned earlier cannot be underestimated, given the plethora of research papers attempting to study intracellular signaling through DCF fluorescence, even in the recent literature.

The global change of cellular OS under intra- or extracellular stimulus could be reflected by DCFH-DA, based on the careful setting of control groups and reasonable analysis of results. Care should be taken to validate the experiments using other agents such as enzyme inhibitors. DCF assay can be used while keeping the mentioned limitations under consideration. Used carefully, this probe can be a powerful tool for analyzing RS production in cells.

11.2.1.3 DCFH-DA Derivatives

CM-DCFH-DA

The chloromethyl derivative of DCFH-DA (CM-DCFH-DA; 5-(and 6-) chloromethyl-2′,7′-dichlorodihydrofluorescein diacetate, acetyl ester) (Fig. 11.5) is sensitive to ROS and exhibits much better retention than DCFH-DA in living cells [38, 39]. Such probe is necessary, especially to avoid side effects due to generation of ROS by the applied probe, to monitor RS outside the cells in solution due to unspecific localization and/or photo-oxidation of the probe by illumination. CM-DCFH-DA passively diffuses into cells, where its acetate groups are cleaved by intracellular esterases and its thiol-reactive chloromethyl group reacts with intracellular GSH and other thiols. Subsequent oxidation yields a fluorescent adduct that is trapped inside the cell, thus facilitating long-term studies. Among its many applications, this probe has been used to perform the followings:

- Analyze FOXO3 transcriptional control of OS [40, 41].
- Assess RS-mediated cytotoxicity and apoptosis (Fig. 11.6) [42–46].
- Detect $OH^{\bullet}$ associated with estrogen-induced DNA damage [47].
- Monitoring RS generation in neurons and brain slices [38, 39, 48].
- Analyze RS production in chromosomally unstable human–hamster hybrid [49].

Carboxylated and Di(acetoxymethyl Ester) Analogs

To enhance the retention of DCFH-DA, the carboxylated DCFH-DA analog (6-carboxy-DCFH-DA), which has two negative charges at physiological pH, and its di(acetoxymethyl ester) analog, 6-carboxy-DCFH-DA, di(acetoxymethyl ester) were developed (Fig. 11.7) [50, 51]. These probes have better cellular retention and

Fig. 11.5 Structure of CM-DCFH-DA

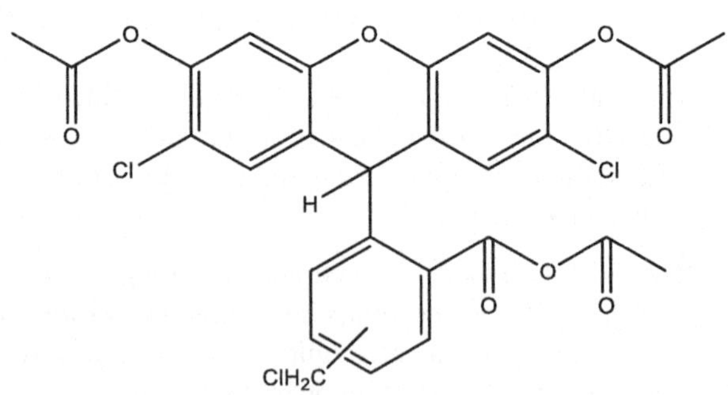

CM-DCFH-DA

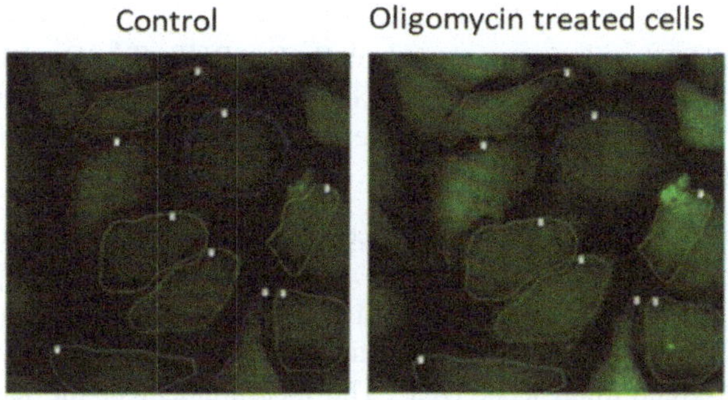

Fig. 11.6 Effect of oligomycin and FCCP on mitochondrial membrane potential and hydrogen peroxide production in HeLa cells
Hydrogen peroxide production was measured with the use of CM-H2DCFDA in a confocal microscope. The addition of oligomycin to the intact cells led to an increase of the rate of hydrogen peroxide production [46]

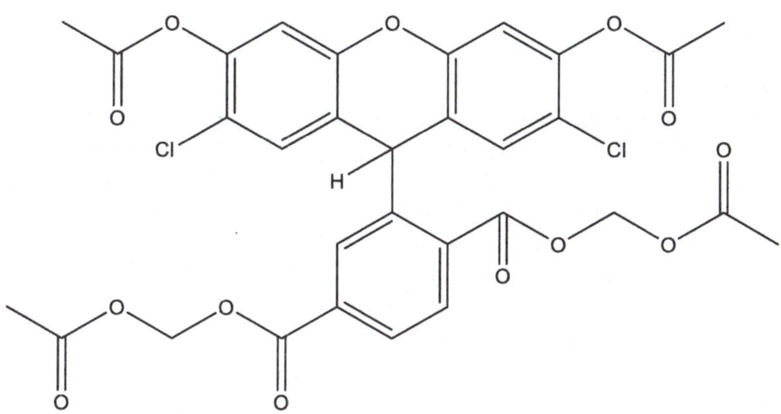

6-carboxy-2',7'-dichlorodihydrofluorescein diacetate, di(acetoxymethyl ester)

Fig. 11.7 Structure of di(acetoxymethyl ester) analog, 6-carboxy-DCFH-DA, di(acetoxymethyl ester)

are commercially available. Upon cleavage of the acetate and ester groups by intracellular esterases and oxidation, both analogs form carboxydichlorofluorescein, with additional negative charges that impede its leakage out of the cell.

Fluorinated Analog

The fluorinated analog, 5-(and 6-)carboxy-2′,7′-difluorodihydrofluorescein diacetate (carboxy-DFFH-DA), is useful for visualizing inflammatory responses and oxidative bursts [52]. When used for intracellular applications, acetate groups are cleaved by endogenous esterases, releasing the corresponding DFFH derivative.

OxyBURST®

The OxyBURST® family is composed of Fc OxyBURST®, OxyBURST® Green H2HFF BSA, and OxyBURST® Green DCFH-DA succinimidyl ester. OxyBURST® conjugates are nonfluorescent until they are oxidized to the corresponding fluorescein derivatives:

- The Fc OxyBURST® Green probe comprises bovine serum albumin (BSA) covalently linked to a DCFH moiety and then complexed with purified rabbit polyclonal anti-BSA antibodies. When this immune complex stimulates the FC receptor, DCFH is internalized within the phagolysosome and subsequently oxidized to green-fluorescent DCF. The measurement of the later reflects the oxidative burst in cells. As FC OxyBurst is based on DCF moiety, it possesses similar inherent problems; however, owing to the specific internalization, it may be a useful tool to study NAD(P)H oxidase (NOX)-induced ROS burst.
- OxyBURST® Green H2HFF BSA is a sensitive fluorescent reagent for detecting extracellular release of oxidative products in a fluorometer or a fluorescence microscope. This probe comprises BSA that is covalently linked to dihydro-2′,4,5,6,7,7′-hexafluorofluorescein (H2HFF), a reduced probe with improved stability. Unlike Fc OxyBURST® Green probe, OxyBURST® Green H2HFF BSA is not complexed with immunoglobulins. OxyBURST® Green H2HFF BSA provides higher sensitivity than conventional assays [53, 54]. OxyBURST® Green H2HFF BSA does not allow the probe to translocate into the cell, which makes the probe useful for ROS measurement in the extracellular space. In fact, the probe was used for the detection of the extracellular release of the ROS in coronary arterial myocytes generated by NOX1 [55].
- The amine-reactive OxyBURST® Green DCFH-DA succinimidyl ester (Fig. 11.8) can be used to prepare oxidation-sensitive conjugates of a wide variety of biomolecules and particles including peptides, proteins, antigens, antibodies, dextrans, bacteria, yeast, and polystyrene microspheres [56, 57]. Following conjugation to amines, the acetate groups of OxyBURST® Green DCFH-DA can be removed by treatment with hydroxylamine at neutral pH to yield the dihydrofluorescein conjugate.

Fig. 11.8 Structure of OxyBURST® Green DCFH-DA succinimidyl ester

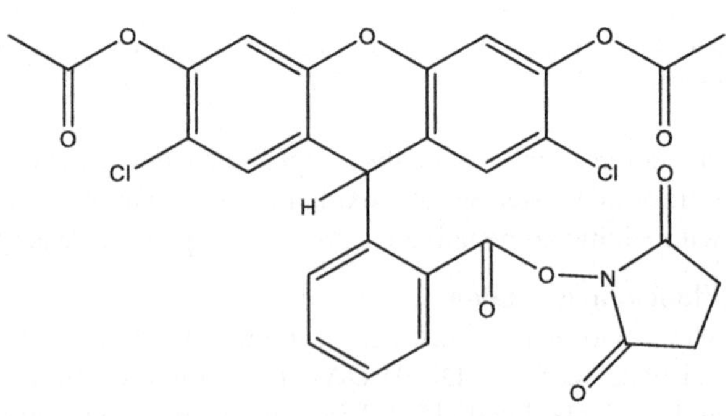

11.2.2 Dihydrorhodamine Compounds

11.2.2.1 Dihydrorhodamine 123

Dihydrorhodamine 123 (DHR) is a cell-permeant, uncharged, and mitochondrial-avid analog of 2,7 dichlorodihydrofluorescein that can undergo two-electron oxidation to form the fluorophore rhodamine 123 (Eq. 11.1), which is highly fluorescent at 536 nm when excited at 500 nm. The oxidized rhodamine is positively charged, lipophilic, and tends to be retained within the cell and mitochondria after tautomerization of its equivalent amino groups [37]. At high levels, DHR can detect $^1O_2$ in the mitochondria and cause NAD(P)H oxidation. This probe has been used to detect $OH^•$, $ONOO^-$, and $NO_2^•$, but it is poorly responsive to $O_2^{•-}$, H_2O_2, and $NO^•$. In addition, DHR is more sensitive than DCFH-DA for the detection of HOCl [58].

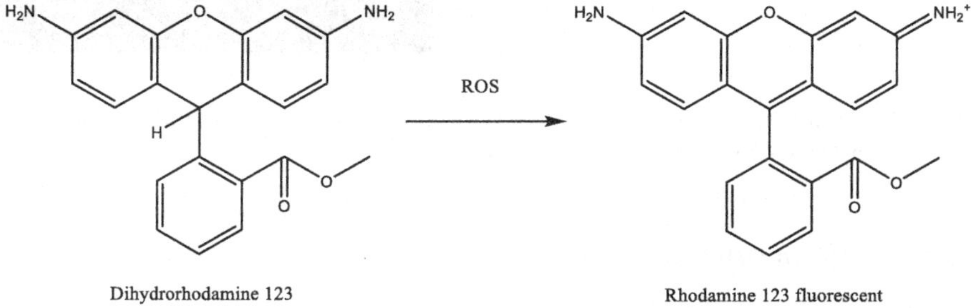

(11.1)

The direct reaction of DHR with H_2O_2 does not occur, but heme-containing peroxidases, such as HRP or other heme compounds including cytochrome c, can catalyze the hydrogen peroxide-dependent oxidation of DHR to rhodamine [16, 59]. Selective application of SOD, CAT, various NOS inhibitors, and peroxynitrite scavengers is required to provide for more precise identification of the RS responsible for DHR oxidation. Thus, cellular oxidation of DHR to rhodamine is at best a qualitative probe for peroxynitrite detection.

Rhodamine can be ejected from cells by drug conjugate efflux pumps; therefore, the activity of these membrane transport systems in the cells being studied is a key factor that needs to be considered [58]. It is noteworthy to mention that the oxidative conversion of DHR to rhodamine can be mediated by an intermediate DHR radical that can be reduced by thiols and ascorbic acid, leading to false-negative data.

DHR has been used to investigate reactive oxygen intermediates produced by phagocytes [60], activated mast cells [61], and vascular endothelial tissues [62, 63]. It has also been employed to study Alzheimer's disease (Fig. 11.9) [64], and the role of the CD14 cell-surface marker in hydrogen peroxide production by human monocytes [65]. However, because of the various biologic substances that can lead to rhodamine fluorescence coupled to the inherent uncertainty relating to endogenous versus artifactual oxidant generation, this fluorescent probe is best

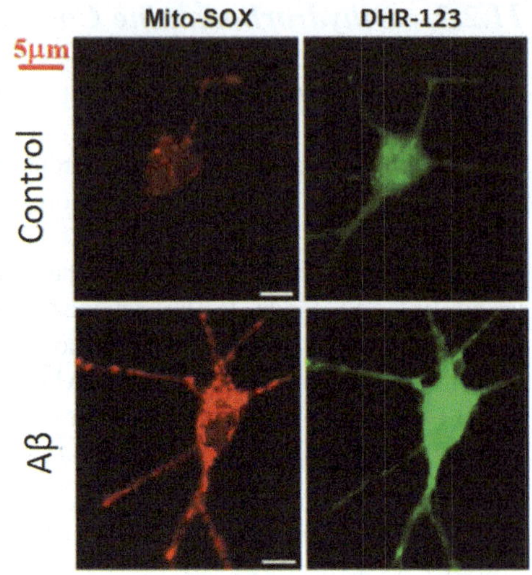

Fig. 11.9 The Aβ treatment-induced increase of mitoROSs (Mito-SOX) and cytROSs (DHR-123) (model cells: isolated of Hippocampal Neurons) [64]

applied as qualitative marker of cellular OS, rather than as precise indicator of rates of a particular RS formation.

11.2.2.2 Dihydrorhodamine 6G

Intracellular oxidation of dihydrorhodamine 6G yields rhodamine 6G, which localizes in the mitochondria of live cells (Eq. 11.2). As compared with rhodamine 123, this cationic oxidation product has longer-wavelength spectrum, making it very useful in multicolor applications and in auto-fluorescent cells and tissues. Dihydrorhodamine 6G has been used for microplate assays of granulocyte activation [66] and for the analysis of ROS levels in human umbilical vein endothelial cells by flow cytometry [41].

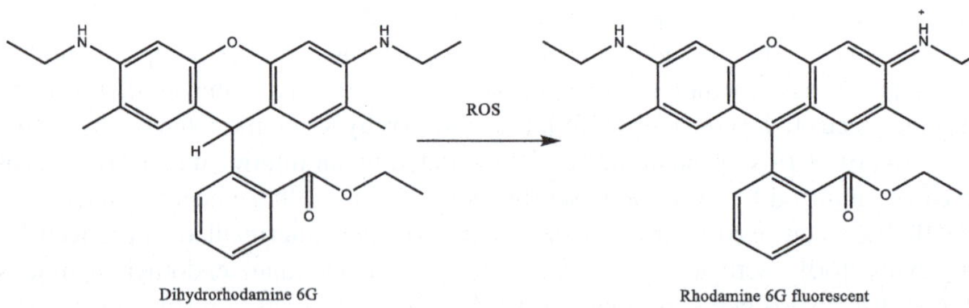

(11.2)

11.2.2.3 MitoTracker® Probes

Two MitoTracker® probes, MitoTracker® Orange CM-H2TMRos and MitoTracker® Red CM-H2XRos, are chemically reactive reduced rosamines (Fig. 11.10).

These probes do not fluoresce until they enter an actively respiring cell, where they are oxidized by ROS to fluorescent mitochondrion-selective probes. Although CM-H2TMRos and CM-H2XRos are widely used as indicators of mitochondrial ROS, their fluorescence cannot be unambiguously associated with the site of oxidant production as the cationic charge that drives their electrophoretic sequestration in active mitochondria is only present after the oxidation of the probe.

MitoTracker has been tested on young satellite cells exposed to bafilomycin A1, as shown in Fig. 11.11 [69]. It has also been used to evaluate mitochondrial ROS production in some pathological conditions such as diabetes [70], obesity [70], doxorubicin exposure (Fig. 11.11) [67], and SARS-COV2 infection (Fig. 11.11) [68].

11.2.2.4 RedoxSensor™ Red CC-1

RedoxSensor™ Red CC-1 probe (2,3,4,5,6-pentafluorotetramethyldihydrorosamine; Fig. 11.10) passively enters living cells and is subsequently oxidized in the cytosol to a red-fluorescent product (excitation/emission maxima ~540/600 nm), which then accumulates in the mitochondria. Alternatively, this nonfluorescent probe may be transported to the lysosomes where it can be oxidized. The differential distribution of the oxidized product between lysosomes and mitochondria seems to depend on the redox potential of the cytosol. In contact-inhibited cells, the staining is primarily lysosomal, whereas in proliferating cells, mitochondrial staining predominates.

RedoxSensor™ Red CC-1 probe has been used in human sperm cells (Fig. 11.22) [71], and some pathological situation such as ethanol-induced mitochondrial T cell ROS generation (Fig. 11.12) [72], and indomethacin-induced ROS production in the mitochondria of intestinal epithelial cell line, RIE-1 (Fig. 11.12) [73].

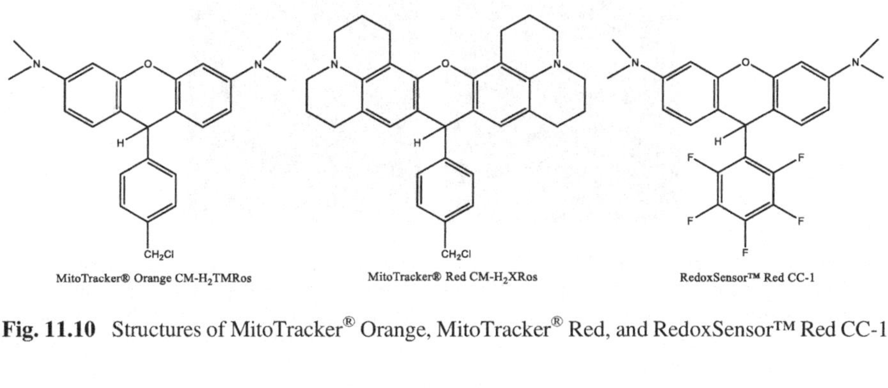

MitoTracker® Orange CM-H_2TMRos　MitoTracker® Red CM-H_2XRos　RedoxSensor™ Red CC-1

Fig. 11.10 Structures of MitoTracker® Orange, MitoTracker® Red, and RedoxSensor™ Red CC-1

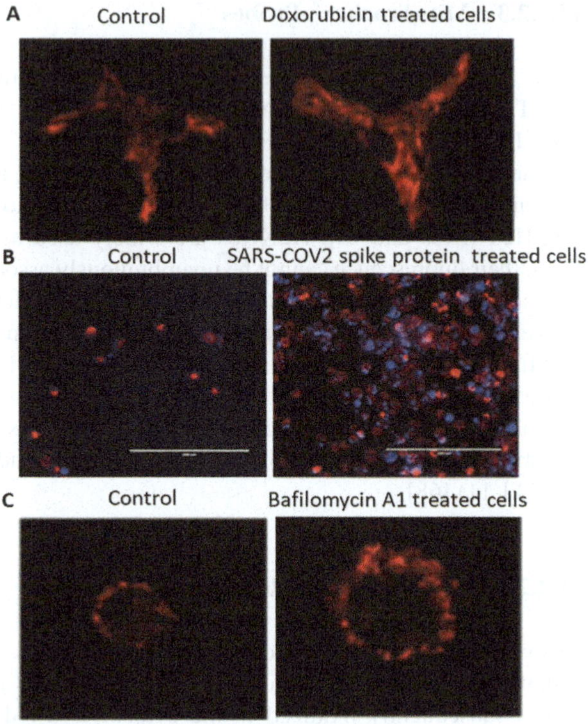

Fig. 11.11 MitoTracker probe for the assessment of mitochondrial production of ROS in pathological conditions. (**a**) Model cells: ventricular cardiomyocytes from neonatal rats (*NRVMs*). NRVMs were exposed either to vehicle or 5 μM doxorubicin for 1 h, and cells were loaded with fluorescent probe MitoTracker Red CM-H2XRos for 30 min. Images were captured with inverted fluorescence microscope using 579 nm excitation and 599 nm emission, wavelengths [67]. (**b**) Human Microglia (HMC3) [68]. (**c**) Isolated quiescent satellite cells. Staining of young satellite cells treated with bafilomycin A1 or vehicle for 24 h [69]

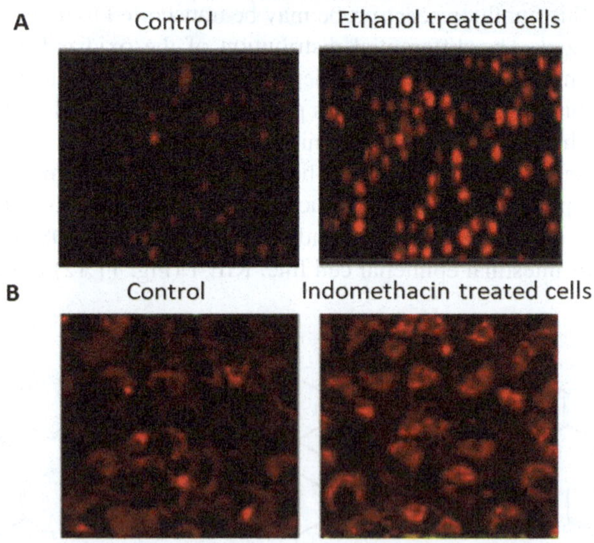

Fig. 11.12 RedoxSensor™ Red CC-1 probe for the assessment of mitochondrial production of ROS in pathological conditions. (**a**) Model cells: Human T cells (Jurkat cells). Ethanol-treated Jurkat cells displayed mitochondrial generation of ROS (indicated by bright red fluorescence in REDCC-1) [72]. (**b**) RIE-1 cells. RedoxSensor Red CC-1-stained RIE-1 cells were observed using a laser scanning confocal microscope. RIE-1 cells were incubated in the absence or presence of 200 μM indomethacin [73]

11.2.3 Dihydroethidium/DHE (Hydroethidine/HE)

Dihydroethidium (DHE) is a cell-permeable probe that undergoes a two-electron oxidation to form ethidium (E), providing fluorescence at an excitation wavelength of 520 nm and an emission wavelength of 610 nm [74]. The reaction is relatively specific for $O_2^{\bullet -}$, with minimal oxidation induced by $OH^{\bullet}$, H_2O_2, $ONOO^-$, or HOCl. DHE can also undergo a direct redox reaction with ferricytochrome c [75].

The treatment of cells with DHE and the monitoring of the generated fluorescence are frequently used to assess $O_2^{\bullet -}$ generation. While the oxidation of DHE appears specific for $O_2^{\bullet -}$, cellular alteration such as changes in the mitochondria membrane potential can affect the fluorescence, and thus, caution must be exercised when using this probe in biological systems [76]. Moreover, some authors suggest that DHE can form two fluorescent products. One is ethidium (E^+), which is formed by nonspecific redox reactions, while the other is 2-hydroxyethidium (2-OHE$^+$), a specific adduct of $O_2^{\bullet -}$ (Fig. 11.13) [77]. The fluorescent spectra of ethidium and 2-OH-E$^+$ overlap, and it is, therefore, difficult to use simple fluorescence detection with methods like confocal microscopy to accurately measure only 2-OH-E$^+$. To differentiate between 2-OH-E$^+$ and E^+, the use of HPLC/FL analysis has been suggested and provides unequivocal differentiation of the two products. Adapted HPLC to directly quantify 2-OH-E$^+$ and validate this approach is a way to quantify $O_2^{\bullet -}$ in biological systems [77]. Careful use of specific wavelengths of excitation might allow separation of these signals, so that confocal imaging or other fluorescence-based approaches could be used [78, 79]. Several assays were developed by taking advantage of this fact either using specific emission wavelengths of the two probes to separate the signals coming from $O_2^{\bullet -}$ and those generated from other oxidants [79]. The specificity of DHE for $O_2^{\bullet -}$ certainly reduces the influence of the observer in the interpretation of the microscopy images.

The intracellular oxidation of DHE by $O_2^{\bullet -}$ has been analyzed with flow cytometry [80], and visualization of cells (e.g., sperm cells (Fig. 11.22)), tissue slices, isolated blood vessels, and anatomic regions displaying increased rates of oxidant production with digital imaging microfluorometry has been described in vivo [71, 81–83].

Several models of OS-related diseases have been studied including ischemic stroke (Fig. 11.14) [84], menadione exposure (Fig. 11.14) [20], salt-induced hypertension (Fig. 11.14) [85], and SARS-CoV-2 Spike Protein S1 exposure [86] (Fig. 11.14).

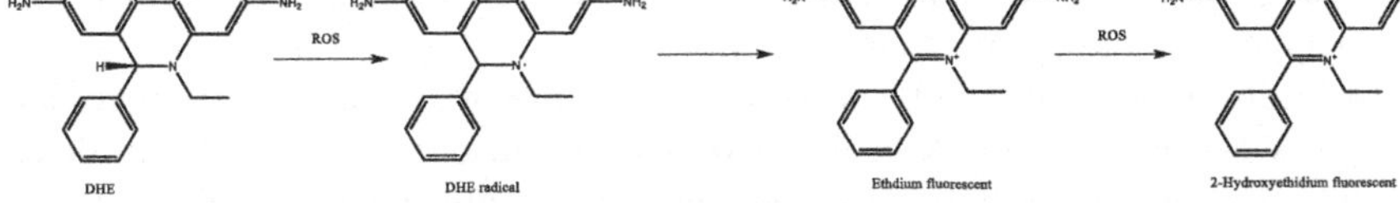

Fig. 11.13 Conversion of DHE to Ethidium and 2-hydroxyethidium (2-OHE$^+$)

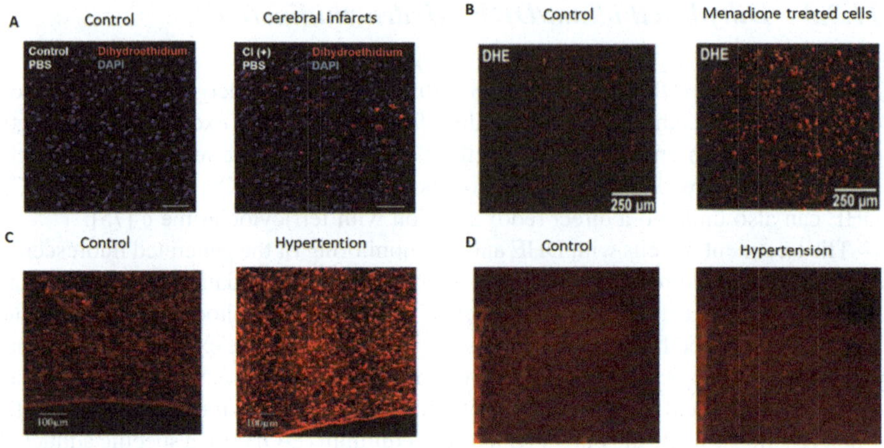

Fig. 11.14 DHE probe for the assessment of ROS in pathological conditions. (**a**) model: brain of mice. $O_2^{\bullet-}$ generation in the penumbra surrounding cerebral infarcts as indicated by DHE microfluorometry [84]. (**b**) model cells: human-induced pluripotent stem cell-derived cardiomyocytes (hiPSC-CMs) [20]. (**c**) model: hypothalamic paraventricular nucleus (PVN) from hypertensive rats and normal salt rats [85]. (**d**) model: hypothalamic paraventricular nucleus (PVN) from SARS-CoV-2 Spike Protein S1 subunit exposed rats and vehicle exposed rats [86]

In contrast to other intracellular probes for $O_2^{\bullet-}$, there is little capacity for artifactual formation of $O_2^{\bullet-}$ by DHE attributable to redox cycling. However, there are three main reactions that may limit the use of DHE in the detection of $O_2^{\bullet-}$ production:

- First, cytochrome *c* is also capable of oxidizing DHE. This may be of importance when mitochondria are the primary source of $O_2^{\bullet-}$ production or when cytochrome *c* is released into the cytosol under conditions leading to apoptosis [87]. Because OS has been suggested to initiate apoptotic processes in several types of cells [88], it would be difficult to ascribe DHE conversion to 2-OHE$^+$ solely to $O_2^{\bullet-}$ and localization of cellular sites of $O_2^{\bullet-}$ generation would be suspect under these circumstances.
- Second, quantitation of $O_2^{\bullet-}$ production using DHE will be inaccurate because of its capacity to enhance rates of $O_2^{\bullet-}$ dismutation to H_2O_2 [75]. Under these conditions, the rate of $O_2^{\bullet-}$ formation will be underestimated.
- Third, DHE fluorescent signals can be falsely enhanced by intercalation of the probe with DNA giving a nuclear fluorescence especially when high concentrations of DHE are used [76].

Because of these considerations, the use of HE for the determination of $O_2^{\bullet-}$ in vivo appears to have significant limitations.

The localized detection of $O_2^{\bullet-}$ can be realized through various DHE-fluorescent derivatives designed for the subcellular organelles. In order to specifically detect $O_2^{\bullet-}$ in mitochondria, DHE was conjugated to triphenylphosphonium moiety (TPP$^+$). This modified DHE analog is referred to as mito-SOX and has become

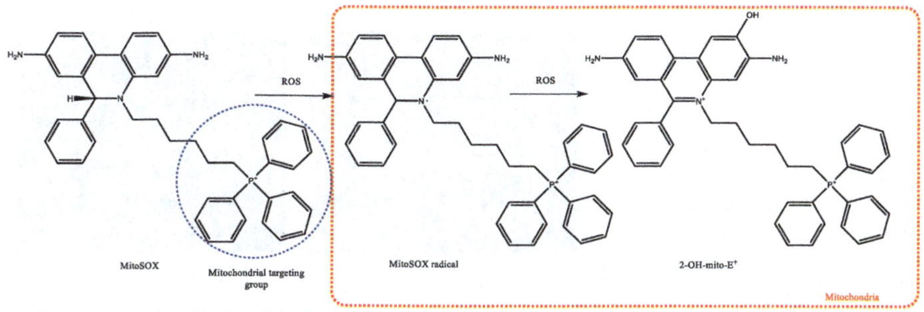

Fig. 11.15 Mitochondrial-targeted accumulation of mito-SOX and detection of mitochondrial $O_2^{\bullet-}$

commonly used for the detection of $O_2^{\bullet-}$ within the mitochondria. Cationic TPP^+ allows the probe to move through the mitochondrial membrane and accumulate in the mitochondrial matrix. Analogous to DHE, mito-SOX reacts with $O_2^{\bullet-}$ to form 2-hydroxy-mito-ethidium (2-OH-mito-E^+) (Fig. 11.15) [89].

A caveat to use mito-SOX is that high concentrations can impair the mitochondrial function [78]. Furthermore, concentrations of mito-SOX exceeding 2 μM can lead to substantial cytoplasmic accumulation of mito-SOX and thus can compromise the probe-mitochondrial specificity of $O_2^{\bullet-}$. Some authors suggest using mito-SOX at concentrations of 2 μM or less to avoid these complications [90].

While measurements of 2-OH-Mito-E^+ are most accurately achieved by HPLC [91], Robinson et al. have reported that mitochondrial $O_2^{\bullet-}$ can be accurately quantified in living cells using selective excitation at 395 nm and detection at an emission of 560 nm [78]. These parameters seem to reduce signals coming from unspecific fluorescent products. Therefore, fluorescence spectroscopy can be used for rapid and specific visualization and measurement of mitochondrial $O_2^{\bullet-}$, and confirmation with HPLC analysis is advisable.

In the study of fenretinide-induced apoptosis of neuroblastoma, mito-SOX was applied to detect the mitochondrial production of $O_2^{\bullet-}$. The results showed the inhibitory effect of the mitochondrial $O_2^{\bullet-}$ on the respiratory chain as well as the anion superoxide cytotoxicity [92].

Mito-SOX probe has also been used in physiological conditions (e.g., human sperm cells (Fig. 11.22)) [71, 93], in pathological condition (e.g., Alzheimer's disease) (Fig. 11.9) [64], and in multiple toxicological conditions related to OS such as menadione exposure (Fig. 11.16) [20], oligomycin exposure (Fig. 11.16) [46], MPP^+-induced cytotoxicity (Fig. 11.16) [94], and doxorubicin exposure [95].

Similar to DHE, the sources of red fluorescence from mito-SOX are complex. In addition to the specific product 2-OH-Mito-E^+ from the oxidation of $O_2^{\bullet-}$, Mito-E^+ generated from other oxidants such as H_2O_2, cytochrome c, and peroxidase can be found, and therefore assistant analysis is necessary [96].

Overall, limitations of DHE and mito-SOX include their complex chemistry and potential interference with heme enzymes [89]. These probes are known for their

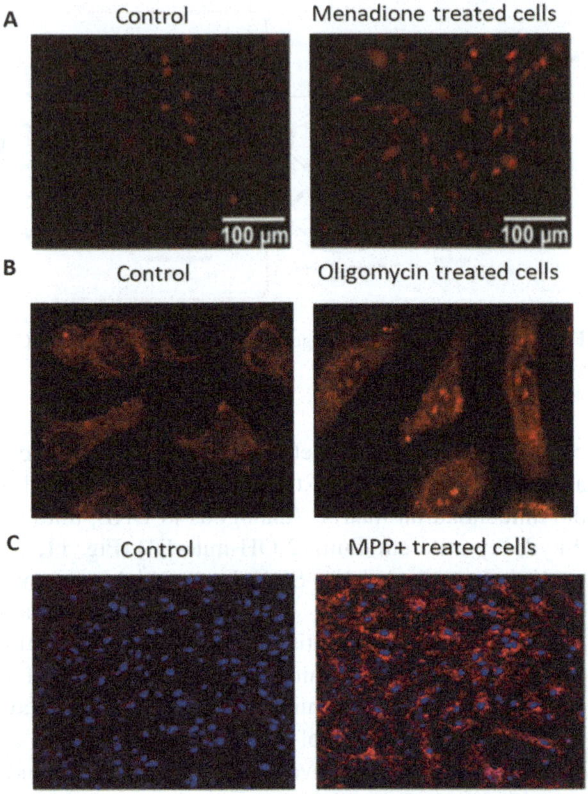

Fig. 11.16 Mito-SOX probe for the assessment of mitochondrial production of ROS in toxicological conditions. (**a**) Model cells: BAEC cells [20]. (**b**) HeLa cells. The mitochondrial $O_2^{\bullet-}$ was estimated using Mito-SOX Red in a confocal microscope. Addition of oligomycin (right panel) to the HeLa cells increased mitochondrial $O_2^{\bullet-}$ production represented by increased fluorescence [46]. C: SH-SY5Y cells. SH-SY5Y cells were treated with 500 µM MPP⁺ for 24 h (representative confocal images of mitochondrial ROS-dependent fluorescence in SH-SY5Y cells) [94]

instability as they are light sensitive and prone to auto-oxidation. They require a two-step reaction to detect $O_2^{\bullet-}$ involving a FR intermediate that can be potentially a subject for reaction with antioxidants [97]. Finally, formation of $O_2^{\bullet-}$specific product can be affected by peroxidase catalysis, which could compromise quantification of $O_2^{\bullet-}$ [98]. Although many research studies have reported the application of DHE and mito-SOX in detecting $O_2^{\bullet-}$ in cells and biological systems, the interference from other oxidative radicals prevented them from quantifying $O_2^{\bullet-}$ accurately.

11.2.4 Amplex Red

Amplex Red (10-acetyl-3,7-dihydroxyphenoxazine) is an oxazine derivative used for the detection of H_2O_2. The latter reacts in the presence of HRP with the probe in a one-electron pathway to form the fluorescent resorufin ($\lambda_{excitation} = 563$ nm and $\lambda_{emission} = 587$ nm). HRP reacts with H_2O_2, generating peroxidase compound I, which produces resorufin on an equimolar basis (Fig. 11.17).

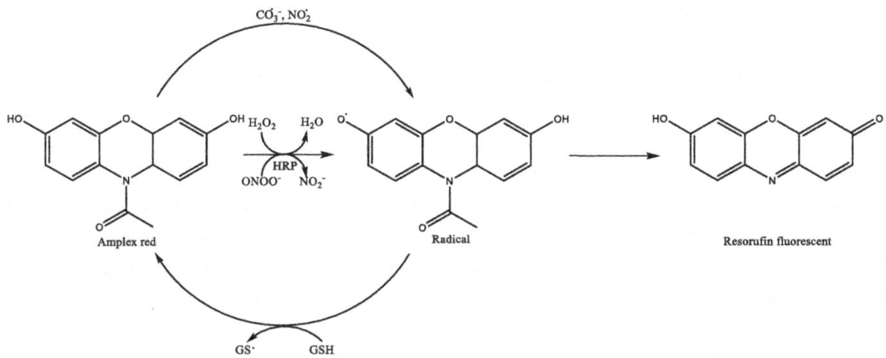

Fig. 11.17 Detection of H_2O_2 with Amplex Red

The product resorufin is stable, allowing the detection of H_2O_2 under both oxidative and reductive conditions, offering scope for a wide range of applications. Amplex Red can be used to detect H_2O_2 released from isolated mitochondria [99]. As mentioned earlier, $O_2^{\bullet -}$ produced in the mitochondrial matrix is quickly dismutated by SOD2 to H_2O_2, which readily diffuses through mitochondrial membranes to the surrounding medium. Amplex Red has been used to detect the role of mitochondria-sourced H_2O_2 in the process of mitochondrial gene damage [100]. This assay is a reliable method to measure the extracellular formation of H_2O_2.

However, confounding factors of this assay should be taken into consideration. The selectivity for H_2O_2 is challenged in the presence of $ONOO^-$, which forms resorufin at a faster rate than H_2O_2, so specific inhibitors are necessary [101]. Another caveat for using Amplex Red is that $O_2^{\bullet -}$ can react with peroxidases either in its basal state or when peroxidases are in compound I or II state and thus can lead to formation of compound III. These reactions can impair the stoichiometry of hydrogen peroxide detection. It is therefore impossible to measure H_2O_2 by Amplex Red in the presence of substantial fluxes of $O_2^{\bullet -}$. This problem can be remedied by addition of Cu, Zn-SOD to the assay. This manipulation eliminates this problem and likely minimally affects the ultimate result, as $O_2^{\bullet -}$ spontaneously dismutates to H_2O_2 even in the absence of SOD at a slightly slower rate.

Besides, HRP can react with reducing agents such as NADPH and GSH, to generate H_2O_2, resulting in the production of additional resorufin. It is also reported that resorufin could be reduced to a colorless, nonfluorescent compound by NADPH-CYP450 reductases, accompanied by the generation of $O_2^{\bullet -}$ or even H_2O_2, posing problems for the identification of the latter [102]. Moreover, Amplex Red shows low permeability through cell membranes, resulting in its inability to assess precisely intracellular production of H_2O_2 due to the diffusing nature of H_2O_2 [103]. Because H_2O_2 is diffusible, it reaches the equilibrium with the tissue's surrounding buffer; therefore, the values measured in the buffer should provide an estimation of the amount of H_2O_2 originally produced by the tissue.

In addition, Amplex Red is light sensitive [104], and at high concentrations (50 μM), it can be affected by auto-oxidation, which results in the production of $O_2^{\bullet -}$ and H_2O_2. Low concentrations of Amplex Red (10 μM) minimize this problem.

Last but not least, in some tissues (e.g., liver and kidney) and cultured cells, Amplex Red could be transformed into resorufin by carboxylesterase without the requirement of H_2O_2 or HRP [105].

To conclude, because of multiple interference reactions induced by confounding compounds and enzymes, Amplex Red is a problematic probe to measure intracellular H_2O_2, but still it could be used in cell-free systems in the presence of HRP, where H_2O_2 is released from the cell or separated mitochondrial particles.

11.2.5 Diphenyl-1-Pyrenylphosphine

Lipid peroxidation in living cells can be visualized using fluorescent derivatives that localize to membranes. Several membrane-partitioning probes have been introduced to measure lipid peroxidation within membranes. One is diphenyl-1-pyrenylphosphine (DPPP). This probe reacts with peroxides to generate a fluorescent product which can be detected at 380 nm when excited at 352 nm (Eq. 11.3). Lipid-soluble hydroxides like methyl linoleate hydroperoxide react with DPPP, while H_2O_2 located in the cell membrane does not react with DPPP. The fluorescent product of DPPP is quite stable in living cells and can remain up to 2 days. Nevertheless, this probe is light and air sensitive, and it is difficult to obtain reproducible results using it under laboratory conditions [106].

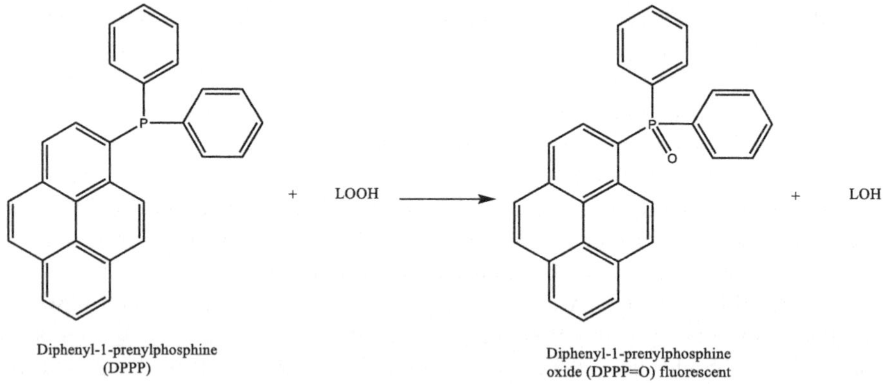

(11.3)

11.2.6 Cis-Parinaric Acid

Cis-Parinaric acid is a fluorescent polyunsaturated fatty acid (PUFA) that is not found endogenously. This probe can rapidly oxidize and lose its fluorescence ($\lambda_{\text{emission}}$ 413 nm and $\lambda_{\text{excitation}}$ 324 nm), when incorporated into lipids undergoing peroxidation (Eq. 11.4) [107]. A decrease in its fluorescence suggests destruction of the fatty acid revealing lipid peroxidation taking place. It can be added as a free acid and used to study its susceptibility to oxidative damage within membranes [108].

Cis-Parinaric acid needs careful handling because its polyunsaturated structure makes it highly susceptible to nonspecific oxidation, so it should be stored in the dark under N_2 atmosphere. The need to use UV range is another disadvantage. For example, the required wavelengths are not always available on microplate readers or even confocal/flow cytometry systems.

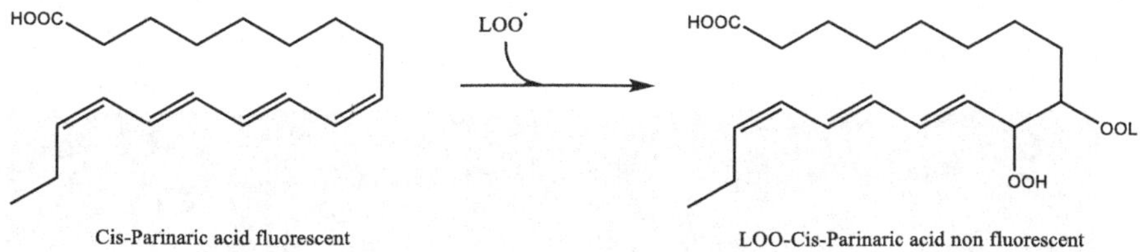

(11.4)

11.2.7 BODIPY

Another fatty acid probe becoming widely used is C-11-BODIPY581/591 (undecanoic acid fluorophore), which upon oxidation of its polyunsaturated butadienyl portion, shifts its fluorescence excitation and emission from red (590 nm) to green (510 nm) (Eq. 11.5) [109].

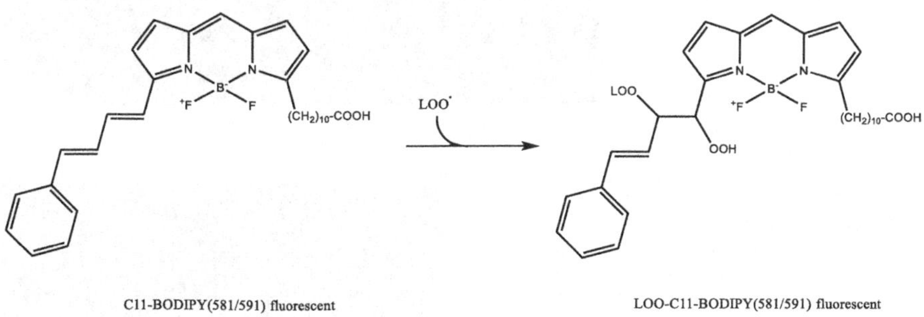

(11.5)

Therefore, the green/red ratio can be used as an estimate of oxidative degradation of membrane lipids in living cells [110]. Simultaneous dual wavelength excitation and emission measurements are particularly useful in determining the fluorescence ratio changes of C-11-BODIPY581/591. Use of the ratio provides a measure of lipid peroxidation that is independent of heterogeneous probe distributions such as lipid density cell thickness, uneven dye loading, or compartmentalization that may influence measurement with single emission [111]. Because this reagent is compatible with living cells, measurements can take place in real time without fixation. Moreover, it can be used to determine RS within membranes. This probe responds to a range of RS ($OH^•$, peroxyl, alkoxyl, and $ONOO^-$) but not to $O_2^{•-}$, $NO^•$, H_2O_2, $^1O_2$, or hydroperoxides [112]. It has been used for demonstrating the antioxidant capacity of plasma and lipid vesicles [113, 114], as well as the pathogenesis of ferroptosis (Fig. 11.18) [115, 116], ischemia–reperfusion injury (Fig. 11.19) [19], and sperm lipid peroxidation (Fig. 11.20) [117].

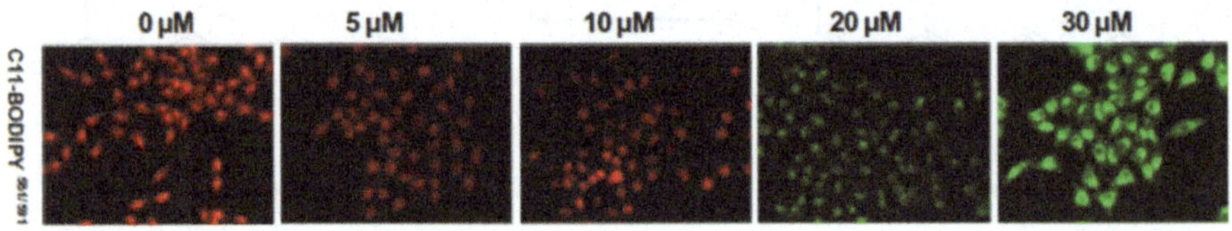

Fig. 11.18 Fluorescence images of HepG2 cells incubated with different concentrations of erastin (ferroptosis activator) for 4 h [115]

Fig. 11.19 Revelation of lipid peroxidation in ischemia-reperfusion (H/R)-induced H9c2 cardiomyocytes. The level of lipid peroxidation was detected by C11 BODIPY 581/591 probe [19]

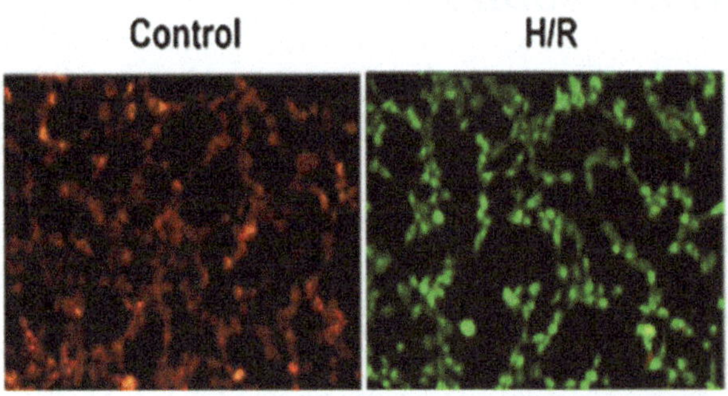

Fig. 11.20 Fluorescent microscopic picture of BODIPY staining for lipid peroxidation in spermatozoa [117]

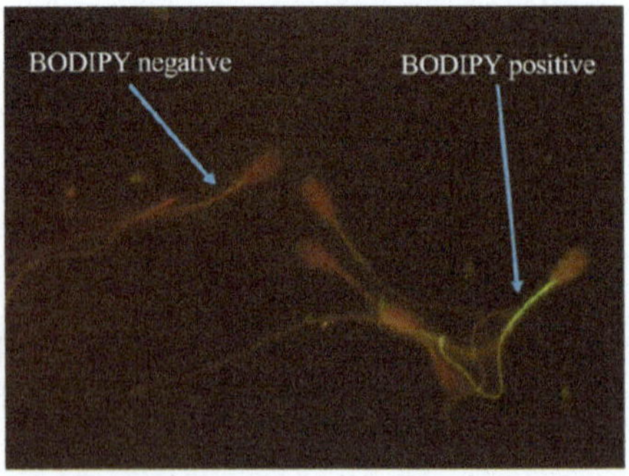

11.2.8 Coumarin-3-Carboxylic Acid (3-CCA)

One of the most commonly used fluorescent probes for the detection of OH· is coumarin-3-carboxylic acid (3-CCA). Following the reaction with OH•, 3-CCA becomes hydroxylated forming a fluorescent 7-OHCCA product (Eq. 11.6). However, due to high pH sensitivity of 3-CCA, pH must be carefully monitored [118].

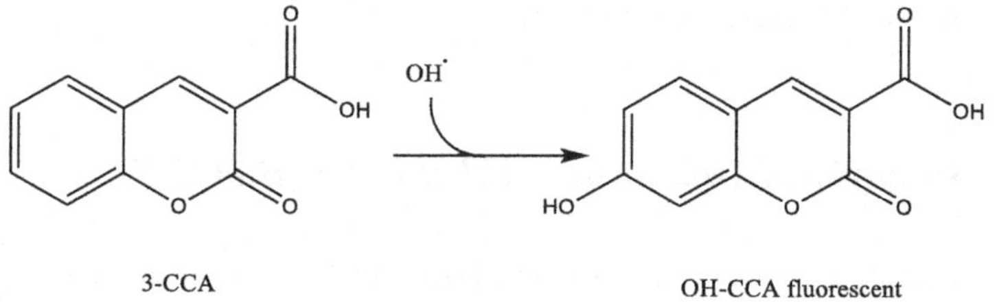

(11.6)

11.2.9 3′-(P-Hydroxyphenyl Fluorescein (HPF) and 3′-(P-Aminophenyl) Fluorescein (APF)

3′-(p-aminophenyl) fluorescein (APF) and 3′-(p-hydroxyphenyl) fluorescein (HPF) are cell-permeable probes, offering greater selectivity and stability than DCFH-DA for ROS detection (Eq. 11.7) [119]. APF and HPF show much more limited reactivity and greater resistance to photo-oxidation. In addition, APF and HPF are resistant to auto-oxidation and can be used in the presence of an enzymatic system [119]. However, similar to other nonratiometric fluorescent probes, the fluorescence of these probes is dependent on the pH, concentration of the probe, and other environmental factors.

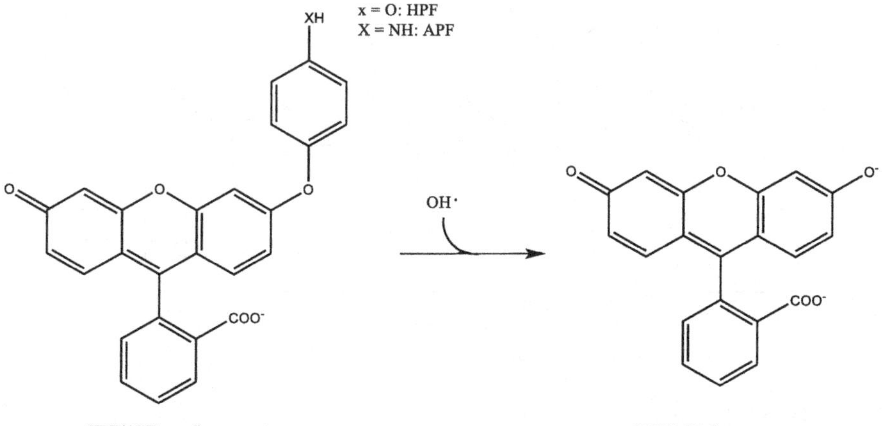

(11.7)

Both probes are nonfluorescent until reaction with $OH^{\bullet}$, $ONOO^{-}$, or $^{1}O_2$ causing the O-dearylation of the aminophenyl or hydroxyphenyl rings. In addition, APF can react with HOCl, making it possible to use APF and HPF together to selectively detect HOCl. Using APF, researchers have been able to detect HOCl generation by activated neutrophils, a feat that has not been possible with traditional RS sensors [119].

In the presence of RS, both APF and HPF yield a bright green-fluorescent product (excitation/emission maxima ~490/515 nm).

11.2.10 Biotinylated Glutathione Ethyl Ester (BioGEE)

Biotinylated glutathione ethyl ester (BioGEE) (Fig. 11.21) is a cell-permeant probe used for the detection of glutathiolation. Under OS conditions, cells may transiently incorporate GSH into proteins. Stressed cells incubated with BioGEE will also incorporate this biotinylated glutathione derivative into proteins, facilitating the identification of oxidation-sensitive proteins [120, 121]. Once these cells are fixed and permeabilized, glutathiolation levels can be detected and measured with a fluorescent streptavidin conjugate using either fluorescence microscopy or flow

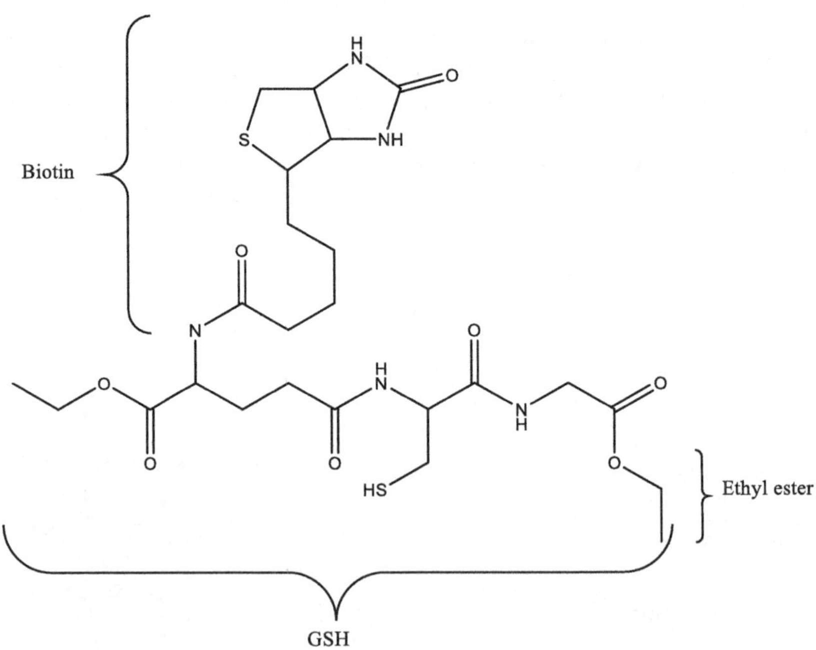

Fig. 11.21 Structure of BioGEE

cytometry. Proteins glutathiolated with BioGEE can be captured using streptavidin agarose and analyzed by mass spectrometry or by western blotting methods [121–123].

11.2.11 Boronates

The propensity of H_2O_2 to interact with boronate groups was used in the synthesis of a family of fluorescent probes that allow the precise investigation of ROS in intracellular signaling [124]. They have been designed as new effective probes through attachment with a fluorophore, which generates a fluorescent product upon reaction with H_2O_2, providing more selective and sensitive detection than the original prototypes [125].

Boronates have also been shown to react with $ONOO^-$ and HOCl with varying rates of reaction [126, 127]. The second-order rate constants showed $ONOO^-$ to be the most reactive ($\sim 10^5$–10^6 M^{-1} s^{-1}) followed by HOCl ($\sim 10^3$–10^4 M^{-1} s^{-1}), and by H_2O_2 (~ 2 M^{-1} s^{-1}) [127].

Boronates have been applied as one of the most effective probes for the monitoring and quantification of ROS and RNS in cells. The general mechanism of oxidative reaction to boronates involves nucleophilic addition of the oxidant to the boron atom followed by the heterolytic cleavage of the X–O bond to form the respective ion. Intramolecular rearrangement of the B–O yields the final fluorescent phenolic products (Eq. 11.8).

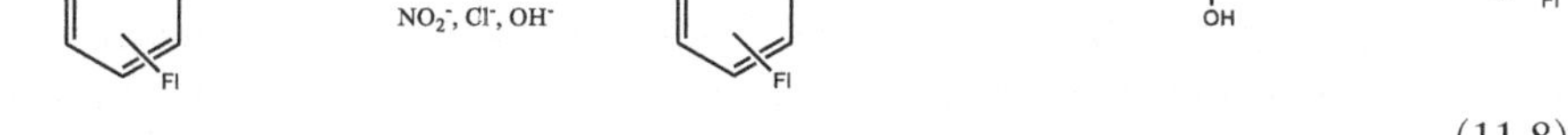

(11.8)

The first developed probes (green fluorescent peroxyfluor 1 (PF1), red fluorescent peroxyresorufin 1, and blue fluorescent peroxyxanthone 1) were based on the fluorescence-emitting opening of the bis-boronate-originated lactone structure [128, 129]. These boronate probes are cell permeable and can detect changes in H_2O_2 in the micromolar range in living cells [129]. As mentioned earlier, these probes possess a fluorophore that is protected by boronate. Upon exposure to H_2O_2, the boronate undergoes a nucleophilic attack, leading to its removal from the fluorophore, allowing light emission (Eq. 11.9).

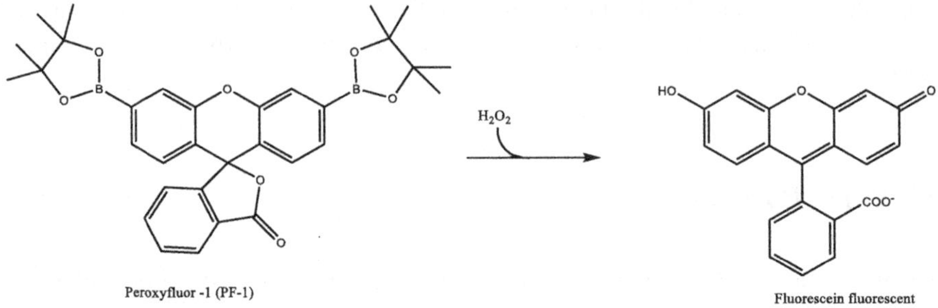

(11.9)

Further optimization of the probes relying on the removal of one of the boronates revealed substantial increase in hydrogen peroxide sensitivity. This chemical modification allowed monoboronate-based probes (Peroxyfluor-3, Peroxy Orange 1, and Peroxy Yellow 1) to detect physiological changes in hydrogen peroxide concentration (Eq. 11.10) [130].

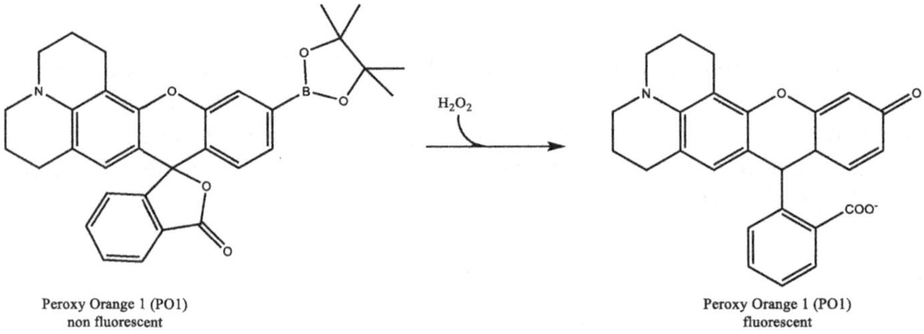

(11.10)

Researchers have synthesized other boronate-based fluorescent probes such as peroxy green 1 (PG1) and peroxy crimson 1 (PC1), all of which show higher selectivity for H_2O_2. These probes were applied in the fluorescence imaging of H_2O_2 in many tissues and cells [128, 131]. PG1 and PC1 have also been reported to be membrane permeable, along with visible excitation and emission wavelengths [74]. Reacting with H_2O_2, results in a 10-fold and 40-fold increase in fluorescence for PG1 and PC1, respectively, compared to initial boronate probes. PG1 features an excitation wavelength of 460 nm with emission maxima at 510 nm (Eq. 11.11). PC1 demonstrates improved characteristics of red-shifted excitation and larger stokes shift which reduces auto-fluorescence (excitation: 480 nm; emission: 584 nm) (Eq. 11.12) [132].

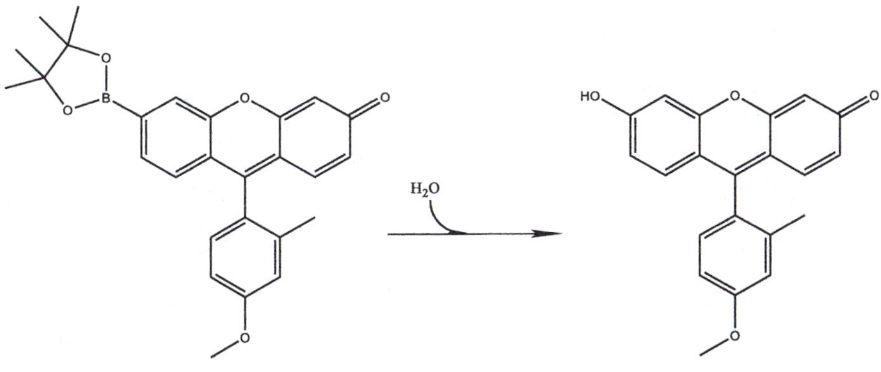

PG1 non fluorescent PG1 fluorescent

(11.11)

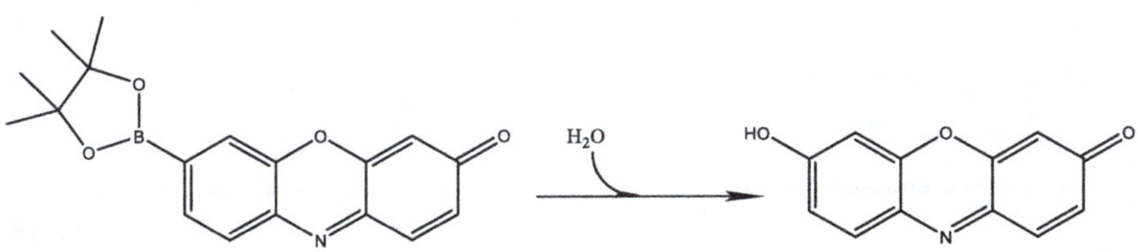

PC1 non fluorescent Resorufin

(11.12)

The regional detection of H_2O_2 in cells has also been achieved by the boronate-based probes. This has been accomplished by combining boronate chemistry with organelle-targetable functioning groups. Several mitochondria-targeted probes have been generated including Mitochondrial Peroxy Yellow 1 (MitoPY1), MitoBoronic acid (MitoB), and SHP-Mito [133–135]. These probes utilize the commonly used positively charged phosphonium moiety for mitochondrial targeting, which allows selective detection of H_2O_2 within the mitochondria (Eqs. 11.13, 11.14 and 11.15). These bifunctional fluorescent probes permit imaging H_2O_2 within the mitochondria using imaging techniques such as flow cytometry and confocal microscopy [134].

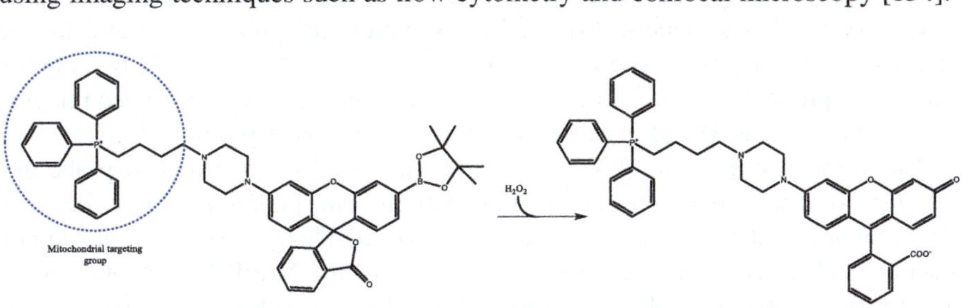

MitoPY1 MitoPY1ox fluorescent

(11.13)

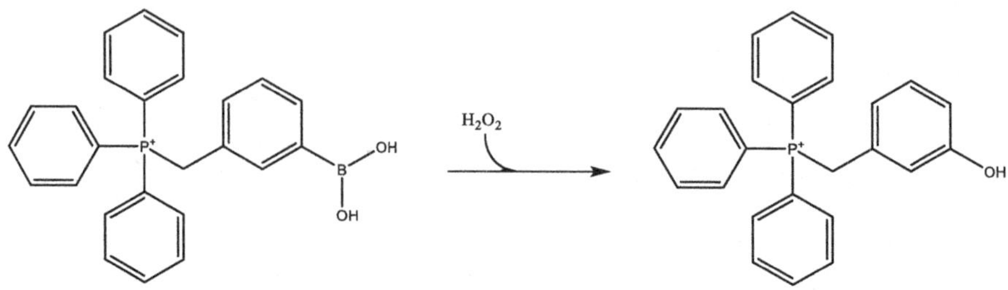

(11.14)

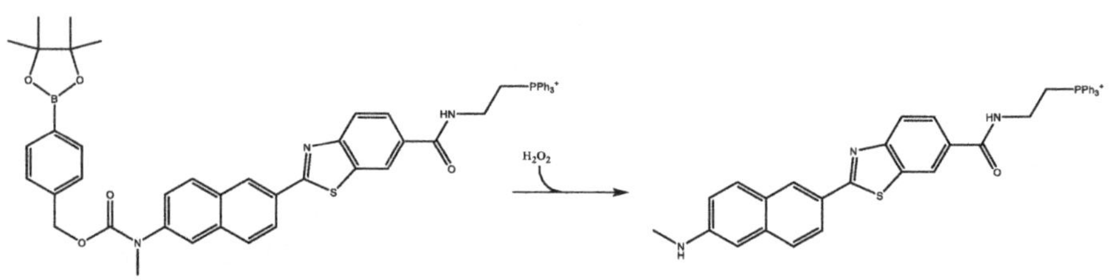

(11.15)

Taking mitochondria peroxy yellow 1 (MitoPY1), for example, it is composed of a chemoselective boronate-based switch and a mitochondria-targeting phosphonium moiety, and the latter targets the MitoPY1 probe into mitochondria, facilitating its reaction with H_2O_2 to produce the highly fluorescent MitoPY1ox, which fluorescence is detectable at 528 nm [134]. This probe has been used to detect H_2O_2 production in human sperm cells (Fig. 11.22) [71, 93].

By combining the boronate chemoselectivity and the cationic triphenylphosphonium affinity for mitochondria (TPP^+), Masanta et al. synthesized a two-photon hydrogen peroxide probe targeted to mitochondria called, SHP-Mito [135]. The addition of 6-(benzo[d]thiazol-20-yl)-2-(N,N-dimethylamino)naphthalene (BTDAN) as a reporter allows for the ratiometric analysis of the probe oxidation. The probe was demonstrated to be unresponsive to other ROS/RNS than H_2O_2 as well as pH change. Due to two-photon excitation, the probe was useful for the deep-tissue imaging ranging from 100 to 180 μm [135]. SHP-Mito is a two-photon ratiometric probe that allows for increased penetration depth and prolonged imaging time [135], whereas MitoB is a ratiometric mass spectrometry probe [133].

The boronic component of MitoB is a ratiometric mass spectroscopy probe that uses the boronate group conjugated to the cationic phosphonium moiety for mitochondrial accumulation. It is rapidly converted to the phenol product (MitoP) upon reaction with H_2O_2. Mass spectrometry analysis of the MitoB/MitoP ratio allows accurate hydrogen peroxide measurements in whole organisms, cells and tissues. The probe offers the measurement of hydrogen peroxide concentration in the mitochondria with a nanomolar accuracy. Combination of specific mitochondrial

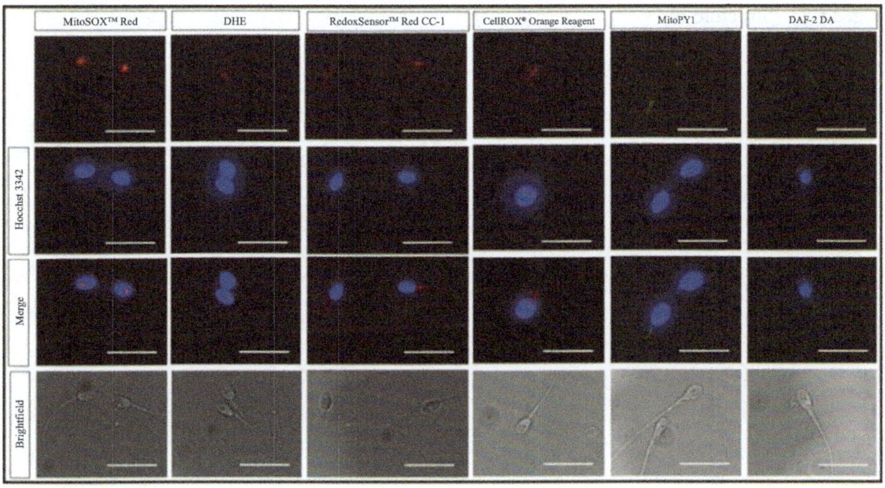

Fig. 11.22 Representative images of human sperm cells labeled with fluorescent probes for ROS/RNS detection
Mito-SOX and DHE labeled the sperm head, while RedoxSensor™ Red CC-1, CellROX® Orange, MitoPY1, and DAF-2-DA labeled the midpiece. Sperm nuclei were labeled with Hoechst 33342 (blue fluorescence) and phase contrast images in gray show the entire sperm cell. Scale bar 100 μm [71]

targeting and hydrogen peroxide-chemoselective switch with mass spectroscopy analysis in MitoB proposes a unique candidate for H_2O_2 measurement in the mitochondria of cells and animals.

Besides, the Nuclear Peroxy Emerald 1 (NucPE1) was designed for the targeting of H_2O_2 in nuclear, enriching the application of boronate probes in organelles (Fig. 11.23) [136]. However, NucPE1 suffers from a relatively poor signal-to-noise ratio, exhibiting only 20% increase in fluorescence in HEK293 cells treated with 100 μM H_2O_2 [136]. Furthermore, the mechanism of trafficking to the nucleus is unclear, possibly involving DNA binding. Thus, NucPE1 may induce genotoxic stress, resulting in radical cycling and false amplification of the signal. In addition, the fluorescence of the probe itself may change upon DNA binding and have an impact on the signal. The probe was also demonstrated to respond to nuclear hydrogen peroxide alterations in various leukemic cell lines [137]. Because of its excellent nuclear localization in a variety of cell, the probe offers a promising tool to study nuclear H_2O_2 [138]. However, further studies need to be carried out to determine the mechanism of its nuclear accumulation, as well as its sensitivity/dynamic range properties.

The further modification of monoborate-based probes by adding methyl-ester group increased cellular retention following cleavage by esterases, as recorded with Peroxy Yellow 1 Methyl-Ester (PY1-ME) probe [139]. This probe has been used in the investigation of hydrogen peroxide uptake by aquaporins in HEK 293 cells [139].

The latest version of PF1, Peroxyfluor 6 acetoxymethyl ester (PF6-AM) (Fig. 11.24), was synthesized in order to explore the mechanism of NOX regulation

Fig. 11.23 Structure of NucPE1

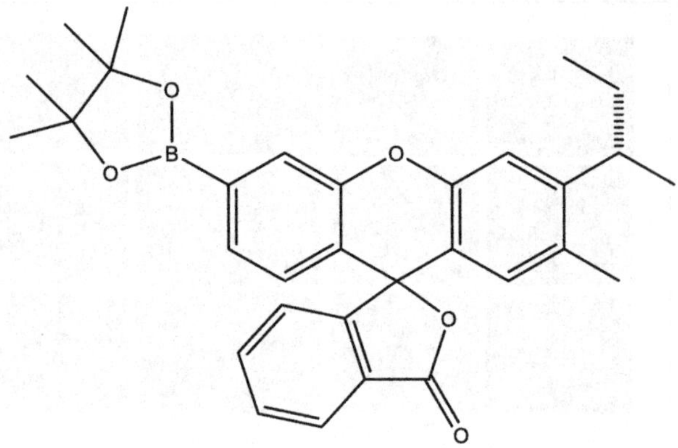

NucPE1

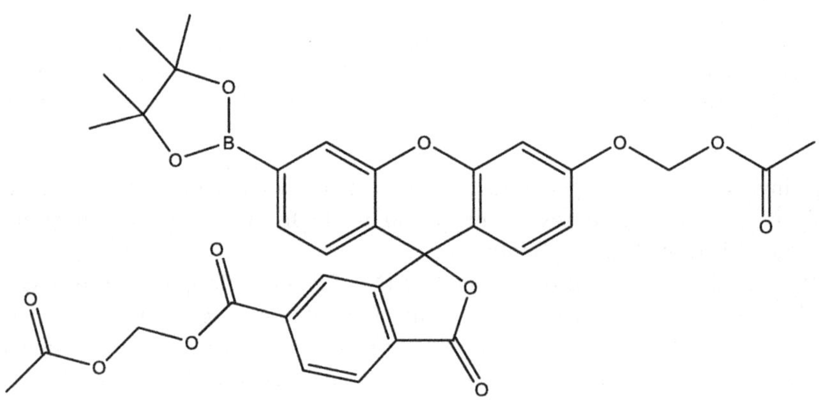

PF6-AM

Fig. 11.24 Structure of PF6-AM

in the brain [140]. Addition of the acetoxymethyl group gives the probe increased cellular retention and sensitivity to H_2O_2. The AM-esters aid in the translocation of the probe through the membrane and its accumulation in the cytoplasm following the release of the AM groups due to cellular esterase catalysis. This structural modification made the PF6-AM probe more sensitive than previous members of the Peroxy family. Indeed, PF6-AM responded to signaling levels of H_2O_2 produced upon FGF-2 stimulation at concentrations as low as 10 μM, which was not possible with PG1 [140]. The increased cellular retention of the probe inside the cells also facilitates prolonged imaging.

Quantitative analysis of hydrogen peroxide generation using fluorescent probes remains problematic as the signal from the single-wavelength emitting probes can be affected by the concentration of the probe. To address this issue, a ratiometric monoboronate-based probe was synthesized, Ratio Peroxyfluor (RPF)1 [141]. This probe overcomes these issues as the quantification of the change in the fluorescence is calculated as a ratio of the fluorescence upon reaction with H_2O_2 (λ517) to the internal fluorescence of the probe (λ464) [141]. Upon reacting with H_2O_2, fluorescence resonance energy transfer (FRET) occurs between the coumarin moiety (donor) and boronate group (acceptor) in the two-fluorophore cassette (Eq. 11.16).

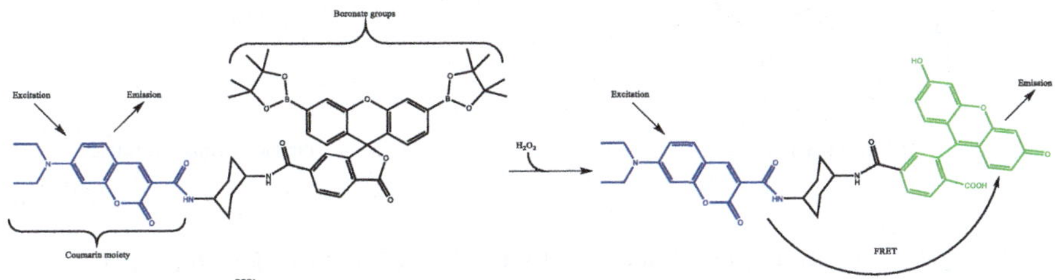

RPF1

(11.16)

Addition of the boronate moiety to the click-modified coumarin platform combines hydrogen peroxide-chemoselectivity with the highly fluorescent characteristics of the coumarin probe (Eq. 11.17) [142].

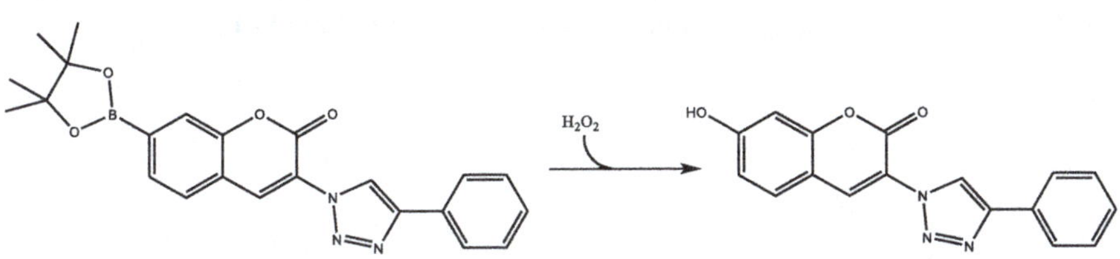

Boronate-Coumarin probe

Fluorescent product (λabs 400 nm; λemis 475 nm)

(11.17)

The coumarin moiety also allows possible tethering of the compound to other functional groups, to target it to different organelles or to improve some physicochemical properties such as sensitivity, photostability, and permeability. The probe responded with a five-fold increase in the fluorescence intensity following 5 μM hydrogen peroxide treatment [142]. Furthermore, the probe's emission intensity was shown to be dependent specifically on hydrogen peroxide concentration, and the concentration dependence does not occur in the presence of other ROS [142].

Further development of FRET concept has yielded another ratiometric hydrogen peroxide probe, Peroxy Lucifer (PL)1 (Eq. 11.18) [143]. Upon reacting with H_2O_2, internal charge transfers within the Lucifer platform result in a change in the emission color of the dye. The key advantage of PL-1 is the fact that it allows the measurement of localized hydrogen peroxide generation with simultaneous monitoring of the hydrogen peroxide variation across the rest of the cell. The relative ratios of the green versus blue fluorescence intensities revealed changes in the local hydrogen peroxide concentration in vesicles versus cytoplasm upon PMA stimulation of RAW 264.7 macrophages [143].

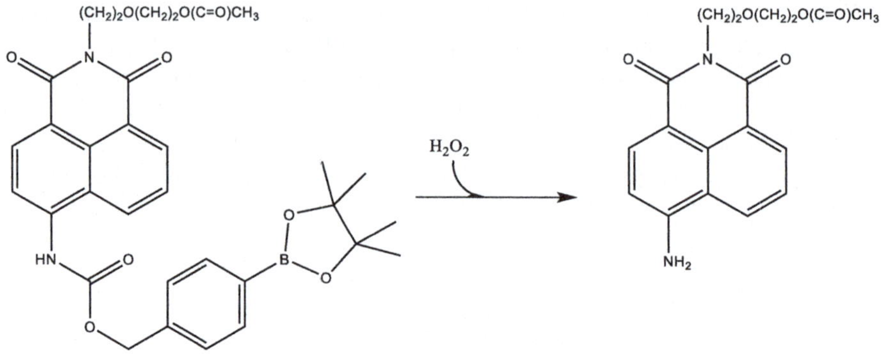

PL1 (λemission 475 nm) Product (λemission 540 nm)

(11.18)

Similarly, dimethylaminocinnamaldehyde (DMACA) linked arylboronate probe, a ratiometric hydrogen peroxide probe, demonstrates intramolecular charge transfer upon the reaction with H_2O_2 (Eq. 11.19) [144]. The transfer involves a change in the ratio of the absorption/emission at two wavelengths. Interestingly, the change in color is also visible to the naked eye, which could be useful in the synthesis of a simple hydrogen peroxide sensor in solutions. The probe demonstrated a significant responsiveness to 2 μM hydrogen peroxide treatment in the prostate cancer cell line (PC3), resulting in a change in a probe color from red to blue [144].

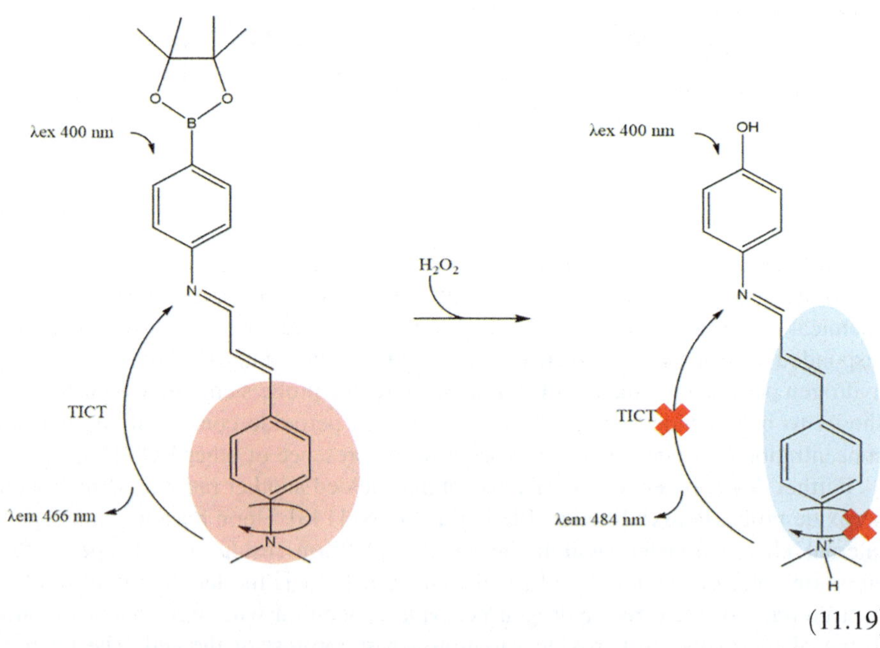

(11.19)

Kalyanaraman et al. have described boronate probes that react with $ONOO^-$ yielding corresponding phenols [145, 146]. One example is coumarin-7-boronic acid (CBA), which readily reacts with ONOO– (Eq. 11.20). Using CBA and fluorescein dimethylamide boronate (FlAmBE) (Eq. 11.21), this group of researchers studied peroxynitrite production in endothelial cells and macrophages [147]. It seemed that the reaction of $ONOO^-$ with boronate led to a site-specific nitration of an aromatic moiety that can be followed by HPLC and mass spectrometry.

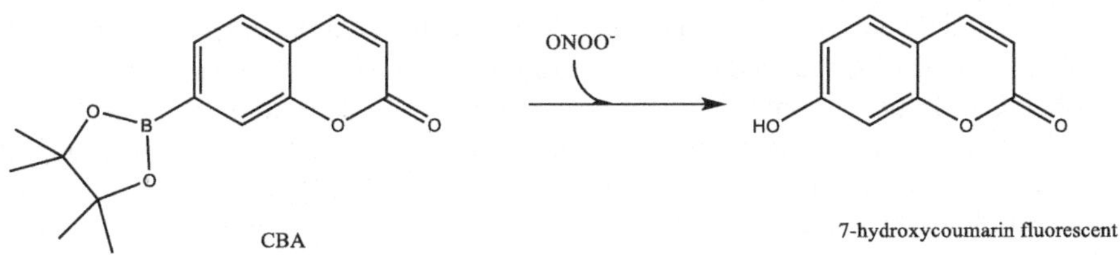

(11.20)

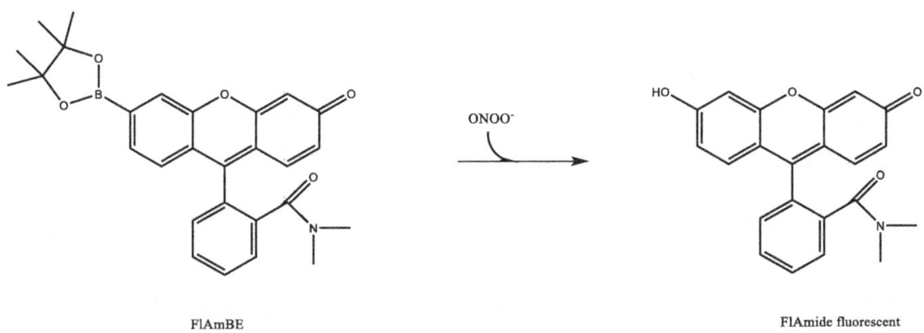

(11.21)

All of the aforementioned probes irreversibly undergo oxidation; therefore, they emit a fluorescent signal, based on the accumulated RS. In contrast, to detect multiple reversible redox reactions in the cellular environment a variant type of fluorescent probes called Redoxfluor (RF)1 was synthesized ($\lambda_{absorption}$ 490 nm, $\lambda_{emission}$ 503 nm) (Eq. 11.22) [148].

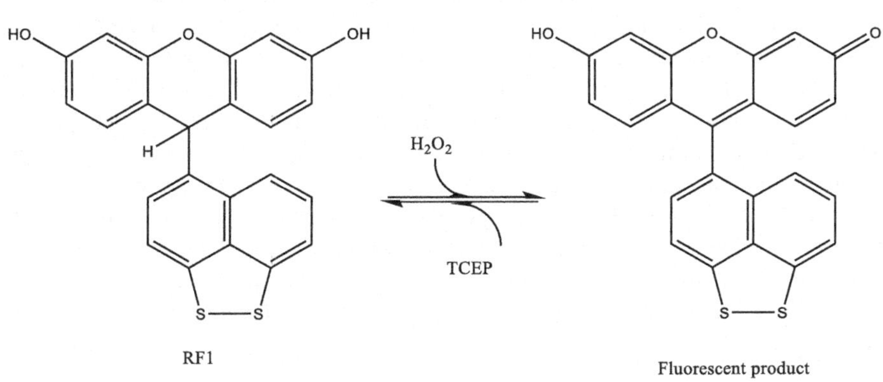

(11.22)

The authors showed that the emission intensity of the probe decreased with treatment of the mild reductant tris(2-carboxyethyl)phosphine (TCEP) and increased with 100 μM hydrogen peroxide treatment. Thus, the fluorescence of this probe reflects the current redox status of its environment and is not an accumulated signal dependent on irreversible oxidation reactions.

One important characteristic of RS fluorescent probes is their compatibility with animal model studies. However, imaging the probe in an animal introduces problems in probe diffusion, tissue and organ penetration, as well as subsequent imaging. To combat these issues, a combination of near-infrared (NIR) detection of the cyanine-7 (Cy-7) with the chemoselectivity of phenylboronic acid was developed, resulting in the formation of boronic sulfo-QCy7 probe (Eq. 11.23) [149]. Sulfonation increases the solubility of the compound.

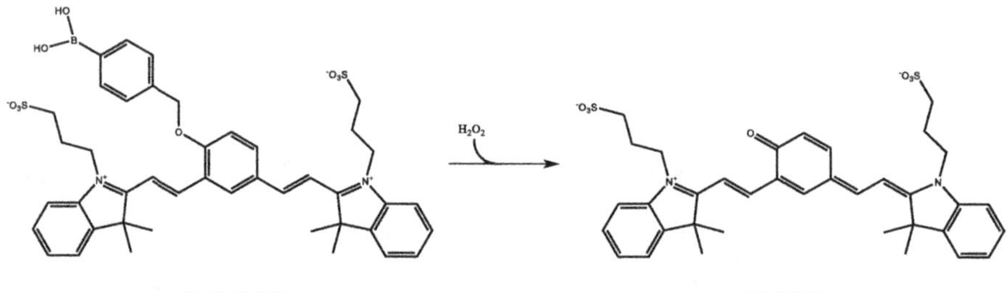

(11.23)

NIR imaging of the cyanine is an attractive tool in animal studies due to the deep penetration of the NIR photons and low background fluorescence of the tissue. Sulfo-QCy7 probe was used to noninvasively image exogenously administered H_2O_2 into mice. The change in the concentration of H_2O_2 following injection resulted in a change in the fluorescent signal. The dye enabled monitoring of hydrogen peroxide signaling upon injection of LPS into mice. The intensity of the fluorescence signal also showed minimal change over the first 5 h, indicating relative chemical stability. Similarly, Peroxy Caged Luciferin (PCL)-1 has been utilized for noninvasive in vivo hydrogen peroxide detection throughout whole mice using boronate protection and Luciferin chemiluminescence (560 nm) (Fig. 11.25) [150]. While maintaining the hydrogen peroxide selectivity, the probe has an ability to noninvasively measure H_2O_2 throughout the whole mouse body without removal of the fur or skin. Moreover, the signal received from the luciferin depends linearly on the concentration of H_2O_2 within a range from 5 to 250 μM in cell culture environment. The probe was used to detect increases in H_2O_2 upon testosterone stimulation that was attenuated by the combination of testosterone with N-acetylcysteine (NAC), a radical scavenger, in the FVB-luc$^+$ mice. While PCL-1 offers an interesting tool to study the redox changes in animals, the animal needs to ubiquitously express the firefly luciferase in order to visualize the reaction of the probe [151].

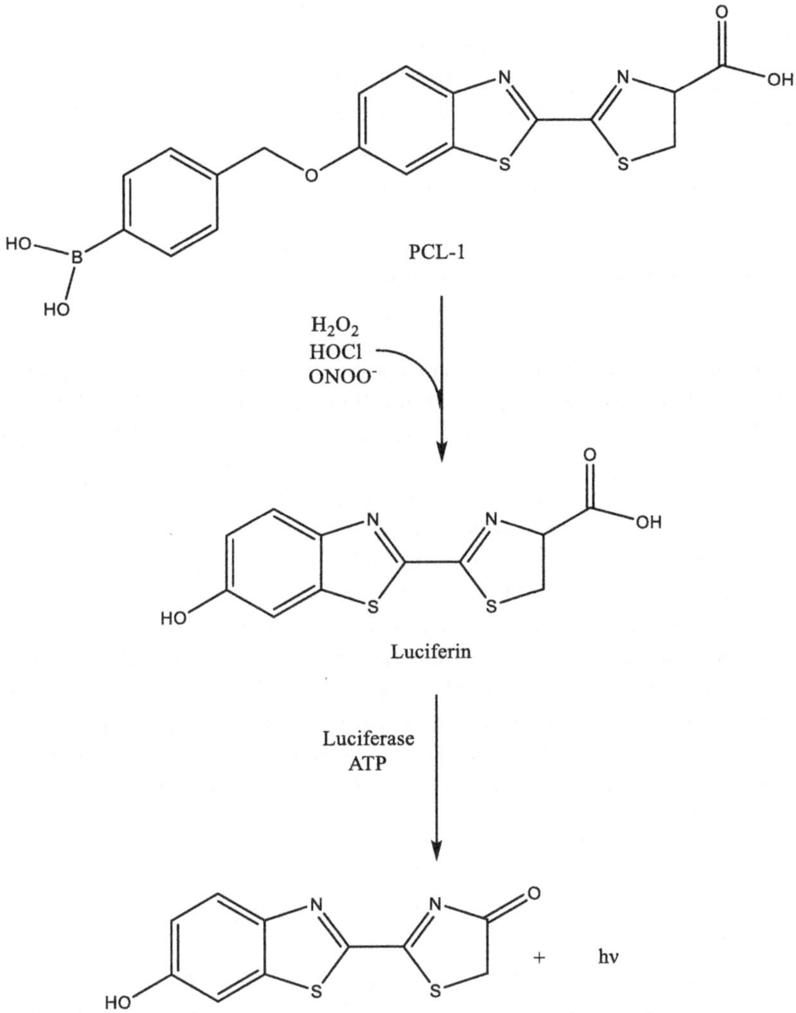

Fig. 11.25 Peroxy Caged Luciferin (PCL)-1 and release of Luciferin chemiluminescence

The boronate chemistry was also utilized in the synthesis of RS-BE probe that fluoresces upon presence of both H_2O_2 and Fe^{3+}/Cu^{2+} ions (Fig. 11.26) [152]. The reaction with H_2O_2 unmasks the specific chelating region of the RS-BE that can react specifically with metal ions, yielding a fluorescent metal-coordinated product. RS-BE was demonstrated to specifically fluoresce in the presence of Fe/Cu ions and H_2O_2 in neuroblastoma cells [152].

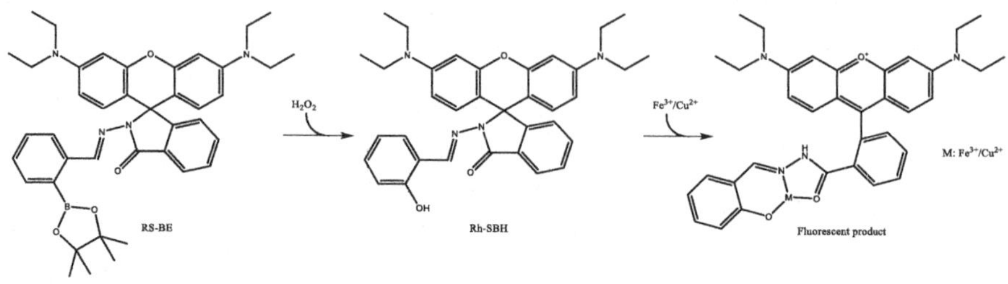

Fig. 11.26 Activation and metal reaction of RS-BE

All of the aforementioned probes are suitable for the detection and measurement of H_2O_2 in cell samples. However, in clinical environments it may be important to measure changes in the redox state of tissue samples, for example, following patient biopsy.

There have been few probes reported utilizing the boronate-phenol chemistry and allowing for hydrogen peroxide imaging in thick tissues. Naphto-Peroxyfluor-1 (NPF1) is one such probes, in which the addition of a naphtofluorescein moiety results in emission in the visible far-red region of the spectrum (Eq. 11.24) [131]. This change in the fluorescence profile has several benefits. First, it avoids problems with background fluorescence from the specimen. Second, it enables imaging of thick samples (>80 μm), for example, tissue slices. While this probe presents the aforementioned advantages, it utilizes one-photon microscopy, which carries a number of drawbacks. The short excitation wavelengths are quite damaging to live specimens, substantially shortening the time for accurate imaging.

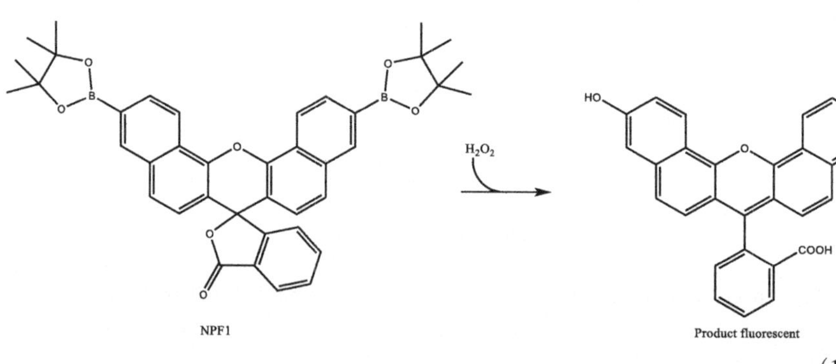

(11.24)

These shortcomings were overcome in Peroxy Naphthalene 1 (PN1), the first hydrogen peroxide-specific two-photon fluorescent probe with the excitation profile in the near-infrared region (Eq. 11.25) [153]. PN1 probe is applicable to deeper tissue imaging as it facilitates prolonged imaging without causing much damage to the specimen.

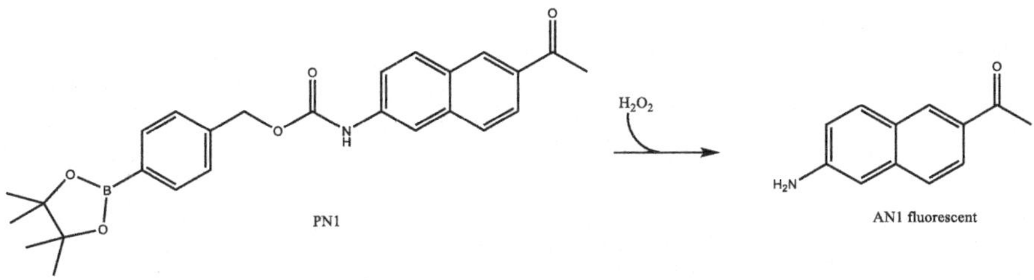

(11.25)

Additionally, PN1 is a ratiometric dye that can be used for quantitative analysis of hydrogen peroxide generation. The probe's fluorescence is not affected by changes in biological range of pH and does not cause cytotoxicity [153]. The combination of two-photon microscopy with ratiometric analysis makes PN1 a good candidate for examining hydrogen peroxide levels in tissue.

The creation of multifunctional or dual signal sensors that react with two or more cellular species and could elucidate their relationship under a stressed system are of high interest. Few multifunctional sensors have been developed which probe for the redox active metals Cu or Fe in the presence of H_2O_2. In each case, the sensors rely on a prochelator strategy that utilizes a boronate ester protecting group to prevent metal coordination prior to oxidation by H_2O_2. Thus, the selectivity of this strategy resides in the reactivity-dependent activation of a prochelator to a chelator, and not on any inherent metal preference of the chelator itself. The Guo laboratory first introduced SBH-B-AN (Fig. 11.27), which is an anthracene version of its boronic acid prochelator, SBH [154]. The protected version didn't show any iron-binding capabilities and any changes to the fluorescence output. However, after reaction with H_2O_2, the emission bands of the chelator at 400 and 480 nm were ~30–40% quenched by the addition of Fe^{3+}.

As a separate approach to this strategy, a group of researchers introduced FlamB [159]. Built upon the spirolactam framework found in several metal sensors described in previous sections, nonfluorescent FlamB is oxidized by H_2O_2 to its

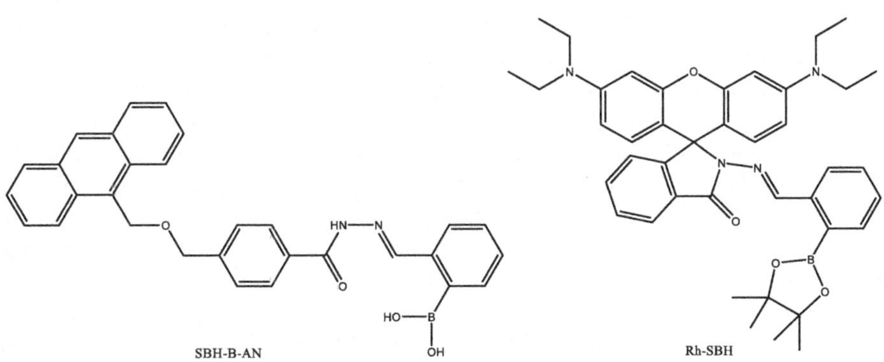

Fig. 11.27 Structures of SBH-B-AN and Rh-SBH

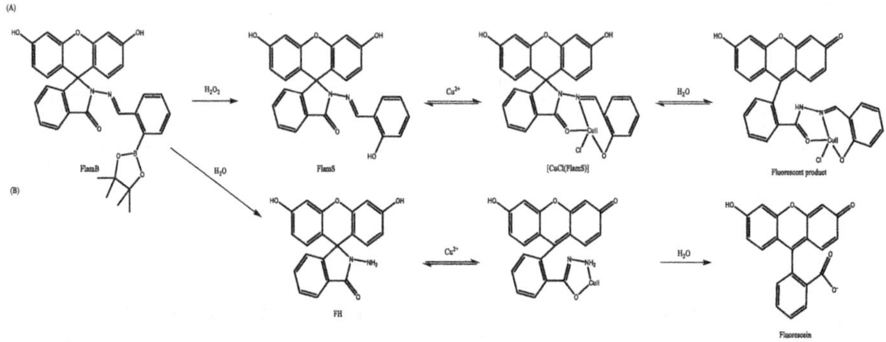

Fig. 11.28 FlamB reaction with H_2O_2 and alternate pathways to fluorescence enhancement. (**a**) FlamB oxidation followed by Cu^{2+} binding and spirolactam ring opening to give modestly fluorescent product [CuCl(FlamS)]. (**b**) Hydrolysis of FlamB to FH followed by its reaction with Cu^{2+} to yield highly fluorescent fluorescein

chelator form, FlamS, which itself exhibits a modest Cu^{2+}-selective fluorescence turn-on response (Fig. 11.28). However, it was found that a competing mechanism occurs in aqueous solution because of the hydrolytic instability of the prochelator's hydrazone backbone. One of FlamB's hydrolysis products, fluorescein hydrazine, reacts with Cu^{2+} (Fig. 11.28) or H_2O_2 to produce the highly fluorescent fluorescein. A very small amount of hydrolytic decomposition could therefore produce a fluorescence response that could be easily misinterpreted. Due to this alternative pathway of fluorescence turn-on, cell studies were not performed.

Despite the instability of FlamB, the Guo laboratory reported a rhodamine derivative containing the same hydrazone prochelator backbone [152]. They showed that Rh-SBH (Fig. 11.27) produced a 5- or 7.5-fold turn-on in the presence of Fe^{3+} or Cu^{2+} and excess H_2O_2, respectively. In turn, Rh-SBH was applied to living SH-SY5Y neuroblastoma cells and showed increased fluorescence when incubated with excess Fe^{3+} or Cu^{2+} and H_2O_2.

The final example in this class of multifunctional sensors is FloB-SI (Eq. 11.26) [160]. This fluorescein prochelator is resistant to hydrolysis in aqueous solution and cell culture media. It undergoes fast deprotection with H_2O_2 through loss of its self-immolative pinacol boronic ester protecting group. Once converted to its chelator form, FloS, emission intensity decreases by 70% with Fe^{3+} or Cu^{2+}, with more modest quenching in the presence of Zn^{2+}, Co^{2+}, or Ni^{2+} and no change with Na^{+}, Ca^{2+}, Mg^{2+}, or Mn^{2+}. Fluorescence microscopy studies in HeLa cells using FloB-SI showed that the sensor's fluorescence intensity remained unchanged until incubation with exogenous H_2O_2, which caused the cellular fluorescence to diminish. Incubation with a competitive chelator, SIH, restored the emission signal, suggesting that FloB-SI is effectively reporting a hydrogen peroxide-induced increase in intracellular labelled metal with the most likely candidate being iron [160]. While these results are encouraging, the fluorescence quenching mechanism and the poor photostability of the compounds indicate that more work remains to be done to optimize this type of multifunctional probes and use them to study the role of metals in OS-associated diseases.

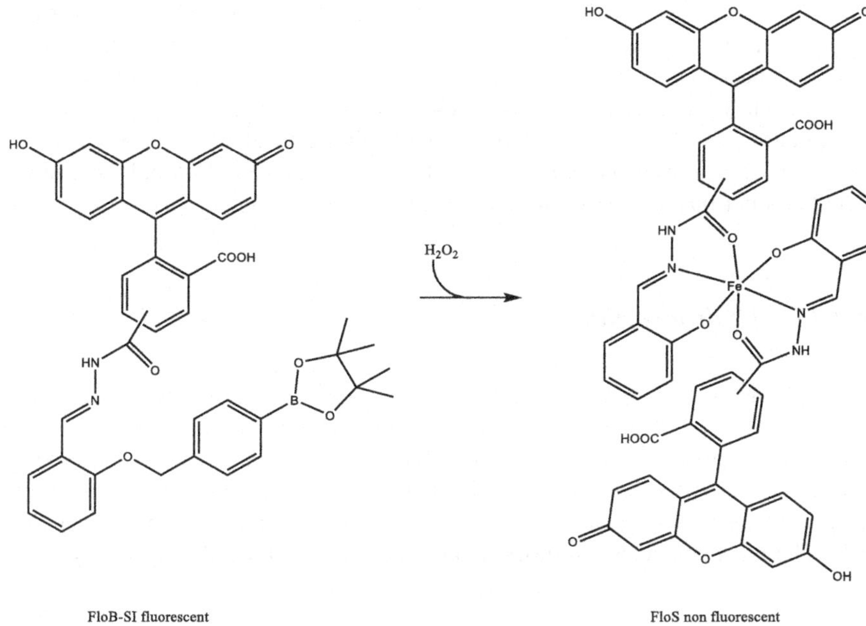

(11.26)

Overall, it is quite likely that boronate probes provide a reliable approach for the detection of H_2O_2 and $ONOO^-$. As suggested earlier, there is a potential problem related to the lack of specificity of some boronate probes that can detect either $ONOO^-$ or H_2O_2. The use of a hydrogen peroxide scavenger such as PEG-CAT or a peroxynitrite scavenger such as uric acid may help increase specificity of boronate probes, but further studies are required to validate such approaches.

The design and synthesis of the combined probes make the boronate-based assay suitable for the specific detection and monitoring of H_2O_2. It provides versatile possibilities to create novel multifunctional probes for better targeting and detection of FR in living systems and cells. They allow for an unattainable level of accuracy in measuring important second messengers represented by RS.

11.2.12 Calcein-Acetoxymethylester

Calcein-acetoxymethylester (Calcein-AM) has been reported as a detector of intracellular OS [161]. Calcein-AM is a fluorogenic cell permeable compound that is converted by intracellular esterases into the cell impermeant anion calcein, which is fluorescent. Historically, intracellular calcein production has been used in both microscopy and fluorometry as an indicator of viable cells. The ROS-oxidized product of Calcein-AM has been shown to still retain the ability to cross cell membranes and has similar spectral properties to the fluorescent calcein [161]. Thus, it is important to keep the compound in the media rather than washing it away after cell loading. Removal of the dye results in movement of the reacted compound back out of the cell on a concentration gradient. This probe tends to aggregate and form intensely fluorescent localizations within the cell. While it has possibilities with confocal microscopy where cellular localization is possible, the inability to distinguish spectrally between RS reacted Calcein-AM and the esterase reacted product makes the use of this probe problematic (Fig. 11.29).

11.2.13 Hydrocyanine

Hydrocyanine probes are fluorogenic sensors for $O_2^{\bullet-}$ and $OH^{\bullet}$ (Eq. 11.27). They are synthesized by reducing the iminium cation of the cyanine (Cy) dyes with sodium borohydride. While weakly fluorescent, upon oxidation their fluorescence intensity increases 100 fold. In addition to being fluorescent, oxidation also converts the molecule from being membrane permeable to an ionic impermeable moiety [162]. The most characterized of these probes are Hydro-Cy3 and Hydro-Cy5.

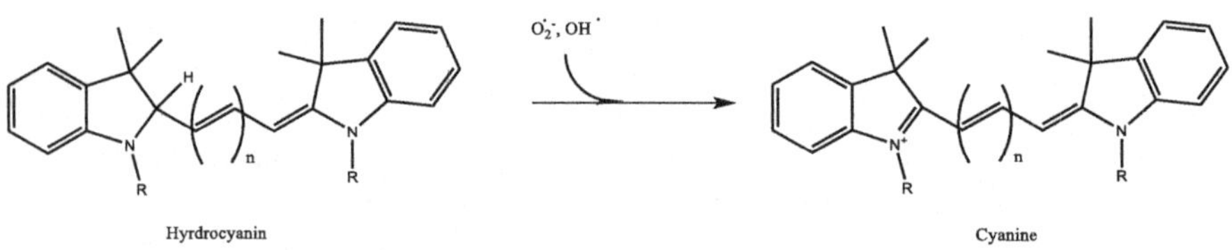

(11.27)

Hydro-Cy3 probe has been used to evaluate ROS production in wounded intestinal epithelium after *Lactobacillus rhamnosus GG* infection (Figs. 11.30 and 11.31) [163].

11.2.14 DanePy

The fluorescence quenching of DanePy reflects the formation of nonfluorescent DanePyO due to reaction of $^1O_2$ with DanePy. As DanePy is detectable by fluorescence quenching due to ROS formation, it is more often used as a typical and highly selective probe for $^1O_2$ detection [164]. DanePy which is quenched in the presence of $^1O_2$ is suitable for an optical measurement of the interaction between RS and spin trap molecules.

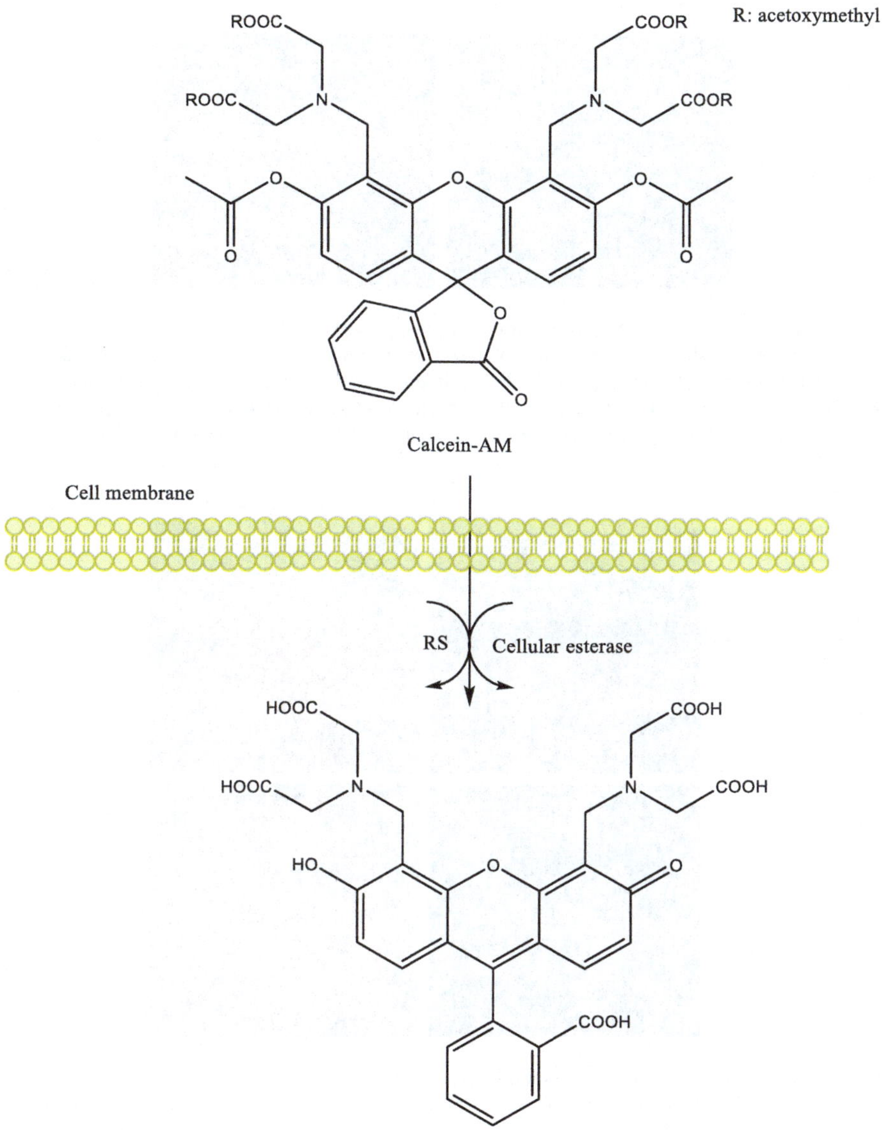

Fig. 11.29 Reaction of Calcein-AM with cellular peroxidases and RS

11.2.15 DAF-2-DA

When measuring real-time changes in nitric oxide production in cells, diaminofluorescein diacetate (DAF-2DA) is usually required. DAF-2-DA is cell permeable and reacts with $NO^{\bullet}$. Like DCFH-DA, esterase activity is required

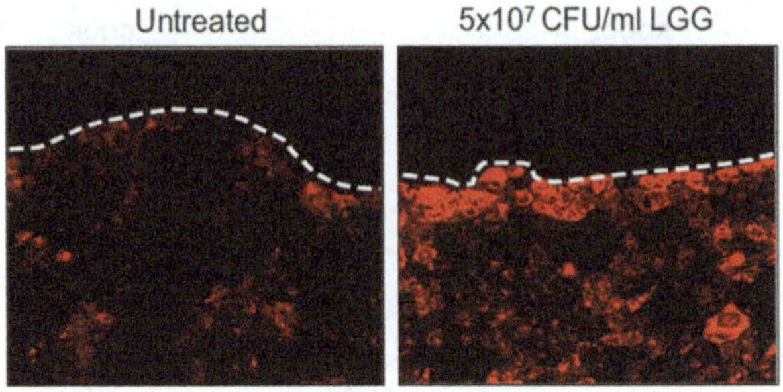

Fig. 11.30 Detection of ROS generation in vitro using hydro-Cy3 probe
Fluorescent images show scratch-wounded epithelium in culture. Dashed line denotes leading edge of scratch wound. Untreated cells exhibit low hydro-Cy3 fluorescence (upper left). Addition of 5×10^7 colony-forming units (CFU)/mL of the bacterium *Lactobacillus rhamnosus* GG results in a robust increase in fluorescence, that is, ROS generation 5 min after treatment [163]

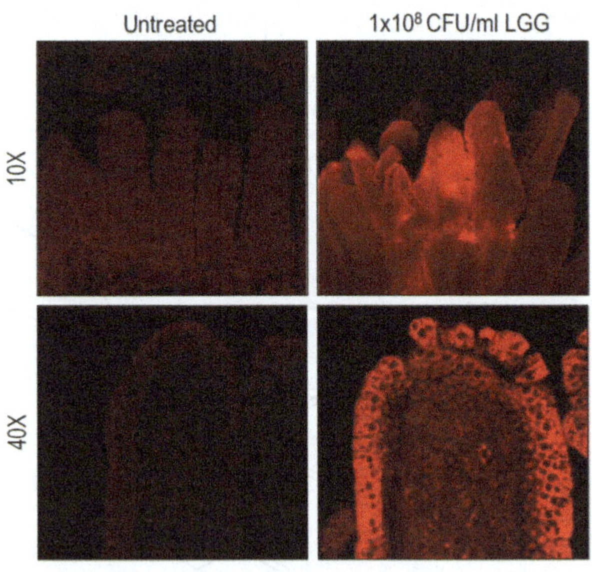

Fig. 11.31 Detection of ROS generation in vivo using hydro-Cy3 probe
Images of hydro-Cy3 fluorescence in whole-mount preparation of small intestine from untreated mice (left) and mice gavaged with 1×10^7 colony-forming units of *Lactobacillus rhamnosus* GG (right). Images taken 1 h after administration of LGG at 10X (upper row) and 40X (lower row) magnifications [163]

before the probe can be oxidized. DAF-2-DA is a nonfluorescent probe which fluoresces on reaction with $NO^{\bullet}$ and can be used to measure intracellular production of $NO^{\bullet}$ (Fig. 11.32) [165]. DAF-2 DA fluorescence can be visualized using confocal microscopy.

DAF-2DA fluoresces in organelles with high Ca^{2+} concentration including mitochondria and ER. However, using DAF-2-DA in systems where intracellular Ca^{2+} increases may produce misleading results of nitric oxide production [166]. Changes

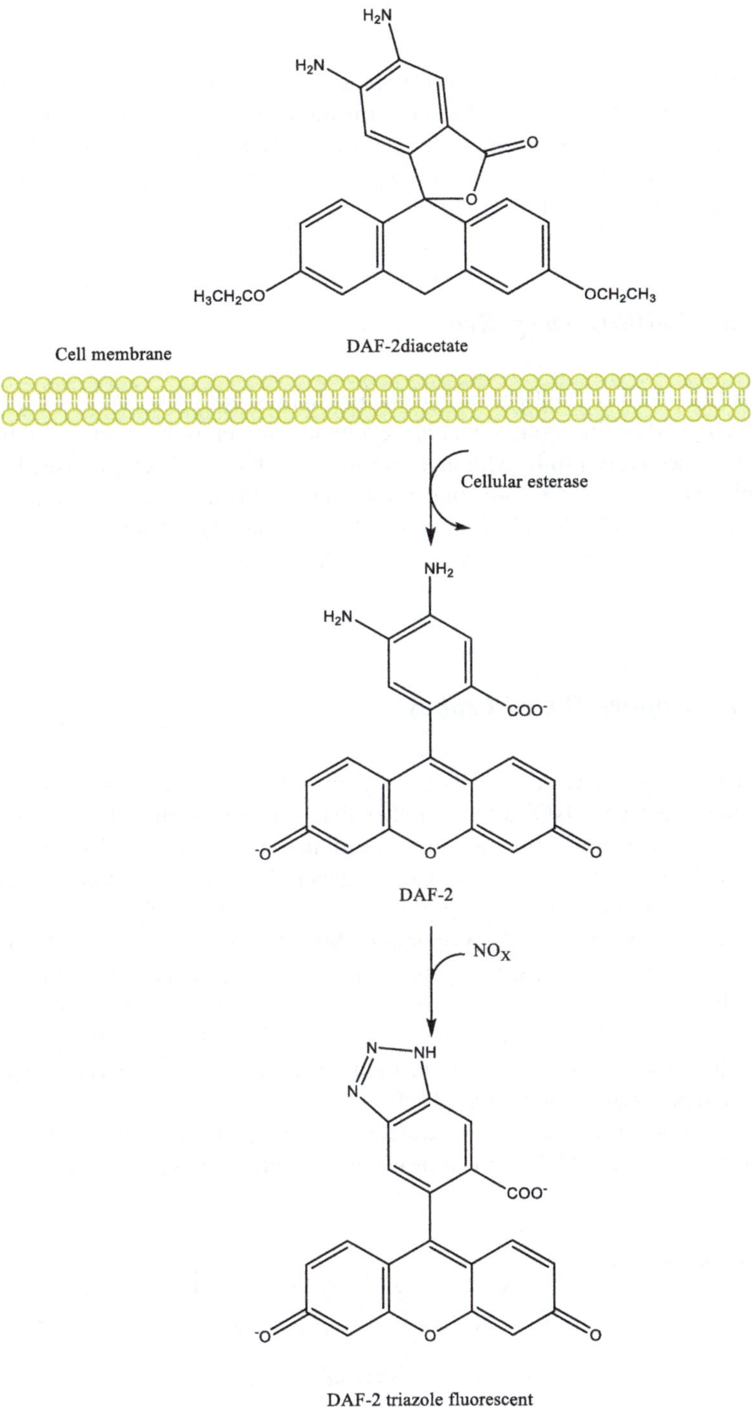

Fig. 11.32 Fluorometric detection of NOx using DAF-2-DA
DAF-2DA diffuses into cells and tissue where esterases hydrolyze the diacetate residues thereby accumulating DAF-2 within the intracellular space. NO-derived nitrosating agents nitrosate DAF-2 to yield its highly fluorescent product DAF-2 triazole

in intracellular Ca^{2+} occurs in several apoptotic systems [167]. When measuring $NO^{\bullet}$ in such systems, it is necessary to confirm nitric oxide production using another probe such as 2,3-diaminonaphthalene or specific NOS inhibitors [167].

Finally, DAF-2 DA probe has been used to assess nitric oxide production in human sperm cells (Fig. 11.22) [71].

11.2.16 CellRox Deep Red

CellRox Deep Red is a cell-permeable and nonfluorescent probe until oxidation with RS, which yields a fluorophore that has excitation and emission maxima at 640 and 665 nm, respectively [168]. Although the structure has not been published yet, the probe claims to offer several advantages over others. Most noticeably, CellRox Deep Red Reagent shows higher photostability when compared to DCFH. One of the most interesting application of this probe is visualization of RS in human sperm cells (Fig. 11.22) [71].

11.2.17 Copper-Based Probes

In the last two decades, there has been great interest in developing NO-specific probes that react with $NO^{\bullet}$ directly rather than react with other RNS. One of the approaches was the utilization of cupric coordination compounds. The reaction with $NO^{\bullet}$ caused reduction of Cu (II) to Cu (I), generating a highly fluorescent product from the ligand. In one study, cupric ions were associated with 4-methoxy-2-(1H-naphtho [2,3-d]imidazol-2-yl)phenol (MNIP) creating an MNIP-Cu probe (Fig. 11.33) that specifically reacts with $NO^{\bullet}$ to generate blue fluorescence [169]. This probe is capable of crossing cell membranes and allows for nitric oxide imaging in vitro and in tissue sections of mice. However, the probe is not suitable for assessing the localization of nitric oxide generation because the dye diffuses quickly across membranes [169].

Another study focused on two strategies utilizing FL1 as a ligand for Cu (II), forming CuFL1 (Eq. 11.28), to improve its potential for subcellular localization studies [170].

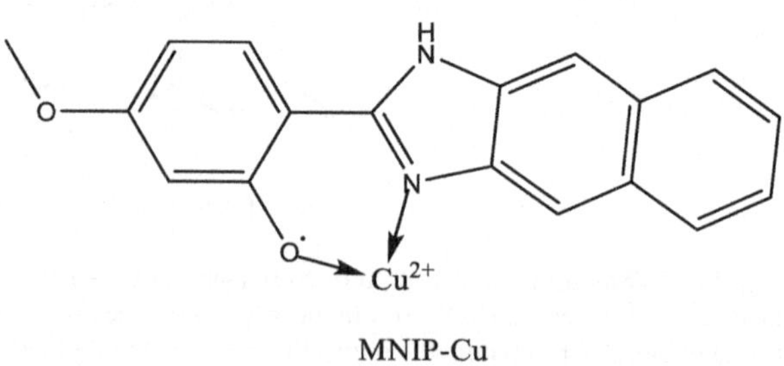

MNIP-Cu

Fig. 11.33 Structure of MNIP-Cu

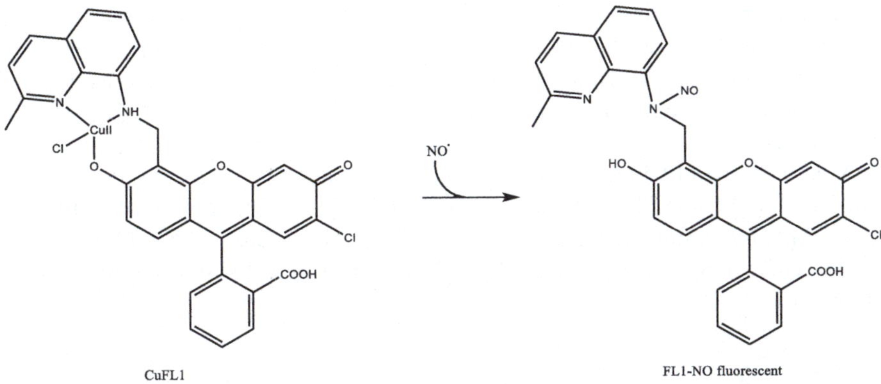

(11.28)

First, the authors added an ester moiety, yielding CuFL1E (Fig. 11.34). Intracellular esterases cleaved the ester moiety, resulting in the corresponding negatively

Fig. 11.34 Structure of CuFL1E and CuFLDex

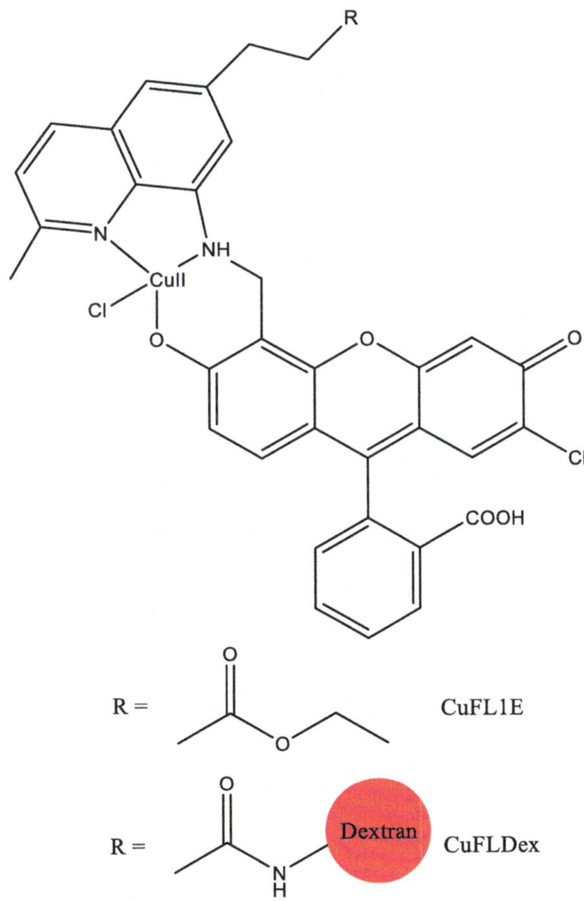

charged, membrane-non-diffusible acids, FL1A. Second, they used dextrans as a vehicle to translocate the probe inside the cell, yielding the FL1 derivative CuFLDEX (Fig. 11.34). Both of these probes are NO$^{\bullet}$ or NO-donor selective and were shown to improve cell localization, with CuFL1E localizing to the mitochondria.

Modification of FL1 and FL1E yielded the seminaphtho-fluorescein-based probes CuSNFL1 and CuSNFL1E (Eqs. 11.29 and 11.30) [171]. These probes emit longer wavelength lights suitable for deep-tissue imaging, undergo a 20–45-fold increase in fluorescence upon reacting with NO$^{\bullet}$, and can detect ligand-stimulated nitric oxide production.

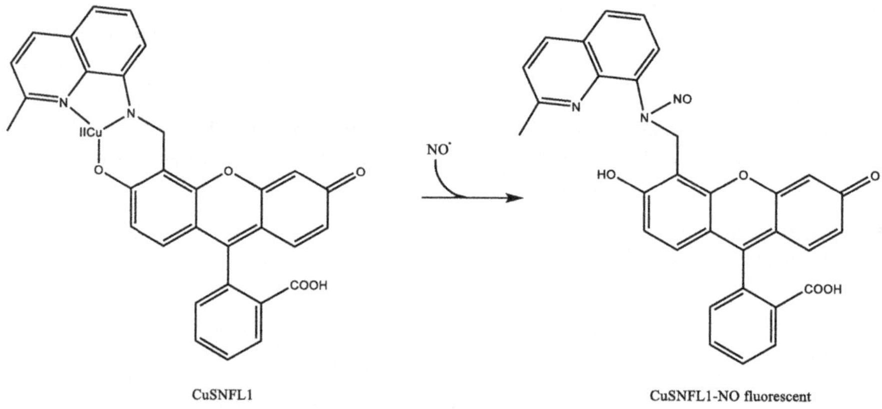

(11.29)

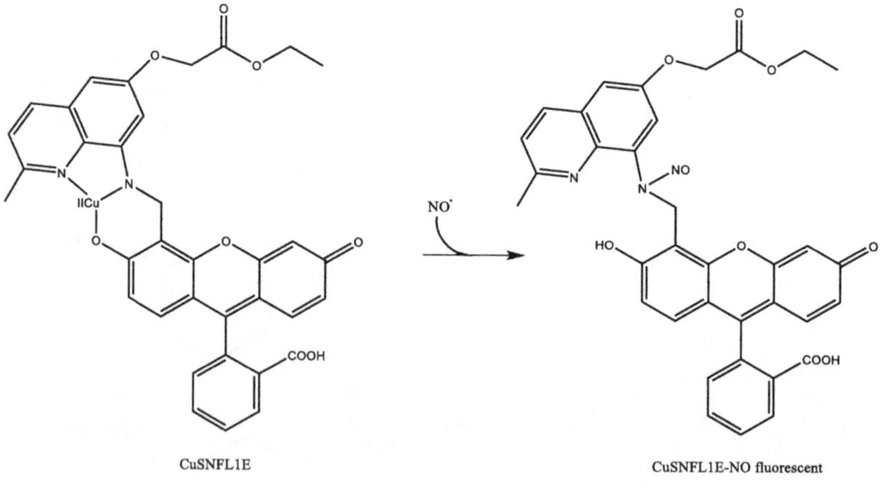

(11.30)

The sensitivity of cupric ions to $NO^•$ was also used in the synthesis of the CuRBT probe (Eq. 11.31), based on a rhodamine B platform [172]. Upon reaction with $NO^•$, Cu(II) undergoes reduction and the subsequent change in this platform exhibits a large increase (700-fold) in fluorescence. CuRBT also offers one of the lowest nitric oxide detection limits (1 nM). As a result of the possible side effects of introduction of heavy metal into cells, other strategies are being developed to detect $NO^•$.

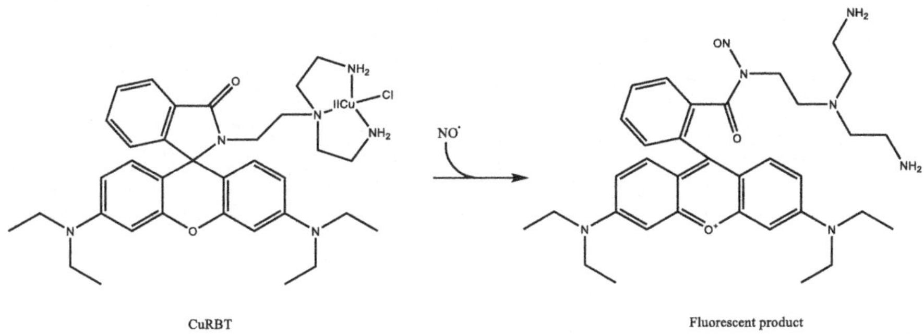

(11.31)

11.2.18 Other Molecular Probes

The NO550 probe utilizes emission of fluorescence following the internal charge transfer between dimethylamino and cyano groups upon reaction with $NO^•$ (Eq. 11.32) [173]. NO550 probe monitors endogenous changes in nitric oxide fluxes and displays a good signal-to-noise ratio, showing a 1500-fold increase in fluorescence emission upon nitric oxide detection. NO550 is also insensitive to changes in pH over the range 4.5–9.

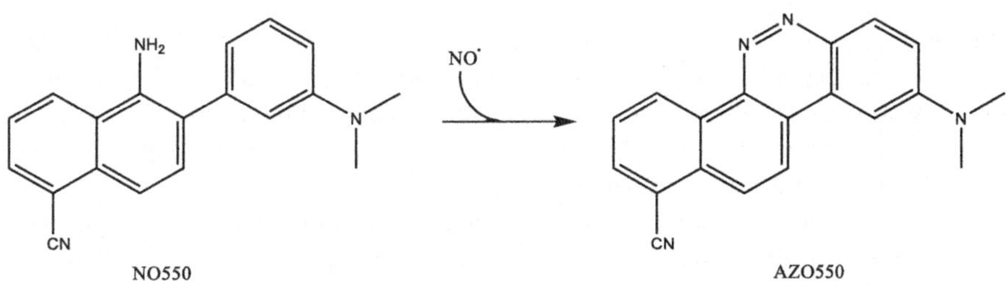

(11.32)

The ratiometric NO-specific probe, Cou-Rho-NO, is composed of rhodamine and coumarin moieties as a FRET (Fig. 11.35) [174]. This probe is selective for $NO^•$, and its limit of detection is 30 nM. It also works over the 3–10 pH range.

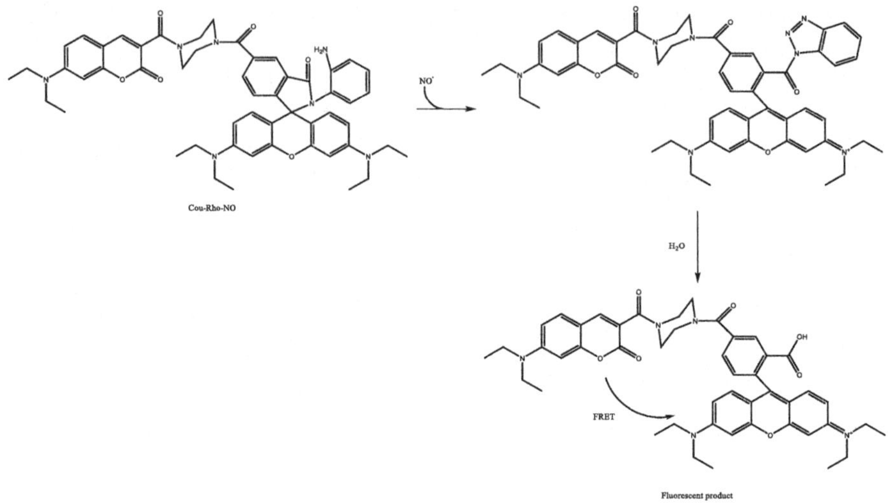

Fig. 11.35 Reaction of Cou-Rho-NO with NO˙

Finally, in order to examine the interplay of different ROS/RNS in signaling reactions, a dual specificity probe, FP-H_2O_2-NO, was developed. It can detect H_2O_2, NO˙, and H_2O_2/NO at once, emitting distinct fluorescent signals upon contact with each species (Fig. 11.36) [175], and with FRET occurring in the presence of both together. A great deal of improvement has been made to fluorescent ROS probes in recent years, mostly in the areas of specificity and sensitivity. Although these improvements allowed for more accurate assessments of cellular redox dynamics, the probes are generally irreversibly oxidized and so are limited in their usefulness.

11.2.19 Nanoparticles

During the past two decades, sustained efforts have been made to develop sensors based on nanoparticles (NPs). The term "nanoparticles" covers an unlimited diversity of nano-objects, which have at least one dimension ranging from a few to several hundred nanometers and a very high surface-to-volume ratio. The core of NPs can be a metal, a polymer, or a semiconductor [157, 176].

Some drawbacks of traditional fluorescent probes can be overcome by encapsulating them in a NP delivery system (Fig. 11.37) [177]. Fluorescent probes may preferentially target cellular organelles, as discussed earlier, but they may undergo nonspecific reactions during delivery to those organelles. NPs can be seen as a platform for custom functionalization (Fig. 11.37), and their surface can be modified with targeting moieties that address them to particular cells such as tumor cells that express specific membrane receptors.

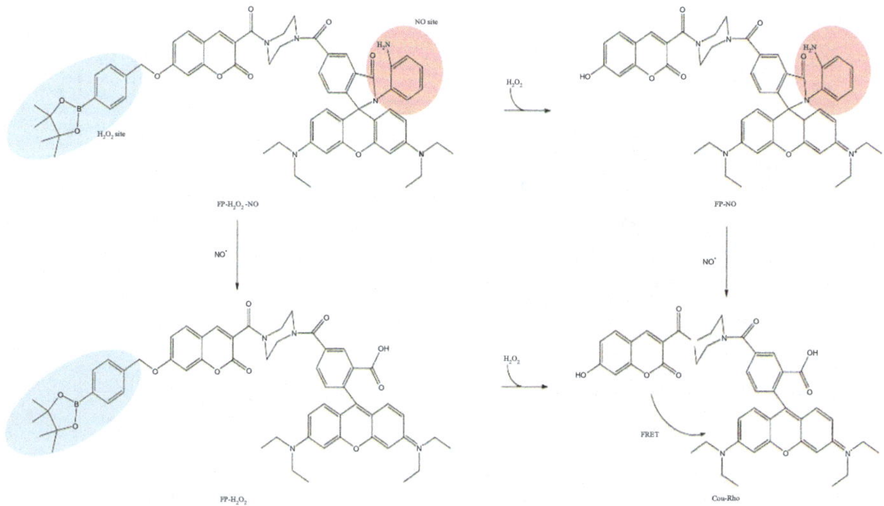

Reaction with ROS/RNS	Excitation 400 nm		Excitation 550 nm	Fluorescence signal sequence
	Emission 460 nm	Emission 580 nm	Emission 580 nm	
None	✸	✸	✸	Black → Black → Black
NO	✸	✸	✸	Black → Black → Red
H_2O_2	✸	✸	✸	Blue → Black → Black
NO + H_2O_2	✸	✸	✸	Black → Red → Red

Fig. 11.36 Reaction of FP-H_2O_2-NO with NO$^{•}$ and/or H_2O_2

Chemically inert NP delivery systems are sufficiently small so as to be introducible into cells by standard mechanisms (microinjection, lipofection, and TAT-protein delivery) [178]. NPs can be either loaded inside, mainly for polymer NPs, or decorated outside with one or several probes or sensing agents (Fig. 11.37). NPs are large enough to encapsulate perfectly relatively large volumes of one or multiple probes, which facilitates measurement of several fluorescent responses. Thus, they can offer a ratiometric fluorescence detection and simultaneous detection of several RS at the same time. At the same time, the NP can protect the probe from interferences with the biological medium, which improves its chemical stability, photostability, and even selectivity.

One of the first approaches to deliver fluorescent probes into cells was the Photonic Explorer for Biomedical use with Biologically Localized Embedding (PEBBLE) [179]. The PEBBLEs consist of a reporter probe encapsulated in a matrix of inert material (e.g., polyacrylamide, PVC, decyl methacrylate, or gold colloid) and are typically 20–600 nm in diameter. Thus, PEBBLEs have a volume relative to cell size of the order of 1 ppb-1 ppm, and so their introduction into the cell will have negligible effects. PEBBLEs also have the benefit of minimizing photo-bleaching of ROS probes [180]. Measurement of PMA-stimulated ROS production via PEBBLEs

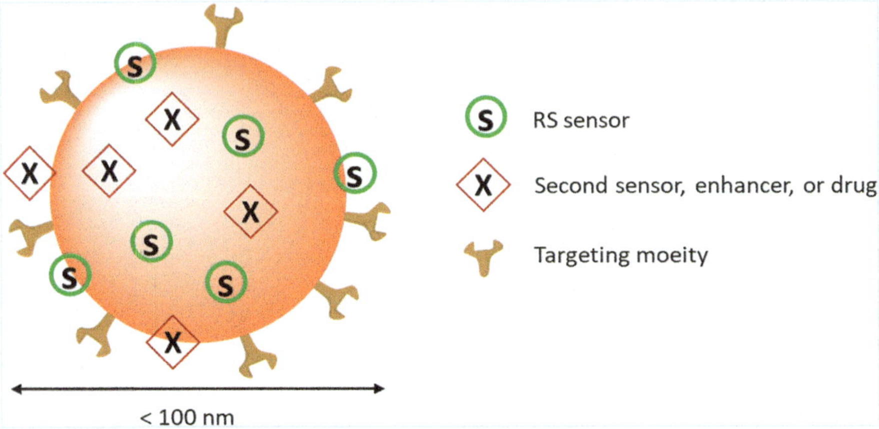

Fig. 11.37 Design of NPs-base probe for RS detection
RS sensors (S) or other compounds (X) may be attached to the NP or introduced inside porous NPs. In the latter case, RS need to penetrate the NP to reach the sensor. The surface can be modified to target the particles to specific cells

has been achieved in macrophages [181]. A good example of delivered fluorescent probes using NP technology is the hydroxyl radical-sensitive 3-CCA, which has been attached to amine-functionalized polyacrylamide NPs, along with Red-Dextran as a reference dye, to allow ratiometric analysis of $OH^{•}$ in vitro [182]. Another example is the COumarin-NEutral Red (CONER), which is a ratiometric nanoprobe composed of biocompatible NPs and encapsulated neutral red with 3-CCA derivative [183]. CONER probe quantifies the hydroxyl radical levels as a fluorescence ratio of the reference Texas Red dye and the hydroxyl radical-sensing 3-CCA moiety. It was demonstrated that CONER nanoprobe responded specifically to $OH^{•}$ in viable cancer cells exposed to OS [183].

An ingenious idea was put forward by Tan et al. (2017), which consists of combining ROS detection with ROS-induced drug delivery. The proposed design was the use of mesoporous silica NPs able simultaneously to detect H_2O_2 using a boronate derivative of fluorescein (FL2) and to release Captopril, a drug efficient against heart failure [184]. For this probe, the cleavage of the boronate by H_2O_2 freed the fluorescent fluorescein. The boronates were "covered" with α-cyclodextrine. The volume of the α-cyclodextrine limits the escape of captopril from the pores of the NP. After reaction with H_2O_2, the fluorescence of FL2 appeared at 520 nm on the NPs. The α-cyclodextrines were released simultaneously with the boronates thereby increasing the diffusion of captopril out of the NPs. These NPs were tested using a zebrafish model of heart failure [185]. The increase of FL2 fluorescence intensity in the heart was proportional to the severity of heart failure. Zebrafish treated with captopril-loaded NPs recovered heartbeat rate and cardiac output comparable to healthy animals. The benefit due to captopril released from NPs seemed to be higher than the one due to soluble captopril.

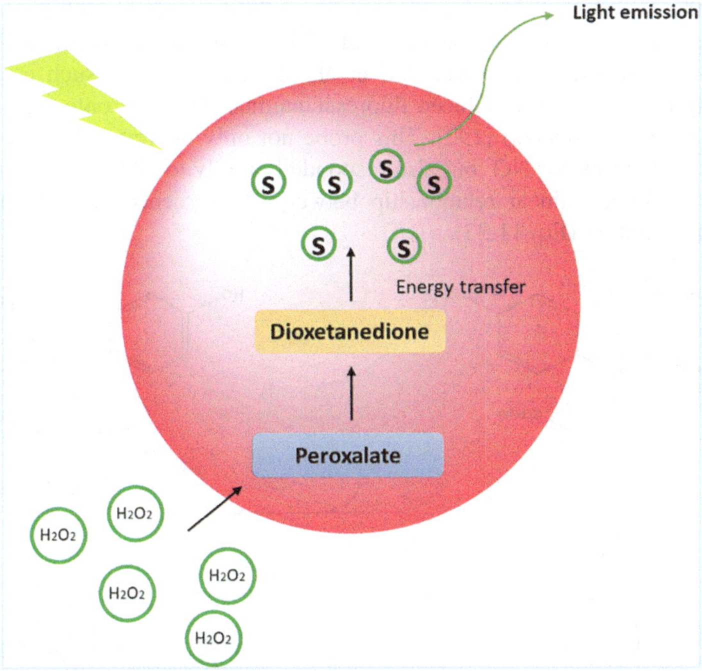

Fig. 11.38 Peroxalate-based NPs for the detection of H_2O_2 in vivo

Peroxalate-based NPs were also developed for the detection of H_2O_2 (Fig. 11.38) [158]. In this system, NPs were loaded with a fluorescent probe and then chemically excited by the high-energy dioxetanedione produced by the reaction of peroxalate ester with H_2O_2. This design has been improved by introducing an enhancer between the dioxetanedione and a near IR probe. With this improved version, nanomolar concentrations of H_2O_2 were detected [186]. These NPs were used to detect H_2O_2 even in a mouse model of peritonitis and arthritis [185]. This approach based on the initial chemiluminescence of the dioxetanedione has a very-low background. There was no noise due to excitation of tissue auto-fluorescence as in the case of fluorescence-based detection.

Sensitivity of HPF for $OH^{\bullet}$ has been exploited in the development of AuNC@HPF, a nanoprobe where the gold nanocluster (AuNC) protected by bovine serum albumin was employed as a reference fluorophore, and HPF functions as a hydroxyl radical sensor [187]. The probe can be detected by a single emission peak at 637 nm, originating from AuNC reference fluorescence [187]. Upon addition of $OH^{\bullet}$, a second HPF emission peak can be observed at 515 nm—which is strictly dependent on the concentration of $OH^{\bullet}$. The probe demonstrated a high selectivity for $OH^{\bullet}$ over other ROS, as well as good water solubility, cell permeability, and long-term stability [187].

Oxazine and hydroethidine have been combined with NPs for the detection of HOCl and $ONOO^-$ or $O_2^{\bullet -}$, respectively [188, 189]. Red probes based on protected cyanines coupled to NPs were also developed [190, 191].

An effective strategy for improving the stability and the selectivity of the fluorophore required for the detection of $NO^{\bullet}$ is through immobilization of the reactive molecules into a matrix. Shi et al. (2012) developed such a sensor by covalently immobilizing a reduced fluoresceinamine molecule onto the surface of silicon nanowires (SiNWs) [192]. This probe not only proved to have high sensitivity and selectivity for $NO^{\bullet}$ but also responded rapidly to $NO^{\bullet}$ released from liver extract exhibiting a linear relationship between fluorescence intensity and nitric oxide concentration (Eq. 11.33).

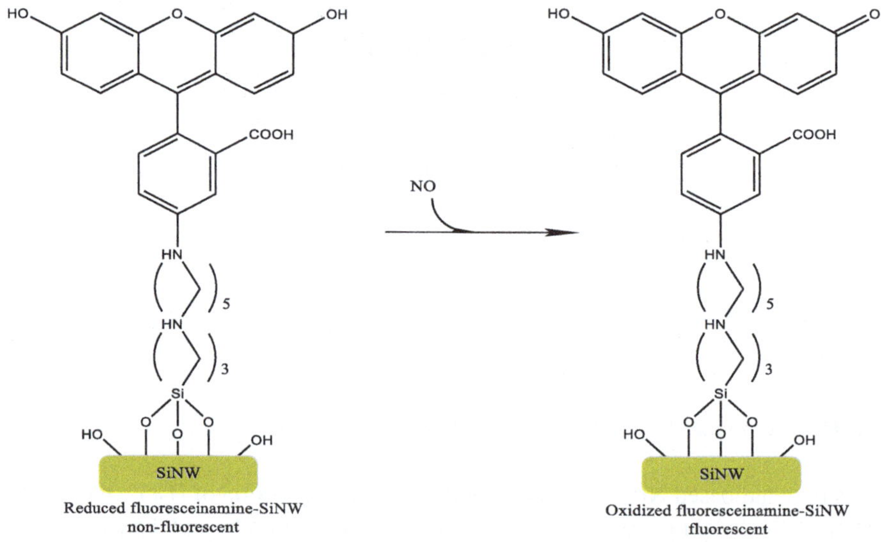

(11.33)

Cy3-functionalized fluorescent Silver NPs were developed by Swanner and Singh (2018) in order to detect ROS production in cancer cells. MDA-MB-231 breast cancer cells were used to validate the model, as shown in Fig. 11.39 [193].

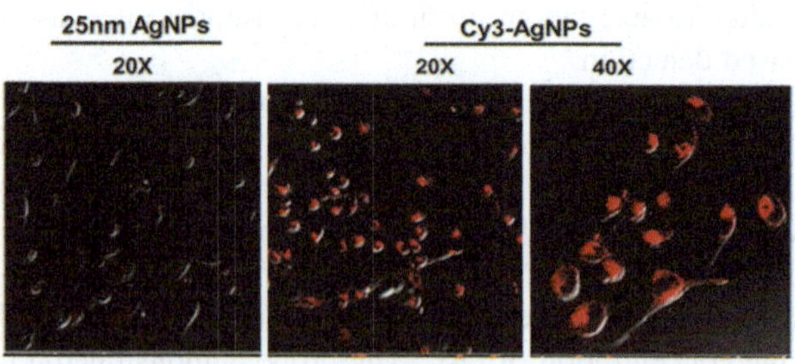

Fig. 11.39 Confocal images of MDA-MB-231 breast cancer cells after 1 h treatment with 25 nm AgNPs or Cy3-AgNPs
The first panel shows cells after 1 h treatment with 25 nm AgNPs. The second and third panels show cells after a 1 h treatment with Cy3-AgNPs [193]

Finally, NP technology has been used to overcome other issues linked to the potential cytotoxicity of the traditional probes and nonspecific interactions with multiple RS. Encapsulation of these probes by inert NPs, which protect them from nonspecific interactions, provides an elegant solution for delivery into cells. Although NPs are very promising tools for imaging, diagnosis, and even therapy, they can also be toxic. Many studies report that NP toxicity is related to an inflammatory response and/or OS, in particular for the particles based on metals, metal oxides, and semi-conductors [194, 195]. This point should be carefully addressed for each new NP design [196]. Furthermore, NPs enter cells by various mechanisms depending on the nature of the NP and the cell type and may end up in the cytosol, endosomes, and other organelles [197, 198]. Thus, the actual location of NPs should be verified before drawing conclusions about the location of RS generation when using NP technology.

11.2.20 Genetically Encoded Biosensors Based on Fluorescent Proteins

Although the aforementioned fluorescent probes measure the endogenously redox status, an exciting new field of research utilizes properties of known redox sensing proteins for the assessment of OS. Fluorescent protein-based biosensors have been developed for the investigation of RS in situ in real time. These sensors are genetically encoded and do not require the addition of any other reagents or cell lysis, making them very amenable to multiplexing.

Fluorescent proteins (FPs) are widely used tools to study a large variety of cellular processes including the assessment of local redox environments in cells and organelles [199]. For more than 20 years, these sensors have been developed for redox imaging in cell culture, tissues, organs, and live animals [200–202]. The application of genetically encoded FPs in fluorescence microscopy provides new insights into the complicated and fascinating world of OS.

FPs are specific markers that can be fused as tags to selected proteins (combination of FPs and prokaryotic redox-sensitive proteins). The cells then produce chimeric proteins that are capable of detecting RS or changes in the redox status [203]. It became then possible to follow the dynamics of a certain proteins and enzymes in living cells. These indicator systems can be genetically encoded at sites in cells and organs to monitor changes in redox states during metabolic activity. They can be targeted to specific subcellular locations such as plasma membrane or mitochondria [204, 205]. Genetically encoded FPs are therefore one approach to overcome problems regarding the specificity of localization when monitoring ROS. Indeed, one of their main advantages is their biocompatibility and targeting to a specific location.

The whole arsenal for tissue-specific genetic engineering is available for site-specific detection, and light-RS-sensitive proteins provide a new optogenetic tool to assess redox biology locally [206, 207].

11.2.20.1 Production

All FPs have a similar structure. They are composed of a β-barrel crossed by an α-helix that bears the chromophore formed usually by the autocatalytic cyclisation of three residues in positions 65, 66, and 67. The residue 66 is an aromatic residue. In roGFP and HyPer discussed below, it is a tyrosine [185].

The gene coding for a recombinant protein is introduced into cells through adenovirus or plasmids. It is targeted to subcellular organelles, reporting the redox states of some regions in the cell due to their combination with specific targeting signal peptides or retention sequences, which target FPs to different organelles [208]. The redox-dependent fluorescence spectral changes of the recombinant protein translate the structural changes in disulfide bonds and partial main chain under oxidized conditions. Several colors of redox-sensitive targeting proteins for redox detection have been generally applied (see the following text).

Figure 11.40 gives an overview on genetically encoded fluorescence proteins and their basic properties of selectivity and applicability in living cells.

11.2.20.2 Types

Green Fluorescent Proteins (GFPs)

GFPs are used as biosensors for the assessment of local redox environments in cells and organelles. They can be detected using highly resolved fluorescence nanoscope [209–211]. Novel microscopic techniques like laser switching contrast microscopy (LSCM) allows for the imaging of the dynamics of GFPs in single cell compartments [212].

To exploit the potential of GFPs for sensing local redox reactions, extensive studies have been undertaken to develop GFP-based biosensors using targeted mutations and directed evolution. The optical properties of these biosensors mainly depend on selective binding of protons (pH), oxygen atoms, cofactors (e.g., NADH) and electro-transfer [210, 213].

Progress in engineering of redox-sensitive fluorescence proteins led to the development of several derivatives of GFPs containing redox-active cysteines forming a disulfide bridge as redox switch [214]. These proteins can be selectively targeted or fused to specific target proteins for precise subcellular localization, enabling a specific and localized monitoring of RS in vivo [215]. It is noteworthy to mention that the fluorescence of GFPs is extremely sensitive to tyrosine nitration, as confirmed by correlated anti-nitrotyrosine immunoreactivity [216].

One of the most known GFPs are roGFPs (redox-sensitive green fluorescent proteins) (Fig. 11.40a) [217]. roGFPs are created by inserting a redox-active pair of cysteines into the structure of the GFP from *Aequorea victoria* at positions S147 and Q204 into the surface of the β-strands 7 and 10 in front of the chromophore, so that the disulfide bond between the two positions is formed in a reducing environment. This enables the probes to respond to redox changes in intra- and extracellular systems, accompanied by reversible ratiometric changes in fluorescence intensity [218].

roGFPs have two fluorescence excitation maxima at 405 and 488 nm with a common emission maximum at 515 nm, which display rapid and reversible ratiometric changes in fluorescence in response to alterations in redox potential in vitro and in vivo (Fig. 11.41). Disulfide formation results in the protonation of GFPs with increases 405 excitation signals at the expense of 488 nm excitation signals when the emission output at 515 nm is determined. Thus, the ratio of fluorescence from excitation at 405 and 488 nm indicates the extent of oxidation.

The reaction speed and pH are the key factors determining the usage of roGFP biosensors:

- The fluorescence excitation ratio of roGFPs displays a negligible pH sensitivity at physiological pHs from 5.5 to 8.5. Indeed, due to the identical modulation of the two excitation bands by pH, their ratio stays unchanged despite the pH change. In the case of single-wavelength excitation, the normalization by the ratio is not possible and the pH sensitivity should be carefully addressed [202].
- As suggested by Dooley et al., the fluorescence intensity of roGFPs does not vary as rapidly as the changes of the redox status, indicating that these proteins are more suitable for monitoring long-lasting redox changes [208]. It should be noted that roGFP probes could neither report the concentration nor the net production of RS, they are only applicable to detect the changes in ROS levels caused by external factors.

Numerous probes have been developed based on roGFP:

roGFP1 and 2 The first two members of roGFPs, roGFP 1 and 2, with difference in one amino acid (T65S of roGFP 1), reflect the conversion of dithiol/disulfide induced by the accumulation of ROS. These GFPs are used to monitor changes in the global thiol/disulfide equilibrium [219, 220]. As roGFP1 and 2 are fully oxidized in certain organelles (endosomes, lysosomes, and ER), they are more suitable for reducing environments such as cytoplasm, peroxisomes, mitochondria, and nucleus [218, 221, 222]. In terms of pH, roGFP2 tends to be more easily affected by change of pH than roGFP1 in the range of 6.0–8.0, so it is necessary to take special care of the detection environment.

roGFP1-iX roGFP1-iX was developed based on roGFP1 for application under oxidizing conditions. In roGFP1-iX, the amino acids inserted adjacent to cysteine 147 as well as the mutation of H148S in β-strand 7 together result in faster reaction kinetics and lower sensitivity to pH than roGFP1 in the range of 6.0–8.0 [223]. These

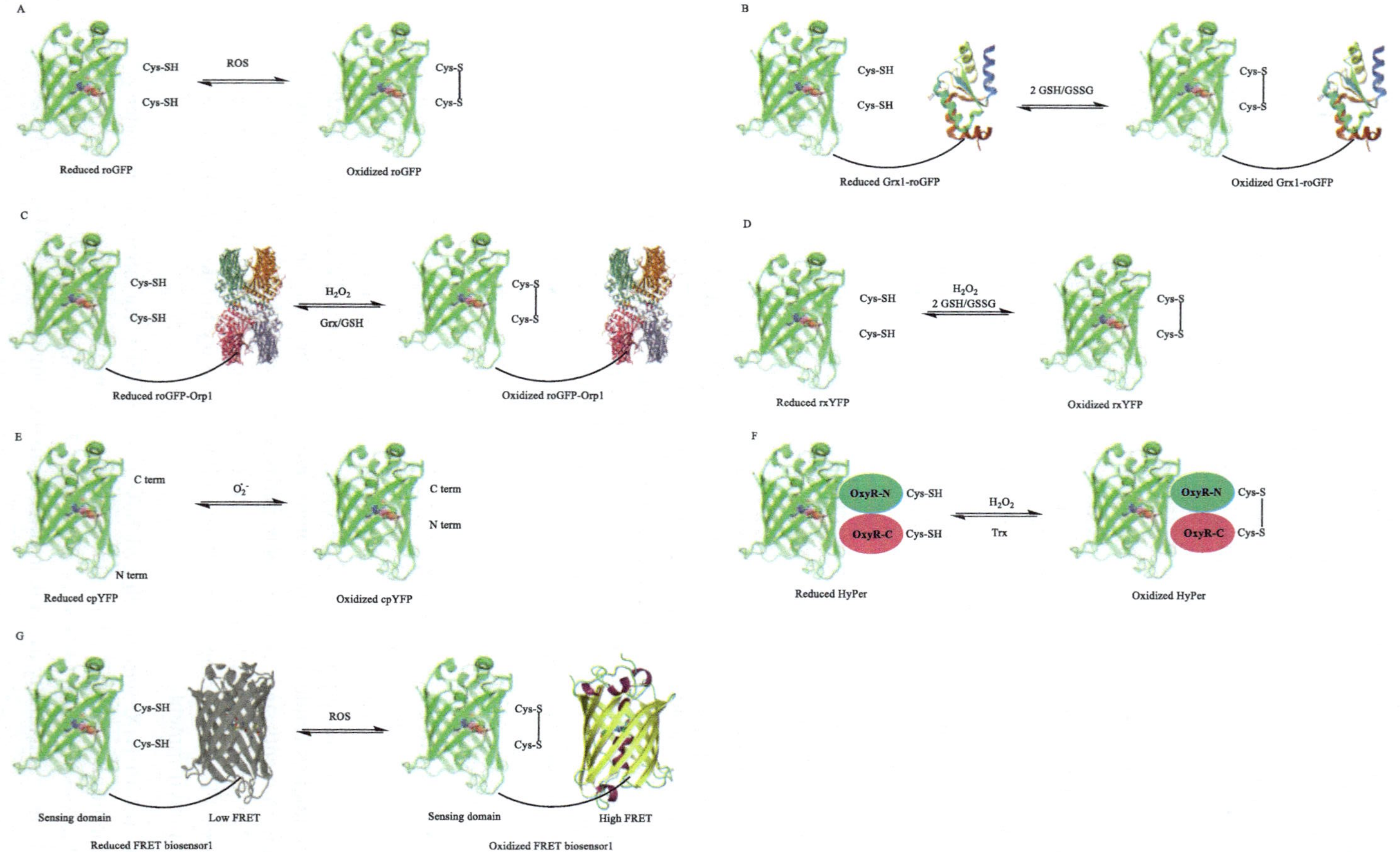

Fig. 11.40 Fluorescent protein (FP)-based biosensors. (**a**) roGFPs were developed by introducing 2 cysteine residues able to form an intramolecular disulfide bond upon reaction with ROS. (**b**) The fusion of roGFPs with glutaredoxin1 (Grx1) led to a family of biosensors dedicated to the measurement of GSH/GSSG status. (**c**) The coupling with the peroxidase Orp1 with roGFPs forms a dynamic hydrogen peroxide biosensor. (**d**) rxYFP is used to assess both H_2O_2 and GSH/GSSG status. (**e**) cpYFP is obtained by circular permutation of the protein sequence. As a result, NH_2 and COOH termini of the FP are positioned on the edge of the barrel, in front of the chromophore. (**f**) The fusion of cpYFP with NH_2- and COOH-terminal domains of OxyR leads to the HyPer family. (**g**) FRET-based sensors, a pair of fluorescent proteins (an energy donor and an energy acceptor) sandwiches a ROS-sensitive protein domain (black line). The changes of conformation in the sensing domain modulate the FRET

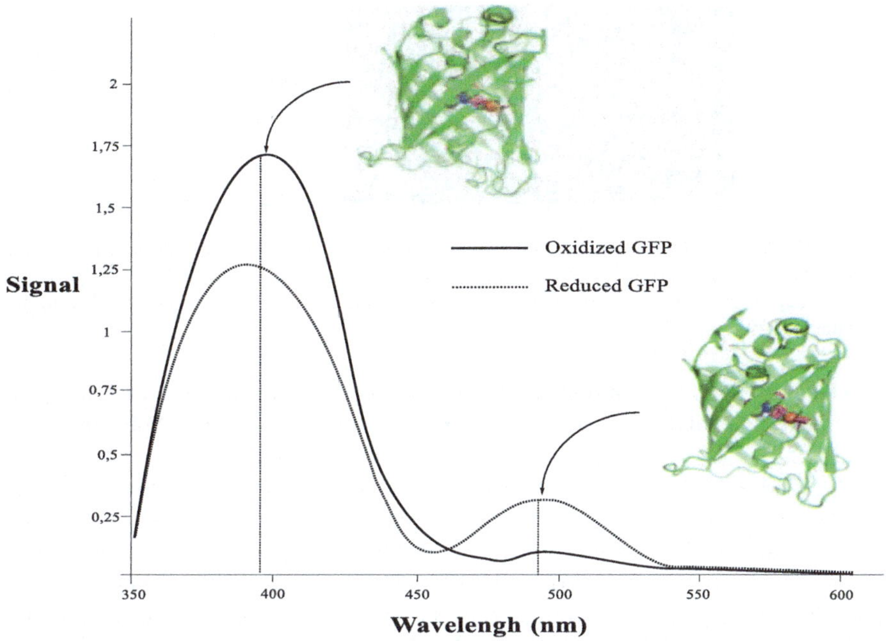

Fig. 11.41 Principle of ratiometric GFPs monitoring
The relative fluorescence intensity of the two excitation maxima of GFPs shifts depending on the redox state: reduction causes a decrease in the excitation at 405 nm and an increase in the excitation at 488 nm (arrows)

factors of roGFP1-iX make it more competent for redox monitoring in oxidizing organelles like ER [224].

roGFP1-iL roGFP1-iL has been modified for redox detection in the ER, where disulfide bonds of roGFP1-iL are partially oxidized, resulting in a relatively high level of reduction potential for monitoring the change of redox status [225]. The combination of roGFP1-iL and Grx1 (glutaredoxin-1) is more sensitive to GSH/GSSG ratio. Indeed, glutaredoxins are small enzymes that are oxidized by substrates and reduced nonenzymatically by glutathione.

roGFP1-RX roGFP1-RX offers higher reaction rate and dynamic range than roGFP1, by introducing three positively charged amino acids next to the key cysteines [226].

Mito-roGFP The mitochondrially targeted redox-sensitive GFP termed "mito-roGFP" does not specifically react in response to certain ROS, but it is used to globally evaluate OS in mitochondria [227]. Guzman et al. reported measurements of mitochondrial OS on dopaminergic neurons in transgenic mice expressing mito-roGFP [228]. Real-time assessment of the redox state of mitochondria in dopaminergic neurons has been achieved using this probe (Fig. 11.42) [229].

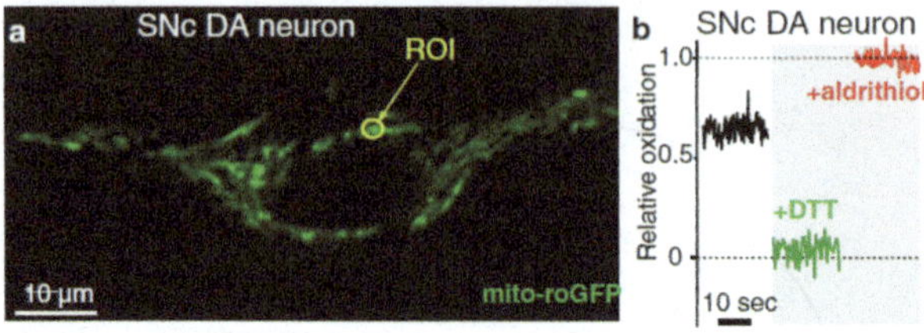

Fig. 11.42 Assessment of mitochondrial redox state. (**a**) SNc Dopaminergic neuron in a brain slice from a mito-roGFP mouse. (**b**) Representative fluorescence time series, showing fluorescence before (control, *black trace*) and after application of dithiothreitol (DTT; green trace) and Aldrithiol (red trace) [229]

Grx1-roGFP2-iL The modified probe Grx1-roGFP2-iL has broader midpoint potentials than roGFP2, therefore, being more suitable for usage both in cytosol and ER.

By expressing Grx1-roGFP fusion sensors in the biological system of interest and targeting the protein to a cellular compartment, it is possible to measure the dynamics of glutathione redox potential in real time, which is a significant advantage over invasive static methods (Fig. 11.40b). This probe does not directly measure ROS compounds. It, however, detects the shift in the GSH/GSSG equilibrium and is more specific than roGFP2 for the dynamic assessment of this ratio [230]. This probe has been used to assess GSH/GSSG changes during PMA induced-inflammation (Fig. 11.43) [231].

A targeted version of this probe, mito-Grx1-roGFP2, has recently been developed [232].

roGFP2-Orp1 To avoid reduction of the GFP by Trx, new probes have been designed in which a hydrogen peroxide sensor (Orp1; yeast peroxidase) is conjugated to a redox-sensitive reporter (roGFP2) (Fig. 11.40c) [203]. Orp1 is oxidized by H_2O_2, forms disulfides, and then accepts an electron from cysteine residues on the roGFP2 transferring disulfides to roGFP2 through a thiol-disulfide exchange mechanism. Afterward, roGFP2 emits fluorescence. The reaction is reversible through the action of cellular Trx or Grx/GSH systems. Therefore, the reversibility and the dynamic of its response rely on the presence of endogenous reductants in cells and may vary depending on the cell type and organism.

The redox-relaying equivalents between these roGFP2-Orp1 parts enable this probe to reflect the level of hydrogen peroxide effectively [203]. roGFP2-Orp1 redox probe represents one of the most advanced and promising tools for specific, quantitative, dynamic, and compartment-specific hydrogen peroxide detection. roGFP2-Orp1 is pH stable, enabling it to be successfully applied to monitor hydrogen peroxide alteration in cytosol and mitochondria, with a sensitivity of micromoles [233, 234]. Indeed, a specific sensor for H_2O_2, targeted to the mitochondrial matrix (Mt-roGFP2-Orp1) has been developed and tested in insulin-secreting cell line (BRIN-BD11) [235]. roGFP2-Orp1 shows higher selectivity toward H_2O_2 than $O_2^{\bullet -}$ and other oxidants.

Others po-roGFP2 targeted to peroxisome and roGFP2 fused to peroxiredoxin [236].

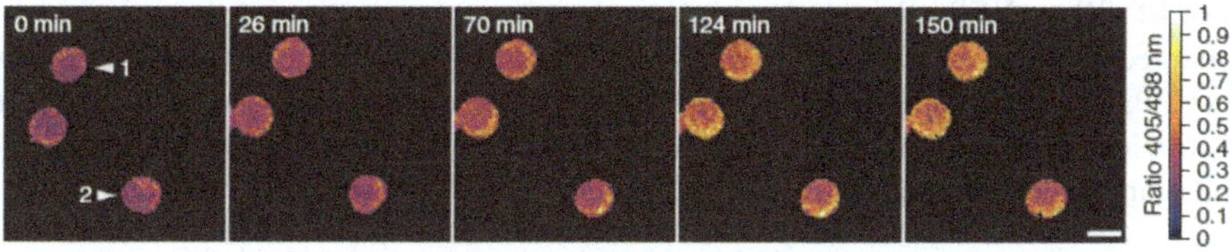

Fig. 11.43 Grx1-roGFP2 visualizes physiological redox changes associated with immune-cell activation
Grx1-roGFP2–positive mouse RAW264.7 cells were stimulated with 1 mM PMA after 6 min. False-color ratio pictures of PMA-treated cells are depicted at representative time points (top). Scale bar, 10 μm. The sensor ratio of two stimulated cells (arrowheads) and one control cell was calculated and plotted over time (bottom) [231]

rxYFPs

The redox-sensitive yellow fluorescent protein (rxYFP) is engineered from yellow fluorescent protein (YFP). A pair of cysteines is introduced near the chromophore region to form a reversible disulfide bond (introduction of two redox-active cysteines in constructs termed rxYFP149 202). The formation of disulfide bridge under oxidative conditions leads to a distortion of the typical β-barrel structure, thus changing the fluorescence properties of rxYFP [237]. Hence, the fluorescence intensity is decreased around 527 nm, enabling it to visualize the cellular redox status in vivo [238].

rxYFP can be used to examine the redox state of the cell in its entirety, producing an intensity map of the thiol-based redox state across the cell [238]. This high-resolution technique can determine subcellular locations of high and low protein oxidation. Due to the appropriate midpoint redox potential of rxYFP, it is applied to detect the intracellular dynamics of GSH/GSSG and thiol/disulfide changes in different parts of cells, especially in reducing compartments such as the cytosol, mitochondria, and the nucleus (Fig. 11.40d) [239, 240].

rxYFP probes have also been developed into sensors that are specifically targeted to the mitochondria and display reversible ratiometric fluorescence in response to redox changes [218].

It should be noted that rxYFP probes are pH sensitive, and the pKa of their chromophores varies according to the pH of the used system. Under neutral conditions, the fluorescence quenching effect would occur on the protonated form of rxYFP probes, resulting in the decrease of fluorescence signal at about 2.2 fold, without inducing significant shift at the excitation wavelength [237].

The following are some examples of rxYFP derivatives:

rxYFP-Grx1P rxYFP-Grx1P is obtained by fusing the glutaredoxin-1 of yeast (Grx1P) with rxYFP, making it suitable for real-time imaging of the intracellular glutathione redox system [241]. This probe reacts more specifically with glutathione than other oxidants like H_2O_2 [8].

rxYFP 3R rxYFP 3R is designed through further introduction of three positively charged residues (200R/204R/227R) in addition to the existing two cysteines in rxYFP, resulting in higher reactivity for glutathione than rxYFP (~13-fold) [242]. It is obvious that the modified probe shows improved specificity toward GSH/GSSG ratio and higher stability under physiological pH conditions.

cpYFPs

The circularly permuted green fluorescent protein (cpYFP) is a probe for $O_2^{\bullet-}$ detection and imaging (Fig. 11.40e) [243, 244]. The binding of mitochondrial targeting sequences with cpYFP allows this probe for subcellular ROS sensing [244]. It has been suggested that pH sensitivity of cpYFP is a possible cause of the apparent "superoxide flashes" [244]. The use of cpYFP as selective superoxide sensor was afterward questioned and a matter of debate [245–247]. In fact, some researchers are questioning the validity of cpYFP in detecting mitochondrial $O_2^{\bullet-}$, either based on the unchanged fluorescence intensity of isolated mitochondria particles at elevated $O_2^{\bullet-}$ or on fluctuations attributed to pH oscillations [248]. The fluorescence intensity of cpYFP is prone to be influenced by the alkaline environment of the mitochondrial matrix, therefore it is necessary to exercise caution during the imaging of mitochondrial $O_2^{\bullet-}$.

HyPer

HyPer sensors were designed as specific detectors of H_2O_2. HyPer (cpYFP-OxyR) consists of cpYFP inserted into the regulatory domain of the *E. coli* hydrogen peroxide-sensing protein, OxyR, a hydrogen peroxide-inducible gene regulator (Fig. 11.40f) [249]. A disulfide bridge is formed between the key reactive cysteines [155, 156] of the regulatory domain during the reaction with H_2O_2, resulting in the critical conformational change of OxyR, which further leads to changes of both conformation and fluorescence intensity of the whole protein, thus correlating the changes of hydrogen peroxide levels with the fluorescence intensity.

One major advantage of this probe is its ratiometric sensitivity to H_2O_2. This is a key property of the probe, allowing comparisons of relative quantification measurements between experiments. Another major advantage of HyPer is that its fluorescence is reversible. Therefore, once this probe is loaded into cells, it can be used to follow varying fluxes of H_2O_2 in real time. However, the fluorescence of HyPer can be affected by reduction of the GFP by Trx [250].

HyPer is a powerful tool for intracellular dynamic changes and real time studies of H_2O_2 in many kinds of cell systems and living animals such as *Arabidopsis*, *Caenorhabditis elegans*, yeast, and mouse [251, 252]. This probe has been successfully targeted to several cellular compartments including the cytoplasm, the nucleus, peroxisomes, the mitochondrial intermembrane space, and the mitochondrial matrix [249, 253]. Mitochondrial-targeted redox-sensitive yellow fluorescent protein pHyPer-dMito has been used to analyze H_2O_2 in the mitochondria of apoptotic HeLa cells [249]. However, the temporal dynamics of reducing agents acting on the probe in various organelles varied greatly.

HyPer demonstrates submicromolar affinity to H_2O_2 and has been shown to be insensitive to other oxidants, such as $O_2^{\bullet-}$, $NO^{\bullet}$, and $ONOO^-$. Using HyPer constructs, Malinouski et al. proved the presence of low levels of H_2O_2 in these subcellular compartments. An exception to this is the ER, where HyPer was found to be predominantly oxidized [253].

The sensitivity and reacting speed of HyPer to H_2O_2 is higher than that of roGFP2-Orp1. The key reason seems to be the position of the redox-sensitive cysteine pair in HyPer: different from rxYFPs and roGFPs, they are located on the redox sensing domain rather than the chromophore [249].

In order to reflect the changes of hydrogen peroxide level more effectively, the oxidized HyPer proteins required to be reduced in time, which seems to be mediated by intracellular glutathione systems [254], so this probe may also be used to report the equilibrium between H_2O_2 and glutathione.

Improved mutations have been applied on HyPer yielding probes called HyPer-2 and HyPer-3, through mutation (A406V) of HyPer and a further mutation of H34Y, respectively [255, 256]. Unlike the roGFP-Orp1 chimera, there is no thiol-disulfide transfer, but rather two cysteines of OxyR form a reversible disulfide bond which induces a conformational change that is transferred to the cpYFP moiety [257].

HyPer-2 is more stable between monomers in OxyR dimers, while probe HyPer-3 shows improved performance with respect to response time and speed in the reaction with H_2O_2. Both HyPer-2 and HyPer-3 have a broader dynamic range than HyPer. The two probes have been successfully used in the fluorescence lifetime imaging of H_2O_2 in vivo.

The major limitation of HyPer sensors is the change of the pH environment. HyPer sensors display high pH sensitivity with pKs1/2 at ~8 [258], and pH variations influence their ratiometric response. It was suggested to use a pH-sensitive FP as a control, for example, SypHer, which only differs from HyPer by a single cysteine [259]. SypHer does not respond to H_2O_2, and the experimental setup to monitor its response is identical to the one needed for HyPer.

rxRFP/HyPerRed

The family of Hyper has been enriched with a red variant called rxRFP [260, 261]. This red fluorescent protein (RFP) has been developed for the detection of H_2O_2. Chromophore transformations in RFPs offer tools for designing suitable red-shifted probes, which are advantageous for imaging studies due to the strong absorption in the green spectral range. Excitation with longer wavelengths also leads to reduced auto-fluorescence [262]. It is indeed a major challenge to broaden the color palette of available FP biosensors in order to monitor in the same cell or organism different ROS in parallel in different compartments.

rxRFP was obtained through replacing the sensing domain of the calcium probe R-GECO1 by OxyR, and the formation of a disulfide bond between Cys199 and

Cys208 making rxRFP specific for the recognition of H_2O_2 [261]. rxRFP has similar kinetics and sensitivity toward H_2O_2 to HyPer, and it is competent enough to detect low levels of H_2O_2 even with a lower dynamic range.

rxRFP is sensitive to the change of pH, so parallel experiments are recommended. For example, HyPerRed-C199S or HyPer-C199S could be applied as control probes for pH change [261]. Additionally, the reaction between rxRFP and H_2O_2 is nonratiometric, limiting its usage in performing quantitative assays, so other GFP sensors are needed for quantitative measurement.

Targeted rxRFP probes (i.e., mitochondria: mito-rxRFP, ER: ER-rxRFP, and nucleus: Nuc-rxRFP) were developed by Fan and Ai and tested in HEK 293 T cells [263].

Many other fluorescent biosensors have been developed based on redox RFP. For example, cpRFP and TrxRFP1 were designed for monitoring the redox dynamics involved in thiol/disulfide and thioredoxin conversions, respectively [264, 265].

OHSer

An analog probe to HyPer has been developed for the detection of organic hydroperoxides. The probe, called OHSer, combines OhrR, a transcriptional regulator of the cellular response to organic hydroperoxide detoxification, with cpVenus, a yellow variant of GFP [266]. The probe gives a fluorescent signal upon interaction with organic hydroperoxides, as the OhrR portion of the chimera undergoes a conformational change and the cpVenus portion is highly conformation sensitive.

FRET-Based Biosensors

The best option to get rid of the pH sensitivity of FPs is to develop a FRET-based biosensor based on pH-insensitive FPs. In a pioneering work, Hammerling's laboratory used the cysteine-containing regulatory domain from the heat-shock protein HSP-33 as a sensing module [267]. More recently, two fluorescent biosensors for H_2O_2, PerFRET and OxyFRET, were built using as sensing module the cysteine-rich domains (NH_2- and COOH-terminal CRD domains) of the yeast activator protein 1 (Yap1) involved in a redox relay of *Saccharomyces cerevisiae*. In PerFRET, one CRD domain is coupled with the peroxidase Orp1. The sensing part of the biosensor that undergoes conformational changes upon oxidation (Fig. 11.40g) is sandwiched between two FPs that form a FRET pair, Cerulean (Δ11) and cp173Venus [204, 268, 269]. OxyFRET expressed in the cytoplasm of neutrophil-like PLB985 cells allowed the measurement of NOX activation stimulated by 1 μM PMA [204]. Nevertheless, further engineering may be needed to decrease pH-sensitivity and improve the brightness and the dynamic range of the biosensor response.

FRET technology has been applied to other important proteins including metallothionein and Yap1. Metallothionein is a small cysteine-rich metal-binding protein target of $NO^•$ that has been used to create a fluorescent nitric oxide probe through chimeric sandwiching between two GFPs [270]. This allows live-cell FRET to be used to measure conformational changes upon metal release from metallothionein, which is dependent upon endogenously generated $NO^{\bullet}$ when cytoplasmic Ca^{2+} levels are elevated. Similarly, Redoxfluor probes utilize the redox-sensitive yeast Yap1 transcription factor to drive FRET between cerulean (CFP) and citrine (YFP) allowing for visualization of redox changes [221].

11.2.20.3 Advantages and Inconveniences

Genetically encoded biosensors have several advantages over other commonly used probes. They provide real-time and dynamic detection of redox-status changes through reversible reactions with RS and radicals, without special requirements for permeability into target cells. Some FPs can stay in cells and allow for a long-term detection of cellular radicals. This allows the temporal study of FR generation and removal within the cell.

These biosensors can be targeted to a specific cellular compartment by utilizing cellular trafficking sequences as part of the chimeric molecule. Peptides can be targeted to the nucleus by inserting repeats of the nuclear localization signal (NLS). This particular signal consists of short sequences of positively charged lysines and arginines expressed on the protein surface. Similar sequences can be used to direct ROS sensors to mitochondria, peroxisomes, or cytoplasm. Given the wide variety of different RS probes available, each with a different combination of properties, it is critical to determine the most useful probe for each experiment. A logic diagram has been proposed by Wooley et al. to help scientists chose the proper probe according to the experimental design [138].

On the other hand, the limitations of fluorescent protein-based probes are as follows:

- First, the preparation of these probes is rather complex, and the possible influencing factors during experiments should be strictly controlled. For example, the nature of target recipient cells would influence the expression level of FPs, which then impacts the ability of probes to detect ROS. Moreover, some FPs like roGFPs have been found to be reversibly oxidized once exposed to air or oxygen, a frequent situation in in vitro studies. In addition, roGFP probes showed lower sensitivity to oxidants in vivo than in vitro [231], which may be the result of the presence of antioxidants.
- Second, compared to other types of probes for radical detection, the reaction rates of these proteins with redox substances are quite low, resulting in low immediacy in reflecting the intracellular radicals.
- Third, as for probes HyPer and roGFP2-Orp1, their selectivity to H_2O_2 is based on the reversible redox reactions mediated by H_2O_2 and the reductive GSH/GSSG couple and Trx. This may lead to inaccurate analysis of data. Therefore, quantitative determination of ROS by protein-based biosensors is difficult, requiring careful calibration and reliable verification.

- Fourth, fluorescence collecting tools may also have an impact on the data generation of protein-based biosensors. For example, microscopy is used in the fluorescence pickup of many kinds of cells, while it is not suitable for imaging large amounts of samples. Flow cytometer is popular in the analysis of mammalian cells, and laser scanning plate readers are applied in high-throughput screening of tissue samples, where rapid dual-color fluorescence imaging of proteins is possible [271]. Fluorescent microplate readers can also report fluorescence data, but they may not be suitable in this case, since they cannot provide information at cellular and subcellular levels.
- Fifth, GFP derivatives often act as singlet oxygen sensitizers themselves. The generation of new and specific GFP mutants is a promising approach to overcome this problem, which again votes for the importance of developing improved genetically FPs for ROS sensing.
- Sixth, most of FP-based biosensors are pH sensitive. In the context of ROS production, this question is particularly relevant. In fact, the initial production of $O_2^{\bullet-}$ by NOX leads to the oxidation of NADPH and the release of protons on the cytosolic side of the oxidase (one proton for two $O_2^{\bullet-}$). The protons may be transported across the membrane by H^+ pumps and channels, and consumed during the dismutation of $O_2^{\bullet-}$ into H_2O_2 [272]. Consequently, the local balance of proton concentration can vary, and ideally, the response of the ROS sensor should be pH insensitive to limit any interferences with ROS measurements.
- Seventh, some FP-based biosensors have low signal-to-noise ratio and further optimization is required to improve their performances in ROS sensing experiments. Indeed, the detection of FP responses at the lowest expression levels limits the interplay among the sensing module, the ROS concentration, and the overall redox homeostasis of the organism. This can be achieved by using FPs with the maximum brightness, that is, with the best extinction coefficient, quantum yield, and photostability. The photostability of HyPer-2 has been compared with the stability of related yellow FPs. Interestingly, it was found to be very high, enough to perform STED microscopy, a super-resolved imaging technique that stresses the fluorophores with high-excitation power [205].

References

1. Collins AR. Assays for oxidative stress and antioxidant status: applications to research into the biological effectiveness of polyphenols2,3. The American journal of clinical nutrition. 2005;81(1):261S–7S.
2. Griffiths HR, Møller L, Bartosz G, Bast A, Bertoni-Freddari C, Collins A, et al. Biomarkers. Molecular Aspects of Medicine. 2002;23(1):101–208.
3. Forkink M, Smeitink JAM, Brock R, Willems PHGM, Koopman WJH. Detection and manipulation of mitochondrial reactive oxygen species in mammalian cells. Biochimica et Biophysica Acta (BBA)—Bioenergetics. 2010;1797(6):1034–44.
4. Owen JB, Butterfield DA. Measurement of Oxidized/Reduced Glutathione Ratio. In: Bross P, Gregersen N, editors. Protein Misfolding and Cellular Stress in Disease and Aging: Concepts and Protocols. Totowa, NJ: Humana Press; 2010. p. 269–77.

5. Murphy ME, Kolvenbach R, Aleksis M, Hansen R, Sies H. Antioxidant depletion in aortic crossclamping ischemia: Increase of the plasma α-tocopheryl quinone/α-tocopherol ratio. Free Radical Biology and Medicine. 1992;13(2):95–100.
6. Halliwell B, Whiteman M. Measuring reactive species and oxidative damage in vivo and in cell culture: how should you do it and what do the results mean? Br J Pharmacol. 2004;142(2): 231–55.
7. Afzal M, Matsugo S, Sasai M, Xu B, Aoyama K, Takeuchi T. Method to overcome photoreaction, a serious drawback to the use of dichlorofluorescin in evaluation of reactive oxygen species. Biochem Biophys Res Commun. 2003;304(4):619–24.
8. Zhang Y, Dai M, Yuan Z. Methods for the detection of reactive oxygen species. Analytical Methods. 2018;10(38):4625–38.
9. Tarpey MM, Fridovich I. Methods of detection of vascular reactive species: nitric oxide, superoxide, hydrogen peroxide, and peroxynitrite. Circulation research. 2001;89(3):224–36.
10. Crow JP. Dichlorodihydrofluorescein and dihydrorhodamine 123 are sensitive indicators of peroxynitrite in vitro: implications for intracellular measurement of reactive nitrogen and oxygen species. Nitric oxide: biology and chemistry. 1997;1(2):145–57.
11. Kooy NW, Royall JA, Ischiropoulos H. Oxidation of 2',7'-dichlorofluorescin by peroxynitrite. Free radical research. 1997;27(3):245–54.
12. Cathcart R, Schwiers E, Ames BN. Detection of picomole levels of hydroperoxides using a fluorescent dichlorofluorescein assay. Anal Biochem. 1983;134(1):111–6.
13. Myhre O, Andersen JM, Aarnes H, Fonnum F. Evaluation of the probes 2',7'--dichlorofluorescin diacetate, luminol, and lucigenin as indicators of reactive species formation. Biochemical pharmacology. 2003;65(10):1575–82.
14. Bilski P, Belanger AG, Chignell CF. Photosensitized oxidation of 2',7'-dichlorofluorescin: singlet oxygen does not contribute to the formation of fluorescent oxidation product 2',7'--dichlorofluorescein. Free Radic Biol Med. 2002;33(7):938–46.
15. Zhu H, Bannenberg GL, Moldéus P, Shertzer HG. Oxidation pathways for the intracellular probe 2',7'-dichlorofluorescein. Arch Toxicol. 1994;68(9):582–7.
16. LeBel CP, Ischiropoulos H, Bondy SC. Evaluation of the probe 2',7'-dichlorofluorescin as an indicator of reactive oxygen species formation and oxidative stress. Chem Res Toxicol. 1992;5 (2):227–31.
17. Tampo Y, Kotamraju S, Chitambar CR, Kalivendi SV, Keszler A, Joseph J, et al. Oxidative stress-induced iron signaling is responsible for peroxide-dependent oxidation of dichlorodihydrofluorescein in endothelial cells: role of transferrin receptor-dependent iron uptake in apoptosis. Circulation research. 2003;92(1):56–63.
18. Reiniers MJ, de Haan LR, Reeskamp LF, Broekgaarden M, Hoekstra R, van Golen RF, et al. Optimal Use of 2',7'-Dichlorofluorescein Diacetate in Cultured Hepatocytes. Methods Mol Biol. 2022;2451:721–47.
19. Zhao L, Shi H, Zhang F, Xue H, Han Q. Hederagenin protects against myocardial ischemia-reperfusion injury via attenuating ALOX5-mediated ferroptosis. Naunyn-Schmiedeberg's archives of pharmacology. 2023.
20. Gardiner B, Dougherty JA, Ponnalagu D, Singh H, Angelos M, Chen CA, et al. Measurement of Oxidative Stress Markers In Vitro Using Commercially Available Kits. In: Berliner LJ, Parinandi NL, editors. Measuring Oxidants and Oxidative Stress in Biological Systems. Cham (CH): Springer; Copyright 2020, Springer Nature Switzerland AG.; 2020. p. 39–60.
21. Rao KM, Padmanabhan J, Kilby DL, Cohen HJ, Currie MS, Weinberg JB. Flow cytometric analysis of nitric oxide production in human neutrophils using dichlorofluorescein diacetate in the presence of a calmodulin inhibitor. J Leukoc Biol. 1992;51(5):496–500.
22. Possel H, Noack H, Augustin W, Keilhoff G, Wolf G. 2,7-Dihydrodichlorofluorescein diacetate as a fluorescent marker for peroxynitrite formation. FEBS Lett. 1997;416(2):175–8.
23. Suematsu M, Schmid-Schönbein GW, Chavez-Chavez RH, Yee TT, Tamatani T, Miyasaka M, et al. In vivo visualization of oxidative changes in microvessels during neutrophil activation. The American journal of physiology. 1993;264(3 Pt 2):H881–91.

24. al-Mehdi A, Shuman H, Fisher AB. Fluorescence microtopography of oxidative stress in lung ischemia-reperfusion. Laboratory investigation; a journal of technical methods and pathology. 1994;70(4):579–87.
25. Kehrer JP, Paraidathathu T. The use of fluorescent probes to assess oxidative processes in isolated-perfused rat heart tissue. Free radical research communications. 1992;16(4):217–25.
26. Sundaresan M, Yu ZX, Ferrans VJ, Irani K, Finkel T. Requirement for generation of H2O2 for platelet-derived growth factor signal transduction. Science. 1995;270(5234):296–9.
27. Behl C, Davis JB, Lesley R, Schubert D. Hydrogen peroxide mediates amyloid beta protein toxicity. Cell. 1994;77(6):817–27.
28. Ubezio P, Civoli F. Flow cytometric detection of hydrogen peroxide production induced by doxorubicin in cancer cells. Free Radic Biol Med. 1994;16(4):509–16.
29. Curtin JF, Donovan M, Cotter TG. Regulation and measurement of oxidative stress in apoptosis. Journal of Immunological Methods. 2002;265(1):49–72.
30. Zielonka J, Kalyanaraman B. "ROS-generating mitochondrial DNA mutations can regulate tumor cell metastasis"—a critical commentary. Free Radical Biology and Medicine. 2008;45 (9):1217–9.
31. Folkes LK, Patel KB, Wardman P, Wrona M. Kinetics of reaction of nitrogen dioxide with dihydrorhodamine and the reaction of the dihydrorhodamine radical with oxygen: implications for quantifying peroxynitrite formation in cells. Arch Biochem Biophys. 2009;484(2):122–6.
32. Qian SY, Buettner GR. Iron and dioxygen chemistry is an important route to initiation of biological free radical oxidations: an electron paramagnetic resonance spin trapping study. Free Radic Biol Med. 1999;26(11–12):1447–56.
33. Tarpey MM, Wink DA, Grisham MB. Methods for detection of reactive metabolites of oxygen and nitrogen: in vitro and in vivo considerations. 2004;286(3):R431–R44.
34. Ohashi T, Mizutani A, Murakami A, Kojo S, Ishii T, Taketani S. Rapid oxidation of dichlorodihydrofluorescin with heme and hemoproteins: formation of the fluorescein is independent of the generation of reactive oxygen species. FEBS Lett. 2002;511(1–3):21–7.
35. Rota C, Fann YC, Mason RP. Phenoxyl free radical formation during the oxidation of the fluorescent dye 2',7'-dichlorofluorescein by horseradish peroxidase. Possible consequences for oxidative stress measurements. J Biol Chem. 1999;274(40):28161–8.
36. O'Malley YQ, Reszka KJ, Britigan BE. Direct oxidation of 2',7'-dichlorodihydrofluorescein by pyocyanin and other redox-active compounds independent of reactive oxygen species production. Free Radic Biol Med. 2004;36(1):90–100.
37. Royall JA, Ischiropoulos H. Evaluation of 2',7'-dichlorofluorescin and dihydrorhodamine 123 as fluorescent probes for intracellular H_2O_2 in cultured endothelial cells. Arch Biochem Biophys. 1993;302(2):348–55.
38. Kirkland RA, Franklin JL. Evidence for redox regulation of cytochrome C release during programmed neuronal death: antioxidant effects of protein synthesis and caspase inhibition. J Neurosci. 2001;21(6):1949–63.
39. Kirkland RA, Saavedra GM, Franklin JL. Rapid activation of antioxidant defenses by nerve growth factor suppresses reactive oxygen species during neuronal apoptosis: evidence for a role in cytochrome c redistribution. J Neurosci. 2007;27(42):11315–26.
40. Yalcin S, Zhang X, Luciano JP, Mungamuri SK, Marinkovic D, Vercherat C, et al. Foxo3 is essential for the regulation of ataxia telangiectasia mutated and oxidative stress-mediated homeostasis of hematopoietic stem cells. J Biol Chem. 2008;283(37):25692–705.
41. Olmos Y, Valle I, Borniquel S, Tierrez A, Soria E, Lamas S, et al. Mutual dependence of Foxo3a and PGC-1alpha in the induction of oxidative stress genes. J Biol Chem. 2009;284 (21):14476–84.
42. Pan J, Chang Q, Wang X, Son Y, Zhang Z, Chen G, et al. Reactive oxygen species-activated Akt/ASK1/p38 signaling pathway in nickel compound-induced apoptosis in BEAS 2B cells. Chem Res Toxicol. 2010;23(3):568–77.
43. Meissner F, Molawi K, Zychlinsky A. Superoxide dismutase 1 regulates caspase-1 and endotoxic shock. Nat Immunol. 2008;9(8):866–72.

44. Kaur P, Schulz K, Heggland I, Aschner M, Syversen T. The use of fluorescence for detecting MeHg-induced ROS in cell cultures. Toxicol In Vitro. 2008;22(5):1392–8.
45. Dykens JA, Jamieson JD, Marroquin LD, Nadanaciva S, Xu JJ, Dunn MC, et al. In vitro assessment of mitochondrial dysfunction and cytotoxicity of nefazodone, trazodone, and buspirone. Toxicological sciences: an official journal of the Society of Toxicology. 2008;103(2):335–45.
46. Suski JM, Lebiedzinska M, Bonora M, Pinton P, Duszynski J, Wieckowski MR. Relation between mitochondrial membrane potential and ROS formation. Methods Mol Biol. 2012;810:183–205.
47. Wang Z, Wijewickrama GT, Peng KW, Dietz BM, Yuan L, van Breemen RB, et al. Estrogen Receptor {alpha} Enhances the Rate of Oxidative DNA Damage by Targeting an Equine Estrogen Catechol Metabolite to the Nucleus. J Biol Chem. 2009;284(13):8633–42.
48. Gerich FJ, Funke F, Hildebrandt B, Fasshauer M, Müller M. H(2)O(2)-mediated modulation of cytosolic signaling and organelle function in rat hippocampus. Pflugers Arch. 2009;458(5): 937–52.
49. Limoli CL, Giedzinski E, Morgan WF, Swarts SG, Jones GD, Hyun W. Persistent oxidative stress in chromosomally unstable cells. Cancer research. 2003;63(12):3107–11.
50. Schwarzer C, Illek B, Suh JH, Remington SJ, Fischer H, Machen TE. Organelle redox of CF and CFTR-corrected airway epithelia. Free Radic Biol Med. 2007;43(2):300–16.
51. Thornber K, Colomba A, Ceccato L, Delsol G, Payrastre B, Gaits-Iacovoni F. Reactive oxygen species and lipoxygenases regulate the oncogenicity of NPM-ALK-positive anaplastic large cell lymphomas. Oncogene. 2009;28(29):2690–6.
52. Kim SH, Castro F, Paterson Y, Gravekamp C. High efficacy of a Listeria-based vaccine against metastatic breast cancer reveals a dual mode of action. Cancer research. 2009;69(14): 5860–6.
53. Cohen HJ, Chovaniec ME. Superoxide generation by digitonin-stimulated guinea pig granulocytes. A basis for a continuous assay for monitoring superoxide production and for the study of the activation of the generating system. J Clin Invest. 1978;61(4):1081–7.
54. Whitin JC, Chapman CE, Simons ER, Chovaniec ME, Cohen HJ. Correlation between membrane potential changes and superoxide production in human granulocytes stimulated by phorbol myristate acetate. Evidence for defective activation in chronic granulomatous disease. J Biol Chem. 1980;255(5):1874–8.
55. Zhang G, Zhang F, Muh R, Yi F, Chalupsky K, Cai H, et al. Autocrine/paracrine pattern of superoxide production through NAD(P)H oxidase in coronary arterial myocytes. Am J Physiol Heart Circ Physiol. 2007;292(1):H483–95.
56. Cao D, Boxer LA, Petty HR. Deposition of reactive oxygen metabolites onto and within living tumor cells during neutrophil-mediated antibody-dependent cellular cytotoxicity. J Cell Physiol. 1993;156(2):428–36.
57. Ryan TC, Weil GJ, Newburger PE, Haugland R, Simons ER. Measurement of superoxide release in the phagovacuoles of immune complex-stimulated human neutrophils. J Immunol Methods. 1990;130(2):223–33.
58. Buxser SE, Sawada G, Raub TJ. Analytical and numerical techniques for evaluation of free radical damage in cultured cells using imaging cytometry and fluorescent indicators. Methods in enzymology. 1999;300:256–75.
59. Kalyanaraman B, Darley-Usmar V, Davies KJ, Dennery PA, Forman HJ, Grisham MB, et al. Measuring reactive oxygen and nitrogen species with fluorescent probes: challenges and limitations. Free Radic Biol Med. 2012;52(1):1–6.
60. Elbim C, Lizard G. Flow cytometric investigation of neutrophil oxidative burst and apoptosis in physiological and pathological situations. Cytometry A. 2009;75(6):475–81.
61. Tsinkalovsky OR, Laerum OD. Flow cytometric measurement of the production of reactive oxygen intermediate in activated rat mast cells. Apmis. 1994;102(6):474–80.

62. Parker RA, Huang Q, Tesfamariam B. Influence of 3-hydroxy-3-methylglutaryl-CoA (HMG-CoA) reductase inhibitors on endothelial nitric oxide synthase and the formation of oxidants in the vasculature. Atherosclerosis. 2003;169(1):19–29.
63. Ishikawa M, Stokes KY, Zhang JH, Nanda A, Granger DN. Cerebral microvascular responses to hypercholesterolemia: roles of NADPH oxidase and P-selectin. Circulation research. 2004;94(2):239–44.
64. Çınar R, Nazıroğlu M. TRPM2 Channel Inhibition Attenuates Amyloid β42-Induced Apoptosis and Oxidative Stress in the Hippocampus of Mice. Cell Mol Neurobiol. 2023;43(3): 1335–53.
65. Zurgil N, Shafran Y, Afrimzon E, Fixler D, Shainberg A, Deutsch M. Concomitant real-time monitoring of intracellular reactive oxygen species and mitochondrial membrane potential in individual living promonocytic cells. J Immunol Methods. 2006;316(1–2):27–41.
66. Moser M, Bauer M, Schmid S, Ruppert R, Schmidt S, Sixt M, et al. Kindlin-3 is required for beta2 integrin-mediated leukocyte adhesion to endothelial cells. Nature medicine. 2009;15(3): 300–5.
67. Deshwal S, Antonucci S, Kaludercic N, Di Lisa F. Measurement of Mitochondrial ROS Formation. Methods Mol Biol. 2018;1782:403–18.
68. Clough E, Inigo J, Chandra D, Chaves L, Reynolds JL, Aalinkeel R, et al. Mitochondrial Dynamics in SARS-COV2 Spike Protein Treated Human Microglia: Implications for Neuro-COVID. J Neuroimmune Pharmacol. 2021;16(4):770–84.
69. García-Prat L, Martínez-Vicente M, Muñoz-Cánoves P. Methods for Mitochondria and Mitophagy Flux Analyses in Stem Cells of Resting and Regenerating Skeletal Muscle. Methods Mol Biol. 2016;1460:223–40.
70. Choo HJ, Kim JH, Kwon OB, Lee CS, Mun JY, Han SS, et al. Mitochondria are impaired in the adipocytes of type 2 diabetic mice. Diabetologia. 2006;49(4):784–91.
71. Escada-Rebelo S, Ramalho-Santos J. Oxidative and Nitrosative Stress Detection in Human Sperm Using Fluorescent Probes. Methods Mol Biol. 2023;2566:45–52.
72. Rehman S, Chandel N, Salhan D, Rai P, Sharma B, Singh T, et al. Ethanol and vitamin D receptor in T cell apoptosis. J Neuroimmune Pharmacol. 2013;8(1):251–61.
73. Omatsu T, Naito Y, Handa O, Mizushima K, Hayashi N, Qin Y, et al. Reactive oxygen species-quenching and anti-apoptotic effect of polaprezinc on indomethacin-induced small intestinal epithelial cell injury. J Gastroenterol. 2010;45(7):692–702.
74. Rothe G, Valet G. Flow cytometric analysis of respiratory burst activity in phagocytes with hydroethidine and 2',7'-dichlorofluorescin. J Leukoc Biol. 1990;47(5):440–8.
75. Benov L, Sztejnberg L, Fridovich I. Critical evaluation of the use of hydroethidine as a measure of superoxide anion radical. Free Radic Biol Med. 1998;25(7):826–31.
76. Budd SL, Castilho RF, Nicholls DG. Mitochondrial membrane potential and hydroethidine-monitored superoxide generation in cultured cerebellar granule cells. FEBS Lett. 1997;415(1): 21–4.
77. Zielonka J, Kalyanaraman B. Hydroethidine- and MitoSOX-derived red fluorescence is not a reliable indicator of intracellular superoxide formation: another inconvenient truth. Free Radic Biol Med. 2010;48(8):983–1001.
78. Robinson KM, Janes MS, Pehar M, Monette JS, Ross MF, Hagen TM, et al. Selective fluorescent imaging of superoxide in vivo using ethidium-based probes. Proc Natl Acad Sci U S A. 2006;103(41):15038–43.
79. Zhao H, Kalivendi S, Zhang H, Joseph J, Nithipatikom K, Vásquez-Vivar J, et al. Superoxide reacts with hydroethidine but forms a fluorescent product that is distinctly different from ethidium: potential implications in intracellular fluorescence detection of superoxide. Free Radic Biol Med. 2003;34(11):1359–68.
80. Carter WO, Narayanan PK, Robinson JP. Intracellular hydrogen peroxide and superoxide anion detection in endothelial cells. J Leukoc Biol. 1994;55(2):253–8.

81. Suzuki H, Swei A, Zweifach BW, Schmid-Schönbein GW. In vivo evidence for microvascular oxidative stress in spontaneously hypertensive rats. Hydroethidine microfluorography. Hypertension. 1995;25(5):1083–9.
82. Bindokas VP, Jordán J, Lee CC, Miller RJ. Superoxide production in rat hippocampal neurons: selective imaging with hydroethidine. J Neurosci. 1996;16(4):1324–36.
83. Miller FJ, Gutterman DD, Rios CD, Heistad DD, Davidson BL. Superoxide Production in Vascular Smooth Muscle Contributes to Oxidative Stress and Impaired Relaxation in Atherosclerosis. 1998;82(12):1298–305.
84. Hirata K, Marushima A, Nagasaki Y, Ishikawa H, Matsumura H, Mujagić A, et al. Efficacy of redox nanoparticles for improving survival of transplanted cells in a mouse model of ischemic stroke. Hum Cell. 2023;36(5):1703–15.
85. Niu LG, Sun N, Liu KL, Su Q, Qi J, Fu LY, et al. Genistein Alleviates Oxidative Stress and Inflammation in the Hypothalamic Paraventricular Nucleus by Activating the Sirt1/Nrf2 Pathway in High Salt-Induced Hypertension. Cardiovasc Toxicol. 2022;22(10–11):898–909.
86. Sun Q, Li L, Jin F, Liu Y, Yang B, Meng W, et al. SARS-CoV-2 Spike Protein S1 Exposure Increases Susceptibility to Angiotensin II-Induced Hypertension in Rats by Promoting Central Neuroinflammation and Oxidative Stress. Neurochemical research. 2023;48(10):3016–26.
87. Green DR, Reed JC. Mitochondria and apoptosis. Science. 1998;281(5381):1309–12.
88. Cai J, Yang J, Jones DP. Mitochondrial control of apoptosis: the role of cytochrome c. Biochim Biophys Acta. 1998;1366(1–2):139–49.
89. Zielonka J, Srinivasan S, Hardy M, Ouari O, Lopez M, Vasquez-Vivar J, et al. Cytochrome c-mediated oxidation of hydroethidine and mito-hydroethidine in mitochondria: identification of homo- and heterodimers. Free Radic Biol Med. 2008;44(5):835–46.
90. Dikalov SI, Harrison DG. Methods for detection of mitochondrial and cellular reactive oxygen species. Antioxid Redox Signal. 2014;20(2):372–82.
91. Zielonka J, Vasquez-Vivar J, Kalyanaraman B. Detection of 2-hydroxyethidium in cellular systems: a unique marker product of superoxide and hydroethidine. Nat Protoc. 2008;3(1): 8–21.
92. Cuperus R, Leen R, Tytgat GAM, Caron HN, van Kuilenburg ABP. Fenretinide induces mitochondrial ROS and inhibits the mitochondrial respiratory chain in neuroblastoma. Cellular and Molecular Life Sciences. 2010;67(5):807–16.
93. Tavares RS, Escada-Rebelo S, Soares MM, Silva AF, Almeida-Santos T, Amaral S, et al. Mitochondrial Functional Assessment in Mammalian Gametes Using Fluorescent Probes. Methods Mol Biol. 2021;2310:57–68.
94. Yi F, He X, Wang D. Lycopene protects against MPP(+)-induced cytotoxicity by maintaining mitochondrial function in SH-SY5Y cells. Neurochemical research. 2013;38(8):1747–57.
95. Lin R, Peng X, Li Y, Wang X, Liu X, Jia X, et al. Empagliflozin attenuates doxorubicin-impaired cardiac contractility by suppressing reactive oxygen species in isolated myocytes. Molecular and cellular biochemistry. 2023.
96. Zielonka J, Hardy M, Kalyanaraman B. HPLC study of oxidation products of hydroethidine in chemical and biological systems: ramifications in superoxide measurements. Free Radical Biology and Medicine. 2009;46(3):329–38.
97. Zielonka J, Zhao H, Xu Y, Kalyanaraman B. Mechanistic similarities between oxidation of hydroethidine by Fremy's salt and superoxide: stopped-flow optical and EPR studies. Free Radic Biol Med. 2005;39(7):853–63.
98. Fernandes DC, Wosniak J, Jr., Pescatore LA, Bertoline MA, Liberman M, Laurindo FR, et al. Analysis of DHE-derived oxidation products by HPLC in the assessment of superoxide production and NADPH oxidase activity in vascular systems. American journal of physiology Cell physiology. 2007;292(1):C413–22.
99. Panov A, Dikalov S, Shalbuyeva N, Taylor G, Sherer T, Greenamyre JT. Rotenone model of Parkinson disease: multiple brain mitochondria dysfunctions after short term systemic rotenone intoxication. J Biol Chem. 2005;280(51):42026–35.

100. Lu B, Yadav S, Shah PG, Liu T, Tian B, Pukszta S, et al. Roles for the human ATP-dependent Lon protease in mitochondrial DNA maintenance. J Biol Chem. 2007;282(24):17363–74.
101. Dębski D, Smulik R, Zielonka J, Michałowski B, Jakubowska M, Dębowska K, et al. Mechanism of oxidative conversion of Amplex® Red to resorufin: Pulse radiolysis and enzymatic studies. Free Radic Biol Med. 2016;95:323–32.
102. Votyakova TV, Reynolds IJ. Detection of hydrogen peroxide with Amplex Red: interference by NADH and reduced glutathione auto-oxidation. Arch Biochem Biophys. 2004;431(1): 138–44.
103. Dikalov S, Griendling KK, Harrison DG. Measurement of reactive oxygen species in cardiovascular studies. Hypertension. 2007;49(4):717–27.
104. Zhao B, Ranguelova K, Jiang J, Mason RP. Studies on the photosensitized reduction of resorufin and implications for the detection of oxidative stress with Amplex Red. Free Radic Biol Med. 2011;51(1):153–9.
105. Miwa S, Treumann A, Bell A, Vistoli G, Nelson G, Hay S, et al. Carboxylesterase converts Amplex red to resorufin: Implications for mitochondrial H2O2 release assays. Free radical biology & medicine. 2016;90:173–83.
106. Takahashi M, Shibata M, Niki E. Estimation of lipid peroxidation of live cells using a fluorescent probe, diphenyl-1-pyrenylphosphine. Free Radic Biol Med. 2001;31(2):164–74.
107. Drummen GP, Op den Kamp JA, Post JA. Validation of the peroxidative indicators, cis-parinaric acid and parinaroyl-phospholipids, in a model system and cultured cardiac myocytes. Biochim Biophys Acta. 1999;1436(3):370–82.
108. Shvedova AA, Tyurina JY, Kawai K, Tyurin VA, Kommineni C, Castranova V, et al. Selective peroxidation and externalization of phosphatidylserine in normal human epidermal keratinocytes during oxidative stress induced by cumene hydroperoxide. The Journal of investigative dermatology. 2002;118(6):1008–18.
109. Zhao YY, Yang YQ, Sheng HH, Tang Q, Han L, Wang SM, et al. GPX4 Plays a Crucial Role in Fuzheng Kang'ai Decoction-Induced Non-Small Cell Lung Cancer Cell Ferroptosis. Front Pharmacol. 2022;13:851680.
110. Pap EH, Drummen GP, Winter VJ, Kooij TW, Rijken P, Wirtz KW, et al. Ratio-fluorescence microscopy of lipid oxidation in living cells using C11-BODIPY(581/591). FEBS Lett. 1999;453(3):278–82.
111. Drummen GP, van Liebergen LC, Op den Kamp JA, Post JA. C11-BODIPY(581/591), an oxidation-sensitive fluorescent lipid peroxidation probe: (micro)spectroscopic characterization and validation of methodology. Free Radic Biol Med. 2002;33(4):473–90.
112. Yoshida Y, Shimakawa S, Itoh N, Niki E. Action of DCFH and BODIPY as a probe for radical oxidation in hydrophilic and lipophilic domain. Free radical research. 2003;37(8):861–72.
113. Beretta G, Aldini G, Facino RM, Russell RM, Krinsky NI, Yeum KJ. Total antioxidant performance: a validated fluorescence assay for the measurement of plasma oxidizability. Anal Biochem. 2006;354(2):290–8.
114. Zhu M, Qin ZJ, Hu D, Munishkina LA, Fink AL. Alpha-synuclein can function as an antioxidant preventing oxidation of unsaturated lipid in vesicles. Biochemistry. 2006;45(26): 8135–42.
115. Xu L, Du X, Liu T, Sun D. In situ and dynamic SERS monitoring of glutathione levels during cellular ferroptosis metabolism. Analytical and bioanalytical chemistry. 2023;415(25): 6145–53.
116. Wang F, Naowarojna N, Zou Y. Stratifying ferroptosis sensitivity in cells and mouse tissues by photochemical activation of lipid peroxidation and fluorescent imaging. STAR Protoc. 2022;3 (2):101189.
117. Elango K, Layek SS, Kumaresan A. Advances in Bovine Sperm Quality Assessment: From Motility to Fertility. In: Kumaresan A, Srivastava AK, editors. Current Concepts in Bovine Reproduction. Singapore: Springer Nature Singapore; 2022. p. 263–91.
118. Manevich Y, Held KD, Biaglow JE. Coumarin-3-carboxylic acid as a detector for hydroxyl radicals generated chemically and by gamma radiation. Radiat Res. 1997;148(6):580–91.

119. Setsukinai K, Urano Y, Kakinuma K, Majima HJ, Nagano T. Development of novel fluorescence probes that can reliably detect reactive oxygen species and distinguish specific species. J Biol Chem. 2003;278(5):3170–5.
120. Edenbaum H, Han J. Assessment of S-Glutathionylated Rac1 in Cells Using Biotin-Labeled Glutathione. In: Rivero F, editor. Rho GTPases: Methods and Protocols. New York, NY: Springer New York; 2018. p. 155–63.
121. Mullen L, Seavill M, Hammouz R, Bottazzi B, Chan P, Vaudry D, et al. Development of 'Redox Arrays' for identifying novel glutathionylated proteins in the secretome. Scientific Reports. 2015;5(1):14630.
122. Chen M, Wang J, Yang Y, Zhong T, Zhou P, Ma H, et al. Redox-dependent regulation of end-binding protein 1 activity by glutathionylation. Science China Life Sciences. 2021;64(4): 575–83.
123. Müller-Schüssele SJ, Bohle F, Rossi J, Trost P, Meyer AJ, Zaffagnini M. Plasticity in plastid redox networks: evolution of glutathione-dependent redox cascades and glutathionylation sites. BMC Plant Biology. 2021;21(1):322.
124. Lippert AR, Van de Bittner GC, Chang CJ. Boronate Oxidation as a Bioorthogonal Reaction Approach for Studying the Chemistry of Hydrogen Peroxide in Living Systems. Acc Chem Res. 2011;44(9):793–804.
125. Zielonka J, Sikora A, Hardy M, Joseph J, Dranka BP, Kalyanaraman B. Boronate probes as diagnostic tools for real time monitoring of peroxynitrite and hydroperoxides. Chem Res Toxicol. 2012;25(9):1793–9.
126. Miller EW, Chang CJ. Fluorescent probes for nitric oxide and hydrogen peroxide in cell signaling. Curr Opin Chem Biol. 2007;11(6):620–5.
127. Sikora A, Zielonka J, Lopez M, Joseph J, Kalyanaraman B. Direct oxidation of boronates by peroxynitrite: mechanism and implications in fluorescence imaging of peroxynitrite. Free Radic Biol Med. 2009;47(10):1401–7.
128. Chang MC, Pralle A, Isacoff EY, Chang CJ. A selective, cell-permeable optical probe for hydrogen peroxide in living cells. J Am Chem Soc. 2004;126(47):15392–3.
129. Miller EW, Albers AE, Pralle A, Isacoff EY, Chang CJ. Boronate-based fluorescent probes for imaging cellular hydrogen peroxide. J Am Chem Soc. 2005;127(47):16652–9.
130. Dickinson BC, Huynh C, Chang CJ. A Palette of Fluorescent Probes with Varying Emission Colors for Imaging Hydrogen Peroxide Signaling in Living Cells. Journal of the American Chemical Society. 2010;132(16):5906–15.
131. Albers AE, Dickinson BC, Miller EW, Chang CJ. A red-emitting naphthofluorescein-based fluorescent probe for selective detection of hydrogen peroxide in living cells. Bioorganic & Medicinal Chemistry Letters. 2008;18(22):5948–50.
132. Miller EW, Tulyathan O, Isacoff EY, Chang CJ. Molecular imaging of hydrogen peroxide produced for cell signaling. Nature chemical biology. 2007;3(5):263–7.
133. Cochemé Helena M, Quin C, McQuaker Stephen J, Cabreiro F, Logan A, Prime Tracy A, et al. Measurement of H2O2 within Living Drosophila during Aging Using a Ratiometric Mass Spectrometry Probe Targeted to the Mitochondrial Matrix. Cell Metabolism. 2011;13(3): 340–50.
134. Dickinson BC, Chang CJ. A targetable fluorescent probe for imaging hydrogen peroxide in the mitochondria of living cells. J Am Chem Soc. 2008;130(30):9638–9.
135. Masanta G, Heo CH, Lim CS, Bae SK, Cho BR, Kim HM. A mitochondria-localized two-photon fluorescent probe for ratiometric imaging of hydrogen peroxide in live tissue. Chemical Communications. 2012;48(29):3518–20.
136. Dickinson Bryan C, Tang Y, Chang Z, Chang Christopher J. A Nuclear-Localized Fluorescent Hydrogen Peroxide Probe for Monitoring Sirtuin-Mediated Oxidative Stress Responses In Vivo. Chemistry & Biology. 2011;18(8):943–8.
137. Guida M, Maraldi T, Beretti F, Follo MY, Manzoli L, De Pol A. Nuclear Nox4-Derived Reactive Oxygen Species in Myelodysplastic Syndromes. BioMed research international. 2014;2014:456937.

138. Woolley JF, Stanicka J, Cotter TG. Recent advances in reactive oxygen species measurement in biological systems. Trends in biochemical sciences. 2013;38(11):556–65.
139. Miller EW, Dickinson BC, Chang CJ. Aquaporin-3 mediates hydrogen peroxide uptake to regulate downstream intracellular signaling. Proc Natl Acad Sci U S A. 2010;107(36): 15681–6.
140. Dickinson BC, Peltier J, Stone D, Schaffer DV, Chang CJ. Nox2 redox signaling maintains essential cell populations in the brain. Nature chemical biology. 2011;7(2):106–12.
141. Albers AE, Okreglak VS, Chang CJ. A FRET-Based Approach to Ratiometric Fluorescence Detection of Hydrogen Peroxide. Journal of the American Chemical Society. 2006;128(30): 9640–1.
142. Du L, Ni N, Li M, Wang B. A Fluorescent Hydrogen Peroxide Probe Based on a 'Click' Modified Coumarin Fluorophore. Tetrahedron letters. 2010;51(8):1152–4.
143. Srikun D, Miller EW, Domaille DW, Chang CJ. An ICT-Based Approach to Ratiometric Fluorescence Imaging of Hydrogen Peroxide Produced in Living Cells. Journal of the American Chemical Society. 2008;130(14):4596–7.
144. Kumar M, Kumar N, Bhalla V, Sharma PR, Qurishi Y. A charge transfer assisted fluorescent probe for selective detection of hydrogen peroxide among different reactive oxygen species. Chemical Communications. 2012;48(39):4719–21.
145. Sikora A, Zielonka J, Lopez M, Dybala-Defratyka A, Joseph J, Marcinek A, et al. Reaction between peroxynitrite and boronates: EPR spin-trapping, HPLC Analyses, and quantum mechanical study of the free radical pathway. Chem Res Toxicol. 2011;24(5):687–97.
146. Zielonka J, Sikora A, Joseph J, Kalyanaraman B. Peroxynitrite is the major species formed from different flux ratios of co-generated nitric oxide and superoxide: direct reaction with boronate-based fluorescent probe. J Biol Chem. 2010;285(19):14210–6.
147. Zielonka J, Zielonka M, Sikora A, Adamus J, Joseph J, Hardy M, et al. Global profiling of reactive oxygen and nitrogen species in biological systems: high-throughput real-time analyses. J Biol Chem. 2012;287(5):2984–95.
148. Miller EW, Bian SX, Chang CJ. A Fluorescent Sensor for Imaging Reversible Redox Cycles in Living Cells. Journal of the American Chemical Society. 2007;129(12):3458–9.
149. Karton-Lifshin N, Segal E, Omer L, Portnoy M, Satchi-Fainaro R, Shabat D. A Unique Paradigm for a Turn-ON Near-Infrared Cyanine-Based Probe: Noninvasive Intravital Optical Imaging of Hydrogen Peroxide. Journal of the American Chemical Society. 2011;133(28): 10960–5.
150. Van de Bittner GC, Dubikovskaya EA, Bertozzi CR, Chang CJ. In vivo imaging of hydrogen peroxide production in a murine tumor model with a chemoselective bioluminescent reporter. Proc Natl Acad Sci U S A. 2010;107(50):21316–21.
151. Stanicka J, Landry W, Cotter TG. Oxidative Stress Biomarkers and ROS Molecular Probes. Oxidative Stress: Diagnostics, Prevention, and Therapy Volume 2. ACS Symposium Series. 1200: American Chemical Society; 2015. p. 353–74.
152. Wei Y, Zhang Y, Liu Z, Guo M. A novel profluorescent probe for detecting oxidative stress induced by metal and H(2)O(2) in living cells. Chem Commun (Camb). 2010;46(25):4472–4.
153. Chung C, Srikun D, Lim CS, Chang CJ, Cho BR. A two-photon fluorescent probe for ratiometric imaging of hydrogen peroxide in live tissue. Chemical Communications. 2011;47(34):9618–20.
154. Wei Y, Guo M. A novel H2O2-triggered anti-Fenton fluorescent pro-chelator excitable with visible light. Chem Commun (Camb). 2009(11):1413–5.
155. Banni S, Lucchi L, Baraldi A, Botti B, Cappelli G, Corongiu F, et al. No direct evidence of increased lipid peroxidation in hemodialysis patients. Nephron. 1996;72(2):177–83.
156. Cao G, Cutler RG. Protein oxidation and aging. I. Difficulties in measuring reactive protein carbonyls in tissues using 2,4-dinitrophenylhydrazine. Arch Biochem Biophys. 1995;320(1): 106–14.

157. Uusitalo LM, Hempel N. Recent advances in intracellular and in vivo ROS sensing: focus on nanoparticle and nanotube applications. International journal of molecular sciences. 2012;13 (9):10660–79.
158. Lee D, Khaja S, Velasquez-Castano JC, Dasari M, Sun C, Petros J, et al. In vivo imaging of hydrogen peroxide with chemiluminescent nanoparticles. Nat Mater. 2007;6(10):765–9.
159. Hyman LM, Stephenson CJ, Dickens MG, Shimizu KD, Franz KJ. Toward the development of prochelators as fluorescent probes of copper-mediated oxidative stress. Dalton Transactions. 2010;39(2):568–76.
160. Hyman LM, Franz KJ. A Cell-Permeable Fluorescent Prochelator Responds to Hydrogen Peroxide and Metal Ions by Decreasing Fluorescence. Inorganica Chim Acta. 2012;380:125–34.
161. Uggeri J, Gatti R, Belletti S, Scandroglio R, Corradini R, Rotoli BM, et al. Calcein-AM is a detector of intracellular oxidative activity. Histochem Cell Biol. 2004;122(5):499–505.
162. Kundu K, Knight SF, Willett N, Lee S, Taylor WR, Murthy N. Hydrocyanines: a class of fluorescent sensors that can image reactive oxygen species in cell culture, tissue, and in vivo. Angew Chem Int Ed Engl. 2009;48(2):299–303.
163. Saeedi BJ, Chandrasekharan B, Neish AS. Hydro-Cy3-Mediated Detection of Reactive Oxygen Species In Vitro and In Vivo. Methods Mol Biol. 2019;1982:329–37.
164. Szabó M, Larkum AWD, Vass I. A Review: The Role of Reactive Oxygen Species in Mass Coral Bleaching. In: Larkum AWD, Grossman AR, Raven JA, editors. Photosynthesis in Algae: Biochemical and Physiological Mechanisms. Cham: Springer International Publishing; 2020. p. 459–88.
165. Sarkar A, Saha P, Mandal G, Mukhopadhyay D, Roy S, Singh SK, et al. Monitoring of intracellular nitric oxide in leishmaniasis: its applicability in patients with visceral leishmaniasis. Cytometry A. 2011;79(1):35–45.
166. Broillet M, Randin O, Chatton J. Photoactivation and calcium sensitivity of the fluorescent NO indicator 4,5-diaminofluorescein (DAF-2): implications for cellular NO imaging. FEBS Lett. 2001;491(3):227–32.
167. Donovan M, Carmody RJ, Cotter TG. Light-induced photoreceptor apoptosis in vivo requires neuronal nitric-oxide synthase and guanylate cyclase activity and is caspase-3-independent. J Biol Chem. 2001;276(25):23000–8.
168. Haugland RP. The handbook: a guide to fluorescent probes and labeling technologies: Molecular probes; 2005.
169. Ouyang J, Hong H, Shen C, Zhao Y, Ouyang C, Dong L, et al. A novel fluorescent probe for the detection of nitric oxide in vitro and in vivo. Free Radical Biology and Medicine. 2008;45 (10):1426–36.
170. Pluth MD, McQuade LE, Lippard SJ. Cell-Trappable Fluorescent Probes for Nitric Oxide Visualization in Living Cells. Organic Letters. 2010;12(10):2318–21.
171. Pluth MD, Chan MR, McQuade LE, Lippard SJ. Seminaphthofluorescein-Based Fluorescent Probes for Imaging Nitric Oxide in Live Cells. Inorganic Chemistry. 2011;50(19):9385–92.
172. Hu X, Wang J, Zhu X, Dong D, Zhang X, Wu S, et al. A copper(ii) rhodamine complex with a tripodal ligand as a highly selective fluorescence imaging agent for nitric oxide. Chemical Communications. 2011;47(41):11507–9.
173. Yang Y, Seidlits SK, Adams MM, Lynch VM, Schmidt CE, Anslyn EV, et al. A Highly Selective Low-Background Fluorescent Imaging Agent for Nitric Oxide. Journal of the American Chemical Society. 2010;132(38):13114–6.
174. Yuan L, Lin W, Xie Y, Chen B, Song J. Development of a ratiometric fluorescent sensor for ratiometric imaging of endogenously produced nitric oxide in macrophage cells. Chemical Communications. 2011;47(33):9372–4.
175. Yuan L, Lin W, Xie Y, Chen B, Zhu S. Single Fluorescent Probe Responds to H2O2, NO, and H2O2/NO with Three Different Sets of Fluorescence Signals. Journal of the American Chemical Society. 2012;134(2):1305–15.

176. Guo H, Aleyasin H, Dickinson BC, Haskew-Layton RE, Ratan RR. Recent advances in hydrogen peroxide imaging for biological applications. Cell Biosci. 2014;4(1):64.
177. Koo YE, Fan W, Hah H, Xu H, Orringer D, Ross B, et al. Photonic explorers based on multifunctional nanoplatforms for biosensing and photodynamic therapy. Appl Opt. 2007;46 (10):1924–30.
178. Lee YE, Smith R, Kopelman R. Nanoparticle PEBBLE sensors in live cells and in vivo. Annu Rev Anal Chem (Palo Alto Calif). 2009;2:57–76.
179. Clark HA, Barker SLR, Brasuel M, Miller MT, Monson E, Parus S, et al. Subcellular optochemical nanobiosensors: probes encapsulated by biologically localised embedding (PEBBLEs). Sensors and Actuators B: Chemical. 1998;51(1):12–6.
180. Hammond VJ, Aylott JW, Greenway GM, Watts P, Webster A, Wiles C. An optical sensor for reactive oxygen species: encapsulation of functionalised silica nanoparticles into silicate nanoprobes to reduce fluorophore leaching. Analyst. 2007;133(1):71–5.
181. Henderson JR, Fulton DA, McNeil CJ, Manning P. The development and in vitro characterisation of an intracellular nanosensor responsive to reactive oxygen species. Biosensors and Bioelectronics. 2009;24(12):3608–14.
182. King M, Kopelman R. Development of a hydroxyl radical ratiometric nanoprobe. Sensors and Actuators B: Chemical. 2003;90(1):76–81.
183. Ganea GM, Kolic PE, El-Zahab B, Warner IM. Ratiometric coumarin-neutral red (CONER) nanoprobe for detection of hydroxyl radicals. Anal Chem. 2011;83(7):2576–81.
184. Tan SY, Teh C, Ang CY, Li M, Li P, Korzh V, et al. Responsive mesoporous silica nanoparticles for sensing of hydrogen peroxide and simultaneous treatment toward heart failure. Nanoscale. 2017;9(6):2253–61.
185. Erard M, Dupré-Crochet S, Nüße O. Biosensors for spatiotemporal detection of reactive oxygen species in cells and tissues. Am J Physiol Regul Integr Comp Physiol. 2018;314(5): R667-r83.
186. Seo YH, Singh A, Cho HJ, Kim Y, Heo J, Lim CK, et al. Rational design for enhancing inflammation-responsive in vivo chemiluminescence via nanophotonic energy relay to near-infrared AIE-active conjugated polymer. Biomaterials. 2016;84:111–8.
187. Zhuang M, Ding C, Zhu A, Tian Y. Ratiometric fluorescence probe for monitoring hydroxyl radical in live cells based on gold nanoclusters. Anal Chem. 2014;86(3):1829–36.
188. Gao X, Ding C, Zhu A, Tian Y. Carbon-dot-based ratiometric fluorescent probe for imaging and biosensing of superoxide anion in live cells. Anal Chem. 2014;86(14):7071–8.
189. Panizzi P, Nahrendorf M, Wildgruber M, Waterman P, Figueiredo J-L, Aikawa E, et al. Oxazine Conjugated Nanoparticle Detects in Vivo Hypochlorous Acid and Peroxynitrite Generation. Journal of the American Chemical Society. 2009;131(43):15739–44.
190. Goto A, Nakahara I, Yamaguchi T, Kamioka Y, Sumiyama K, Matsuda M, et al. Circuit-dependent striatal PKA and ERK signaling underlies rapid behavioral shift in mating reaction of male mice. Proc Natl Acad Sci U S A. 2015;112(21):6718–23.
191. Kim JY, Choi WI, Kim YH, Tae G. Highly selective in-vivo imaging of tumor as an inflammation site by ROS detection using hydrocyanine-conjugated, functional nano-carriers. J Control Release. 2011;156(3):398–405.
192. Miao R, Mu L, Zhang H, Xu H, She G, Wang P, et al. Modified silicon nanowires: a fluorescent nitric oxide biosensor with enhanced selectivity and stability. Journal of Materials Chemistry. 2012;22(8):3348–53.
193. Swanner J, Singh R. Synthesis, Purification, Characterization, and Imaging of Cy3-Functionalized Fluorescent Silver Nanoparticles in 2D and 3D Tumor Models. Methods Mol Biol. 2018;1790:209–18.
194. Abdal Dayem A, Hossain MK, Lee SB, Kim K, Saha SK, Yang GM, et al. The Role of Reactive Oxygen Species (ROS) in the Biological Activities of Metallic Nanoparticles. International journal of molecular sciences. 2017;18(1).
195. Roy R, Kumar S, Tripathi A, Das M, Dwivedi PD. Interactive threats of nanoparticles to the biological system. Immunol Lett. 2014;158(1–2):79–87.

196. Florence AT. "Targeting" nanoparticles: the constraints of physical laws and physical barriers. J Control Release. 2012;164(2):115–24.
197. Beddoes CM, Case CP, Briscoe WH. Understanding nanoparticle cellular entry: A physicochemical perspective. Adv Colloid Interface Sci. 2015;218:48–68.
198. Tammam SN, Azzazy HME, Lamprecht A. How successful is nuclear targeting by nanocarriers? J Control Release. 2016;229:140–53.
199. Nicolas M-T, Moreau M. Prix Nobel de Chimie 2008 (Osumo Shimomura, Martin Chalfie et Roger Y. Tsien) : Osamu Shimomura. Med Sci (Paris). 2008;24(11):983–4.
200. Breckwoldt MO, Wittmann C, Misgeld T, Kerschensteiner M, Grabher C. Redox imaging using genetically encoded redox indicators in zebrafish and mice. Biological chemistry. 2015;396(5):511–22.
201. Ezeriņa D, Morgan B, Dick TP. Imaging dynamic redox processes with genetically encoded probes. Journal of molecular and cellular cardiology. 2014;73:43–9.
202. Schwarzländer M, Dick TP, Meyer AJ, Morgan B. Dissecting Redox Biology Using Fluorescent Protein Sensors. Antioxid Redox Signal. 2016;24(13):680–712.
203. Gutscher M, Sobotta MC, Wabnitz GH, Ballikaya S, Meyer AJ, Samstag Y, et al. Proximity-based protein thiol oxidation by H2O2-scavenging peroxidases. J Biol Chem. 2009;284(46):31532–40.
204. Enyedi B, Zana M, Donkó Á, Geiszt M. Spatial and temporal analysis of NADPH oxidase-generated hydrogen peroxide signals by novel fluorescent reporter proteins. Antioxid Redox Signal. 2013;19(6):523–34.
205. Mishina NM, Mishin AS, Belyaev Y, Bogdanova EA, Lukyanov S, Schultz C, et al. Live-Cell STED Microscopy with Genetically Encoded Biosensor. Nano Lett. 2015;15(5):2928–32.
206. Zeng H, Madisen L. Mouse transgenic approaches in optogenetics. Prog Brain Res. 2012;196:193–213.
207. Wojtovich AP, Foster TH. Optogenetic control of ROS production. Redox Biol. 2014;2:368–76.
208. Dooley CT, Dore TM, Hanson GT, Jackson WC, Remington SJ, Tsien RY. Imaging dynamic redox changes in mammalian cells with green fluorescent protein indicators. J Biol Chem. 2004;279(21):22284–93.
209. Eggeling C, Hilbert M, Bock H, Ringemann C, Hofmann M, Stiel AC, et al. Reversible photoswitching enables single-molecule fluorescence fluctuation spectroscopy at high molecular concentration. Microsc Res Tech. 2007;70(12):1003–9.
210. Brakemann T, Stiel AC, Weber G, Andresen M, Testa I, Grotjohann T, et al. A reversibly photoswitchable GFP-like protein with fluorescence excitation decoupled from switching. Nat Biotechnol [Internet]. 2011 2011/09//; 29(10):[942–7 pp.]. Available from: http://europepmc.org/abstract/MED/21909082, http://pubman.mpdl.mpg.de/pubman/item/escidoc:1222582/component/escidoc:1454610/1222582.pdf, https://doi.org/10.1038/nbt.1952.
211. Westphal V, Seeger J, Salditt T, Hell SW. Stimulated emission depletion microscopy on lithographic nanostructures. Journal of Physics B: Atomic, Molecular and Optical Physics. 2005;38(9):S695.
212. Junghans C, Schmitt F-J, Vukojević V, Friedrich T. Monitoring the diffusion behavior of Na, K-ATPase by fluorescence correlation spectroscopy (FCS) upon fluorescence labelling with eGFP or Dreiklang. Optofluidics, Microfluidics and Nanofluidics. 2015;2(1):1–14.
213. Kremers GJ, Gilbert SG, Cranfill PJ, Davidson MW, Piston DW. Fluorescent proteins at a glance. J Cell Sci. 2011;124(Pt 2):157–60.
214. Jiménez-Banzo A, Nonell S, Hofkens J, Flors C. Singlet oxygen photosensitization by EGFP and its chromophore HBDI. Biophys J. 2008;94(1):168–72.
215. Swanson SJ, Choi WG, Chanoca A, Gilroy S. In vivo imaging of Ca2+, pH, and reactive oxygen species using fluorescent probes in plants. Annu Rev Plant Biol. 2011;62:273–97.
216. Espey MG, Xavier S, Thomas DD, Miranda KM, Wink DA. Direct real-time evaluation of nitration with green fluorescent protein in solution and within human cells reveals the impact of nitrogen dioxide vs. peroxynitrite mechanisms. 2002;99(6):3481–6.

217. Schwarzländer M, Fricker MD, Müller C, Marty L, Brach T, Novak J, et al. Confocal imaging of glutathione redox potential in living plant cells. Journal of microscopy. 2008;231(2): 299–316.
218. Hanson GT, Aggeler R, Oglesbee D, Cannon M, Capaldi RA, Tsien RY, et al. Investigating mitochondrial redox potential with redox-sensitive green fluorescent protein indicators. J Biol Chem. 2004;279(13):13044–53.
219. Arias-Barreiro CR, Okazaki K, Koutsaftis A, Inayat-Hussain SH, Tani A, Katsuhara M, et al. A Bacterial Biosensor for Oxidative Stress Using the Constitutively Expressed Redox-Sensitive Protein roGFP2. Sensors. 2010;10(7):6290–306.
220. Liu X, Wu J, Liu H, Zong N, Zhao J. RoGFP1 is a quantitative biosensor in maize cells for cellular redox changes caused by environmental and endogenous stimuli. Biochem Biophys Res Commun. 2014;452(3):503–8.
221. Yano T, Oku M, Akeyama N, Itoyama A, Yurimoto H, Kuge S, et al. A novel fluorescent sensor protein for visualization of redox states in the cytoplasm and in peroxisomes. Mol Cell Biol. 2010;30(15):3758–66.
222. Austin CD, Wen X, Gazzard L, Nelson C, Scheller RH, Scales SJ. Oxidizing potential of endosomes and lysosomes limits intracellular cleavage of disulfide-based antibody-drug conjugates. Proc Natl Acad Sci U S A. 2005;102(50):17987–92.
223. Lohman JR, Remington SJ. Development of a family of redox-sensitive green fluorescent protein indicators for use in relatively oxidizing subcellular environments. Biochemistry. 2008;47(33):8678–88.
224. Birk J, Meyer M, Aller I, Hansen HG, Odermatt A, Dick TP, et al. Endoplasmic reticulum: reduced and oxidized glutathione revisited. J Cell Sci. 2013;126(Pt 7):1604–17.
225. van Lith M, Tiwari S, Pediani J, Milligan G, Bulleid NJ. Real-time monitoring of redox changes in the mammalian endoplasmic reticulum. J Cell Sci. 2011;124(Pt 14):2349–56.
226. Cannon MB, Remington SJ. Re-engineering redox-sensitive green fluorescent protein for improved response rate. Protein Sci. 2006;15(1):45–57.
227. Schwarzländer M, Fricker MD, Sweetlove LJ. Monitoring the in vivo redox state of plant mitochondria: Effect of respiratory inhibitors, abiotic stress and assessment of recovery from oxidative challenge. Biochimica et Biophysica Acta (BBA) - Bioenergetics. 2009;1787(5): 468–75.
228. Guzman JN, Sanchez-Padilla J, Wokosin D, Kondapalli J, Ilijic E, Schumacker PT, et al. Oxidant stress evoked by pacemaking in dopaminergic neurons is attenuated by DJ-1. Nature. 2010;468(7324):696–700.
229. Plotkin JL, Guzman JN, Schwarz N, Kress G, Wokosin DL, Surmeier DJ. Optical Approaches to Studying the Basal Ganglia. In: Lane EL, Dunnett SB, editors. Animal Models of Movement Disorders: Volume I. Totowa, NJ: Humana Press; 2012. p. 191–220.
230. Aller I, Rouhier N, Meyer AJ. Development of roGFP2-derived redox probes for measurement of the glutathione redox potential in the cytosol of severely glutathione-deficient rml1 seedlings. Front Plant Sci. 2013;4:506.
231. Gutscher M, Pauleau AL, Marty L, Brach T, Wabnitz GH, Samstag Y, et al. Real-time imaging of the intracellular glutathione redox potential. Nat Methods. 2008;5(6):553–9.
232. Kurz FT, Breckwoldt MO. Automated Quantification and Network Analysis of Redox Dynamics in Neuronal Mitochondria. Methods Mol Biol. 2022;2399:261–74.
233. Morgan B, Sobotta MC, Dick TP. Measuring E(GSH) and H2O2 with roGFP2-based redox probes. Free Radic Biol Med. 2011;51(11):1943–51.
234. Müller A, Schneider JF, Degrossoli A, Lupilova N, Dick TP, Leichert LI. Systematic in vitro assessment of responses of roGFP2-based probes to physiologically relevant oxidant species. Free Radical Biology and Medicine. 2017;106:329–38.
235. Vilas-Boas EA, Karabacz N, Marsiglio-Librais GN, Valle MMR, Nalbach L, Ampofo E, et al. Chronic activation of GPR40 does not negatively impact upon BRIN-BD11 pancreatic β-cell physiology and function. Pharmacol Rep. 2020;72(6):1725–37.

236. Lismont C, Walton PA, Fransen M. Quantitative Monitoring of Subcellular Redox Dynamics in Living Mammalian Cells Using RoGFP2-Based Probes. Methods Mol Biol. 2017;1595: 151–64.
237. Ostergaard H, Henriksen A, Hansen FG, Winther JR. Shedding light on disulfide bond formation: engineering a redox switch in green fluorescent protein. The EMBO journal. 2001;20(21):5853–62.
238. Maulucci G, Labate V, Mele M, Panieri E, Arcovito G, Galeotti T, et al. High-resolution imaging of redox signaling in live cells through an oxidation-sensitive yellow fluorescent protein. Sci Signal. 2008;1(43):pl3.
239. Miyawaki A, Griesbeck O, Heim R, Tsien RY. Dynamic and quantitative Ca2+ measurements using improved cameleons. Proc Natl Acad Sci U S A. 1999;96(5):2135–40.
240. Dardalhon M, Kumar C, Iraqui I, Vernis L, Kienda G, Banach-Latapy A, et al. Redox-sensitive YFP sensors monitor dynamic nuclear and cytosolic glutathione redox changes. Free Radic Biol Med. 2012;52(11–12):2254–65.
241. Björnberg O, Ostergaard H, Winther JR. Measuring intracellular redox conditions using GFP-based sensors. Antioxid Redox Signal. 2006;8(3–4):354–61.
242. Hansen RE, Østergaard H, Winther JR. Increasing the Reactivity of an Artificial Dithiol–Disulfide Pair through Modification of the Electrostatic Milieu. Biochemistry. 2005;44(15): 5899–906.
243. Pouvreau S. Superoxide flashes in mouse skeletal muscle are produced by discrete arrays of active mitochondria operating coherently. PLoS One. 2010;5(9).
244. Wang W, Fang H, Groom L, Cheng A, Zhang W, Liu J, et al. Superoxide flashes in single mitochondria. Cell. 2008;134(2):279–90.
245. Pouvreau S. Genetically encoded reactive oxygen species (ROS) and redox indicators. Biotechnol J. 2014;9(2):282–93.
246. Schwarzländer M, Wagner S, Ermakova YG, Belousov VV, Radi R, Beckman JS, et al. The ‘mitoflash’ probe cpYFP does not respond to superoxide. Nature. 2014;514(7523):E12–4.
247. Quatresous E, Legrand C, Pouvreau S. Mitochondria-targeted cpYFP: pH or superoxide sensor? J Gen Physiol. 2012;140(5):567–70.
248. Schwarzländer M, Logan DC, Fricker MD, Sweetlove LJ. The circularly permuted yellow fluorescent protein cpYFP that has been used as a superoxide probe is highly responsive to pH but not superoxide in mitochondria: implications for the existence of superoxide ‘flashes’. The Biochemical journal. 2011;437(3):381–7.
249. Belousov VV, Fradkov AF, Lukyanov KA, Staroverov DB, Shakhbazov KS, Terskikh AV, et al. Genetically encoded fluorescent indicator for intracellular hydrogen peroxide. Nat Methods. 2006;3(4):281–6.
250. Hanschmann EM, Lönn ME, Schütte LD, Funke M, Godoy JR, Eitner S, et al. Both thioredoxin 2 and glutaredoxin 2 contribute to the reduction of the mitochondrial 2-Cys peroxiredoxin Prx3. J Biol Chem. 2010;285(52):40699–705.
251. Hernández-Barrera A, Quinto C, Johnson EA, Wu HM, Cheung AY, Cárdenas L. Using hyper as a molecular probe to visualize hydrogen peroxide in living plant cells: a method with virtually unlimited potential in plant biology. Methods in enzymology. 2013;527:275–90.
252. Labuschagne CF, Brenkman AB. Current methods in quantifying ROS and oxidative damage in Caenorhabditis elegans and other model organism of aging. Ageing research reviews. 2013;12(4):918–30.
253. Malinouski M, Zhou Y, Belousov VV, Hatfield DL, Gladyshev VN. Hydrogen peroxide probes directed to different cellular compartments. PLoS One. 2011;6(1):e14564.
254. Lukyanov KA, Belousov VV. Genetically encoded fluorescent redox sensors. Biochim Biophys Acta. 2014;1840(2):745–56.
255. Markvicheva KN, Bilan DS, Mishina NM, Gorokhovatsky AY, Vinokurov LM, Lukyanov S, et al. A genetically encoded sensor for H2O2 with expanded dynamic range. Bioorg Med Chem. 2011;19(3):1079–84.

256. Bilan DS, Pase L, Joosen L, Gorokhovatsky AY, Ermakova YG, Gadella TW, et al. HyPer-3: a genetically encoded H(2)O(2) probe with improved performance for ratiometric and fluorescence lifetime imaging. ACS Chem Biol. 2013;8(3):535–42.
257. Choi H, Kim S, Mukhopadhyay P, Cho S, Woo J, Storz G, et al. Structural basis of the redox switch in the OxyR transcription factor. Cell. 2001;105(1):103–13.
258. Bilan DS, Belousov VV. HyPer Family Probes: State of the Art. Antioxid Redox Signal. 2016;24(13):731–51.
259. Shirmanova MV, Druzhkova IN, Lukina MM, Matlashov ME, Belousov VV, Snopova LB, et al. Intracellular pH imaging in cancer cells in vitro and tumors in vivo using the new genetically encoded sensor SypHer2. Biochim Biophys Acta. 2015;1850(9):1905–11.
260. Bilan D, Lukyanov S, Belousov V. Genetically encoded fluorescent sensors for redox processes. Russian Journal of Bioorganic Chemistry. 2015;41:231–44.
261. Ermakova YG, Bilan DS, Matlashov ME, Mishina NM, Markvicheva KN, Subach OM, et al. Red fluorescent genetically encoded indicator for intracellular hydrogen peroxide. Nat Commun. 2014;5:5222.
262. Subach FV, Verkhusha VV. Chromophore Transformations in Red Fluorescent Proteins. Chemical reviews. 2012;112(7):4308–27.
263. Fan Y, Ai HW. Development of redox-sensitive red fluorescent proteins for imaging redox dynamics in cellular compartments. Analytical and bioanalytical chemistry. 2016;408(11): 2901–11.
264. Fan Y, Chen Z, Ai HW. Monitoring redox dynamics in living cells with a redox-sensitive red fluorescent protein. Anal Chem. 2015;87(5):2802–10.
265. Fan Y, Makar M, Wang MX, Ai HW. Monitoring thioredoxin redox with a genetically encoded red fluorescent biosensor. Nature chemical biology. 2017;13(9):1045–52.
266. Simen Zhao B, Liang Y, Song Y, Zheng C, Hao Z, Chen PR. A highly selective fluorescent probe for visualization of organic hydroperoxides in living cells. Journal of the American chemical society. 2010;132(48):17065–7.
267. Guzy RD, Hoyos B, Robin E, Chen H, Liu L, Mansfield KD, et al. Mitochondrial complex III is required for hypoxia-induced ROS production and cellular oxygen sensing. Cell Metab. 2005;1(6):401–8.
268. Cranfill PJ, Sell BR, Baird MA, Allen JR, Lavagnino Z, de Gruiter HM, et al. Quantitative assessment of fluorescent proteins. Nat Methods. 2016;13(7):557–62.
269. Mérola F, Fredj A, Betolngar DB, Ziegler C, Erard M, Pasquier H. Newly engineered cyan fluorescent proteins with enhanced performances for live cell FRET imaging. Biotechnol J. 2014;9(2):180–91.
270. Pearce LL, Gandley RE, Han W, Wasserloos K, Stitt M, Kanai AJ, et al. Role of metallothionein in nitric oxide signaling as revealed by a green fluorescent fusion protein. Proc Natl Acad Sci U S A. 2000;97(1):477–82.
271. Pasin F, Kulasekaran S, Natale P, Simón-Mateo C, García JA. Rapid fluorescent reporter quantification by leaf disc analysis and its application in plant-virus studies. Plant Methods. 2014;10(1):22.
272. Nunes P, Demaurex N, Dinauer MC. Regulation of the NADPH oxidase and associated ion fluxes during phagocytosis. Traffic. 2013;14(11):1118–31.

Chapter 12
Electron Paramagnetic Resonance/Electron Spin Resonance

A. Khelfi

Abstract Electron spin resonance (ESR), also known as electron proton resonance (EPR), is a technique used to study the chemical species exhibiting unpaired electrons. EPR spectroscopy allows for the understanding of the function of organic and inorganic radicals as well as transition metal complexes. It can be particularly useful for studying oxidative stress (OS) in biological samples. Reactive species (RS) can be directly measured by EPR most often coupled to sensors to increase their lifetime and the sensitivity of the method. EPR techniques are known to be powerful in detecting primarily RS with properties similar to radicals such as $O_2^{\bullet-}$, $OH^{\bullet}$, and $NO^{\bullet}$. EPR techniques have been developed for non-invasive detection of these RS in accessible sites such as human skin. They have been successfully applied in several other models using specific probes. There are three major approaches for RS detection using EPR technology with probes: spin trapping, spin formation, and spin quenching. In this chapter, we describe these approaches for the detection of RS. We capture the state of development of this expanding field, covering many important topics from immune-spin trapping to nanosensors. Recent advances in the application of sensitive and noninvasive EPR to intact cells and organs are brought into focus here.

Keywords Electron spin resonance (ESR) · Electron proton resonance (EPR) · Spin trapping · Spin formation · Spin quenching · Immune-spin traps · Nanosensors

---

A. Khelfi (✉)
Faculty of Pharmacy, University of Algiers1, Algiers, Algeria

National Centre of Toxicology, Algiers, Algeria

Department of Chemistry and Biomolecular Science, Clarkson University, Potsdam, NY, USA
e-mail: ab.khelfi@univ-alger.dz; akhelfi@clarkson.edu

S. Andreescu et al. (eds.), *Biomarkers of Oxidative Stress*,
https://doi.org/10.1007/978-3-031-60738-7_12

12.1 General Principle

EPR/ESR technologies are based on the properties of electrons as charged particles that spin around their axis, creating a magnetic moment. Electrons exist in two possible spin states, +1/2 and −1/2, and pair with neighboring electrons of the opposite spin. This results in a net spin of zero. On the other hand, unpaired electrons, which have charge and spin, act as tiny magnets that align themselves in two possible directions, either parallel or antiparallel with respect to an applied magnetic field. The energy differences of these states correspond to the microwave region of the electromagnetic spectrum.

The unpaired electrons, which are parallel to the magnetic field, exist in a lower energy state than do electrons aligned antiparallel to the applied field. When an appropriate amount of energy is applied in the form of electromagnetic radiation (usually in the microwave range) using two magnets, electrons move from the lower energy state (parallel to the magnetic field) to the higher energy state (antiparallel to the magnetic field). This resonance or "flipping" of electron depends on the energy and the frequency used as well as the applied magnetic field. The field at which electron flipping spin states occurs is known as the "field of resonance."

EPR/ESR spectroscopy exploits the magnetic moment of the unpaired electron through absorption of microwave radiation energy in the presence of the external magnetic field. The microwave frequency is kept constant, and the magnetic field is measured and scanned. The absorbed energy during resonance can be visualized, analyzed, and even quantified since the height of the signal is directly proportional to the concentration of FR present in the sample. EPR/ESR spectroscopy can also identify the generated FR since each radical has a specific hyperfine splitting pattern for a given set of EPR conditions (EM frequency and magnetic field strength) [1]. Like electrons, atomic nuclei have charge and spin; therefore, they behave like magnets. The nuclei affect the strength of the applied magnetic field on unpaired electrons in a manner called "shielding." Since nuclei can also align either parallel or antiparallel to the applied magnetic field, unpaired electrons experience two slightly different magnetic fields that result in the hyperfine splitting of the original spectrum. As a result of the particular nature of each radical, this interaction yields typical ESR spectra and the hyperfine splitting gives information about compound's structure. It can then be used to check the compound's identity. Indeed, each FR exhibits unique hyperfine splitting patterns that can be easily identified by their EPR spectra. For this reason, EPR/ESR spectroscopy is considered the gold standard for the detection and identification of radicals in biological systems [2].

EPR/ESR techniques are known to be powerful in detecting primarily RS with properties similar to radicals such as $O_2^{\bullet -}$, $OH^{\bullet}$, and $NO^{\bullet}$ [3]. These techniques can also be used to detect secondary radicals with unpaired electron spin states; for example, transition metal complexes implicated in reactive oxygen species (ROS) generation, like ferryl species as well as lipid-derived radicals (alkoxyl, peroxyl, etc.) and protein radicals produced following reaction of RS with the surrounding biomolecules [4].

EPR/ESR techniques have been developed for noninvasive detection of these RS, for example, in accessible sites such as human skin. They have been successfully applied in several models using specific probes [5–8]. Exceptions are ascorbyl, tocopheroxyl radicals, and heme-nitrosyl complexes, which are more stable FR that do not need any probes for their detection and are highly produced in vascular tissues during OS and inflammation [9]. They can be directly measured in plasma as well, but with limited sensitivity. Ascorbyl radical has a longer lifetime than $OH^{\bullet}$ and presents a characteristic doublet in ESR (AH4 = 1.8 G) [10]. The variations of the intensity of its EPR signal have been proposed as a marker of OS [11].

12.2 Advantages and Inconveniences

The major advantage of EPR spectroscopy is its ability to measure and localize FR in vivo. When this technology is applied to complex biological systems and rodent models, it enables information regarding redox reactions, which is evidence of FR production. It permits to measure the kinetics of decomposition of FR as well. However, overlap between different signals in the EPR spectrum makes it difficult to measure individual FR in biological materials. This is the main reason behind using EPR spectroscopy to assess total FR formation [12]. Besides, there are many common factors influencing the detection of FR by EPR, among which the pH value has a considerable influence on radical formation. Simultaneous changes in temperature could affect the energy level of radicals as the liquid environment of analytes is responsible for the resonant state of radicals [13]. Moreover, considering the high cost of EPR spectroscopy, experiments must be performed rationally. Finally, another drawback with EPR is that this technique is not able to efficiently detect short-lived FR involved in oxidation of other biomolecules (half-lives of $OH^{\bullet}$, $O_2^{\bullet -}$, peroxyl radicals are 10^{-9}, 10^{-6}, and few seconds, respectively).

12.3 Spin Trapping, Spin Quenching, and Spin Formation

There are three major approaches for FR detection using EPR technology with probes (Fig. 12.1):

- Spin trapping using nitrones
- Spin formation using hydroxylamine
- Spin quenching (or spin-loss) using trityl

The first and the second involve signal enhancement, and the third involves signal loss.

It is important to distinct spin traps and other spin probes. Spin traps form covalent bond with the analyzed radical by addition reaction, while the other probes are oxidized by ROS without binding.

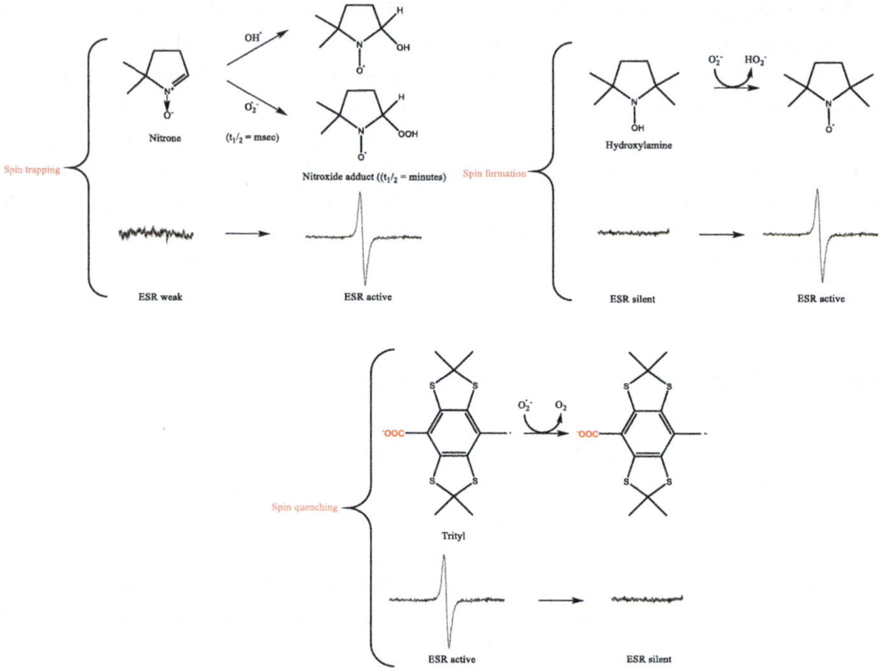

Fig. 12.1 Major approaches for the FR detection using EPR

12.3.1 Spin Trapping

12.3.1.1 Principle

ESR technique allows for unambiguous measurement of the overall FR production, but the major limitation of this method in the study of FR in cells and tissues is related to its poor sensitivity to detect very low concentrations of FR (1 nmol to 1 pmol). Indeed, reactive radicals generated under physiological conditions do not accumulate to enough levels that can be measured, especially highly reactive radicals. As stated above, FR are short lived species and it is extremely difficult to observe the appearance of these highly reactive species with conventional ESR directly in in vivo studies. Only three radicals can be detected directly by ESR in biological samples: ascorbyl radical, tocopheroxyl radicals, and heme-nitrosyl complexes. The ascorbyl radical resulting from one-electron reduction of ascorbate has a prolonged lifetime because of the delocalization of its unpaired electron on the whole structure. Its detection in plasma using ESR at room temperature has been proposed as a noninvasive marker for OS.

To overcome sensitivity limitation and to obtain steady-state concentration for other FR in the detectable range, several approaches have been used such as continuous flow systems, pulse radiolysis, UV photolysis, extremely low temperature, and the so-called spin trapping. Among these, the spin trapping strategy is the most commonly used for biological samples.

The technique of spin trapping is based on the use of an additional compound called "spin trap," which reacts specifically with a radical to produce a measurable product with a characteristic ESR signature. The primary FR is usually added across the double bond of a diamagnetic spin trap to form a radical adduct more stable, less reactive, and longer-lasting than the primary product [14, 15]. This adduct accumulates to levels that can be detected using ESR spectroscopy and provides information that enables the identification of the originating FR species.

Spin trap radicals exhibit unique hyperfine splitting patterns that can be easily identified by their ESR spectra. The hyperfine splitting constants of the spin adduct provide information to identify the primary FR, whereas the amplitude of the emitted signals enables the quantification of the generated FR. The most frequently used unit for the hyperfine splitting constants is Gauss (G) and that for SI unit for magnetic flux density is Tesla (T). Besides, some spin trap agents also act by oxidizing the primary FR to a relatively stable radical, easily detected by electron spin resonators. Finally, the latest progress in EPR technology combined with the development of new spin traps allows for noninvasive mapping of FR in the whole animal [8, 16, 17].

12.3.1.2 Types of Probes

A wide range of diamagnetic spin traps is available for use both in cell cultures and whole animal studies. They vary in chemical structure and membrane permeability. Two main groups of compounds are used: nitroso and nitrone derivatives, which involve addition reactions or electron transfer with RS, forming an EPR-detectable paramagnetic aminoxyl (or commonly called nitroxides) species (Fig. 12.1). In this approach, the nitrone/nitroso probe reacts with a short-lived FR to produce a nitroxide (a spin-radical adduct) having a lifetime considerably greater than that of the parent FR and, therefore, detectable by conventional ESR instruments [18]. The greater the stability of the adduct, the lower the detected concentration of a given FR. As spin traps react specifically with ROS/reactive nitrogen species (RNS)/FR, ESR spectroscopy is able to distinguish between some ROS and RNS [19]. Due to the relatively rapid decomposition of nitroxides that results from the reaction of nitroso compounds with oxygen-centered FR (i.e., $O_2^{\bullet-}$ and $OH^{\bullet}$), nitrones have been considered as the principal spin traps [20].

Nitroso Traps

In nitroso spin traps, the nitrogen atom reacts directly with the FR, giving distinctive spectral features. Common examples of nitroso spin traps, commercially available and used in biological investigation, are 2-methyl-2-nitrosopropane (MNP)

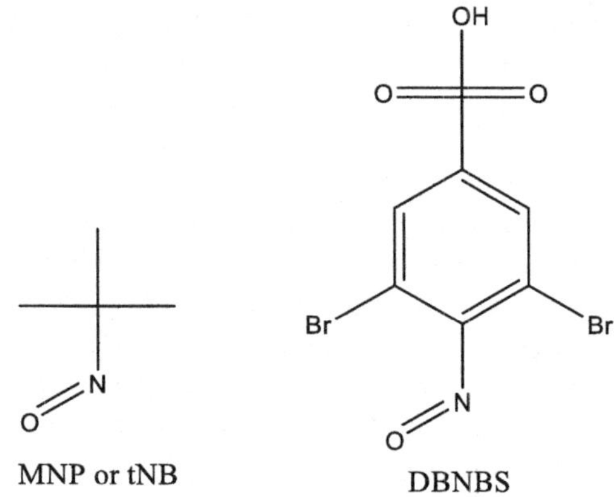

Fig. 12.2 Structure of nitroso traps

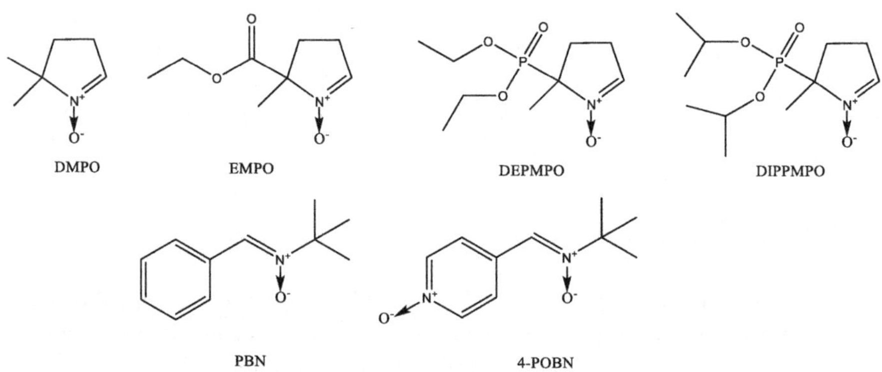

Fig. 12.3 Structure of nitrone traps

(*tert*-nitrosobutane (tNB)) and 3,5-dibromo-4-nitrosobenzenesulfonate (DBNBS) (Fig. 12.2).

Nitrone Traps

Commonly used nitrone traps are 5,5-dimethyl-1-pyrroline N-oxide (DMPO), 5-(ethoxycarbonyl)-5-methyl-1-pyrroline N-oxide (EMPO), 5-diethoxyphosphoryl-5-methyl-1-pyrroline N-oxide (DEPMPO), 5-(Diisopropoxyphosphoryl)-5-methyl-1-pyrroline N-oxide (DIPPMPO), phenyl-tert-butylnitrone/α-phenyl-*tert*-butylnitrone/N-tertbutyl-p-phenylnitrone (PBN), and α-(4-pyridyl-1-oxide)-*N*-*tert*-butylnitrone (4-POBN) (Fig. 12.3). The lack of spectral information about the generated adducts is the major drawback of this class of spin traps.

A demonstration of the mechanism of nitrone spin traps is given in Fig. 12.1.

The most widely used nitrone is DMPO, which produces spin-trapped adducts with characteristic EPR spectra as a result of its reaction with $O_2^{\bullet-}$ and $OH^{\bullet}$. DMPO adducts with these species are DMPO/OOH$^•$ and DMPO/OH$^•$, respectively, both of which could be detected using a standard EPR spectrometer [21].

DMPO-based methods are usually applied to analyze radicals produced by cellular or organelle systems. For instance, the release of $O_2^{•-}$ from the mitochondrial respiratory chain into the inter-membrane space was evidenced by the results of DMPO–EPR assay [22]. DMPO is also useful in studies on isolated enzymes.

Nevertheless, several factors must be taken into account:

- First, in the presence of some reducing agents like ascorbate for instance, adducts of DMPO/OOH$^•$ and DMPO/OH$^•$ could be reduced back to ESR silent species, yielding false negative results, which severely limits the use of DMPO in biological systems [21].
- Second, DMPO reacts with $O_2^{•-}$ at very slow rate constants (30–70 M^{-1} s^{-1}) and has difficulty in detecting $O_2^{•-}$ in most biological systems due to competition with the enzymatic activity of superoxide dismutase (SOD) [23]. These slow rate constants imply the used of DMPO at concentrations ranging from 20 to 100 mM. These high concentrations can affect cell viability.
- Third, certain factors in cells and tissues obstruct the recognition of DMPO adducts like further reaction of $O_2^{•-}$ with DMPO/OOH$^•$ or DMPO/OH$^•$, and the possibility that DMPO/OOH$^•$ decays to form DMPO/OH$^•$ [24]. Besides, DMPO is susceptible to metal ion-catalyzed air oxidation, which may result in the production of unwanted products.
- Fourth, DMPO is inefficient for the reaction with $O_2^{•-}$ in biological systems since it has a partition coefficient of only 0.09, indicating a preference for water over lipid environments [20].

For all these reasons, it is wise to take the above factors into account when interpreting EPR data with DMPO.

The structural analogs of DMPO are frequently used as alternatives or supplements in experimental studies. DEPMPO (Fig. 12.3), an alkylated form of DMPO, has been shown to be a good trap for $O_2^{•-}$ [25], but its secondary radical adducts influence the interpretation of the spectra [26]. Using these traps, OH$^•$ can also be detected in samples at room temperature with a fair amount of sensitivity [27]. DEPMPO has been modified by adding a triphenylphosphonium group, which promotes its selective uptake by mitochondria [28]. Unfortunately, this triphenylphosphonium conjugate of DEPMPO, called "mito-DEPMPO" (Fig. 12.4) must be used at high concentrations (50 mM), which leads to the inhibition of the mitochondrial machinery of ATP synthesis due to the high accumulation of lipophilic cations in the mitochondrial matrix and the alteration of the mitochondrial membrane potential [29].

The use of mito-DEPMPO spin trap is further limited by its slow rate constant for reaction with $O_2^{•-}$ [30] and its potential toxicity [31]. Its use in biological systems is, thus, unlikely. Because of all these limitations, other sensors have been developed to detect FR in cells. They include 5-tert-butyl-5-methyl-1-pyrroline 1-oxide (BMPO) and 5,5-dipropyl-1-pyrroline 1-oxide (DPPO) with more lipophilic and more stable adducts compared to DMPO, showing more potential for wide applications

Fig. 12.4 Structure of mito-DEPMPO probe

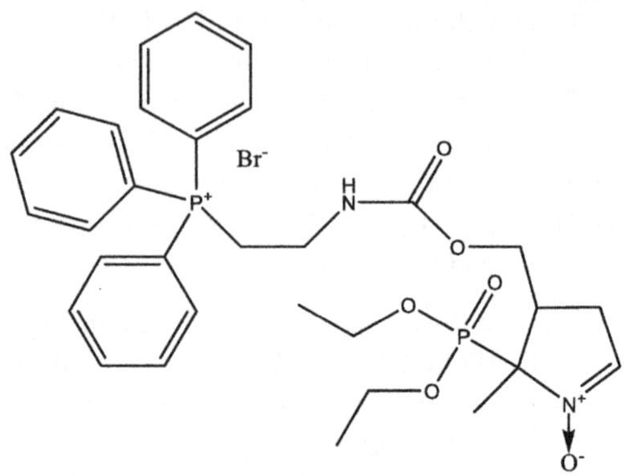

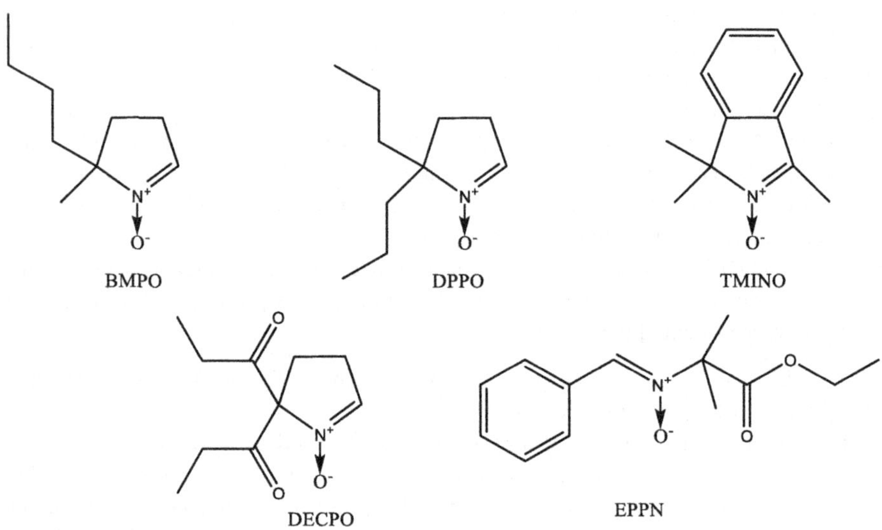

Fig. 12.5 Structure of other nitrone traps

(Fig. 12.5) [21]. In addition, unlike DMPO, the BMPO-$O_2^{\bullet-}$ adduct does not undergo nonenzymatic transformation to BMPO-$OH^{\bullet}$ adducts, making its identification easier, providing more accurate results than that of DMPO since the spectra of DMPO/$OOH^{\bullet}$ and DMPO/$OH^{\bullet}$ do overlap a lot. Moreover, BMPO-radical adducts give much higher signal-to-noise ratios than those of DMPO, showing the potential of BMPO as a promising spin trap for the detection of $O_2^{\bullet-}$ and $OH^{\bullet}$ [32].

Finally, other relative newer nitrones were developed including 1,1,3-trimethyl-isoindole N-oxide (TMINO), 5,5-diethylcarbonyl-1-pyrroline N-oxide (DECPO), and N-2-(2-ethoxycarbonyl-propyl)-α-phenylnitrone (EPPN) (Fig. 12.5) [33, 34].

Immuno-Spin Traps

During the last two decades, a new immuno-spin-trapping approach has been developed based on the reaction of DMPO with protein radicals, forming epitopes that can be specifically detected by immunoglobulins [35]. Both monoclonal and polyclonal antibodies are now available to detect DMPO. Mason et al. have developed multiple antibodies that react with DMPO–protein radical adducts and have demonstrated that they work well for immune-fluorescence, immuno-staining, Western blotting, and even flow cytometry [36]. This approach makes use of DMPO and eventually other spin traps possible without EPR analysis, which, as discussed earlier, is of restricted accessibility, has limited applicability, and lacks sensitivity for biological materials, [37]. In initial studies, anti-DMPO has been used to detect DMPO–protein adducts of hemoglobin and myoglobin that happened as a result of self-peroxidation by H_2O_2 [38]. Immuno-spin trapping has also been used to detect FR adduct formation in the mitochondria, cells, and tissues [39]. Of note, DMPO adducts of components of the mitochondrial electron transfer have been identified using immuno-spin trapping methods [40]. It is likely that this approach can be used to examine specific targets of oxidative attacks within cells and organelles. For instance, immune-spin trapping has been used by Pan et al. who have demonstrated ROS production in hypoxic astrocytes during cerebral ischemia (Fig. 12.6) [41].

12.3.1.3 Specificity Toward Free Radicals

Spin trapping enables to investigate the mechanism of action of FR and hence offers the opportunity to address fundamental questions on their role in biological systems. An important factor, which greatly influences the outcome of EPR results, is the spin trap–FR interaction. It has been found that each FR has a specific affinity for a given spin trap. This highly influences the final outcome of reactions and hence the reliability of the technique. For example, DMPO, as hydrophilic agents, is useful for trapping $O_2^{\bullet-}$ and $OH^{\bullet}$, while PBN, a lipophilic compound, preferentially reacts with lipid-derived radicals. However, uncertainties about influences caused by PBN on nitric oxide metabolism and the ability of ferric ion to oxidize DMPO resulting in

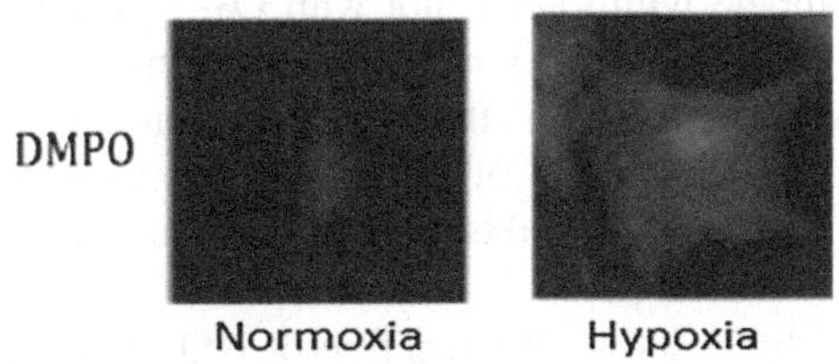

Fig. 12.6 ROS/RNS generation in hypoxic astrocytes
Primary astrocytes collected from Sprague–Dawley rats' cortices were exposed to 3-h hypoxia or normoxia. Levels of DMPO–protein-radical adducts in astrocytes were measured by immuno-spin trapping (green). Scale bar: 30 μm [41]

a classic four-line spectrum associated with hydroxyl radical metabolism, led to the development of more selective probes [42].

Compounds such as TMINO [43] and EPPN [34] have been reported to be more selective to $O_2^{\bullet-}$ [31]. A more specific and stable spin trap, DEPMPO [25, 44], has been proven to be useful in measuring extents of cell and tissue $O_2^{\bullet-}$ production [45, 46].

The half-lives of spin adducts vary significantly, from seconds to hours depending on the type of nitrone and the targeted radical. The least stable adducts are those formed from $O_2^{\bullet-}$ on one side and DMPO or PBN on the other side with a half-life less than a minute in aqueous solution. However, C-5 spin traps like DEPMPO have longer $O_2^{\bullet-}$ adduct half-life of nearly 14 min [47].

Aside from $O_2^{\bullet-}$ and $OH^{\bullet}$, other radicals that can also be identified using spin trapping approach are $NO^{\bullet}$, $RO^{\bullet}$, $NO_2^{\bullet}$, $CO_3^{\bullet-}$, and $RS^{\bullet}$. S- and carbon-centered radicals have been detected in blood and bile samples after systemic injection of some xenobiotics [48–50].

Interestingly, spin-trap approach can also be used to measure oxygen consumption of tissues and to spatially map oxygen concentrations in living tissues [51, 52]. With a sensitivity limit of 10^{-9} mol/L, EPR spectroscopy is also capable of detecting more stable FR-derived species produced in the vascular compartment during oxidative insult or inflammatory injuries such as tocopheroxyl radical, ascorbyl radical, and heme-nitrosyl complexes [9]. Spin traps are also used to label biomolecules and probe basal and oxidative events occurring in protein and lipid microenvironments [53].

Regarding RNS, $NO^{\bullet}$, for instance, does not add to nitrone spin traps, but it undergoes redox reactions with nitronyl nitroxides (NNs) and addition reactions with iron-thiocarbamate complexes which products are detectable by EPR. There are two commonly used NNs: 2-PTIO (phenyl-4,4,5,5-tetramethylimidazoline-l-oxyl 3-oxide) and its analog C-PTIO (2-(4-carboxyphenyl)-4,4,5,5-tetramethylimidazoline-1-oxyl 3-oxide).

As illustrated in Fig. 12.7, $NO^{\bullet}$ reacts with NN via addition reaction to the nitroxyl-O. This results in a subsequent release of $NO_2^{\bullet}$ and the formation of an imino nitroxide (IN). The use of NN probes provides a clear distinction between the spectral profiles imparted by the NN versus the IN product.

The rate constant for the reaction of $NO^{\bullet}$ with NN is around $10^3\ M^{-1}\ s^{-1}$, which is fast enough to compete with O_2 but not with $O_2^{\bullet-}$. Nevertheless, $NO_2^{\bullet}$ formed during the process can participate with other reactions involving NN and, thus, requires careful consideration in the interpretation of the obtained signal [54, 55]. One major inconvenient of NN probes is their possible reduction to EPR-silent hydroxylamines due to the presence of some reductants like ascorbate or metal ions.

Some experiences have shown the reaction of $NO^{\bullet}$ with endogenous heme and non-heme iron proteins, which led to the formation of iron-nitrosyl complexes with characteristic spectra [56–58]. This led to the development of exogenous iron-dithiocarbamates to detect nitric oxide formation [59]. Several Fe^{2+}-dithiocarbamate complexes were successful in nitric oxide detection using EPR (Fig. 12.8) [60–63].

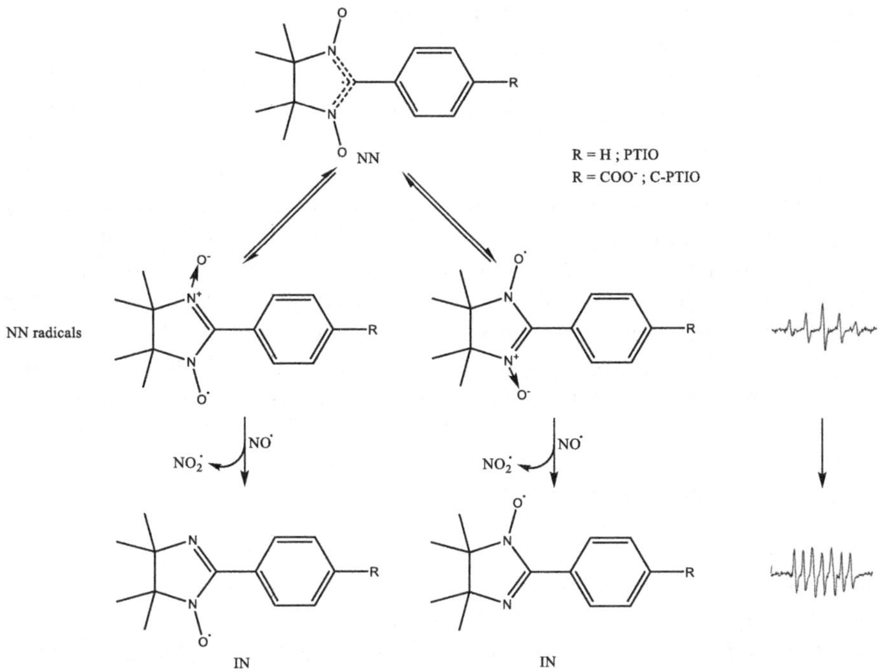

Fig. 12.7 Reaction of NO˙ with nitronyl nitroxides (NN; PTIO or C-PTIO), forming the imino nitroxide (IN)

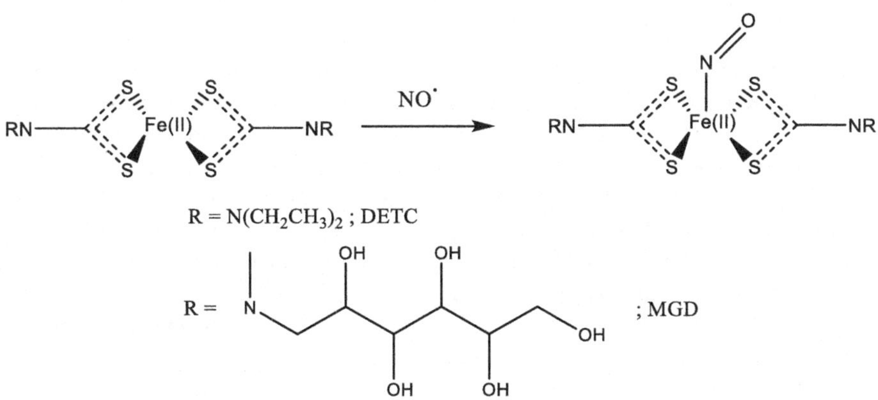

Fig. 12.8 Complexation of NO˙ with Fe^{2+}-dithiocarbamates

The redox state of Fe^{2+}-dithiocarbamates plays a critical role in the detection of NO˙. Under aerobic condition, Fe^{2+} complex readily oxidizes to form Fe^{3+}-dithiocarbamates. Reaction of ferric complex with NO˙ forms the EPR-silent NO-Fe^{3+}-dithiocarbamate complex but can be converted to an EPR-detectable

$NO\text{-}Fe^{2+}$-dithiocarbamates by $NO^{\bullet}$ itself or by reductants such as hydroquinone, ascorbate, or cysteine.

This type of probes was used in multiple systems and models. For example, Fe $(DETC)_2$ (Ferrous-diethyldithiocarbamate) has been reported to measure $NO^{\bullet}$ in porcine aorta with high sensitivity (10 pmol/mL) [64]. Detection of $NO^{\bullet}$ in blood vessels and human umbilical endothelial cells has been successfully achieved using colloidal Fe^{2+}-DETC [65, 66]. Compared to NN probes, iron-dithiocarbamate complexes are better sensors for the detection of $NO^{\bullet}$ due to the stability of the formed adducts. However, cautions should be made since iron complexes have also been shown to detect nitroxyl (HNO), NO_2^{-}, and S-nitrosothiols. Additionally, dithiocarbamate complexes have the potential to chelate metals and act as enzyme inhibitors, which may disturb the analyzed system. Through the use of NOS inhibitors, the obtained signals can be integrated to identify NOS-derived $NO^{\bullet}$ [67].

12.3.1.4 Advantages and Inconveniences

Spin traps are widely applied to EPR-detectable species like $O_2^{\bullet -}$, $OH^{\bullet}$, and $NO^{\bullet}$. As mentioned earlier, spin trapping enables a more sensitive technique for FR monitoring [68]. They give specific response to the external magnetic field and even if one spin trap binds different RS, the complex will typically exhibit specific EPR resonance for different bound RS, which makes spin trapping a selective RS sensing approach.

Although detection through spin trapping offers some distinct advantages in its high sensitivity and specificity toward radical species, there are also several drawbacks and potential issues using this strategy:

- Changes in the spin trap/adduct (chemical, non-radical, photo-chemical, or enzymatic reactions): radical adducts giving the ESR signal can be removed quickly in vivo and in vitro as a result of enzymatic metabolism and direct reductive action of endogenous antioxidants (e.g., ascorbate). For example, nitroxide probes can be reduced enzymatically to hydroxylamines, products that cannot be detected by EPR. When DMPO is used to trap the $OH^{\bullet}$, ascorbate can directly reduce DMPO-$OH^{\bullet}$ adduct to EPR-silent species [69]. Due to the susceptibility of the produced adducts to reduction into diamagnetic hydroxylamine species, post-treatments of the samples are needed in order to oxidize hydroxylamines back to the EPR-detectable nitroxides using mild oxidants such as potassium ferricyanide or gaseous O_2.
- Perturbation of the biological system by sensors: some spin trap agents interfere with the radical-generating system. For example, DMPO must be used at high concentrations (millimolar range), which may affect the whole system (e.g., decomposing H_2O_2 or chelating iron in the Fe^{2+}/H_2O_2 system).
- Toxicity of the probes: stable concentrations of spin traps are difficult to establish in vivo. Their diffusion from tissues and breakdown by several biological components require the use of high concentrations needed for in vivo detection. Most

of the probes (e.g., DMPO) if not all cannot be directly administered to human beings because of unknown toxicity at high concentrations that would be required for RS trapping in vivo. This is the main reason behind the limited use of these probes in vivo.

- Slow reactivity: one of the major inconvenient of spin trapping is the slow reactivity of several probes with $O_2^{\bullet -}$ with rate constants ranging from 1 to 10 M^{-1} s^{-1}. This is another reason behind using high concentrations (typically 10–100 mM) of these sensors in solution. Nevertheless, other radicals exhibit fast reactivity and long adduct half-lives with spin traps.
- Instability of some probes: DMPO can be oxidized by ferric ions, which generates the four-line spectrum normally produced by $OH^{\bullet}$ [42].

12.3.1.5 Summary

Overall, spin trapping is a powerful approach that facilitates the detection of FR, even in complex biological systems. Spin trapping is a good example of the complex interaction between the model system and the artificial addition of probes, with a possible recording in complex biological systems. The most commonly used spin traps are DMPO and PBN. With newer spin traps being developed and more intricate techniques such as EPR imaging, the use of EPR in FR research will continue to grow.

Important criteria for radical detection using EPR spectroscopy combined with spin trapping in biological systems are as follows:

- Specificity of the spin trap toward the radical of interest
- Favorable rate constants of reactions between the spin trap and the radical of interest;
- Reasonable stability of the resulting spin trap-radical adducts.

Finally, a tremendous help in disentangling the resulting EPR spectra has been given for spin trapping by Li and Chignell [70], which has been made freely available to those interested in the EPR field.

12.3.2 Spin Formation

This group of probes is represented by cyclic hydroxylamine derivatives. These compounds are not spin traps since they do not trap radicals but are oxidized by FR to form EPR-detectable and stable nitroxides with lifetimes of several hours in cell culture [71].

Figure 12.9 shows the commonly used hydroxylamines, which are N-hydroxy-pyrrole or piperidine derivatives: 2,2,6,6-tetramethylpiperidine-1-oxyl (TEMPO), 1-hydroxy-2,2,6,6-tetramethyl-4-oxo-piperidine (TEMPONE-H), 1,4-dihydroxy-2,2,6,6-tetramethylpiperidine (TPL-H), 4-phosphonooxy-2,2,6,6-

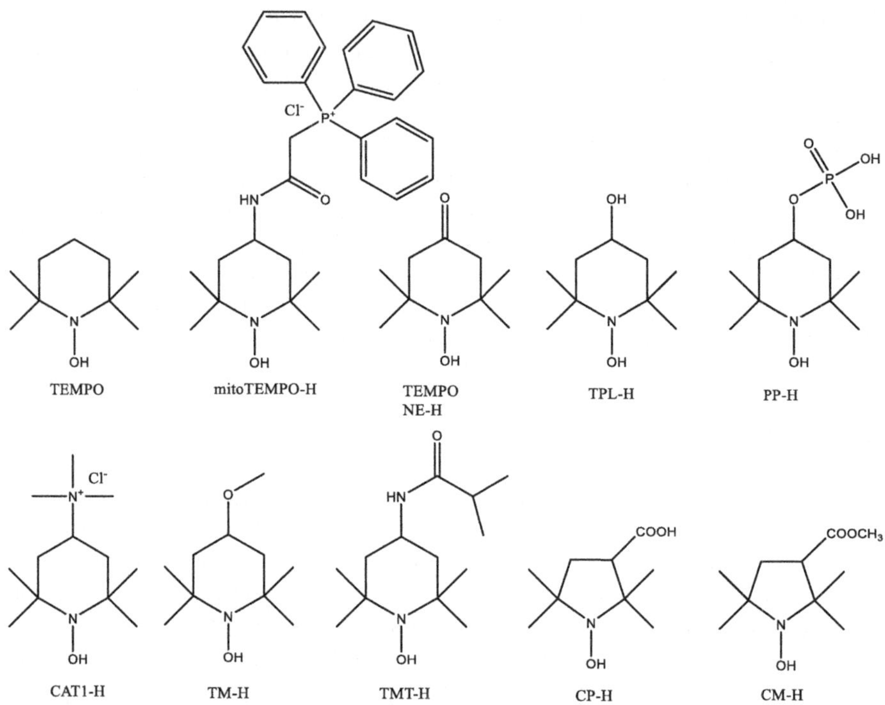

Fig. 12.9 Structure of hydroxylamine probes

tetramethylpiperidine-N-hydroxyl (PP-H), 1-hydroxy-2,2,6,6-tetramethylpiperidin-4-yl-trimethylammonium (CAT1-H), 1-hydroxy-4-methoxy-2,2,6,6-tetramethylpiperidine (TM-H), N-(1-Hydroxy-2,2,6,6-tetramethylpiperidin-4-yl)-2-methylpropanamide (TMT-H), 3-carboxy-2,2,5,5-tetramethylpyrrolin-1-hydroxide (CP-H), and 1-hydroxy-3-methoxycarbonyl-2,2,5,5-tetramethylpyrrolidine (CM-H).

Cyclic hydroxylamines can be used to measure $O_2^{\bullet -}$ in cultured cells, tissues, and whole animals [72–74]. These probes are oxidized by $O_2^{\bullet -}$ via electron transfer to yield a paramagnetic aminoxyl species and HO_2^- (Fig. 12.1). Therefore, they can be considered as an artificial source of ROS. Since other RS and metal ions can also give the exact same EPR triplet signal, caution should be made in data interpretation using these probes.

Number of neutral, cationic, and anionic hydroxylamine spin probes, which have different degrees of lipophilicity and cell permeability, are currently available. By careful selection of these probes, one can separately detect intracellular and extracellular ROS. Combined use of these probes helps identifying different sources of cellular radicals [75]. In fact, intracellular, extracellular, mitochondrial, and cytoplasmic sources of $O_2^{\bullet -}$ were distinguished by using a set of hydroxylamine spin probes simultaneously. Among these probes, the positively charged CAT1-H can only detect extracellular $O_2^{\bullet -}$ released from intact cells and tissues, or extramitochondrial $O_2^{\bullet -}$ released from isolated mitochondria. The mitochondria-targeted

TEMPO-H (mito-TEMPO-H) detects $O_2^{\bullet-}$ production in isolated mitochondrial particles, and the cell-permeable CP-H and CM-H are responsible for detecting $O_2^{\bullet-}$ in the cytoplasm. Moreover, CM-H is highly cell permeable and can detect both intracellular and extracellular ROS. CM-H also accumulates in mitochondria. The negatively charged PP-H accumulates inside cells, probably via active transport.

Of note, mito-TEMPO-H is a modified hydroxylamine probe, which contains a triphenylphosphonium group, allowing its accumulation inside the mitochondria. Upon stimulation of cytoplasmic production of $O_2^{\bullet-}$ with PMA (phorbol 12-myristate 13-acetate), both PP-H and CM-H provide strong signals, while the signal with mito-TEMPO-H is weak. In contrast, stimulation of the mitochondrial production of $O_2^{\bullet-}$ with rotenone only increases the CM-H and mito-TEMPO-H signals [71]. Thus, judicious use of hydroxylamine spin probes and their combinations provides unique information about site-specific production of ROS in the intracellular, extracellular, or mitochondrial compartments [76].

Although hydroxylamine probes exhibit 10- to 1000-fold higher reactivity to $O_2^{\bullet-}$ than nitrones, the paramagnetic species generated from hydroxylamine does not allow for a clear differentiation between the generated radicals [77]. Indeed, spin traps add to a FR at its α-carbon (C-2) position to form aminoxyl adducts, resulting in more complex signals than those obtained from hydroxylamines (Fig. 12.1) [78]. This property is behind the popularity of spin trapping over spin formation, not only in FR detection but also in their identification.

Hydroxylamine probes react with $O_2^{\bullet-}$ with rate constants of 10^3–10^4 M^{-1} s^{-1} (pH = 7.4), which are much more fast that those of the reaction of $O_2^{\bullet-}$ with nitrone spin traps. This favors the competition of hydroxylamine probes with cellular antioxidants and enhances the efficiency for detection of intracellular $O_2^{\bullet-}$ [74]. Despite significant differences in the chemical structures, these hydroxylamine probes have similar rates of oxidation in response to $O_2^{\bullet-}$ generated by xanthine and Xanthine oxidase (XO) [76]. For this reason, hydroxylamine probes can be used at relatively low concentrations (0.05–1 mM), minimizing potential toxicity and interfering effects in biological systems.

Another advantage of hydroxylamines over other probes is that they react with $O_2^{\bullet-}$ in a single chemical reaction. This minimizes potential artifacts that often occur during multistep redox reactions with other probes [79]. Moreover, hydroxylamine probes are incapable of redox cycling; therefore, they have no potential to artificially produce ROS [80].

A distinct advantage of hydroxylamine probes is that they can be used for real-time monitoring of ROS production at room temperature by following the accumulation of nitroxide radical using an EPR spectrometer. For this goal, a field scan is run, the precise G-value of low field peak is detected, and the scanner is set to follow this G-value over time.

Low-temperature EPR is useful for ROS detection in intact tissues. For this purpose, samples are incubated in physiological saline solution containing hydroxylamine probes and subsequently frozen in liquid nitrogen at −80 °C. The specimens are analyzed later at a low temperature in quartz dewar [81].

For the detection of ROS using EPR techniques, it is essential to remove artificial signals generated by transition metals, which are often present in physiological buffers. This is particularly important for hydroxylamine probes. Premixing of buffers with chelex overnight is recommended as is the addition of deferoxamine or DTPA (diethylene triamine penta-acetic acid).

It is noteworthy to know that the conversion of hydroxylamine probes to their respective nitroxide radicals is not specific for $O_2^{\bullet-}$ but can be mediated by $ONOO^-$ and other oxidants. Thus, it is necessary to use specific scavengers/inhibitors to identify the precise nature of the RS behind nitroxide radical signal. For instance, parallel experiments using external SOD confirms whether the signal is derived from $O_2^{\bullet-}$. As a practical example, overexpression of SOD2, targeted to the mitochondria, inhibits the mito-TEMPO-H and CM-H signals caused by the presence of rotenone [76]. In other cases, polyethylene glycol (PEG)-SOD has been found to reduce nitroxide radical production by 70% to 98% [76]. Besides, uric acid can be used as a selective scavenger of $ONOO^-$ [82]. In general, there is no reaction between H_2O_2 and hydroxylamine probes; however, H_2O_2 can be detected in isolated mitochondria and other sub-cellular compartments using HPR mediated co-oxidation of CAT1-H and CP-H [83, 84].

12.3.3 Spin Quenching

Due to their stability, trityl radicals have been used as sensors for the detection of $O_2^{\bullet-}$ in solutions. The most common are the triarylmethyl (trityl)-based radicals, which undergo electro-transfer reaction with $O_2^{\bullet-}$ to form an EPR silent trityl anion and O_2 [85, 86]. Trityl radicals, TAM OX063 and perchlorotriphenylmethyl (PCM-TC) (Fig. 12.10), have been shown to give high reactivity and sensitivity toward $O_2^{\bullet-}$ in aqueous solution [85, 86]. In this case, $O_2^{\bullet-}$ reacts with the trityl probe to cause the loss of the EPR signal.

One major advantage of trityl radicals is that they are not subjected to reduction by endogenous components including ascorbate and GSH. However, they exhibit reactivity with other oxidative species including $HO_2^{\bullet}$, $ROO^{\bullet}$, and $OH^{\bullet}$ [86].

12.4 Direct Detection with EPR/ESR

Ascorbic acid reacts with a wide range of FR. One of its oxidation products, ascorbyl radical, can easily be detected by EPR spectrometers. Measurement of ascorbyl radical has been used as a marker of FR production in blood and organs (e.g., skin) [10, 87, 88]. The assay is only semiquantitative, since ascorbyl radicals quickly react with each other to generate EPR-silent species (ascorbate and dehydroascorbate) [89]. Ascorbyl radicals are also reduced by enzymes in vivo [90].

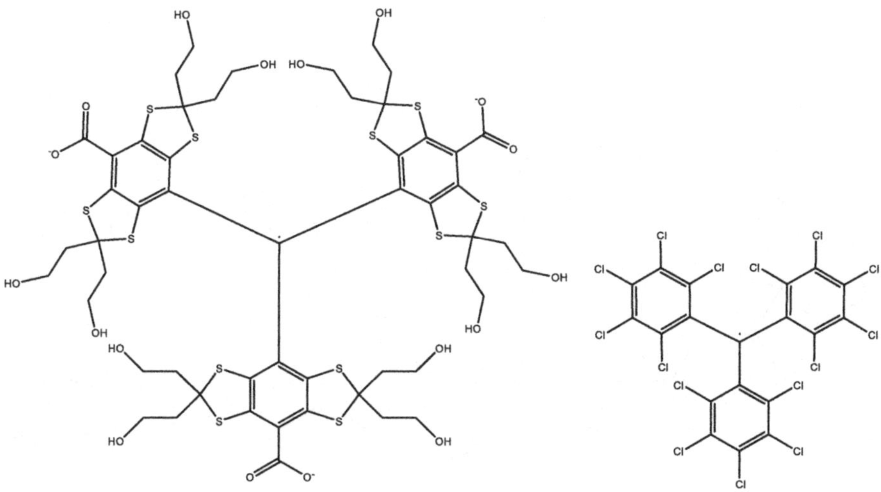

TAM OX063 PCM-TC

Fig. 12.10 Structure of trityl probes

EPR spectrum of $NO^{\cdot}$, which is recognized as an endothelial derived factor, can be directly obtained in whole blood samples frozen at 77°K without using spin trap agents since $NO^{\cdot}$ binds firmly to hemoglobin to form a stable species, $HbNO^{\cdot}$ [91]. Following 5-min period of leg ischemia in human studies, Wennmalm et al. reported increased EPR signals of $HbNO^{\cdot}$ in plasma samples collected within the 5–10 min of reperfusion [92].

12.5 Nanoparticle-Based Sensors

The use of spin traps for EPR detection of radicals is a well-established approach in applied chemistry [2]. However, the low trapping efficiency and the need for high concentrations of spin traps present significant limitations for their use in biological systems. To address this issue, Liu et al. were able to successfully link EMPO spin traps to gold NPs, therefore, exploiting the higher reaction rates of self-assembled monolayers of organic molecules, compared to individual molecules and increasing the reaction rates with FR [93]. The probe (Fig. 12.11) exhibited improved stability and increased molecular rate constants for reaction with $OH^{\bullet}$ compared to free EMPO. This approach of using spin traps combined with nanotechnology shows great promise for FR detection and OS assessment in biological systems [94].

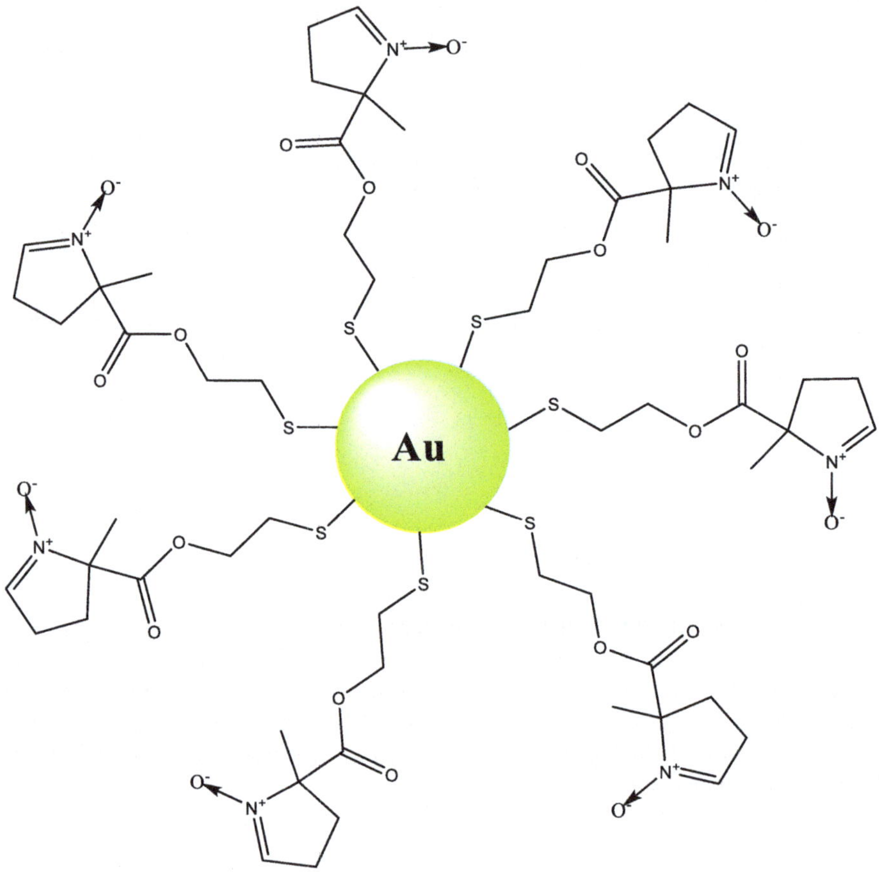

Fig. 12.11 EMPO-gold NPs for FR detection using EPR

12.6 Imaging

Significant advancements have been made in the development of highly sensitive probes, analysis modalities, data acquisition, and imaging systems. The availability of low-frequency EPR instrumentation can provide direct visualization of probe signals to FR production or O_2 consumption in whole animals [95, 96]. The approach involves the use of low-frequency, highly sensitive spectrometers, operating between 200 M MHz and 1.5 GHz and paramagnetic probes.

Since FR production is directly correlated with O_2 consumption, probes that respond to the pO_2 are very desirable like trityl [97], phthalocyanines [98, 99], and charcoal [100], the latter two are stable enough from being metabolized. In vivo imaging of $NO^•$ has also been achieved using iron-dithiocarbamate spin traps [101].

As mentioned earlier, hydroxylamines and trityls probes react with FR and show characteristic spectral behaviors resulting from signal formation or disappearance. In vivo mapping using EPR techniques with hydroxylamine probes, combined with tomographic technology, is an example of ingenious tools available for noninvasive detection and real-time monitoring of RS production and tissue localization as well as analyzing the effects of antioxidants on RS detoxification and prevention of oxidative damage [102, 103]. It also shows spatially resolved oxidative insults in the body [104, 105]. This is achieved by the ability of a biological system to reduce the introduced probe, which intensity is proportional to the body antioxidant activity. Therefore, the disappearance of the EPR signal can be used as an indicator of the antioxidant defense system.

Finally, commonly used spin trapping agents for RS imaging in biological materials are DMPO, PBN, DBNBS, and DIPPMPO. DBNBS is used for hydrogen peroxide sensing resulting in specific spectra, while DIPPMPO is used for mitochondrial production of $O_2^{\bullet-}$.

12.7 Application

The ability of the EPR to monitor different types of FR in different biological systems makes it a versatile tool for research. EPR spectroscopy coupled with spin trap agents has been used to measure radicals directly, in whole animals. In experimental studies, spin trap agents are injected before inducing OS conditions. Several examples can be found in the literature such as ischemia reperfusion [106], hyperoxia [107], septic shock [108], and carbon tetrachloride exposure [109]. Unfortunately, spin trap agents have some toxic properties and cannot, therefore, be injected safely to humans. This problem can be avoided by using a noninvasive spin-trapping approach in which ex vivo tissues or body fluids, for example, blood samples, are mixed with the spin-trap agent as soon as possible. For example, valuable data on FR production has been obtained after reperfusion during coronary angioplasty [110]. Another example is detection of carbon-centered and alkoxyl FR in coronary sinus blood from patients undergoing elective cardioplegia [111]. PBN has been employed to demonstrate FR overproduction by adding it to coronary sinus blood drawn during coronary bypass surgery [112, 113]. Using DMPO spin trap agent, Culcasi et al. found an increase in $DMPO\text{-}OH^{\cdot}$ formation in the coronary sinus blood drawn from patients having undergone cardio-pulmonary bypass during the first minutes following aortic declamping [114]. Pincemail et al. obtained first evidence of an increasing FR production in patients subjected to renal transplantation [115]. They recorded increased FR production within the first minutes after the graft reperfusion. Of course, highly reactive FR such as $OH^{\bullet}$, and highly reactive non-radical species such as $ONOO^{-}$, will simply not remain to be detected in ex vivo materials. Therefore, EPR of ex vivo materials to which traps have been added detects probably secondary radicals resulting from the interaction of RS with macromolecules. These secondary radicals include protein and lipid-derived radicals.

Using PBN, Davison et al. demonstrated high levels of FR in blood samples of patients with type 1 diabetes [116]. The hyperfine coupling constants recorded from ESR spectra suggested the presence of carbon- and oxygen-centered radicals in patients with diabetes. High levels of lipid hydroperoxides were also observed in diabetic patients, which was consistent with the observation of high levels of FR in these patients as measured by EPR [117].

Besides, EPR measurements can be used to detect FR in human tissue obtained ex vivo. For example, analysis of frozen samples taken from patients with pathological conditions such as fibroleiomyoma and carcinoma resulted in the detection of a signal, which was believed to be that of lipid peroxyl radicals in human uterine cervix [118]. DMPO and hydroxylamine probes (e.g. bis(1-hydroxy-2,2,6,6-tetramethyl-4-piperidinyl)decandioate) were used to measure FR in the skin and liver biopsies, respectively [88, 119].

Spin trapping has been employed to detect $O_2^{\bullet -}$ formed from XO [85], the mitochondrial electron transport chain (ETC) [120], and NAD(P)H oxidase (NOX) [121]. Nitrones have also been successfully used to detect $O_2^{\bullet -}$ generation in human neutrophils [122], human epithelial cells [123], and reperfused cardiac tissue [124], using ex vivo samples [48, 125].

EPR spectroscopy was also performed for the evaluation of the capacity of engineered NPs to produce FR in both acellular and cellular systems. DMPO has been successfully used for the assessment of TiO_2-induced OS in human lung cells [126]. A good overview of EPR spectra of TiO_2 with nitrone and nitroso spin traps was provided by Li et al. suggesting that combination of EPR spin label oximetry and immune spin trapping techniques are promising tools for studying oxidative damage caused by TiO_2 NPs [127].

As mentioned earlier, EPR spectroscopy can be used to monitor some antioxidant species in biological systems without the use of spin traps. Ascorbate, tocopherol, and other antioxidant radicals are relatively stable and can be directly detected by EPR at mild temperatures. It was first shown by using EPR that tocopheroxyl radicals are recycled by ascorbic acid [1]. Ascorbyl radical can be easily detected by EPR. This radical has been detected in skin after UV irradiation and was associated with myocardial ischemia [10, 128, 129]. Direct EPR analysis of ascorbyl radical was performed in patients undergoing aortic cross-clamping ischemia [130]. The results indicated a decrease in ascorbyl radical after aortic valve replacement.

The Laboratory of Kanagawa Dental College has used an in vitro EPR technique to evaluate antioxidative capacity and OS in an experimental animal brain model of lifestyle-related diseases in which OS causes stroke and hypertension. The laboratory has further applied this technique to evaluate antioxidative properties of a drug in brain [3, 131, 132]. In the future, assessment strategies using EPR technology may help further identify antioxidative properties of drugs, drinks, foods, and supplements, on the basis of which new products with excellent antioxidative properties would then be developed.

Overall, EPR can be applied in both laboratory and clinical settings where biological fluids derived from patients or whole experimental models can be directly analyzed for FR, antioxidants, and radical scavengers. It is important to note that spin trapping is highly dependent on the employed spin trap and the antioxidant status of biofluids. Also, one should keep in mind the half-lives of FR, the time and conditions for sampling and storage of samples, which play a crucial role in having precise and accurate results.

References

1. Han D, Loukianoff S, McLaughlin L. Part VII Chapter 17 – Oxidative stress indices: analytical aspects and significance. In: Sen CK, Packer L, Hänninen OOP, editors. Handbook of Oxidants and Antioxidants in Exercise. Amsterdam: Elsevier Science B.V.; 2000. p. 433–83.
2. Hawkins CL, Davies MJ. Detection and characterisation of radicals in biological materials using EPR methodology. Biochim Biophys Acta. 2014;1840(2):708-21.
3. Kobayashi K, Yoshino F, Takahashi SS, Todoki K, Maehata Y, Komatsu T, et al. Direct assessments of the antioxidant effects of propofol medium chain triglyceride/long chain triglyceride on the brain of stroke-prone spontaneously hypertensive rats using electron spin resonance spectroscopy. Anesthesiology. 2008;109(3):426-35.
4. Shulaev V, Oliver DJ. Metabolic and proteomic markers for oxidative stress. New tools for reactive oxygen species research. Plant Physiol. 2006;141(2):367-72.
5. Khoo JP, Alp NJ, Bendall JK, Kawashima S, Yokoyama M, Zhang Y-H, et al. EPR quantification of vascular nitric oxide production in genetically modified mouse models. Nitric oxide : biology and chemistry. 2004;10(3):156-61.
6. Yokoyama H, Itoh O, Aoyama M, Obara H, Ohya H, Kamada H. In vivo EPR imaging by using an acyl-protected hydroxylamine to analyze intracerebral oxidative stress in rats after epileptic seizures. Magnetic Resonance Imaging. 2000;18(7):875-9.
7. Menezes LB, Segat BB, Tolentino H, Pires DC, Mattos LMdM, Hottum HM, et al. ROS scavenging of SOD/CAT mimics probed by EPR and reduction of lipid peroxidation in S. cerevisiae and mouse liver, under severe hydroxyl radical stress condition. Journal of inorganic biochemistry. 2023;239:112062.
8. Emoto MC, Sato-Akaba H, Matsuoka Y, Yamada K-i, Fujii HG. Non-invasive mapping of glutathione levels in mouse brains by in vivo electron paramagnetic resonance (EPR) imaging: Applied to a kindling mouse model. Neuroscience Letters. 2019;690:6-10.
9. Laurindo FR, Pedro Mde A, Barbeiro HV, Pileggi F, Carvalho MH, Augusto O, et al. Vascular free radical release. Ex vivo and in vivo evidence for a flow-dependent endothelial mechanism. Circulation research. 1994;74(4):700-9.
10. Buettner GR, Jurkiewicz BA. Ascorbate free radical as a marker of oxidative stress: an EPR study. Free Radic Biol Med. 1993;14(1):49–55.
11. Lemineur T, Deby-Dupont G, Preiser JC. Biomarkers of oxidative stress in critically ill patients: what should be measured, when and how? Current opinion in clinical nutrition and metabolic care. 2006;9(6):704-10.
12. Muckenschnabel I, Goodman BA, Williamson B, Lyon GD, Deighton N. Infection of leaves of Arabidopsis thaliana by Botrytis cinerea: changes in ascorbic acid, free radicals and lipid peroxidation products. Journal of Experimental Botany. 2002;53(367):207-14.
13. Berliner L. The evolution of biomedical EPR (ESR). Biomedical Spectroscopy and Imaging. 2016;5:5-26.
14. Hideg E, Kálai T, Hideg K. Direct detection of free radicals and reactive oxygen species in thylakoids. Methods Mol Biol. 2011;684:187-200.
15. Haywood R. Spin-Trapping Theory and Applications. In: Roberts G, Watts A, editors. Encyclopedia of Biophysics. Berlin, Heidelberg: Springer Berlin Heidelberg; 2020. p. 1-8.

16. Bačić G, Pavićević A, Peyrot F. In vivo evaluation of different alterations of redox status by studying pharmacokinetics of nitroxides using magnetic resonance techniques. Redox Biol. 2016;8:226-42.
17. Lauricella R, Tuccio B. Detection and Characterisation of Free Radicals After Spin Trapping. In: Bertrand P, editor. Electron Paramagnetic Resonance Spectroscopy: Applications. Cham: Springer International Publishing; 2020. p. 51-82.
18. Sharma SK, Hamblin MR. The Use of Fluorescent Probes to Detect ROS in Photodynamic Therapy. In: Espada J, editor. Reactive Oxygen Species: Methods and Protocols. New York, NY: Springer US; 2021. p. 215-29.
19. Freinbichler W, Colivicchi MA, Stefanini C, Bianchi L, Ballini C, Misini B, et al. Highly reactive oxygen species: detection, formation, and possible functions. Cell Mol Life Sci. 2011;68(12):2067-79.
20. Apak R, Özyürek M, Güçlü K, Çapanoğlu E. Antioxidant Activity/Capacity Measurement. 1. Classification, Physicochemical Principles, Mechanisms, and Electron Transfer (ET)-Based Assays. Journal of Agricultural and Food Chemistry. 2016;64(5):997-1027.
21. Zhang Y, Dai M, Yuan Z. Methods for the detection of reactive oxygen species. Analytical Methods. 2018;10(38):4625–38.
22. Han D, Williams E, Cadenas E. Mitochondrial respiratory chain-dependent generation of superoxide anion and its release into the intermembrane space. The Biochemical journal. 2001;353(Pt 2):411-6.
23. Zhao H, Joseph J, Fales HM, Sokoloski EA, Levine RL, Vasquez-Vivar J, et al. Detection and characterization of the product of hydroethidine and intracellular superoxide by HPLC and limitations of fluorescence. Proc Natl Acad Sci U S A. 2005;102(16):5727–32.
24. Villamena FA, Zweier JL. Superoxide radical trapping and spin adduct decay of 5-tert-butoxycarbonyl-5-methyl-1-pyrroline N-oxide (BocMPO): kinetics and theoretical analysis. Journal of the Chemical Society, Perkin Transactions 2. 2002(7):1340–4.
25. Roubaud V, Sankarapandi S, Kuppusamy P, Tordo P, Zweier JL. Quantitative measurement of superoxide generation using the spin trap 5-(diethoxyphosphoryl)-5-methyl-1-pyrroline-N-oxide. Anal Biochem. 1997;247(2):404-11.
26. Dikalov S, Jiang J, Mason RP. Characterization of the high-resolution ESR spectra of superoxide radical adducts of 5-(diethoxyphosphoryl)-5-methyl-1-pyrroline N-oxide (DEPMPO) and 5,5-dimethyl-1-pyrroline N-oxide (DMPO). Analysis of conformational exchange. Free radical research. 2005;39(8):825-36.
27. Mason RP. In vitro and in vivo detection of free radical metabolites with electron spin resonance: Oxford University Press, Oxford; 1996.
28. Hardy M, Rockenbauer A, Vásquez-Vivar J, Felix C, Lopez M, Srinivasan S, et al. Detection, characterization, and decay kinetics of ROS and thiyl adducts of mito-DEPMPO spin trap. Chem Res Toxicol. 2007;20(7):1053-60.
29. Kurtoglu M, Lampidis TJ. From delocalized lipophilic cations to hypoxia: blocking tumor cell mitochondrial function leads to therapeutic gain with glycolytic inhibitors. Mol Nutr Food Res. 2009;53(1):68-75.
30. Keszler A, Kalyanaraman B, Hogg N. Comparative investigation of superoxide trapping by cyclic nitrone spin traps: the use of singular value decomposition and multiple linear regression analysis. Free Radic Biol Med. 2003;35(9):1149-57.
31. Khan N, Wilmot CM, Rosen GM, Demidenko E, Sun J, Joseph J, et al. Spin traps: in vitro toxicity and stability of radical adducts. Free Radic Biol Med. 2003;34(11):1473-81.
32. Zhao H, Joseph J, Zhang H, Karoui H, Kalyanaraman B. Synthesis and biochemical applications of a solid cyclic nitrone spin trap: a relatively superior trap for detecting superoxide anions and glutathiyl radicals. Free Radic Biol Med. 2001;31(5):599-606.
33. Karoui H, Clément J-L, Rockenbauer A, Siri D, Tordo P. Synthesis and structure of 5,5-diethoxycarbonyl-1-pyrroline N-oxide (DECPO). Application to superoxide radical trapping. Tetrahedron Letters. 2004;45(1):149-52.

34. Stolze K, Udilova N, Rosenau T, Hofinger A, Nohl H. Spin trapping of superoxide, alkyl- and lipid-derived radicals with derivatives of the spin trap EPPN. Biochemical pharmacology. 2003;66(9):1717-26.
35. Deletioglu V, Tuncay E, Toy A, Atalay M, Turan B. Immuno-spin trapping detection of antioxidant/pro-oxidant properties of zinc or selenium on DNA and protein radical formation via hydrogen peroxide. Molecular and cellular biochemistry. 2015;409(1):23-31.
36. Ramirez DC, Chen YR, Mason RP. Immunochemical detection of hemoglobin-derived radicals formed by reaction with hydrogen peroxide: involvement of a protein-tyrosyl radical. Free Radic Biol Med. 2003;34(7):830-9.
37. Miller EW, Albers AE, Pralle A, Isacoff EY, Chang CJ. Boronate-based fluorescent probes for imaging cellular hydrogen peroxide. J Am Chem Soc. 2005;127(47):16652–9.
38. Deterding LJ, Ramirez DC, Dubin JR, Mason RP, Tomer KB. Identification of free radicals on hemoglobin from its self-peroxidation using mass spectrometry and immuno-spin trapping: observation of a histidinyl radical. J Biol Chem. 2004;279(12):11600-7.
39. Gomez-Mejiba SE, Zhai Z, Akram H, Deterding LJ, Hensley K, Smith N, et al. Immuno-spin trapping of protein and DNA radicals: "tagging" free radicals to locate and understand the redox process. Free Radic Biol Med. 2009;46(7):853-65.
40. Chen YR. EPR spin-trapping and nano LC MS/MS techniques for DEPMPO/OOH and immunospin-trapping with anti-DMPO antibody in mitochondrial electron transfer system. Methods Mol Biol. 2008;477:75-88.
41. Pan R, Liu KJ, Qi Z. Zinc causes the death of hypoxic astrocytes by inducing ROS production through mitochondria dysfunction. Biophysics Reports. 2019;5(4):209-17.
42. Makino K, Hagiwara T, Hagi A, Nishi M, Murakami A. Cautionary note for DMPO spin trapping in the presence of iron ion. Biochemical and Biophysical Research Communications. 1990;172(3):1073-80.
43. Bottle SE, Hanson GR, Micallef AS. Application of the new EPR spin trap 1,1,3-trimethylisoindole N-oxide (TMINO) in trapping HO. and related biologically important radicals. Org Biomol Chem. 2003;1(14):2585-9.
44. Frejaville C, Karoui H, Tuccio B, Le Moigne F, Culcasi M, Pietri S, et al. 5-(Diethoxyphosphoryl)-5-methyl-1-pyrroline N-oxide: a new efficient phosphorylated nitrone for the in vitro and in vivo spin trapping of oxygen-centered radicals. Journal of medicinal chemistry. 1995;38(2):258-65.
45. Vasquez-Vivar J, Kalyanaraman B, Kennedy MC. Mitochondrial aconitase is a source of hydroxyl radical. An electron spin resonance investigation. J Biol Chem. 2000;275(19): 14064-9.
46. Dambrova M, Baumane L, Kalvinsh I, Wikberg JE. Improved method for EPR detection of DEPMPO-superoxide radicals by liquid nitrogen freezing. Biochem Biophys Res Commun. 2000;275(3):895-8.
47. Villamena F. Chemistry of Reactive Species. 2013. p. 1–48.
48. Jiang JJ, Liu KJ, Jordan SJ, Swartz HM, Mason RP. Detection of free radical metabolite formation using in vivo EPR spectroscopy: evidence of rat hemoglobin thiyl radical formation following administration of phenylhydrazine. Arch Biochem Biophys. 1996;330(2):266-70.
49. Sentjurc M, Mason RP. Inhibition of radical adduct reduction and reoxidation of the corresponding hydroxylamines in in vivo spin trapping of carbon tetrachloride-derived radicals. Free Radic Biol Med. 1992;13(2):151-60.
50. Chamulitrat W, Jordan SJ, Mason RP. Fatty acid radical formation in rats administered oxidized fatty acids: in vivo spin trapping investigation. Arch Biochem Biophys. 1992;299 (2):361-7.
51. Baker JE, Froncisz W, Joseph J, Kalyanaraman B. Spin label oximetry to assess extracellular oxygen during myocardial ischemia. Free Radic Biol Med. 1997;22(1–2):109–15.
52. Velan SS, Spencer RG, Zweier JL, Kuppusamy P. Electron paramagnetic resonance oxygen mapping (EPROM): direct visualization of oxygen concentration in tissue. Magn Reson Med. 2000;43(6):804–9.

53. Borbat PP, Costa-Filho AJ, Earle KA, Moscicki JK, Freed JH. Electron spin resonance in studies of membranes and proteins. Science. 2001;291(5502):266–9.
54. Goldstein S, Russo A, Samuni A. Reactions of PTIO and carboxy-PTIO with *NO, *NO2, and O2-*. J Biol Chem. 2003;278(51):50949–55.
55. Samuni U, Samuni Y, Goldstein S. On the distinction between nitroxyl and nitric oxide using nitronyl nitroxides. J Am Chem Soc. 2010;132(24):8428–32.
56. Henry Y, Lepoivre M, Drapier JC, Ducrocq C, Boucher JL, Guissani A. EPR characterization of molecular targets for NO in mammalian cells and organelles. FASEB journal : official publication of the Federation of American Societies for Experimental Biology. 1993;7(12): 1124–34.
57. Lancaster JR, Jr., Langrehr JM, Bergonia HA, Murase N, Simmons RL, Hoffman RA. EPR detection of heme and nonheme iron-containing protein nitrosylation by nitric oxide during rejection of rat heart allograft. J Biol Chem. 1992;267(16):10994–8.
58. Glover RE, Ivy ED, Orringer EP, Maeda H, Mason RP. Detection of nitrosyl hemoglobin in venous blood in the treatment of sickle cell anemia with hydroxyurea. Mol Pharmacol. 1999;55(6):1006–10.
59. Kalyanaraman B. Detection of nitric oxide by electron spin resonance in chemical, photochemical, cellular, physiological, and pathophysiological systems. Methods in enzymology. 1996;268:168–87.
60. Mordvintcev P, Mülsch A, Busse R, Vanin A. On-line detection of nitric oxide formation in liquid aqueous phase by electron paramagnetic resonance spectroscopy. Anal Biochem. 1991;199(1):142–6.
61. Vanin AF, Huisman A, van Faassen EE. Iron dithiocarbamate as spin trap for nitric oxide detection: pitfalls and successes. Methods in enzymology. 2002;359:27–42.
62. Vanin AF. Iron diethyldithiocarbamate as spin trap for nitric oxide detection. Methods in enzymology. 1999;301:269–79.
63. Maia LB, Moura JJG. Detection of Nitric Oxide by Electron Paramagnetic Resonance Spectroscopy: Spin-Trapping with Iron-Dithiocarbamates. In: Gupta KJ, editor. Plant Nitric Oxide: Methods and Protocols. New York, NY: Springer New York; 2016. p. 81–102.
64. Tsuchiya K, Takasugi M, Minakuchi K, Fukuzawa K. Sensitive quantitation of nitric oxide by EPR spectroscopy. Free Radical Biology and Medicine. 1996;21(5):733–7.
65. Kleschyov AL, Münzel T. Advanced spin trapping of vascular nitric oxide using colloid iron diethyldithiocarbamate. Methods in enzymology. 2002;359:42–51.
66. Kleschyov AL, Mollnau H, Oelze M, Meinertz T, Huang Y, Harrison DG, et al. Spin Trapping of Vascular Nitric Oxide Using Colloid Fe(II)-Diethyldithiocarbamate. Biochemical and Biophysical Research Communications. 2000;275(2):672–7.
67. Hogg N. Detection of nitric oxide by electron paramagnetic resonance spectroscopy. Free Radic Biol Med. 2010;49(2):122–9.
68. Zulfugarov I, Tovuu A, Kim J-H. Detection of Reactive Oxygen Species in Higher Plants. Journal of Plant Biology. 2011;54:351–7.
69. Halliwell B, Gutteridge JM. Free radicals in biology and medicine: Oxford University Press, USA; 2015.
70. Li AS, Chignell CF. The NoH value in EPR spin trapping: a new parameter for the identification of 5,5-dimethyl-1-pyrroline-N-oxide spin adducts. J Biochem Biophys Methods. 1991;22(1):83–7.
71. Dikalov SI, Harrison DG. Methods for detection of mitochondrial and cellular reactive oxygen species. Antioxid Redox Signal. 2014;20(2):372–82.
72. Dikalov S, Fink B, Skatchkov M, Bassenge E. Comparison of glyceryl trinitrate-induced with pentaerythrityl tetranitrate-induced in vivo formation of superoxide radicals: effect of vitamin C. Free Radic Biol Med. 1999;27(1–2):170–6.
73. Dikalov S, Skatchkov M, Fink B, Bassenge E. Quantification of superoxide radicals and peroxynitrite in vascular cells using oxidation of sterically hindered hydroxylamines and electron spin resonance. Nitric oxide : biology and chemistry. 1997;1(5):423–31.

74. Dikalov SI, Dikalova AE, Mason RP. Noninvasive diagnostic tool for inflammation-induced oxidative stress using electron spin resonance spectroscopy and an extracellular cyclic hydroxylamine. Arch Biochem Biophys. 2002;402(2):218–26.
75. Kuznetsov AV, Kehrer I, Kozlov AV, Haller M, Redl H, Hermann M, et al. Mitochondrial ROS production under cellular stress: comparison of different detection methods. Analytical and bioanalytical chemistry. 2011;400(8):2383–90.
76. Dikalov SI, Kirilyuk IA, Voinov M, Grigor'ev IA. EPR detection of cellular and mitochondrial superoxide using cyclic hydroxylamines. Free radical research. 2011;45(4):417–30.
77. Dikalov S, Grigor'ev IA, Voinov M, Bassenge E. Detection of Superoxide Radicals and Peroxynitrite by 1-Hydroxy-4-phosphonooxy-2,2,6,6-tetramethylpiperidine: Quantification of Extracellular Superoxide Radicals Formation. Biochemical and Biophysical Research Communications. 1998;248(2):211–5.
78. Villamena FA, Das A, Nash KM. Potential implication of the chemical properties and bioactivity of nitrone spin traps for therapeutics. Future Med Chem. 2012;4(9):1171–207.
79. Fernandes DC, Wosniak J, Jr., Pescatore LA, Bertoline MA, Liberman M, Laurindo FR, et al. Analysis of DHE-derived oxidation products by HPLC in the assessment of superoxide production and NADPH oxidase activity in vascular systems. American journal of physiology Cell physiology. 2007;292(1):C413–22.
80. Marchesi E, Rota C, Fann YC, Chignell CF, Mason RP. Photoreduction of the fluorescent dye 2'-7'-dichlorofluorescein: a spin trapping and direct electron spin resonance study with implications for oxidative stress measurements. Free Radic Biol Med. 1999;26(1–2):148–61.
81. Dikalov S, Griendling KK, Harrison DG. Measurement of reactive oxygen species in cardiovascular studies. Hypertension. 2007;49(4):717–27.
82. Kuzkaya N, Weissmann N, Harrison DG, Dikalov S. Interactions of peroxynitrite with uric acid in the presence of ascorbate and thiols: implications for uncoupling endothelial nitric oxide synthase. Biochemical pharmacology. 2005;70(3):343–54.
83. Doughan AK, Dikalov SI. Mitochondrial redox cycling of mitoquinone leads to superoxide production and cellular apoptosis. Antioxid Redox Signal. 2007;9(11):1825–36.
84. Dikalov SI, Dikalova AE, Bikineyeva AT, Schmidt HH, Harrison DG, Griendling KK. Distinct roles of Nox1 and Nox4 in basal and angiotensin II-stimulated superoxide and hydrogen peroxide production. Free Radic Biol Med. 2008;45(9):1340–51.
85. Kutala VK, Villamena FA, Ilangovan G, Maspoch D, Roques N, Veciana J, et al. Reactivity of superoxide anion radical with a perchlorotriphenylmethyl (trityl) radical. J Phys Chem B. 2008;112(1):158–67.
86. Rizzi C, Samouilov A, Kutala VK, Parinandi NL, Zweier JL, Kuppusamy P. Application of a trityl-based radical probe for measuring superoxide. Free Radic Biol Med. 2003;35(12): 1608–18.
87. Jurkiewicz BA, Buettner GR. EPR detection of free radicals in UV-irradiated skin: mouse versus human. Photochem Photobiol. 1996;64(6):918–22.
88. Haywood RM, Wardman P, Gault DT, Linge C. Ruby laser irradiation (694 nm) of human skin biopsies: assessment by electron spin resonance spectroscopy of free radical production and oxidative stress during laser depilation. Photochem Photobiol. 1999;70(3):348–52.
89. Kobayashi K, Harada Y, Hayashi K. Kinetic behavior of the monodehydroascorbate radical studied by pulse radiolysis. Biochemistry. 1991;30(34):8310–5.
90. May JM, Qu Z, Cobb CE. Recycling of the ascorbate free radical by human erythrocyte membranes. Free Radic Biol Med. 2001;31(1):117–24.
91. Kruszyna H, Kruszyna R, Smith RP, Wilcox DE. Red blood cells generate nitric oxide from directly acting, nitrogenous vasodilators. Toxicol Appl Pharmacol. 1987;91(3):429–38.
92. Wennmalm Å, Lanne B, Petersson A-S. Detection of endothelial-derived relaxing factor in human plasma in the basal state and following ischemia using electron paramagnetic resonance spectrometry. Analytical Biochemistry. 1990;187(2):359–63.
93. Du L, Huang S, Zhuang Q, Jia H, Rockenbauer A, Liu Y, et al. Highly sensitive free radical detection by nitrone-functionalized gold nanoparticles. Nanoscale. 2014;6(3):1646–52.

94. Jeong MS, Yu K-N, Chung HH, Park SJ, Lee AY, Song MR, et al. Methodological considerations of electron spin resonance spin trapping techniques for measuring reactive oxygen species generated from metal oxide nanomaterials. Scientific Reports. 2016;6(1):26347.
95. Kuppusamy P, Zweier JL. Cardiac applications of EPR imaging. NMR Biomed. 2004;17(5): 226–39.
96. Jackson SK, Thomas MP, Smith S, Madhani M, Rogers SC, James PE. In vivo EPR spectroscopy: biomedical and potential diagnostic applications. Faraday Discuss. 2004;126: 103–17; discussion 69–83.
97. Liu Y, Villamena FA, Sun J, Wang TY, Zweier JL. Esterified trityl radicals as intracellular oxygen probes. Free Radic Biol Med. 2009;46(7):876–83.
98. Pandian RP, Chacko SM, Kuppusamy ML, Rivera BK, Kuppusamy P. Evaluation of lithium naphthalocyanine (LiNc) microcrystals for biological EPR oximetry. Advances in experimental medicine and biology. 2011;701:29–36.
99. Khan N, Williams BB, Hou H, Li H, Swartz HM. Repetitive tissue pO2 measurements by electron paramagnetic resonance oximetry: current status and future potential for experimental and clinical studies. Antioxid Redox Signal. 2007;9(8):1169–82.
100. He G, Shankar RA, Chzhan M, Samouilov A, Kuppusamy P, Zweier JL. Noninvasive measurement of anatomic structure and intraluminal oxygenation in the gastrointestinal tract of living mice with spatial and spectral EPR imaging. Proc Natl Acad Sci U S A. 1999;96(8): 4586–91.
101. Berliner LJ, Fujii H. In vivo spin trapping of nitric oxide. Antioxid Redox Signal. 2004;6(3): 649–56.
102. Utsumi H, Yamada K. In vivo electron spin resonance-computed tomography/nitroxyl probe technique for non-invasive analysis of oxidative injuries. Arch Biochem Biophys. 2003;416 (1):1–8.
103. Utsumi H, Yamada K, Ichikawa K, Sakai K, Kinoshita Y, Matsumoto S, et al. Simultaneous molecular imaging of redox reactions monitored by Overhauser-enhanced MRI with 14N- and 15N-labeled nitroxyl radicals. Proc Natl Acad Sci U S A. 2006;103(5):1463–8.
104. Kopáni M, Celec P, Danisovic L, Michalka P, Biró C. Oxidative stress and electron spin resonance. Clin Chim Acta. 2006;364(1–2):61–6.
105. Utsumi H, Yasukawa K, Soeda T, Yamada K, Shigemi R, Yao T, et al. Noninvasive mapping of reactive oxygen species by in vivo electron spin resonance spectroscopy in indomethacin-induced gastric ulcers in rats. The Journal of pharmacology and experimental therapeutics. 2006;317(1):228–35.
106. Connor HD, Gao W, Mason RP, Thurman RG. New reactive oxidizing species causes formation of carbon-centered radical adducts in organic extracts of blood following liver transplantation. Free Radic Biol Med. 1994;16(6):871–5.
107. Torbati D, Church DF, Keller JM, Pryor WA. Free radical generation in the brain precedes hyperbaric oxygen-induced convulsions. Free Radic Biol Med. 1992;13(2):101–6.
108. Lloyd SS, Chang AK, Taylor FB, Janzen EG, McCay PB. Free radicals and septic shock in primates: The role of tumor necrosis factor. Free Radical Biology and Medicine. 1993;14(3): 233–42.
109. McCay PB, Lai EK, Poyer JL, DuBose CM, Janzen EG. Oxygen- and carbon-centered free radical formation during carbon tetrachloride metabolism. Observation of lipid radicals in vivo and in vitro. J Biol Chem. 1984;259(4):2135–43.
110. Coghlan JG, Flitter WD, Holley AE, Norell M, Mitchell AG, Ilsley CD, et al. Detection of free radicals and cholesterol hydroperoxides in blood taken from the coronary sinus of man during percutaneous transluminal coronary angioplasty. Free radical research communications. 1991;14(5–6):409–17.
111. Tortolani AJ, Powell SR, Misík V, Weglicki WB, Pogo GJ, Kramer JH. Detection of alkoxyl and carbon-centered free radicals in coronary sinus blood from patients undergoing elective cardioplegia. Free Radic Biol Med. 1993;14(4):421–6.

112. Tarkka MR, Vuolle M, Kaukinen S, Holm P, Eloranta J, Kaukinen U, et al. Effect of allopurinol on myocardial oxygen free radical production in coronary bypass surgery. Scand Cardiovasc J. 2000;34(6):593–6.
113. Clermont G, Vergely C, Jazayeri S, Lahet JJ, Goudeau JJ, Lecour S, et al. Systemic free radical activation is a major event involved in myocardial oxidative stress related to cardiopulmonary bypass. Anesthesiology. 2002;96(1):80–7.
114. Culcasi M, Pietri S, Carriere I, d'Arbigny P, Drieu K. Electron-spin-resonance study of the protective effects of Ginkgo biloba extract (EGb 761) on reperfusion-induced free-radical generation associated with plasma ascorbate consumption during open-heart surgery in man. Advances in Ginkgo Biloba Extract Research. 1993;2:153–62.
115. Pincemail J, Defraigne JO, Franssen C, Bonnet P, Deby-Dupont G, Pirenne J, et al. Evidence for free radical formation during human kidney transplantation. Free Radical Biology and Medicine. 1993;15(3):343–8.
116. Davison GW, George L, Jackson SK, Young IS, Davies B, Bailey DM, et al. Exercise, free radicals, and lipid peroxidation in type 1 diabetes mellitus. Free Radical Biology and Medicine. 2002;33(11):1543–51.
117. Dasgupta A, Klein K. Methods for Measuring Oxidative Stress in the Laboratory. 2014. p. 19–40.
118. Benedetto C, Bocci A, Dianzani MU, Ghiringhello B, Slater TF, Tomasi A, et al. Electron Spin Resonance Studies on Normal Human Uterus and Cervix and on Benign and Malignant Uterine Tumors1. Cancer research. 1981;41(7):2936–42.
119. Valgimigli M, Valgimigli L, Trerè D, Gaiani S, Pedulli GF, Gramantieri L, et al. Oxidative stress EPR measurement in human liver by radical-probe technique. Correlation with etiology, histology and cell proliferation. Free radical research. 2002;36(9):939–48.
120. Du G, Mouithys-Mickalad A, Sluse FE. Generation of superoxide anion by mitochondria and impairment of their functions during anoxia and reoxygenation in vitro. Free Radic Biol Med. 1998;25(9):1066–74.
121. Bannister JV, Bellavite P, Serra MC, Thornalley PJ, Rossi F. An EPR study of the production of superoxide radicals by neutrophil NADPH oxidase. FEBS Lett. 1982;145(2):323–6.
122. Britigan BE, Cohen MS, Rosen GM. Detection of the production of oxygen-centered free radicals by human neutrophils using spin trapping techniques: a critical perspective. J Leukoc Biol. 1987;41(4):349–62.
123. Shi H, Timmins G, Monske M, Burdick A, Kalyanaraman B, Liu Y, et al. Evaluation of spin trapping agents and trapping conditions for detection of cell-generated reactive oxygen species. Arch Biochem Biophys. 2005;437(1):59–68.
124. Hirata H, He G, Deng Y, Salikhov I, Petryakov S, Zweier JL. A loop resonator for slice-selective in vivo EPR imaging in rats. J Magn Reson. 2008;190(1):124–34.
125. Kadiiska MB, Burkitt MJ, Xiang QH, Mason RP. Iron supplementation generates hydroxyl radical in vivo. An ESR spin-trapping investigation. J Clin Invest. 1995;96(3):1653–7.
126. Bhattacharya K, Davoren M, Boertz J, Schins RP, Hoffmann E, Dopp E. Titanium dioxide nanoparticles induce oxidative stress and DNA-adduct formation but not DNA-breakage in human lung cells. Particle and fibre toxicology. 2009;6:1–11.
127. Li M, Yin J-J, Wamer WG, Lo YM. Mechanistic characterization of titanium dioxide nanoparticle-induced toxicity using electron spin resonance. Journal of Food and Drug Analysis. 2014;22(1):76–85.
128. Mori A, Wang X, Liu J. Electron spin resonance assay of ascorbate free radicals in vivo. Methods in enzymology. 1994;233:149–54.
129. Pietri S, Culcasi M, Stella L, Cozzone PJ. Ascorbyl free radical as a reliable indicator of free-radical-mediated myocardial ischemic and post-ischemic injury: A real-time continuous-flow ESR study. European journal of biochemistry. 1990;193(3):845–54.
130. Pietri S, Séguin JR, d'Arbigny PD, Culcasi M. Ascorbyl free radical: a noninvasive marker of oxidative stress in human open-heart surgery. Free Radic Biol Med. 1994;16(4):523–8.

131. Lee MC, Shoji H, Miyazaki H, Yoshino F, Hori N, Toyoda M, et al. Assessment of oxidative stress in the spontaneously hypertensive rat brain using electron spin resonance (ESR) imaging and in vivo L-Band ESR. Hypertens Res. 2004;27(7):485–92.
132. Lee MC, Shoji H, Miyazaki H, Yoshino F, Hori N, Miyake S, et al. Measurement of oxidative stress in the rodent brain using computerized electron spin resonance tomography. Magn Reson Med Sci. 2003;2(2):79–84.

Chapter 13
Electrochemical Methods and Sensors

E. DeVoe, T. Bechard, A. S. Deshpande, and S. Andreescu

Abstract With the increasing evidence of the involvement of oxidative stress in many diseases, interest in developing methods that can measure oxidative stress biomarkers in real time has continuously increased. Electrochemical methods and sensors are ideally suited for monitoring ROS and RNS with very high sensitivity and real-time resolution, crucial for quantifying these short-lived reactive species (RS) *in situ*. This chapter describes the development status of electrochemical methods, sensors, and biosensors for *in vivo/in vitro* monitoring of ROS/RNS and provides examples of measurement capabilities in pathophysiological conditions associated with oxidative stress for disease monitoring, pharmacological screening, and other relevant biomedical applications. Recently reported microelectrode designs and methodologies based on chemical and enzyme-based approaches as well as their advantages and limitations are discussed. Examples of possible implementation and translation of this technology to evaluate oxidative stress in cellular, tissues, and whole animal models are also provided to demonstrate potential of this technology in biology and biomedical fields.

Keywords Oxidative stress · Electrochemical sensors · Biosensors · ROS/RNS · Measurements · Microelectrodes

13.1 Introduction

In this chapter, we describe the principle, performance, and application of electrochemical methods and sensors for the detection of oxidative stress biomarkers, focusing on ROS and RNS, formed by reactions involving oxygen or nitrogen species. These include H_2O_2, $O_2^{\bullet -}$, $OH^{\bullet}$, $ROO^{\bullet}$, $HOCl/OCl^-$, and $^1O_2$ [1, 2], as examples of ROS, as well as $NO^{\bullet}$ and $ONOO^-$ as representative RNS. These species are produced during cellular processes such as oxidative phosphorylation,

E. DeVoe · T. Bechard · A. S. Deshpande · S. Andreescu (✉)
Department of Chemistry and Biomolecular Science, Clarkson University, Potsdam, NY, USA
e-mail: devoeel@clarkson.edu; becharta@clarkson.edu; deshpaas@clarkson.edu; eandrees@clarkson.edu

S. Andreescu et al. (eds.), *Biomarkers of Oxidative Stress*,
https://doi.org/10.1007/978-3-031-60738-7_13

catabolism, and phagocytosis and have different lifetime, bioactivity, and cellular function [3, 4]. ROS and/or RNS are responsible for the overall oxidative load in biological systems [2, 4]. In normal physiological conditions, enzymatic (e.g., SOD, CAT, and GPx) and nonenzymatic (e.g., glutathione, Trx, and ascorbate) antioxidants regulate oxidative levels and prevent oxidative overload [5, 6]. ROS is known to play a critical role in redox signaling, allowing cells to adapt to environmental changes [3, 5, 7]. However, when these species are present in excess, they can damage cells, proteins, and DNA and can lead to an overall state of OS [8]. This redox imbalance contributes to the development and progression of a myriad of diseases [9, 10] and if not corrected could result in severe conditions, including organ failure [11–13]. Because ROS/RNS are highly reactive and short-lived, with lifetimes as low as in the nanoseconds, their direct measurement in cells, tissues, and organisms is extremely difficult. Therefore, in order to fully understand the contribution of ROS/RNS and the relationship between OS in physiological and pathological conditions [14, 15], there is a need to develop analytical methods and tools that can monitor ROS/RNS in real time and in situ and investigate disease mechanisms and pharmacological efficacy. Their accurate measurements in situ could contribute to the understanding of their physiological functions.

Conventional methods to measure ROS/RNS include electron paramagnetic resonance and photon emission spectroscopy [16, 17] and fluorescence spectroscopy with ROS/RNS-specific dyes [18]. These methods have limited ability to measure these species in biological tissues and do not provide accurate quantification of the level of these species in biological samples. Due to the high sensitivity and the ability to fabricate small probes that can be inserted in tissues and be used in cellular cultures, electrochemistry provides unique opportunities to detect ROS/RNS release in real time and with high spatial resolution [19–23]. Many different configurations of electrochemical sensors and detection methods have been reported in literature, with the most common examples for the detection of species such as H_2O_2, $NO^{\bullet}$ and $ONOO^-$ and $O_2^{\bullet-}$ [24]. This chapter describes the functioning and provides several examples of electrochemical microbiosensors for ROS/RNS measurements in biological systems. We also provide a discussion of the remaining challenges and translational aspects of this technology for biology and biomedical applications.

13.2 Electrochemical Properties of ROS/RNS

ROS/RNS have characteristic redox potentials enabling measurement by electrochemical means. For example, the reduction potential of $O_2^{\bullet-}$ to H_2O_2 is ~0.94V, while that of $NO^{\bullet}$ as a representative RNS is ~ +0.8V depending on the electrode and electrolyte conditions. Table 13.1 provides representative examples of redox reactions and oxidation potentials of common RS.

Nitric oxide, as a representative RNS, is a diffusible gaseous radical with a 1–10 s biological half-life playing an essential physiological regulator in immunology and cell signaling [36, 37]. Depending on the pH and the environment, $NO^{\bullet}$ can be found

Table 13.1 Electrode reactions and standard potentials for powerful oxidants formed in aqueous sulfate and chloride solutions [25]

Oxidant	Reaction	E°, V vs. NHE	Ref.
Ozonide radical anion	$O_3{}^{\bullet -}{}_{(aq)} + 6H^+ + 5e^- = 3H_2O$	1.575	[26][a]
Hydroxyl radical	$OH^{\bullet}{}_{(g)} + H^+ + e^- = H_2O$ $OH^{\bullet}{}_{(aq)} + H^+ + e^- = H_2O$ $O^{\bullet -}{}_{(aq)} + 2H^+ + e^- = H_2O$	2.82 2.722 3.426	[27, 28][g] [26, 29, 30][b] [26, 31][c]
Atomic oxygen	$O_{(g)} + 2H^+ + e^- = H_2O$	2.421	[29]
Ozone	$O_{3(g)} + 2H^+ + 2e^- = O_2 + H_2O$ $O_{3(g)} + 6H^+ + 6e^- = 3H_2O$	2.076 1.511	[29]
Hydrogen peroxyl radical	$HO_2{}^{\bullet} + 3H^+ + 3e^- = 2H_2O$	1.684	[29][h]
Hydrogen peroxide	$H_2O_2 + 2H^+ + 2e^- = 2H_2O$ $HO_2{}^- + 3H^+ + 2e^- = 2H_2O$	1.776 2.119	[29]
Oxygen	$O_2 + 4H^+ + 4e^- = 2H_2O$	1.228	[29]
Atomic chlorine	$Cl^{\bullet} + e^- = Cl^-$	2.410	[30, 32] [33, 34][d]
Dichlorine radical anion	$Cl_2{}^{\bullet -} + e^- = 2Cl^-$	2.090	[32]
Hydroxyl chloride radical anion	$ClOH^{\bullet -} + H^+ + e^- = Cl^- + H_2O$ $ClOH^{\bullet -} + e^- = Cl^- + OH^-$	2.729 1.90	[32][e]
Chlorine monoxide	$Cl_2O + 2H^+ + 2e^- = 2Cl^- + H_2O$	2.153	[29]
Chlorine dioxide	$ClO_2 + 4H^+ + 5e^- = Cl^- + 2H_2O$	1.511	[29]
Hypochlorous acid	$HOCl + H^+ + e^- = Cl^- + H_2O$ $OCl^- + 2H^+ + e^- = Cl^- + H_2O$	1.494 1.715	[29]
Chlorine	$Cl_2 + 2e^- = 2Cl^-$	1.395	[29]
Persulfate radical	$HSO_4{}^{\bullet} + e^- = HSO_4{}^-$	2.60	[26, 33]
Peroxysulfate	$S_2O_8{}^{-2} + 2e^- = 2SO_4{}^{-2}$	1.940	[33][f] [33][f] [26]

[a]On the basis of $\Delta G^0{}_f$ ($O_3{}^{\bullet -}$ (aq)) = 11.25 obtained from $E^0(O_2/O_3{}^{\bullet -})$ [34]
[b]Using $\Delta G(OH^{\bullet}{}_{aq})$ = +6.0 [30]; Koppenol-Liebman [34] obtained 2.59 V; calculated from Pourbaix data [29]: 2.832 V
[c]On the basis of pK of $OH^{\bullet}$: pK = 11.9 [31]; E^0 = 3.308 calculated from [34]
[d]Berdnikov and Bazhin [27] obtained 2.55 V; Gratzel: 2.55 V [33]
[e]Using $\Delta G(ClOH^{\bullet -}{}_{aq}) = -25.18$ kcal/mol obtained from data of Ref. [32]
[f]2.01 V given by Pourbaix [29]
[g]From [28, 35]
[h]From [29]

in equilibrium with $NO_2{}^-$ and dinitrogen oxide (N_2O) [38], and under oxygen conditions can form highly reactive peroxynitrate, $ONOO^-$, which is known to react with tyrosine residues in protein phosphorylation [39]. Superoxide is a by-product of mitochondrial respiration [40] and known to be involved in redox regulatory signaling [41, 42]. Superoxide can induce formation of more toxic radicals such as $OH^{\bullet}$, that can damage cells, DNA, and intracellular molecules, following mechanism as shown in Eq. 2.29.

It is well known that SOD converts $O_2^{\bullet -}$ to H_2O_2 [43, 44] which can be assayed to estimate the amount and rate of production of $O_2^{\bullet -}$. H_2O_2 can also result from the degradation of adenosine monophosphate (AMP), which is converted to hypoxanthine and then further oxidized to xanthine and uric acid by an enzyme XO, as shown in Eq. 13.1. H_2O_2 is also present in organelles, peroxisomes, during the reduction of dioxygen with fatty acids, polyamines, and other compounds, catalyzed by flavin adenine dinucleotide (FAD).

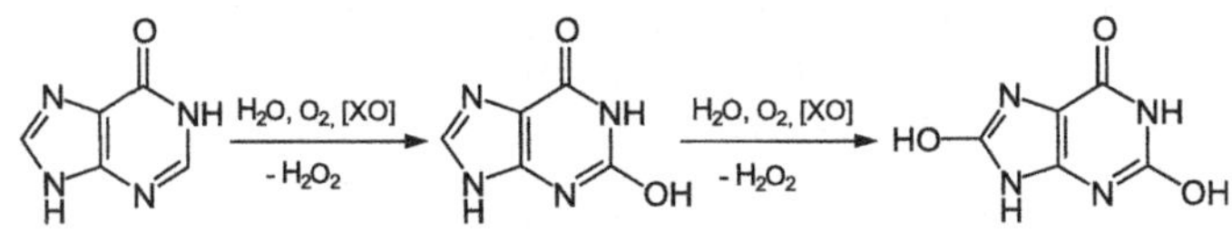

(13.1)

Hypoxanthine xanthine uric acid

13.3 Electrochemical Sensors and Biosensors for ROS/RNS Detection

Electrochemistry is a valuable tool for the quantification of ROS/RNS species in biological settings. Using small size microelectrodes, specially designed probes coated with ROS/RNS-specific receptors and coatings can be used to measure RS directly in biological media, cell populations, be inserted in tissues, measure single cell activity, and ultimately measure inside live cells [45]. ROS/RNS detection in biological models has been reviewed [19, 46, 47], but only a few reported studies demonstrate measurements in vivo due to issues such as short lifetime, calibration drift, cross-reactivity, and interferences.

13.3.1 Materials and Fabrication of Electrochemical (Bio)sensors

While many studies demonstrate usefulness of electrochemical measurements of ROS/RNS, in most cases, the probes used have micro-size dimensions such as glassy carbon electrodes (GCEs) which have restricted usability for cells or live tissues work. Therefore, for these to be used in these environments, electrode probes need to be designed to match the biological model of study to minimize perturbation of the environment. Parameters such as electrode size, coatings, and morphology play an important role. Most microelectrodes are fabricated from carbon fiber microelectrodes (CFMEs) using a capillary puller. These CFMEs have sizes ranging from

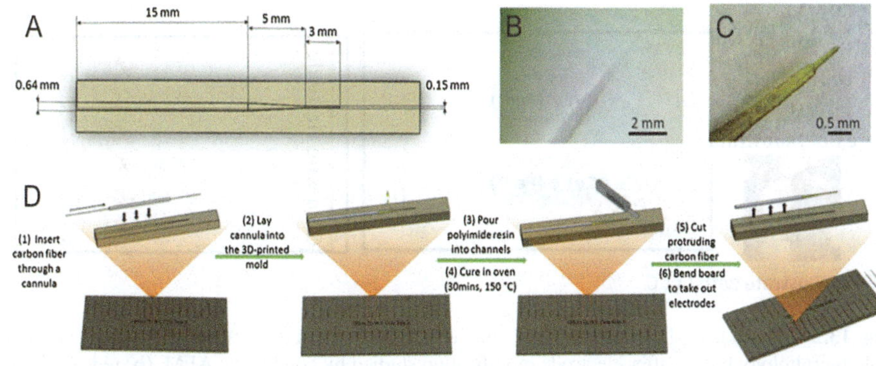

Fig. 13.1 Example of fabrication process of CFMEs using 3D printed mold on polyimide: (**a**) Close-up of the design of one channel for 3D printed microelectrode fabrication. (**b**) Image of the tip from a 3D printed mold, scale bar: 2 mm. (**c**) Image of a polyimide-insulated carbon fiber microelectrode fabricated using the mold, scale bar: 0.5 mm. (**d**) Scheme of the process of microelectrode fabrication. (Reprinted with permission from Ref. [53])

5 to 10 μm. Gold or platinum wire microelectrodes with a diameter of ~100 μm are also frequently used, particularly for measurements of $O_2^{\bullet-}$ that involve bioimmobilization of biological receptors [48]. Efforts to develop miniaturized probes [49] involved procedures such as flame-etching, which was shown to form micrometer-sized carbon fibers with diameter of 50-100 nm at the top and a length of 30-100 μm [50]; electrochemical pre-treatment [51]; pyrolytic deposition of carbon within pre-pulled nanopipettes with diameter varying in the range 5-200 nm [52]; or the use of a 3D printed mold, as shown in Fig. 13.1 [53], a procedure enabling high volume production.

To achieve sensitive and selective measurements, electrodes are functionalized with chemical or biological coatings. ROS/RNS can be measured directly at their characteristic potentials, such as measurements of nitric oxide oxidation at ~ 0.8 V vs. Ag/AgCl, or the $O_2/O_2^{\bullet-}$ redox couple at ~-0.33 V vs. NHE [22]. Use of platinized carbon MEs (~10 μm diameter) was demonstrated to be particularly suitable for single cell measurements [22, 31, 54, 55] and enable studies of species inside single phagolysosomes of living macrophages by chronoamperometry [55]. Commonly, MEs are functionalized with catalytic materials and nanomaterials to enhance the sensitivity and selectivity. Carbon nanomaterials such as carbon nanotubes [56] or graphene [57] are the most commonly used due to their high surface area and conductivity, facilitating MEs with increased electrochemically active surface [58, 59] and improved detection performance, for example, for H_2O_2 [60]. Catalytic materials based on metallic layers or polymerized electrocatalytic mediators are also common. Examples include modification with composites made with graphene, Prussian Blue [61], or hemin [62, 63]. Bimetallic [64], metal–polymer [65], or metal–metal oxides [66, 67] are common examples of electrocatalytic materials for hydrogen peroxide detection. For example, CuO nanostructures improved performance of hydrogen peroxide sensing [68], which

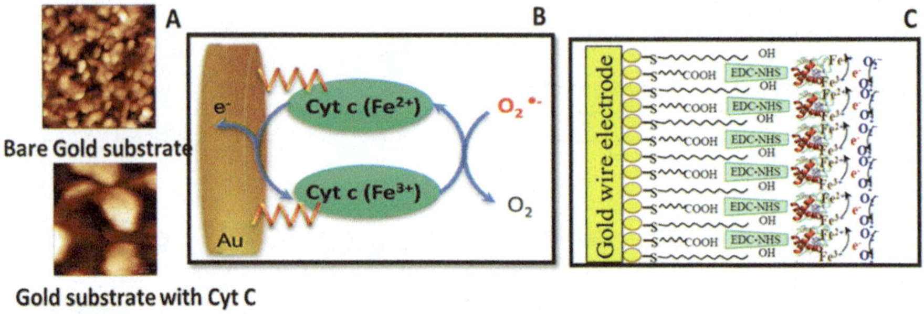

Fig. 13.2 Electrode design and immobilization of a redox protein (Cyt c) showing: (**a**) changes in gold morphology before/after electrode modification studied by conductive AFM, (**b**) redox wiring and electrode transfer process, and (**c**) immobilization of Cyt c using a layer of mixed thiols on a gold wire electrode modified with Au nanoparticles

can be used to develop miniaturized sensors [69]. Common electrocatalytic mediators are materials such as Prussian Blue [70] and Meldola Blue [71, 72].

Because of the complexity of the biological environment and the presence of redox active compounds, such as uric acid and ascorbic acid in the media, a common difficulty for translating their use in "real" environments is the issue of interferences [23]. To address this issue, MEs are chemically coated with perm-selective or electrostatic layers to reject interfering species through either size or charge. The use of materials such as Nafion and chitosan [73] has been shown to be effective at removing ascorbic acid interferences. Another problem is biofouling due to protein adsorption or deposition of reaction by-products. Some studies have shown that the preparation of CFMEs with carbon nanotubes improves the resistance to fouling [74].

Biological sensors or biosensors are functionalized electrodes that contain a biological recognition layer, typically a redox protein, immobilized at the surface of the electrode to specifically recognize the targeted species and translate recognition into an electrochemical signal. For example, cytochrome c (Cyt C) provides biomolecular recognition for $O_2^{\bullet-}$ through an electron transfer process at the surface of MEs [75]. To fabricate the biosensor, Cyt C is immobilized onto the surface of a gold microwire and the immobilized Cyt C reacts with $O_2^{\bullet-}$; the Cyt C is then oxidized to/from the electrode, generating a current proportional to the $O_2^{\bullet-}$ concentration [48, 76, 77]. A general schematic of the immobilization procedure using self-assembled layers of mixed thiols is shown in Fig. 13.2. Since biosensors utilize biomolecules for recognition, they tend to be more selective then chemical sensors. Moreover, because they require immobilization of the biomolecule, the long-term stability of these probes might pose challenges for translation and field implementation. Table 13.2 provides examples of some common microelectrode platforms for measuring ROS/RNS in biological systems. Performance characteristics for representative sensors and examples of applications are discussed in the next sections.

Table 13.2 Examples of electrochemical sensors and biosensors and electrode designs for ROS/RNS detection in biological conditions

#	ROS/ RNS	Electrochemical technique	Electrode materials	Working electrode	LOD	Biological system	Refs.
1	H_2O_2	Fast scan cyclic voltammetry	1,3-phenylenediamine	CFME	20 μM	Rat brain	[78]
2	H_2O_2	Chronoamperometry	Pt-Pd bimetallic nanocoral	CFME	0.42 μM	A549 living cells, milk	[79]
3	H_2O_2	Amperometry, CV	Hemoglobin, SWCNT	CFME	0.23 μM	HePG2 cancer cells	[80]
4	H_2O_2	Amperometry	Au-Pd alloy NPs, Graphene Quantum Dots	CFME	500 nM	Clinical breast cancer tissue	[81]
5	H_2O_2	Amperometry	Pt NPs, Nafion, PPD	CFME	0.53 μM	In vitro	[82]
6	H_2O_2	Amperometry	Platinized silica nanoporous membranes	CFME Or ITO	0.01mM	Rat brain	[83]
7	H_2O_2	Chronoamperometry	Heat treatment to create nanopores to improve catalytic performance	Heat-treated CFME	1 μM	In vitro	[84]
8	H_2O_2	Chronoamperometry	Core–shell 2D VS_2,@VC@N-doped carbon sheets decorated by Pd NPs	CFME	50 nM	MCS-7 cancer cells, and breast cancer tissue	[85]
9	H_2O_2	Chronoamperometry	Pt–Pd NPs, graphene oxide	CFME	0.3 μM	Raw 264.7 cells secretion	[86]
10	H_2O_2	Chronoamperometry	Au–Ag bimetallic NPs/polydopamine	CFME	0.12 μM	HepG2 cells	[87]
11	HClO/ ClO^-	DPV	Graphene oxide, carbon nanotubes, MBS	CFME	0.5 μM	Body fluids	[88]
12	$O_2^{\cdot-}$	Chronoamperometry	MWCNTs, ionic Liquid-Br, SOD, Prussian Blue NPs	CFME	0.42 μM	Alzheimer rat brains	[89]
13	$O_2^{\cdot-}$	DPV with ratiometric signal output	Diphenylphosphonate-2-naphthol ester, methylene blue SWCNTs	CFME	2 μM	Rat brain	[90]
14	NO	DPV	NiTSPc/Nafion	CFME	0.34 μM	Zebrafish intestine	[21]

Reproduced with permission from Ref. [24]

MWCNTs multi-walled carbon nanotubes, *CV* cyclic voltammetry, *BBY* Bismarck Brown Y, *rGO* reduced graphene oxide, *FTO* fluorine doped tin oxide, *AgNPs* silver nanoparticles, *CNT* carbon nano tubes, *DNA* deoxyribose nucleic acid, *APTES* (3-aminopropyl) triethoxysilane, *Cyt C* cytochrome C, *Poly(5A1N)* electropolymerized 5-amino-1-naphthol, *XG* Xero gel, *PSS* polystyrene sulfonic acid, *BA* 5-(1,2-dithiolan-3-yl)-N-(4-(4,4,5,5-tetramethyl-1,3,2-dioxaborolan-2-yl)phenyl)pent-anamide, *CFME* carbon fiber micro electrode, *AuNP* gold nanoparticles, *PDDA* poly(diallyl dimethyl ammonium chloride), *D-cell* Plastic disposable carbon-based electrochemical cell, *a-NSGF* taurine-functionalized graphene foam, *ITO-PET* indium tin oxide supported on poly-ethylene-terephthalate foil, *CTS* chitosan, *MPNS* microporous polymeric nanospheres, *HTCMP* hollow tubular conjugated organic microporous polymer

13.3.2 Electrochemical Sensors for H_2O_2 and $O_2^{\bullet-}$

Hydrogen peroxide and superoxide species are characterized by short half-life in the extra- and intracellular space and are present at low and variable concentrations due to the complex and dynamic oxidant/antioxidant system. Another radical species, $OH^{\bullet}$ has the lowest lifetime among the ROS and its formation is limited to the Fenton reaction. Methods for the detection of $OH^{\bullet}$ are typically based on its strong oxidation properties after exposure to $OH^{\bullet}$ during the Fenton reaction [91] or by indirectly measuring the oxidation properties of $OH^{\bullet}$ toward DNA strands [92], which is quantified by electrochemistry using methylene blue as a mediator.

Among ROS, H_2O_2 is the easiest to detect electrochemically [19]. A CFME modified with a synthetic receptor with affinity toward H_2O_2 enabled its determination in a single drop of blood using differential pulse voltammetry (DPV) in the potential range −0.5 to −0.6 V vs. Ag/AgCl and a linearity range of 0.5–400 μM [93]. A carbon-based electrochemical electrode modified with Ag nanoparticles and δ-FeOOH enabled hydrogen peroxide detection in fetal bovine serum [94]. Electrodes functionalized with chemical mediators such as Prussian blue (PB) are a preferred modality for ROS, especially H_2O_2 detection at concentration levels down to 1.3 μM [95]. Strategies to increase selectivity of CFMEs include modification with catalytic materials like Pt, often combined with Au, Ag, or carbon nanostructures for enhancing performance. A CFME functionalized with VS_2@VC@N-doped carbon sheets and Pd nanoparticles facilitated detection of H_2O_2 at −0.05 V with no major interferences from electroactive species [88].

Detection of $O_2^{\bullet-}$ is complicated by the fact that, in aqueous solutions, $O_2^{\bullet-}$ undergoes disproportionation to H_2O_2. The disproportionation is a very fast second order reaction with respect to $O_2^{\bullet-}$ ($k = 2 \times 10^5\ M^{-1}\ s^{-1}$ at 25 °C and a pH of 7.4 [96]). The standard redox potential of the $O_2/O_2^{\bullet-}$ redox couple is between 330 and 140 mV vs. NHE. Superoxide detection can be achieved by direct electrochemistry using platinized CFME or MEs modified with N-doped carbon Ag NPs [97] or porous carbon [98]. A common approach to improve selectivity is to immobilize Cyt C or SOD onto a functionalized MEs [19] or use enzyme mimetic materials such as $MnTiO_3$ microdiscs [99], manganese phosphate [100] or nanostructured MXenes [101], and graphene/Ag NP/CeO_2/TiO_2 [102] as alternative to natural enzymes. Cyt c-based nanoporous gold electrodes with a detection limit of 1.9 nM and a sensitivity of 1.9 $nM.nM^{-1}cm^{-2}$ enabled the detection of $O_2^{\bullet-}$ in skeletal muscle tissue [103]. Figure 13.3 shows an example of Cyt c-based electrochemical biosensor fabricated on a Au ME wire for $O_2^{\bullet-}$ detection, the electrode processes, and detection mechanism. The sensitivity of the biosensor was 42.4 nA/μM, and the LOD was 2.4 nM. The biosensor was tested in human blood platelets and skeletal rat limb muscles following ischemia reperfusion injury (IRI), confirming the possibility of these measurements in complex biological systems [48]. A similar configuration also demonstrated performance for real-time measurements of $O_2^{\bullet-}$ in the intestine (Fig. 13.4) [104].

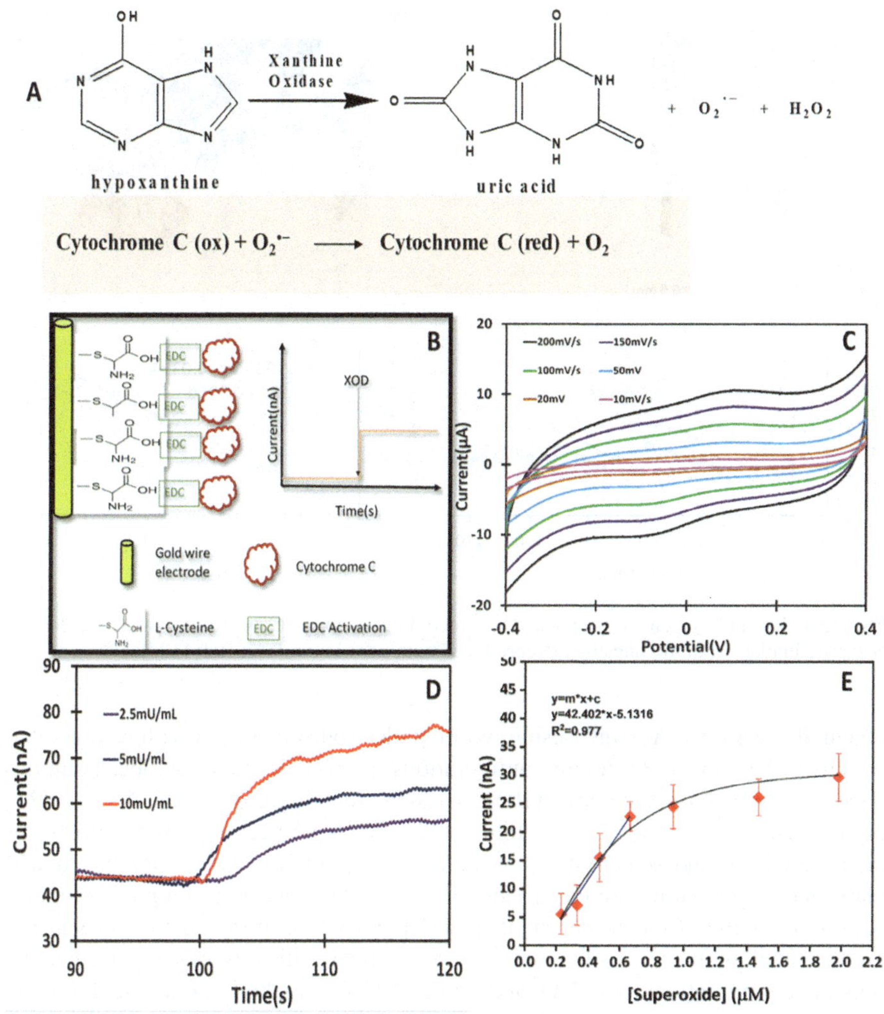

Fig. 13.3 Example of electrochemical biosensors for $O_2^{\bullet-}$ using a Cyt c-modified electrode showing: (**a**) formation of $O_2^{\bullet-}$ using hypoxanthine/XO, (**b**) immobilization process using L-cysteine and EDC chemistry, (**c**) effect of scan rate indicating a surface confined process, (**d**) effect of different concentrations of enzymatically generated and $O_2^{\bullet-}$ obtained from different XOD levels, and (**e**) calibration curve. (Reproduced with permission Ref. [48])

13.3.3 Electrochemical Measurements of Nitric Oxide ($NO^{\bullet}$) and Peroxynitrite ($ONOO^{-}$)

Nitric oxide is a highly reactive diffusible species, which can combine with $O_2^{\bullet-}$ to form the more toxic $ONOO^{-}$. Electrochemical detection of $NO^{\bullet}$ can be accomplished based on the electrochemical oxidation of $NO^{\bullet}$, which takes place at a

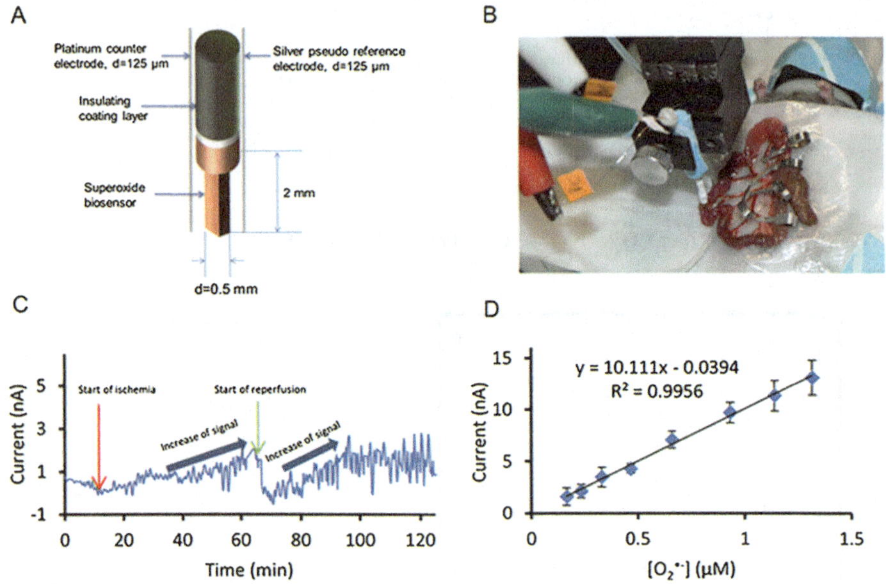

Fig. 13.4 Example of electrochemical $O_2^{\bullet -}$ recordings using a Cyt-c thiol-modified gold wire electrode implanted in the intestine. (Reproduced with permission from Ref. [104])

potential >0.8V vs. Ag/AgCl using working electrodes modified with catalytic and selective membranes. Different configurations of $NO^{\bullet}$ sensors have been summarized in an extensive review [105]. CFMEs functionalized with Nafion [21], o-phenylene diamine (o-PD), or chitosan [21, 73] have been reported that showed high selectivity and sensitivity. In vivo measurements can be accomplished with platinized or platinum wire electrodes. An electrode consisting of a glass nanopore platinized platinum electrode enabled in-depth $NO^{\bullet}$ profiling in the cortical and hippocampal brain areas of a rodent model [106]. Other works reported multi-sensing configurations for $^{\bullet}NO$ and oxygen [107], carbon monoxide [108], or potassium ions [109], which enabled real-time monitoring of dynamic concentration changes and signaling of these species. A Pt disk electrode modified with an electropolymerized permselective 5-amino-1-naphtol (Poly(5A1N)) and fluorinated xerogel enabled measurement of $NO^{\bullet}$ in proinflammatory macrophages [110], achieving a detection limit (LOD) of 1 nM and a linearity range (LR) from 0.01 to 10 μM. An electrode coated with reduced graphene oxide and platinum nanoparticles enabled nitric oxide detection in human serum, with a LOD of 52 nM and a LR from 0.25 to 40 μM [111]. A microsensor with poly(eugenol) coating enabled high selectivity detection in the presence of H_2O_2 [112]. Other sensors integrate a flexible implantable microelectrode coated with polylactic acid and poly (trimethylene carbonate), an ultrathin Au membrane, and a poly(eugenol) film, with wireless monitoring to track release over a period of 5 day in the heart and joint cavity of rabbits [113]. The sensor had a LOD of 0.97 nm and a LR of 0.01–100 μM.

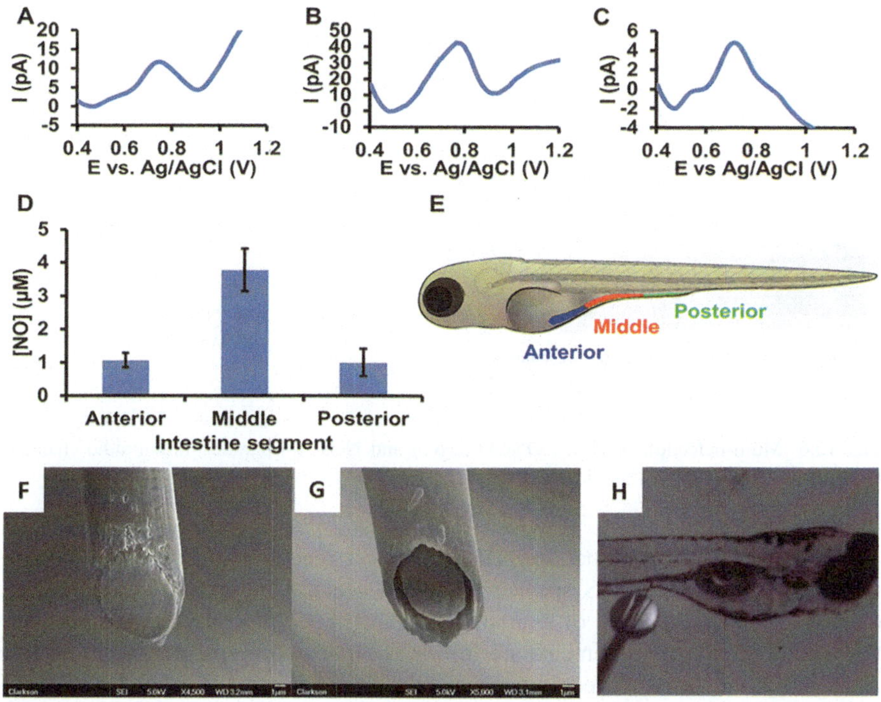

Fig. 13.5 In vivo DPV recordings with the microelectrode inserted in the intestine of 5 dpf zebrafish embryos in the: anterior (**a**), middle (**b**), and posterior (**c**) segments. Due to the measurement being made in live embryos without the use of anesthetics, small shoulders can be seen in the voltammograms recorded in vivo, which may be due to the movements of the embryo. Nitric oxide concentrations measured in vivo along the intestine of 5 dpf zebrafish embryos; error bars represent standard deviations for $n = 3$ measurements in different embryos using individual microelectrodes (**d**). Schematic representation of the three regions of the zebrafish embryo intestine where the sensor was implanted (**e**). The length of the embryo at 5 dpf is approximately 4 mm. (**f**, **g**) Scanning electron micrograph images of the carbon fiber microelectrode, before (**f**) and after (**g**) modification with NiTSPc and Nafion. Electrochemical deposition of NiTSPc takes place with etching of the carbon fiber microelectrode tip. The scale bar is 1 µm. (Panels A–C, with permission from Ref. [51])

Our group developed CFMEs modified with selective membranes to track the evolution of $NO^{\bullet}$ in different locations of the intestine of zebrafish embryos [21, 114]. The sensor was used as a measurement tool to evaluate changes in nitric oxide levels in response to drugs and to assess oxidative stress response as a result of nanoparticle exposure. Examples of electrochemical recordings in the zebrafish intestine, location of the electrode during implantation, and scanning electron microscope (SEM) micrographs of the microelectrode before and after electrochemical treatment to increase sensitivity are shown in Fig. 13.5. Measurements showed good correlation with results obtained by fluorescence imaging using the 4-amino-5-methylamino-2',7'-difluorofluorescein diacetate (DAF-FM-DA) dye [22].

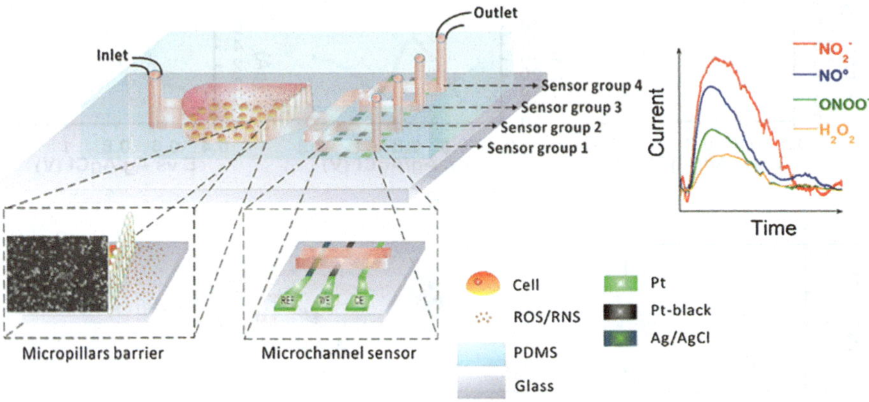

Fig. 13.6 Multi-detection of H_2O_2, $ONOO^-$, $NO^•$, and NO_2^- on parallel microfluidic channels. (With permission from Ref. [121])

Peroxynitrite, the product formed in the fast reaction of $O_2^{•-}$ with $NO^•$, is relatively newer ROS/RNS species important because of its high reactivity and damaging effects. Because of its extremely short time, $ONOO^-$ has been difficult to measure [115]. Measurements are further complicated by the myriad of cellular reaction and the dependence of pH and cellular environments. The $ONOO^-/ONOO^•$ formal potential is 0.27 V vs. SSCE [116], which makes possible direct electrochemical sensing of these species. Most commonly used electrodes are based on platinized MEs modified with electrocatalysts such as conjugated Mn complexes (e.g., tetraaminophthalocyanine manganese (II) [117], MnO_2-Hemin [118], and PEDOT-Hemin [119] layers, and polymeric nanospheres [120]). Simultaneous detection of ROS/RNS is of great interest due to inter-relation between these species, changing concentrations dynamically over time, and their high reactivity. An example of multi-ROS/RNS microfluidic-based detection has been reported for ROS/RNS released by cells [121] and ratiometric measurements [122], shown in Fig. 13.6.

The development of electrochemical sensors for $ONOO^-$ detection is a big challenge due to its instability and short lifetime, as well as challenges related to the preparation and handling of $ONOO^-$ to perform appropriate calibration. Similar to $O_2^{•-}$, the validation of peroxynitrite sensors is often questionable. Because of a lack of selective catalysts for its detection, only a few publications report electrode modifications for peroxynitrite sensing using electrochemical mediators. Hemin, an iron containing porphyrin, was shown to possess electrocatalytic activity for $ONOO^-$ [123]. The sensing performance of hemin toward $ONOO^-$ was improved by synthesis of composite materials with graphene [124] or PEDOT [119]. The hemin–PEDOT material is particularly attractive for use in biological studies due to its facile deposition on carbon fiber electrodes [125]. Cobalt-containing mediators have also been used for peroxynitrite detection. The performance of a cobalt porphyrin–graphene composite material has been demonstrated for simultaneous detection of $ONOO^-$ and H_2O_2 [126]. Catalytic activity for peroxynitrite detection was also reported for a cobalt center-containing complex, cyanocobalamin [127]. Its use is advantageous due to possibility of forming thin films at the electrode surface by electrochemical polymerization from monomer solutions.

13.4 Biological Applications and Translational Aspects

Accurate ROS/RNS measurements are important to understand the relationship between OS and disease as well as in monitoring single cell events [128]. Because OS is implicated in the progression of many conditions, such measurements can be used to provide insights into the development of diseases, evaluate effectiveness of therapies, and quantify ROS at subcellular levels such as, for example, in fibroblast and cancer cell lines [128], or in single cells to evaluate effect of treatment with chemotherapeutics [129]. The translational utility of these sensors was evaluated in biological fluids like human platelets [48] and several animal models, e.g., ROS/RNS during intestinal ischemia-reperfusion injury [104].

Despite the desirable characteristics of these sensors, several challenges remain before they can be translated to clinical work. Limitations include: (1) the risk of biofouling, or adsorption of biomolecules onto the probe, which reduces detection capability, particularly for probes used in vivo [130], (2) interference from co-existing electroactive compounds [131], (3) quantitative data interpretation, (4) selectivity, (5) calibration drifts and (6) accuracy of measurements. In particular, ROS/RNS concentrations are harder to estimate due to limited stability and high reactivity of these species. Biological media are complex and difficult to replicate for external calibration purpose, or calibration is unreliable due to lack of stability of RS. For these reasons, validation using alternative methods should be conducted along with calibration during measurements in biological models [114].

Current development status of electrochemical sensors for ROS/RNS measurements demonstrates the potential of this technology for investigating complex biochemical mechanisms in biology and medicine. Their capabilities enable rapid, sensitive, and real-time detection of RS in live organisms, with potential to provide detailed investigation of biochemical mechanisms at organ levels when implanted, and at fundamental level, in cell populations or at single cells. These studies can elucidate biological processes and study mechanisms of disease development or injury states. The potential of these methods has been demonstrated in cells and biological media, and future work can further advance this technology for broad adoption by the biological, clinical, and biomedical communities. Work demonstrating intracellular measurements is also expected and can make significant contributions to the field. Development of methodologies for simultaneous detection of multiple ROS/RNS is underway, and examples of measurements in different biological models can decipher complex oxidative stress mechanisms and processes. The applicability of this approach for measurements in living organisms has been demonstrated, and recent works provide evidence highlighting the potential of this methodology in human subjects. Such studies could provide significant information on the biochemistry of living organisms and advance clinical measurements and disease diagnostics.

13.5 Conclusions and Trends

ROS/RNS measurements are important because OS is implicated in the progression of many pathological conditions; sensors for ROS/RNS can be used to understand the relationship between OS and disease. When carefully designed, electrochemical sensors and biosensors can provide continuous monitoring of the appearance and evolution of ROS/RNS species and the study of the dynamics of reactions and processes involving ROS/RNS species. However, electrochemical sensors are relatively seldom used to investigate processes in whole organisms. Their capabilities as biomedical tools in the clinical field are not well known by the biological and clinical communities, and their reliability for measurements in live organisms and clinical applications is still to be proved.

Although sensors for ROS/RNS are well developed, most work was done in standard solutions or synthetically generated species. While electrochemical sensors for ROS/RNS have demonstrated improved capabilities in recent years, their implementation requires further refinement to improve robustness, selectivity, and sensitivity along with reducing fouling. Improving the selectivity toward individual radicals or developing multianalyte sensors for simultaneous quantification of radicals could increase capabilities and advance applications. Manufacturing MEs using advanced methods is also needed to accelerate adoption. The developmental work paved the way in designing future clinical-ready probes for detection of ROS/RNS in vivo and enabling monitoring physiological and pathological events in conditions such as cancer, traumatic brain injury, hypovolemic shock, and many others.

References

1. Chen XQ, Wang F, Hyun JY, Wei TW, Qiang J, Ren XT, et al. Recent progress in the development of fluorescent, luminescent and colorimetric probes for detection of reactive oxygen and nitrogen species. Chem Soc Rev. 2016;45(10):2976–3016.
2. Murphy MP, Holmgren A, Larsson NG, Halliwell B, Chang CJ, Kalyanaraman B, et al. Unraveling the Biological Roles of Reactive Oxygen Species. Cell Metab. 2011;13(4):361–6.
3. Sies H, Jones DP. Reactive oxygen species (ROS) as pleiotropic physiological signalling agents. Nat Rev Mol Cell Bio. 2020;21(7):363–83.
4. Schrader M, Fahimi HD. Peroxisomes and oxidative stress. Bba-Mol Cell Res. 2006;1763(12):1755–66.
5. Checa J, Aran JM. Reactive Oxygen Species: Drivers of Physiological and Pathological Processes. J Inflamm Res. 2020;13:1057–73.
6. Gorlach A, Dimova EY, Petry A, Martinez-Ruiz A, Hernansanz-Agustin P, Rolo AP, et al. Reactive oxygen species, nutrition, hypoxia and diseases: Problems solved? Redox Biol. 2015;6:372–85.

7. Chen Y, Azad MB, Gibson SB. Superoxide is the major reactive oxygen species regulating autophagy. Cell Death Differ. 2009;16(7):1040–52.
8. Nash KM, Rockenbauer A, Villamena FA. Reactive Nitrogen Species Reactivities with Nitrones: Theoretical and Experimental Studies. Chem Res Toxicol. 2012;25(8):1581–97.
9. Elbatreek MH, Pachado MP, Cuadrado A, Jandeleit-Dahm K, Schmidt H. Reactive Oxygen Comes of Age: Mechanism-Based Therapy of Diabetic End-Organ Damage. Trends Endocrinol Metab. 2019;30(5):312–27.
10. Senoner T, Dichtl W. Oxidative Stress in Cardiovascular Diseases: Still a Therapeutic Target? Nutrients. 2019;11(9).
11. Burmeister DM, Gomez BI, Dubick MA. Molecular mechanisms of trauma-induced acute kidney injury: Inflammatory and metabolic insights from animal models. Biochim Biophys Acta Mol Basis Dis. 2017;1863(10 Pt B):2661–71.
12. Collard CD, Gelman S. Pathophysiology, clinical manifestations, and prevention of ischemia-reperfusion injury. Anesthesiology. 2001;94(6):1133–8.
13. Eltzschig HK, Eckle T. Ischemia and reperfusion-from mechanism to translation. Nat Med. 2011;17(11):1391–401.
14. Clanton TL. Hypoxia-induced reactive oxygen species formation in skeletal muscle. J Appl Physiol (1985). 2007;102(6):2379–88.
15. Sies H, Jones DP. Reactive oxygen species (ROS) as pleiotropic physiological signalling agents. Nat Rev Mol Cell Biol. 2020;21(7):363–83.
16. Prasad A, Balukova A, Pospisil P. Triplet Excited Carbonyls and Singlet Oxygen Formation During Oxidative Radical Reaction in Skin. Front Physiol. 2018;9:1109.
17. Togashi H, Aoyama M, Oikawa K. Imaging of reactive oxygen species generated in vivo. Magn Reson Med. 2016;75(3):1375–9.
18. Wang HS. Development of fluorescent and luminescent probes for reactive oxygen species. Trac-Trend Anal Chem. 2016;85:181–202.
19. Calas-Blanchard C, Catanante G, Noguer T. Electrochemical Sensor and Biosensor Strategies for ROS/RNS Detection in Biological Systems. Electroanal. 2014;26(6):1277–86.
20. Malferrari M, Becconi M, Rapino S. Electrochemical monitoring of reactive oxygen/nitrogen species and redox balance in living cells. Anal Bioanal Chem. 2019;411(19):4365–74.
21. Dumitrescu E, Wallace KN, Andreescu S. Real time electrochemical investigation of the release, distribution and modulation of nitric oxide in the intestine of individual zebrafish embryos. Nitric oxide : biology and chemistry. 2018;74:32–8.
22. Amatore C, Arbault S. Oxidative Stress at the Single Cell Level. Front Neuroeng. 2007;1:261–83.
23. Ozel RE, Hayat A, Andreescu S. Recent Developments in Electrochemical Sensors for the Detection of Neurotransmitters for Applications in Biomedicine. Analytical letters. 2015;48 (7):1044–69.
24. Deshpande AM, W.; Andreescu, S. Electrochemical sensors for oxidative stress monitoring. Curr Opin Electroche. 2021;29:100809.
25. Hepel M, Luo J. Photoelectrochemical mineralization of textile diazo dye pollutants using nanocrystalline WO_3 electrodes. Electrochim Acta. 2001;47(5):729–40.
26. Shafiee SAA, J.; Hamzah, H.H. Review—Electroreduction of Peroxodisulfate. J Electrochem Soc. 2018 165 H785.
27. Berdnikov VM, Bazin NM. Russ J Phys Chem. 1970;44:395.
28. JANAF Thermochemical Tables. 2nd ed. Stull DR, Prophet H, editors. Washington DC: US Natl. Bur. Stand.; 1971.
29. Pourbaix M. Atlas of Electrochemical Equilibria in Aqueous Solutions: Pergamon Press; 1966.
30. Schwarz HA, Dodson RW. J Phys Chem. 1984;88:3643.
31. Amatore C, Arbault S, Guille M, Lemaitre F. Electrochemical monitoring of single cell secretion: Vesicular exocytosis and oxidative stress. Chemical Reviews. 2008;108(7): 2585–621.

32. Jayson GG, Parsons BJ, Swallow AJ. J Chem Soc, Faraday Soc Trans. 1973;69:1597.
33. Desilvestro J, Gratzel M. J Electroanal Chem. 1987;238:129.
34. Koppenol WH, Liebman JF. J Phys Chem. 1984;88:99.
35. Wagman DD, Evans WH, Parker VB, Schum RH, Halow I, Bailey SM, et al. J Phys Chem Ref Data Suppl. 1982;2.
36. Dejam A, Hunter CJ, Schechter AN, Gladwin MT. Emerging role of nitrite in human biology. Blood Cells Mol Dis. 2004;32(3):423–9.
37. Sies H. Strategies of antioxidant defense. Eur J Biochem. 1993;215(2):213–9.
38. Gladwin MT, Grubina R, Doyle MP. The new chemical biology of nitrite reactions with hemoglobin: R-state catalysis, oxidative denitrosylation, and nitrite reductase/anhydrase. Acc Chem Res. 2009;42(1):157–67.
39. Brennan ML. A tale of two controversies: defining both the role of peroxidases in nitrotyrosine formation in vivo using eosinophil peroxidase and myeloperoxidase-deficient mice, and the nature of peroxidase-generated reactive nitrogen species. J Biol Chem. 2002;277:17415–27.
40. Hirst J. Mitochondrial complex I. Annu Rev Biochem. 2013;82:551–75.
41. Bae YS, Oh H, Rhee SG, Yoo YD. Regulation of reactive oxygen species generation in cell signaling. Mol Cells. 2011;32(6):491–509.
42. Nickel A, Kohlhaas M, Maack C. Mitochondrial reactive oxygen species production and elimination. J Mol Cell Cardiol. 2014;73:26–33.
43. Fridovich I. Superoxide dismutases. Adv Enzymol Relat Areas Mol Biol. 1974;41(0):35–97.
44. Saltzman HA, Fridovich I. Editorial: Oxygen toxicity. Introduction to a protective enzyme: superoxide dismutase. Circulation. 1973;48(5):921–3.
45. Yum K, Wang N, Yu M-F. Nanoneedle: A multifunctional tool for biological studies in living cells. Nanoscale. 2010;2(3):363–72.
46. Chen W, Ren Q-Q, Yang Q, Wen W, Zhao Y-D. In Vivo Electrochemical Biosensors for Reactive Oxygen Species Detection: A Mini-Review. Analytical Letters. 2012;45(2–3): 156–67.
47. Liu X, Dumitrescu E, Andreescu S. Electrochemical Biosensors for Real-Time Monitoring of Reactive Oxygen and Nitrogen Species. Oxidative Stress: Diagnostics, Prevention, and Therapy Volume 2. ACS Symposium Series. 1200: American Chemical Society; 2015. p. 301–27.
48. Deshpande AS, Muraoka W, Wait J, Colak A, Andreescu S. Direct real-time measurements of superoxide release from skeletal muscles in rat limbs and human blood platelets using an implantable Cytochrome C microbiosensor. Biosens Bioelectron. 2023;240:115664.
49. Yeh JI, Shi H. Nanoelectrodes for biological measurements. Wiley Interdisciplinary Reviews: Nanomedicine and Nanobiotechnology. 2010;2(2):176–88.
50. Li X, Majdi S, Dunevall J, Fathali H, Ewing AG. Quantitative Measurement of Transmitters in Individual Vesicles in the Cytoplasm of Single Cells with Nanotip Electrodes. Angewandte Chemie International Edition. 2015;54(41):11978–82.
51. Dumitrescu E, Deshpande A, Wallace KN, Andreescu S. Time-Dependent Monitoring of Dopamine in the Brain of Live Embryonic Zebrafish Using Electrochemically Pretreated Carbon Fiber Microelectrodes. ACS measurement science au. 2022;2(3):261–70.
52. Actis P, Tokar S, Clausmeyer J, Babakinejad B, Mikhaleva S, Cornut R, et al. Electrochemical Nanoprobes for Single-Cell Analysis. ACS Nano. 2014;8(1):875–84.
53. Trikantzopoulos E, Yang C, Ganesana M, Wang Y, Venton BJ. Novel carbon-fiber microelectrode batch fabrication using a 3D-printed mold and polyimide resin. Analyst. 2016.
54. Hu KK, Liu YL, Oleinick A, Mirkin MV, Huang WH, Amatore C. Nanoelectrodes for intracellular measurements of reactive oxygen and nitrogen species in single living cells. Curr Opin Electroche. 2020;22:44–50.
55. Hu KK, Li Y, Rotenberg SA, Amatore C, Mirkin MV. Electrochemical Measurements of Reactive Oxygen and Nitrogen Species inside Single Phagolysosomes of Living Macrophages. J Am Chem Soc. 2019;141(11):4564–8.

56. Merkoçi A, Pumera M, Llopis X, Pérez B, del Valle M, Alegret S. New materials for electrochemical sensing VI: Carbon nanotubes. TrAC Trends in Analytical Chemistry. 2005;24(9):826–38.
57. Pumera M, Ambrosi A, Bonanni A, Chng ELK, Poh HL. Graphene for electrochemical sensing and biosensing. TrAC Trends in Analytical Chemistry. 2010;29(9):954–65.
58. Jacobs CB, Ivanov IN, Nguyen MD, Zestos AG, Venton BJ. High Temporal Resolution Measurements of Dopamine with Carbon Nanotube Yarn Microelectrodes. Analytical Chemistry. 2014;86(12):5721–7.
59. Mercante LA, Pavinatto A, Iwaki LEO, Scagion VP, Zucolotto V, Oliveira ON, et al. Electrospun Polyamide 6/Poly(allylamine hydrochloride) Nanofibers Functionalized with Carbon Nanotubes for Electrochemical Detection of Dopamine. ACS Applied Materials & Interfaces. 2015;7(8):4784–90.
60. Cao X, Zeng Z, Shi W, Yep P, Yan Q, Zhang H. Three-Dimensional Graphene Network Composites for Detection of Hydrogen Peroxide. Small. 2013;9(9–10):1703–7.
61. Yang J-H, Myoung N, Hong H-G. Facile and controllable synthesis of Prussian blue on chitosan-functionalized graphene nanosheets for the electrochemical detection of hydrogen peroxide. Electrochimica Acta. 2012;81:37–43.
62. Chen J, Zhao L, Bai H, Shi G. Electrochemical detection of dioxygen and hydrogen peroxide by hemin immobilized on chemically converted graphene. Journal of Electroanalytical Chemistry. 2011;657(1–2):34–8.
63. Guo Y, Li J, Dong S. Hemin functionalized graphene nanosheets-based dual biosensor platforms for hydrogen peroxide and glucose. Sensors and Actuators B: Chemical. 2011;160 (1):295–300.
64. Wang N, Han Y, Xu Y, Gao C, Cao X. Detection of H2O2 at the Nanomolar Level by Electrode Modified with Ultrathin AuCu Nanowires. Analytical Chemistry. 2015;87(1): 457–63.
65. Li Y, Zhang J-J, Xuan J, Jiang L-P, Zhu J-J. Fabrication of a novel nonenzymatic hydrogen peroxide sensor based on Se/Pt nanocomposites. Electrochemistry Communications. 2010;12 (6):777–80.
66. Liu Z, Zhao B, Shi Y, Guo C, Yang H, Li Z. Novel nonenzymatic hydrogen peroxide sensor based on iron oxide–silver hybrid submicrospheres. Talanta. 2010;81(4–5):1650–4.
67. Miao Y-E, He S, Zhong Y, Yang Z, Tjiu WW, Liu T. A novel hydrogen peroxide sensor based on Ag/SnO2 composite nanotubes by electrospinning. Electrochimica Acta. 2013;99:117–23.
68. Ping J, Ru S, Fan K, Wu J, Ying Y. Copper oxide nanoparticles and ionic liquid modified carbon electrode for the non-enzymatic electrochemical sensing of hydrogen peroxide. Microchimica Acta. 2010;171(1):117–23.
69. Song M-J, Hwang SW, Whang D. Non-enzymatic electrochemical CuO nanoflowers sensor for hydrogen peroxide detection. Talanta. 2010;80(5):1648–52.
70. Karyakin AA. Prussian Blue and Its Analogues: Electrochemistry and Analytical Applications. Electroanalysis. 2001;13(10):813–9.
71. Wring SA, Hart JP. Chemically modified, carbon-based electrodes and their application as electrochemical sensors for the analysis of biologically important compounds. A review. Analyst. 1992;117(8):1215–29.
72. Njagi JE, JS.; Aston, JW.; Leiter, JC.; Andreescu, S. A sensitive electrochemical sensor based on chitosan and electropolymerized Meldola blue for monitoring NO in brain slices. Sensors and Actuators B: Chemical. 2010;143(2):673–80.
73. Ozel RE, Wallace KN, Andreescu S. Chitosan coated carbon fiber microelectrode for selective in vivo detection of neurotransmitters in live zebrafish embryos. Anal Chim Acta. 2011;695 (1–2):89–95.
74. Harreither W, Trouillon R, Poulin P, Neri W, Ewing AG, Safina G. Carbon Nanotube Fiber Microelectrodes Show a Higher Resistance to Dopamine Fouling. Analytical Chemistry. 2013;85(15):7447–53.

75. Beissenhirtz MK, Scheller FW, Lisdat F. A superoxide sensor based on a multilayer cytochrome c electrode. Anal Chem. 2004;76(16):4665–71.
76. Ganesana M, Erlichman JS, Andreescu S. Real-time monitoring of superoxide accumulation and antioxidant activity in a brain slice model using an electrochemical cytochrome c biosensor. Free radical biology & medicine. 2012;53(12):2240–9.
77. Liu X, Marrakchi M, Jahne M, Rogers S, Andreescu S. Real-time investigation of antibiotics-induced oxidative stress and superoxide release in bacteria using an electrochemical biosensor. Free radical biology & medicine. 2016;91:25–33.
78. Wilson LR, Panda S, Schmidt AC, Sombers LA. Selective and Mechanically Robust Sensors for Electrochemical Measurements of Real-Time Hydrogen Peroxide Dynamics in Vivo. Anal Chem. 2018;90(1):888–95.
79. Li H, Zhao H, He H, Shi L, Cai X, Lan M. Pt-Pd bimetallic nanocoral modified carbon fiber microelectrode as a sensitive hydrogen peroxide sensor for cellular detection. Sensors and Actuators B: Chemical. 2018;260:174–82.
80. Zhang H, Ruan J, Liu W, Jiang X, Du T, Jiang H, et al. Monitoring dynamic release of intracellular hydrogen peroxide through a microelectrode based enzymatic biosensor. Anal Bioanal Chem. 2018;410(18):4509–17.
81. Xu Q, Yuan H, Dong X, Zhang Y, Asif M, Dong Z, et al. Dual nanoenzyme modified microelectrode based on carbon fiber coated with AuPd alloy nanoparticles decorated graphene quantum dots assembly for electrochemical detection in clinic cancer samples. Biosensors and Bioelectronics. 2018;107:153–62.
82. Wang B, Wen X, Chiou P-Y, Maidment NT. Pt Nanoparticle-modified Carbon Fiber Microelectrode for Selective Electrochemical Sensing of Hydrogen Peroxide. Electroanal. 2019;31 (9):1641–5.
83. Li X, Zhou L, Ding J, Sun L, Su B. Platinized Silica Nanoporous Membrane Electrodes for Low-Fouling Hydrogen Peroxide Detection. ChemElectroChem. 2020;7(9):2081–6.
84. Seven F, Gölcez T, ŞEn M. Nanoporous carbon-fiber microelectrodes for sensitive detection of H2O2 and dopamine. Journal of Electroanalytical Chemistry. 2020;864:114104.
85. Yuan H, Zhao J, Wang Q, Manoj D, Zhao A, Chi K, et al. Hierarchical Core–Shell Structure of 2D VS2@VC@N-Doped Carbon Sheets Decorated by Ultrafine Pd Nanoparticles: Assembled in a 3D Rosette-like Array on Carbon Fiber Microelectrode for Electrochemical Sensing. ACS Applied Materials & Interfaces. 2020;12(13):15507–16.
86. Qi H, Song J, Fu Y, Wu X, Qi H. Highly dispersive Pt–Pd nanoparticles on graphene oxide sheathed carbon fiber microelectrodes for electrochemical detection of H2O2 released from living cells. Nanotechnology. 2020;31(13):135503.
87. Sun W, Cai X, Wang Z, Zhao H, Lan M. A novel modification method via in-situ reduction of AuAg bimetallic nanoparticles by polydopamine on carbon fiber microelectrode for H2O2 detection. Microchemical Journal. 2020;154(January):104595-.
88. Dong H, Zhou Y, Zhao L, Hao Y, Zhang Y, Ye B, et al. Dual-Response Ratiometric Electrochemical Microsensor for Effective Simultaneous Monitoring of Hypochlorous Acid and Ascorbic Acid in Human Body Fluids. Anal Chem. 2020;92(22):15079–86.
89. Peng Q, Yan X, Shi X, Ou S, Gu H, Yin X, et al. In vivo monitoring of superoxide anion from Alzheimer's rat brains with functionalized ionic liquid polymer decorated microsensor. Biosensors and Bioelectronics. 2019;144:111665-.
90. Huang S, Zhang L, Dai L, Wang Y, Tian Y. Nonenzymatic Electrochemical Sensor with Ratiometric Signal Output for Selective Determination of Superoxide Anion in Rat Brain. Anal Chem. 2021;93(13):5570–6.
91. Gualandi I, Guadagnini L, Zappoli S, Tonelli D. A Polypyrrole Based Sensor for the Electrochemical Detection of OH Radicals. Electroanalysis. 2014;26(7):1544–50.
92. Wu L, Yang Y, Zhang H, Zhu G, Zhang X, Chen J. Sensitive electrochemical detection of hydroxyl radical with biobarcode amplification. Analytica Chimica Acta. 2012;756:1–6.

93. Dong H, Zhou Y, Hao Y, Zhao L, Sun S, Zhang Y, et al. "Turn-on" ratiometric electrochemical detection of H2O2 in one drop of whole blood sample via a novel microelectrode sensor. Biosensors and Bioelectronics. 2020;165:112402-.
94. de Meira FHA, Resende SF, Monteiro DS, Pereira MC, Mattoso LHC, Faria RC, et al. A Non-enzymatic Ag/delta-FeOOH Sensor for Hydrogen Peroxide Determination using Disposable Carbon-based Electrochemical Cells. Electroanal. 2020;32(10):2231–6.
95. Liu X, Zhang X, Zheng J. One-pot fabrication of AuNPs-Prussian blue-Graphene oxide hybrid nanomaterials for non-enzymatic hydrogen peroxide electrochemical detection. Microchemical Journal. 2021;160:105595-.
96. Amatore C, Arbault S. Oxidative Stress at the Single Cell Level. In: Michael AC, Borland LM, editors. Electrochemical Methods for Neuroscience. Frontiers in Neuroengineering. Boca Raton (FL): CRC Press; 2007.
97. Wu TD, Li L, Song GJ, Ran MM, Lu XQ, Liu XH. An ultrasensitive electrochemical sensor based on cotton carbon fiber composites for the determination of superoxide anion release from cells. Microchim Acta. 2019;186(3).
98. Gao QM, Zhao HL, Wang ZX, Cai X, Zhou LF, Lan MB. Fabrication of hierarchically porous carbon networks for the electrochemical determination of superoxide anion released from living cells. Sensor Actuat B-Chem. 2021;330.
99. Zhao SF, Shi ZZ, Guo CX, Li CM. A high-energy-state biomimetic enzyme of oxygen-deficient MnTiO3 nanodiscs for sensitive electrochemical sensing of the superoxide anion. Chem Commun. 2019;55(54):7836–9.
100. Cai X, Shi LB, Sun WQ, Zhao HL, Li H, He HY, et al. A facile way to fabricate manganese phosphate self-assembled carbon networks as efficient electrochemical catalysts for real-time monitoring of superoxide anions released from HepG2 cells. Biosens Bioelectron. 2018;102: 171–8.
101. Zheng JS, Wang B, Jin YZ, Weng B, Chen JC. Nanostructured MXene-based biomimetic enzymes for amperometric detection of superoxide anions from HepG2 cells. Microchim Acta. 2019;186(2).
102. Wang ZX, Zhao HL, Gao QM, Chen KC, Lan MB. Facile synthesis of ultrathin two-dimensional graphene-like CeO2 & ndash;TiO2 & nbsp;mesoporous nanosheet loaded with Ag nanoparticles for non-enzymatic electrochemical detection of superoxide anions in HepG2 cells. Biosens Bioelectron. 2021;184.
103. Sadeghian RB, Ostrovidov S, Han JH, Salehi S, Bahraminejad B, Bae H, et al. Online Monitoring of Superoxide Anions Released from Skeletal Muscle Cells Using an Electrochemical Biosensor Based on Thick-Film Nanoporous Gold. Acs Sensors. 2016;1(7):921–8.
104. Gubernatorova EO, Liu X, Othman A, Muraoka WT, Koroleva EP, Andreescu S, et al. Europium-Doped Cerium Oxide Nanoparticles Limit Reactive Oxygen Species Formation and Ameliorate Intestinal Ischemia-Reperfusion Injury. Adv Healthc Mater. 2017;6(14).
105. Brown MD, Schoenfisch MH. Electrochemical Nitric Oxide Sensors: Principles of Design and Characterization. Chem Rev. 2019;119(22):11551–75.
106. Jo A, Do H, Jhon G-J, Suh M, Lee Y. Real-time evaluation of nitric oxide (NO) levels in cortical and hippocampal areas with a nanopore-based electrochemical NO sensor. Neuroscience Letters. 2011;498(1):22–5.
107. Park SS, Hong M, Song C-K, Jhon G-J, Lee Y, Suh M. Real-Time in Vivo Simultaneous Measurements of Nitric Oxide and Oxygen Using an Amperometric Dual Microsensor. Analytical Chemistry. 2010;82(18):7618–24.
108. Park SS, Hong M, Ha Y, Sim J, Jhon G-J, Lee Y, et al. The real-time in vivo electrochemical measurement of nitric oxide and carbon monoxide release upon direct epidural electrical stimulation of the rat neocortex. Analyst. 2015;140(10):3415–21.
109. Moon J, Ha Y, Kim M, Sim J, Lee Y, Suh M. Dual Electrochemical Microsensor for Real-Time Simultaneous Monitoring of Nitric Oxide and Potassium Ion Changes in a Rat Brain during Spontaneous Neocortical Epileptic Seizure. Analytical Chemistry. 2016.

110. Brown MD, Schoenfisch MH. Selective and Sensocompatible Electrochemical Nitric Oxide Sensor with a Bilaminar Design. Acs Sensors. 2019;4(7):1766–73.
111. Mathew G, Narayanan N, Abraham DA, De M, Neppolian B. Facile Green Approach for Developing Electrochemically Reduced Graphene Oxide-Embedded Platinum Nanoparticles for Ultrasensitive Detection of Nitric Oxide. ACS omega. 2021;6(12):8068–80.
112. Oliveira R, Sella C, Souprayen C, Ait-Yahiatene E, Slim C, Griveau S, et al. Development of a flow microsensor for selective detection of nitric oxide in the presence of hydrogen peroxide. Electrochim Acta. 2018;286:365–73.
113. Jin ZH, Liu YL, Fan WT, Huang WH. Integrating Flexible Electrochemical Sensor into Microfluidic Chip for Simulating and Monitoring Vascular Mechanotransduction. Small. 2020;16(9):e1903204.
114. Özel RE, Alkasir RSJ, Ray K, Wallace KN, Andreescu S. Comparative Evaluation of Intestinal Nitric Oxide in Embryonic Zebrafish Exposed to Metal Oxide Nanoparticles. Small. 2013;9(24):4250–61.
115. Ferrer-Sueta G, Radi R. Chemical Biology of Peroxynitrite: Kinetics, Diffusion, and Radicals. Acs Chem Biol. 2009;4(3):161–77.
116. Amatore C, Arbault S, Bruce D, de Oliveira P, Erard M, Vuillaume M. Characterization of the electrochemical oxidation of peroxynitrite: Relevance to oxidative stress bursts measured at the single cell level. Chem-Eur J. 2001;7(19):4171–9.
117. Xue J, Ying XY, Chen JS, Xian YH, Jin LT, Jin J. Amperometric ultramicrosensors for peroxynitrite detection and its application toward single myocardial cells. Anal Chem. 2000;72 (21):5313–21.
118. Kalil H, Maher S, Bose T, Bayachou M. Manganese Oxide/Hemin-Functionalized Graphene as a Platform for Peroxynitrite Sensing. J Electrochem Soc. 2018;165(12):G3133–G40.
119. Peteu SF, Whitman BW, Galligan JJ, Swain GM. Electrochemical detection of peroxynitrite using hemin-PEDOT functionalized boron-doped diamond microelectrode. Analyst. 2016;141 (5):1796–806.
120. Liu FX, Li L, Zhang BY, Fan WZ, Zhang RJ, Liu GA, et al. A novel electrochemical sensor based on microporous polymeric nanospheres for measuring peroxynitrite anion released by living cells and studying the synergistic effect of antioxidants. Analyst. 2019;144(23): 6905–13.
121. Li Y, Sella C, Lemaitre F, Guille-Collignon M, Amatore C, Thouin L. Downstream Simultaneous Electrochemical Detection of Primary Reactive Oxygen and Nitrogen Species Released by Cell Populations in an Integrated Microfluidic Device. Anal Chem. 2018;90(15):9386–94.
122. Lin L, Qian L, Weizhou F, Lan D, Xue Z, Xiuhui L, et al. A novel ratiometric electrochemical sensing strategy for monitoring of peroxynitrite anion released from high glucose-induced cells. Sensors and Actuators, B: Chemical. 2021;328:925–4005.
123. Peteu SF, Bose T, Bayachou M. Polymerized hemin as an electrocatalytic platform for peroxynitrite's oxidation and detection. Analytica Chimica Acta. 2013;780:81–8.
124. Oprea R, Peteu SF, Subramanian P, Qi W, Pichonat E, Happy H, et al. Peroxynitrite activity of hemin-functionalized reduced graphene oxide. Analyst. 2013;138(15):4345–52.
125. Peteu S, Peiris P, Gebremichael E, Bayachou M. Nanostructured poly (3,4-ethylenedioxythiophene)–metalloporphyrin films: Improved catalytic detection of peroxynitrite. Biosensors and Bioelectronics. 2010;25(8):1914–21.
126. Hosu IS, Wang Q, Vasilescu A, Peteu SF, Raditoiu V, Railian S, et al. Cobalt phthalocyanine tetracarboxylic acid modified reduced graphene oxide: a sensitive matrix for the electrocatalytic detection of peroxynitrite and hydrogen peroxide. RSC Advances. 2015;5 (2):1474–84.
127. Wang Y, Chen Z-z. A novel poly(cyanocobalamin) modified glassy carbon electrode as electrochemical sensor for voltammetric determination of peroxynitrite. Talanta. 2010;82(2): 534–9.

128. Jiang H, Zhang XW, Liao QL, Wu WT, Liu YL, Huang WH. Electrochemical Monitoring of Paclitaxel-Induced ROS Release from Mitochondria inside Single Cells. Small. 2019;15(48): e1901787.
129. Vaneev AN, Gorelkin PV, Garanina AS, Lopatukhina HV, Vodopyanov SS, Alova AV, et al. In Vitro and In Vivo Electrochemical Measurement of Reactive Oxygen Species After Treatment with Anticancer Drugs. Anal Chem. 2020;92(12):8010–4.
130. Peng Q, Yan X, Shi X, Ou S, Gu H, Yin X, et al. In vivo monitoring of superoxide anion from Alzheimer's rat brains with functionalized ionic liquid polymer decorated microsensor. Biosens Bioelectron. 2019;144:111665.
131. Wilson LR, Panda S, Schmidt AC, Sombers LA. Selective and Mechanically Robust Sensors for Electrochemical Measurements of Real-Time Hydrogen Peroxide Dynamics in Vivo. Anal Chem. 2018;90(1):888–95.

Chapter 14
Enzymatic Methods

A. Khelfi

Abstract Enzymes are important entities in oxidative stress and play multiple functions at different levels. Some pro-oxidant/anti-oxidant enzymes, normally present intracellularly, are found in the blood circulation and thus can be assessed and used as biomarkers of oxidative stress (e.g., myeloperoxidase (MPO), xanthine oxidase (XO), nitric oxide synthases (NOS), glutathione peroxidase (GPx), glutathione reductase (GR), superoxide dismutase (SOD), thioredoxin reductase (TrxR), catalase (CAT)). Assays for activity assessment of these enzymes are now available on the market and are widely used in research. Besides, certain enzymes can be used as reagents to assess the level of some reactive species (RS) (e.g., hydrogen peroxide and glutathione (GSH)) in situ.

This chapter provides a valuable information and a laboratory resource for oxidative stress assessment using simple and low-cost methodology. We examine in this chapter the molecular basis of each enzymatic reaction with especial emphasis on its limitations and advantages.

Keywords Myeloperoxidase (MPO) · Xanthine oxidase (XO) · Nitric oxide synthases (NOS) · Glutathione peroxidase (GPx) · Glutathione reductase (GR) · Superoxide dismutase (SOD) · Thioredoxin reductase (TrxR) · Catalase (CAT) · Hydrogen peroxide · Glutathione (GSH)

14.1 Determination of Enzyme Activity

Enzymes are important actors of OS and play pleiotropic functions at different levels. Some ROS-generating enzymes, normally present intracellularly, are found in blood circulation and thus can be assessed and used as markers of OS (e.g., MPO,

A. Khelfi (✉)
Faculty of Pharmacy, University of Algiers1, Algiers, Algeria

National Centre of Toxicology, Algiers, Algeria

Department of Chemistry and Biomolecular Science, Clarkson University, Potsdam, NY, USA
e-mail: ab.khelfi@univ-alger.dz; akhelfi@clarkson.edu

S. Andreescu et al. (eds.), *Biomarkers of Oxidative Stress*,
https://doi.org/10.1007/978-3-031-60738-7_14

EPO, XO, and NOS involved in oxidative burst). Assay for activity assessment of these enzymes in their characteristic location is also an important indication of increased production of various ROS in situ.

On the other hand, OS has been speculated to cause consumption of antioxidant enzymes and their assessment has known a continuous development over time. Antioxidant enzymes that can be analyzed kinetically include SOD, CAT, and glutathione-related enzymes (GPx, GR, and GST), and TrxR.

14.1.1 Myeloperoxidase (MPO)

Myeloperoxidase is a unique heme peroxidase that catalyzes the conversion of H_2O_2 and chloride to HOCl (Eq. 2.75), a strong oxidant with powerful antimicrobial activity and broad-spectrum reactivity with biomolecules. Myeloperoxidase also oxidizes a variety of compounds including phenols and anilines, via the peroxidation cycle. It is a lysosomal hemoprotein located in the azurophilic granules of PMNs and monocytes.

Commercial kits are available for the determination of MPO activity. They use either the chlorination or the peroxidase activity. Assays based on measurement of chlorination catalysis are more specific than those based on peroxidase substrates.

In one chlorination assay, the non-fluorescent substrate 3′-(p-aminophenyl)-fluorescein is cleaved by HOCl generated after MPO catalysis, yielding fluorescein, which is quantified fluorimetrically (excitation 485 nm, emission 530 nm) as provided by Invitrogen. Another chlorination assay uses the trapping of HOCl by taurine to produce taurine chloramines. The latter is reacted with the colored 5-thio-2-nitrobenzoic acid (TNB) to generate a colorless product, dithionitrobenzoic acid (NWLSS). The decrease in absorbance at 412 nm is proportional to MPO activity of the sample.

In peroxidase activity-based assays, o-dianisidine and tetramethylbenzidine (TMB) are general peroxidase substrates, lacking specificity toward MPO [1]. In an assay provided by invitrogen, peroxidation is detected using non-fluorescent Amplex® UltraRed reagent, which is oxidized by the hydrogen peroxide-generated intermediates MPO-I and MPO-II to form a fluorescent product.

Even with availability of several ways of assessment, a limitation of MPO as a biomarker is that the current protocols are not standardized.

Myeloperoxidase is not only important in the innate immune system regarding its antimicrobial activities but also contributes to inflammatory diseases via multiple mechanisms. The suitability of MPO as a potential independent prognostic biomarker of inflammation has been proposed in numerous pathologies. Myeloperoxidase is considered as an important marker for inflammatory and autoimmune diseases as well as certain cancers [2, 3]. Besides, it has been shown that high levels of MPO in circulation are associated with CVD (myocardial infarction and acute coronary syndrome), chronic obstructive pulmonary disease, preeclampsia, Parkinson's, and Alzheimer's diseases [4–8]. Several evidence have shown that

MPO-catalyzed reactions contribute to atherogenesis, and significant amounts of MPO have been demonstrated to be present in atherosclerotic lesions [9, 10].

Myeloperoxidase has emerged as an independent marker and predictor of CVD. For example, Brennan et al. reported that MPO levels allow better identification of patients at risk for future cardiac events than other known risk factors like C-reactive protein [11]. Baldus et al. similarly reported that elevated levels of MPO (>350 mg/L) are a powerful predictor of myocardial infarction in patients with acute coronary syndrome [12]. Finally, MPO levels in plasma have also been shown to be an independent indicator of EC dysfunction [13].

Overall, the determination of MPO activity or its products may provide an important diagnostic tool for heart diseases. It has been advocated as a prognostic marker of CVD due to its role in the pathology of atherogenesis.

14.1.2 Eosinoperoxidase (EPO)

Eosinoperoxidase is an enzyme found within the eosinophil granulocytes. The major function of this enzyme is to catalyze the formation of HOBr from H_2O_2 and bromide (Eq. 2.76). Eosinoperoxidase has been implicated in promoting oxidative tissue injury in conditions ranging from asthma and other allergic inflammatory disorders to cancer and parasitic/helminthic infections [14].

The amount of EPO activity is usually determined in tissue homogenate and is assessed by an endpoint kinetic assay. The most common method is based on measuring the enzymatic activity of EPO using o-Phenylenediamine (OPD) as a substrate and H_2O_2 as an oxidant in alkaline buffer (Tris-HCl pH 8.0) [15]. The reaction is usually stopped by the addition of sulfuric acid. The absorbance at 490 nm is measured as indicator of EPO activity.

14.1.3 Xanthine Oxidase (XO)

Xanthine oxidase plays a key role in the production of RS. This enzyme exists in two interconvertible forms, XO (that catalyzes the oxidation of xanthine to uric acid using oxygen as the electron acceptor and produces $O_2^{\bullet -}$ and H_2O_2) and xanthine dehydrogenase (XDH) (that carries out the same reaction but uses NAD^+ and generates NADH). While the product is a known antioxidant, the enzyme is particularly known as an important source of $O_2^{\bullet -}$ (Eq. 2.11).

XDH is the predominant form in oxygenized tissues, but it can be converted to XO under various conditions. Inflammatory or hypoxic conditions promote XDH expression in tissues and EC, and stimulate its release into the circulation [16, 17].

The commercial Amplite™ XO assay provides a quick and ultrasensitive method for the determination of XO activity. In this assay, XO catalyzes the oxidation of purine bases, hypoxanthine, or xanthine, to uric acid and $O_2^{\bullet -}$. Subsequently, $O_2^{\bullet -}$

spontaneously dismutates to H_2O_2, which in the presence of HRP reacts stoichiometrically with Amplex® Red reagent (N-acetyl-3,7-dihydroxyphenoxazine) to generate the red-fluorescent oxidation product. The fluorescence can be easily detected using a fluorescence microplate reader at Ex/Em = 540/590 nm.

Biovision has developed another version of this protocol to evaluate XO activity. In this assay, H_2O_2 reacts stoichiometrically with OxiRed™ Probe to generate a colored (at $\lambda = 570$ nm) and fluorescent compound (at Ex/Em = 535/587 nm), which is proportional to XO content.

During OS, XO activity will increase due to the dehydrogenase-to-oxidase conversion following thiol oxidation [18]. XO-induced OS has been detected in renal diseases as well as COPD, pulmonary hypertension, sickle cell disease, and DM [17]. An increase in XO activity has been suggested in patients with heart failure [19]. The most important breakthrough was the hypothesis of the key role of XO in ischemia–reperfusion injury [20].

In another context, Amplex® Red xanthine/XO assay has been used for XO determination as a marker of recovery from exercise stress [21].

14.1.4 Nitric Oxide Synthases (NOS)

Nitric oxide synthases are a family of heme-flavoenzymes that produce the radical $NO^{\bullet}$ from *l*-arginine using NADPH and molecular oxygen. The level of different forms of NOS enzymes can be monitored and can indicate an increased production of $NO^{\bullet}$ in certain physiological and pathological conditions (see Sect. 4.1).

A general biochemical assay to assess the level of NOS activity in tissues such as muscle has been described [22]. In this assay, muscle homogenates were incubated with [^{3}H]arginine (NOS substrate), and the amount of [^{3}H]citrulline produced by NOS and separated on Dowex AG SOWX-S was quantified by liquid scintillation counting. However, as the constitutive and inducible forms of NOS may be present in the same organ and tissues, it is important to use specific inhibitors of different isoforms of NOS. Such inhibitors can distinguish quite clearly between nNOS, eNOS, and iNOS [23].

14.1.5 Glutathione Peroxidase (GPx) and Glutathione Reductase (GR)

Glutathione peroxidase is essentially intracellular enzymes, and its activity outside cells is relatively low. Besides the plasmatic GPx (GPx-3), the occurrence of this enzyme in the extracellular environment may also result from their delivery from damaged cells [24].

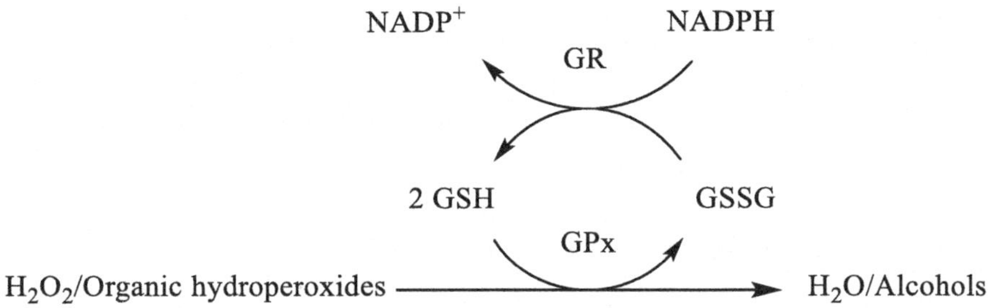

Fig. 14.1 Overview on GPx/GR system

Glutathione peroxidase catalyzes the reduction of H_2O_2 and ROOH such as lipid hydroperoxides to water and the corresponding alcohol, respectively, using GSH as the reducing agent. GR then reconverts GSSG to GSH using the reducing power of NADPH.

Glutathione peroxidase activity can be measured by following GSH consumption using polarography [25]. However, this assay suffers from many interferences. The most widely used method is based on the oxidation of reduced glutathione to the oxidized form by the enzyme in the presence of H_2O_2, and the subsequent reduction of GSSG by GR which uses NADPH as the cofactor [26]. The oxidation of NADPH to $NADP^+$ is accompanied by a decrease in absorbance at 340 nm, which is proportional to GPx activity (Fig. 14.1).

The activity is calculated from a calibration curve for NADPH after subtraction of the appropriate blanks. Commercial kits based on this principle are now available (e.g., NWLSS).

Since larger amounts of the enzyme are present in cells, hemolysis of blood can give rise to spuriously high values in plasma samples. GR activity, as a cellular biomarker, can be measured in a similar manner using NADPH and GSSG as substrates [27].

Biological samples contain various proportion of GPx which are not all selenoproteins but could act as peroxidases (called non-selenium GPx or GST). For example, erythrocytes contain only the selenoprotein form of GPx, whereas plasma contains both the selenium-dependent (80%) and the selenium-independent forms (20%) [25]. In tissues, the distribution of both forms is variable depending on the organ. As a function of the substrate used, the activity of these enzymes can be measured. For example, GST (called non-selenium GPx), which shows a glutathione peroxidase-like activity, catalyzes reaction of ROOH (cumene hydroperoxide) but not of H_2O_2. Therefore, H_2O_2 (or t butyl hydroperoxide) is used as a substrate to assay the selenium-dependent GPx [28]. On the other hand, GPx and GST activities can be measured using cumene hydroperoxide and GSH as substrates in a coupled reaction with GR.

We can also mention that a fluorimetric microassay has been developed by Martinez et al. using o-phthalaldehyde to measure GPx activity into human blood platelets [29]. This assay can be applied to many biological samples including plasma, whole blood, erythrocytes, platelets, and tissues.

Finally, globular GPx activity is significantly decreased in COPD [30–32]. Some researchers have shown that GPx-1 is a high predictor of cardiovascular risk [33, 34].

14.1.6 Superoxide Dismutase (SOD)

Superoxide dismutase is a two-edged sword. It catalyzes the dismutation of $O_2^{\bullet-}$ to molecular oxygen and H_2O_2, which freely passes through the membranes and possibly act as an oxidant and produce a more reactive and cytotoxic species ($ONOO^-$) by reaction with $NO^{\bullet}$. Hydrogen peroxide, generated by SOD catalysis, is rapidly eliminated by GPx and CAT. Therefore, SOD activity is a critical part of the antioxidant system as it participates in $O_2^{\bullet-}$ clearance. Imbalance in SOD and GPx/CAT ratio may be dangerous and favors untolerated conditions of OS.

Superoxide dismutase is essentially intracellular enzymes. Its activity outside cells is relatively low compared to the intracellular ones, and its occurrence in the extracellular environment may simply result from their release from damaged cells.

Three isoenzymes of SOD have been described in humans with different cellular localization and metal cofactor [35]:

- Homodimeric Cu/ZnSOD (SOD1), which is mainly located in the cytosol but also present in lysosome, nuclei, and the space between inner and outer mitochondrial membranes. It contains Cu and Zn at the active site.
- Homotetrameric MnSOD (SOD2), which is mitochondrial and has Mn at the active site. Note that mammalian erythrocytes (which have no mitochondria) contain no MnSOD.
- Homotetrameric Cu/ZnSOD (SOD3), which is extracellular and contains Cu and Zn at the active site. The majority of SOD present in plasma is the extracellular form (SOD3).

Measurements of SOD activity employ both direct and indirect approaches. One can directly measure the loss of UV absorbance of $O_2^{\bullet-}$ when KOH is added to aqueous solution at basic pH. However, this approach is difficult to execute because of the alkaline conditions that are incompatible with the physiological pH of 7.4 [36]. Most assays measure the activity indirectly by generating a constant flux of $O_2^{\bullet-}$ either chemically or enzymatically with a constant flux, which reduces a detector molecule and produces a colored dye.

Many indirect methods have been published in the literature using various detector molecules such as cytochrome c, luminol, nitroblue tetrazolium, lucigen, pyrogallol, NO_2^-, adrenalin, and p-iodonitrotetnazolium [37, 38]. Several commercial kits are now available to assess SOD activity in plasma and tissues using this approach. Most of the commercial kits use nitroblue tetrazolium salts (NBT) to generate a colored product (NBT-formazan) with consumption of $O_2^{\bullet-}$ (Fig. 14.2).

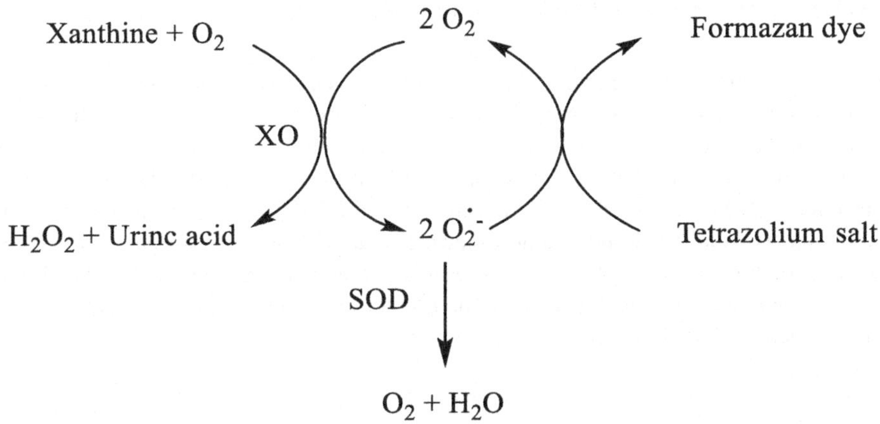

Fig. 14.2 Overview on the principle of SOD assay
XO system is used to generate $O_2^{\bullet -}$. SOD catalyzes the transformation of $O_2^{\bullet -}$ and enters in competition with tetrazolium salt, which reacts with $O_2^{\bullet -}$ to produce the colored Formazan dye

In the presence of SOD, $O_2^{\bullet -}$ reacts preferably with this enzyme resulting in a decrease in the color formation. Quantification of reduced formation of formazan by SOD is used as a means of defining SOD activity. One unit of SOD activity is defined as the amount of enzyme needed to catalyze 50% dismutation of $O_2^{\bullet -}$.

The original method developed by McCord and Fridoivich used XO system as a source of $O_2^{\bullet -}$ and cytochrome c to produce the colored product (Fig. 14.2) [39]. By removing $O_2^{\bullet -}$, SOD inhibits reaction with the detector, which causes a drop in absorbance. As mentioned earlier, an improved tetrazolium compound that generates a water-soluble product (WST-1 formazan) is currently used in almost all commercial kits (e.g., Biovision, Dojindo, Fluka, and Cayman).

In addition, some methods include CAT in the assay system to remove the H_2O_2 produced during the reaction, which can interfere in the assay by re-oxidizing the reduced dye.

Overall, SOD activity is usually measured indirectly by analyzing the inhibition in the rate of reduction of a tetrazolium salt by $O_2^{\bullet -}$ generated through XO enzymatic system. It is possible to differentiate Cu/Zn from MnSOD by adding cyanide to the reaction mixture (Cu/ZnSOD are inhibited by cyanide) [40]. Otherwise, to distinguish between the different isoforms of SOD, there have been reports on a traditional approach using specific antibodies and inhibitors against undesirable isoforms [41]. Cu/ZnSOD is the major SOD in extracellular fluids. However, extracellular SOD activity is low and its measurement requires a specific protein concentration using heparin or concanavalin A and highly sensitive SOD assay [42].

14.1.7 Thioredoxin Reductase (TrxR)

Kits are commercially available to assess TrxR activity. They provide a convenient method for quantifying TrxR activity in tissue homogenates and cell lysates. For example, the Trx assay kit available from Sigma Aldrich and Cayman is based on TrxR-mediated reduction of 5,5′-dithio-bis(2-nitrobenzoic acid) (DTNB) in the presence of NADPH to 5-thio-2-nitrobenzoic acid (TNB), which produces a yellow color that is monitored spectrophotometrically at 412 nm. Measurement of TrxR activity in the absence and in the presence of aurothiomalate, a specific TrxR inhibitor included in the kit, allows for correction of non-thioredoxin reductase-independent DTNB reduction.

14.1.8 Catalase (CAT)

Catalase is a common antioxidant enzyme found in nearly all living organisms. Catalase, which catalyzes the dismutation of H_2O_2 into water and oxygen, is a homotetrameric protein containing four iron hemes (Eqs. 5.8, 5.9 and 5.10). This enzyme is largely located in the peroxisomes and prevents excessive buildup of H_2O_2. Thus, CAT allows important cellular processes that produce H_2O_2 as a by-product to take place safely. In addition, CAT exhibits peroxidase activity, whereby it catalyzes the conversion of low molecular weight alcohols to form aldehydes. In fact, the enzyme has been found to oxidize hydrogen donors like methanol into formaldehyde when methanol concentration is in excess compared to hydrogen peroxide concentration.

Catalase is an intracellular enzyme that is present in plasma at low concentration. It is, therefore, preferable to determine its activity in erythrocyte lysates rather than in plasma in order to assess the antioxidant capacity.

Catalase activity is usually determined by its catalytic function. There are three major ways of activity measurement. They involve either the examination of H_2O_2 remaining after short incubation of the sample with known amount of H_2O_2, the determination of molecular oxygen appearance using electrochemical detection, or the determination of aldehydes generated by its peroxidase activity with alcohols.

1. In the first approach, which is rather straightforward determination, the decrease of H_2O_2 is measured by a spectrophotometer. If the level of H_2O_2 is in excess, the initial decomposition of H_2O_2 is proportional to the presence of CAT in a biological sample. Removal of H_2O_2 can be monitored by the loss of absorption at 240 nm. This method is available for measurements in whole blood and erythrocytes. Alternatively, the remaining H_2O_2 is reacted with 4-aminoantipyrene and 3,5-dichloro-2-hydroxybenzene sulfonic acid (DHBS) (Trinder reagent) in the presence of HRP. The resulting quinoneimine compound is measured at 520 nm. The decrease in the color is proportional to the activity of

CAT in the sample. This principle (Eqs. 14.1 and 14.2) is found in the commercial kits marketed by Cell Biolabs, Oxis, Sigma, and others:

$$2H_2O_2 \xrightarrow{\text{CAT}} O_2 + 2H_2O \tag{14.1}$$

$$\text{Remaining}\, H_2O_2 + \text{DHBS} + \text{AAP} \xrightarrow{\text{Peroxidase}} \text{Quinineimine dye} \tag{14.2}$$

Another method is available where the remaining H_2O_2 is reacted with Amplex Red and peroxidase to produce a fluorescent product (resorufin), which is measured fluorometrically (Ex 570 nm; Em, 590 nm). This method is employed in the kits sold by Molecular Probes, Assay Designs, Invitrogen, and Bachem.

Similarly, the assay offered by aatbio uses Amplite™ Red, which reacts with H_2O_2 to generate a red-fluorescent product. Decay of fluorescence intensity is proportional to CAT activity. Amplite™ Red enables a dual recordable mode as the fluorescent signal can be easily recorded by either a fluorescence microplate reader at Ex/Em = 540/590 nm or an absorbance microplate reader at 576 nm.

2. In the second approach, the liberation of oxygen is determined using an oxygen electrode or by polarography [25]
3. The third and more recent approach is based on the peroxidative activity of CAT. It is based on two steps mechanism of CAT reaction, and the assay kits based on this approach are sold by Calbiochem and Cayman [43]. Here the sample containing CAT is reacted with methanol in the presence of an optimal amount of H_2O_2. The resulting formaldehyde is reacted with 4-amino 3-hydrazino-5-mercapto-1,2,4 triazole (Purpald) to give a dark pink color which is measured at 540 nm. The use of the dual function ensures good reaction specificity. This method is suitable for use on blood elements (white and red blood cells) with a satisfactory sensitivity.

14.2 Determination of Reactive Species and Antioxidant Molecules

14.2.1 Determination of H_2O_2

14.2.1.1 Determination of H_2O_2 Using Catalase Activity

One of the most traditional ways of measuring hydrogen peroxide production is represented by the inhibition of its clearance by CAT. The principle is based on the rapid formation of an intermediate complex called Compound I ($^{+\bullet}$CAT-Fe^{+4}) during CAT-dependent metabolism of H_2O_2 (Fig. 14.3). Because CAT is specific for H_2O_2, monitoring of the CAT intermediate is a highly selective method to assess hydrogen peroxide levels. Briefly, intracellular production of H_2O_2 can be assessed by

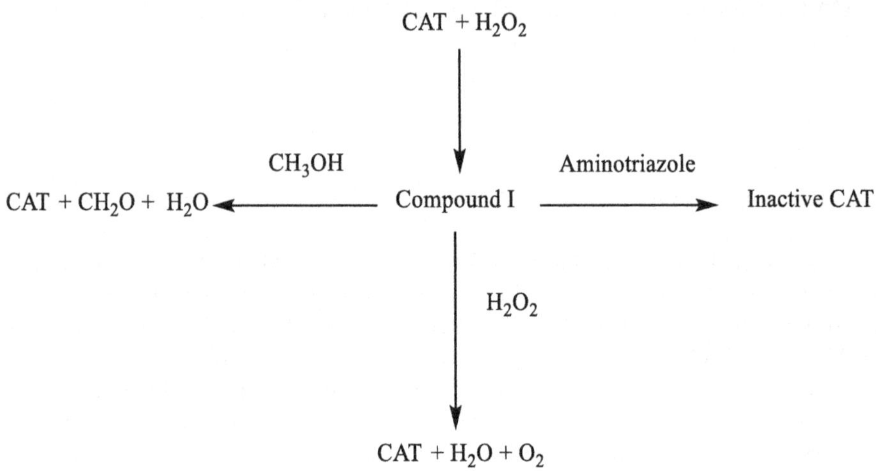

Fig. 14.3 Principle of hydrogen peroxide determination using catalase method

introducing a specific inhibitor for CAT, aminotriazole (at concentrations between 1 and 10 mM), which reacts with the Compound I intermediate causing irreversible inhibition of the enzyme only in the presence of H_2O_2. As the inhibition is dependent on the presence of H_2O_2, the amount of H_2O_2 generated in the unknown sample can be calculated from the rate of inhibition. The intermediate complex that is formed in catalase-dependent metabolism of H_2O_2 is stable; thus, steady-state monitoring of a hydrogen peroxide-catalase complex can be followed with dual-wavelength absorption meter at 640–660 nm or time-dependent elimination of H2O2, determined by the DCF substrate as described earlier. Intracellular rates of production of H_2O_2 can be estimated by determining the half time of aminotriazole-dependent inactivation of CAT.

This method has in fact been used to assess rates of hydrogen peroxide formation in different cell types as well as in tissue homogenates [44–46].

14.2.1.2 Determination of H_2O_2 Using Horse Radish Peroxidase Activity

Peroxidase-based assays have been developed to specifically quantify hydrogen peroxide generation in some biological specimens. The HRP-linked assays are particularly useful for the quantification of H_2O_2 in cultured cells and isolated buffer-perfused tissue preparations. However, these assays are not suitable for H_2O_2 quantification in serum and plasma since many reducing agents present in the extracellular fluids will interfere with the assay.

HRP-linked assays are based on the HRP-catalyzed hydrogen peroxide oxidation of a wide range of electron donating compounds (AH):

$$\text{HRPH-Fe}^{+3} + H_2O_2 \rightarrow {}^{\bullet +}\text{HRPH-Fe}^{+4} \quad (14.3)$$

$$ {}^{\bullet +}\text{HRPH-Fe}^{+4} + \text{AH} \rightarrow \text{HRPH-Fe}^{+3} + \text{A}^{+} \quad (14.4)$$

${}^{+\bullet}$HRP-Fe^{+4} represents the ferryl porphyrin cation radical intermediate of HRP.

The amount of H_2O_2 is estimated by following the decay of fluorescence of initially fluorescent probes, reacting as electron donors, such as scopoletin (7-hydroxy-6-methoxy-coumarin; the most commonly used electron donor), or by monitoring the increase in fluorescence from non-fluorescent electron donors, which become fluorescent upon oxidation such as diacetyldichlorofluorescin ($DCFH_2$-DA), p-hydroxyphenylacetate (p-HPA), homovanillic acid (3-methoxy-4-hydroxyphenylacetic acid; HVA), 4-aminoantipyrine, and Amplex Red (N-acetyl-3,7-dihydroxyphenoxazine) [47–50]. For example, Amplex Red is oxidized by H_2O_2 in the presence of HRP and in doing so converted to resorufin (see Chaps. 10 and 11). Unlike Amplex Red, resorufin is a highly colored compound that can be detected colorimetrically at 570 nm or by fluorescence using excitation of 560 nm and emission of 590 nm. Homovanillic acid dimerizes when oxidized by H_2O_2 through HRP catalysis. As with Amplex red, homovanillic acid monomer is non-fluorescent, but as a dimmer, it possesses a peak excitation wavelength of 315 nm, with an emission wavelength of 425 nm (see Chap. 1). Care should be taken when using these compounds to assess hydrogen peroxide production. Several peroxidase-like metalloporphoryns have been shown to be catalytic for the reaction in addition to HRP.

Oxidation of non-fluorescent dyes such as tetramethylbenzidine or phenol red in the presence of H_2O_2 and HRP is followed spectrophotometrically using the same principle [51]. The phenol red-HRP assay is pH-dependent and less sensitive than fluorescence-based methods, reducing its utility when attempting to detect H_2O_2 in cells with low endogenous rates of oxidant formation.

The major advantages of using HRP are its specificity toward H_2O_2 and its commercial availability at low cost. The major disadvantages are its relatively low specificity toward hydrogen-donating substrates:

- First, numerous biological substances, including thiol compounds and vitamin C, can serve as substrates for HRP and thus compete with the detector molecule for oxidation, leading to underestimation of hydrogen peroxide formation.
- Second, competition with HRP by endogenous CAT for H_2O_2 can artificially reduce the amount of H_2O_2 and lead to underestimation of H_2O_2. For example, in the scopoletin-HRP-coupled assay, rates of H_2O_2 formation in isolated mitochondria or submitochondrial particles ranged from 2% to 76% of those detected by either oxygen consumption or by hydrogen peroxide-induced cytochrome c peroxidase Compound I formation [52].
- Third, quenching of fluorescent signals by cell and tissue components can lead to overestimation of H_2O_2 in the case of scopoletin or underestimation, as with homovanillic acid. When measuring cellular or subcellular hydrogen peroxide accumulation, accuracy of the assay can be improved by separation of the

incubation media from cellular components before addition of the detection system thus limiting confounding interactions of the detection system with cellular elements [53].

Use of HRP-coupled reactions with spectrophotometric detection can also have limitations. Direct reduction of the oxidized detector molecule by electron transport components can limit the utility of tetramethylbenzidine in hydrogen peroxide determination, for example, in mitochondria [53]. Although spectrophotometric detection of H_2O_2 with phenol red has been used successfully in purified enzyme–substrate mixtures and to detect hydrogen peroxide release from activated leukocytes [54], the phenol red-HRP assay is pH-dependent and less sensitive than fluorescent methods, reducing its utility when attempting to detect H_2O_2 from vascular tissue [52].

Overall, HRP-linked assays are usually performed in solutions in which electron donors and other interfering substances are at minimum. In complex systems such as plasma, the high amount of competing hydrogen donors makes this assay impractical. Tarpley et al. summarized the utility of HRP-linked assays quite succinctly by stating that the method(s) are quite useful for the quantification of hydrogen peroxide levels in cultured cells, organ cultures, and isolated buffer-perfused tissue preparations [52].

14.2.1.3 Determination of H_2O_2 Using GPx and GR Activities

The use of GR and GPx allows for the specific measurement of hydrogen peroxide efflux. This assay is based on the reaction of H_2O_2 with GSH and the regeneration of the latter by GR (Eqs. 5.2 and 5.4). Disappearance of NADPH is monitored spectrophotometrically to estimate hydrogen peroxide levels. This method is easy to run but not very sensitive and needs to inactivate endogenous enzymes present in biological fluids. The addition of cyanide is used to inactivate many NOX that may interfere with the assay.

14.2.1.4 Determination of H_2O_2 Using Cytochrome c Peroxidase Activity

In this assay, cytochrome c peroxidase forms a stable complex with H_2O_2 that can be measured with a dual-wavelength spectrophotometer. This assay is not specific and cannot be used in biological samples.

14.2.2 Determination of $O_2^{\bullet -}$ Using Aconitase Activity

Aconitase is a citric acid cycle enzyme belonging to the family of dehydrogenases containing iron–sulfur (4Fe–4S) centers that catalyze the conversion of citrate to isocitrate. The mitochondrial and the cytosolic forms of aconitase are inactivated by $O_2^{\bullet -}$. Intracellular generation of $O_2^{\bullet -}$ can be estimated by quantifying the superoxide-dependent inactivation of aconitase, with low levels of enzyme activity reflecting high levels of $O_2^{\bullet -}$. The latter is a cytosolic and mitochondrial enzyme that catalyzes the interconversion of citrate to isocitrate. This enzyme contains iron–sulfur cluster [^{4}Fe–^{4}S] in its active site. Superoxide inactivates aconitase due to the reversible oxidation and loss of iron from enzyme clusters [55]. This results in aconitase absorbance changes between 240 nm and 340 nm.

Because of ongoing basal production of $O_2^{\bullet -}$, a fraction of aconitase at any moment is inactivated but capable of reactivation. This ratio of inactive/active aconitase increases as rates of $O_2^{\bullet -}$ production are augmented [52]. Therefore, aconitase assay is used as a sensitive index of changes in steady-state levels of $O_2^{\bullet -}$, and relative rates of $O_2^{\bullet -}$ production within the cytosol and/or mitochondria can be determined. Aconitase activity can be measured in cells and tissue homogenates by monitoring spectrophotometrically the conversion of isocitrate to cis-aconitate at 240 nm or 340 nm.

Aconitase assay suffers from several background interferences. Other possibilities are available such as the coupled assay where citrate serves as substrate for aconitase, and the isocitrate formed is then converted to α-ketoglutarate by NADP-dependent isocitrate dehydrogenase. This allows monitoring of NADPH formation at 340 nm with diminished background interference. Also, because inactivation of the enzyme occurs over several hours or days, it is impossible to intervene with specific scavengers to restore its activity. Aconitase activity is restored by the addition of reducing agent dithiothreitol, Na_2S, or ferrous ammonium sulfate (thiol reagents and reduced iron) to cellular extracts and remeasuring aconitase activity [56].

The reaction between $O_2^{\bullet -}$ and aconitase is quite rapid and sensitive, so this assay has been used to determine relative rates of $O_2^{\bullet -}$ formation in a variety of mammalian cells including cultured neuronal cells, fibroblasts, and macrophages, as well as heart, liver, lung, and brain tissues [52].

There are precautions and limitations that need to be considered when performing this assay. In addition to $O_2^{\bullet -}$, other oxidants can modulate aconitase activity. For instance, $ONOO^-$ can oxidize the iron–sulfur cluster of the enzyme with subsequent loss of activity. It has also been reported that $NO^\bullet$ can reduce aconitase activity, but it is still unclear if $NO^\bullet$ itself is responsible for enzyme inactivation or it is the result of ensuing $ONOO^-$ formation. Hydrogen peroxide has also been found to inhibit aconitase activity, albeit with rate constants 3–4 orders of magnitude lower than those recorded with $O_2^{\bullet -}$ and $ONOO^-$ [57]. Thus, a direct contribution of H_2O_2 to oxidant-mediated aconitase inactivation is insignificant [58]. It is wise to avoid the influences from the above oxidants with specific inhibitors when detecting the level of $O_2^{\bullet-}$ by this method.

14.2.3 Determination of Organic Hydroperoxides

14.2.3.1 Determination of Organic Hydroperoxides Using Cyclooxygenase Activity

Organic hydroperoxides have been shown to activate COX, which can be measured by monitoring oxygen consumption using electrodes. Although COX activity is proportional to ROOH concentration in the reactive medium, this assay is not specific and cannot be used for ROOH assessment in biological samples.

14.2.3.2 Determination of Organic Hydroperoxides Using GPx Activity

In this assay, GPx breaks down ROOH to their alcohol constituents by using the reducing power of GSH (Eq. 5.3). Organic hydroperoxides in samples can be measured by the addition of GPx and GSH and monitoring the GSSG formation. Consumption of NADPH (after reduction of oxidized glutathione with GR) can be used, followed by UV-spectroscopy [59].

The possible interference for this assay is high amount of H_2O_2 as GPx catalyzes GSH oxidation in the presence of this compound. This assay is not specific and cannot be used for ROOH detection in biological samples. To overcome the problem, various extraction procedures have been introduced, but these lead to extensive pretreatment of biological samples and to artifacts in vitro. In enzyme-based assays, the activity of the enzyme used varies to some extent with the type of lipid peroxide present in the sample. Despite this issue, this assay yield results comparable to the colorimetric assays cited above when applied to human plasma [60].

14.2.4 Determination of Ascorbic Acid Using Ascorbate Oxidase Activity

Great specificity for ascorbate can be achieved if ascorbate oxidase is coupled with an assay for dehydroascorbate. Ascorbic acid oxidase catalyzes the following reaction:

$$\text{Ascorbate} + 1/2O_2 \xrightarrow{\text{Ascorbate Oxidase}} \text{Dehydroascorbate} + H_2O \quad (14.5)$$

In this assay, dehydroascorbic acid production is monitored spectrophotometrically. Ascorbic acid is determined as the difference between the initial dehydroascorbate value and dehydroascorbate value after ascorbate oxidase addition. Absorbance changes are monitored at 346 nm [61].

This enzyme-linked method is specific for ascorbic acid due to the high specificity that ascorbate oxidase possesses for its natural substrate, ascorbic acid. The detection limit is less than 0.5 μM.

14.2.5 Determination of GSH Using Glutathione S-Transferase

By using a luciferin derivative in conjunction with glutathione S-transferase enzyme, the amount of GSH would be proportional to the luminescent signal generated when luciferase is added in a subsequent step.

References

1. Pulli B, Ali M, Forghani R, Schob S, Hsieh KLC, Wojtkiewicz G, et al. Measuring Myeloperoxidase Activity in Biological Samples. PLOS ONE. 2013;8(7):e67976.
2. Strzepa A, Pritchard KA, Dittel BN. Myeloperoxidase: A new player in autoimmunity. Cellular immunology. 2017;317:1–8.
3. Qian S, Golubnitschaja O, Zhan X. Chronic inflammation: key player and biomarker-set to predict and prevent cancer development and progression based on individualized patient profiles. EPMA Journal. 2019;10(4):365–81.
4. Frijhoff J, Winyard PG, Zarkovic N, Davies SS, Stocker R, Cheng D, et al. Clinical Relevance of Biomarkers of Oxidative Stress. Antioxid Redox Signal. 2015;23(14):1144–70.
5. Kimak E, Zięba B, Duma D, Solski J. Myeloperoxidase level and inflammatory markers and lipid and lipoprotein parameters in stable coronary artery disease. Lipids Health Dis. 2018;17 (1):71.
6. Scharnagl H, Kleber ME, Genser B, Kickmaier S, Renner W, Weihrauch G, et al. Association of myeloperoxidase with total and cardiovascular mortality in individuals undergoing coronary angiography–the LURIC study. International journal of cardiology. 2014;174(1):96–105.
7. Morrow J, Qiu W, Chhabra D, Rennard S, Belloni P, Belousov A, et al. Identifying a gene expression signature of frequent COPD exacerbations in peripheral blood using network methods. BMC medical genomics. 2015;8:1.
8. Tzikas S, Schlak D, Sopova K, Gatsiou K, Stakos D, Stamatelopoulos K, et al. Increased Myeloperoxidase Plasma Levels in Patients with Alzheimer's Disease. Journal of Alzheimer's disease : JAD. 2013;39.
9. Daugherty A, Dunn JL, Rateri DL, Heinecke JW. Myeloperoxidase, a catalyst for lipoprotein oxidation, is expressed in human atherosclerotic lesions. J Clin Invest. 1994;94(1):437–44.
10. Senders ML, Mulder WJM. Targeting myeloperoxidase in inflammatory atherosclerosis. European Heart Journal. 2018;39(35):3311–3.
11. Brennan M-L, Penn MS, Van Lente F, Nambi V, Shishehbor MH, Aviles RJ, et al. Prognostic Value of Myeloperoxidase in Patients with Chest Pain. 2003;349(17):1595–604.

12. Baldus S, Heeschen C, Meinertz T, Zeiher AM, Eiserich JP, Münzel T, et al. Myeloperoxidase Serum Levels Predict Risk in Patients With Acute Coronary Syndromes. 2003;108(12):1440–5.
13. Vita JA, Brennan M-L, Gokce N, Mann SA, Goormastic M, Shishehbor MH, et al. Serum Myeloperoxidase Levels Independently Predict Endothelial Dysfunction in Humans. 2004;110 (9):1134–9.
14. Wu W, Chen Y, Hazen SL. Eosinophil peroxidase nitrates protein tyrosyl residues. Implications for oxidative damage by nitrating intermediates in eosinophilic inflammatory disorders. J Biol Chem. 1999;274(36):25933–44.
15. Yamaguchi S, Tomomatsu N, Komatsu H. Effect of Y-24180, a receptor antagonist to platelet-activating factor (PAF), on allergic cutaneous reactions in actively sensitized mice. Inflammation Research. 2000;49(11):584–90.
16. Nishino T, Okamoto K, Eger BT, Pai EF, Nishino T. Mammalian xanthine oxidoreductase – mechanism of transition from xanthine dehydrogenase to xanthine oxidase. 2008;275(13): 3278–89.
17. Cantu-Medellin N, Kelley EE. Xanthine oxidoreductase-catalyzed reactive species generation: A process in critical need of reevaluation. Redox Biol. 2013;1(1):353–8.
18. Nishino T, Okamoto K, Kawaguchi Y, Hori H, Matsumura T, Eger BT, et al. Mechanism of the conversion of xanthine dehydrogenase to xanthine oxidase: identification of the two cysteine disulfide bonds and crystal structure of a non-convertible rat liver xanthine dehydrogenase mutant. J Biol Chem. 2005;280(26):24888–94.
19. Doehner W, Jankowska EA, Springer J, Lainscak M, Anker SD. Uric acid and xanthine oxidase in heart failure – Emerging data and therapeutic implications. Int J Cardiol. 2016;213:15–9.
20. Wu MY, Yiang GT, Liao WT, Tsai APY, Cheng YL, Cheng PW, et al. Current Mechanistic Concepts in Ischemia and Reperfusion Injury. Cellular Physiology and Biochemistry. 2018;46 (4):1650–67.
21. Probes for Reactive Oxygen Species, Including Nitric Oxide: ThermoFisher scientific;
22. Kobzik L, Reid MB, Bredt DS, Stamler JS. Nitric oxide in skeletal muscle. Nature. 1994;372 (6506):546–8.
23. Garvey EP, Furfine ES, Sherman PA. Purification and inhibitor screening of human nitric oxide synthase isozymes. Methods in enzymology. 268: Academic Press; 1996. p. 339–49.
24. Giustarini D, Dalle-Donne I, Tsikas D, Rossi R. Oxidative stress and human diseases: Origin, link, measurement, mechanisms, and biomarkers. Critical reviews in clinical laboratory sciences. 2009;46(5–6):241–81.
25. Thérond P, Bonnefont-Rousselot D, Davit-Spraul A, Conti M, Legrand A. Biomarkers of oxidative stress: an analytical approach. Current opinion in clinical nutrition and metabolic care. 2000;3(5):373–84.
26. Massarsky A, Kozal JS, Di Giulio RT. Glutathione and zebrafish: Old assays to address a current issue. Chemosphere. 2017;168:707–15.
27. Marrocco I, Altieri F. Measurement and Clinical Significance of Biomarkers of Oxidative Stress in Humans. 2017;2017:6501046.
28. Kulak MV, Cyr AR, Woodfield GW, Bogachek M, Spanheimer PM, Li T, et al. Transcriptional regulation of the GPX1 gene by TFAP2C and aberrant CpG methylation in human breast cancer. Oncogene. 2013;32(34):4043–51.
29. Martinez JI, Launay JM, Dreux C. A sensitive fluorimetric microassay for the determination of glutathione peroxidase activity. Application to human blood platelets. Anal Biochem. 1979;98 (1):154–9.
30. Mohammed A, Gutta V, Ansari MS, Saladi Venkata R, Jamil K. Altered antioxidant enzyme activity with severity and comorbidities of chronic obstructive pulmonary disease (COPD) in South Indian population. COPD Research and Practice. 2017;3(1):4.
31. Vlahos R, Bozinovski S. Glutathione peroxidase-1 as a novel therapeutic target for COPD. Redox Report. 2013;18(4):142–9.
32. Singh S, Verma SK, Kumar S, Ahmad MK, Nischal A, Singh SK, et al. Evaluation of Oxidative Stress and Antioxidant Status in Chronic Obstructive Pulmonary Disease. 2017;85(2):130–7.

33. Schnabel R, Lackner KJ, Rupprecht HJ, Espinola-Klein C, Torzewski M, Lubos E, et al. Glutathione Peroxidase-1 and Homocysteine for Cardiovascular Risk Prediction: Results from the AtheroGeneStudy. Journal of the American College of Cardiology. 2005;45(10): 1631–7.
34. Souiden Y, Mallouli H, Meskhi S, Chaabouni Y, Rebai A, Chéour F, et al. MnSOD and GPx1 polymorphism relationship with coronary heart disease risk and severity. Biological Research. 2016;49(1):22.
35. Che M, Wang R, Li X, Wang H-Y, Zheng XFS. Expanding roles of superoxide dismutases in cell regulation and cancer. Drug Discov Today. 2016;21(1):143–9.
36. Reznick AZ, Packer L, Sen CK. Strategies to assess oxidative stress. In: Reznick AZ, Packer L, Sen CK, Holloszy JO, Jackson MJ, editors. Oxidative Stress in Skeletal Muscle. Basel: Birkhäuser Basel; 1998. p. 43–58.
37. Xu C, Liu S, Liu Z, Song F, Liu S. Superoxide generated by pyrogallol reduces highly water-soluble tetrazolium salt to produce a soluble formazan: a simple assay for measuring superoxide anion radical scavenging activities of biological and abiological samples. Analytica chimica acta. 2013;793:53–60.
38. Arauz J, Ramos-Tovar E, Muriel P. Redox state and methods to evaluate oxidative stress in liver damage: From bench to bedside. Annals of Hepatology. 2016;15(2):160–73.
39. McCord JM, Fridovich I. Superoxide dismutase. An enzymic function for erythrocuprein (hemocuprein). J Biol Chem. 1969;244(22):6049–55.
40. Imai N, Okabe T. Kinetics Differ Between Copper-Zinc and Manganese Superoxide Dismutase Activity With Acute Ischemic Stroke. Journal of Stroke and Cerebrovascular Diseases. 2011;20 (1):75–8.
41. Suzuki K, Ohno H, Oh-ishi S, Kizaki T, Ookawara T, Fujii J, et al. Chapter 11. Superoxide dismutases in exercise and disease. 2000. p. 243–95.
42. Antonyuk S, Strange R, Marklund S, Hasnain S. The Structure of Human Extracellular Copper-Zinc Superoxide Dismutase at 1.7 Å Resolution: Insights into Heparin and Collagen Binding. Journal of molecular biology. 2009;388:310–26.
43. Johansson LH, Håkan Borg LA. A spectrophotometric method for determination of catalase activity in small tissue samples. Analytical Biochemistry. 1988;174(1):331–6.
44. Ruiz-Ojeda FJ, Gomez-Llorente C, Aguilera CM, Gil A, Rupérez AI. Impact of 3-Amino-1,2,4-Triazole (3-AT)-Derived Increase in Hydrogen Peroxide Levels on Inflammation and Metabolism in Human Differentiated Adipocytes. PLOS ONE. 2016;11(3):e0152550.
45. Walton PA, Pizzitelli M. Effects of peroxisomal catalase inhibition on mitochondrial function. Front Physiol. 2012;3:108.
46. Wagner BA, Evig CB, Reszka KJ, Buettner GR, Burns CP. Doxorubicin increases intracellular hydrogen peroxide in PC3 prostate cancer cells. Archives of biochemistry and biophysics. 2005;440(2):181–90.
47. Han D, Loukianoff S, McLaughlin L. Part VII Chapter 17: Oxidative stress indices: analytical aspects and significance. In: Sen CK, Packer L, Hänninen OOP, editors. Handbook of Oxidants and Antioxidants in Exercise. Amsterdam: Elsevier Science B.V.; 2000. p. 433–83.
48. Miwa S, Treumann A, Bell A, Vistoli G, Nelson G, Hay S, et al. Carboxylesterase converts Amplex red to resorufin: Implications for mitochondrial H2O2 release assays. Free radical biology & medicine. 2016;90:173–83.
49. Summers F, Zhao B, Ganini D, Mason R. Photooxidation of Amplex Red to Resorufin: Implications of Exposing the Amplex Red Assay to Light. Methods in enzymology. 2013;526:1–17.
50. Quinlan CL, Perevoschikova IV, Goncalves RLS, Hey-Mogensen M, Brand MD. The determination and analysis of site-specific rates of mitochondrial reactive oxygen species production. Methods in enzymology. 2013;526:189–217.
51. Fernando CD, Soysa P. Optimized enzymatic colorimetric assay for determination of hydrogen peroxide (H2O2) scavenging activity of plant extracts. MethodsX. 2015;2:283–91.

52. Tarpey MM, Wink DA, Grisham MB. Methods for detection of reactive metabolites of oxygen and nitrogen: in vitro and in vivo considerations. Am J Physiol Regul Integr Comp Physiol. 2004;286(3):R431–44.
53. Staniek K, Nohl H. H(2)O(2) detection from intact mitochondria as a measure for one-electron reduction of dioxygen requires a non-invasive assay system. Biochim Biophys Acta. 1999;1413 (2):70–80.
54. Pick E, Keisari Y. A simple colorimetric method for the measurement of hydrogen peroxide produced by cells in culture. J Immunol Methods. 1980;38(1–2):161–70.
55. Grisham MB. Detection of Reactive Metabolites of Oxygen and Nitrogen. Oxidative Stress in Aquatic Ecosystems. p. 327–43.
56. Hausladen A, Fridovich I. Measuring nitric oxide and superoxide: rate constants for aconitase reactivity. Methods in enzymology. 1996;269:37–41.
57. Flint DH, Tuminello JF, Emptage MH. The inactivation of Fe-S cluster containing hydro-lyases by superoxide. J Biol Chem. 1993;268(30):22369–76.
58. Tarpey MM, Fridovich I. Methods of Detection of Vascular Reactive Species. 2001;89(3): 224–36.
59. Ruiz C, Alegría A, Barberá R, Farré R, Lagarda MJ. Determination of plasma lipid hydroperoxides by an NADPH/NADP+ coupled enzyme reaction system. Evaluation of a method. Eur J Clin Chem Clin Biochem. 1997;35(12):893–8.
60. Abuja PM, Albertini R. Methods for monitoring oxidative stress, lipid peroxidation and oxidation resistance of lipoproteins. Clin Chim Acta. 2001;306(1–2):1–17.
61. Moeslinger T, Brunner M, Volf I, Spieckermann PG. Spectrophotometric determination of ascorbic acid and dehydroascorbic acid. Clinical Chemistry. 1995;41(8):1177–81.

Chapter 15
Chromatographic Methods

A. Khelfi

Abstract Chromatography is an analytical technique commonly used for separating a mixture of substances into its individual components. It is a family of separation techniques known for their high resolution and specificity. These techniques are one of the most widely used technologies in oxidative stress assessment in research and clinical practice. Several components of oxidative stress such as protein, lipid, carbohydrate, and DNA oxidative end-products (e.g., malondialdehyde (MDA), 4-hydroxy-2-nonenal (4-HNE), conjugated dienes, isoprostanes (IsoPs), protein carbonyl) can be detected using a variety of chromatographic methods (particularly HPLC and GC equipped with different detectors). Antioxidant compounds can also be detected using this technology with high sensitivity (e.g., glutathione (GSH), coenzyme Q_{10}, and vitamin E).

The present state of knowledge and future trends in oxidative stress research using chromatographic techniques are presented in depth in this chapter. We give a comprehensive review of all available methods allowing for the detection of most of the stress markers with especial emphasis on their limitations and advantages. Many solutions are given and briefly summarized.

Keywords Malondialdehyde (MDA) · 4-Hydroxy-2-nonenal (4-HNE) · Conjugated dienes · isoprostanes (IsoPs) · Protein carbonyl · Carbohydrate · Glutathione (GSH) · Coenzyme Q_{10}, · Vitamin E · HPLC

A. Khelfi (✉)
Faculty of Pharmacy, University of Algiers1, Algiers, Algeria

National Centre of Toxicology, Algiers, Algeria

Department of Chemistry and Biomolecular Science, Clarkson University, Potsdam, NY, USA
e-mail: ab.khelfi@univ-alger.dz; akhelfi@clarkson.edu

S. Andreescu et al. (eds.), *Biomarkers of Oxidative Stress*,
https://doi.org/10.1007/978-3-031-60738-7_15

15.1 Hydroxyl Radical

Hydroxyl radical is formed when H_2O_2 reacts with ferrous ions (Fenton reaction). This radical can be detected using specific reagents to generate stable compounds detectable by chromatography methods, of which liquid chromatography and its combination with convenient detectors are often used. This method is based on the ability of $OH^•$ to attack the benzene rings of aromatic molecules used as probes and to produce hydroxylated products that can be measured by chromatographic procedures.

For in vivo studies, two important conditions are required for the probe; the hydroxylated product must not be metabolized and must be different from possible enzyme-produced hydroxylated metabolites.

To measure $OH^•$ in biological systems, exogenously aromatic probes are added to stabilize the radical. When small amounts are added to a biological system, both salicylate and phenylalanine undergo addition reactions with $OH^•$ to give hydroxylated products, which can be determined by HPLC (UV and EC) or GC instruments.

The success of aromatic compounds in detecting $OH^•$ depends on their concentration at sites of FR generation in relation to that of other $OH^•$ scavengers (virtually all other biomolecules react fast with $OH^•$ and will compete). The following are aromatic traps, which are useful in in vivo studies as some them are known to be acceptable for human consumption:

15.1.1 Salicylate Method

Grootveld and Halliwell have first discovered the aromatic hydroxylation products of 2,3- and 2,5-dihydroxybenzoic acids (2,3-DHBA and 2,5-DHBA with 49% and 40% yields, respectively), along with a minor metabolite, catechol (11%), all formed from a salicylate probe upon $OH^•$ attack (Eq. 15.1) [1]. 2,3-DHBA is excited at 305 nm and is highly fluorescent at 407 nm.

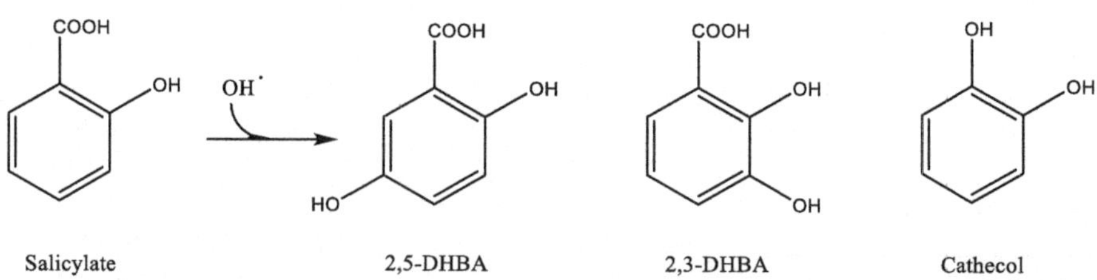

(15.1)

As $OH^•$ half-life is very short (Table 1.1), a real-time detection of $OH^•$ has been proposed. As salicylate reacts quite fast to yield stable products, this has proved to be a successful technique for measuring the appearance of $OH^•$ in living organisms. In vivo studies in humans have shown that salicylic acid in its acetylated form (aspirin) is also a suitable trapping agent. Therefore, oral administration of the drug was proposed for OS assessment in many reaction systems, and the amount of 2,3-DHBA appearing in the urine after an aspirin challenge appears to be a direct reflection of $OH^{\bullet}$ concentration [2–4].

It is not surprising that 2,3-DHBA has been reported to be a sensitive indicator of oxidative damage in many diseases. After oral administration of aspirin, elevated levels of 2,3-DHBA were found in the plasma of patients undergoing chronic OS and RA, suggesting the involvement of $OH^{\bullet}$ in such pathologies [1, 5]. The research by Ghiselli suggests that $OH^{\bullet}$ is involved in the pathogenesis of late complications in diabetes mellitus (DM) [6]. By using salicylate hydroxylation probes, it has been shown that $OH^{\bullet}$ is generated during myocardial and cerebral ischemia in experimental models [7–9]. Salicylate has been used to measure ex vivo radical formation in humans exposed to ozone [10]. Other examples include alcoholism [11], and myocardial infarction [12].

Although widely used, this method has several fundamental disadvantages. Essentially, the drawbacks of this method are the hydroxylation behavior and bioactivity of salicylate. It has been reported that 2,5-DHBA may be produced enzymatically via the CYP450 system. Therefore, this compound may not be a specific marker of $OH^{\bullet}$ generation and its measurement can give misleading results [13]. Because 2,5-DHBA can also be produced enzymatically [14], only 2,3-DHBA can be used for hydroxyl radical determination. 2,3-DHBA is not generated by any known enzyme [15] and can be considered a specific marker for $OH^{\bullet}$ production. Moreover, 2,3-DHBA is a highly stable product, which can be easily assayed by HPLC. Catechol is believed to correlate with 2,3-DHB levels and can be used as an additional indicator of $OH^{\bullet}$ generation. Accurate determination of DHBA by HPLC is accomplished with an electrochemical detector (amperometric or coulometric electrochemical detection) [2, 16–18]. A GC-MS method also exists for the determination of DHBA and offers better sensitivity [19]. The 2,3-DHBA/2,5-DHBA ratio has often been used as a measure of $OH^{\bullet}$, but the ratio of these two varies dependent on the chemical environment and thus a reliable quantification cannot be performed using this approach [20]. In vitro work also demonstrated that even in a simple Fenton system, the ratio showed a time dependence, which again raised doubt on its reliability [21–23]. Furthermore, salicylate may directly perturb the responses to ROS as it has been shown to affect inflammatory responses [24, 25].

Besides, 2,3-DHBA can be metabolized in the liver, and possibly other organs, causing its loss. It follows that salicylate method is not a quantitative measure of hydroxyl radical generation, especially if the end products are measured in the plasma where their origin is unclear. Moreover, specificity of the trap for $OH^{\bullet}$ is uncertain; it has been shown that $ONOO^{-}$ can hydroxylate salicylate possibly by direct reaction with ONOOH [26]. Simultaneous determination of nitration products can be used to distinguish the two products [27].

15.1.2 Phenylalanine Method

The hydroxylation reaction of phenylalanine has been proposed as an alternative to salicylate, mainly because the administration of phenylalanine has fewer identified side effects [17]. Hydroxyl radical hydroxylates L- or D-phenylalanine to produce ortho-, meta-, and para-tyrosines (Eq. 15.2).

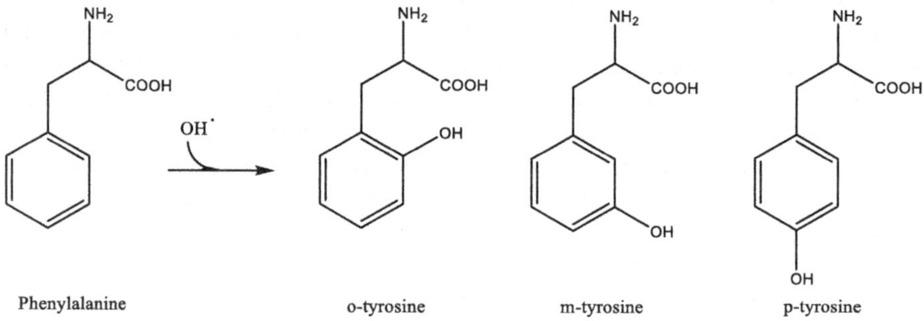

(15.2)

The quantification of the phenylalanine hydroxylation products, o-, m-, and p-tyrosine, can be achieved by either UV, FL, or EC detection coupled with HPLC or by GC-MS [28, 29]. These metabolites seem not to be enzymatically produced in humans except p-tyrosine, which is an endogenous compound formed from L-phenylalanine by the enzyme phenylalanine hydroxylase. Therefore, it cannot be used for detection of $OH^{\bullet}$. Since D-phenylalanine is not a substrate for this enzymatic hydroxylation, it has been proposed as a more suitable compound for hydroxyl radical detection [30].

Phenylalanine has been used to measure hydroxyl radical formation in saliva [31]. Phenylalanine has also been used to measure ex vivo radical formation in blood from patients with rheumatoid arthritis [5], and in humans exposed to ozone [32]. By using phenylalanine hydroxylation probes, it has been shown that $OH^{\bullet}$ is generated during myocardial and cerebral ischemia in experimental models [7–9].

A major problem with phenylalanine probe is that HPLC analysis of meta-tyrosine in human samples can be confused by the frequent presence of a coeluting peak of unknown identity [33]. Be suspicious if you see an apparent meta-tyrosine peak on HPLC much greater than that of ortho-tyrosine: this is suggestive of an artifact since both products should be formed in comparable amounts when $OH^{\bullet}$ attacks phenylalanine. Mass spectrometry can be used to avoid this problem [34].

Chemically, this system shows the same drawbacks as the salicylate method since D-phenylalanine is a substrate for D-amino acid oxidase, which is present in brain [33] and produces H_2O_2 as a product. Using the same arguments as above, it is doubtful whether the sum of o- and m-tyrosine can be taken for an artifact-free ROS quantitation in vivo [27, 35]. It should be mentioned that this method does have the potential to detected nitration simultaneously, allowing concurrent monitoring of peroxynitrite formation [35].

15.1.3 4-Hydroxybenzoic Acid (4-HBA) Method

Detection of ROS with 4-hydroxybenzoic acid (4-HBA) is less complicated, because only one hydroxylation product, 3,4-dihydroxybenzoic acid (3,4-DHBA) is formed in significant amounts (Eq. 15.3). Thus, from a chemical viewpoint, this method should provide a much more reliable possibility for hydroxyl radical quantitation. It has been used in microdialysis experiments, using HPLC separation with electrochemical detection [36, 37].

Although 4-HBA can be hydroxylated by monooxygenases from some microorganisms, this apparently does not occur in mammals, suggesting its suitability as a chemical trap for hydroxyl radical determination in vivo. As 4-HBA is an endogenous compound that shows little or no apparent toxicity, it has been claimed that it could be also used for human studies [37]. However, it is necessary to use high (mM range) concentrations of 4-HBA and further work is necessary to assess whether any biochemical processes are affected by 4-HBA, or its metabolites, at such levels, particularly in view of it being a substrate for 4-hydroxybenzoate polyprenyltransferase, which is involved in the biosynthesis of ubiquinone [37].

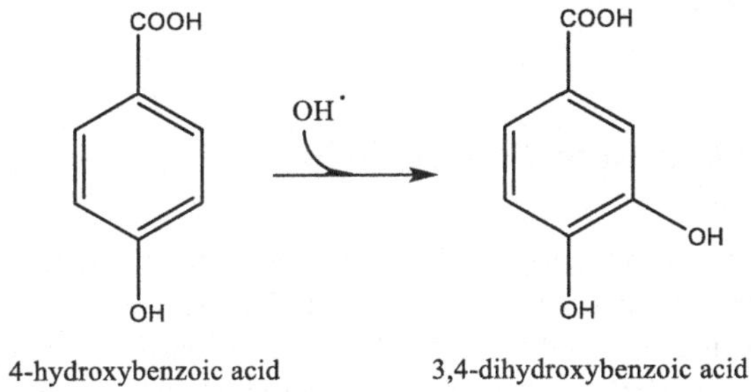

(15.3)

15.1.4 Terephthalic Acid Method

The terephthalic acid (TA) assay does not have the drawbacks of the salicylate and phenylalanine methods, as the symmetry of the molecule allows only one hydroxylated product to be formed (2-hydroxyterephthalic acid (OH-TA)), as shown in reaction 15.4. TA is not present endogenously and is poorly metabolized. Moreover, TA has low toxicity and is non-accumulative.

OH-TA shows a brilliant fluorescence and the method allows the detection down to 0.5 nM OHTA using a commonly available fluorescence detector. This reaction has been used for many years for in vitro studies [38, 39]. Its ease of use and its good performance, especially in the low-dose range [40]. This has led to the suggestion that this system could be used in vivo [41]. Yan et al. reported its application for detecting ROS during brain microdialysis in sheep [14].

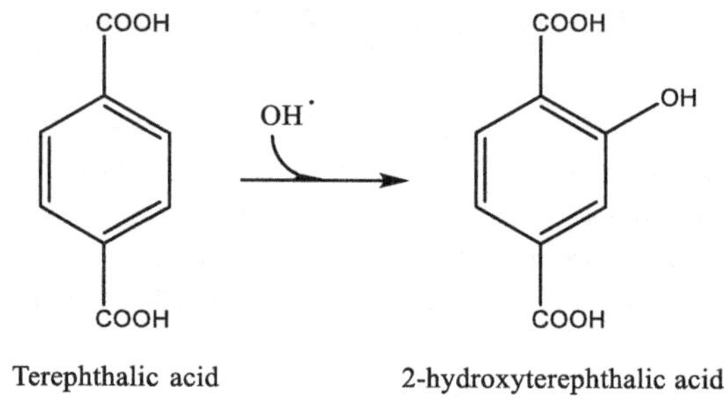

(15.4)

15.1.5 *Antipyrene Method*

The compound antipyrene is hydroxylated by OH˙ into p- and o-hydroxylation products, the latter again not apparently formed enzymatically in humans. Antipyrene has been used to measure hydroxyl radical production during exercise [42].

15.2 Nitric Oxide and Peroxynitrite

There are various methods reported in the literature for the determination of nitric oxide radical end products such as NO_2^- and NO_3^-. Several chromatographic methods have been developed for their detection in plasma, including HPLC-UV/FL (with or without Griess derivatization), anion-exchange chromatography, and GC-MS [43, 44].

Urate, being an endogenous antioxidant, may be used as a non-toxic trap. Urate is readily oxidized by a range of RS including $ONOO^-$ [45]. Several groups have used urate as a "selective" scavenger of $ONOO^-$ in animal studies, neglecting the fact that it reacts with many species. One of the urate's oxidation products, allantoin, can be measured in human body fluids using HPLC technique. Its plasma, urine, or cerebral microdialysis fluid levels are elevated in conditions associated with OS such as chronic inflammation, iron overload, diabetes, chronic heart failure, premature birth, and exercise [46–53]. Levels of allantoin rise in the human muscle during exhaustive exercise, presumably due to oxidation of urate by RS generated during exercise [54]. Allantoin measurement may be one of the more promising methods for human use, since human urate levels in vivo are high and urate reacts with a wide range of RS.

15.3 Lipid End products

15.3.1 Malondialdehyde

The determination of MDA using the TBARS spectrophotometric methods has been severely criticized, with some authors dismissing it as unacceptable in modern research and not credible to monitor lipid peroxidation, simply because most TBA-reactive materials in human body fluids are not exclusively related to MDA generation [55, 56]. These methods overestimate the extent of lipid peroxidation and the lack of specificity has rendered them rather obsolete. Today, most of the researchers use an HPLC modification of the TBARS method with well-defined MDA standards [56, 57]. The chromatographic approach uses HPLC-UV to separate the TBA-MDA adduct from interfering chromophores, thereby resulting in improved specificity. Up to a three-fold difference was found between TBARS and MDA values measured in human plasma using specific chromatographic methods [55].

One of the most important preparation of samples is to add BHT during the TBA-MDA conjugation process to prevent lipid peroxidation during heating. The interlaboratory study carried out by 16 research teams in 2010 showed that determination of MDA by HPLC will be the method of choice for the assessment of lipid peroxidation in human plasma samples [58]. Note that both free MDA and protein-bound MDA have been successfully measured in plasma by HPLC-UV, after derivatization with 2,4-dinitrophenylhydrazine (DNPH) or diaminonaphthalene [59, 60]. More reliable and reproducible methods based on synchronous fluorimetry or TBA-MDA conjugate analysis using fluorescence detector were described, giving a similar specificity and a greater sensitivity [61–63].

Recently, GC-FID has been used to detect MDA in human plasma [82], however, much lower submicromolar concentrations of MDA are detected using GC-MS [64, 65]. This indicates that the problem of interference can be further overcome and GC can provide more accurate estimate for MDA levels in biological samples. The derivatization reagents usually used for GC-MS analysis include methyl hydrazine, phenylhydrazine, pentafluorophenylhydrazine, and pentafluoro benzylhydroxylamine [65, 66]. Since the majority of MDA is believed to form adducts with proteins, the MDA-TBA assay would greatly underestimate total MDA concentrations. Derivatization of MDA with compounds such as 2-hydroxypyrimidine can be used to detect total MDA, whereas the other reagents are used to determine free MDA [67].

It is noteworthy that the values obtained for plasma with mass spectrometric methods are up to 10-fold lower than those obtained with the MDA levels in urine as reported by the BOSS where the effect of CCl_4 poisoning in rats was studied, and MDA levels were found to correlate well with the extent of OS [68].

MDA is also excreted in breath and can be analyzed as a marker of OS preferably using GC or HPLC coupled with mass or tandem mass spectrometry. For instance, Peroni et al. analyzed MDA in breath using LC-MS after derivatization with DNPH [69].

Some artifacts are still present and make the application of these chromatographic methods difficult at present:

- MDA analysis by HPLC and GC is time-consuming involving complex sample preparation to remove contaminants and sample extraction into organic solvents to improve sensitivity and peak separation.
- The validity of MDA as a marker of OS remains uncertain in some conditions, making it less applicable for routine clinical use.
- Measurements of MDA in plasma and urine can be confounded by diet and should not be used as an index of whole-body lipoxidation until diet is strictly monitored.

Overall, chromatographic methods are sensitive and selective systems. They have been used for the measurement of MDA in many clinical situations: CF, Duchenne's myopathy, breast cancer, AIDS, RA, DM, coronary atherosclerosis, and leukemia [70–77]. Besides, determination of this marker allows the demonstration of the efficiency of antioxidant supplementation when treating patients, for example, for phenylketonuria, CF, or patients undergoing hemodialysis [78].

Reaction of MDA with nucleotides is a subject that has received considerable attention in the 90th because of its mutagenicity. GC-MS method using pentafluorobenzyl derivatization has been developed to detect endogenous MDA-deoxyguanosine adducts (M1G-deoxyribose), which has been implicated in G to T transversions [79].

15.3.2 4-Hydroxy-2-Nonenal

4-HNE is a significant aldehyde produced as a cleavage product from the oxidation of polyunsaturated fatty acids (PUFAs). Various chromatographic methods have been applied for the determination of free 4-HNE in biological samples. GC and HPLC with different detectors are accurate techniques for its determination.

In HPLC procedures, 4-HNE and other aldehydes (e.g., alkanals, hydroxyalkenals, and alkenals) are derivatized with DNPH to hydrazones. After a step of purification on solid thin-layer chromatography (TCL), the DNPH derivatives are analyzed and resolved using the HPLC system with UV detection. The HNE-DNPH adduct can also be measured by HPLC-EC with greater sensitivity. Note that formation of derivatives with DNPH is time-consuming. A less time-consuming alternative uses 1,3-cyclohexanedione to form fluorescent dihydroacridine derivatives, which can be detected by HPLC-FL, giving a better sensitivity. At present, this HPLC technique is unlikely to be used as routine measures. It can be used where it is necessary to know the range of aldehydes produced in a particular condition.

4-HNE and other aldehydes have also been measured by LC-MS as their pentafluorobenzyloxime (PFB) derivative. Using ESI-MS, Gioacchini et al. were able to detect 4-HNE in cell extracts directly without derivatization [80].

Greater sensitivity and selectivity can be achieved with GC-MS instruments. They have been greatly developed and are now used to quantitate 4-HNE in plasma, urine, and cerebrospinal fluid with high efficiency [81, 82]. Some methods are based on the use of o-pentafluorobenzyl hydroxylamine to form the pentafluorobenzyl-oxime derivatives. The hydroxynonenal pentafluorobenzyl-oxime is then analyzed by GC-MS (NICI) using deuterated hydroxynonenal as internal standard [83]. Overall, GC-MS can now detect as low as 1 picomole of HNE in biological samples.

4-HNE can be analyzed in EBC as marker of OS preferably using GC or LC coupled with tandem mass spectrometry. Peroni et al. analyzed 4-HNE in EBC samples using LC-MS/MS after derivatization with DNPH [69].

During heavy stress (e.g., patients with severe RA, lupus erythematosus, systemic sclerosis, chronic lymphedema, or chronic renal failure), HNE is increased in serum to concentrations up to 3- to 10-fold higher than physiological ones [84].

Finally, a urinary stable metabolite of 4-HNE, 1,4-dihydroxynonane-mercapturic acid, can be used as biomarker of OS and assessed using chromatographic methods. Other compounds such as 4-oxo-2-nonenal are less studied than 4-HNE as indicators of lipid peroxidation because 4-oxo-2-nonenal is more reactive and unstable in human serum.

15.3.3 Acrolein

Acrolein is one of the most reactive of all aldehydes generated during lipid peroxidation. Attempts have been made to use this reactive compound as a potential biomarker of OS [85].

Acrolein can be directly measured by headspace GC-MS and HPLC-FL or as its metabolic product 3-hydroxypropyl mercapturic acid by LC-MS/MS [86–88]. Concentrations of acrolein were found to be increased in CVD and RA [89].

15.3.4 Conjugated Dienes

Chromatographic methods allow the separation of different diene conjugates, which are double-single-double bond compounds. They are formed during early steps of lipid peroxidation. As primary products of lipid oxidation, they have been expected to be interesting substances to measure.

Among other products of lipid peroxidation, both endoperoxides and hydroperoxides possess a conjugated diene structure with a characteristic UV absorption around 234 nm. Consequently, conjugated dienes are frequently measured by spectrophotometry and HPLC-UV as biomarkers of lipid peroxidation [67]. Lipids are extracted using the standard Folch extraction method and the conjugated dienes are read at 234 nm. As many other molecules also absorb at 234 nm, a problem with spectrophotometric measurement of conjugated dienes is interference by other substances leading to a lack of specificity and tendency to give artificially high values. HPLC-UV allows selective identification of lipids containing conjugated dienes. Such methods have been used to identify phospholipid peroxides and cholesterol ester peroxides formed during LDL oxidation [90]. The most common usage is a dynamic measurement of the oxidation of LDL in vitro [91].

For other applications, HPLC-UV suffers from several artifacts. Indeed, a non-oxygen-containing isomer of linoleic acid (9*cis*, 11*trans*-octadecadienoic acid) and the most (90%) available of the conjugated diene in human plasma have been shown to strongly absorb at 234 nm and interfere with conjugated diene values [92, 93]. This product is not found in the plasma of animals subjected to OS and may be of dietary origin or acquired from gut bacteria. Therefore, application of conjugated diene methods to human body fluids is not probably measuring lipid peroxidation and cannot be recommended for OS assessment. Measurement is also complicated by the relatively low levels of conjugated dienes normally present in blood circulation. This is due to the fact that conjugated dienes readily break down to other lipid oxidation products. Finally, conjugated dienes can be measured more accurately with GC-MS and LC-MS but have not really been convincing as markers.

15.3.5 Organic and Lipid Hydroperoxides

Organic and lipid hydroperoxides are most frequently analyzed by using a number of chromatographic methods coupled with appropriate detection systems (HPLC and GC-MS). They reflect an early stage of lipid peroxidation and can be used as biomarkers of oxidative damage of membrane lipids in vivo. Because lipid hydroperoxides are rapidly cleared in cells, these compounds do not build up to high levels and are usually difficult to detect.

Since analysis of lipid hydroperoxides requires HPLC-UV, separation of individual lipids can be used for the determination of positional isomers of lipid peroxides. Lipid hydroperoxides can be measured as cholesteryl ester hydroperoxides, which are the most abundant plasma lipid hydroperoxides in vivo. Determination of 7-hydroperoxycholesterol and its products by the normal-phase HPLC separation and UV detection at 205 nm has been proposed [94]. Chromatographic methods have been also used for determining phospholipid hydroperoxides isolated from LDL, phosphatidylcholine and cholesteryl ester hydroperoxides extracted from LDL, phosphatidylcholine and triacylglycerol ester hydroperoxides extracted from plasma, cholesteryl linoleate hydroperoxides extracted from LDL, fatty acid hydroperoxides in intact native or copper ion oxidized LDL on a single run [94].

HPLC chemiluminescence has been frequently used to measure lipid hydroperoxides [95]. Processing of chromatograms is performed, after extraction of lipids, and extensive isolation of lipid peroxides with a very specific and selective separation using HPLC. This assay can reach high sensitivity (picomole range). It is based on the decomposition of lipid peroxides in the presence of heme compounds (e.g., cytochrome c and microperoxidase), and subsequent reaction with luminol or isoluminol, which generates oxidized products that emit chemiluminescence. This property can be used to measure individual lipid peroxides after separation by HPLC, employing post-column derivatization [96]. Chemiluminescence assay is not specific for lipid hydroperoxides because it also detects ascorbic acid, ubiquinols, hydroquinones, and consequently overestimates lipid hydroperoxide content in biological samples. Inclusion of controls with reducing agents such as $NaBH_4$ or triphenylphosphine specifically eliminates chemiluminescence produced by constant hydroperoxides [97].

HPLC-chemiluminescence methods, although sensitive, are time-consuming and are best used only when information on individual hydroperoxides is required. Using these methods, cholesterol-hydroperoxides have been detected in patients with adult respiratory distress syndrome or undergoing angioplasty [98].

HPLC-FL can be used for the determination of lipid hydroperoxides. The method is based on the specific reaction of diphenyl pyrenyl phosphine (DPP) with lipid hydroperoxides, producing DPP oxide, which can be measured by its fluorescence (excitation 352 nm and emission 380 nm). Since the reaction takes place in different types of solvents, it can be used both in reverse and normal phase systems. Therefore, HPLC-FL assay is more versatile for post-column detection of a variety of compounds, and positional isomers of hydroperoxides [99]. Unlike chemiluminescence that is subject to interferences, the fluorescence method appears to be less susceptible to interferences.

There are other methods where lipid hydroperoxides can be measured such as HPLC-EC and LC-MS, which yield values up to two orders of magnitude. LC-MS is considered a more specific analytical method for analysis of lipid hydroperoxide in human serum [100]. All of these methods are distinctly research tools (gold standard) rather than a convenient routine clinical assays.

GC-MS is also a sensitive method to measure lipid hydroperoxides and involves reduction of lipid hydroperoxides (LOOH) to the corresponding hydroxyl derivative (LOH), usually with sodium borohydride or triphenylphosphine. The hydroxy-fatty acid is then derivatized, separated by GC and detected by mass spectrometry.

Accurate measurement of lipid hydroperoxides is difficult due to their rapid degradation in vitro. It is extremely important to limit this by adding antioxidants and quick processing of samples at 4 °C, which is likely not possible in a clinical situation. It is noteworthy to mention that there is a possibility that the results may be confounded by the absorption of lipid peroxides present in dietary fat. Indeed, lipid hydroperoxides and aldehydes such as MDA can be absorbed from the diet and then excreted in urine. It follows that measurements of these compounds can be confounded by diet and should not be used as an index of human body lipoxidation unless diet is strictly controlled.

Finally, HETE, known for their effects in the immune system and their involvement in OS, can be measured by GC-MS, operated in the negative ion chemical ionization mode.

15.3.6 Alkane Gases

Another possibility with chromatographic methods is the measurement of volatile hydrocarbon end products such as ethane and pentane, which are minor components of lipid peroxidation released into the breath. These are not easy measurements (GC-FID and GC-MS) and do depend on the decomposition of peroxides, which varies between tissues. They can be analyzed in EBC (commercially instruments are available for alkane collection). Such methods can detect these markers in picomolar concentrations. Short-chain alkanes and other compounds in EBC are regarded as one of the least invasive methods to measure OS. Alkane determination by chromatographic methods seems perfect as it does not need any reagent and is sensitive to changes in oxidative status. Moreover, they are insensitive to fasting or meals.

Alkane levels have been associated with some pathological conditions. For instance, patients with chronic heart failure present an increase in pentane inside expired air [101]. According to Humad et al., pentane excreted by RA patients is proportional to the severity of the disease [102].

15.3.7 Lysophophosholipids and Cholesterol Oxides

Some lipid-oxidized products such as cholesterol oxides and lysophosphatidylcholine have been considered for detection using chromatographic methods, but they are rapidly metabolized and do not accumulate in the blood as they do in atherosclerotic tissue. Nevertheless, these may prove to be useful markers using LC-MS [103]. Oxidized derivatives of cholesterol can be measured by thin-layer chromatography followed by GC. They showed an increase in their levels in diabetic patients [78]. Several specific HPLC assays for determining other degradation products of cholesterol, for example 9-oxononanoyl cholesterol (cholesterol core-aldehyde formed during lipid peroxidation) in isolated LDL or atherosclerotic lesion have been proposed. 7-Ketocholesterol was also identified as the main cholesterol peroxidation product, using an HPLC assay with UV detection [104–106].

15.3.8 Isoprostanes

Several IsoPs can be measured with great sensitivity using chromatographic techniques. They can be measured accurately down to picomolar concentrations with GC-MS, GC-MS/MS, LC-MS, and LC-MS/MS.

Isoprostanes are chemically stable in vivo and ex vivo, but once produced and released into blood circulation, they are readily metabolized and eliminated. Their fast disappearance from plasma may prevent practical application. Despite the fact that IsoPs have a short half-life, they seem to be produced at a steady rate, which is increased under OS in several diseases [107].

One has to distinguish between two fractions of IsoPs in human blood, that is free and esterified forms. Analysis of the esterified molecules requires hydrolysis to yield free derivatives. Isoprostanes are ideally measured in blood, 24-h urine output, and EBC. Because urine samples contain a very low lipid content, auto-oxidation is not a drawback. Urinary IsoPs are increased in conditions that predispose to CVD, such as smoking, hypercholesterolemia, hypertension, obesity, and DM [108].

By far the most convincing measurement to be made is that of the F_2-IsoPs which arise from the oxidation of PUFAs [109]. F_2-isoprostanes are reliable marker of lipid peroxidation and are mainly formed from the arachidonic acid oxidation.

The preferred method for measurement of multiple products is LC-MS, but the more successful approach is the measurement of IsoPs by GC(NICI)MS. This technique is considered as the reference method as it is highly accurate, reliable, and sensitive [110, 111]. Typical GC-MS assays involve the use of deuterated IsoPs, the conversion of the IsoPs to pentafluorobenzylesters (PFBE), and the derivatization to trimethylsilyl (TMS) ethers [112]. Precautions are taken to prevent artifact generation and adequate purification steps are included. The assays are, however, labor-intensive, involve considerable sample preparation, and require sophisticated instrumentation.

Following the introduction of electrospray ionization (ESI) techniques, several LC-(ESI)MS assays have been published for IsoPs analysis. Some of the reported limits of detection (LOD) were down to 0.5 pg/mL in plasma and the intra- and inter-assay imprecision were under 5%. The range of IsoPs values in normal human plasma (31–220 pg/mL) is in the range of published values of other methods (31–166 pg/mL). The sample volume required is usually less than 0.3 mL, which compares quite favorably with the GC-MS procedures that usually require up to 3.0 mL. A significant advantage of the LC-MS methods over the GC-MS methods is that the sample preparation does not require derivatization steps and greater recovery can be achieved. They are not only simpler and faster but also are less prone to artifact generation. However, the sensitivity of LC-MS methods is somewhat lower than that of GC-MS.

At present, the most useful methodology to measure F_2-IsoPs in biological samples is the tandem mass spectrometry, that is, GC-MS/MS and LC-MS/MS [113, 114].

A simplified LC-MS/MS method for the quantification of several IsoPs has been described [115]. By combining a solid phase extraction (SPE) and LC-MS/MS using negative mode, five different IsoPs were quantified in human urine. A semi-automated high-throughput LC-(APCI (Atmospheric Pressure Chemical Ionization))MS/MS method was developed by Haschke et al. for F_2-IsoPs quantitation in human urine and plasma [116]. Quantification of IsoPs by GC-MS/MS is considered the "gold standard" as it offers the highest sensitivity and specificity for several IsoP compounds. GC-MS/MS is also considered as the most accurate method with a marked advantage over immunoassays. It must be taken into account that measures of IsoPs by GC-MS/MS in plasma may be markedly compromised by hemolysis [117]. One challenge to widespread adaptation of F_2-IsoPs in clinical trials is that these reliable methods are labor-intensive and require expensive and specialized instrumentation.

A key product that can be measured is 8-iso-PGF2α (15-F_{2t}-IsoP) because of its known biological activity. The commercial availability of immunoaffinity chromatography columns renders possible rapid, specific, and accurate analysis of this IsoPs in human plasma and urine even by GC-MS and LC-MS. Indeed, a reliable method was developed to overcome the contamination problem in plasma and urine samples by using a single-step extraction method taking advantage of a commercially available immunoaffinity column (IAC) (Cayman) which was primarily developed for ELISA [118–120]. Finally, it is noteworthy to know that several IsoPs can measured in EBC as illustrated in Table 7.2.

15.3.9 Isolevuglandins

Same as IsoPs, IsoLGs are produced during lipid peroxidation from arachidonic acid. They are highly sensitive to changes to oxidative modifications occurring during OS. While F_2-IsoPs are stable compounds, IsoLGs are highly reactive and readily and irreversibly react with primary amines in cells to form pyrrole (lactam) and oxidized pyrrole (hydroxylactam) adducts [121]. Therefore, only IsoLG adducts, and not unreacted IsoLGs, are detected in biological fluids, cells, and tissues.

IsoLG adducts may eventually prove to have great utility as OS biomarkers because they appear to directly participate in some pathological processes (e.g., hypertension, arrhythmia, atherosclerosis, and neurodegenerative diseases) [121, 122]. Using LC-MS, IsoLG-phosphatidyl-ethanolamine adducts have been found to be increased in plasma from patients with macular degeneration compared to healthy volunteers [123]. However, the utility of measuring IsoLG adducts in plasma or urine as reliable biomarkers remains to be established. The current findings provide strong background to further investigate the potential use of IsoLG adducts as clinical biomarkers to identify individuals at high risk and estimate the efficacy of treatments targeting IsoLGs.

15.4 Protein End products

15.4.1 Protein Carbonyl

The number of protein-bound aldehyde molecules is measured via determination of protein carbonyl groups. A major problem with simple-based measurements to detect protein carbonyls is that they do not distinguish well between oxidation and glyco-modification of proteins. Although, spectrophotometric techniques are simple and give reasonable estimate of the total carbonyl content in a mixture of proteins, such techniques measuring optical absorption at 360 nm following reaction with diphenylhydrazine and production of DNP (dinitrophenyl) derivative will be subject to significant interference. Interference from unmodified proteins, such as hemoglobin and myoglobin which also absorb at 360 nm, may occur as well. Therefore, these methods are not suitable for use in biological fluids and tissues. Currently, more complex methods are applied. One of the most commonly used methods in research to detect protein oxidation is HPLC. Separation of the DNP-proteins by HPLC significantly improves the sensitivity and specificity of the assay. Unfortunately, even with HPLC-UV improvements, criticism still exists against this method for being non-specific and unreliable. Currently, protein carbonyls can be separated on HPLC and detected by highly sensitive instruments such as EC detectors. Newly developed procedures for mass spectrometry seem promising in the analysis of protein carbonyls, but the diversity in protocols and labeling procedures is too broad to combine and adequately compare results from different studies [124].

15.4.2 Oxidized Methionine

Chromatographic analysis of oxidized methionine such as methionine sulfoxide in proteins has been described using a cyanogen bromide procedure. In this procedure, amino acids are separated from acid hydrolysates of proteins treated with cyanogen bromide using reverse-phase HPLC. This approach enables the simultaneous determination of homoserine, corresponding to converted non-oxidized methionine, and methionine oxidized derivatives [125].

15.4.3 Oxidized Tyrosine

15.4.3.1 Nitrotyrosine

3-Nitrotyrosine, as a biomarker of protein nitration, can be detected in biological fluids and tissues by various chromatographic methods. Among the quantitative approaches, both LC and GC have been described. The same holds for small

NO_2Tyr-containing peptides, which, like free NO_2Tyr, can be measured. HPLC method using UV absorbance detection does not provide adequate sensitivity and specificity for the quantification of NO_2Tyr and NO_2TyrProt in biological materials. HPLC-FL following derivatization improves sensitivity and selectivity as compared to HPLC-UV. HPLC-FL revealed clearly elevated concentrations of NO_2Tyr and NO_2TyrProt in several human diseases investigated so far, suggesting the involvement of nitrosative stress in virtually a large number of human diseases.

HPLC with electrochemical detection is the technique of choice for sensitive and selective determination of 3-NO_2Tyr in different biological fluids and tissues.

One of the most advanced and precise techniques is mass spectrometry for either free nitrotyrosine or nitrotyrosine from nitrated proteins that have been hydrolyzed or digested to their constituent amino acids. Accurate determination requires prior separation using gas or liquid chromatographic techniques.

Some GC-MS methods involve isolation of 3-NO_2Tyr by ion exchange chromatography, reduction to 3-aminotyrosine, and derivatization to per-heptafluorobutyryl, which is finally quantified by MS detector [126]. However, evidence of interference despite the use of a HPLC step for sample purification suggested that the reported values of free NO_2Tyr, provided by GC-MS methods, are not accurate [127, 128].

A serious pitfall in chromatographic techniques is the possible artificial nitration of tyrosine residues in the sample by the presence of NO_2^- and acid conditions during protein precipitation and hydrolysis [129]. Nevertheless, some mass spectrometric methods allow the evaluation and quantification of artifactual production of NO_2Tyr. Besides, the high-throughput analysis of clinical samples needs to be addressed as mass spectrometry, while considered the gold standard, is not yet feasible for a large number of analysis [130].

At present, only tandem mass spectrometry (LC-MS/MS and GC-MS/MS) provides reliable values and accurate, sensitive, and specific quantification of NO_2Tyr and NO_2TyrProt in various biological samples. This is due to the fact that only tandem mass spectrometry can eliminate considerable interference by coeluting molecules. The high selectivity and sensitivity of LC-MS/MS and GC-MS/MS were convenient for quantifying the low levels of 3-NO_2Tyr found in healthy human plasma (1nM). LC-MS/MS technology made a derivatization step of 3-NO_2Tyr unnecessary, thus minimizing the artifactual formation of 3-NO_2Tyr. However, LC-MS/MS is less sensitive than GC-MS/MS, and the basal concentrations obtained by this approach may serve as reference values.

Ryberg and Caidahl reviewed chromatographic and mass spectrometric methods for quantitative determination of 3-NO_2Tyr in different biological samples including plasma and urine. The authors emphasized the importance of 3-NO_2Tyr as a biomarker for over 40 different diseases [127].

Paradoxical results can be obtained as the use of the most reliable techniques (GC-MS/MS and LC-MS/MS) may reveal that NO_2Tyr and NO_2TyrProt levels are not elevated in various human diseases as it was expected with simple techniques.

15.4.3.2 Dityrosine

The appearance of a dityrosine adduct in proteins as a result of tyrosyl radical reaction and oxidative modification has been described in the literature. As a marker, dityrosine has the advantage of being biochemically stable. The carbon–carbon bond created between two molecules of tyrosine is resistant to enzymatic hydrolysis.

A useful method for detection of dityrosine based on HPLC-FL has been described by Colombo et al. [131]. An isotope dilution LC-MS/MS method utilizing [2D6]dityrosine was developed by Orhan et al. for the quantitative determination of o,o′-dityrosine in urine [132]. GC-MS can be used for dityrosine detection in LDL isolated from human atherosclerotic plaques [133].

15.4.3.3 Chlorotyrosine

Very sensitive GC-MS and LC-MS methods are available for the detection of Cl-Tyr and other chlorination products [134]. Free 3-Cl-Tyr is a highly specific biomarker of the MPO-OS pathway and has been determined in isolated LDL and atherosclerotic tissue after delipidation and hydrolysis of proteins [135].

15.4.4 Other Derivatives

p,m,o-tyrosine are interesting derivatives of protein oxidation because they can only be formed by a non-enzymatic free radical oxidation of phenylalanine. They can be detected by HPLC-FL and GC-MS, and their concentrations in tissues often increase during OS [136]. Oxidized tryptophan can also be identified using HPLC using UV detector.

15.4.5 Simultaneous Determination of Several Oxidized Amino Acids

Several research groups have used HPLC methods to simultaneously quantify tyrosine congeners in serum, cerebrospinal fluid, saliva, brain, and even eye lens. For instance, HPLC-EC array can simultaneously monitor at least seven distinct tyrosine derivatives and isomers in a biological sample after enzymatic or acid hydrolysis [31, 137–141]. Simultaneous detection of tyrosine oxidation products (3,4 DOPA; 3-Cl-Tyr, 3,3′-dityrosine; 3-NO_2Tyr, m- and o-tyrosine) has been described using several coulometric cells in series following HPLC separation [142]. LC(APCI)MS/MS for the simultaneous analysis of several modified tyrosine has been described by Orhan et al. [143]. The method is sufficiently sensitive and highly selective for direct determination of phenylalanine, *p*-tyrosine, *o*-tyrosine, and o,o′-dityrosine in urine samples. The LOD ranges between 0.01 and 0.05 μM for the oxidized products, while they are 0.5 μM for the parent amino acids, phenylalanine, and p-tyrosine. The authors recommended freeze-drying and SPE of urine samples in order to bring the amount of m-tyrosine, 3-Cl-Tyr, 3-NO_2Tyr in extracts within the range of the method's LOD. All of these studies aptly demonstrate the potential of simultaneous analysis of tyrosine congeners as biomarkers of OS.

Several studies have demonstrated a clear relationship between inflammatory pathologies and elevation of tyrosine oxidation products. Nitrotyrosine has been detected at elevated levels in serum and synovial fluid from patients with RA [144]. Hensley et al. have reported that protein-bound NO_2Tyr and dityrosine, liberated by proteolytic digestion prior to HPLC-EC analysis, increase to 3-7-fold in some regions of brain in patients with AD relative to age-matched normal human brain. Extending upon this work, the authors have detected multiple tyrosine congeners in cerebral microvessels using a 12-channel electrochemical array [138].

15.5 Carbohydrate End Products

Chromatographic-based techniques represent a key method in detecting AGEs and the specific site of protein glycation, but their use is still limited in routine clinical analysis. They allow the identification and measurement of specific AGEs such as pentosidine (HPLC-FL), CML (LC-MS and GC-MS), CEL (LC-MS), glyoxal (HPLC-FL, GC-MS, and LC-MS), and methylglyoxal (HPLC-FL, GC-MS, and LC-MS) [145, 146]. However, the use of AGEs as biomarkers and the development of specific methods for clinical application are hampered by the structural heterogeneity of these compounds, due to different ways of generation.

15.6 DNA End Products

Different methods have been proposed for the measurement of oxidative damage to DNA. Some of the earliest methods include chromatography to study products of RS action on nucleic acids, with the hope of better understanding apoptosis and oncogenesis. These methods have been widely used for the analysis of one of the classical biomarkers of oxidative damage to DNA, which is 8-OHdG.

15.6.1 8-OHdG

Analysis of samples, particularly urine, requires SPE or chromatographic pre-purification by normal, reverse phase, or immunoaffinity columns to prevent interference by many urinary compounds [147]. Cells used for DNA analysis are usually leukocytes and isolated lymphocytes from the peripheral blood. The general procedure involves the extraction of DNA, followed by its enzymatic hydrolysis to release the 8-OHdG. HPLC-UV was used to measure 8-OHdG but great volumes of samples were required to reach sufficient sensitivity.

HPLC-EC was the most common means of determining the levels of 8-OHdG for at least two decades. Although susceptible to adventitious oxidation of deoxyguanosine during sample preparation (see below), HPLC-EC was chosen as the method of choice, affording satisfactory accuracy and great sensitivity when quantifying 8-OHdG levels. Coulometric electrochemical detection provides an excellent choice for measurement and gives three orders of magnitude superior over UV detection, both in selectivity and sensitivity (sensitivity at subpicomolar level) [148]. Since 8-OHdG is a rare compound, normally existing in a ratio of approximately on 8-OHdG residue/10^5 deoxyguanosine residues, the selectivity and sensitivity of HPLC-EC are a necessity for accurate 8-OHdG determination [149]. The concentration is usually expressed as moles of 8-OHdG per 10^6 moles of undamaged deoxyguanosine.

Other methods for 8-OHdG quantification include GC-MS and LC-MS/MS [150]. The former is a powerful technique and very advantageous because it enables the simultaneous determination of a wide variety of oxidized DNA bases. It can be applied to DNA–protein complexes as well. GC-MS method involves derivatization of the 8-OHdG by trimethylsilylation before detection. GC-MS has been used to measure 8-OHdG in urine, and the LOD was extremely low (1.8 pmol) [151]. Although this method is highly sensitive and provides information on individual products of oxidized bases, the values obtained with this technique are 10- to 50 fold higher than with HPLC-EC [152]. Some papers have shown that the high values of oxidized base pairs obtained by GC-MS were due, in part, to the artificial oxidation of base pairs that occurs during the derivatization step [153]. To prevent oxidation, it has been suggested that normal bases should be separated from the oxidized ones by chromatography before derivatization process [154]. This would guarantee that only oxidized bases are present during derivatization, which prevents artificially high values. These improvements in GC-MS techniques resulted in an agreement between GC-MS and HPLC measurements of cellular 8-OHdG.

LC-MS/MS methods have great potential because derivatization modification is not necessary and extraction of samples may not be needed, especially for urine samples, although a SPE is commonly employed. Because of great advances in LC-MS/MS instruments in recent years, it is expected that LC-MS/MS will be increasingly used for quantitative determination of 8-OHdG and may become the method of choice in near future. Relative to HPLC-EC methods, LC-MS/MS provides greater specificity and allows analysis of multiple analytes.

The completed BOSS, using acute CCl_4 exposure of rodents as a model for OS, has demonstrated that 8-OHdG in urine is a potential candidate and a general biomarker of OS [68]. Indeed, elevations in 8-OHdG (which probably exists as the tautomer 8-oxo-dG in vivo) have been noted in diverse disease states and conditions that involve OS including urine of cancer patients [155], urine of patients with AD [156], and urine and leukocytes of smokers [157, 158].

While measurement of 8-OHdG has proved a valuable tool in biomedical research, analysis of this analyte is prone to extreme interlaboratory variability. Regardless of the employed detection method, artifactual oxidation of guanine is a major confounding issue for the determination of the correct basal amounts of 8-OHdG in DNA [159]. 8-OHdG quantification subsequent to DNA extraction and hydrolysis, which gives inaccurate representation of the level of oxidative DNA damage and introduction of artifactual 8-OHdG, brings into question the validity of results determined by chromatographic methods. As a result of this issue, the European Standards Committee on Oxidative DNA Damage (ESCODD) was set up to identify and resolve methodological problems, which occur during isolation, purification, hydrolysis, and handling of DNA products [159]. This committee has been testing the ability of different laboratories using a variety of existing methods to measure 8-OHdG in standard and identical samples (untreated or treated to induce formation of different amounts of 8-OHdG within a DNA molecule). The performance of different methods was compared, and the putatively best analytical approach was evaluated. Although, no sign of consensus was found over the measured concentrations of 8-OHdG using various techniques, HPLC-EC was able to measure 8-OHdG in cellular DNA with great accuracy, not only with respect to the other chromatographic techniques but also in comparison with enzymatic methods [160, 161]. Additionally, the ESCODD reports confirmed that several methods used for detecting and quantifying 8-OHdG damages in extracted DNA from cells gave rise to incorrect conclusions due to the generation of oxidative artifacts, with large discrepancies noted between the levels of 8-OHdG documented using different quantification methods. The ESCODD provided more reliable methods for sample preparation and analysis. Later, a series of refinements were made to reduce artifactual 8-OHdG formation, which should diminish this concern. In particular, DNA extraction methods, which do not involve phenol, chloroform, or thiocyanate, seem to significantly reduce artifactual 8-OHdG formation. One possibility is the use of chaotropic NaI method for DNA extraction. Other techniques used to minimize oxidation of guanine in samples include the use of antioxidants (2,2,6,6-tetramethylpiperidine-N-oxyl (TEMPO), DTT, GSH, and CAT), metal chelators (EDTA and deferoxamine mesylate), purging of solutions with argon, and storage of samples at −80 °C until analysis [162]. As mentioned earlier, if 8-OHdG determination is planned using GC-MS instrument, guanine bases should be removed by chromatography before oxidized guanine is derivatized. This ensures that no guanine bases are present during the derivatization procedure, where it can oxidize and artificially inflate oxidized guanine values. All these adjustments highly improve the background level of 8-OHdG.

15.6.2 8-oxo-dG

An extension of chromatographic methods has been made to detect another DNA damage marker, 8-oxo-dG, in biological fluids (urine, plasma, and DNA extract). Analysis of 8-oxo-dG has been hampered for a long period of time by the occurrence of several drawbacks associated with their measurement, which led to a wide discrepancy in results. Even when multiple methods have been used, different values of 8-oxo-G (RNA form of 8-oxo-dG) were reported in biological fluids, cells, and tissues [163].

Floyd et al. first demonstrated that 8-oxo-G could be determined using HPLC-EC [164]. This method was later adapted to simultaneously measure 8-OHdG [165]. HPLC values have generally yielded higher cellular 8-oxo-G values than those of enzymatic assays. HPLC measurements have given values ranging from 1.7 to 32 8-oxo-G/8-oxo-dG per 10^6 base pairs, while enzyme-based assays provided the lowest values, ranging from 0.08 to 0.24 8-oxo-G/8-oxo-dG per 10^6 base pairs [163].

As for 8-OHdG, the variation in values is attributed to nucleic acid extraction procedures. Extraction of oxidized nucleic acids using phenol-chloroform has been suggested to cause artificial oxidation [166]. Collins et al. used a milder extraction procedure for 8-oxo-G, which resulted in values that were closer to the values given by enzymatic assays [167]. With the improved nucleic acid extraction procedures, measurement of 8-oxo-G by HPLC remains a reliable and sensitive method.

GC-MS provides generally the highest values (ranging from 40 to 660 8-oxo-G or 8-oxo-dG per 10^6 base pairs). Artifacts generated during extraction and especially derivatization are responsible for most of the deviations in values among different protocols.

Optimized assays are now available, and the most reliable ones are represented by liquid chromatography coupled with mass spectroscopy. Topic et al. described LC-MS/MS method for analysis of 8-oxo-dG in urine samples to establish a gender-specific reference range of this marker in Serbian population [168]. Determination of 8-oxo-G in urine can be performed by LC-MS/MS as well, as described by Jorgensen et al. [169].

15.6.3 8-NO_2dG

One of the most devastating oxidants in terms of DNA damage is $ONOO^-$, which can attack guanine at its C-8 position, yielding 8-NO_2dG and 8-oxo-dG. 8-NO_2dG can be determined by HPLC-EC after enzymatic or acidic digestion of DNA and its reduction to 8-aminoguanine [170]. 8-NO_2dG can be measured LC-MS GC-MS as well.

15.6.4 Other Bases and Compounds

Other DNA damage products can be measured by protocols involving chromatographic methods. They can be measured directly in DNA, after release of oxidized derivatives by glycosylase in plasma or urine.

Since the popularization of HPLC-EC assays for the quantification of 8-OHdG, 8-oxo-dG, and 8-NO_2dG, the technology has been brought to bear on other rare and potentially mutagenic DNA bases. For instance, several methods have been reported for HPLC-EC analysis of FAPY, 5-hydroxyuracil, 5-hydroxy-deoxycytidine, hydroxyadenine, and 5-hydroxyuridine as well as methylated guanine residues [171–173].

Faure et al. have developed a precise method to measure 5-hydroxymethyluracile in urine by GC-MS using a stable isotopic internal standard that gives reproducible results [174]. Thereafter, over 20 different modified DNA bases have been identified using GC-MS method [175]. These modified bases can also be measured using post-labeling CG-MS and LC-MS as well [176]. Cadet et al. have observed, using post-labeling techniques, the formation of adenine Nl-oxide in fibroblasts subjected to a stress with H_2O_2 [177]. Besides, LC-MS/MS-based methods have quantified many different oxidized DNA products from various biological matrices.

Modified bases may also result from the attack of reactive products related to the oxidation of other molecules such as lipids. In this context, etheno-DNA adducts such as 1,N(6)-etheno-2′-deoxyadenosine (εdA) and 3,N(4)-etheno-2′--deoxycytidine (εdC) are formed among others and can be used as biomarkers of OS. They may be used to assess progression of inflammatory and cancer diseases as well [178]. Etheno-DNA adducts can be measured by LC/MS-based methods [179, 180]. Elevated levels of etheno-DNA adducts were found in patients suffering from chronic inflammatory diseases, whereas increased levels of urinary εdA were observed in workers exposed to diesel engine exhaust [181, 182]. It is noteworthy to know that adducts between MDA and deoxyguanosine can also be measured by HPLC-FL and GC-MS methods [183, 184].

15.6.5 Simultaneous Determination of Several Oxidized Bases

Because exposure to pro-oxidant agents can lead to different kinds of oxidative DNA damage, it is unreliable to measure a single product as an index of oxidative modification. A better approach is to measure multiple modified bases at the same time.

For instance, GC-MS can detect a wide range of modified bases in a single DNA sample. Although, this instrument can simultaneously quantify modified nucleotide bases, the risk of overestimation of DNA damage due to artifact oxidation of bases during sample processing is still of major concern. Such analysis requires highly specialized personnel and expensive instrumentation.

Finally, all indications are that measurement of DNA and RNA oxidation products will increase in popularity and utility as a means of studying OS and chemical mutagenesis. Improvements in sample preparation and cleanup, particularly the introduction of multicolumn preparatory systems, may favor the use of HPLC in monitoring DNA oxidation in urine or plasma within a clinical setting. Each of the mentioned methods has advantages and disadvantages, and results cannot be compared at present due to the lack of standardization, differences in specificity and sensitivity, variations in DNA extraction protocols, DNA digestion, and storage conditions.

15.7 Glutathione

Several chromatographic methods have been optimized in order to identify and quantify GSH and its disulfides (GSSG and GSSP) in human samples. The most common procedures for the simultaneous determination of these compounds involve HPLC separation of the derivatized molecules followed by UV or fluorescence detection. The HPLC-UV methods rely on the derivatization of GSH and GSSG with various chromophores (iodoacetic acid/dinitrobenzene, 4,4′-dithiodipyridine, and DTNB), resulting in a sensitivity range of 10–100 picomoles [67]. For fluorescence detection, the derivatization is generally carried out with dansyl chloride, N-chlorodansylamide, monobromobimanes, and o-phthalaldehyde [185].

A major problem with derivatization is the possibility of auto-oxidation of GSH. Most derivatization procedures require high pH, where auto-oxidation of glutathione is likely to occur. The commonly used method to limit the amount of auto-oxidation of GSH was developed by Reed et al. This method has been reported to generate 0.7% oxidation of GSH to GSSG [67]. Some researchers suggest measurement of GSSG separately to avoid artificial oxidation. It is also recommended that GSH be neutralized prior to derivatizing GSSG for measurement. In a method described by Asensi et al., the thiol group of GSH is neutralized by N-ethylmaleimide (NEM; an S-alkylating agent), followed by the derivatization of GSSG [186]. This method ensures that the measured product is endogenous GSSG and not artificial GSSG generated during the derivatization process. Such a procedure limits auto-oxidization of GSH and is recommended for GSSG measurements. In other studies, pre-treatment with NEM indicated that blood GSH concentration in healthy people is about 1.5 mmol/L, whereas GSSG and GSSP concentrations are only about 2–3 μmol/L, or about three orders of magnitude lower [187, 188].

Svardal et al. developed a HPLC-FL method for the determination of GSH and GSSG, in which thiol derivatization using monobromobimane is performed [189]. GSSG is first converted to GSH by $NaBH_4$, followed by derivatization of thiol groups and chromatographic separation. GSSG is calculated as the difference between free GSH (without reduction) and total GSH (after reduction).

Ion-exchange chromatography method using EC detector without derivatization procedures can be performed for the determination of GSH and GSSG in biological samples. The two analytes, as a function of the potential applied to the electrochemical cell (coulometric electrodes), are transformed into oxidized form. The current generated by this redox reaction is measured. This method is one of the most sensitive methods for GSH/GSSG detection (LOD $\approx$ 10 picomoles) and may be used in a complex matrix without the need of extraction and purification. Glassy carbon/gold-mercury coupled with HPLC can be advantageous as it offers high selectivity for thiols and disulfides.

As a consequence, HPLC-EC was regarded as the most attractive method of detection due to its simplicity. However, GC-MS and LC-(ESI)MS/MS have been shown to have improved precision and accuracy as well as a higher selectivity and sensitivity while still demonstrating a good agreement with previous studies that used HPLC-EC [190]. These added benefits of GC-MS and LC-MS-MS over HPLC-EC resulted in mass spectrometric methods being more widely used for analysis of GSH and its related compounds in recent years [191–194].

LC-MS/MS analytical methods for reduced and oxidized glutathione as well as their derivatives include separation by reverse-phase or ion-exchange chromatography, and quantitation by MS/MS using SRM.

Reduced glutathione is very unstable, and GSSG and GSSP concentrations may be overestimated if samples are not properly processed. It is first necessary to take precautions to prevent auto-oxidation of GSH to GSSG after collection and during sample preparation. Because the GSH/GSSG ratio is greater than 100:1, even a tiny amount of GSH oxidation can underestimate its redox state. There is an eventual oxidation of GSH in blood samples during blood withdrawal, centrifugation, and further handling. The samples are usually prepared under standardized conditions in order to avoid this artifact and to provide satisfactory relative quantitation among different biological samples. This is usually done by blocking GSH with thiol-reagent in the sampling tube such as N-ethyl maleimide, iodoacetate, or 2-vinyl pyridine [195]. Another way to prevent GSH oxidation is to perform an acidification of samples during or immediately after sample collection. Acids such as perchloric acid, m-phosphoric acid, 5-sulfosalicylic acid, and picric acid are commonly used. Acidification commonly used for protein acidification can paradoxally lead to large overestimation of GSSG and protein-glutathione mixed disulfides (GSSP), due to artifactual oxidation. These effects were confirmed in samples treated with trichloroacetic acid at different conditions of dilution. It is also necessary to minimize hemolysis of the sample during blood draw because of very high levels of GSH in erythrocytes [196].

The measurement of different forms of oxidized GSH and the estimation of the molar ratios (GSH/GSSG) may give fundamental information on oxidative reactions. This ratio is often used to examine redox status and is a well-established marker of OS. It has been also suggested as a reliable marker for whole-body antioxidant index of diseases [197]. Indeed, levels of GSH decrease during OS due to the oxidation and subsequent formation of GSH disulfides, resulting in a decrease in the ratio of GSH to GSSG and PSSG.

Many published reports evaluated blood levels of GSH, GSSG, and GSSP both in healthy controls and in subjects affected by various pathologies. The ratio GSH/GSSG has been shown to be decreased in several conditions including smoking, DM, chemotherapy, and CVD [198–200].

GSH determination is important to identify patients who may benefit from GSH repletion by thiol derivatives or precursors, for instance, in clinical trials. Therefore, accurate determination of GSH provides valuable information and may represent one of the best choices to monitor antioxidant treatments based on this compound.

Besides GSH, other thiol compounds have been analyzed in peripheral blood, lymphocytes, and tissues especially hepatocytes using different types of mass spectrometers [193]. An efficient LC-MS/MS SRM method has been developed for quantitation of multiple biologically important thiols obtained from liver homogenates [201]. This method allows simultaneous quantification of GSH, GSSG, homocysteine, and homocystine using glutathione ethyl ester as a single internal standard. The sensitivity is suitable for these endogenous compounds.

15.8 Homocysteine

Chromatographic methods represent the method of choice for the determination of homocysteine and its oxidized form homocysteine. HPLC protocols with multiple detectors (UC, FL, EC, and MS) have been developed for this purpose and provide high specificity and sensitivity for quantitative analysis of homocysteine in biological samples (plasma and urine). LC-MS/MS SRM method allows accurate determination of very low levels of homocystine in plasma using d^8-homocystine as an internal standard [202]. These methods have been used to study the involvement of homocysteine in certain diseases linked to OS such as breast cancer [203].

15.9 Coenzyme Q_{10}

Coenzyme Q_{10} exists in oxidized (ubiquinone) and reduced (ubiquinol) forms in circulation. Ubiquinol importance in mitochondrial metabolism is well established, and the ratio of the oxidized/reduced form in cells and plasma is an important indicator of OS.

Methods for simultaneous measurement of ubiquinone and ubiquinol involve HPLC separation, followed by UV or electrochemical detection, with the latter being more sensitive [204, 205]. Mainly, there are three possible configurations for HPLC methods: UV, dual UV-electrochemical system, and dual electrode electrochemical system.

Ubiquinone and ubiquinol can be detected using HPLC-UV detection (290 nm for ubiquinol and 275 nm for ubiquinone). Although the use of HPLC-UV to determine ubiquinone and ubiquinol in tissue specimens has been useful, this method is not sensitive, particularly for ubiquinol detection at 290 nm. Oxidation of ubiquinol to ubiquinone by treatment with parabenzoquinone before extraction and separation enables the UV detection with high sensitivity [206]. Precolumn reduction with borohydride can be performed and it is useful for determination of total coenzyme Q_{10}.

Electrochemical detection is much more sensitive and because ubiquinone has a fairly strong extinction coefficient ($s = 14{,}200$) at 275 nm, several methods have coupled electrochemical with HPLC for ubiquinol/ubiquinone detection [207, 208]. Most biological samples fall within this sensitivity limits and so this analytical system is frequently used.

Dual UV/electrochemical detection is based on UV detection of ubiquinone and electrochemical detection of ubiquinol. Probably the most sensitive way to simultaneously determine ubiquinol and ubiquinone is by this dual system. In this configuration, ubiquinone is reduced to ubiquinol by a high negative potential at the upstream electrode, and the newly converted ubiquinol is detected at the downstream electrode [209]. Consequently, both compounds are detected at the downstream electrode as ubiquinol, but because HPLC separation of ubiquinol and ubiquinone occurs upstream of the electrode, the two compounds are individually measured. An alternative HPLC system can be performed, which uses oxidation of serum samples by $K_3Fe(CN)_6$ before analysis, so that the results are expressed as ubiquinone [210]. There is a possibility to construct a chromatographic system that comprises two revered-phase columns and a three-electrode coulometric detector coupled in series. Note that most of the described methods for ubiquinol and ubiquinone can simultaneously detect α-tocopherol, γ-tocopherol, and β-carotene in plasma samples [211, 212].

Finally, LC-MS/MS is considered the "gold standard" for ubiquinone/ubiquinol detection as it offers the highest sensitivity and specificity in biological fluids. For all described methods, sample handling must be performed in shaded light to avoid photochemical decomposition of coenzyme Q_{10}. It is noteworthy to know that the ratio of ubiquinone to ubiquinol is 5/95 in normal human plasma, but it increases significantly in several pathological states including DM, acute respiratory distress syndrome (ARDS), and coronary heart disease [213, 214].

15.10 Vitamin A

The most commonly used method for determining plasma concentrations of vitamin A–related compounds is HPLC (UV and MS), which allows separation of several carotenoids. All steps of the assay require protection from light. Simultaneous measurement of seven carotenoids (i.e., lutein, zeaxanthine, canthaxanthine, β-cryptoxanthine, lycopene, α- and β-carotene), and retinol has been performed in serum by HPLC [215].

In chronic stress during HIV infection, it has been found that the most easily depletable antioxidant is β-carotene, probably because this vitamin cannot be recycled after scavenging [216].

15.11 Vitamin C

Vitamin C is the most potent intra- and extracellular antioxidant in the organism. This compound scavenges $O_2^{\bullet-}$, $OH^{\bullet}$, and $ROO^{\bullet}$, and can react with OCl^- and $^1O_2$.

Determination of vitamin C concentration in plasma requires caution, given the great oxidizability of this compound and the instability of the dehydroascorbic acid. Due to the rapid oxidation of ascorbic acid at neutral or alkaline pH, plasma should be immediately acidified after its separation by centrifugation. Different types of acids are used such as metaphosphoric and trichloroacetic acids, oxalic acid alone, or in association with trichloroacetic acid. Another precaution to take is that plasma samples should be analyzed as soon as possible to avoid redistribution of ascorbic acid from plasma to erythrocytes and oxidation of ascorbic acid.

Chromatographic methods to measure ascorbate in biological preparations are numerous and several detectors are proposed. They range from spectroscopic methods where ascorbate has strong absorption in the UV region, to fluorometric, electrochemical, and mass spectrometric methods. These methods are commonly used and have the advantage of allowing detection of both vitamin C and its degradation products.

Ascorbate is primarily detected by reverse-phase HPLC-UV. Although ascorbate is a highly polar molecule, the use of ion couplers allows its separation via reverse-phase-ion-pair chromatography. The UV determination of ascorbate at 254 nm is usually used to measure ascorbic acid in urine and plasma [217]. This method is rather insensitive and cannot be used for some biological samples. Moreover, interference by other absorbing substances, such as hypoxanthine, xanthine, and uric acid, makes the use of UV detection impractical for most biological samples. On the other hand, electrochemical detection (either amperometric or coulometric detection) of ascorbic acid is extremely selective and sensitive and is considered as the method of choice for vitamin C determination in biological samples.

The detection of dehydroascorbic acid by HPLC is more troublesome and indirect. Because this oxidized form cannot be directly detected, two strategies have been proposed:

- Derivatization of dehydroascorbate with a strong fluorescent compound (e.g., o-phenylenediamine) followed by separation using HPLC and detection by fluorescence [218].
- Reduction of dehydroascorbate to ascorbate using 2,3- dimercapto-1-propanol or DTT and estimation of dehydroascorbate as ascorbate using electrochemical detection [219, 220]. In order to use electrochemistry, samples must be run twice. First, samples must be analyzed without 2,3-dimercapto-l-propanol or

DTT treatment to obtain concentration values of ascorbate and then measured again after reduction to obtain the total ascorbate value. Concentration of dehydroascorbate is estimated by subtracting the ascorbate value from the total ascorbate value. Note that 2,3-dimercapto-l-propanol can interfere with ascorbate and must be removed from the sample in advance. The use of DTT is advantageous because it does not need to be extracted, saving therefore time and reducing sample variability.

Finally, Deutsch and Kolhouse have published a reference method for vitamin C determination in biological samples, associating gas chromatography with mass spectrometry [221].

The importance of vitamin C as an antioxidant in the physiology of health and disease (e.g., DM) has been widely studied using chromatographic methods [222].

15.12 Vitamin E

Chromatographic techniques provide specific and accurate determination of all tocopherol species and many references clinical laboratories provide chromatographic analysis of these species in human serum and plasma.

Several methods exist for the simultaneous detection of the vitamin E family, especially α- and γ-tocopherol since they represent the major forms of this vitamin in cells. Indeed, plasma essentially contains 80–90% of α-tocopherol, the rest being other isomers (γ and β-tocopherol). They can be detected by GC-MS or HPLC, the latter being widely used. Vitamin E species are separated by reverse-phase HPLC and detected with either UV, FL, or EC detectors [223–227]. While all these detectors have been used with HPLC to determine tocopherols, there are analytical differences between detection methods. By taking into consideration GC-MS, chromatography sensitivity for vitamin E is as follows: GC-MS > HPLC-EC > HPLC-FL > HPLC-UV.

HPLC-UV can be problematic for two reasons. First, the extinction coefficient for tocopherol is not high enough ($\varepsilon = 3058$); therefore, this method lacks sensitivity. Secondly, the maximum absorption of tocopherol is located at 294 nm, a wavelength at which many other biomolecules interfere. Despite these limitations, HPLC-UV with proper and optimized chromatographic conditions can be applicable for the measurement of vitamin E in some biological systems.

Electrochemical determination of vitamin E is one of the most frequently used methods because of its high sensitivity. For samples that need extremely sensitive detection methods, electrochemical detection remains one of the best choices available.

Although fluorometric detection is less sensitive than electrochemical detection, for most biological samples, fluorescence detection represents a trouble-free method for tocopherol determination. Overall, electrochemical and fluorescent detections are recommended over UV detection for analysis of biological samples due to their sensitivity and selectivity.

For all mentioned methods, the preanalytical step should be performed in dark and under refrigerated conditions. The addition of antioxidants such as pyrogallol, vitamin C, and butylated hydroxytoluene is useful to protect vitamin E from oxidation reactions. Tocopherols are lipophilic molecules that can be extracted by organic solvents like hexane. Because α-tocopherol loss during extraction procedures can be a source of error, internal standards like tocopherol acetate and tocol can be used to increase reproducibility.

Plasma vitamin E assays can be useful for the follow-up of an oral supplementation with vitamin E when the management of OS is required in some diseases.

Several methods were published for simultaneous detection of several vitamins involved in OS. Faure et al. used HPLC-UV for analysis of α-tocopherol, retinol, and β-carotene [228]. In another work, simultaneous measurement of α-tocopherol, retinol, and five carotenoids (lutein, cryptoxanthin, lycopene, and α- and β-carotene) in plasma and isolated LDL have been performed using reverse-phase HPLC-UV [229].

Probably one of the most efficient and time-saving procedures for identifying and quantifying lipid-soluble antioxidants is to use electrochemical detection and thereto separate these antioxidants on HPLC [230]. In this procedure, it is possible to perform simultaneous detection of α- and β-tocopherols, β-carotene, lycopene, ubiquinol, and ubiquinone in human plasma. The sensitivity of this procedure ranges from about 1–50 μM.

Finally, Andreoli et al. described LC-MS/MS method for simultaneous determination of α-tocopherol, retinol, and β-carotene in human serum. The authors used atmospheric pressure chemical ionization for ionization in mass spectrometric analysis [231].

References

1. Grootveld M, Halliwell B. Aromatic hydroxylation as a potential measure of hydroxyl-radical formation in vivo. Identification of hydroxylated derivatives of salicylate in human body fluids. The Biochemical journal. 1986;237(2):499-504.
2. Coudray C, Favier A. Determination of salicylate hydroxylation products as an in vivo oxidative stress marker. Free Radical Biology and Medicine. 2000;29(11):1064-70.
3. Piszcz P, Żurawski K, Głód BK. Application of HPLC to Study the Reaction of Free Radicals with Antioxidants and/or Toxins. Journal of Chemistry. 2014;2014:385908.
4. Głód BK, Czapski GA, Haddad PR. Application of high-performance liquid chromatography to the investigation of free radical reactions in biological systems. TrAC Trends in Analytical Chemistry. 2000;19(8):492-7.
5. Kaur H, Edmonds SE, Blake DR, Halliwell B. Hydroxyl radical generation by rheumatoid blood and knee joint synovial fluid. Annals of the rheumatic diseases. 1996;55(12):915-20.
6. Ghiselli A, Laurenti O, De Mattia G, Maiani G, Ferro-Luzzi A. Salicylate hydroxylation as an early marker of in vivo oxidative stress in diabetic patients. Free Radic Biol Med. 1992;13(6):621-6.

7. Sun JZ, Kaur H, Halliwell B, Li XY, Bolli R. Use of aromatic hydroxylation of phenylalanine to measure production of hydroxyl radicals after myocardial ischemia in vivo. Direct evidence for a pathogenetic role of the hydroxyl radical in myocardial stunning. 1993;73(3):534-49.
8. O'Neill CA, Fu LW, Halliwell B, Longhurst JC. Hydroxyl radical production during myocardial ischemia and reperfusion in cats. The American journal of physiology. 1996;271(2 Pt 2): H660-7.
9. Piantadosi CA, Zhang J. Mitochondrial generation of reactive oxygen species after brain ischemia in the rat. Stroke. 1996;27(2):327–31; discussion 32.
10. Liu L, Leech JA, Urch RB, Silverman FS. In vivo salicylate hydroxylation: a potential biomarker for assessing acute ozone exposure and effects in humans. American journal of respiratory and critical care medicine. 1997;156(5):1405-12.
11. Thome J, Zhang J, Davids E, Foley P, Weijers HG, Wiesbeck GA, et al. Evidence for increased oxidative stress in alcohol-dependent patients provided by quantification of in vivo salicylate hydroxylation products. Alcohol Clin Exp Res. 1997;21(1):82-5.
12. Tubaro M, Cavallo G, Pensa V, Chessa MA, Natale E, Ricci R, et al. Demonstration of the formation of hydroxyl radicals in acute myocardial infarction in man using salicylate as probe. Cardiology. 1992;80(3–4):246–51.
13. Sloot WN, Gramsbergen JBP. Detection of salicylate and its hydroxylated adducts 2,3- and 2,5-dihydroxybenzoic acids as possible indices for in vivo hydroxyl radical formation in combination with catechol- and indoleamines and their metabolites in cerebrospinal fluid and brain tissue. Journal of Neuroscience Methods. 1995;60(1):141–9.
14. Yan EB, Unthank JK, Castillo-Melendez M, Miller SL, Langford SJ, Walker DW. Novel method for in vivo hydroxyl radical measurement by microdialysis in fetal sheep brain in utero. J Appl Physiol (1985). 2005;98(6):2304–10.
15. Dupont I, Berthou F, Bodenez P, Bardou L, Guirriec C, Stephan N, et al. Involvement of cytochromes P-450 2E1 and 3A4 in the 5-hydroxylation of salicylate in humans. Drug Metab Dispos. 1999;27(3):322–6.
16. Diez L, Livertoux M-H, Stark A-A, Wellman-Rousseau M, Leroy P. High-performance liquid chromatographic assay of hydroxyl free radical using salicylic acid hydroxylation during in vitro experiments involving thiols. Journal of Chromatography B: Biomedical Sciences and Applications. 2001;763(1):185–93.
17. Themann C, Teismann P, Kuschinsky K, Ferger B. Comparison of two independent aromatic hydroxylation assays in combination with intracerebral microdialysis to determine hydroxyl free radicals. Journal of Neuroscience Methods. 2001;108(1):57–64.
18. McCabe DR, Maher TJ, Acworth IN. Improved method for the estimation of hydroxyl free radical levels in vivo based on liquid chromatography with electrochemical detection. Journal of Chromatography B: Biomedical Sciences and Applications. 1997;691(1):23–32.
19. Luo X, Lehotay DC. Determination of hydroxyl radicals using salicylate as a trapping agent by gas chromatography-mass spectrometry. Clinical Biochemistry. 1997;30(1):41–6.
20. Freinbichler W, Tipton KF, Corte LD, Linert W. Mechanistic aspects of the Fenton reaction under conditions approximated to the extracellular fluid. Journal of inorganic biochemistry. 2009;103(1):28–34.
21. Lang K, Wagnerová D, Brodilová J. The Role of Hydrogen Peroxide in Dioxygen Induced Hydroxylation of Salicylic Acid. Collection of Czechoslovak Chemical Communications. 1994;59.
22. Luňák S, Muzart J, Brodilová J. Photochemical Hydroxylation of Salicylic Acid Derivatives with Hydrogen Peroxide, Catalyzed with Fe(III) and Sensitized with Methylene Blue. Collection of Czechoslovak Chemical Communications. 1994;59.
23. Maskos Z, Rush JD, Koppenol WH. The hydroxylation of the salicylate anion by a fenton reaction and Γ-radiolysis: A consideration of the respective mechanisms. Free Radical Biology and Medicine. 1990;8(2):153–62.
24. Salzberg-Brenhouse HC, Chen EY, Emerich DF, Baldwin S, Hogeland K, Ranelli S, et al. Inhibitors of cyclooxygenase-2, but not cyclooxygenase-1 provide structural and functional

protection against quinolinic acid-induced neurodegeneration. The Journal of pharmacology and experimental therapeutics. 2003;306(1):218–28.
25. Borthwick GM, Johnson AS, Partington M, Burn J, Wilson R, Arthur HM. Therapeutic levels of aspirin and salicylate directly inhibit a model of angiogenesis through a Cox-independent mechanism. FASEB journal : official publication of the Federation of American Societies for Experimental Biology. 2006;20(12):2009–16.
26. Kaur H, Whiteman M, Halliwell B. Peroxynitrite-dependent aromatic hydroxylation and nitration of salicylate and phenylalanine. Is hydroxyl radical involved? Free radical research. 1997;26(1):71–82.
27. Ferger B, Themann C, Rose S, Halliwell B, Jenner P. 6-hydroxydopamine increases the hydroxylation and nitration of phenylalanine in vivo: implication of peroxynitrite formation. Journal of neurochemistry. 2001;78(3):509–14.
28. H. K, B. H. Salicyclic acid and phenylalanine as probes to detect hydroxyl radicals. In: NA. P, FJ. K, editors. Free Radicals: A pratical approach: Oxford: Oxford University Press; 1996.
29. Leeuwenburgh C, Hansen P, Shaish A, Holloszy JO, Heinecke JW. Markers of protein oxidation by hydroxyl radical and reactive nitrogen species in tissues of aging rats. The American journal of physiology. 1998;274(2):R453–61.
30. Moreno S, Nardacci R, Cimini A, Cerù MP. Immunocytochemical localization of D-amino acid oxidase in rat brain. J Neurocytol. 1999;28(3):169–85.
31. Nair UJ, Nair J, Friesen MD, Bartsch H, Ohshima H. Ortho- and meta-tyrosine formation from phenylalanine in human saliva as a marker of hydroxyl radical generation during betel quid chewing. Carcinogenesis. 1995;16(5):1195–8.
32. Frischer T, Pullwitt A, Kühr J, Meinert R, Haschke N, Studnicka M, et al. Aromatic hydroxylation in nasal lavage fluid following ambient ozone exposure. Free Radic Biol Med. 1997;22 (1–2):201–7.
33. Reddy S, Halliwell B, Jones AD, Longhurst JC. The use of phenylalanine to detect hydroxyl radical production in vivo: a cautionary note. Free Radic Biol Med. 1999;27(11–12):1465.
34. Pennathur S, Wagner JD, Leeuwenburgh C, Litwak KN, Heinecke JW. A hydroxyl radical-like species oxidizes cynomolgus monkey artery wall proteins in early diabetic vascular disease. J Clin Invest. 2001;107(7):853–60.
35. Liu M, Liu S, Peterson SL, Miyake M, Liu KJ. On the application of 4-hydroxybenzoic acid as a trapping agent to study hydroxyl radical generation during cerebral ischemia and reperfusion. Molecular and cellular biochemistry. 2002;234–235(1–2):379–85.
36. Marklund N, Clausen F, Lewander T, Hillered L. Monitoring of reactive oxygen species production after traumatic brain injury in rats with microdialysis and the 4-hydroxybenzoic acid trapping method. J Neurotrauma. 2001;18(11):1217–27.
37. Kalén A, Appelkvist EL, Chojnacki T, Dallner G. Nonaprenyl-4-hydroxybenzoate transferase, an enzyme involved in ubiquinone biosynthesis, in the endoplasmic reticulum-Golgi system of rat liver. Journal of Biological Chemistry. 1990;265(2):1158–64.
38. Soto-Otero R, Méndez-Alvarez E, Hermida-Ameijeiras A, Muñoz-Patiño AM, Labandeira-Garcia JL. Autoxidation and neurotoxicity of 6-hydroxydopamine in the presence of some antioxidants: potential implication in relation to the pathogenesis of Parkinson's disease. Journal of neurochemistry. 2000;74(4):1605–12.
39. Saran M, Summer KH. Assaying for hydroxyl radicals: hydroxylated terephthalate is a superior fluorescence marker than hydroxylated benzoate. Free radical research. 1999;31(5): 429–36.
40. Dai GD, Cui LB, Song L, Zhao RZ, Chen JF, Wang YB, et al. Metabolism of terephthalic acid and its effects on CYP4B1 induction. Biomed Environ Sci. 2006;19(1):8–14.
41. Schimmel M, Bauer G. Proapoptotic and redox state-related signaling of reactive oxygen species generated by transformed fibroblasts. Oncogene. 2002;21(38):5886–96.
42. Meijer EP, Coolen SA, Bast A, Westerterp KR. Exercise training and oxidative stress in the elderly as measured by antipyrine hydroxylation products. Free radical research. 2001;35(4): 435–43.

43. Romitelli F, Santini SA, Chierici E, Pitocco D, Tavazzi B, Amorini AM, et al. Comparison of nitrite/nitrate concentration in human plasma and serum samples measured by the enzymatic batch Griess assay, ion-pairing HPLC and ion-trap GC–MS: The importance of a correct removal of proteins in the Griess assay. Journal of Chromatography B. 2007;851(1):257-67.
44. Fang Y-I, Ohata H, Honda K. Fluorometric determination of nitrite with 2,3-diaminonaphthalene by reverse phase HPLC under alkaline conditions. Journal of Pharmacological and Toxicological Methods. 2009;59(3):153–5.
45. Whiteman M, Ketsawatsakul U, Halliwell B. A reassessment of the peroxynitrite scavenging activity of uric acid. Annals of the New York Academy of Sciences. 2002;962:242–59.
46. Moison RM, de Beaufort AJ, Haasnoot AA, Dubbelman TM, van Zoeren-Grobben D, Berger HM. Uric acid and ascorbic acid redox ratios in plasma and tracheal aspirate of preterm babies with acute and chronic lung disease. Free Radic Biol Med. 1997;23(2):226–34.
47. Ogihara T, Kim HS, Hirano K, Imanishi M, Ogihara H, Tamai H, et al. Oxidation products of uric acid and ascorbic acid in preterm infants with chronic lung disease. Biol Neonate. 1998;73 (1):24–33.
48. Benzie IF, Chung W, Tomlinson B. Simultaneous measurement of allantoin and urate in plasma: analytical evaluation and potential clinical application in oxidant:antioxidant balance studies. Clin Chem. 1999;45(6 Pt 1):901–4.
49. Mikami T, Kita K, Tomita S, Qu GJ, Tasaki Y, Ito A. Is allantoin in serum and urine a useful indicator of exercise-induced oxidative stress in humans? Free radical research. 2000;32(3): 235–44.
50. Doehner W, Schoene N, Rauchhaus M, Leyva-Leon F, Pavitt DV, Reaveley DA, et al. Effects of xanthine oxidase inhibition with allopurinol on endothelial function and peripheral blood flow in hyperuricemic patients with chronic heart failure: results from 2 placebo-controlled studies. Circulation. 2002;105(22):2619–24.
51. Pavitt DV, de Fonseka S, Al-Khalaf N, Cam JM, Reaveley DA. Assay of serum allantoin in humans by gas chromatography-mass spectrometry. Clin Chim Acta. 2002;318(1–2):63–70.
52. Yardim-Akaydin S, Sepici A, Ozkan Y, Torun M, Simşek B, Sepici V. Oxidation of uric acid in rheumatoid arthritis: is allantoin a marker of oxidative stress? Free radical research. 2004;38 (6):623–8.
53. Marklund N, Ostman B, Nalmo L, Persson L, Hillered L. Hypoxanthine, uric acid and allantoin as indicators of in vivo free radical reactions. Description of a HPLC method and human brain microdialysis data. Acta Neurochir (Wien). 2000;142(10):1135–41; discussion 41–2.
54. Hellsten Y, Svensson M, Sjödin B, Smith S, Christensen A, Richter EA, et al. Allantoin formation and urate and glutathione exchange in human muscle during submaximal exercise. Free Radic Biol Med. 2001;31(11):1313–22.
55. Moselhy HF, Reid RG, Yousef S, Boyle SP. A specific, accurate, and sensitive measure of total plasma malondialdehyde by HPLC. J Lipid Res. 2013;54(3):852–8.
56. Bertolín JR, Joy M, Blanco M. Malondialdehyde determination in raw and processed meat products by UPLC-DAD and UPLC-FLD. Food Chemistry. 2019;298:125009.
57. Bastos AS, Loureiro APdM, Oliveira TFd, Corbi SCT, Caminaga RMS, Júnior CR, et al. Quantitation of malondialdehyde in gingival crevicular fluid by a high-performance liquid chromatography-based method. Analytical Biochemistry. 2012;423(1):141–6.
58. Breusing N, Grune T, Andrisic L, Atalay M, Bartosz G, Biasi F, et al. An inter-laboratory validation of methods of lipid peroxidation measurement in UVA-treated human plasma samples. Free radical research. 2010;44(10):1203–15.
59. Kim B, Jung W, Kho Y. Quantification of Malondialdehyde in Human Urine by HPLC-DAD and Derivatization with 2,4-Dinitrophenylhydrazine: Quantification of MDA in human urine by DNPH derivatization. Bulletin of the Korean Chemical Society. 2017;38.
60. Steghens JP, Kappel A, Denis I, Collombel C. Diaminonaphtalene, a new highly specific regent for HPLC-UV measurement of total and free malondialdehyde in human plasma or serum. Free radical biology & medicine. 2001;31:242–9.

61. Cheng G-W, Wu H-L, Huang Y-L. Simultaneous determination of malondialdehyde and ofloxacin in plasma using an isocratic high-performance liquid chromatography/fluorescence detection system. Analytica chimica acta. 2008;616(2):230-4.
62. Domijan AM, Ralić J, Radić Brkanac S, Rumora L, Žanić-Grubišić T. Quantification of malondialdehyde by HPLC-FL – application to various biological samples. Biomedical chromatography : BMC. 2015;29(1):41–6.
63. Dong X, Tang J, Chen X. Sensitive determination of malondialdehyde in rat prostate by high performance liquid chromatography with fluorescence detection. Scientific Reports. 2020;10 (1):3990.
64. Zelzer S, Oberreither R, Bernecker C, Stelzer I, Truschnig-Wilders M, Fauler G. Measurement of total and free malondialdehyde by gas-chromatography mass spectrometry--comparison with high-performance liquid chromatography methology. Free radical research. 2013;47(8): 651–6.
65. Tsikas D, Rothmann S, Schneider JY, Gutzki FM, Beckmann B, Frölich JC. Simultaneous GC-MS/MS measurement of malondialdehyde and 4-hydroxy-2-nonenal in human plasma: Effects of long-term L-arginine administration. Anal Biochem. 2017;524:31–44.
66. Giera M, Lingeman H, Niessen WMA. Recent Advancements in the LC- and GC-Based Analysis of Malondialdehyde (MDA): A Brief Overview. Chromatographia. 2012;75(9–10): 433–40.
67. Han D, Loukianoff S, McLaughlin L. Part VII • Chapter 17 – Oxidative stress indices: analytical aspects and significance. In: Sen CK, Packer L, Hänninen OOP, editors. Handbook of Oxidants and Antioxidants in Exercise. Amsterdam: Elsevier Science B.V.; 2000. p. 433–83.
68. Kadiiska M, Gladen BC, Baird DD, Germolec D, Graham LB, Parker C, et al. Biomarkers of Oxidative Stress Study II. Are oxidation products of lipids, proteins, and DNA markers of CCl4 poisoning? Free radical biology & medicine. 2005;38:698–710.
69. Peroni D, Bodini A, Corradi M, Coghi A, Boner AL, Piacentini G. Markers of oxidative stress are increased in exhaled breath condensates of children with atopic dermatitis. The British journal of dermatology. 2011;166:839–43.
70. Antus B, Barta I. Response to the Letter Entitled "Comments Concerning Comparison of Airway and Systemic Malondialdehyde Levels for Assessment of Oxidative Stress in Cystic Fibrosis". Lung. 2015;193(5):861–2.
71. do Val Carneiro JL, Nixdorf SL, Mantovani MS, da Silva do Amaral Herrera AC, Aoki MN, Amarante MK, et al. Plasma malondialdehyde levels and CXCR4 expression in peripheral blood cells of breast cancer patients. Journal of cancer research and clinical oncology. 2009;135(8):997–1004.
72. Luczaj W, Jarocka-Karpinska I, Sierakowski S, Andrisic L, Zarkovic N, Skrzydlewska E. Lipid peroxidation in Rheumatoid arthritis; consequences and monitoring. Free Radic Biol Med. 2014;75 Suppl 1:S49.
73. Arribas L, Almansa I, Miranda M, Muriach M, Romero FJ, Villar VM. Serum Malondialdehyde Concentration and Glutathione Peroxidase Activity in a Longitudinal Study of Gestational Diabetes. PLOS ONE. 2016;11(5):e0155353.
74. Ito F, Sono Y, Ito T. Measurement and Clinical Significance of Lipid Peroxidation as a Biomarker of Oxidative Stress: Oxidative Stress in Diabetes, Atherosclerosis, and Chronic Inflammation. Antioxidants (Basel). 2019;8(3):72.
75. Tsamesidis I, Pantaleo A, Pekou A, Gusani A, Iliadis S, Makedou K, et al. Correlation of Oxidative Stress Biomarkers and Hematological Parameters in Blood Cancer Patients from Sardinia, Italy. Int J Hematol Oncol Stem Cell Res. 2019;13(2):49–57.
76. Abdel SE, Abdel-Meguid I, Korraa S. Markers of oxidative stress and aging in Duchene muscular dystrophy patients and the possible ameliorating effect of He:Ne laser. Acta myologica : myopathies and cardiomyopathies : official journal of the Mediterranean Society of Myology. 2007;26(1):14–21.

77. Mebrat Y, Amogne W, Mekasha A, Gleason RL, Jr., Seifu D. Lipid Peroxidation and Altered Antioxidant Profiles with Pediatric HIV Infection and Antiretroviral Therapy in Addis Ababa, Ethiopia. Journal of Tropical Pediatrics. 2016;63(3):196–202.
78. Favier AE. How to demonstrate the occurrence of an oxidative stress in human? In: Favier AE, Cadet J, Kalyanaraman B, Fontecave M, Pierre JL, editors. Analysis of Free Radicals in Biological Systems. Basel: Birkhäuser Basel; 1995. p. 99–117.
79. Chaudhary AK, Nokubo M, Reddy GR, Yeola SN, Morrow JD, Blair IA, et al. Detection of Endogenous Malondialdehyde-Deoxyguanosine Adducts in Human Liver. Science. 1994;265 (5178):1580–2.
80. Gioacchini AM, Calonghi N, Boga C, Cappadone C, Masotti L, Roda A, et al. Determination of 4-hydroxy-2-nonenal at cellular levels by means of electrospray mass spectrometry. 1999;13(15):1573–9.
81. Łuczaj W, Moniuszko A, Jarocka-Karpowicz I, Pancewicz S, Andrisic L, Zarkovic N, et al. Tick-borne encephalitis--lipid peroxidation and its consequences. Scandinavian journal of clinical and laboratory investigation. 2016;76(1):1–9.
82. Zelzer S, Mangge H, Oberreither R, Bernecker C, Gruber H-J, Prüller F, et al. Oxidative stress: Determination of 4-hydroxy-2-nonenal by gas-chromatography/mass spectrometry in human and rat plasma. Free radical research. 2015;49:1–19.
83. Brahmbhatt VV, Nold C, Albert CJ, Ford DA. Quantification of pentafluorobenzyl oxime derivatives of long chain aldehydes by GC-MS analysis. Lipids. 2008;43(3):275–80.
84. Ratcliffe N, Wieczorek T, Drabińska N, Gould O, Osborne A, De Lacy Costello B. A mechanistic study and review of volatile products from peroxidation of unsaturated fatty acids: an aid to understanding the origins of volatile organic compounds from the human body. Journal of breath research. 2020;14(3):034001.
85. Pizzimenti S, Ciamporcero E, Daga M, Pettazzoni P, Arcaro A, Cetrangolo G, et al. Interaction of aldehydes derived from lipid peroxidation and membrane proteins. Front Physiol. 2013;4: 242-.
86. Paci A, Rieutord A, Guillaume D, Traoré F, Ropenga J, Husson HP, et al. Quantitative high-performance liquid chromatographic determination of acrolein in plasma after derivatization with Luminarin® 3. Journal of Chromatography B: Biomedical Sciences and Applications. 2000;739(2):239–46.
87. Takamoto S, Sakura N, Yashiki M, Kojima T. Determination of acrolein by headspace solid-phase microextraction gas chromatography and mass spectrometry. Journal of Chromatography B: Biomedical Sciences and Applications. 2001;758(1):123–8.
88. Zhang X, Xiong W, Shi L, Hou H, Hu Q. Simultaneous determination of five mercapturic acid derived from volatile organic compounds in human urine by LC–MS/MS and its application to relationship study. Journal of Chromatography B. 2014;967:102–9.
89. Kawabata C, Nagasawa T, Ono M, Tarumoto N, Katoh N, Hotta Y, et al. Plasma acrolein level in rheumatoid arthritis increases independently of the disease characteristics. Modern Rheumatology. 2020:1–8.
90. Handelman GJ, Frankel EN, Fenz R, German JB. Chemical changes during the early phase of in vitro oxidative damage to human LDL. Biochemistry and molecular biology international. 1993;31(4):777–88.
91. Esterbauer H, Puhl H, Dieber-Rotheneder M, Waeg G, Rabl H. Effect of antioxidants on oxidative modification of LDL. Annals of medicine. 1991;23(5):573–81.
92. Gutteridge JM, Halliwell B. The measurement and mechanism of lipid peroxidation in biological systems. Trends in biochemical sciences. 1990;15(4):129–35.
93. Corongiu FP, Poli G, Dianzani MU, Cheeseman KH, Slater TF. Lipid peroxidation and molecular damage to polyunsaturated fatty acids in rat liver. Recognition of two classes of hydroperoxides formed under conditions in vivo. Chem Biol Interact. 1986;59(2):147–55.
94. Cerne D, Lukac-Bajalo J. Oxidative stress assays for disease risk stratification. Acta pharmaceutica (Zagreb, Croatia). 2006;56(1):1–17.

95. Henderson DE, Slickman AM, Henderson SK. Quantitative HPLC Determination of the Antioxidant Activity of Capsaicin on the Formation of Lipid Hydroperoxides of Linoleic Acid: A Comparative Study against BHT and Melatonin. Journal of Agricultural and Food Chemistry. 1999;47(7):2563–70.
96. Miyazawa T. Determination of phospholipid hydroperoxides in human blood plasma by a chemiluminescence-HPLC assay. Free Radic Biol Med. 1989;7(2):209–17.
97. Frei B, Yamamoto Y, Niclas D, Ames BN. Evaluation of an isoluminol chemiluminescence assay for the detection of hydroperoxides in human blood plasma. Analytical biochemistry. 1988;175(1):120–30.
98. Yamamoto Y, Frei B, Ames BN. [38] Assay of lipid hydroperoxides using high-performance liquid chromatography with isoluminal chemiluminescence detection. Methods in enzymology. 186: Academic Press; 1990. p. 371–80.
99. Ohshima T, Hopia A, German JB, Frankel EN. Determination of hydroperoxides and structures by high-performance liquid chromatography with post-column detection with diphenyl-1-pyrenylphosphine. Lipids. 1996;31(10):1091–6.
100. Ferreiro-Vera C, Ribeiro JP, Mata-Granados JM, Priego-Capote F, Luque de Castro MD. Standard operation protocol for analysis of lipid hydroperoxides in human serum using a fully automated method based on solid-phase extraction and liquid chromatography-mass spectrometry in selected reaction monitoring. Journal of chromatography A. 2011;1218(38): 6720–6.
101. Sobotka PA, Brottman MD, Weitz Z, Birnbaum AJ, Skosey JL, Zarling EJ. Elevated breath pentane in heart failure reduced by free radical scavenger. Free Radic Biol Med. 1993;14(6): 643–7.
102. Humad S, Zarling E, Clapper M, Skosey JL. Breath pentane excretion as a marker of disease activity in rheumatoid arthritis. Free radical research communications. 1988;5(2):101–6.
103. Goodenough AK, Onorato JM, Ouyang Z, Chang S, Rodrigues AD, Kasichayanula S, et al. Quantification of 4-Beta-Hydroxycholesterol in Human Plasma Using Automated Sample Preparation and LC-ESI-MS/MS Analysis. Chemical Research in Toxicology. 2011;24(9): 1575–85.
104. Karten B, Boechzelt H, Abuja PM, Mittelbach M, Oettl K, Sattler W. Femtomole analysis of 9-oxononanoyl cholesterol by high performance liquid chromatography. J Lipid Res. 1998;39 (7):1508–19.
105. Bailey AL, Wortley G, Southon S. Measurement of Aldehydes in Low Density Lipoprotein by High Performance Liquid Chromatography. Free Radical Biology and Medicine. 1997;23(7): 1078–85.
106. Kritharides L, Jessup W, Gifford J, Dean RT. A method for defining the stages of low-density lipoprotein oxidation by the separation of cholesterol- and cholesteryl ester-oxidation products using HPLC. Anal Biochem. 1993;213(1):79–89.
107. Basu S. The enigma of in vivo oxidative stress assessment: isoprostanes as an emerging target. Scandinavian Journal of Food and Nutrition. 2007;51(2):48–61.
108. Oxidative Stress. In: Stanner S, editor. Cardiovascular Disease Diet, Nutrition and Emerging Risk Factors: Blackwell Publishing; 2005. p. 78–99.
109. Nikolaidis M, Kyparos A, Vrabas I. F2-isoprostane formation, measurement and interpretation: The role of exercise. Progress in lipid research. 2010;50:89–103.
110. Liu W, Morrow J, Yin H. Quantification of F2-Isoprostanes as a Reliable Index of Oxidative Stress in vivo Using Gas Chromatography – Mass Spectrometry (GC-MS) Method. Free Radical Biology and Medicine. 2009;47:1101–7.
111. Yen H-C, Wei H-J, Chen T-W. Analytical Variables Affecting Analysis of F_2-Isoprostanes and F_4-Neuroprostanes in Human Cerebrospinal Fluid by Gas Chromatography/Mass Spectrometry. BioMed research international. 2013;2013:810915.
112. Milne GL, Gao B, Terry ES, Zackert WE, Sanchez SC. Measurement of F2- isoprostanes and isofurans using gas chromatography-mass spectrometry. Free radical biology & medicine. 2013;59:36–44.

113. Briskey DR, Wilson GR, Fassett RG, Coombes JS. Optimized method for quantification of total F2-isoprostanes using gas chromatography–tandem mass spectrometry. Journal of Pharmaceutical and Biomedical Analysis. 2014;90:161–6.
114. Holder C, Adams A, McGahee E, Xia B, Blount BC, Wang L. High-Throughput and Sensitive Analysis of Free and Total 8-Isoprostane in Urine with Isotope-Dilution Liquid Chromatography-Tandem Mass Spectrometry. ACS Omega. 2020;5(19):10919–26.
115. Yan W, Byrd GD, Ogden MW. Quantitation of isoprostane isomers in human urine from smokers and nonsmokers by LC-MS/MS. J Lipid Res. 2007;48(7):1607–17.
116. Haschke M, Zhang YL, Kahle C, Klawitter J, Korecka M, Shaw LM, et al. HPLC-atmospheric pressure chemical ionization MS/MS for quantification of 15-F2t-isoprostane in human urine and plasma. Clin Chem. 2007;53(3):489–97.
117. Dreissigacker U, Suchy MT, Maassen N, Tsikas D. Human plasma concentrations of malondialdehyde (MDA) and the F2-isoprostane 15(S)-8-iso-PGF(2alpha) may be markedly compromised by hemolysis: evidence by GC-MS/MS and potential analytical and biological ramifications. Clin Biochem. 2010;43(1–2):159–67.
118. Sircar D, Subbaiah PV. Isoprostane Measurement in Plasma and Urine by Liquid Chromatography–Mass Spectrometry with One-Step Sample Preparation. Clinical Chemistry. 2007;53(2):251–8.
119. Schwedhelm E, Benndorf R, Boeger R, Tsikas D. Mass Spectrometric Analysis of F2-Isoprostanes: Markers and Mediators in Human Disease. Current Pharmaceutical Analysis – CURR PHARM ANAL. 2007;3.
120. Tsikas D, Schwedhelm E, Suchy MT, Niemann J, Gutzki FM, Erpenbeck VJ, et al. Divergence in urinary 8-iso-PGF(2alpha) (iPF(2alpha)-III, 15-F(2t)-IsoP) levels from gas chromatography-tandem mass spectrometry quantification after thin-layer chromatography and immunoaffinity column chromatography reveals heterogeneity of 8-iso-PGF(2alpha). Possible methodological, mechanistic and clinical implications. Journal of chromatography B, Analytical technologies in the biomedical and life sciences. 2003;794 (2):237–55.
121. Salomon RG, Bi W. Isolevuglandin adducts in disease. Antioxid Redox Signal. 2015;22(18): 1703–18.
122. Kirabo A, Fontana V, de Faria AP, Loperena R, Galindo CL, Wu J, et al. DC isoketal-modified proteins activate T cells and promote hypertension. J Clin Invest. 2014;124(10):4642–56.
123. Li W, Laird JM, Lu L, Roychowdhury S, Nagy LE, Zhou R, et al. Isolevuglandins covalently modify phosphatidylethanolamines in vivo: detection and quantitative analysis of hydroxylactam adducts. Free Radic Biol Med. 2009;47(11):1539–52.
124. Fedorova M, Bollineni RC, Hoffmann R. Protein carbonylation as a major hallmark of oxidative damage: Update of analytical strategies. 2014;33(2):79–97.
125. Maier KL, Lenz A-G, Beck-Speier I, Costabel U. [44] Analysis of methionine sulfoxide in proteins. Methods in enzymology. 251: Academic Press; 1995. p. 455–61.
126. Crowley JR, Yarasheski K, Leeuwenburgh C, Turk J, Heinecke JW. Isotope Dilution Mass Spectrometric Quantification of 3-Nitrotyrosine in Proteins and Tissues Is Facilitated by Reduction to 3-Aminotyrosine. Analytical Biochemistry. 1998;259(1):127–35.
127. Ryberg H, Caidahl K. Chromatographic and mass spectrometric methods for quantitative determination of 3-nitrotyrosine in biological samples and their application to human samples. Journal of Chromatography B. 2007;851(1):160–71.
128. Tsikas D, Caidahl K. Recent methodological advances in the mass spectrometric analysis of free and protein-associated 3-nitrotyrosine in human plasma. J Chromatogr B Analyt Technol Biomed Life Sci. 2005;814(1):1–9.
129. Tsikas D, Duncan MW. Mass spectrometry and 3-nitrotyrosine: Strategies, controversies, and our current perspective. 2014;33(4):237–76.
130. Tsikas D. Analytical methods for 3-nitrotyrosine quantification in biological samples: the unique role of tandem mass spectrometry. Amino Acids. 2012;42(1):45–63.

131. Colombo G, Clerici M, Giustarini D, Portinaro N, Badalamenti S, Rossi R, et al. A central role for intermolecular dityrosine cross-linking of fibrinogen in high molecular weight advanced oxidation protein product (AOPP) formation. Biochimica et Biophysica Acta (BBA) – General Subjects. 2015;1850(1):1–12.
132. Orhan H, Coolen S, Meerman J. Quantification of urinary o,o′-dityrosine, a biomarker for oxidative damage to proteins, by high performance liquid chromatography with triple quadrupole tandem mass spectrometry: A comparison with ion-trap tandem mass spectrometry. Journal of chromatography B, Analytical technologies in the biomedical and life sciences. 2005;827:104–8.
133. Leeuwenburgh C, Rasmussen JE, Hsu FF, Mueller DM, Pennathur S, Heinecke JW. Mass Spectrometric Quantification of Markers for Protein Oxidation by Tyrosyl Radical, Copper, and Hydroxyl Radical in Low Density Lipoprotein Isolated from Human Atherosclerotic Plaques*. Journal of Biological Chemistry. 1997;272(6):3520–6.
134. Song Y, Liao J, Zha C, Wang B, Liu CC. Simultaneous determination of 3-chlorotyrosine and 3-nitrotyrosine in human plasma by direct analysis in real time–tandem mass spectrometry. Acta Pharmaceutica Sinica B. 2015;5(5):482–6.
135. Hazen SL, Heinecke JW. 3-Chlorotyrosine, a specific marker of myeloperoxidase-catalyzed oxidation, is markedly elevated in low density lipoprotein isolated from human atherosclerotic intima. J Clin Invest. 1997;99(9):2075–81.
136. Matayatsuk C, Poljak A, Bustamante S, Smythe GA, Kalpravidh RW, Sirankapracha P, et al. Quantitative determination of ortho- and meta-tyrosine as biomarkers of protein oxidative damage in β-thalassemia. Redox Report. 2007;12(5):219–28.
137. Hensley K, Maidt ML, Pye QN, Stewart CA, Wack M, Tabatabaie T, et al. Quantitation of protein-bound 3-nitrotyrosine and 3,4-dihydroxyphenylalanine by high-performance liquid chromatography with electrochemical array detection. Anal Biochem. 1997;251(2):187–95.
138. Hensley K, Maidt ML, Yu Z, Sang H, Markesbery WR, Floyd RA. Electrochemical analysis of protein nitrotyrosine and dityrosine in the Alzheimer brain indicates region-specific accumulation. J Neurosci. 1998;18(20):8126–32.
139. Salman-Tabcheh S, Guérin MC, Torreilles J. Nitration of tyrosyl-residues from extra- and intracellular proteins in human whole blood. Free Radic Biol Med. 1995;19(5):695–8.
140. Abdelrahim M, Morris E, Carver J, Facchina S, White A, Verma A. Liquid chromatographic assay of dityrosine in human cerebrospinal fluid. J Chromatogr B Biomed Sci Appl. 1997;696 (2):175–82.
141. Maruyama W, Hashizume Y, Matsubara K, Naoi M. Identification of 3-nitro-L-tyrosine, a product of nitric oxide and superoxide, as an indicator of oxidative stress in the human brain. Journal of chromatography B, Biomedical applications. 1996;676(1):153–8.
142. Hensley K, Williamson KS, Floyd RA. Measurement of 3-nitrotyrosine and 5-nitro-gamma-tocopherol by high-performance liquid chromatography with electrochemical detection. Free Radic Biol Med. 2000;28(4):520–8.
143. Orhan H, Vermeulen NPE, Tump C, Zappey H, Meerman JHN. Simultaneous determination of tyrosine, phenylalanine and deoxyguanosine oxidation products by liquid chromatography–tandem mass spectrometry as non-invasive biomarkers for oxidative damage. Journal of Chromatography B. 2004;799(2):245–54.
144. Misko TP, Radabaugh MR, Highkin M, Abrams M, Friese O, Gallavan R, et al. Characterization of nitrotyrosine as a biomarker for arthritis and joint injury. Osteoarthritis and Cartilage. 2013;21(1):151–6.
145. Palma-Duran SA, Kontogianni MD, Vlassopoulos A, Zhao S, Margariti A, Georgoulis M, et al. Serum levels of advanced glycation end-products (AGEs) and the decoy soluble receptor for AGEs (sRAGE) can identify non-alcoholic fatty liver disease in age-, sex- and BMI-matched normo-glycemic adults. Metabolism. 2018;83:120-7.
146. Lee JS, Chung Y-S, Chang SY, Jung Y-S, Kim SH. Simple Quantification of Pentosidine in Human Urine and Plasma by High-Performance Liquid Chromatography. International Journal of Analytical Chemistry. 2017;2017:1389807.

147. Palmieri B, Sblendorio V. Oxidative stress tests: overview on reliability and use. Part I. European review for medical and pharmacological sciences. 2007;11(5):309–42.
148. Thérond P, Bonnefont-Rousselot D, Davit-Spraul A, Conti M, Legrand A. Biomarkers of oxidative stress: an analytical approach. Current opinion in clinical nutrition and metabolic care. 2000;3(5):373–84.
149. Hensley K, Williamson KS, Maidt ML, Prasad Gabbita S, Grammas P, Floyd RA. Determination of Biological Oxidative Stress Using High Performance Liquid Chromatography with Electrochemical Detection (HPLC-ECD). 1999;22(8):429–37.
150. Khelfi A, Djemili I, Bellouard M-M, Hedhili A, Nebbali M, Hoummani H, et al. CHAPITRE 8 – Toxicologie des autres agents/produits toxiques. In: Khelfi A, Djemili I, Bellouard M-M, Hedhili A, Nebbali M, Hoummani H, et al., editors. Tout-En-un de Toxicologie. Paris: Elsevier Masson; 2023. p. 623–86.
151. Teixeira AJR, Gommersampt JH, Vandewerken G, Westra JG, Stavenuiter JFC, Dejong APJM. Method for the Analysis of Oxidized Nucleosides by Gas Chromatography Mass Spectrometry. Analytical Biochemistry. 1993;214(2):474–83.
152. Biomarkers of Oxidative Stress in Plasma and Urine. Analytical Techniques for Clinical Chemistry. p. 555–94.
153. Ravanat J-L, Turesky RJ, Gremaud E, Trudel LJ, Stadler RH. Determination of 8-Oxoguanine in DNA by Gas Chromatography-Mass Spectrometry and HPLC-Electrochemical Detection: Overestimation of the Background Level of the Oxidized Base by the Gas Chromatography-Mass Spectrometry Assay. Chemical Research in Toxicology. 1995;8(8):1039–45.
154. Douki T, Dealtour T, Bianchini F, Cadet J. Observation and prevention of an artefactual formation of oxidized DNA bases and nucleosides in the GC-EMS method. Carcinogenesis. 1996;17(2):347–53.
155. Guo C, Li X, Wang R, Yu J, Ye M, Mao L, et al. Association between Oxidative DNA Damage and Risk of Colorectal Cancer: Sensitive Determination of Urinary 8-Hydroxy-2′-deoxyguanosine by UPLC-MS/MS Analysis. Scientific Reports. 2016;6(1):32581.
156. Barbagallo M, Marotta F, Dominguez LJ. Oxidative stress in patients with Alzheimer's disease: effect of extracts of fermented papaya powder. Mediators Inflamm. 2015;2015: 624801-.
157. Kulikowska-Karpińska E, Czerw K. [Estimation of 8-hydroxy-2′-deoxyguanosine (8-OHdG) concentration in the urine of cigarette smokers]. Wiadomosci lekarskie (Warsaw, Poland : 1960). 2015;68(1):32–8.
158. Lodovici M, Casalini C, Cariaggi R, Michelucci L, Dolara P. Levels of 8-hydroxydeoxyguanosine as a marker of DNA damage in human leukocytes. Free Radic Biol Med. 2000;28(1):13–7.
159. Inter-laboratory validation of procedures for measuring 8-oxo-7,8-dihydroguanine/8-oxo-7,8-dihydro-2′-deoxyguanosine in DNA. Free radical research. 2002;36(3):239–45.
160. Collins AR, Cadet J, Möller L, Poulsen HE, Viña J. Are we sure we know how to measure 8-oxo-7,8-dihydroguanine in DNA from human cells? Arch Biochem Biophys. 2004;423(1): 57–65.
161. Measurement of DNA oxidation in human cells by chromatographic and enzymic methods. Free Radic Biol Med. 2003;34(8):1089–99.
162. Gedik CM, Collins A. Establishing the background level of base oxidation in human lymphocyte DNA: results of an interlaboratory validation study. FASEB journal : official publication of the Federation of American Societies for Experimental Biology. 2005;19(1): 82–4.
163. Collins A, Cadet J, Epe B, Gedik C. Problems in the measurement of 8-oxoguanine in human DNA. Report of a workshop, DNA oxidation, held in Aberdeen, UK, 19–21 January, 1997. Carcinogenesis. 1997;18(9):1833–6.
164. Floyd RA, Watson JJ, Wong PK, Altmiller DH, Rickard RCJFrrc. Hydroxyl free radical adduct of deoxyguanosine: sensitive detection and mechanisms of formation. 1986;1(3): 163–72.

165. Shigenaga MK, Gimeno CJ, Ames BN. Urinary 8-hydroxy-2′-deoxyguanosine as a biological marker of in vivo oxidative DNA damage. Proc Natl Acad Sci U S A. 1989;86(24):9697–701.
166. Finnegan MTV, Herbert KE, Evans MD, Griffiths HR, Lunec J. Evidence for sensitisation of DNA to oxidative damage during isolation. Free Radical Biology and Medicine. 1996;20(1): 93–8.
167. Collins AR, Dusinská M, Gedik CM, Stětina R. Oxidative damage to DNA: do we have a reliable biomarker? Environ Health Perspect. 1996;104 Suppl 3(Suppl 3):465–9.
168. Topic A, Francuski D, Markovic B, Stankovic M, Dobrivojevic S, Drca S, et al. Gender-related reference intervals of urinary 8-oxo-7,8-dihydro-2′-deoxyguanosine determined by liquid chromatography-tandem mass spectrometry in Serbian population. Clinical Biochemistry. 2013;46(4):321–6.
169. Jorgensen A, Broedbaek K, Fink-Jensen A, Knorr U, Greisen Soendergaard M, Henriksen T, et al. Increased systemic oxidatively generated DNA and RNA damage in schizophrenia. Psychiatry Research. 2013;209(3):417–23.
170. Ohshima H, Virág L, Souza J, Yermilov V, Pignatelli B, Masuda M, et al. Detection of Certain Peroxynitrite-Induced DNA Modifications. In: Armstrong D, editor. Oxidative Stress Biomarkers and Antioxidant Protocols. Totowa, NJ: Humana Press; 2002. p. 77–88.
171. Kaur H, Halliwell B. Measurement of oxidized and methylated DNA bases by HPLC with eletrochemical detection. The Biochemical journal. 1996;318 (Pt 1):21–3.
172. Wagner JR, Blount BC, Weinfeld M. Excision of oxidative cytosine modifications from gamma-irradiated DNA by Escherichia coli endonuclease III and human whole-cell extracts. Anal Biochem. 1996;233(1):76–86.
173. Wagner JR, Hu CC, Ames BN. Endogenous oxidative damage of deoxycytidine in DNA. 1992;89(8):3380–4.
174. Faure H, Incardona M-F, Boujet C, Cadet J, Ducros V, Favier A. Gas chromatographic—mass spectrometric determination of 5-hydroxymethyluracil in human urine by stable isotope dilution. Journal of Chromatography B: Biomedical Sciences and Applications. 1993;616 (1):1–7.
175. Dizdaroglu M. [1] Chemical determination of oxidative DNA damage by gas chromatography-mass spectrometry. Methods in enzymology. 234: Academic Press; 1994. p. 3–16.
176. Cadet J, Incardona MF, Odin F, Molko D, Mouret JF, Polverelli M, et al. Measurement of oxidative base damage to DNA by using HPLC-32P-postlabelling and GC/MS-selective ion monitoring assays. IARC scientific publications. 1993(124):271–6.
177. Cadet J, Odin F, Mouret J-F, Polverelli M, Audic A, Giacomoni P, et al. Chemical and biochemical postlabeling methods for singling out specific oxidative DNA lesions. Mutation Research/DNAging. 1992;275(3):343–54.
178. Bartsch H, Nair J. Accumulation of lipid peroxidation-derived DNA lesions: potential lead markers for chemoprevention of inflammation-driven malignancies. Mutation research. 2005;591(1–2):34–44.
179. Li H, Cui S, Wang S, Jiang X, Zhang S, Zhang R, et al. Ultrasensitive UPLC–MS/MS method for analysis of etheno-DNA adducts in human white blood cells. Free radical research. 2015;49(9):1049–54.
180. Cui S, Li H, Wang S, Jiang X, Zhang S, Zhang R, et al. Ultrasensitive UPLC-MS-MS method for the quantitation of etheno-DNA adducts in human urine. Int J Environ Res Public Health. 2014;11(10):10902–14.
181. Increased Etheno-DNA Adducts in Affected Tissues of Patients Suffering from Crohn's Disease, Ulcerative Colitis, and Chronic Pancreatitis. 2006;8(5-–6):1003–10.
182. Bin P, Shen M, Li H, Sun X, Niu Y, Meng T, et al. Increased levels of urinary biomarkers of lipid peroxidation products among workers occupationally exposed to diesel engine exhaust. Free radical research. 2016;50(8):820–30.
183. Agarwal S, Draper HH. Isolation of a malondialdehyde-deoxyguanosine adduct from rat liver DNA. Free Radical Biology and Medicine. 1992;13(6):695–9.

184. Rouzer CA, Chaudhary AK, Nokubo M, Ferguson DM, Reddy GR, Blair IA, et al. Analysis of the Malondialdehyde−2′- Deoxyguanosine Adduct Pyrimidopurinone in Human Leukocyte DNA by Gas Chromatography/Electron Capture/Negative Chemical Ionization/Mass Spectrometry. Chemical Research in Toxicology. 1997;10(2):181–8.
185. Martin J, White INH. Fluorimetric determination of oxidised and reduced glutathione in cells and tissues by high-performance liquid chromatography following derivatization with dansyl chloride. Journal of Chromatography B: Biomedical Sciences and Applications. 1991;568(1): 219–25.
186. Asensi M, Sastre J, Pallardo FV, Estrela JM, Viña J. [35] Determination of oxidized glutathione in blood: High-performance liquid chromatography. Methods in enzymology. 234: Academic Press; 1994. p. 367–71.
187. Rossi R, Dalle-Donne I, Milzani A, Giustarini D. Oxidized forms of glutathione in peripheral blood as biomarkers of oxidative stress. Clin Chem. 2006;52(7):1406–14.
188. Rossi R, Milzani A, Dalle-Donne I, Giustarini D, Lusini L, Colombo R, et al. Blood glutathione disulfide: in vivo factor or in vitro artifact? Clin Chem. 2002;48(5):742–53.
189. Svardal AM, Mansoor MA, Ueland PM. Determination of reduced, oxidized, and protein-bound glutathione in human plasma with precolumn derivatization with monobromobimane and liquid chromatography. Anal Biochem. 1990;184(2):338–46.
190. Radabaugh MR, Nemirovskiy OV, Misko TP, Aggarwal P, Mathews WR. Immunoaffinity liquid chromatography–tandem mass spectrometry detection of nitrotyrosine in biological fluids: Development of a clinically translatable biomarker. Analytical Biochemistry. 2008;380(1):68–76.
191. Squellerio I, Caruso D, Porro B, Veglia F, Tremoli E, Cavalca V. Direct glutathione quantification in human blood by LC–MS/MS: comparison with HPLC with electrochemical detection. Journal of Pharmaceutical and Biomedical Analysis. 2012;71:111–8.
192. Yu J, Li C, Shen S, Liu X, Peng Y, Zheng J. Mass spectrometry based detection of glutathione with sensitivity for single-cell analysis. 2015;29(7):681–9.
193. Carroll D, Howard D, Zhu H, Paumi CM, Vore M, Bondada S, et al. Simultaneous quantitation of oxidized and reduced glutathione via LC-MS/MS: An insight into the redox state of hematopoietic stem cells. Free radical biology & medicine. 2016;97:85–94.
194. Tsikas D, Hanff E, Kayacelebi AA, Böhmer A. Gas chromatographic-mass spectrometric analysis of the tripeptide glutathione in the electron-capture negative-ion chemical ionization mode. Amino Acids. 2016;48(2):593–8.
195. Ochtrop P, Hackenberger CPR. Recent advances of thiol-selective bioconjugation reactions. Current Opinion in Chemical Biology. 2020;58:28–36.
196. Jones DP, Carlson JL, Samiec PS, Sternberg P, Mody VC, Reed RL, et al. Glutathione measurement in human plasma: Evaluation of sample collection, storage and derivatization conditions for analysis of dansyl derivatives by HPLC. Clinica Chimica Acta. 1998;275(2): 175–84.
197. Zitka O, Skalickova S, Gumulec J, Masarik M, Adam V, Hubalek J, et al. Redox status expressed as GSH:GSSG ratio as a marker for oxidative stress in paediatric tumour patients. Oncol Lett. 2012;4(6):1247–53.
198. Paolisso G, Di Maro G, Pizza G, D'Amore A, Sgambato S, Tesauro P, et al. Plasma GSH/GSSG affects glucose homeostasis in healthy subjects and non-insulin-dependent diabetics. The American journal of physiology. 1992;263(3 Pt 1):E435–40.
199. Rodgers C, Sanborn C, Taylor O, Gundy P, Pasvogel A, Moore IMK, et al. Fatigue and Oxidative Stress in Children Undergoing Leukemia Treatment. Biol Res Nurs. 2016;18(5): 515–20.
200. Bajic VP, Van Neste C, Obradovic M, Zafirovic S, Radak D, Bajic VB, et al. Glutathione "Redox Homeostasis" and Its Relation to Cardiovascular Disease. Oxidative medicine and cellular longevity. 2019;2019:5028181-.
201. Bouligand J, Deroussent A, Paci A, Morizet J, Vassal G. Liquid chromatography-tandem mass spectrometry assay of reduced and oxidized glutathione and main precursors in mice liver.

Journal of chromatography B, Analytical technologies in the biomedical and life sciences. 2006;832(1):67–74.
202. Tomaiuolo M, Vecchione G, Grandone E, Cocomazzi N, Casetta B, Minno G, et al. A new method for determination of plasma homocystine by isotope dilution and electrospray tandem mass spectrometry. Journal of chromatography B, Analytical technologies in the biomedical and life sciences. 2006;842:64–9.
203. Stachniuk J, Kubalczyk P, Furmaniak P, Głowacki R. A versatile method for analysis of saliva, plasma and urine for total thiols using HPLC with UV detection. Talanta. 2016;155:70–7.
204. Hirayama A, Kubo H, Mita M, Shirota O, Yamamoto Y. High-Sensitivity Simultaneous Analysis of Ubiquinol-10 and Ubiquinone-10 in Human Plasma. Journal of chromatographic science. 2008;46:717–21.
205. Yamashita S, Yamamoto Y. Simultaneous detection of ubiquinol and ubiquinone in human plasma as a marker of oxidative stress. Anal Biochem. 1997;250(1):66–73.
206. Mosca F, Fattorini D, Bompadre S, Littarru GP. Assay of Coenzyme Q10 in Plasma by a Single Dilution Step. Analytical Biochemistry. 2002;305(1):49–54.
207. Franke AA, Morrison CM, Bakke JL, Custer LJ, Li X, Cooney RV. Coenzyme Q10 in human blood: native levels and determinants of oxidation during processing and storage. Free radical biology & medicine. 2010;48(12):1610–7.
208. Abdul-Rasheed O, Farid Y. Development of a new high performance liquid chromatography method for measurement of coenzyme Q10 in healthy blood plasma. Saudi medical journal. 2009;30:1138–43.
209. Tang PH, Miles MV. Measurement of oxidized and reduced coenzyme Q in biological fluids, cells, and tissues: an HPLC-EC method. Methods Mol Biol. 2012;837:149–68.
210. Laaksonen R, Riihimäki A, Laitila J, Mårtensson K, Tikkanen MJ, Himberg JJ. Serum and muscle tissue ubiquinone levels in healthy subjects. The Journal of laboratory and clinical medicine. 1995;125(4):517–21.
211. Karpinska J, Mikołuć B, Motkowski R, Piotrowska-Jastrzebska J. HPLC method for simultaneous determination of retinol, α-tocopherol and coenzyme Q10 in human plasma. Journal of pharmaceutical and biomedical analysis. 2006;42:232–6.
212. Kand'ár R, Novotná P, Drábková P. Determination of Retinol, α-Tocopherol, Lycopene, and β-Carotene in Human Plasma Using HPLC with UV-Vis Detection: Application to a Clinical Study. Journal of Chemistry. 2013;2013:460242.
213. Lim SC, Tan HH, Goh SK, Subramaniam T, Sum CF, Tan IK, et al. Oxidative burden in prediabetic and diabetic individuals: evidence from plasma coenzyme Q(10). Diabetic medicine : a journal of the British Diabetic Association. 2006;23(12):1344–9.
214. Lagendijk J, Ubbink JB, Delport R, Vermaak WJ, Human JA. Ubiquinol/ubiquinone ratio as marker of oxidative stress in coronary artery disease. Res Commun Mol Pathol Pharmacol. 1997;95(1):11–20.
215. Steghens JP, van Kappel AL, Riboli E, Collombel C. Simultaneous measurement of seven carotenoids, retinol and alpha-tocopherol in serum by high-performance liquid chromatography. J Chromatogr B Biomed Sci Appl. 1997;694(1):71–81.
216. Favier A, Sappey C, Leclerc P, Faure P, Micoud M. Antioxidant status and lipid peroxidation in patients infected with HIV. Chem Biol Interact. 1994;91(2–3):165–80.
217. Kim Y, Kim M-G. HPLC-UV method for the simultaneous determinations of ascorbic acid and dehydroascorbic acid in human plasma. 2016;24:37–42.
218. Tessier F, Birlouez-Aragon I, Tjani C, Guilland JC. Validation of a micromethod for determining oxidized and reduced vitamin C in plasma by HPLC-fluorescence. Int J Vitam Nutr Res. 1996;66(2):166–70.
219. Pullar JM, Bayer S, Carr AC. Appropriate Handling, Processing and Analysis of Blood Samples Is Essential to Avoid Oxidation of Vitamin C to Dehydroascorbic Acid. Antioxidants (Basel). 2018;7(2):29.
220. Mochnik PA, Frei B, Ames BN. [23] Measurement of antioxidants in human blood plasma. Methods in enzymology. 234: Academic Press; 1994. p. 269–79.

221. Deutsch JC, Kolhouse JF. Ascorbate and dehydroascorbate measurements in aqueous solutions and plasma determined by gas chromatography-mass spectrometry. Analytical Chemistry. 1993;65(4):321–6.
222. Khan A, Khan MI, Iqbal Z, Shah Y, Ahmad L, Nazir S, et al. A new HPLC method for the simultaneous determination of ascorbic acid and aminothiols in human plasma and erythrocytes using electrochemical detection. Talanta. 2011;84(3):789–801.
223. Zerbinati C, Galli F, Regolanti R, Poli G, Iuliano L. Gas chromatography–mass spectrometry microanalysis of alpha- and gamma-tocopherol in plasma and whole blood. Clinica Chimica Acta. 2015;446:156–62.
224. Marco R, Federico R, Carla Q, Afro Q, Alberto B. Simplified HPLC-UV method for the determination of α-tocopherol in plasma. Italian Journal of Animal Science. 2010;4.
225. Galli F, Lee R, Dunster C, Kelly FJ. Gas chromatography mass spectrometry analysis of carboxyethyl-hydroxychroman metabolites of alpha- and gamma-tocopherol in human plasma. Free Radic Biol Med. 2002;32(4):333–40.
226. Che H-L, Tan DM-Y, Meganathan P, Gan Y-L, Abdul Razak G, Fu J-Y. Validation of a HPLC/FLD Method for Quantification of Tocotrienols in Human Plasma. International Journal of Analytical Chemistry. 2015;2015:357609.
227. Yamauchi R, Noro H, Shimoyamada M, Kato K. Analysis of vitamin E and its oxidation products by HPLC with electrochemical detection. Lipids. 2002;37(5):515–22.
228. Faure H, Preziosi P, Roussel AM, Bertrais S, Galan P, Hercberg S, et al. Factors influencing blood concentration of retinol, α-tocopherol, vitamin C, and β-carotene in the French participants of the SU.VI.MAX trial. European Journal of Clinical Nutrition. 2006;60(6):706–17.
229. Arnaud J, Fortis I, Blachier S, Kia D, Favier A. Simultaneous determination of retinol, alpha-tocopherol and beta-carotene in serum by isocratic high-performance liquid chromatography. Journal of chromatography. 1991;572(1–2):103–16.
230. Motchnik PA, Frei B, Ames BN. Measurement of antioxidants in human blood plasma. Methods in enzymology. 1994;234:269–79.
231. Andreoli R, Manini P, Poli D, Bergamaschi E, Mutti A, Niessen WM. Development of a simplified method for the simultaneous determination of retinol, alpha-tocopherol, and beta-carotene in serum by liquid chromatography-tandem mass spectrometry with atmospheric pressure chemical ionization. Analytical and bioanalytical chemistry. 2004;378(4):987–94.

Chapter 16
Immunoassays/ELISA

A. Khelfi

Abstract Immunoassays are a class of analytical techniques where the reaction between an antigen and an antibody is used for the identification and quantification of the antigen in biological samples. Immunoassays are one the most commonly used techniques in clinical practice for oxidative stress assessment. The assays are often standardized and do not require expensive instrumentation. They are easy to perform and fast and provides a sensitive, automated, and high-throughput alternative for measuring a variety of oxidative stress biomarkers (e.g., protein carbonyl, nitrotyrosine, glutathionylated proteins, carbohydrate end-products, MDA adducts, 4-HNE-Protein adducts, OxLDL, isoprostanes (IsoPs), and DNA end-products). They are cost-effective with sensitivity and specificity comparable to conventional methods. More important, in situ analysis is possible with immunoassays. The versatility in the applications of immunoassays has also made these methods more popular in recent years. For these reasons, immunoassays play an important role in various bioanalytical settings such as clinical diagnostics, research, and therapeutic assessment. In fact, many tests have already been converted into commercial kits for this purpose and are reproducible and user-friendly. A detailed and thorough review of these assays is presented in this chapter with their challenges and applications.

Keywords Protein carbonyl · Nitrotyrosine · Glutathionylated proteins · Carbohydrate end-products · MDA adducts · 4-HNE-Protein adducts · OxLDL · isoprostanes (IsoPs) · DNA end-products · Antibody

A. Khelfi (✉)
Faculty of Pharmacy, University of Algiers 1, Algiers, Algeria

National Centre of Toxicology, Algiers, Algeria

Department of Chemistry and Biomolecular Science, Clarkson University, Potsdam, NY, USA
e-mail: ab.khelfi@univ-alger.dz; akhelfi@clarkson.edu

S. Andreescu et al. (eds.), *Biomarkers of Oxidative Stress*,
https://doi.org/10.1007/978-3-031-60738-7_16

16.1 Protein End Products

16.1.1 Protein Carbonyl

Immunoassays have wide application for analyzing protein carbonyls in experimental and clinical samples, especially where concentrations are low and only small amounts of sample are required. The main advantage of protein carbonyl analysis is the relative stability of these compounds compared with other markers of OS such as MDA and oxidized glutathione. Among the first methods, there was an ELISA for protein carbonyl detection that gave semi-quantitative data in biological samples [1]. Protein carbonyls having either ketone and/or aldehyde moieties are commonly detected by first derivatizing these proteins using DNPH and formation of stable carbonyl-2,4-dinitrophenylhydrazone adducts followed by detection using 2,4-dinitrophenyl antibodies commercially available. The sensitivity is now greatly improved, employing enhanced antidinitrophenylhydrazone antibodies. These antibodies are commercially available from Sigma-Aldrich (St. Louis, MO). Polac et al. used ELISA to evaluate carbonyl groups and commented that these groups in protein coupled to DNPH are good markers of OS [2].

The results of ELISA assays correlate well with colorimetric methods. They are more convenient to analyze a larger number of samples within one run and require significantly less sample volume [1]. Indeed, the only method that seems to be applicable in clinical settings is ELISA (kits are available) as it enables high throughput, involves external standards, and provides comparison of samples under constant conditions. ELISA assays are sufficiently sensitive to detect carbonyls in plasmatic proteins and detect significant increases of these markers in critically ill patients. They were applied as markers of OS during coronary heart surgery [3]. Following coronary surgery in the reperfused human heart, a two- to three-fold increase of protein carbonyls was recorded in plasma collected from the venous coronary sinus. Protein carbonyls remained increased in blood for up to 18 h, which is an important criterion for being a stable marker of OS.

16.1.2 Nitrotyrosine and Nitroproteins

The most of the available data on NO_2Tyr in biological fluids and tissues are derived from antibody-based techniques, which, however, are often not rigorously validated. It is worth mentioning that ELISA for NO_2Tyr detection lacks specificity and sensitivity [77]. Such immunoassays should be considered, at best, semi-quantitative, and the results should be interpreted accordingly. Quantification by ELISA is limited by several factors:

- The employed antibodies have different affinities for different nitrated proteins [4], although this issue can be addressed by methods that measure a single nitrated protein [5].

- Some NO_2Tyr ELISA may be prone to interference.
- The sensitivity of some assays may be too low to detect NO_2Tyr in healthy human plasma.

Increases in NO_2Tyr in smokers and patients with various pathologies have been noted using ELISA and blotting techniques such as coeliac disease, children with chronic renal failure, patients with DM or acute respiratory distress [6–10].

Semi-quantitative ELISA assays based on specific antibodies are available for nitro-protein detection [4]. A convenient semi-quantitative method has been developed to assess nitrated proteins in biological fluids and tissue homogenates using a competitive ELISA [4]. This assay selectively detected 3-NO_2Tyr residues in a variety of peptides and proteins treated with $ONOO^-$ but had a low affinity to free 3-NO_2Tyr. The main problem with these assays is low sensitivity. Furthermore, the epitopes may be different from a nitration site to another and their use is limited by the variable affinity of antibodies for different nitrated proteins [11].

A variety of nitroproteins have been identified in plasma but so far the only immunoassay that has been developed and clinically validated is ELISA for nitrated albumin [12, 13]. This ELISA has been validated previously in studies on perinatal asphyxia where the results were shown to correlate with the severity of neonatal encephalopathy [5], a condition that has been previously reported to be associated with increased protein nitration in the human brain [14]. In neonatal hypoglycemia, protein nitration was shown to correlate with the number and severity of hypoglycemic events, which possibly impair psychomotor development [13].

Although monoclonal antibodies (mAb) against chlorinated proteins are available, the nature of the epitope is uncertain as for nitroproteins, and the antibody does not react to free Cl-Tyr [15].

16.1.3 Glutathionylated Proteins

Immunoassay methods including ELISA have been used to quantify the level of glutathiolation of proteins, with some success demonstrated for its use for glutathiolated actin [16]. In addition, ELISA for other glutathionylated proteins (e.g., erythrocyte Na^+/K^+ Pump) using monoclonal anti-glutathione antibody has been developed [17].

16.2 Carbohydrate End Products

A number of ELISA methods have been reported to measure AGEs. The first methods were introduced in 1992 when Araki et al. used ELISA with both polyclonal and monoclonal antibodies (mAb) to demonstrate an age-dependent increase in AGEs in lens proteins [18]. CML was later identified as the epitope recognized by these antibodies [19]. Makita et al. used an ELISA assay with a polyclonal anti-AGE antibody of unknown specificity to demonstrate an increase of AGEs in circulating proteins of diabetic and nondiabetic patients with end-stage renal disease [20]. Makita et al. reported that AGEs on plasma proteins were decreased during hemodialysis and following renal transplantation. These results suggest a role for the kidney in removal of AGEs proteins or AGEs precursors from plasma [21]. Using polyclonal antibodies, Beisswenger et al. were able to measure increases in AGEs in skin collagen during the early stages of retinopathy and nephropathy in diabetic patients [22]. Using an anti-CML-adduct antibody, Schleicher et al. also reported that CML levels were increased twofold in plasma of diabetic patients, independent of the presence of nephropathy [23].

Papanastasiou et al. introduced a time-resolved fluorescence ELISA assay for AGEs in serum of diabetic patients, which significantly increases the sensitivity of absorbance-based assays [24]. Degenhardt et al. reported that the antibodies of Papanastasiou et al. were monospecific for CML [25].

Hayase et al. used an antipyrraline antibody in a RIA procedure to demonstrate increased levels of pyrraline in plasma proteins from diabetic patients [26]. In another work, an antibody anti-pentosidine hapten was used in an ELISA assay and showed that pentosidine was increased in plasma proteins and skin collagen from diabetic patients [27]. Bucala et al. detected both lipid- and protein-bound carbohydrates on human LDL by ELISA in diabetic patients using an anti-AGE antibody [28]. The authors found a strong correlation among levels of AGE-protein, AGE-lipid, and lipid peroxidation aldehydes (thiobarbituric acid assay) in LDL. Since MDA-Lys, CML, and CM lipids are produced together during lipoprotein oxidation in vitro, and levels of MDA-Lys and CML are increased in uremic plasma and in LDL isolated from atherosclerotic plaque, it is not surprising that levels of these compounds and CM lipids would increase in concert with one another in circulating LDL during acute coronary syndrome, DM, and uremia.

Taneda and Monnier used a specific antibody against pentosidine in an ELISA to measure the age-dependent increase of pentosidine in skin collagen in nondiabetic patients as well as an increase of pentosidine in plasma proteins of diabetic and uremic patients [27]. Digestion of AGEs proteins by either enzymatic or acid hydrolysis improved the sensitivity of the assay by 10–100 fold, which is consistent with the expectation that most of the pentosidine is buried in inaccessible intermolecular cross-links between proteins. The enhancement of sensitivity by hydrolysis was observed with both insoluble collagen and soluble plasma proteins, suggesting that hydrolysis of proteins, in general, may enhance the sensitivity of ELISA for AGEs assessment. It is noteworthy to know that alkaline denaturation can be used to enhance sensitivity of the assay, for instance, Araki et al. used alkaline treatment to denature and solubilize lens proteins, whereas others have used enzymatic digestion for the assay [18, 22, 29].

Compared to LC-MS and GC-MS, analysis of CML in proteins by an ELISA technique greatly simplifies its detection and provides a valuable tool for comparative analysis of the CML content of proteins. However, some anti-AGE antibodies are clearly specific for AGEs, other than CML. Moreover, the fact that CML is formed during both lipoxidation and glycoxidation reactions raises questions about the specificity of some “anti-AGE” antibodies. While commercial ELISA kits for AGEs measurement provide ease of use, many of them do not specify the antibody used. This inconvenient may lead to differences in the given values between commercial kits.

16.3 Lipid End Products

16.3.1 MDA Adducts

Several commercial ELISA kits are available to detect TBARS as free MDA equivalents. Because the majority MDA is bound to protein, several antibodies were developed against MDA-protein adducts. Some of these are used for MDA–LDL detection and have been validated against HPLC methods with good performance and high specificity [30]. Using antibodies against MDA-DNA adducts, various immunological techniques have been used to detect such adducts [31].

16.3.2 4-HNE-Protein Adducts

4-HNE can be detected by means of immunological techniques using specific anti-HNE antibodies. Because the majority of 4-HNE is bound to protein, several research groups developed antibodies against HNE-protein adducts.

The availability of monoclonal and polyclonal antibodies directed against different HNE-protein adducts (involving lysine, cysteine, or histidine residues) allowed the development of immunoassays that are currently available. The reliability of immunoassays is mostly dependent on the specificity of the antibody used and may lead to discrepancies between the available commercial kits. Highly specific antibodies against HNE-His adducts are widely used for the analysis of biological samples.

The simultaneous analyses of several human plasma samples of HNE-His and MDA using HPLC and ELISA gave good correlations, indicating a favorable use of ELISA for human plasma samples [78].

In 2013, modified versions of the HNE-His ELISA assay were described to be highly convenient for human samples. Interestingly, the modified versions used not only noncommercial but also commercial mAb, which were tenfold less sensitive than the noncommercial ones. Nevertheless, in both cases, the 50% increase of the HNE-His products in the blood was detected for apparently healthy but obese, young to mid-aged men [79]. It is important to mention that further modifications of the HNE-His ELISA were introduced, enabling to measure HNE-His in murine tissue homogenates, which were found 50% increased in liver of rats fed with PUFA-enriched diet [32]. On the top of plasma and serum samples, the specificity of

HNE-His mAb allows identification of the corresponding adducts in broad spectrum of samples such as cell cultures and human tissue homogenates, which is of primary importance for the studies on lipid peroxidation end products during OS [80]. Accordingly, HNE-His adducts are often referred to as one of the major biomarkers of lipid peroxidation, implying the applicability of methods for their detection in various samples as well as their role in crucial cellular processes, in particular redox signaling. The relevance of 4-HNE-linked His adducts as biomarkers of OS is also supported by the fact that these adducts dominate in Cu^{2+}-oxLDL, while the other major 4-HNE adducts (4-HNE-lysine and 4-HNE-cysteine) appeared in negligible amounts in the oxLDL, as analyzed by the fluorescent probe 2-aminopyridine [33]. Such immunological techniques were used to detect these adducts in pathological situations such as CVD and particularly atherosclerotic lesions of human aorta [34]. In another type of diseases, HNE-modified tau proteins were found to be associated with neurofibrillary tangles in AD [35].

16.3.3 OxLDL and Autoantibodies

OxLDL can be detected using ELISA methods. Several configurations for determining oxLDL concentration have been proposed (usually sandwich and competitive ELISA). Most of the assays are validated and employ mAb that recognize unique oxidation-specific epitopes present on oxLDL [36, 37]. The specificity of these assays relies on the primary (capturing) mAb used. There are three well-known antibodies that are employed for this purpose, namely, 4E6, FOH1a/DLH3, and E06. This can be explained by the fact that oxLDL does not describe a unique chemical entity but represents many different chemical, biological, and immunologic changes in LDL. Figure 16.1 shows the methodology for several well-described oxLDL assays.

These assays are not necessarily similar because they either use antibodies detecting different epitopes or have methodologies that do not represent the same measurements.

***The mAb (4E6), which was generated by immunizing mice with Cu^{2+}-oxidized LDL, recognizes the MDA-lysine-modified protein epitope [38, 39]. The epitope recognized by mAb 4E6 is generated only after the modification of at least 60 lysine residues on apoB-100 of the oxLDL, as fewer lysine modifications cannot be detected in the assay (Fig. 16.1a). A general limitation of the most commonly used 4E6-based immunoassay is that native LDL is also detected. As a result, the concentrations of oxLDL closely reflect the levels of LDL, and the predictive value of the assays is dependent on the amount of apoB [40]. This leads to a serious doubt over the usefulness of oxLDL as a measure of OS and its clinical utility to predict CVD.

***The murine mAb FOH1a/DLH3 recognizes specifically phosphocholine epitopes of the oxidized phosphatidylcholine (PC) on oxLDL [41, 42]. Since the oxidized phospholipids are also present on other plasma lipoproteins, it is usually

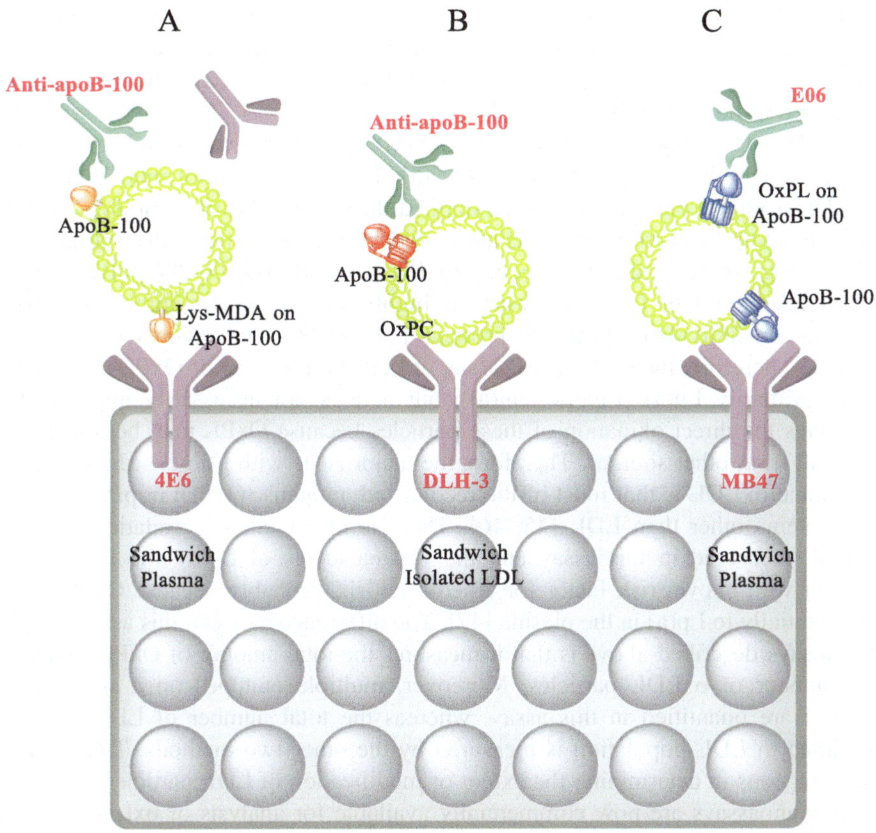

Fig. 16.1 Overview on different immunoassays used to assess oxLDL in blood. (**a**) Sandwich ELISA using 4E6 antibody to assess oxLDL in plasma (4E6 recognizes Lys-MDA epitopes on ApoB-100); (**b**) Sandwich ELISA using DLH-3 antibody to assess isolated oxLDL (DLH-3 recognizes oxidized PC on isolated LDL); and (**c**) Sandwich ELISA using E06 antibody to assess oxLDL in plasma (E06 recognizes oxidized PL epitopes on ApoB-100)

necessary to use isolated LDL collected by sequential ultracentrifugation rather than whole plasma to measure oxLDL (Fig. 16.1b). This minimizes the artifactual influence of aAb to oxLDL as well, but an extensive isolation procedure increases the possibility of LDL oxidation in vitro. In addition, this assay is prone to ex vivo oxidation when stored at 4 °C or after coating on plates. Besides, the assay is difficult to implement in large-scale studies because the extraction of LDL is time-consuming and may take up to 4 days to prepare.

A commercial kit has been developed based on this assay (Kyowa Medex, Japan). The advantage of this kit is that it can be used with the whole plasma instead of isolated LDL. Since the mAb used in these assays does not react with MDA–LDL, it provides complementary information to the above method that uses mAb 4E6.

The LOD of the assay is 0.5 ng of Cu^{2+}-oxidized LDL. The oxidation unit is defined as the reactivity of 1 ng of Cu^{2+}-oxidized LDL standard. Using this method, the level of oxLDL was shown to be eightfold higher in hemodialysis patients compared to normal individuals [43].

***The OxPL/apoB assays use the IgM mAb E06, which recognizes the phosphorylcholine moiety of OxPL [44, 45]. The assay does not require isolation of LDL from the plasma and can be applied directly to plasma samples. The E06 assay is set up by capturing plasma apoB in microtiter wells using the murine mAb MB47 (Fig. 16.1c) and then detecting OxPL on apoB-100 (OxPL/apoB) with the biotinylated form of E06. This assay is highly specific to the number of OxPL epitopes on individual apoB-100 proteins but not the total amount of OxPL in plasma. Therefore, the OxPL/apoB assay detects the presence of OxPL on lipoproteins, primarily Lp(a) (Lipoprotein(a)), but it does not necessarily imply that it measures the direct oxidation of these particles because OxPL may be transferred to Lp(a) from other sources. Therefore, the major issue with E06-based methods to determine oxLDL is that most of the recognized antigens (OxPL) in plasma lie on lipoproteins other than LDL [45, 46]. There is an excellent correlation between oxLDL measured in this assay and the plasma levels of Lp(a), the latter being another well-known risk factor for atherosclerosis, probably because OxPC binds preferentially to Lp(a) in the plasma [47]. The difference between this assay and the two assays described above is that it measures the total amount of OxPC, not only the number of oxLDL particles. Moreover, multiple epitopes on a single LDL particle are quantified in this assay, whereas the total number of LDL particles modified in LDL population is measured by the other two methods. This may be advantageous in determining the extent of modification of LDL particles.

Immunoassays are now commercially available for analysis of oxLDL and are widely used to study OS in a variety of diseases. Pawlak et al. used a commercially available sandwich ELISA (Mercodia, Uppsala, Sweden) for analysis of oxLDL in dialysis patients [37]. The assay is specific for oxLDL and based on murine mAb directed against an epitope in the apoB-100 moiety generated following to reaction of lysine residues with aldehydes.

Measurements have linked oxLDL with the risk of CVD. The oxidation of LDL is not only an important toxic mechanism involved in atheroma, but it represents a major marker of OS. Plasma oxLDL levels have been consistently found elevated in patients with CVD, independently of the assay. However, conflicting results have been found on the association of oxLDL with atherosclerosis severity and the usefulness of oxLDL for CVD prediction.

Van der Zwan et al. reported that the oxLDL/apo B100 ratio correlated inversely with flow-mediated dilation of brachial artery even after adjusting known risk factors (age, sex, and glucose tolerance), showing that this is a clinically useful marker and an independent contributor to CVD [48]. In addition to CVD, plasma levels of oxLDL have been found to be increased in patients with insulin resistance, DM, and obesity [49].

The use of oxLDL as a biomarker of OS has been criticized because of the heterogeneity of oxidized products, the low specificity of several commercial antibodies, and the contrasting results obtained depending on the assay utilized [50, 80]. Indeed, the results obtained with different immunoassays cannot be compared and often do not correlate with each other, which is inconsistent with oxLDL being a quantitative measure of OS or representing a relevant tool to predict CVD.

Physicochemical changes in oxLDL can trigger the synthesis of aAb. Indeed, the oxLDL particles are particularly immunogenic, and antibodies directed against them have been located in the circulation. These aAb can also be used as a measure of OS in vivo. Several assays for determining plasma concentration of aAb to oxidatively modified LDL have been proposed. Most of the assays are based on comparison of the reactivity of samples with immobilized oxLDL and with fixed native LDL. The results are expressed in arbitrary units, either as a difference or as a ratio that reflects the increased binding to oxLDL.

Koskinen et al. proposed a competitive assay that allows calculation of specific binding by highlighting the difference in binding between unabsorbed and oxidized LDL-absorbed aliquots [51]. In addition, the authors suggest the preparation of a standard with a known antibody concentration, allowing expression of antibody concentration in standard mass unit rather than in arbitrary units. Further attempts to standardize the assay will clarify the pathogenic significance of aAb in the pathogenesis of atherosclerosis and the usefulness of aAb determination in atherosclerotic risk stratification. aAb to oxLDL have been detected in healthy individuals but appear to increase in CVD [52].

Finally, Pawlak et al. proposed oxLDL/oxLDL-Ab ratio as a new biomarker that is significantly and independently associated with carotid atherosclerosis and cardiovascular complications in dialyzed patients [53].

16.3.4 Isoprostanes

F_2-Isoprostanes are considered the "gold standard" biomarker of OS. Several immunoassays have been developed for the measurement of these compounds and some of them are commercially available. ELISA has expanded research in this area because of its low cost and relative ease of use.

While commercial immunoassays allow high-throughput analysis of samples and represent an easier and cheaper alternative to mass spectrometry, multiple studies have demonstrated variable performance between the commercially available F_2-IsoP kits and poor correlation with mass spectrometric results [54, 55]. Il'yasova et al. found a weak correlation between ELISA and GC-MS values for the urinary metabolite of IsoPs [55]. Moreover, examination of the published data for ELISA and LC-MS/MS shows a wide variation in IsoP values for normal human plasma [81]. Besides, the sensitivity and specificity of immunoassays may vary substantially among manufacturers. These discrepancies are suggested in part to be a consequence of the polyclonal antibodies used to bind to one specific isomer (8-isoPG_{F2}), and which demonstrate cross-reactivity with other isomers and molecules of a similar structure [55]. The results of Proudfoot et al. show that the ELISA method underestimated the total IsoP compounds compared to the GC-MS method. This is due to the low immunoreactivity of some isomers [56]. Another drawback of these techniques is that limited information is currently available regarding their precision and accuracy. These methods are also less sensitive and less specific compared to chromatography combined with mass spectrometry.

In general, all these factors demonstrate that immunoassay-based methods for IsoP level determination is a poor choice. However, due to ease of use, ELISA still remains a popular method for F_2-IsoP quantification.

In addition to commercial immunoassays, some researchers have generated polyclonal antibodies and developed specific assays for F_2-IsoPs. It appears that there is good correlation between these techniques and mass spectrometry [57]. It is highly recommended that samples should be partially purified before analysis because of considerable interference by structurally related metabolites found in biological fluids and tissue extracts. Indeed, some comparative studies showed that the commercially available enzyme immunoassays for 15-F_{2t}-IsoP are not applicable without prior and proper extraction of 15-F_{2t}-IsoP from the biological sample [58]. Antibody cross-reactivity can inflate quantification of IsoP concentrations, and the presence of some biological impurities can interfere further with antibody binding [55]. Besides, the study of Tsikas et al. showed that only about 50% of the urinary IsoPs react with the antibody [82].

In a typical ELISA assay, the plasma sample is first hydrolyzed in basic conditions to release the esterified IsoPs, and the sample is extracted by either SPE cartridges, silicic acid column, or immunoaffinity columns containing an antibody for 8-iso-PGF_2. The sample is then subjected to competitive ELISA, in which the competition of 8-IsoP in the unknown sample with the tracer (8-IsoP-PGF-linked acetylcholinesterase or HRP) to bind to the isoprostane-specific polyclonal antibody on the plate is measured. All these limitations and time-consuming preparations can be avoided by analyzing IsoPs in noninvasive samples. For instance, Caballero et al. measured both leukotrienes and IsoPs in EBC using specific EIA capable of determining concentration at the picogram level [59].

It is important to know that artifactual formation of F_2-IsoPs during sample storage may readily occur [83], and post-prandial levels of 15-F_{2t}-IsoP may rise following a fatty meal, demonstrating the confounding effect of diet [60]. Thus, the results from immunoassays, particularly for individual patients must be used with extreme caution, only with appropriate sample cleanup and should be validated whenever possible. For some of immunoassay methods, notably for RIA and EIA methods for 15-F_{2t}-IsoP, a potential drawback may be limited information about accuracy and selectivity.

Overall, immunoassay techniques, based on specific antibodies against IsoPs, are still under development and their application is limited since the obtained results do not correlate well with mass spectrometry determination [54]. Alternative methods have also been developed to quantify F_2-IsoPs by immunologic techniques (e.g., RIA) and a few immunoassay sets are now commercially available. However, many of these assays are not able to distinguish between prostanoids and IsoPs, much less between the different types of IsoPs.

16.3.5 Other Lipid Peroxidation End Products

Few studies have focused on the development of immunoassays for the detection of IsoLG and acrolein. This is because these compounds are not widely studied in OS context and we may see their assessment increased in the next few years. Acrolein is usually measured as protein adducts using ELISA and commercial kits based on the reactivity of specific antibody with lysine adducts of acrolein are now available (Genox, Biovendor) [61]. Quantitative immunoassays using polyclonal antibodies detected increased levels of IsoLG-protein adducts in plasma from patients with atherosclerosis or end-stage renal disease, as well as in the glaucomatous trabecular meshwork [84].

16.4 DNA End Products

Many laboratories use ELISA for measuring 8-OHdG in urine and plasma samples. Although ELISA methods are less specific compared to HPLC-EC and LC/GC-MS, some kits with specific antibodies are appropriate for urine samples. Examining urinary 8-OHdG levels has multiple benefits over methods that require DNA extraction, allowing quicker high-throughput studies, reducing the need for expensive chromatography or mass spectrometry equipment, and trained personnel.

An ELISA system for urinary 8-OHdG using a mAb has been developed by Erhola et al. [62]. Another mAb more specific for 8-OHdG has been shown not to cross-react with the original versions of deoxyribonucleosides, other DNA base-modified products such as 8-OHdA and O-methyl-2′-deoxyguanosine, as well as urine components such as uric acid, creatine, and creatinine [63]. This antibody called N45.1 was established using 8-OHdG–keyhole limpet hemocyanin as an antigen. Moreover, N45.1 is scarcely reactive with DNA adducts that the authors have examined (21 molecules) and thus highly specific for 8-OHdG. Several commercial kits are currently available for the assay of 8-OHdG by ELISA (Cayman, NWLSS, Neogen, Oxis, CosmoBio, Cell Biolabs, Northwest Life Sciences, and others).

This approach continues to receive widespread attention with multiple studies demonstrating that urinary levels of 8-OHdG are good biomarkers for oxidative DNA damage. This resulted in its extensive use when examining some degenerative diseases and several cancers [64–66].

Although good correlations of the results obtained by ELISA and HPLC-EC methods have been reported, the concentrations of 8-OHdG obtained with some ELISA kits were up to sixfold higher, probably because of the nonselective detection of some oligonucleotides or other modified bases by ELISA [67]. Unfortunately, ELISA methods are not without drawbacks, as levels of urinary 8-OHdG may not be entirely DNA-derived but rather be from the deoxyGTP precursor pool, which can also give a false estimation of the level of DNA oxidative damage [68].

Some studies reported discrepancies between ELISA and different methods utilized for 8-OHdG/8-oxo-dG quantification, demonstrating a high degree of variability when compared to the more accurate chromatography-based methods [69]. Given the discrepancies between urinary 8-oxo-dG quantification methods, the European Standards Committee on Urinary Lesion Analysis (ESCULA) was established to perform analytical-based selection of the most appropriate methods for quantification. The committee identified LC-MS/MS as the most appropriate technique for urinary 8-oxo-dG quantification [69]. However, due to the high-throughput and inexpensive nature of ELISA, this method is still a more favorable quantification method. As a result, there has been much work to optimize this technique as the study of Rossner et al. which demonstrated that purification of 8-oxo-dG in urine using SPE substantially improves ELISA quantification of this analyte and provides better accuracy and more consistency with LC-MS/MS quantification [70]. Although not as accurate as LC-MS/MS, this study showed the potential of ELISA to be an accurate method for 8-OHdG quantification, providing a more accessible approach to assess DNA oxidative damage.

Besides, ELISA methods can be used for analysis of 8-oxo-dG and 8-NO_2dG in serum and urine. Several commercial assays are now available; however, the clinical significance of these methods has been questioned [69]. Analysis of 8-oxo-dG has been hampered for a long time by the occurrence of several drawbacks related to their measurement. Optimized assays were developed and commercial ELISA assays based on specific antibodies are now available [71, 85]. Other methods to assess oxidative DNA damage such as the ARP assays in combination with an ELISA-like assay are also available.

Overall, methods based on ELISA for DNA oxidation assessment are more convenient in clinical assessments compared to chromatographic methods because of high throughput and standardization. Although ELISA methods are less specific compared to HPLC-EC or LC/GC-MS, some kits with highly specific antibodies are appropriate for urine samples [66].

Oxidation of DNA can result in the generation of aAb against oxidized DNA. An increase in anti-hydroxy-methyl-deoxyuridine antibodies of IgM type was recorded in serum of patients with systemic lupus erythematosus and other autoimmune diseases [72]. Among other oxidized derivatives, deoxyguanosine-malondialdehyde adducts (MDAdG) are possibly analyzed immunochemically [73].

16.5 Myeloperoxidase (MPO)

Myeloperoxidase content of biological samples can be measured by sandwich ELISA assay. Myeloperoxidase is first captured on the ELISA plate with murine mAb. After washing, a rabbit polyclonal antibody directed against MPO is added to bind to the captured MPO. Thereafter, a goat antibody directed against the rabbit IgG and conjugated with HRP is added and after washing and catalysis of a specific chromogen, the color is developed and read at the corresponding wavelength

(450 nm). The amount of MPO is calculated with the help of a calibrated standard of human MPO.

Commercial ELISA Kit usually provides an accurate and sensitive quantitation of human MPO in a variety of biological samples, including human serum. In the case of Cell Sciences kit, the captured MPO is bound to a biotin conjugated antibody and streptavidin-peroxidase is used for color development. The Calbiochem assay is based on the capture of MPO on ELISA plate with a polyclonal antibody and the activity of the captured enzyme itself is measured colorimetrically by reaction with tetramethyl benzidine and H_2O_2.

Invitrogen utilizes the Amplex® UltraRed reagent, a fluorogenic substrate for HRP, which reacts with H_2O_2 to produce the fluorescent Amplex® UltraRed oxidation product (excitation/ emission maxima ≈ 568/581 nm). With good quantum efficiency, high extinction coefficient, and resistance to auto-oxidation, the fluorogenic Amplex® UltraRed reagent delivers overall a better sensitivity and a broader assay range than colorimetric assays.

16.6 Thioredoxin (Trx)

ELISA methods have been used to demonstrate elevated levels of Trx in AIDS patients, where it paradoxically and negatively correlates with survival of patients [74].

Various reports have suggested that Trx concentrations in blood are diagnostically relevant in a range of diseases. For instance, Qi et al. found that determination of Trx concentrations in serum using ELISA reflected disease severity in acute ischemic stroke [75], and the addition of Trx levels to an established disease severity score provided a significant improvement in diagnostic capacity.

16.7 Peroxiredoxin (Prx)

Several ELISA and ELISA-like methods are currently available for the detection of Prx isoforms (Prx 1, 2, 3, 4, and 6) in human plasma and serum. They have been used to explore these antioxidant agents in several diseases linked to OS including pancreatic cancer, hepatocellular carcinoma, and T2DM [76, 86–90].

16.8 Other Biomarkers

Immunoassays are used for quantitative detection of other important biomarkers of OS including SOD, GPx, aldehyde dehydrogenase, NOX, and nNOS. However, little data are available on the specificity of the corresponding antibodies used for their detection.

References

1. Buss H, Chan TP, Sluis KB, Domigan NM, Winterbourn CC. Protein carbonyl measurement by a sensitive ELISA method. Free Radic Biol Med. 1997;23(3):361–6.
2. Polac I, Borowiecka M, Wilamowska A, Nowak P. Oxidative stress measured by carbonyl groups level in postmenopausal women after oral and transdermal hormone therapy. The journal of obstetrics and gynaecology research. 2012;38:1177–81.
3. Pantke U, Volk T, Schmutzler M, Kox WJ, Sitte N, Grune T. Oxidized proteins as a marker of oxidative stress during coronary heart surgery. Free Radical Biology and Medicine. 1999;27(9): 1080–6.
4. Khan J, Brennand DM, Bradley N, Gao B, Bruckdorfer R, Jacobs M. 3-Nitrotyrosine in the proteins of human plasma determined by an ELISA method. The Biochemical journal. 1998;330 (Pt 2)(Pt 2):795–801.
5. Wayenberg J-L, Ransy V, Vermeylen D, Damis E, Bottari S. Nitrated plasma albumin as a marker of nitrative stress and neonatal encephalopathy in perinatal asphyxia. Free radical biology & medicine. 2009;47:975–82.
6. ter Steege JC, Koster-Kamphuis L, van Straaten EA, Forget PP, Buurman WA. Nitrotyrosine in plasma of celiac disease patients as detected by a new sandwich ELISA. Free Radic Biol Med. 1998;25(8):953–63.
7. Kari JA, Donald AE, Vallance DT, Bruckdorfer KR, Leone A, Mullen MJ, et al. Physiology and biochemistry of endothelial function in children with chronic renal failure. Kidney Int. 1997;52 (2):468–72.
8. Ceriello A, Mercuri F, Quagliaro L, Assaloni R, Motz E, Tonutti L, et al. Detection of nitrotyrosine in the diabetic plasma: Evidence of oxidative stress. Diabetologia. 2001;44:834–8.
9. Gole MD, Souza JM, Choi I, Hertkorn C, Malcolm S, Foust RF, 3rd, et al. Plasma proteins modified by tyrosine nitration in acute respiratory distress syndrome. Am J Physiol Lung Cell Mol Physiol. 2000;278(5):L961–7.
10. Pignatelli B, Li CQ, Boffetta P, Chen Q, Ahrens W, Nyberg F, et al. Nitrated and oxidized plasma proteins in smokers and lung cancer patients. Cancer research. 2001;61(2):778–84.
11. Xu S, Ying J, Jiang B, Guo W, Adachi T, Sharov V, et al. Detection of sequence-specific tyrosine nitration of manganese SOD and SERCA in cardiovascular disease and aging. Am J Physiol Heart Circ Physiol. 2006;290:H2220–7.
12. Piroddi M, Palmese A, Pilolli F, Amoresano A, Pucci P, Ronco C, et al. Plasma nitroproteome of kidney disease patients. Amino Acids. 2011;40(2):653–67.
13. Wayenberg JL, Cavedon C, Ghaddhab C, Lefèvre N, Bottari SP. Early transient hypoglycemia is associated with increased albumin nitration in the preterm infant. Neonatology. 2011;100(4): 387–97.
14. Groenendaal F, Lammers H, Smit D, Nikkels PGJ. Nitrotyrosine in brain tissue of neonates after perinatal asphyxia. Arch Dis Child Fetal Neonatal Ed. 2006;91(6):F429–F33.
15. Malle E, Hazell L, Stocker R, Sattler W, Esterbauer H, Waeg G. Immunologic detection and measurement of hypochlorite-modified LDL with specific monoclonal antibodies. Arteriosclerosis, thrombosis, and vascular biology. 1995;15(7):982–9.

16. Johansson M, Lundberg M. Glutathionylation of beta-actin via a cysteinyl sulfenic acid intermediary. BMC Biochemistry. 2007;8(1):26.
17. Liu CC, Fry N, Karimi Galougahi K, Rasmussen H, Figtree G. Glutathionylation of erythrocyte Na-K pump in heart failure: a novel biomarker that reflects a key oxidative abnormality in the heart. Circulation. 2012;126:A12793.
18. Araki N, Ueno N, Chakrabarti B, Morino Y, Horiuchi S. Immunochemical evidence for the presence of advanced glycation end products in human lens proteins and its positive correlation with aging. The Journal of biological chemistry. 1992;267:10211–4.
19. Ikeda K, Higashi T, Sano H, Jinnouchi Y, Yoshida M, Araki T, et al. Nε-(Carboxymethyl)lysine Protein Adduct Is a Major Immunological Epitope in Proteins Modified with Advanced Glycation End Products of the Maillard Reaction. Biochemistry. 1996;35(24):8075–83.
20. Makita Z, Vlassara H, Cerami A, Bucala R. Immunochemical detection of advanced glycosylation end products in vivo. J Biol Chem. 1992;267(8):5133–8.
21. Makita Z, Radoff S, Rayfield EJ, Yang Z, Skolnik E, Delaney V, et al. Advanced glycosylation end products in patients with diabetic nephropathy. The New England journal of medicine. 1991;325(12):836–42.
22. Beisswenger PJ, Makita Z, Curphey TJ, Moore LL, Jean S, Brinck-Johnsen T, et al. Formation of immunochemical advanced glycosylation end products precedes and correlates with early manifestations of renal and retinal disease in diabetes. Diabetes. 1995;44(7):824–9.
23. Schleicher ED, Wagner E, Nerlich AG. Increased accumulation of the glycoxidation product N (epsilon)-(carboxymethyl)lysine in human tissues in diabetes and aging. J Clin Invest. 1997;99 (3):457–68.
24. Papanastasiou P, Grass L, Rodela H, Patrikarea A, Oreopoulos D, Diamandis EP. Immunological quantification of advanced glycosylation end-products in the serum of patients on hemodialysis or CAPD. Kidney International. 1994;46(1):216–22.
25. Degenhardt TP, Grass L, Reddy S, Thorpe SR, Diamandis EP, Baynes JW. The serum concentration of the advanced glycation end-product Nε-(carboxymethyl)lysine is increased in uremia: Technical Note. Kidney International. 1997;52(4):1064–7.
26. Hayase F, Nagaraj R, Miyata S, Njoroge G, Monnier V. Aging of proteins: Immunological detection of a glucose-derived pyrrole formed during Maillard reaction in vivo. The Journal of biological chemistry. 1989;264:3758–64.
27. Taneda S, Monnier VM. ELISA of pentosidine, an advanced glycation end product, in biological specimens. Clinical Chemistry. 1994;40(9):1766–73.
28. Bucala R, Makita Z, Koschinsky T, Cerami A, Vlassara H. Lipid advanced glycosylation: pathway for lipid oxidation in vivo. Proc Natl Acad Sci U S A. 1993;90(14):6434–8.
29. Rumble JR, Cooper ME, Soulis T, Cox A, Wu L, Youssef S, et al. Vascular hypertrophy in experimental diabetes. Role of advanced glycation end products. J Clin Invest. 1997;99(5): 1016–27.
30. Bevan RJ, Durand MF, Hickenbotham PT, Kitas GD, Patel PR, Podmore ID, et al. Validation of a novel ELISA for measurement of MDA-LDL in human plasma. Free Radical Biology and Medicine. 2003;35(5):517–27.
31. Sevilla CL, Mahle NH, Eliezer N, Uzieblo A, O'Hara SM, Nokubo M, et al. Development of monoclonal antibodies to the malondialdehyde-deoxyguanosine adduct, pyrimidopurinone. Chem Res Toxicol. 1997;10(2):172–80.
32. Guéraud F, Taché S, Steghens JP, Milkovic L, Borovic-Sunjic S, Zarkovic N, et al. Dietary polyunsaturated fatty acids and heme iron induce oxidative stress biomarkers and a cancer promoting environment in the colon of rats. Free Radic Biol Med. 2015;83:192–200.
33. Wakita C, Honda K, Shibata T, Akagawa M, Uchida K. A method for detection of 4-hydroxy-2-nonenal adducts in proteins. Free Radical Biology and Medicine. 2011;51(1):1–4.
34. Uchida K, Itakura K, Kawakishi S, Hiai H, Toyokuni S, Stadtman ER. Characterization of epitopes recognized by 4-hydroxy-2-nonenal specific antibodies. Arch Biochem Biophys. 1995;324(2):241–8.

35. Liu Q, Smith MA, Avilá J, DeBernardis J, Kansal M, Takeda A, et al. Alzheimer-specific epitopes of tau represent lipid peroxidation-induced conformations. Free Radic Biol Med. 2005;38(6):746–54.
36. Kuppan K, Mohanlal J, Mohammad AM, Babu KA, Sen P, Undurti ND, et al. Elevated serum OxLDL is associated with progression of type 2 Diabetes Mellitus to diabetic retinopathy. Experimental eye research. 2019;186:107668.
37. Pawlak K, Mysliwiec M, Pawlak D. Oxidized low-density lipoprotein (oxLDL) plasma levels and oxLDL to LDL ratio — Are they real oxidative stress markers in dialyzed patients? Life Sciences. 2013;92(4):253–8.
38. Holvoet P, Donck J, Landeloos M, Brouwers E, Luijtens K, Arnout J, et al. Correlation between oxidized low density lipoproteins and von Willebrand factor in chronic renal failure. Thrombosis and haemostasis. 1996;76(5):663–9.
39. Holvoet P, Cleemput JV, Collen D, Vanhaecke J. Oxidized Low Density Lipoprotein Is a Prognostic Marker of Transplant-Associated Coronary Artery Disease. 2000;20(3):698–702.
40. Wu T, Willett WC, Rifai N, Shai I, Manson JE, Rimm EB. Is Plasma Oxidized Low-Density Lipoprotein, Measured With the Widely Used Antibody 4E6, an Independent Predictor of Coronary Heart Disease Among U.S. Men and Women? Journal of the American College of Cardiology. 2006;48(5):973–9.
41. Itabe H, Yamamoto H, Suzuki M, Kawai Y, Nakagawa Y, Suzuki A, et al. Oxidized phosphatidylcholines that modify proteins. Analysis by monoclonal antibody against oxidized low density lipoprotein. The Journal of biological chemistry. 1997;271:33208–17.
42. Itabe H, Takeshima E, Iwasaki H, Kimura J, Yoshida Y, Imanaka T, et al. A monoclonal antibody against oxidized lipoprotein recognizes foam cells in atherosclerotic lesions. Complex formation of oxidized phosphatidylcholines and polypeptides. J Biol Chem. 1994;269(21): 15274–9.
43. Itabe H, Yamamoto H, Imanaka T, Shimamura K, Uchiyama H, Kimura J, et al. Sensitive detection of oxidatively modified low density lipoprotein using a monoclonal antibody. J Lipid Res. 1996;37(1):45–53.
44. Hörkkö S, Bird DA, Miller E, Itabe H, Leitinger N, Subbanagounder G, et al. Monoclonal autoantibodies specific for oxidized phospholipids or oxidized phospholipid-protein adducts inhibit macrophage uptake of oxidized low-density lipoproteins. J Clin Invest. 1999;103(1): 117–28.
45. Friedman P, Horkko S, Steinberg D, Witztum JL, Dennis EA. Correlation of antiphospholipid antibody recognition with the structure of synthetic oxidized phospholipids. Importance of Schiff base formation and aldol condensation. J Biol Chem. 2002;277(9):7010–20.
46. Tsimikas S, Lau HK, Han K-R, Shortal B, Miller ER, Segev A, et al. Percutaneous Coronary Intervention Results in Acute Increases in Oxidized Phospholipids and Lipoprotein(a). 2004;109(25):3164–70.
47. Tsimikas S. Measures of oxidative stress. Clinics in laboratory medicine. 2006;26(3):571–90, v–vi.
48. van der Zwan LP, Teerlink T, Dekker JM, Henry RM, Stehouwer CD, Jakobs C, et al. Circulating oxidized LDL: determinants and association with brachial flow-mediated dilation. J Lipid Res. 2009;50(2):342–9.
49. Trpkovic A, Resanovic I, Stanimirovic J, Radak D, Mousa SA, Cenic-Milosevic D, et al. Oxidized low-density lipoprotein as a biomarker of cardiovascular diseases. Critical reviews in clinical laboratory sciences. 2015;52(2):70–85.
50. Maiolino G, Rossitto G, Caielli P, Bisogni V, Rossi GP, Calò LA. The role of oxidized low-density lipoproteins in atherosclerosis: the myths and the facts. Mediators Inflamm. 2013;2013:714653-.
51. Koskinen S, Enockson C, Lopes-Virella MF, Virella G. Preparation of a human standard for determination of the levels of antibodies to oxidatively modified low-density lipoproteins. Clin Diagn Lab Immunol. 1998;5(6):817–22.

52. van den Berg VJ, Vroegindewey MM, Kardys I, Boersma E, Haskard D, Hartley A, et al. Anti-Oxidized LDL Antibodies and Coronary Artery Disease: A Systematic Review. Antioxidants (Basel). 2019;8(10):484.
53. Pawlak K, Mysliwiec M, Pawlak D. Oxidized LDL to autoantibodies against oxLDL ratio—The new biomarker associated with carotid atherosclerosis and cardiovascular complications in dialyzed patients. Atherosclerosis. 2012;224(1):252–7.
54. Smith KA, Shepherd J, Wakil A, Kilpatrick ES. A comparison of methods for the measurement of 8-isoPGF(2α): a marker of oxidative stress. Annals of clinical biochemistry. 2011;48(Pt 2): 147–54.
55. Il'yasova D, Morrow JD, Ivanova A, Wagenknecht LE. Epidemiological marker for oxidant status: comparison of the ELISA and the gas chromatography/mass spectrometry assay for urine 2,3-dinor-5,6-dihydro-15-F2t-isoprostane. Annals of epidemiology. 2004;14(10):793–7.
56. Proudfoot J, Barden A, Mori TA, Burke V, Croft KD, Beilin LJ, et al. Measurement of urinary F (2)-isoprostanes as markers of in vivo lipid peroxidation-A comparison of enzyme immunoassay with gas chromatography/mass spectrometry. Anal Biochem. 1999;272(2):209–15.
57. Wang Z, Ciabattoni G, Créminon C, Lawson J, Fitzgerald GA, Patrono C, et al. Immunological characterization of urinary 8-epi-prostaglandin F2 alpha excretion in man. The Journal of pharmacology and experimental therapeutics. 1995;275(1):94–100.
58. Klawitter J, Haschke M, Shokati T, Klawitter J, Christians U. Quantification of 15-F2t-isoprostane in human plasma and urine: results from enzyme-linked immunoassay and liquid chromatography/tandem mass spectrometry cannot be compared. Rapid communications in mass spectrometry: RCM. 2011;25(4):463–8.
59. Caballero S, Martorell A, Escribano A, Belda J. Markers of airway inflammation in the exhaled breath condensate of preschool wheezers. Journal of investigational allergology & clinical immunology. 2013;23(1):7–13.
60. Gopaul NK, Zacharowski K, Halliwell B, Änggård EE. Evaluation of the postprandial effects of a fast-food meal on human plasma F2-isoprostane levels. Free Radical Biology and Medicine. 2000;28(5):806–14.
61. Satoh K, Yamada S, Koike Y, Igarashi Y, Toyokuni S, Kumano T, et al. A 1-Hour Enzyme-Linked Immunosorbent Assay for Quantitation of Acrolein- and Hydroxynonenal-Modified Proteins by Epitope-Bound Casein Matrix Method. Analytical Biochemistry. 1999;270(2): 323–8.
62. Erhola M, Toyokuni S, Okada K, Tanaka T, Hiai H, Ochi H, et al. Biomarker evidence of DNA oxidation in lung cancer patients: association of urinary 8-hydroxy-2'-deoxyguanosine excretion with radiotherapy, chemotherapy, and response to treatment. FEBS Lett. 1997;409(2): 287–91.
63. Toyokuni S, Tanaka T, Hattori Y, Nishiyama Y, Yoshida A, Uchida K, et al. Quantitative immunohistochemical determination of 8-hydroxy-2'-deoxyguanosine by a monoclonal antibody N45.1: its application to ferric nitrilotriacetate-induced renal carcinogenesis model. Laboratory investigation; a journal of technical methods and pathology. 1997;76(3):365–74.
64. Valavanidis A, Vlachogianni T, Fiotakis C. 8-hydroxy-2' -deoxyguanosine (8-OHdG): A critical biomarker of oxidative stress and carcinogenesis. Journal of environmental science and health Part C, Environmental carcinogenesis & ecotoxicology reviews. 2009;27(2):120–39.
65. Qing X, Shi D, Lv X, Wang B, Chen S, Shao Z. Prognostic significance of 8-hydroxy-2′-deoxyguanosine in solid tumors: a meta-analysis. BMC Cancer. 2019;19(1):997.
66. Cooke MS, Olinski R, Loft S. Measurement and meaning of oxidatively modified DNA lesions in urine. Cancer epidemiology, biomarkers & prevention: a publication of the American Association for Cancer Research, cosponsored by the American Society of Preventive Oncology. 2008;17(1):3–14.
67. Yin B, Whyatt RM, Perera FP, Randall MC, Cooper TB, Santella RM. Determination of 8-hydroxydeoxyguanosine by an immunoaffinity chromatography-monoclonal antibody-based ELISA. Free Radical Biology and Medicine. 1995;18(6):1023–32.

68. Sakumi K, Furuichi M, Tsuzuki T, Kakuma T, Kawabata S, Maki H, et al. Cloning and expression of cDNA for a human enzyme that hydrolyzes 8-oxo-dGTP, a mutagenic substrate for DNA synthesis. 1993;268(31):23524–30.
69. Barregard L, Møller P, Henriksen T, Mistry V, Koppen G, Rossner P, Jr., et al. Human and methodological sources of variability in the measurement of urinary 8-oxo-7,8-dihydro-2'-deoxyguanosine. Antioxid Redox Signal. 2013;18(18):2377–91.
70. Rossner P, Jr., Mistry V, Singh R, Sram RJ, Cooke MS. Urinary 8-oxo-7,8-dihydro-2'-deoxyguanosine values determined by a modified ELISA improves agreement with HPLC-MS/MS. Biochem Biophys Res Commun. 2013;440(4):725–30.
71. Evans MD, Olinski R, Loft S, Cooke MS. Toward consensus in the analysis of urinary 8-oxo-7,8-dihydro-2'-deoxyguanosine as a noninvasive biomarker of oxidative stress. FASEB journal: official publication of the Federation of American Societies for Experimental Biology. 2010;24(4):1249–60.
72. Frenkel K, Karkoszka J, Kim E, Taioli E. Recognition of oxidized DNA bases by sera of patients with inflammatory diseases. Free Radical Biology and Medicine. 1993;14(5):483–94.
73. Leuratti C, Singh R, Lagneau C, Farmer PB, Plastaras JP, Marnett LJ, et al. Determination of malondialdehyde-induced DNA damage in human tissues using an immunoslot blot assay. Carcinogenesis. 1998;19(11):1919–24.
74. Nakamura H, De Rosa SC, Yodoi J, Holmgren A, Ghezzi P, Herzenberg LA, et al. Chronic elevation of plasma thioredoxin: Inhibition of chemotaxis and curtailment of life expectancy in AIDS. 2001;98(5):2688–93.
75. Qi A-q, Li Y, Liu Q, Si J-Z, Tang X-M, Zhang Z-Q, et al. Thioredoxin is a novel diagnostic and prognostic marker in patients with ischemic stroke. Free Radical Biology and Medicine. 2015;80:129–35.
76. Abbasi A, Corpeleijn E, Gansevoort RT, Gans ROB, Struck J, Schulte J, et al. Circulating peroxiredoxin 4 and type 2 diabetes risk: the Prevention of Renal and Vascular Endstage Disease (PREVEND) study. Diabetologia. 2014;57(9):1842–9.
77. Tsikas D. Analytical methods for 3-nitrotyrosine quantification in biological samples: the unique role of tandem mass spectrometry. Amino Acids. 2012;42(1):45–63.
78. Breusing N, Grune T, Andrisic L, Atalay M, Bartosz G, Biasi F, et al. An inter-laboratory validation of methods of lipid peroxidation measurement in UVA-treated human plasma samples. Free radical research. 2010;44(10):1203–15.
79. Weber D, Milkovic L, Bennett SJ, Griffiths HR, Zarkovic N, Grune T. Measurement of HNE-protein adducts in human plasma and serum by ELISA—Comparison of two primary antibodies. Redox Biol. 2013;1(1):226–33.
80. Frijhoff J, Winyard PG, Zarkovic N, Davies SS, Stocker R, Cheng D, et al. Clinical Relevance of Biomarkers of Oxidative Stress. Antioxid Redox Signal. 2015;23(14):1144–70.
81. Sircar D, Subbaiah PV. Isoprostane Measurement in Plasma and Urine by Liquid Chromatography–Mass Spectrometry with One-Step Sample Preparation. Clinical Chemistry. 2007;53(2):251–8.
82. Tsikas D, Schwedhelm E, Suchy MT, Niemann J, Gutzki FM, Erpenbeck VJ, et al. Divergence in urinary 8-iso-PGF(2alpha) (iPF(2alpha)-III, 15-F(2t)-IsoP) levels from gas chromatography-tandem mass spectrometry quantification after thin-layer chromatography and immunoaffinity column chromatography reveals heterogeneity of 8-iso-PGF(2alpha). Possible methodological, mechanistic and clinical implications. Journal of chromatography B, Analytical technologies in the biomedical and life sciences. 2003;794(2):237–55.
83. Giustarini D, Dalle-Donne I, Tsikas D, Rossi R. Oxidative stress and human diseases: Origin, link, measurement, mechanisms, and biomarkers. Critical reviews in clinical laboratory sciences. 2009;46(5–6):241–81.
84. Salomon RG, Bi W. Isolevuglandin adducts in disease. Antioxid Redox Signal. 2015;22(18):1703–18.

85. Gedik CM, Collins A. Establishing the background level of base oxidation in human lymphocyte DNA: results of an interlaboratory validation study. FASEB journal: official publication of the Federation of American Societies for Experimental Biology. 2005;19(1):82–4
86. Cai CY, Zhai LL, Wu Y, Tang ZG. Expression and clinical value of peroxiredoxin-1 in patients with pancreatic cancer. European Journal of Surgical Oncology (EJSO). 2015;41(2):228–35.
87. Al-Masri AA, El Eter E, Tayel S, Zamil H. Differential associations of circulating peroxiredoxins levels with indicators of glycemic control in type 2 diabetes mellitus. European review for medical and pharmacological sciences. 2014;18(5):710–6.
88. Zhang Y, Ying X, Zhao Q, Ma J, Zhang D, He C, et al. Identification of Protein Expression Changes in Hepatocellular Carcinoma through iTRAQ. Dis Markers. 2020;2020:2632716.
89. Ismail S, Mayah W, Battia HE, Gaballah H, Jiman-Fatani A, Hamouda H, et al. Plasma nuclear factor kappa B and serum peroxiredoxin 3 in early diagnosis of hepatocellular carcinoma. Asian Pacific journal of cancer prevention : APJCP. 2015;16(4):1657–63.
90. Gateva A, Assyov Y, Velikova T, Kamenov Z. Increased peroxiredoxin 4 levels in patients with prediabetes compared to normal glucose tolerance subjects. Clinical endocrinology. 2016;85 (4):551–5.

Chapter 17
Immunohistochemistry

A. Khelfi

Abstract Several biomarkers of protein oxidation, glycoxidation, and lipid peroxidation are now identified and can be detected in situ by immunohistochemistry. In situ measurements allow for the detection of specific cells and specific localization of oxidative stress biomarkers (e.g., protein carbonyls, nitrotyrosine, chlorotyrosine, carbohydrate end-products, malondialdehyde (MDA), 4-hydroxy-2-nonenal (4-HNE), iNOS, and COX) affecting only a small fraction of tissue and cells. Because tissues are usually fixed in formalin, investigators must focus on well-defined products that are stable in this fixative. However, a major problem may emerge, which is the difficulty in measuring trace concentration of individual oxidation end-products in tissue by relatively sophisticated chemical assay procedures. Protein, lipid, and DNA damage can be visualized in tissues using monoclonal or polyclonal antibodies. In this chapter, we feature several cases in which immunohistochemical techniques directed to damage products and pro-oxidant enzymes are used to support investigation of the pathogenesis of several diseases. The chapter also gives a perspective on the diversity of applications of immunochemistry in the ever-emerging field of oxidative stress.

Keywords Protein carbonyls · Nitrotyrosine · Chlorotyrosine · Carbohydrate end-products · Malondialdehyde (MDA) · 4-Hydroxy-2-nonenal (4-HNE) · iNOS · COX

A. Khelfi (✉)
Faculty of Pharmacy, University of Algiers 1, Algiers, Algeria

National Centre of Toxicology, Algiers, Algeria

Department of Chemistry and Biomolecular Science, Clarkson University, Potsdam, NY, USA
e-mail: ab.khelfi@univ-alger.dz; akhelfi@clarkson.edu

S. Andreescu et al. (eds.), *Biomarkers of Oxidative Stress*,
https://doi.org/10.1007/978-3-031-60738-7_17

17.1 Protein End Products

17.1.1 Protein Carbonyls

The carbonyl assay is a general measurement of oxidative protein damage, which helps to assess steady-state protein damage in human tissues and body fluids. This assay is based on the ability of several RS to attack histidine, arginine, lysine, and proline residues in proteins to produce carbonyl functions that can be measured after derivatization using DNPH. This reaction forms the protein-bound hydrazine, which can be detected by anti-2,4-dinitrophenyl antibody (Fig. 17.1a). An alternative can be performed through the use of biotin-hydrazide reagent in which the protein-bound acyl hydrazone is detected by the enzyme-linked avidin or streptavidin [1].

Several applications of immunochemical methods involving anti-DNPH antibodies have been published, some of them can be found in the following references [2, 3].

17.1.2 Nitrotyrosine

Nitrotyrosine is a biochemical marker of peroxynitrite-induced damage to proteins. Evidence for protein nitration in cells and tissues has been first provided by immunochemistry using mAb against NO_2Tyr that do not cross-react with 3-aminotyrosine, 3-Cl-Tyr, o-phosphotyrosine, or nitrotryptophan [4]. In order to exclude this cross-reactivity, controls should be performed by reducing NO_2Tyr to aminotyrosine with sodium thiosulfate, which should suppress recognition by the antibody and result in total disappearance of the signal.

Commercial high-activity rabbit polyclonal antibody for detecting nitrotyrosine-containing proteins and peptides is available [5, 6]. Fluorescence of Green Fluorescent Protein (GFP) is extremely sensitive to tyrosine nitration and can be used as a probe of nitrosative stress as confirmed by correlated anti-nitrotyrosine immunoreactivity [63].

Immunohistochemistry of NO_2Tyr has been widely used to assess nitrosative stress in many diseases (Fig. 17.1b). Based on immunohistochemical analysis, NO_2Tyr is increased primarily at sites of inflammation, for example, in foam cells in atherosclerotic plaque, in the eye in experimental autoimmune uveitis, and in the lung following ischemic injury [7, 8, 9]. Leeuwenburgh et al. also confirmed by chemical analysis a nearly 100-fold increase in NO_2Tyr in plaque versus circulating LDL [10].

Specific locations and targets of tyrosine nitration in patients with lung cancer were detailed using histochemical analysis [11]. The authors have recorded an increased NO_2Tyr immunostaining, which is limited to the tumor and not to surrounding healthy tissue, suggesting a unique environment inside the tumor that may contribute to the disease process.

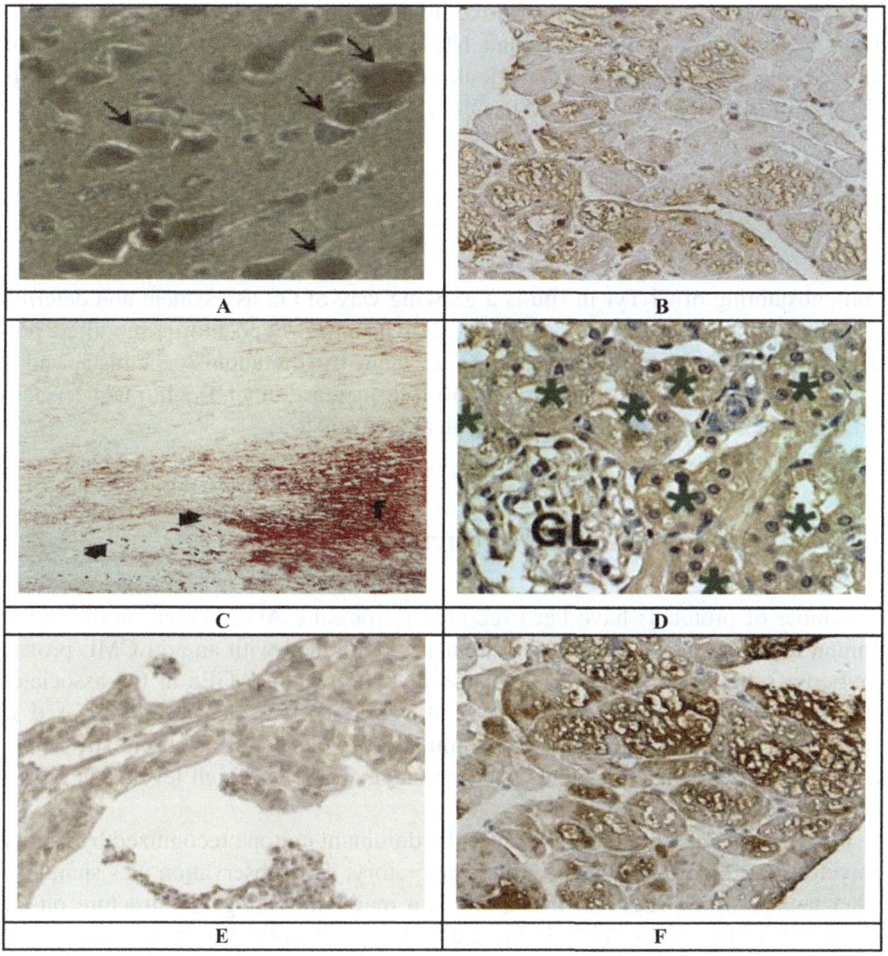

Fig. 17.1 Immunohistochemistry of protein carbonyl, NO_2Tyr, CML, 4-HNE-His adduct, 8-OHdG, and iNOS [68–72]. (**a**) Brain biopsy from patients with hereditary ferritinopathy and immunostaining with anti-DNP antibody. (**b**) Endomyocardial biopsy from Fabry disease patients and immunostaining with anti-NO_2Tyr antibody. (**c**) Aorta biopsy from patients with atherosclerotic plaque and immunostaining with 6D12 antibody. (**d**) Kidney biopsy and immunostaining of 4-HNE-His adduct using HNIJ-2 antibody. GL: Glomerulus. (**e**) Pancreas biopsy from patient with pancreatic ductal adenocarcinoma and immunostaining of 8-OHdG using N45.1 antibody. (**f**) Endomyocardial biopsy from Fabry disease patients and immunostaining with anti-iNOS antibody

The presence of NO_2Tyr has also been demonstrated in models of xenobiotic toxicity. For example, immunostaining for NO_2Tyr has been used to investigate the pathophysiology of acetaminophen-induced hepatotoxicity. Initial formation of NO_2Tyr was noted in the sinusoidal EC, with a later accumulation within hepatocytes, indicating OS in both types of cells [12, 13].

Investigating the mechanism of doxorubicin-induced cardiotoxicity in mice and using immunostaining for iNOS and NO_2Tyr, Weinstein et al. and Mihm et al. showed that the extent of left ventricular dysfunction was highly correlated with NO_2Tyr immunoreactivity [14, 15].

17.1.3 Chlorotyrosine

Immunostaining of Cl-Tyr in situ is a growing way of OS assessment and determination of inflammatory response extent. For instance, hypochlorite-modified proteins have been identified in atherosclerotic plaque by immunohistochemistry and in renal inflammatory lesions [16, 17]. A 30-fold increase in Cl-Tyr in plaque versus circulating LDL was confirmed by chemical analysis [64].

17.2 Carbohydrate End Products

A number of protocols have been reported to measure AGEs based on the use of immunohistochemistry. Using histochemistry combined with an anti-CML-protein antibody, several works have shown the involvement of AGEs in OS-associated pathologies (Fig. 17.1c). Schleicher et al. reported that CML was increased in kidneys of diabetic patients [65]. Inflammation induced by metabolic stress may cause oxidative damage in kidneys and vascular walls with high levels of CML in these tissues.

Reddy et al. reported that CML was the dominant epitope recognized by several polyclonal antibodies prepared in their laboratory. This observation was shared by other teams, which suggests that CML is a major immunogenic structure on the surface of AGEs proteins [18]. Using the CML-specific 6D12 antibody, Kume et al. detected increased levels of AGEs in atherosclerotic plaque of non-diabetic subjects [19].

Anti-AGE antibodies with specificity for CML have also been used in dialysis patient and found increased levels of AGEs in amyloid plaque in dialysis-related amyloidosis. According to several studies, CML seems to be a primary AGEs epitope in Alzheimer's plaque [20, 21]. Horiuchi et al. also reported increased levels of CML in elastin of sun-exposed skin of individuals with actinic elastosis [22].

Niwa et al. described a murine mAb specific for (dihydro)imidazolones, which is formed in reaction of 3DG with the guanidino group of arginine and reported that erythrocyte levels of these adducts were significantly elevated in diabetic patients [23]. This antibody was used in immunohistochemical studies and detected increased levels of 3DG-Arg adducts in vascular and renal tissues from diabetic patients. In addition to preparation of anti-AGE antibodies using AGEs proteins as antigens, Monnier et al. have prepared specific antibodies against other AGEs, for example pyrraline, linked as haptens to carrier macromolecules. As an application, the authors found that the advanced glycosylation reaction may contribute to the decreased turnover and thickening of the extracellular matrix in DM and aging [24].

17.3 Lipid End Products

17.3.1 *MDA/4-HNE-Protein Adducts*

Histochemistry has been used to explore lipoxidation and aldehyde generation in different samples. To measure the presence of aldehydes and particularly 4-HNE in tissues, including protein bound-HNE, immunodetection is preferred. Antibodies against 4-HNE and MDA conjugated to different amino acids have been developed and applied to immunohistochemistry in paraffin-embedded tissues [25–27]. They are used in qualitative and semi-quantitative analyses where their presence and relative abundance are detected in cells and tissues in parallel with evaluation of morphological changes.

Haberland et al. used antibodies against MDA-protein adducts and first reported the detection of MDA-protein epitopes in atherosclerotic plaque [28]. Subsequent studies have shown that both MDA-Lys and 4-HNE-Lys adducts are present in atherosclerotic plaque, MDA-protein epitopes are enriched in Alzheimer's plaque, and 4-HNE epitopes are highly present in amyloid deposits in patients with systemic amyloidosis [29, 30].

In this context, mAb HNEJ-1-5 have been developed against the hemi-acetal structure of Michael adducts [31]. Most of antibodies recognized 4-HNE lysine, cysteine, and histidine adducts equally, while mAb HNEJ-2 selectively recognized histidine adducts (Fig. 17.1d). The most important feature with HNEJ-2 is that these antibodies do not recognize 4-HHE, formaldehyde- or glutaraldehyde-modified proteins [32]. Besides, Pompella et al. and Waeg et al. independently produced antibodies against HNE-modified proteins [33, 34]. These antibodies were successfully applied to immunohistochemistry in paraffin-embedded tissue samples [35].

4-HNE-His adducts are often referred as the major biomarkers of OS, as it has been verified by a series of clinical trials combining HNE-His immunochemistry with metabolomics/lipidomics in patients with RA, neuroborreliosis, encephalitis, metabolic syndrome, and liver cancer [36–40, 66], while in patients with lung malignancies it was verified by lipidomics and PET/CT [41, 42]. These antibodies have been applied to a variety of other pathological conditions including malignant neoplasm (renal cell carcinoma), atherosclerosis (foam cells), PD (Nigral neuron), alcohol liver disease (hepatocytes), and chronic hepatitis (hepatocytes) [43–46].

Besides visualization of the distribution of ALEs obtained by immunohistochemistry, a particular advantage of the HNE-His mAb is that the same epitopes could be detected using several immunoassays. Moreover, while these epitopes remain stable for nearly 1 year at −80 °C, the samples are usually formalin-fixed and well preserved for years, especially in the case of paraffin-embedded tissue specimens.

Zarkovic et al. summarized other benefits as follows [47]:

- Immunohistochemistry specific for HNE-protein adducts, mainly for HNE-histidine epitopes, is a valuable semi-quantitative method to study pathophysiology of OS and lipid peroxidation.
- Immunohistochemical studies of HNE-protein adducts can give a confirmation of the association of lipid peroxidation with major human diseases.
- Formation of HNE-protein adducts is not irreversible and correlates with the age of patients and severity of some degenerative, metabolic, and inflammatory diseases.
- In case of carcinogenesis, 4-HNE can play undesirable pro-carcinogenic role. Conversely, cells in the vicinity of cancer can generate desirable 4-HNE to defend the organism from cancer invasion.

17.3.2 Isolevuglandins

Immunohistochemical staining using a single-chain antibody D11ScFv, which selectively bond to IsoLG-protein adducts, showed increased levels of these determinants in the hippocampus of AD patients [48], in the epicardial border zone of myocardial infarcts [49], and in heart, aorta, and dendritic cells during hypertension [50].

17.3.3 Acrolein

Immunohistochemical studies using mAb against acrolein adducts revealed that the latter colocalized with apoA-I in human atherosclerotic lesions. Some of these studies found an impairment of the ability of acrolein-modified apoA-I to remove cholesterol from cultured cells, suggesting that acrolein-carbonylation may interfere with the normal function of apoA-I as a promoter of cholesterol removal from artery wall cells, thus playing a critical function in atherogenesis [51].

17.4 DNA End Products

Immunostaining is widely used for the detection of oxidized bases, especially 8-OHdG (Fig. 17.1e). Several polyclonal and mAb for 8-OHdG have been reported (e.g., 15A3 and 1F7). According to Toyokuni, most reported polyclonal and mAb used for 8-OHdG detection show a high cross-reactivity with 8-OHG (RNA form), except for the mAb N45.1 [52, 67]. According to Toyokuni et al., N45.1 is scarcely reactive with other DNA adducts (21 molecules) and thus highly specific for 8-OHdG detection when applicable for immunohistochemistry in paraffin-embedded sections [67]. Monoclonal antibody N45.1 was established using 8-OHdG–keyhole limpet hemocyanin as an antigen [53]. Immunohistochemistry using N45.1 has been successfully applied to a variety of pathological conditions including UV irradiation, malignant neoplasms, ischemia-reperfusion injury, and DM [52, 54].

Takahashi et al. confirmed in a model of acute-liver-damage induced by CCl_4 that the anti-8-OHdG antibody detects cells targeted by FR in paraffin-embedded sections [55]. Immunohistochemical detection of markedly high levels of 8-OHdG was reported to occur in various human tumors like high-grade glioma, compared to adjacent, normal tissues, or low-grade glioma [56].

17.5 Inducible Nitric Oxide Synthase (iNOS)

One of the most interesting approaches for the localization of NOS in tissues and highlighting the involvement of a precise isoform of NOS in a specific pathology is to make immunohistochemical evaluations. Immuunohistochemical techniques employ specific antibodies against each individual isoform of NOS (Fig. 17.1f). This exploration has gained considerable popularity in recent years and helped to understand several pathophysiological mechanisms.

The inducible form, iNOS, is found predominantly in mononuclear phagocytes where it may be induced by different stimulators such as endotoxins and cytokines, resulting in the production of high levels of $NO^•$. The use of immunohistochemical techniques with specific antibodies indicates that iNOS promotes tumor progression. For example, in the prostate where high-grade prostatic intraepithelial neoplasia (PIN) and carcinoma display more intense iNOS immunoreactivity than benign prostatic hyperplasia and low-grade PIN samples [57, 58].

17.6 Cyclooxygenase (COX)

Cyclooxygenase catalyzes the formation of prostaglandins and other eicosanoids from arachidonic acid. Cyclooxygenase-2 is induced at the site of inflammation following stimulation with pro-inflammatory agents such as IL-1, TNF-α, and lipopolysaccharides. COX-2 overexpression was found to be involved in cellular proliferation and carcinogenesis in different organs [59–61]. Investigators suggested that this enzyme may increase the production of FR in situ. Indeed, oxygenation of arachidonic acid by COX-1 via FR mechanism has been described to produce peroxyl radicals [62].

17.7 Other Biomarkers of OS

Immunohistochemistry is used for in situ detection of other important biomarkers of OS including NOX, OGG-1, and Nrf-2. However, little data are available on the specificity of the corresponding antibodies used for their detection.

References

1. Moreira PI, Sayre LM, Zhu X, Nunomura A, Smith MA, Perry G. Detection and localization of markers of oxidative stress by in situ methods: application in the study of Alzheimer disease. Methods Mol Biol. 2010;610:419–34.
2. Zheng J, Bizzozero OA. Accumulation of protein carbonyls within cerebellar astrocytes in murine experimental autoimmune encephalomyelitis. J Neurosci Res. 2010;88(15):3376–85.
3. Luo S, Wehr NB. Protein carbonylation: avoiding pitfalls in the 2,4-dinitrophenylhydrazine assay. Redox report : communications in free radical research. 2009;14(4):159–66.
4. Beckmann JS, Ye YZ, Anderson PG, Chen J, Accavitti MA, Tarpey MM, et al. Extensive nitration of protein tyrosines in human atherosclerosis detected by immunohistochemistry. Biological chemistry Hoppe-Seyler. 1994;375(2):81–8.
5. Ijsselmuiden AJ, Musters RJ, de Ruiter G, van Heerebeek L, Alderse-Baas F, van Schilfgaarde M, et al. Circulating white blood cells and platelets amplify oxidative stress in heart failure. Nature clinical practice Cardiovascular medicine. 2008;5(12):811–20.
6. Ellis EA. Correlative transmission microscopy: cytochemical localization and immunocytochemical localization in studies of oxidative and nitrosative stress. Methods Mol Biol. 2008;477:41–8.
7. Wu GS, Zhang J, Rao NA. Peroxynitrite and oxidative damage in experimental autoimmune uveitis. Investigative Ophthalmology & Visual Science. 1997;38(7):1333–9.
8. Ischiropoulos H, al-Mehdi AB, Fisher AB. Reactive species in ischemic rat lung injury: contribution of peroxynitrite. The American journal of physiology. 1995;269(2 Pt 1):L158–64.
9. Thomson L. 3-nitrotyrosine modified proteins in atherosclerosis. Dis Markers. 2015;2015: 708282.
10. Leeuwenburgh C, Hardy MM, Hazen SL, Wagner P, Oh-ishi S, Steinbrecher UP, et al. Reactive nitrogen intermediates promote low density lipoprotein oxidation in human atherosclerotic intima. J Biol Chem. 1997;272(3):1433–6.
11. Masri FA, Comhair SAA, Koeck T, Xu W, Janocha A, Ghosh S, et al. Abnormalities in nitric oxide and its derivatives in lung cancer. American journal of respiratory and critical care medicine. 2005;172(5):597–605.
12. Knight TR, Jaeschke H. Peroxynitrite formation and sinusoidal endothelial cell injury during acetaminophen-induced hepatotoxicity in mice. Comparative hepatology. 2004;3 Suppl 1 (Suppl 1):S46-S.
13. Knight TR, Kurtz A, Bajt ML, Hinson JA, Jaeschke H. Vascular and hepatocellular peroxynitrite formation during acetaminophen toxicity: role of mitochondrial oxidant stress. Toxicological sciences : an official journal of the Society of Toxicology. 2001;62(2):212–20.
14. Weinstein DM, Mihm MJ, Bauer JA. Cardiac peroxynitrite formation and left ventricular dysfunction following doxorubicin treatment in mice. The Journal of pharmacology and experimental therapeutics. 2000;294(1):396–401.
15. Mihm MJ, Yu F, Weinstein DM, Reiser PJ, Bauer JA. Intracellular distribution of peroxynitrite during doxorubicin cardiomyopathy: evidence for selective impairment of myofibrillar creatine kinase. Br J Pharmacol. 2002;135(3):581–8.

16. Hazell LJ, Arnold L, Flowers D, Waeg G, Malle E, Stocker R. Presence of hypochlorite-modified proteins in human atherosclerotic lesions. J Clin Invest. 1996;97(6):1535–44.
17. Malle E, Woenckhaus C, Waeg G, Esterbauer H, Gröne EF, Gröne HJ. Immunological evidence for hypochlorite-modified proteins in human kidney. The American journal of pathology. 1997;150(2):603–15.
18. Reddy S, Bichler J, Wells-Knecht KJ, Thorpe SR, Baynes JW. N epsilon-(carboxymethyl)-lysine is a dominant advanced glycation end product (AGE) antigen in tissue proteins. Biochemistry. 1995;34(34):10872–8.
19. Kume S, Takeya M, Mori T, Araki N, Suzuki H, Horiuchi S, et al. Immunohistochemical and ultrastructural detection of advanced glycation end products in atherosclerotic lesions of human aorta with a novel specific monoclonal antibody. The American journal of pathology. 1995;147 (3):654–67.
20. Miyata T, Oda O, Inagi R, Iida Y, Araki N, Yamada N, et al. beta 2-Microglobulin modified with advanced glycation end products is a major component of hemodialysis-associated amyloidosis. J Clin Invest. 1993;92(3):1243–52.
21. Vitek MP, Bhattacharya K, Glendening JM, Stopa E, Vlassara H, Bucala R, et al. Advanced glycation end products contribute to amyloidosis in Alzheimer disease. Proc Natl Acad Sci U S A. 1994;91(11):4766–70.
22. Mizutari K, Ono T, Ikeda K, Kayashima K, Horiuchi S. Photo-enhanced modification of human skin elastin in actinic elastosis by N(epsilon)-(carboxymethyl)lysine, one of the glycoxidation products of the Maillard reaction. The Journal of investigative dermatology. 1997;108(5): 797–802.
23. Niwa T, Katsuzaki T, Miyazaki S, Miyazaki T, Ishizaki Y, Hayase F, et al. Immunohistochemical detection of imidazolone, a novel advanced glycation end product, in kidneys and aortas of diabetic patients. J Clin Invest. 1997;99(6):1272–80.
24. Miyata S, Monnier V. Immunohistochemical detection of advanced glycosylation end products in diabetic tissues using monoclonal antibody to pyrraline. J Clin Invest. 1992;89(4):1102–12.
25. Sampey BP, Korourian S, Ronis MJ, Badger TM, Petersen DR. Immunohistochemical characterization of hepatic malondialdehyde and 4-hydroxynonenal modified proteins during early stages of ethanol-induced liver injury. Alcohol Clin Exp Res. 2003;27(6):1015–22.
26. Jang EJ, Kim DH, Lee B, Lee EK, Chung KW, Moon KM, et al. Activation of proinflammatory signaling by 4-hydroxynonenal-Src adducts in aged kidneys. Oncotarget. 2016;7(32): 50864–74.
27. Podszun MC, Chung J-Y, Ylaya K, Kleiner DE, Hewitt SM, Rotman Y. 4-HNE Immunohistochemistry and Image Analysis for Detection of Lipid Peroxidation in Human Liver Samples Using Vitamin E Treatment in NAFLD as a Proof of Concept. 2020;68(9):635–43.
28. Haberland ME, Fong D, Cheng L. Malondialdehyde-altered protein occurs in atheroma of Watanabe heritable hyperlipidemic rabbits. Science. 1988;241(4862):215–8.
29. Yan S, Chen X, Schmidt A, Brett J, Godman G, Zou Y, et al. Glycated tau protein in Alzheimer disease: a mechanism for induction of oxidant stress. 1994;91(16):7787–91.
30. Ando Y, Nyhlin N, Suhr O, Holmgren G, Uchida K, el Sahly M, et al. Oxidative stress is found in amyloid deposits in systemic amyloidosis. Biochem Biophys Res Commun. 1997;232(2): 497–502.
31. Uchida K, Szweda LI, Chae HZ, Stadtman ER. Immunochemical detection of 4-hydroxynonenal protein adducts in oxidized hepatocytes. Proc Natl Acad Sci U S A. 1993;90(18):8742–6.
32. Toyokuni S, Miyake N, Hiai H, Hagiwara M, Kawakishi S, Osawa T, et al. The monoclonal antibody specific for the 4-hydroxy-2-nonenal histidine adduct. FEBS Lett. 1995;359(2–3): 189–91.
33. Pompella A, Cambiaggi C, Dominici S, Paolicchi A, Tongiani R, Comporti M. Single-cell investigation by laser scanning confocal microscopy of cytochemical alterations resulting from extracellular oxidant challenge. Histochemistry and Cell Biology. 1996;105(3):173–8.

34. Waeg G, Dimsity G, Esterbauer H. Monoclonal Antibodies for Detection of 4-Hydroxynonenal Modified Proteins. Free radical research. 1996;25(2):149–59.
35. Tanaka T, Nishiyama Y, Okada K, Hirota K, Matsui M, Yodoi J, et al. Induction and nuclear translocation of thioredoxin by oxidative damage in the mouse kidney: independence of tubular necrosis and sulfhydryl depletion. Laboratory investigation; a journal of technical methods and pathology. 1997;77(2):145–55.
36. Łuczaj W, Gindzienska-Sieskiewicz E, Jarocka-Karpowicz I, Andrisic L, Sierakowski S, Zarkovic N, et al. The onset of lipid peroxidation in rheumatoid arthritis: consequences and monitoring. Free radical research. 2016;50(3):304–13.
37. Moniuszko-Malinowska A, Łuczaj W, Jarocka-Karpowicz I, Pancewicz S, Zajkowska J, Andrisic L, et al. Lipid peroxidation in the pathogenesis of neuroborreliosis. Free Radical Biology and Medicine. 2016;96:255–63.
38. Elrayess MA, Almuraikhy S, Kafienah W, Al-Menhali A, Al-Khelaifi F, Bashah M, et al. 4-hydroxynonenal causes impairment of human subcutaneous adipogenesis and induction of adipocyte insulin resistance. Free Radic Biol Med. 2017;104:129–37.
39. Jaganjac M, Almuraikhy S, Al-Khelaifi F, Al-Jaber M, Bashah M, Mazloum NA, et al. Combined metformin and insulin treatment reverses metabolically impaired omental adipogenesis and accumulation of 4-hydroxynonenal in obese diabetic patients. Redox Biol. 2017;12:483–90.
40. Zhong H, Xiao M, Zarkovic K, Zhu M, Sa R, Lu J, et al. Mitochondrial control of apoptosis through modulation of cardiolipin oxidation in hepatocellular carcinoma: A novel link between oxidative stress and cancer. Free Radical Biology and Medicine. 2017;102:67–76.
41. Gęgotek A, Nikliński J, Žarković N, Žarković K, Waeg G, Łuczaj W, et al. Lipid mediators involved in the oxidative stress and antioxidant defence of human lung cancer cells. Redox Biol. 2016;9:210–9.
42. Živković NP, Petrovečki M, Lončarić Č T, Nikolić I, Waeg G, Jaganjac M, et al. Positron emission tomography-computed tomography and 4-hydroxynonenal-histidine immunohistochemistry reveal differential onset of lipid peroxidation in primary lung cancer and in pulmonary metastasis of remote malignancies. Redox Biol. 2017;11:600–5.
43. Okamoto K, Toyokuni S, Uchida K, Ogawa O, Takenewa J, Kakehi Y, et al. Formation of 8-hydroxy-2'-deoxyguanosine and 4-hydroxy-2-nonenal-modified proteins in human renal-cell carcinoma. International journal of cancer. 1994;58(6):825–9.
44. Yoritaka A, Hattori N, Uchida K, Tanaka M, Stadtman ER, Mizuno Y. Immunohistochemical detection of 4-hydroxynonenal protein adducts in Parkinson disease. Proc Natl Acad Sci U S A. 1996;93(7):2696–701.
45. Ohhira M, Ohtake T, Matsumoto A, Saito H, Ikuta K, Fujimoto Y, et al. Immunohistochemical detection of 4-hydroxy-2-nonenal-modified-protein adducts in human alcoholic liver diseases. Alcohol Clin Exp Res. 1998;22(S3 Pt 1):145s–9s.
46. Kageyama FJJPT. Enhanced lipid peroxidation in chronic hepatitis C. 1998;26:S405–S8.
47. Zarkovic K, Jakovcevic A, Zarkovic N. Contribution of the HNE-immunohistochemistry to modern pathological concepts of major human diseases. Free Radical Biology and Medicine. 2017;111:110–26.
48. Davies SS, Bodine C, Matafonova E, Pantazides BG, Bernoud-Hubac N, Harrison FE, et al. Treatment with a γ-ketoaldehyde scavenger prevents working memory deficits in hApoE4 mice. Journal of Alzheimer's disease : JAD. 2011;27(1):49-59.
49. Fukuda K, Davies SS, Nakajima T, Ong BH, Kupershmidt S, Fessel J, et al. Oxidative mediated lipid peroxidation recapitulates proarrhythmic effects on cardiac sodium channels. Circulation research. 2005;97(12):1262–9.
50. Guo L, Amarnath V, Davies SS. A liquid chromatography-tandem mass spectrometry method for measurement of N-modified phosphatidylethanolamines. Analytical biochemistry. 2010;405 (2):236–45.

51. Shao B, O'Brien K D, McDonald TO, Fu X, Oram JF, Uchida K, et al. Acrolein modifies apolipoprotein A-I in the human artery wall. Annals of the New York Academy of Sciences. 2005;1043:396–403.
52. Toyokuni S. Reactive oxygen species-induced molecular damage and its application in pathology. Pathology international. 1999;49(2):91–102.
53. Osawa T, Yoshida A, Kawakishi S, Yamashita K, Ochi H, editors. Protective role of dietary antioxidants in oxidative stress. Oxidative Stress and Aging; 1995 1995//; Basel: Birkhäuser Basel.
54. Hattori Y, Nishigori C, Tanaka T, Uchida K, Nikaido O, Osawa T, et al. 8-hydroxy-2'--deoxyguanosine is increased in epidermal cells of hairless mice after chronic ultraviolet B exposure. The Journal of investigative dermatology. 1996;107(5):733–7.
55. Takahashi S, Hirose M, Tamano S, Ozaki M, Orita S, Ito T, et al. Immunohistochemical detection of 8-hydroxy-2'-deoxyguanosine in paraffin-embedded sections of rat liver after carbon tetrachloride treatment. Toxicologic pathology. 1998;26(2):247–52.
56. Iida T, Furuta A, Kawashima M, Nishida J, Nakabeppu Y, Iwaki T. Accumulation of 8-oxo-2'--deoxyguanosine and increased expression of hMTH1 protein in brain tumors. Neuro-oncology. 2001;3(2):73–81.
57. Hassan HE, Mohamed AAB, Bakhiet AO, Ahmed HG. Immunohistochemical expression of COX2 and iNOS in bladder cancer and its association with urinary schistosomiasis among Sudanese patients. Infect Agent Cancer. 2013;8(1):9.
58. Baltaci S, Orhan D, Gögüs C, Türkölmez K, Tulunay O, Gögüs O. Inducible nitric oxide synthase expression in benign prostatic hyperplasia, low- and high-grade prostatic intraepithelial neoplasia and prostatic carcinoma. BJU international. 2001;88(1):100–3.
59. Matsumoto G, Muta M, KojiTsuruta, Horiguchi S, Karasawa K, Okamoto A. Tumor Size Significantly Correlates with Postoperative Liver Metastases and COX-2 Expression in Patients with Resectable Pancreatic Cancer. Pancreatology. 2007;7(2):167–73.
60. Bin W, He W, Feng Z, Xiangdong L, Yong C, Lele K, et al. Prognostic relevance of cyclooxygenase-2 (COX-2) expression in Chinese patients with prostate cancer. Acta Histochemica. 2011;113(2):131–6.
61. Mattsson JSM, Bergman B, Grinberg M, Edlund K, Marincevic M, Jirström K, et al. Prognostic impact of COX-2 in non-small cell lung cancer: A comprehensive compartment-specific evaluation of tumor and stromal cell expression. Cancer letters. 2015;356(2, Part B):837–45.
62. Mukherjee A, Brinkley DW, Chang K-M, Roth JP. Molecular Oxygen Dependent Steps in Fatty Acid Oxidation by Cyclooxygenase-1. Biochemistry. 2007;46(13):3975–89.
63. Espey MG, Xavier S, Thomas DD, Miranda KM, Wink DA. Direct real-time evaluation of nitration with green fluorescent protein in solution and within human cells reveals the impact of nitrogen dioxide vs. peroxynitrite mechanisms. 2002;99(6):3481–6.
64. Hazen SL, Heinecke JW. 3-Chlorotyrosine, a specific marker of myeloperoxidase-catalyzed oxidation, is markedly elevated in low density lipoprotein isolated from human atherosclerotic intima. J Clin Invest. 1997;99(9):2075–81.
65. Schleicher ED, Wagner E, Nerlich AG. Increased accumulation of the glycoxidation product N (epsilon)-(carboxymethyl)lysine in human tissues in diabetes and aging. J Clin Invest. 1997;99 (3):457–68.
66. Łuczaj W, Moniuszko A, Jarocka-Karpowicz I, Pancewicz S, Andrisic L, Zarkovic N, et al. Tick-borne encephalitis--lipid peroxidation and its consequences. Scandinavian journal of clinical and laboratory investigation. 2016;76(1):1–9.
67. Toyokuni S, Tanaka T, Hattori Y, Nishiyama Y, Yoshida A, Uchida K, et al. Quantitative immunohistochemical determination of 8-hydroxy-2'-deoxyguanosine by a monoclonal antibody N45.1: its application to ferric nitrilotriacetate-induced renal carcinogenesis model. Laboratory investigation; a journal of technical methods and pathology. 1997;76(3):365–74.
68. Inokuchi S, Itoh S, Yoshizumi T, Yugawa K, Yoshiya S, Toshima T, et al. Mitochondrial expression of the DNA repair enzyme OGG1 improves the prognosis of pancreatic ductal adenocarcinoma. Pancreatology. 2020;20(6):1175–82.

69. Sakata N, Imanaga Y, Meng J, Tachikawa Y, Takebayashi S, Nagai R, et al. Immunohistochemical localization of different epitopes of advanced glycation end products in human atherosclerotic lesions. Atherosclerosis. 1998;141(1):61–75.
70. Baraibar MA, Barbeito AG, Muhoberac BB, Vidal R. A mutant light-chain ferritin that causes neurodegeneration has enhanced propensity toward oxidative damage. Free Radic Biol Med. 2012;52(9):1692–7.
71. Chimenti C, Scopelliti F, Vulpis E, Tafani M, Villanova L, Verardo R, et al. Increased oxidative stress contributes to cardiomyocyte dysfunction and death in patients with Fabry disease cardiomyopathy. Human pathology. 2015;46(11):1760–8.
72. Ozeki M, Miyagawa-Hayashino A, Akatsuka S, Shirase T, Lee WH, Uchida K, et al. Susceptibility of actin to modification by 4-hydroxy-2-nonenal. J Chromatogr B Analyt Technol Biomed Life Sci. 2005;827(1):119–26.

Chapter 18
MiOXSYS

R. Henkel

Abstract Since the landmark paper on the production of hydrogen peroxide by human sperm published by MacLeod in 1943, reports on oxidative stress and its involvement in numerous diseases including cardiovascular diseases, cancers, endometriosis, aging, obesity, or infertility have significantly increased to 315,251 entries on PubMed on 26 September 2023. Subsequently, it became clear that the homeostasis between oxidative stress (OS) and reductive stress (RS) is essential for normal physiology. Therefore, one should rather speak of "redox stress" to cover the whole spectrum. There are five different approaches to detect redox stress, namely, by measurement of (1) antioxidant enzymes, (2) low molecular weight antioxidants, (3) oxidants, (4) products of oxidative damage, and (5) the balance between oxidants and antioxidants. The latter approach includes the recently developed electrochemical technique of the measurement of the oxidation-reduction potential (ORP) using the Male Infertility Oxidative System (MiOXSYS). This methodology is a snapshot of the redox balance at the time of measurement based on the Nernst Equation. Using this technique, elevated oxidative stress levels have been detected in trauma patients, in patients with sepsis, diabetes, and metabolic syndrome, as well as in semen of infertile men. Relevant cut-off values have been identified. Available data indicate this method as a quick and reliable diagnostic method.

Keywords Oxidative stress · Reductive stress · Electrochemistry · Sperm · Infertility · Oxidation-reduction potential

Since the landmark discovery by MacLeod (1943) that human sperm cells are producing small amounts of H_2O_2 and that this oxidant can impair sperm functionality, thus can be implicated in male infertility, research on "oxidative stress"

R. Henkel (✉)
LogixX Pharma Ltd., Theale, Berkshire, UK

Department of Medical Bioscience, University of the Western Cape, Bellville, South Africa

Department of Metabolism, Digestion & Reproduction, Imperial College, London, UK
e-mail: Ralf.henkel@logixxpharma.com; rhenkel@uwc.ac.za; r.henkel@imperial.ac.uk

S. Andreescu et al. (eds.), *Biomarkers of Oxidative Stress*,
https://doi.org/10.1007/978-3-031-60738-7_18

(OS) has significantly increased with 315,251 entries on PubMed on 26.09.2023. Oxidative stress is a situation that is characterized by an imbalance between the available oxidants and antioxidants in favour of the oxidants. Oxidative stress is involved in and characteristic for numerous diseases and conditions including cardiovascular diseases, cancers, endometriosis, ageing, obesity, or infertility. The term "oxidative stress" was coined by Sies and Cadenas in 1985 and defined as "a disturbance in the pro-oxidant-antioxidant balance in favour of the former" [1]. Later, in 2006, Jones (2016) proposed a modification of this definition as "a disruption of redox signalling and control" because it was shown that under certain conditions (such as aging, exercise, or metal ion deficiency/excess) some antioxidant enzymes can cause an increase in reactive oxygen species (ROS) [2]. This suggests that organisms are able to modify their internal response to an oxidative assault thus adapting the redox homeostasis [3].

Oxidative stress, or better said "redox stress" as oxidation and reduction reactions are chemically always linked, can be determined in body fluids using five different approaches, namely by measurement of (1) antioxidant enzymes (e.g. CAT, SOD, and total antioxidant capacity (TAC)), (2) low molecular weight antioxidants (e.g. vitamins C or E, glutathione), (3) oxidants (e.g. H_2O_2 and free radicals), (4) products of oxidative damage (e.g. malondialdehyde, 8-hydroxyguanosine (8-OHdG)), and (5) the balance between oxidants and antioxidants (e.g. ROS-TAC score, electrochemical techniques). The methods used for the approaches 1–4 are spectroscopic, fluorometric, enzymatic, chromatographic, and Electron Paramagnetic Resonance (EPR) Spectroscopy-EPR Imaging methods, which are discussed in the chapters before, determine only one side of the same coin (redox reactions), either oxidants or antioxidants. Contrary, in the electrochemical determination of the balance between oxidants and antioxidants (approach #5), the balance between oxidation and reduction is measured. Whereas for the ROS-TAC score, ROS and TAC still have to be determined separately and the ROS–TAC score is then calculated, electrochemical techniques measure the electron transfer from antioxidants (electron donators) to the oxidants (electron acceptors) in one measurement (Fig. 18.1). Unfortunately, no consensus for routine clinical use of a specific approach or technique has been reached yet. This may be due to different believes in different techniques, cost, time, amount of sample that is used and/or lack of clinical validation. As a new technique, the electrochemical measurement of the static oxidation–reduction potential (sORP) as a measure of OS using the Male Infertility Oxidative System (MiOXSYS) was introduced recently. sORP describes the current standard electrochemical potential between reductants (antioxidants) and oxidants. The potential is measured in milli-Volts (mV) with a platinum-based indicator electrode and a reference electrode made of silver and silver chloride. Both electrodes are connected via a salt bridge and an external small current is applied (Fig. 18.2). In addition, the so-called capacity ORP (cORP) as a measure of the antioxidant reserve with high values indicating a high availability of antioxidants in the system can be measured. The cORP is expressed in micro-Coulombs (μC).

Since information on the ROS levels or the TAC alone is not sufficient to draw conclusions about the actual redox stress [4, 5], the MiOXSYS is providing a

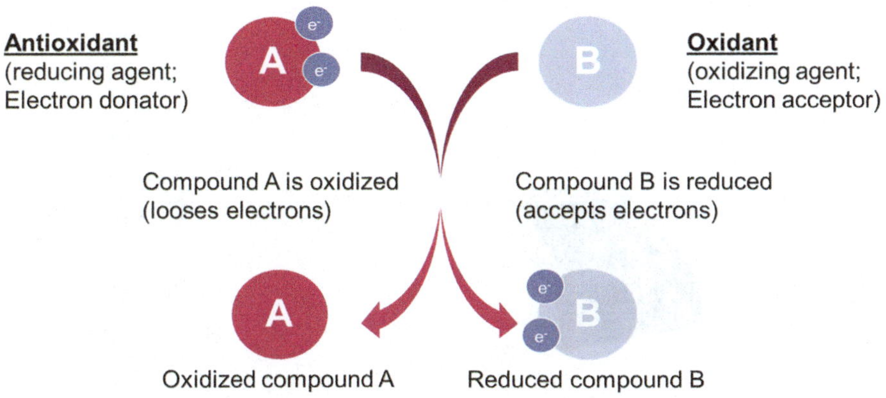

Fig. 18.1 Principle of redox reactions
In redox reactions, the antioxidant (electron donator) is oxidized by the oxidant (electron acceptor) and is losing electrons. In return, the oxidant is reduced by accepting electrons

comprehensive measure of the redox situation (oxidative or reductive stress) in a fluid whereby higher sORP values indicate an imbalance towards the activity of oxidants relative to antioxidants. The MiOXSYS is a novel potentiometric technique based on the electrochemical measurement of electron transfers and provides a measure of the sORP as a snapshot of the redox balance at the time of measurement based on the Nernst Equation [6]:

$$E = E^0 + \frac{RT}{z_e F} \ln\left(\frac{a_{\mathrm{Ox}}}{a_{\mathrm{Red}}}\right)$$

where

E is the cell potential (electromotive force) at the temperature of interest
E^0 is the standard cell potential
R is the universal gas constant: $R = 8.314\ \mathrm{J\ K^{-1}\ mol^{-1}}$
T is the temperature in kelvins
n is the number of electrons transferred in the cell reaction or half-reaction
F is the Faraday constant, the magnitude of charge (in Coulombs) per mole of electrons: $F = 96485.332\ \mathrm{C\ mol^{-1}}$
α is the chemical activity for the relevant species, where α_{Red} is the activity of the reduced form and α_{Ox} is the activity of the oxidized form

ORP measurements have been shown to be a rapid and highly efficient indicator of the quality of a variety of aqueous solutions including water, milk, cheese, blood and blood products, or semen [7–11]. Since the ORP is only correlated with the electron acceptors (oxidants) and electron donators (reductants) and strictly follows the Nernst Equation under equilibrium conditions, it is difficult to interpret the ORP by theoretical reasoning without the addition of a mediator or modifications of the electrode in biological systems. Therefore, a modified Nernst Equation needs to be used [12].

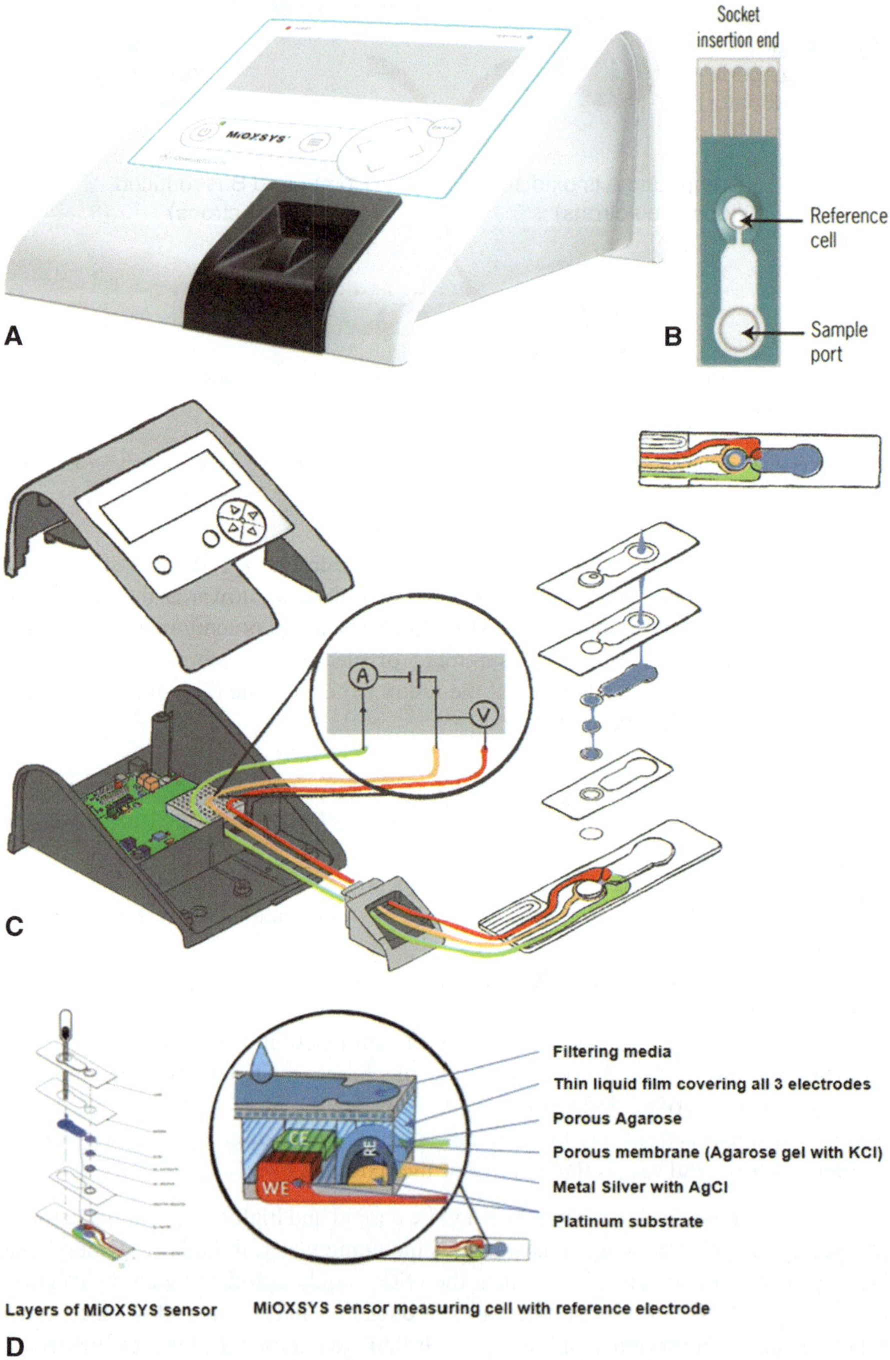

Fig. 18.2 The MiOXSYS system. (**a**) MiOXSYS device, (**b**) MiOXSYS sensor, (**c**) Schematic, drawing of different components of the MiOXSYS system, (**d**) Schematic drawing of the measuring cell with electrodes connected via a salt bridge

In contrast to all other techniques, the MiOXSYS has the major advantage to determine OS and reductive stress (RS), which is as harmful as OS [13]. Furthermore, the MiOXSYS is less time-consuming, cheaper [14, 15], and can also be used for a variety of different specimens including semen, seminal plasma, serum, or culture medium. Hence, this electrochemical testing of the ORP can be used as a point-of-care or bedside diagnostic allowing the physician quick decisions in the patient management and could therefore be very useful in preventive medicine and health care, disease management and the control of relevant therapies, in numerous fields of medicine listed below [16]:

- Testing pharmacological treatments
- Monitoring antioxidant therapies
- Monitoring lifestyle changes
- Sports sector (e.g. during various phases of training or before and after a performance)
- Preventive medicine
- Anti-ageing
- Disease management in several fields
- Diagnosis and treatment of infertility
- Clinical research

As disadvantages, at least for human semen, the age of the sample, high viscosity and repeated centrifugation have been identified as this may alter the result [17, 18]. Clinically, the measurement of ORP using the MiOXSYS or the RedoxSYS™, a similar system to the MiOXSYS based on the same technology, has been used for the detection and monitoring of various diseases and conditions. For the ORP measurement, a 30 μL sample is pipetted on the sample port of the sensor. Then, a low (1 nA) oxidizing current is automatically applied. Following equilibration of the sample on the sensor for about 2 min, the electrochemical potential between the working electrode (silver) and reference electrodes (counter electrode: silver; reference electrode: KCl) in the sensor is measured and the sORP is displayed in milli Volts after about 4 to 5 min. Thereafter, the cORP is measured by applying a linear increasing oxidizing current until a rapid change in the charges between the working and reference electrodes is taking place indicating that no further oxidizable compounds are available. According to the Faraday Law (Q=nFNA), the total charge ("Q") in the electrochemical cell (the ORP sensor) is proportional to the absolute amount of the analyte with "n" being the number of electrons per mole of analyte, "F" Faraday's constant (96487 C mol^{-1}), and NA the number of moles of the analyte. A Coulomb (C) is then equivalent to the amount of analyte oxidized per time unit [19].

A number of studies reported that surgery is associated with a significant increase in OS [20, 21] and increased sORP values [22]. For patients with trauma [23], traumatic brain injury [24], or severe trauma and inflammation [25], the group by Rael and co-workers concluded that the measurement of ORP in plasma could be a useful tool for the assessment of oxidative stress levels in these patients. Whereas the ORP levels in the controls were ranging between −52.0 mV and −34.4 mV, in severely traumatized patients, the plasma ORP levels were with values between −1.1 mV and −22.4 mV significantly higher. Based on measurements of the ratio of reduced to oxidized glutathione (GSH/GSSG), Jones et al. (2000) calculated a redox potential of -137 ± 9 mV in healthy subjects between 25 and 35 years of age [26].

Furthermore, Spanidis et al. (2015) reported that in patients with sepsis the plasma sORP increased significantly by 9.3% and 6.9% in obese patients with sepsis [27]. Similarly, in patients with type-2 diabetes and metabolic syndrome, the sORP together with the conventional oxidative stress parameters GSH decreased significantly while catalase activity increased [28]. Bobe et al. (2017) were able to identify metabolic syndrome with 90% specificity and 80% sensitivity at sORP levels higher than 140 mV [29]. The cORP levels were, however, similar to those found in healthy controls. Likewise, in athletes after strenuous exercise [30], the OS measured as sORP increased significantly by 7%, whereas the cORP decreased by 14.6%, thus indicating a lack of antioxidant capacity. These results were corroborated by decrease of the relevant conventional oxidative stress markers.

In the field of male reproduction, where the MiOXSYS was introduced recently, numerous articles were published on the use and clinical value of sORP to identify patients with seminal OS and poor sperm fertilizing ability [31, 32]. In order to account for variations in sperm counts, the original reading in mV was normalized as mV/10^6 sperm/mL. In a multi-centre study involving 2092 patients in nine countries, Agarwal et al. (2019) established a cut-off value for sORP of 1.34 mV/10^6 sperm/mL [33]. Gill et al. (2021) reported significant differences in the normalized seminal sORP values between healthy normozoospermic volunteers, proven fertile men and infertile patients with and without varicocele [34]. Furthermore, Sallam et al. (2017) and Henkel et al. (2022) were able to predict clinical pregnancy using the normalized sORP in in vitro fertilization (IVF) and intracytoplasmic sperm injection (ICSI), respectively [35, 36]. Hence, indicating sORP as a reliable clinical marker of OS in human semen to predict sperm fertilizing potential. In a meta-analysis, Panner Selvam et al. (2022) indicated the role of seminal OS as measured as sORP in the induction of sperm DNA fragmentation [37], a parameter that has a significant role in the pathogenesis of failed pregnancy and recurrent pregnancy loss [38].

On the other hand, there are also studies that could not find an association between sORP and perioperative OS [39], or sORP expressed as mV/10^6 sperm/mL and sperm function and nitroblue tetrazolium (NBT)-reactivity [40]. Moreover, Joao et al. (2022) reported that the sORP measured in mV showed no difference if sperm were present or not [41]. Moreover, the way in which the MiOXSYS results are expressed by using sperm concentration as denominator appears to have an influence as this approach would overestimate the OS in patients with low sperm concentrations [41, 42]. Furthermore, in a study analysing sperm DNA fragmentation and seminal sORP in the same patients, no correlation between these two parameters could be found [43]. All this indicates that, although the technology is reliable, using the oxidation–reduction potential as diagnostic measure still has its shortcomings and therefore needs to be more investigated.

References

1. Sies H, Cadenas E. Oxidative stress: damage to intact cells and organs. Philos Trans R Soc Lond B Biol Sci. 1985;311(1152):617–31.
2. Jones DP. Redefining oxidative stress. Antioxid Redox Signal. 2006;8(9-10):1865–79.
3. Powers SK, Jackson MJ. Exercise-induced oxidative stress: cellular mechanisms and impact on muscle force production. Physiological reviews. 2008;88(4):1243–76.
4. Dotan Y, Lichtenberg D, Pinchuk I. Lipid peroxidation cannot be used as a universal criterion of oxidative stress. Progress in Lipid Research. 2004;43(3):200–27.
5. Costantini D, Verhulst S. Does high antioxidant capacity indicate low oxidative stress? Functional Ecology. 2009;23(3):506–9.
6. Douglas C, Parekh N, Kahn LG, Henkel R, Agarwal A. A Novel Approach to Improving the Reliability of Manual Semen Analysis: A Paradigm Shift in the Workup of Infertile Men. The world journal of men's health. 2021;39(2):172–85.
7. Noyhouzer T, Kohen R, Mandler D. A new approach for measuring the redox state and redox capacity in milk. Analytical Methods. 2009;1(2):93–9.
8. Caldeo V, McSweeney PLH. Changes in oxidation-reduction potential during the simulated manufacture of different cheese varieties. International Dairy Journal. 2012;25(1):16–20.
9. Copeland A, Lytle DA. Measuring the oxidation–reduction potential of important oxidants in drinking water. Journal AWWA. 2014;106(1):E10–E20.
10. Agarwal A, Sharma R, Roychoudhury S, Du Plessis S, Sabanegh E. MiOXSYS: a novel method of measuring oxidation reduction potential in semen and seminal plasma. Fertility and sterility. 2016;106(3):566–73.e10.
11. Rozemeijer S, Smit B, Elbers PWG, Girbes ARJ, Oudemans-van Straaten HM, de Man AME. Rapid screening of critically ill patients for low plasma vitamin C concentrations using a point-of-care oxidation-reduction potential measurement. Intensive Care Med Exp. 2021;9(1):40.
12. Nagpal S, Chuichulcherm S, Livingston A, Peeva L. Ethanol utilization by sulfate-reducing bacteria: an experimental and modeling study. Biotechnol Bioeng. 2000;70(5):533–43.
13. Castagné V, Lefèvre K, Natero R, Becker DA, Clarke PGH. An optimal redox status for the survival of axotomized ganglion cells in the developing retina. Neuroscience. 1999;93(1): 313–20.
14. Noyhouzer T, Valdinger I, Mandler D. Enhanced potentiometry by metallic nanoparticles. Anal Chem. 2013;85(17):8347–53.
15. Bakker E. Potentiometric Sensors. In: Moretto LM, Kalcher K, editors. Environmental Analysis by Electrochemical Sensors and Biosensors: Fundamentals. New York, NY: Springer New York; 2014. p. 193–238.
16. Palmieri B, Sblendorio V. Oxidative stress tests: overview on reliability and use. Part II. European review for medical and pharmacological sciences. 2007;11(6):383–99.
17. Agarwal A, Bui AD. Chapter 3.4—Oxidation-Reduction Potential Methodology Using the MiOXSYS System. In: Henkel R, Samanta L, Agarwal A, editors. Oxidants, Antioxidants and Impact of the Oxidative Status in Male Reproduction: Academic Press; 2019. p. 217–24.
18. Agarwal A, Roychoudhury S, Bjugstad KB, Cho CL. Oxidation-reduction potential of semen: what is its role in the treatment of male infertility? Ther Adv Urol. 2016;8(5):302–18.
19. Harvey D. Analytical chemistry 2.0. 2010; https://drive.google.com/file/d/1eKRCpmmd8dtr8UA5ztQ4FsVdZY3TUpAB/view. Accessed: 23.05.2024.
20. Misthos P, Katsaragakis S, Theodorou D, Milingos N, Skottis I. The degree of oxidative stress is associated with major adverse effects after lung resection: a prospective study. Eur J Cardiothorac Surg. 2006;29(4):591–5.
21. Koksal GM, Dikmen Y, Erbabacan E, Aydin S, Çakatay U, Sitar ME, et al. Hyperoxic oxidative stress during abdominal surgery: a randomized trial. J Anesth. 2016;30(4):610–9.
22. Schwarz C, Fitschek F, Bar-Or D, Klaus DA, Tudor B, Fleischmann E, et al. Inflammatory response and oxidative stress during liver resection. PLoS One. 2017;12(10):e0185685.
23. Rael LT, Bar-Or R, Aumann RM, Slone DS, Mains CW, Bar-Or D. Oxidation-reduction potential and paraoxonase-arylesterase activity in trauma patients. Biochem Biophys Res Commun. 2007;361(2):561–5.

24. Rael LT, Bar-Or R, Mains CW, Slone DS, Levy AS, Bar-Or D. Plasma oxidation-reduction potential and protein oxidation in traumatic brain injury. J Neurotrauma. 2009;26(8):1203–11.
25. Rael LT, Bar-Or R, Salottolo K, Mains CW, Slone DS, Offner PJ, et al. Injury severity and serum amyloid A correlate with plasma oxidation-reduction potential in multi-trauma patients: a retrospective analysis. Scand J Trauma Resusc Emerg Med. 2009;17:57.
26. Jones DP, Carlson JL, Mody VC, Cai J, Lynn MJ, Sternberg P. Redox state of glutathione in human plasma. Free Radic Biol Med. 2000;28(4):625–35.
27. Spanidis Y, Goutzourelas N, Stagos D, Kolyva AS, Gogos CA, Bar-Or D, et al. Assessment of Oxidative Stress in Septic and Obese Patients Using Markers of Oxidation-reduction Potential. In Vivo. 2015;29(5):595–600.
28. Spanidis Y, Mpesios A, Stagos D, Goutzourelas N, Bar-Or D, Karapetsa M, et al. Assessment of the redox status in patients with metabolic syndrome and type 2 diabetes reveals great variations. Exp Ther Med. 2016;11(3):895–903.
29. Bobe G, Cobb TJ, Leonard SW, Aponso S, Bahro CB, Koley D, et al. Increased static and decreased capacity oxidation-reduction potentials in plasma are predictive of metabolic syndrome. Redox Biol. 2017;12:121–8.
30. Stagos D, Goutzourelas N, Bar-Or D, Ntontou AM, Bella E, Becker AT, et al. Application of a new oxidation-reduction potential assessment method in strenuous exercise-induced oxidative stress. Redox report: communications in free radical research. 2015;20(4):154–62.
31. Agarwal A, Roychoudhury S, Sharma R, Gupta S, Majzoub A, Sabanegh E. Diagnostic application of oxidation-reduction potential assay for measurement of oxidative stress: clinical utility in male factor infertility. Reprod Biomed Online. 2017;34(1):48–57.
32. Takeshima T, Usui K, Mori K, Asai T, Yasuda K, Kuroda S, et al. Oxidative stress and male infertility. Reprod Med Biol. 2021;20(1):41–52.
33. Agarwal A, Panner Selvam MK, Arafa M, Okada H, Homa S, Killeen A, et al. Multi-center evaluation of oxidation-reduction potential by the MiOXSYS in males with abnormal semen. Asian J Androl. 2019;21(6):565–9.
34. Gill K, Kups M, Harasny P, Machalowski T, Grabowska M, Lukaszuk M, et al. The Negative Impact of Varicocele on Basic Semen Parameters, Sperm Nuclear DNA Dispersion and Oxidation-Reduction Potential in Semen. Int J Environ Res Public Health. 2021;18(11).
35. Sallam NH, Hegab M, Mohamed F, El-Kaffash D. Effect of oxidative stress in semen, follicular fluid and embryo culture medium on the outcome of assisted reproduction. Al-Azhar International Medical Journal. 2021;2(7):59–65.
36. Henkel R, Morris A, Vogiatzi P, Saleh R, Sallam H, Boitrelle F, et al. Predictive value of seminal oxidation-reduction potential analysis for reproductive outcomes of ICSI. Reprod Biomed Online. 2022;45(5):1007–20.
37. Panner Selvam MK, Baskaran S, O'Connell S, Almajed W, Hellstrom WJG, Sikka SC. Association between Seminal Oxidation-Reduction Potential and Sperm DNA Fragmentation—A Meta-Analysis. Antioxidants. 2022;11(8):1563.
38. McQueen DB, Zhang J, Robins JC. Sperm DNA fragmentation and recurrent pregnancy loss: a systematic review and meta-analysis. Fertility and sterility. 2019;112(1):54–60.e3.
39. Reiterer C, Fleischmann E, Taschner A, Adamowitsch N, von Sonnenburg MF, Graf A, et al. Perioperative supplemental oxygen and oxidative stress in patients undergoing moderate—to high-risk major abdominal surgery—A subanalysis of randomized clinical trial. J Clin Anesth. 2022;77:110614.
40. Castleton P, Gyawali P, Mathews N, Mutuku SM, Sharkey DJ, McPherson NO. MiOXSYS(®) and OxiSperm(®) II assays appear to provide no clinical utility for determining oxidative stress in human sperm-results from repeated semen collections. Andrology. 2023;11(8):1566–78.
41. Joao F, Duval C, Bélanger MC, Lamoureux J, Xiao CW, Ates S, et al. Reassessing the interpretation of oxidation-reduction potential in male infertility. Reprod Fertil. 2022;3(2): 67–76.
42. Aitken RJ. Oxidative stress and reproductive function. Reproduction. 2022;164(6):E5–E8.
43. Humaidan P, Haahr T, Povlsen BB, Kofod L, Laursen RJ, Alsbjerg B, et al. The combined effect of lifestyle intervention and antioxidant therapy on sperm DNA fragmentation and seminal oxidative stress in IVF patients: a pilot study. Int Braz J Urol. 2022;48(1):131–56.

Chapter 19
TUNEL Assay and SCSA

R. Henkel

Abstract In their recent 2021 manual for the examination of human semen, the World Health Organization recommends the testing for sperm DNA fragmentation (SDF) in the section "Extended Examination." SDF is an endpoint parameter for which one has to distinguish between test systems that measure real DNA damage, that is, actual strand breaks of the DNA, such as 8-hydroxydeoxyguanosine (8-OHdG), single cell gel electrophoresis (COMET), the TdT (terminal deoxynucleotidyl transferase)-mediated dUDP nick-end labelling (TUNEL), and test systems detecting potential DNA damage such as the sperm chromatin dispersion test (SCD) or the sperm chromatin structure assay (SCSA®). The TUNEL assay and the SCSA are the two most commonly used test systems to determine SDF. The principle of the TUNEL assay is based on the binding of labelled the DNA precursor deoxyuridine triphosphate (dUTP) to 3'-OH ends of DNA, thereby labelling single- and double-strand DNA breaks. This test can be done using fluorescence microscopy or flow cytometry. Relevant cut-off values to identify infertile men have been established. The SCSA is a flow cytometric test based on the metachromatic properties of acridine orange where an orange-red fluorescence indicates single-stranded DNA and green fluorescence double-stranded DNA. While different protocols are available for the TUNEL assay, the SCSA is the most standardized test for SDF. Yet, the TUNEL assay is reportedly more predictive of male infertility than the SCSA.

Keywords Sperm DNA fragmentation · Semen · TUNEL assay · Sperm chromatin structure assay (SCSA) · Strand DNA breaks · Flow cytometry

R. Henkel (✉)
LogixX Pharma Ltd., Theale, Berkshire, UK

Department of Medical Bioscience, University of the Western Cape, Bellville, South Africa

Department of Metabolism, Digestion & Reproduction, Imperial College, London, UK
e-mail: Ralf.henkel@logixxpharma.com; rhenkel@uwc.ac.za; r.henkel@imperial.ac.uk

S. Andreescu et al. (eds.), *Biomarkers of Oxidative Stress*,
https://doi.org/10.1007/978-3-031-60738-7_19

19.1 Introduction

During the past years, oxidative stress (OS) has been recognized as a major cause of male infertility and sperm DNA fragmentation which in turn has repeatedly been reported as a significant parameter in the diagnosis of male infertility [1–3] is recommended as a diagnostic test by the World Health Organization in its latest laboratory guidelines (2021) [4]. Oxidative stress in biological systems can be measured with various techniques, directly or indirectly. Among the indirect tests, there are tests that measure endpoints of the damage caused by OS to lipids by lipid peroxidation such as malondialdehyde and damage to the DNA of cells. For the latter, one has to distinguish between test systems that measure real DNA damage, that is, strand break of the DNA, (1) 8-hydroxydeoxyguanosine (8-OHdG), (2) TdT (terminal deoxynucleotidyl transferase)-mediated dUDP nick-end labelling (TUNEL) [5], (3) in situ nick translation [5], and (4) single-cell gel electrophoresis (COMET) under neutral conditions [6], and potential DNA damage (1) sperm chromatin structure assay (SCSA®), (2) acridine orange test [7], (3) sperm chromatin dispersion test (SCD) with its improved version (Halosperm® assay) [8], (4) DNA breakage detection FISH (DBD-FISH) [9], (5) chromomycin A3 (CMA3) [10], and (6) the COMET assay under denaturing conditions [11]. Such tests require an initial denaturation step in the protocol. These tests must not be confused with test systems that assay nuclear maturity and quality of chromatin condensation, such as the aniline blue stain [12], or the toluidine blue stain [13].

In this chapter, the two most commonly used assays to test sperm DNA fragmentation (SDF) [14, 15], namely the TUNEL assay and the SCSA with their features and respective advantages and disadvantages will be described and discussed.

19.2 TUNEL Assay

The TUNEL assay was developed in 1993 by Gorczyca et al. (1993) [5]. The principle of this test is based on the binding of the fluorescence- or biotin-labelled DNA precursor deoxyuridine triphosphate (dUTP) to 3'-OH ends of the DNA and thereby labels single- and double-strand DNA breaks in an enzymatically catalysed reaction using the template-independent terminal deoxynucleotidyl transferase (TdT). The evaluation can then either be performed by analysing the fluorescence using a fluorescence microscope [16] or a flow cytometer [17]. In case of the biotin-labelling, the evaluation is done with a light microscope. However, the fluorescence-based techniques are most commonly used with both techniques correlating very well ($r = 0.720$; $r = 0.783$); results obtained by flow cytometry are higher than those detected with fluorescence microscopy [18, 19]. Considering its advantages, nowadays the flow cytometric assay is used in most laboratories.

For the flow cytometric test, a number of commercial kits comprising a reaction buffer, TdT, fluorescein dithiocyanate (FITC)-labelled dUTP, and propidium iodide/

RNAse (PI/RNAse) to label dead cells are available. Approximately 2.5×10^6 sperm are fixed for 30 min at 4 °C in 7% paraformaldehyde, centrifuged, and permeabilized in ice-cold 70% ethanol. Then, the samples are stained with the reaction mixture, incubated in PI/RNAse and analysed in a flow cytometer using two lasers at a wavelength of 488 nm (FL-1 channel; green fluorescence; FITC) and 640 nm (FL-2 channel; red fluorescence; PI), respectively. The percentage of live, TUNEL-positive, DNA-fragmented sperm, is then calculated by the relevant software after appropriate gating. According to Mitchell et al. (2011), incubation of the sample in 2 mM dithiothreitol (DTT) significantly increases the sensitivity by loosening the chromatin structure by reducing the disulphide bridges between neighbouring protamine molecules and thereby rendering the DNA more accessible to the probe [20].

As compared with other methods to detect DNA damage, the TUNEL assay is more sophisticated, more expensive, and more time-consuming. However, good quality control parameters like low intra-observer and inter-observer variability [21], and inter-laboratory variation [22], have been demonstrated. Its relevance in respect of sperm function as well as of fertilization and pregnancy has repeatedly been proven for intrauterine insemination (IUI), in vitro fertilization (IVF), and intracytoplasmic sperm injection (ICSI) [16, 23–25]. As determined with the TUNEL assay, couples with recurrent pregnancy loss have been shown to have significantly higher SDF values than the fertile controls [26]. In a meta-analysis analysing a total of 16 studies with 2969 couples, Robinson et al. (2012) found that the TUNEL assay is most predictive for miscarriage (RR = 3.94; 95% CI: 2.45, 6.32; P = 0.00001), followed by the SCSA (seven studies: RR = 1.47; 95% CI: 1.04, 2.09; P = 0.03) [27]. Similar results were obtained by Yifu et al. (2020) [28]. In another meta-analysis, Osman et al. (2015) showed that the TUNEL assay was most predictive for live birth after IVF and ICSI (nine studies (four IVF, five ICSI): RR = 1.17; 95% CI: 1.07, 1.28; $p = 0.0005$) [29]. According to an analysis of the IVF/ICSI outcomes in 30 studies, Cissen et al. (2016) demonstrated a fair predictive value of the TUNEL assay [30]. Although SDF is now generally accepted as a major cause of male infertility, and testing for it is now recommended by the WHO (2021) [4], the European Association of Urology (2021) [31], the ESHRE (2023), and Hervas et al. (2022) by analysing 1903 ICSI cycles with 6340 donated oocytes [32] indicate that elevated SDF does not negatively affect the success of cycles with donated oocytes.

19.3 SCSA

The idea that the structure of sperm chromatin is reflecting sperm fertilizing ability was first proposed in 1980 by Evenson et al. [33] and was then introduced as SCSA by Evenson and Jost in 1994 [34]. This assay is based on two principles; first, the stability and acid-resistance of double-stranded DNA, while fragmentated DNA is rather loose and can be denaturated. Second, the metachromatic properties of the intercalating dye acridine orange which fluoresces green if bound to double-stranded DNA, and orange-red if bound to single-stranded DNA [35], and the principle of the acridine orange test [7]. Whereas the original acridine orange test is fluorescence microscope-based and shows high inter- and intra-assay variability, as well as technical problems, the SCSA is using flow cytometry, a widely automated methodology that allows the evaluation of not only 100–200 sperm but of 5000–10,000 sperm in a very short period of time. This procedure, together with its high standardization, renders this assay a very robust, stabile, accurate, and repeatable measure of DNA integrity, proportionate to sperm DNA strand breaks [36–39]. As for the TUNEL assay, freezing of the samples does not affect the result of the SCSA [40].

For this assay, a semen sample is diluted to a sperm concentration of about 1 to 2×10^6 sperm/mL with buffer. Subsequently, the sample is treated with an acid (pH 1.2) 0.1% Triton X-100 solution for 30 s before a solution of purified acridine orange (6 mg/L) in phosphate-citrate buffer is added and flow cytometric analysis is performed. The level of sperm DNA damage is expressed as (1) DNA fragmentation index (DFI) (the percentage of red fluorescing, DNA damaged, sperm), and (2) percentage of highly DNA stainable (HDS) cells, that is the percentage of high DNA stainable sperm. The HDS is associated with the amount of histones not replaced by protamines during spermatogenesis, thus indicating immature sperm [41]. While the DFI has repeatedly been reported of being predictive for in vitro fertilization (IVF) and intracytoplasmic sperm injection (ICSI) [42, 43], the combination of DFI and HDS has been shown to be predictive for the outcome of intrauterine insemination (IUI) [43]. On the other hand, a recent large study by Zhu et al. (2022) evaluating 1500 IUI cycles found no statistical difference between the normal and abnormal DFI groups for the HDS, while the DFI was positively correlated with the HDS, thus indicating that human sperm functionality is multifactorial and the DFI may have only limited predictive value for clinical reproductive outcomes after IUI [44]. This result is corroborated by Luo et al. (2022) reporting that male age is positively associated with DFI but not with HDS [45]. Although both, DFI and HDS, are significantly negatively correlated with most sperm parameters, they are not affecting the reproductive outcome after IUI. Donatti et al. (2023) also found a positive association between male age and DFI in IVF cycles [46]. According to Jerre et al. (2019), an HDS >15% indicates an increased risk of miscarriage in ICSI, but not IVF treatments [47]. On the other hand, Mohammadi et al. (2020) reported that HDS is not a reliable parameter to indicate sperm nuclear compaction and maturity because it correlates poorly with tests that specifically evaluate sperm nuclear compaction such as aniline blue or chromomycin A3 [48].

With regard to the accuracy of the prediction of pregnancy after IVF/ICSI, according to a meta-analysis of 21 studies by Cissen et al. (2016), the predictive power of the SCSA is poor [30]. Simon et al. (2017) reported a similar result confirming a stronger odds ratio for the TUNEL assay (OR: 1.85, 95% CI: 1.52–2.26; $p < 0.0001$) than for the SCSA (OR: 0.88, 95% CI: 0.80–0.96; $p = 0.0041$) after conducting a meta-analysis including 18 studies using the TUNEL assay and 31 using the SCSA [49]. In a prospective cohort study with a total of 234 IVF and ICSI cycles, Green et al. (2020) reported no differences for fertilization (86.3 vs. 84.2%), blastulation (49.5 vs. 48.8%), and euploidy (55.7 vs. 52.1%) between men with a DFI ≤15% and those who had a DFI >15%, respectively [50].

19.4 Advantages and Disadvantages of the TUNEL Assay and SCSA

With the increasing understanding and recognition of the importance of sperm DNA fragmentation in the clinical examination of male infertility, there is a demand for validated practical and clinical guidelines as a global survey revealed that almost 75% of clinicians request SDF testing in cases of idiopathic infertility and 39% as a routine test in the work-up of recurrent pregnancy loss [51]. This testing is especially important with the increasing number of couples that are treated with ICSI as with this technique all natural barriers for sperm entry into the oocyte are bypassed and thereby DNA damaged sperm can fertilize the oocyte. The oocyte has the potential to repair some of the sperm-derived DNA damage. However, if this repair fails, SDF can be fixed in the germ line causing mutations leading to male infertility, early childhood cancer, or genetic imprinting defects [52]. Therefore, clinicians should be provided with clinical indications as well as the advantages and disadvantages of the assays. Table 19.1 summarizes the advantages and disadvantages of the TUNEL assay and SCSA.

Table 19.1 Comparison of advantages and disadvantages of the flow cytometric TUNEL assay and the SCSA

Assay	Advantages	Disadvantages
TUNEL	Commercial assay available Can be performed in fresh or frozen samples Highly accurate, sensitive, specific and reliable Minimal inter-observer variability Direct objective Correlates with fertility and sperm functions Clinically significant	No established standardized protocol Variable clinical thresholds in literature Thresholds not standardized Not specific to oxidative damage
SCSA	Established clinical thresholds Can be performed in fresh or frozen samples Highly comparable data Most published studies reproducible Clinically significant Highly sensitive, specific, and repeatable Correlates with sperm functions and fertility	No commercial assay available Indirect assay only detects ssDNA breaks Involves acid denaturation Lower predictive power Not available in commercial kits Small variations in lab conditions affect results Calculations involve qualitative decisions Very few labs conduct this assay

References

1. Sun TC, Zhang Y, Li HT, Liu XM, Yi DX, Tian L, et al. Sperm DNA fragmentation index, as measured by sperm chromatin dispersion, might not predict assisted reproductive outcome. Taiwan J Obstet Gynecol. 2018;57(4):493–8.
2. Simon L, Emery B, Carrell DT. Sperm DNA Fragmentation: Consequences for Reproduction. Advances in experimental medicine and biology. 2019;1166:87–105.
3. Agarwal A, Farkouh A, Parekh N, Zini A, Arafa M, Kandil H, et al. Sperm DNA Fragmentation: A Critical Assessment of Clinical Practice Guidelines. The world journal of men's health. 2022;40(1):30–7.
4. World Health Organization. WHO Laboratory Manual for the Examination and Processing of Human Semen. Sixth Edition Geneva, Switzerland: Licence: CC BY-NC-SA 3.0 IGO; 2021.
5. Gorczyca W, Gong J, Darzynkiewicz Z. Detection of DNA strand breaks in individual apoptotic cells by the in situ terminal deoxynucleotidyl transferase and nick translation assays. Cancer research. 1993;53(8):1945–51.
6. Ostling O, Johanson KJ. Microelectrophoretic study of radiation-induced DNA damages in individual mammalian cells. Biochem Biophys Res Commun. 1984;123(1):291–8.
7. Tejada RI, Mitchell JC, Norman A, Marik JJ, Friedman S. A test for the practical evaluation of male fertility by acridine orange (AO) fluorescence. Fertility and sterility. 1984;42(1):87–91.
8. Fernández JL, Muriel L, Goyanes V, Segrelles E, Gosálvez J, Enciso M, et al. Simple determination of human sperm DNA fragmentation with an improved sperm chromatin dispersion test. Fertility and sterility. 2005;84(4):833–42.
9. Fernández JL, Goyanes VJ, Ramiro-Díaz J, Gosálvez J. Application of FISH for in situ detection and quantification of DNA breakage. Cytogenet Cell Genet. 1998;82(3-4):251–6.
10. Manicardi GC, Bianchi PG, Pantano S, Azzoni P, Bizzaro D, Bianchi U, et al. Presence of endogenous nicks in DNA of ejaculated human spermatozoa and its relationship to chromomycin A3 accessibility. Biol Reprod. 1995;52(4):864–7.
11. Singh NP, Danner DB, Tice RR, McCoy MT, Collins GD, Schneider EL. Abundant alkali-sensitive sites in DNA of human and mouse sperm. Experimental Cell Research. 1989;184(2): 461–70.
12. Terquem A, Dadoune JP. Aniline Blue Staining of Human Spermatozoon Chromatin. Evaluation of Nuclear Maturation. In: André J, editor. The Sperm Cell: Fertilizing Power, Surface Properties, Motility, Nucleus and Acrosome, Evolutionary Aspects Proceedings of the Fourth International Symposium on Spermatology, Seillac, France, 27 June–1 July 1982. Dordrecht: Springer Netherlands; 1982. p. 249–52.
13. Erenpreisa J, Freivalds T, Roach H, Alston R. Apoptotic cell nuclei favour aggregation and fluorescence quenching of DNA dyes. Histochem Cell Biol. 1997;108(1):67–75.
14. Majzoub A, Agarwal A, Cho CL, Esteves SC. Sperm DNA fragmentation testing: a cross sectional survey on current practices of fertility specialists. Transl Androl Urol. 2017;6(Suppl 4):S710–s9.
15. Baskaran S, Agarwal A, Panner Selvam MK, Finelli R, Robert KA, Iovine C, et al. Tracking research trends and hotspots in sperm DNA fragmentation testing for the evaluation of male infertility: a scientometric analysis. Reprod Biol Endocrinol. 2019;17(1):110.
16. Henkel R, Hajimohammad M, Stalf T, Hoogendijk C, Mehnert C, Menkveld R, et al. Influence of deoxyribonucleic acid damage on fertilization and pregnancy. Fertility and sterility. 2004;81 (4):965–72.
17. Sharma R, Iovine C, Agarwal A, Henkel R. TUNEL assay—Standardized method for testing sperm DNA fragmentation. Andrologia. 2021;53(2):e13738.
18. Muratori M, Piomboni P, Baldi E, Filimberti E, Pecchioli P, Moretti E, et al. Functional and ultrastructural features of DNA-fragmented human sperm. J Androl. 2000;21(6):903–12.
19. Domínguez-Fandos D, Camejo MI, Ballescà JL, Oliva R. Human sperm DNA fragmentation: correlation of TUNEL results as assessed by flow cytometry and optical microscopy. Cytometry A. 2007;71(12):1011–8.

20. Mitchell LA, De Iuliis GN, Aitken RJ. The TUNEL assay consistently underestimates DNA damage in human spermatozoa and is influenced by DNA compaction and cell vitality: development of an improved methodology. Int J Androl. 2011;34(1):2–13.
21. Sharma R, Ahmad G, Esteves SC, Agarwal A. Terminal deoxynucleotidyl transferase dUTP nick end labeling (TUNEL) assay using bench top flow cytometer for evaluation of sperm DNA fragmentation in fertility laboratories: protocol, reference values, and quality control. J Assist Reprod Genet. 2016;33(2):291–300.
22. Ribeiro S, Sharma R, Gupta S, Cakar Z, De Geyter C, Agarwal A. Inter- and intra-laboratory standardization of TUNEL assay for assessment of sperm DNA fragmentation. Andrology. 2017;5(3):477–85.
23. Bakos HW, Thompson JG, Feil D, Lane M. Sperm DNA damage is associated with assisted reproductive technology pregnancy. Int J Androl. 2008;31(5):518–26.
24. Siddhartha N, Reddy NS, Pandurangi M, Muthusamy T, Vembu R, Kasinathan K. The Effect of Sperm DNA Fragmentation Index on the Outcome of Intrauterine Insemination and Intracytoplasmic Sperm Injection. J Hum Reprod Sci. 2019;12(3):189–98.
25. Sugihara A, Punjabi U, Roelant E, De Neubourg D. Is There a Relationship between Sperm DNA Fragmentation and Intra-Uterine Insemination Outcome in Couples with Unexplained or Mild Male Infertility? Results from the ID-Trial. Life (Basel). 2022;13(1).
26. Busnelli A, Garolla A, Di Credico E, D'Ippolito S, Merola AM, Milardi D, et al. Sperm DNA fragmentation and idiopathic recurrent pregnancy loss: Results from a multicenter case-control study. Andrology. 2023;11(8):1673–81.
27. Robinson L, Gallos ID, Conner SJ, Rajkhowa M, Miller D, Lewis S, et al. The effect of sperm DNA fragmentation on miscarriage rates: a systematic review and meta-analysis. Human reproduction (Oxford, England). 2012;27(10):2908–17.
28. Yifu P, Lei Y, Shaoming L, Yujin G, Xingwang Z. Sperm DNA fragmentation index with unexplained recurrent spontaneous abortion: A systematic review and meta-analysis. J Gynecol Obstet Hum Reprod. 2020:101740.
29. Osman A, Alsomait H, Seshadri S, El-Toukhy T, Khalaf Y. The effect of sperm DNA fragmentation on live birth rate after IVF or ICSI: a systematic review and meta-analysis. Reprod Biomed Online. 2015;30(2):120–7.
30. Cissen M, Wely MV, Scholten I, Mansell S, Bruin JP, Mol BW, et al. Measuring Sperm DNA Fragmentation and Clinical Outcomes of Medically Assisted Reproduction: A Systematic Review and Meta-Analysis. PLoS One. 2016;11(11):e0165125.
31. Minhas S BC, Boeri L, Capogrosso P, Carvalho J, Cilesiz NC, Cocci A, Corona G, Dimitropoulos K, Gül M, Hatzichristodoulou G, Jones TH, Kadioglu A, Martínez Salamanca JI, Milenkovic U, Modgil V, Russo GI, Serefoglu EC, Tharakan T, Verze P, Salonia A. EAU Working Group on Male Sexual and Reproductive Health; European Association of Urology Guidelines on Male Sexual and Reproductive Health: 2021 Update on Male Infertility. Eur Urol 80: 603–6202021.
32. Hervás I, Pacheco A, Gil Julia M, Rivera-Egea R, Navarro-Gomezlechon A, Garrido N. Sperm deoxyribonucleic acid fragmentation (by terminal deoxynucleotidyl transferase biotin dUTP nick end labeling assay) does not impair reproductive success measured as cumulative live birth rates per donor metaphase II oocyte used. Fertility and sterility. 2022;118(1):79–89.
33. Evenson DP, Darzynkiewicz Z, Melamed MR. Relation of mammalian sperm chromatin heterogeneity to fertility. Science. 1980;210(4474):1131–3.
34. Evenson D, Jost L. Sperm chromatin structure assay: DNA denaturability. Methods in cell biology. 42: Elsevier; 1994. p. 159–76.
35. Bradley DF, Wolf MK. Aggregation of dyes bound to polyanions. Proc Natl Acad Sci U S A. 1959;45(7):944–52.
36. Evenson DP, Jost LK, Marshall D, Zinaman MJ, Clegg E, Purvis K, et al. Utility of the sperm chromatin structure assay as a diagnostic and prognostic tool in the human fertility clinic. Human reproduction (Oxford, England). 1999;14(4):1039–49.

37. Evenson DP, Larson KL, Jost LK. Sperm chromatin structure assay: its clinical use for detecting sperm DNA fragmentation in male infertility and comparisons with other techniques. J Androl. 2002;23(1):25–43.
38. Virro MR, Larson-Cook KL, Evenson DP. Sperm chromatin structure assay (SCSA) parameters are related to fertilization, blastocyst development, and ongoing pregnancy in in vitro fertilization and intracytoplasmic sperm injection cycles. Fertility and sterility. 2004;81(5):1289–95.
39. Dutta S, Henkel R, Agarwal A. Comparative analysis of tests used to assess sperm chromatin integrity and DNA fragmentation. Andrologia. 2021;53(2):e13718.
40. Sailer BL, Jost LK, Evenson DP. Mammalian sperm DNA susceptibility to in situ denaturation associated with the presence of DNA strand breaks as measured by the terminal deoxynucleotidyl transferase assay. J Androl. 1995;16(1):80–7.
41. Evenson DP. Sperm chromatin structure assay (SCSA®). Methods Mol Biol. 2013;927:147–64.
42. Larson KL, DeJonge CJ, Barnes AM, Jost LK, Evenson DP. Sperm chromatin structure assay parameters as predictors of failed pregnancy following assisted reproductive techniques. Human reproduction (Oxford, England). 2000;15(8):1717–22.
43. Bungum M, Humaidan P, Spano M, Jepson K, Bungum L, Giwercman A. The predictive value of sperm chromatin structure assay (SCSA) parameters for the outcome of intrauterine insemination, IVF and ICSI. Human reproduction (Oxford, England). 2004;19(6):1401–8.
44. Zhu C, Zhang S, Chen F, She H, Ju Y, Wen X, et al. Correlations between elevated basal sperm DNA fragmentation and the clinical outcomes in women undergoing IUI. Front Endocrinol (Lausanne). 2022;13:987812.
45. Luo Y, Wu S, Zhang M, Zhou H, Yuan J, Yang Y, et al. Sperm DNA integrity is critically impacted by male age but does not influence outcomes of artificial insemination by husband in the Chinese infertile couples. Aging (Albany NY). 2022;14(10):4326–35.
46. Donatti LM, Martello CL, Andrade GM, Oliveira NP, Frantz N. Advanced Paternal Age Affects the Sperm DNA Fragmentation Index and May Lead to Lower Good-quality Blastocysts. Reprod Sci. 2023;30(8):2489–94.
47. Jerre E, Bungum M, Evenson D, Giwercman A. Sperm chromatin structure assay high DNA stainability sperm as a marker of early miscarriage after intracytoplasmic sperm injection. Fertility and sterility. 2019;112(1):46–53.e2.
48. Mohammadi Z, Tavalaee M, Gharagozloo P, Drevet JR, Nasr-Esfahani MH. Could high DNA stainability (HDS) be a valuable indicator of sperm nuclear integrity? Basic Clin Androl. 2020;30:12.
49. Simon L, Emery BR, Carrell DT. Review: Diagnosis and impact of sperm DNA alterations in assisted reproduction. Best Practice & Research Clinical Obstetrics & Gynaecology. 2017;44: 38–56.
50. Green KA, Patounakis G, Dougherty MP, Werner MD, Scott RT, Jr., Franasiak JM. Sperm DNA fragmentation on the day of fertilization is not associated with embryologic or clinical outcomes after IVF/ICSI. J Assist Reprod Genet. 2020;37(1):71–6.
51. Agarwal A, Farkouh A, Saleh R, Abdel-Meguid Hamoda TA, Harraz AM, Kavoussi P, et al. Controversy and Consensus on Indications for Sperm DNA Fragmentation Testing in Male Infertility: A Global Survey, Current Guidelines, and Expert Recommendations. The world journal of men's health. 2023;41(3):575–602.
52. Henningsen AA, Gissler M, Rasmussen S, Opdahl S, Wennerholm UB, Spangmose AL, et al. Imprinting disorders in children born after ART: a Nordic study from the CoNARTaS group. Human Reproduction. 2020;35(5):1178–84.

Chapter 20
Proteomics

A. K. Moharana, S. Baskaran, S. C. Sikka, and M. K. Panner Selvam

Abstract Oxidative stress (OS) induces differential expression of proteins and nucleic acids in several disease states. The proteomics approach is used to analyze the involvement of OS-associated proteins in several body fluids, including seminal plasma. In addition, advanced redox mass-proteomic approaches are used to identify the redox-induced modifications of proteins, peptides, and amino acids. Seminal OS affects the functional ability of spermatozoa, including its motility, hyperactivation, capacitation, acrosome reaction, and zona pellucida penetration, all of which are required to fertilize the ovum. This chapter presents an overview of novel proteomics methods and bioinformatics used to analyze proteomics data under conditions of OS. Furthermore, we provide a summary of recent findings from proteomics investigations conducted in males with significantly high seminal OS compared to males with normal levels of free radicals.

Keywords Oxidative stress · Proteomics · Seminal plasma · Spermatozoa · Biomarkers.

20.1 Introduction

Cellular oxidative metabolism is the principal energy source in aerobic organisms, requiring oxygen as a key molecule. However, increased generation of reactive oxygen species (ROS) results in oxidative stress (OS) when their production exceeds the limit of cellular antioxidant defense systems, causing severe cellular damage [1]. ROS, a family of free radicals (FR) such as H_2O_2, $O_2^{\bullet-}$, and $OH^{\bullet}$ in which at

A. K. Moharana
Department of Urology, Tulane University School of Medicine, New Orleans, LA, USA

Redox Biology & Proteomics Laboratory, Department of Zoology, School of Life Sciences, Ravenshaw University, Cuttack, Odisha, India

S. Baskaran · S. C. Sikka · M. K. Panner Selvam (✉)
Department of Urology, Tulane University School of Medicine, New Orleans, LA, USA
e-mail: mpannerselvam@tulane.edu

S. Andreescu et al. (eds.), *Biomarkers of Oxidative Stress*,
https://doi.org/10.1007/978-3-031-60738-7_20

least one electron is missing from their valence shell [2], frequently utilize other molecules to stabilize their structure. Along with the cell's mitochondrial electron transport chain system, several external variables, including alcohol, tobacco smoke, and air pollutants also support ROS generation resulting in OS. Antioxidant enzymes (e.g., CAT, SOD, and glutathione oxidase) and non-enzymes (e.g., vitamins A, E, and zinc) are required to eliminate the excess of FR and to maintain an optimal level of ROS in the cell [3]. However, a minimal level of cellular ROS is vital to the proper function of biological systems that maintain body homeostasis [2].

Molecules like DNA, proteins, and lipids are prone to excessive and unmanageable free radical synthesis in conditions of OS, which compromises cellular health and leads to the onset of disease [4]. Several recent studies have demonstrated that OS greatly affects the normal aging process and a wide range of human diseases, such as cancer, multiple sclerosis, rheumatoid arthritis, neurological diseases, and cardiovascular disorders [5–13]. Therefore, it is crucial to use biomarkers of OS to evaluate the state of illness and to determine the benefits of antioxidants on human health [5].

According to several studies, augmented levels of OS contribute to the prevalence of male subfertility [14–18], which affects 30–80% of couples worldwide [18]. In about 50% of those infertility cases, male factor is the major cause, mainly due to congenital, acquired, or idiopathic factors that impair spermatogenesis [19]. Spermatozoa and seminal plasma are rich in proteins and, therefore, most suitable for proteomics [14]. Since mature spermatozoa are both transcriptionally and translationally inactive, their biological roles are principally achieved by proteins. In addition, seminal OS alters specific protein pathways, and aberrant protein expression may contribute to the impairment of sperm function resulting in male infertility.

Proteomics is the scientific study of how various proteins express themselves in a cell or tissue, temporally and spatially [2]. Approximately thousands of proteins can be identified by the most common bottom-up or shotgun proteomics techniques. The highest number of sperm and seminal plasma proteins are profiled via high-throughput platforms, including liquid chromatography combined with tandem mass spectrometry (LC-MS-MS) and matrix-assisted laser desorption/ionization-time-of-flight MS (MALDI-TOF-MS) [20]. Using the technologies and methodologies of proteomics, researchers can identify important alterations in proteins within the spermatozoa and seminal plasma of infertile men. Most proteomics investigations on male infertility compare, identify, and quantify the proteins expressed differently in spermatozoa or semen from fertile and infertile males [14, 21]. Both traditional proteomics methods such as gel electrophoresis and novel methods such as MALDI-TOF-MS and LC-MS/MS are frequently used to detect proteins.

In this chapter, we discuss the various conventional and advanced proteomics methods and approaches to evaluate the role of OS in males with different types of infertility. Further, we also highlighted the use of various advanced redox mass-proteomic approaches to identify the redox-induced modifications of proteins, peptides, and amino acids.

20.2 Proteomics Techniques

The term "omics" denotes the study of a specific subject. Accordingly, proteomics is the complete profiling of proteins within a cell or tissue [22]. Interactions between proteins and post-translational modifications are responsible for the most varied and complex functions of proteins within cells. The protein composition differs significantly between two distinct cells and within a single cell, depending on the cell's time, space, and disease state. Researchers have examined the protein components relevant to infertility by analyzing tissue and physiological fluids from the reproductive tracts [2]. Additionally, proteomics research may facilitate the discovery of putative "fertility proteins" that could be very helpful in identifying the cause of male infertility [2, 20]. Currently, various techniques are used to separate and identify proteins based on the structural needs of an experiment influenced by physical variables, such as the molecular weight, acidity, charge, and amino acid concentration of proteins. Frequently utilized techniques include systematic two-dimensional gel electrophoresis (2D-GE) for protein segregation, mass spectrometry (MS) for protein identification, and western blot assays for protein validation.

20.2.1 Two-Dimensional Gel Electrophoresis and Difference Gel Electrophoresis

Combined with both isoelectric focusing and sodium dodecyl-sulfate polyacrylamide gel electrophoresis (SDS-PAGE), 2D-GE separates a complex mixture of proteins, which are visualized either by fluorescent or visible dyes. Low reproducibility and limited sensitivity are associated with 2D-GE. However, these limitations were overcome by the development of difference gel electrophoresis (DIGE), where two different protein samples are individually labeled with various fluorescent dyes and electrophoretically separated on the same gel. Using 2D-DIGE, one can perform quantitative as well as comparative proteomics [2, 22, 23].

20.2.2 Mass Spectrometry

MS-based peptide separation depends on the mass-to-charge ratio. The proteins become charged after colliding with hydrogen ions (by an ionizer) while passing through a mass analyzer toward the detector [24]. Time of flight (TOF), triple quadrupole, quadrupole-TOF, ion trap, and hybrid ion trap orbitrap are the most frequently employed mass analyzers in protein applications. Protein identification and sequencing, detection of post-translational modification, and multiprotein complexes characterization are the four major applications of MS-based proteomics [2].

20.2.3 Matrix-Assisted Laser Desorption/Ionization Mass Spectrometry

MALDI and electrospray ionization (ESI) techniques are the most widely utilized ionization methods in experimental biology. These techniques typically generate an initial scanning of peptides liberated by in-gel digestion by proteolytic enzymes in an acrylamide gel. Unlike MALDI, a method in which a solid-state laser repeatedly pulses, ESI is a continuous and liquid-phase method. Capillary electrophoresis and high-pressure liquid chromatography (LC) can be performed prior to concentration and purification of peptides before mass analysis [2, 22, 25].

20.2.4 Surface-Enhanced Laser Desorption/Ionization

Surface-enhanced laser desorption/ionization (SELDI) is used to analyze protein mixtures in clinical samples, such as tissue, blood, and urine, in conjunction with TOF-MS. By combining the two powerful MS technologies and chromatography, SELDI-TOF-MS provides a unique method for finding biomarkers. Unlike the MALDI method, SELDI-TOF-MS uses an altered target to establish biological affinity with the analyte substance, whereas SELDI identifies protein mixtures on a chemically altered surface. SELDI-TOF-MS has some disadvantages, such as having a low resolution, being very time-consuming, and, most often, not allowing for the simple detection of proteins that could serve as biomarkers of disease [2].

20.2.5 Liquid Chromatography-Mass Spectrometry

The analytical chemistry method known as LC-MS integrates the strength of LC to separate individual proteins and the capacity of MS to analyze the masses. The four basic parts of the LC-MS system are the chromatographic column, ionization source, mass analyzer, and detector. LC-MS, a powerful method with great sensitivity and selectivity, is commonly used in several proteomics research, including developing drugs, characterizing target proteins, discovering and screening biomarkers, and conducting toxicological studies. LC-MS/MS shows dominance over MS when masses overlap due to complex samples. Targeted proteomics uses LC-MS/MS to analyze the proteins recognized as biomarkers. However, the augmented use of LC-MS/MS in clinical laboratories has revealed several limitations, such as complicated instrument operation and maintenance as well as inadequate detection for specific analytes [2, 20, 22, 25].

20.3 Protein Identification and Database Searching

Once the molecular weights of peptides are acquired by MS, these peptide masses are compared with DNA sequences, expressed sequence tags, and previously sequenced proteins readily accessible in the global database. Protein analysis software programs, such as Mascot, SEQUEST, and X! Tandem, calculate the spectral counts. Each program uses a different set of algorithms to produce a list of proteins. Gene names matching the discovered proteins within the database are provided with functional annotations, and gene prediction tools identify the undiscovered proteins using their specific functional groups. The selected list of proteins will then undergo gene ontology (GO) analysis, utilizing databases like GO Term Finder and GO Term Mapper. These tools reveal the specific protein structure, function, and biological significance in cellular pathways. Identification of the pathways, cellular interactions, and cellular localizations of the protein groups is achieved via software programs, such as Ingenuity Pathway Analysis (IPA) and Metacore™. The bioinformatics tool STRING (Search Tool for the Retrieval of Interacting Genes/Proteins) provides the functional relationship in protein–protein interactions [22, 26–28].

20.4 Validation of Identified Proteins

The western blot (also known as the protein immunoblot) is an extensively used analytical method for identifying certain proteins in a tissue homogenate or extract sample. Dot-blot analysis and immunohistochemistry are additional similar techniques. In these procedures, antibodies are employed to detect the proteins in tissues and cells via immunostaining and enzyme-linked immunosorbent assay (ELISA). Multiple reaction monitoring for quantifying peptides on an LC-MS/MS platform is another method used to validate biomarker tests [2, 29].

20.5 Types of Proteomics in Redox Research

The powerful strategy offered by mass-proteomic makes it possible to reach high sensitivity and specificity in determining modifications of selected proteins. Redox mass-proteomic, a large-scale study of proteins using bioinformatics tools, can provide information on identifying the oxidized protein. The great resolution of mass spectrometric instruments allows the characterization of redox-induced modifications, identifying not only the susceptible proteins but also the amino acid residues that are modified. A further goal of redox proteomic approaches is to quantify the proportion of modified residues and the extent of oxidative damage occurring at the protein level.

Proteins may become reversibly oxidized in response to redox signaling, but irreversible oxidative modifications are usually associated with disorders and pathologies [30, 31]. Thus, a profile of oxidative modification of proteins from biological fluids or tissue is a promising approach that will help the clinical determination of several pathological states [32, 33]. This will also make the identification of new biomarkers and therapeutic targets for different human diseases possible. For example, it has been shown that infertile patients have differential expression of sperm proteins in relation to ROS levels [34]. The identification of differentially expressed proteins in infertile men using mass spectrometry represents an interesting indicator for ROS-induced male infertility. Understanding the involvement of differentially expressed proteins may reveal the underlying etiological relationship between male infertility and OS.

In the context of redox proteomics, major players are thiol functions (including cysteine residues and GSH system), antioxidant agents (SOD, CAT, GR, and GPx), ROS-generating enzymes (NOX, MPO, XO, and NOS), as well as transcription factors involved in their regulation (Nrf-2 and NF-κB). Various mass spectrometers, including soft ionization ones such as matrix-assisted laser desorption/ionization (MALDI) combined with time-of-flight (TOF) analyzers, have been used for the analysis of these indicators as well as complex products generated by oxidative damage to proteins.

20.5.1 Thiol Proteomics

MS-proteomic can assess thiol groups in proteins, which undergo reversible and irreversible oxidation reactions impacting protein structure and function. These include multiple oxidation states (sulfenic, sulfinic, and sulfonic acids) and adduct forms (glutathionylation, S-nitrosylation, cysteinylation, persulfidation, etc.).

20.5.1.1 Cysteine Proteomics

The location of redox-sensitive cysteine residues within a protein can play an important role with regard to its redox state. Cysteine residues are one of the least abundant amino acids and have the most extreme conservation pattern within proteins. In fact, they are highly conserved when they are located in the active site or involved in co-factor binding [35]. Besides, surface-exposed cysteine residues are particularly sensitive to oxidation by RS and are the most vulnerable among all amino acids.

The cysteine redox proteome has been widely studied using mass instruments. Experimental studies on cultured cells and animal organs without imposed oxidant and/or impaired antioxidant systems showed that the oxidation percentage of cysteine residues in the proteome ranges from 5% to 12%. Cysteine residues within functional pathways share similar percentages of oxidation, as demonstrated by pathway and network studies [36]. This observation has been highlighted in many actin-associated proteins, nuclear import proteins, and mRNA processing enzymes.

Sulfinic and sulfonic acid formation resulting from cysteine oxidation are stable modifications. They can be directly identified by including the change in mass as a variable modification. Due to their low abundance, it can be difficult to characterize under normal conditions, although sensitivity can be increased using strong cation exchange [37].

The application of targeted mass spectrometry approaches increases the sensitivity of redox state determination. It is necessary to have a prior knowledge of the modified peptide that contains cysteine residues for processing approaches, such as selected/multiple reaction monitoring or parallel reaction monitoring (S/MRM or PRM). Thiol proteomic approach has been used to demonstrate elevated levels of thioredoxin (Trx) in patients with cancer [38]. Another related protein, peroxiredoxin (Prx), has been often studied using a proteomic approach and several works identified Prx in the secretome under various disease conditions [39].

20.5.1.2 Methionine Proteomics

The methionine redox proteome is increasingly viewed as part of highly regulated machinery controlling signaling factors during OS. Although enzymes are mostly inhibited or inactivated by methionine oxidation, examples of enzyme activation have also been found [40].

Sulfur in methionine can be reversibly oxidized by some ROS to a sulfoxide. Some studies reported an increase of methionine sulfoxide in plasma proteins and HDL in DM (diabetes mellitus) [41]. Although not as widely studied as a form of thiol oxidation, methionine sulfoxide has potential advantages as a biomarker because it is easily measured and is more stable than thiol oxidized products.

20.5.1.3 Disulfide Bond Proteomics

High-throughput analysis of proteins containing disulfide bonds in complex biological mixtures is based on a top-down mass spectrometry approach. This method compares non-reduced and reduced proteins, where a mass shift of 2 Da is indicative of a disulfide bond. The effect of disulfide bonds is highly dependent on the position of the cysteine residues involved. If it is part of the active site, the disulfide moiety may be part of the catalytic cycle or act as an "on-off" switch for protein activity [42]. Formation of disulfides can also be part of cell signaling, such as the regulation of Nrf-2 during OS [43].

One of the most important RS-induced disulfide bonds is the reversible formation of glutathionylated proteins. Reversible S-glutathionylation of proteins occurs either under physiological conditions within redox signaling pathways or as a result of GSH antioxidant activity against OS lesions through the reduction of oxidized cysteine residues and the formation of mixed disulfide protein-glutathione (PSSG).

Cysteine S-glutathionylation acts as a protective mechanism preventing further irreversible oxidation of the disulfide bond to sulfinic or sulfonic acids. Reduction of PSSG takes place either spontaneously when the GSH/GSSG ratio is high or through enzymatic catalysis via a number of enzymes called protein thiol-disulfide oxidoreductases (e.g., glutaredoxins, GST, protein disulfide isomerases, Trx, Prx, and sulfiredoxins).

Glutathionylation can be directly detected by mass instruments by an increase in the mass of proteins by 305 Da. Recent advances in redox proteomic techniques allowed the identification of many S-glutathionylated proteins and their involvement in several redox-pathways.

Measurement of S-glutathionylation of some functional proteins is a promising way to assess OS. However, evaluation of this parameter is hampered by complexity in analytical methodologies (accessing tissue samples and procedural artifacts), which requires special care in sample handling and preparation [44]. A simpler approach is the analysis of S-glutathionylated proteins in plasma and circulating cells. Reversible S-glutathionylation of proteins in monocytes has emerged as a new signaling paradigm, which provides a molecular basis for the well-established relationship between metabolic disorders, OS, and CVD (cardiovascular diseases) [45].

S-glutathionylation of hemoglobin is also an interesting metabolic reaction as glutathionyl hemoglobin is implicated in different chronic diseases. Glutathionylated hemoglobin can be quantified by LC-MS/MS as a possible intermediate in oxidative damage. Many studies propose glutathionylated hemoglobin, measured in red blood cells, as a marker of OS. An increase of this marker has been reported in patients with hyperlipidemia, DM, cigarette smoking, and renal failure [46–49]. Moreover, levels of glutathionylated hemoglobin were markedly elevated in patients undergoing dialysis and marginally elevated in patients that had received transplanted organs compared to control groups. Glutathionylated hemoglobin levels correlated positively with lipid peroxidation and negatively with erythrocyte GSH content in patients undergoing renal replacement therapy, indicating enhanced OS [46].

20.5.2 Nitroproteome

Proteomic approach has been successful in identifying nitration in proteins. Nitrotyrosine residues can now be identified by proteomic-based mass spectrometry, but this approach requires prior derivatization and subsequent enrichment of nitropeptides as only a reduced fraction of the proteins of interest is usually nitrated [50, 51].

Although the hydrophobicity of tyrosine nitration has limited its detection by MS/MS, some reports have surpassed these limitations and showed some interesting findings. For example, enrichment in NO_2Tyr and Cl-Tyr amounts within ApoA-I, one of the major protein constituents of HDL in human plasma and atherosclerotic lesions, has been demonstrated by proteomic-based mass spectrometric methods. Analysis of serum also showed that concentrations of protein bound NO_2Tyr and Cl-Tyr in ApoA-I are markedly higher in individuals with established coronary artery disease compared to healthy volunteers [52, 53]. According to Bergt et al., these results suggest that the concentrations of NO_2Tyr and Cl-Tyr in circulating HDL using tandem mass spectrometry might represent specific markers for clinically notable atherosclerosis.

Other works have shown that patients with lung cancer and respiratory distress syndrome have significantly higher blood concentrations of nitrated proteins, which supports the presence of oxidative and nitrosative stress [54, 55]. Using mass spectrometry-based proteomic and genomic approaches, authors have identified targeted proteins for nitrosative stress, which include antioxidant enzymes such as SOD. The study carried out by Redondo-Horcajo et al. described the nitration of tyrosine in MnSOD, where cyclosporine-induced $ONOO^-$ formation results in MnSOD inactivation and cell damage [56].

Direct detection of cysteine-SNO residues by mass spectrometry has been described for Trx by Barglow et al. [57]. This approach is obviously of interest when targeting specific proteins but not for complex analysis. An improvement of this approach, called d-switch, has been proposed, allowing simultaneous identification of both S-nitrosated and non-nitrosated cysteines by mass spectrometry following derivatization with NEM [58].

20.5.3 Carbonylation Proteomics

A number of recent proteomic approaches based on the purification and subsequent identification of carbonylated proteins by LC-MS/LC-MS/MS have been described. Most of them take advantage of the reaction of carbonyl groups with hydrazide [59]. Fluorescence-based two-dimensional gel electrophoresis coupled with mass spectrometry has also been used to analyze complex carbonylated proteins [32].

20.6 Proteomics Evaluation of Semen

Semen is a complex physiological fluid that includes numerous peptides and proteins with various functions. Because spermatozoa are unique and neither transcriptionally nor translationally active, it is important to study the sperm proteome and identify the proteins crucial for sperm development, maturation, and function [60]. Proteomics can be used to analyze the entire proteome of sperm as well as the expression of proteins in subcellular compartments [20]. Recently, subcellular proteomics investigations have garnered greater interest than whole sperm proteome, as it offers a complete analysis of the composition and precise distribution of sperm protein [61]. The development of subcellular proteomics led to an understanding of the metabolic enzymes essential for sperm motility. Proteins involved in sperm–oocyte interaction were also revealed by proteomic analysis of sperm tail and different fractions of sperm membrane, respectively [62–65].

20.6.1 Whole Sperm Proteomics

Several studies have focused on analyzing the whole proteome of spermatozoa in normozoospermic men. Martinez-Heredia et al. used 2D-GE and MALDI-TOF-MS analysis and identified 98 proteins in normozoospermic men [66]. These proteins were primarily associated with cell structure or metabolic pathways, including OS-mediated pathways [66]. With the aid of 2D-GE and MS, 73 proteins crucial to energy metabolism were identified [67]. The same investigators identified 25 more sperm proteins, including a nuclear protein (c14orf105) and a membrane protein (ANXA7), which is a Ca^{2+}-dependent, phospholipid-binding protein. Because each subset of spermatozoa has a unique protein profile, different components of mature and immature sperm are subjected to global proteomics research of fertile and infertile individuals [68]. Immature spermatozoa were reported to have increased expression levels of proteins to manage the cellular response to OS. The most matured sperm fraction of subfertile men displayed alterations in the expression of proteins implicated in energy-related signaling cascade and protein folding [69]. Furthermore, proteomic analysis revealed a total of 117 differentially expressed proteins (DEPs) in the spermatozoa of males with primary and secondary infertility compared to those in the spermatozoa of fertile men [70].

20.6.2 Subcellular Proteomics

Proteomic investigations on whole cell lysate can only partially reveal the intended functions of proteins found in sperm subcellular spaces and organelles. However, proteomic analysis of sperm organelles provides additional details regarding the subcellular metabolism. More specifically, the function of the spermatozoa is directly correlated with the differential expression of proteins in its subcellular compartments, such as the nucleus, acrosome, and flagella [61, 62, 71]. For isolation of specific nuclear proteome, sperm are treated with cetyltrimethyl ammonium bromide (CTAB) to evade contamination from the flagella, mitochondria, and acrosomal parts [72]. Nuclear proteins linked to histones and chromatin are important for epigenetic modifications of sperm, fertilization, and embryo development [72]. There have been reports that the head and tail of spermatozoa express a total of 521 and 721 proteins exclusively [71]. According to studies supporting the significance of sperm tail in motility, the flagellar subregion has an abundance of proteins regulating energy-generating pathways, such as glycolysis, gluconeogenesis, the Krebs cycle, and oxidative phosphorylation [62, 73, 74]. Similarly, proteins of signaling pathways, such as proteases and kinases, are restricted to sperm heads [71].

20.6.3 Sperm Surface Proteomics

Sperm require binding proteins expressed on their surface to detect the zona pellucida and to control the acrosome reaction. As a result, the zonal penetration and fertilization processes depend specifically on proteins located on the membrane. Analysis of the membrane's composition, which is the end product of spermatogenesis, can help us understand how sperm differentiate and mature. To purify membrane proteins, sperm cells must first be lysed. Researchers use nitrogen cavitation or sonication methods along with differential centrifugations to isolate the various biological components [75]. The plasma membrane contains lipid drafts where proteins from the identical pathway colocalize with sphingolipids and cholesterol. These proteins play a crucial role in the binding of sperm with the zona pellucida and subsequent fusion with the oolemma [76]. Nano-LC-MS/MS identified 124 sperm membrane proteins, among which arylsulphatase, sperm adhesion molecule 1, β-2 integrin, and angiotensin converting enzymes (ACE) were observed to be involved in intracellular transfer, cellular signaling, and protein folding in addition to sperm–oocyte interaction and adhesion [76]. To find prospective candidates for developing a contraceptive vaccine, Shetty et al. studied the sperm surface proteome [77]. In addition, advanced proteomics techniques have identified all the surface proteins involved in the autoimmune responses of sperm [78].

20.6.4 Seminal Plasma Proteomics

In the past two decades, seminal plasma has been used in 23% of proteomics research focused on male reproductive disorders [79]. In 2006, the very first investigation of the seminal plasma proteome was conducted [80]. A total of 923 proteins, including LF, LM, FN1, SEMG1, and SEMG2, were documented via LC-MS/MS analysis [81]. The following seminal plasma proteins are associated with various functions related to male fertility: SEMG1 and SEMG2 contribute to the formation of coagulum post-ejaculation; lactoferrin acts as an immunomodulator as well as an antibacterial and antioxidant candidate; and clusterin is associated with sperm agglutination [81]. Intasqui et al. presented a group of seminal biomarkers linked to mitochondrial dysfunction, acrosome integrity, and sperm DNA abnormalities [82]. Augmented expression of ANXA7, GSTM3, and ERP44 was linked to changes in mitochondrial activity [74]. Similarly, augmented expression of the phospholipid transfer protein was linked to acrosome defects. CRISPLD1, CRISPLD2, and RARRES1 have been suggested to be indicators of sperm DNA damage [74]. LC-MS/MS analysis revealed that OS impacts the seminal proteome by increasing the expression of proteins involved in the production of unsaturated fatty acids [21]. The 44 differentially expressed proteins (DEPs) reported in fertile males were categorized according to the amount of seminal ROS identified by chemiluminescence, evidence which further strengthened these preliminary observations [21]. Fertile men with high concentrations of seminal ROS generated increased levels of SOD1 and PRDX4, antioxidant defense mechanisms to balance the oxidant levels, and proteins involved in the breakdown of malfunctioning proteins.

20.6.5 Testicular Proteomics

There is a paucity of literature available on proteomic analysis of testicular tissue. More specifically, in the past 20 years, only a few articles have been published regarding testicular proteomics [79]. Cellular signaling pathways are significantly altered in pathological circumstances, including obstructive and non-obstructive azoospermia as well as testicular cancer [83, 84]. The testicular proteome gives considerable insight into these altered pathways. Future proteomics research on testicular tissue will aid in investigating spermatogenesis-related proteins and the discovery of novel therapeutics to treat male infertility [20].

20.7 Proteomics and Semen Oxidative Stress

About 80% of infertile males suffer from seminal OS caused by elevated generation of ROS in semen [85, 86]. The biological activity of spermatozoa is adversely affected by excessive levels of ROS. Prolonged OS and associated increased levels of oxidants can damage DNA and impair sperm parameters (e.g., motility, morphology, count, and viability) [87–90]. In addition to influencing the metabolic and physiological response of spermatozoa to stress, OS also has a similar impact on seminal plasma components. To further understand the impact of oxidative damage on the cellular processes as well as on the sperm proteins related to the reproductive function, numerous proteomics studies on sperm and seminal plasma have been conducted. Several possible biomarkers for both spermatozoa and seminal plasma were reported after conducting a thorough proteomics investigation in sperm of patients with or without ROS [91]. Essential proteins like FN1, MIF, and LGALS3BP were absent in ROS-positive patients but abundantly present in ROS-negative patients. In contrast, the membrane metallo-endopeptidase (MME) protein was absent in the seminal plasma of ROS-depleted patients [92].

20.7.1 Molecular Changes in Sperm

Seminal OS can induce alterations in the expression of sperm proteins. A total of 74 sperm proteins were reported to express differently in conditions of OS [91]. Many of these DEPs were implicated in the CREM pathway, which controls the development of spermatozoa. The specific proteins HSP90B1, MDH2, TGM4, GPX4, GLUL, HIST1H2BA, and HSPA5 have been suggested to be indicators of OS [93]. Other proteins linked to impaired sperm function in patients with increased seminal ROS include ACE, HSPA2, RPS27A, MAP3K3, and APP [94]. Ayaz et al. reported that a unique group of six proteins, CLGN, TPPII, DNAI2, EEA1, HSPA4L, and SERPINA5, were differentially expressed in patients with low, medium, and high seminal OS [34]. These six DEPs impacted the normal biological activity of sperm, thereby diminishing reproductive function [34]. Recent proteomics research, which evaluated fertile men with elevated levels of ROS, reported significant changes in their proteome of sperm and seminal plasma, which contained 371 and 41 DEPs, respectively [21]. Furthermore, antioxidant and proteosomal systems were activated in fertile males with increased levels of ROS [14, 21].

20.7.2 Molecular Changes in Seminal Plasma

To date, only a few proteomics studies have been conducted on the seminal plasma of infertile men with significant levels of OS. Infertile men with high levels of seminal ROS have different proteome profiles compared to that in the seminal plasma of fertile donors with low ROS [91]. In men with ROS-positive seminal plasma, many specific proteins, including ACPP, AZGP1, CLU, KLK3, and PIP were regulated by the androgen receptor, whereas proteins such as MME and FAM3D were suggested as biomarkers in assessing the reproductive status of men with low, medium, and high levels of ROS in their seminal plasma [95]. Additionally, MUC5B is predicted to be a biomarker of seminal lipid peroxidation levels in men with fertility issues [92].

20.8 Oxidative Stress-Associated Biomarkers in Semen

The discovery of biomarkers is the most challenging and critical step in the development of novel diagnostics or therapies. These molecules are naturally found in body fluids, and their altered expression suggests a pathophysiological status. Specific biomarkers predict diagnosis, prognosis, and effectiveness of treatment [87]. Therefore, researchers are looking for protein biomarkers of OS in the semen of males with high levels of ROS. Proteome analysis of semen and sperm from these subjects has furthered the understanding of the cellular and molecular processes involved in the pathophysiology of OS-related infertility. Six proteins, including DJ-1, TBCB, AACT, ALDR, DGK, and PIP, have been proposed as biomarkers of OS in seminal plasma [26]. Except for DJ-1, other proteins are substantially produced in seminal plasma during conditions of acute OS. Additionally, PIP is being proposed as a biomarker of both azoospermia and asthenozoospermia, with its quantities decreasing in both circumstances [26]. Several sperm proteins, including MDH2, TGM4, GPX4, GLUL, HSP90B1, and HSPA5, have been proposed as potential biomarkers because of their elevated expression in ROS-positive samples compared to samples with low OS [26]. Heat shock proteins, such as HSP90B1 and HSPA5, are strongly associated with varicocele, contributing to OS and infertility. Based on OS-induced dysfunction, a range of putative sperm biomarkers, including PARK7, SOD2, NELFE, and PRDX2, was also observed in varicocele patients [96].

20.9 Current Challenges and Future Directions

There are many challenges to overcome on the long, complex journey toward the science of proteomics. Studies describing protein profiles have reported large volumes of data, intricate protein compositions, and significant quantities of isolated proteins, but mostly with unknown activities. These limiting considerations may constrain proteomics techniques from identifying biological markers exclusively for research purposes. With a growing interest in proteomics, researchers and clinicians will face numerous challenges as they utilize this state-of-the-art technology to comprehend the molecular basis of high OS-related male factor infertility. Researchers and physicians should consider that proteome profiles from patients with a specific infertility disorder are subject to significant variation across experimental conditions and among individuals. As thiol-independent activation pathways are being documented more frequently, further global methodologies are needed to uncover chaperones and components of the degradation machinery that employ thiol-independent regulatory pathways during OS. The possibility that these thiol-independent processes could control different signaling pathways is an equally fascinating subject. In any scenario, more methodical techniques are required to identify and study the roles of redox-sensitive proteins in reducing and oxidizing environments.

20.10 Conclusions

Substantial evidence suggests that OS contributes significantly to male infertility. Low sperm function and low semen quality are linked to changes in the proteome of sperm and seminal plasma. Proteomic studies show that OS is associated with significant alterations in protein compositions of both seminal fluid and sperm of infertile men. The development and utilization of proteomics-based techniques offer a better understanding of the fundamental molecular mechanisms and cellular pathways responsible for male infertility. Identifying reliable proteomic biomarkers for various male infertility scenarios, particularly of idiopathic nature, would be extremely beneficial in their clinical evaluation and possible reliable therapy of male infertility.

References

1. Swain N, Moharana AK, Jena SR, Samanta L. Impact of Oxidative Stress on Embryogenesis and Fetal Development. Adv Exp Med Biol. 2022;1391:221–41.
2. Gupta S, Ghulmiyyah J, Sharma R, Halabi J, Agarwal A. Power of proteomics in linking oxidative stress and female infertility. Biomed Res Int. 2014;2014:916212.
3. Borowiecka M, Wojsiat J, Polac I, Radwan M, Radwan P, Zbikowska HM. Oxidative stress markers in follicular fluid of women undergoing in vitro fertilization and embryo transfer. Syst Biol Reprod Med. 2012;58(6):301–5.
4. Birben E, Sahiner UM, Sackesen C, Erzurum S, Kalayci O. Oxidative stress and antioxidant defense. World Allergy Organ J. 2012;5(1):9–19.
5. Katerji M, Filippova M, Duerksen-Hughes P. Approaches and Methods to Measure Oxidative Stress in Clinical Samples: Research Applications in the Cancer Field. Oxid Med Cell Longev. 2019;2019:1279250.
6. Dhalla NS, Temsah RM, Netticadan T. Role of oxidative stress in cardiovascular diseases. J Hypertens. 2000;18(6):655–73.
7. Gelderman KA, Hultqvist M, Olsson LM, Bauer K, Pizzolla A, Olofsson P, et al. Rheumatoid arthritis: the role of reactive oxygen species in disease development and therapeutic strategies. Antioxid Redox Signal. 2007;9(10):1541–67.
8. Hwang O. Role of oxidative stress in Parkinson's disease. Exp Neurobiol. 2013;22(1):11–7.
9. Lau AT, Wang Y, Chiu JF. Reactive oxygen species: current knowledge and applications in cancer research and therapeutic. J Cell Biochem. 2008;104(2):657–67.
10. Lyras L, Cairns NJ, Jenner A, Jenner P, Halliwell B. An assessment of oxidative damage to proteins, lipids, and DNA in brain from patients with Alzheimer's disease. J Neurochem. 1997;68(5):2061–9.
11. Sayre LM, Smith MA, Perry G. Chemistry and biochemistry of oxidative stress in neurodegenerative disease. Curr Med Chem. 2001;8(7):721–38.
12. Sosa V, Moliné T, Somoza R, Paciucci R, Kondoh H, ME LL. Oxidative stress and cancer: an overview. Ageing Res Rev. 2013;12(1):376–90.
13. Toshniwal PK, Zarling EJ. Evidence for increased lipid peroxidation in multiple sclerosis. Neurochem Res. 1992;17(2):205–7.
14. Panner Selvam MK, Durairajanayagam D, Sikka SC. Molecular Interactions Associated with Oxidative Stress-Mediated Male Infertility: Sperm and Seminal Plasma Proteomics. Adv Exp Med Biol. 2022;1358:63–76.
15. Agarwal A, Mulgund A, Alshahrani S, Assidi M, Abuzenadah AM, Sharma R, et al. Reactive oxygen species and sperm DNA damage in infertile men presenting with low level leukocytospermia. Reprod Biol Endocrinol. 2014;12:126.
16. Panner Selvam MK, Agarwal A, Baskaran S. Proteomic analysis of seminal plasma from bilateral varicocele patients indicates an oxidative state and increased inflammatory response. Asian J Androl. 2019;21(6):544–50.
17. Sikka SC. Relative impact of oxidative stress on male reproductive function. Curr Med Chem. 2001;8(7):851–62.
18. Tremellen K. Oxidative stress and male infertility--a clinical perspective. Human reproduction update. 2008;14(3):243–58.

19. Agarwal A, Baskaran S, Parekh N, Cho CL, Henkel R, Vij S, et al. Male infertility. Lancet. 2021;397(10271):319–33.
20. Panner Selvam MK, Durairajanayagam D, Agarwal A. Proteomic and Metabolomic Fingerprinting in Male Infertility. Male Infertility: Contemporary Clinical Approaches, Andrology, ART and Antioxidants. 2020:123–38.
21. Dias TR, Samanta L, Agarwal A, Pushparaj PN, Panner Selvam MK, Sharma R. Proteomic Signatures Reveal Differences in Stress Response, Antioxidant Defense and Proteasomal Activity in Fertile Men with High Seminal ROS Levels. Int J Mol Sci. 2019;20(1).
22. Selvam MKP, Durairajanayagam D, Agarwal A. Proteomics and metabolomics. In Vitro Fertilization: A Textbook of Current and Emerging Methods and Devices. 2019:535–47.
23. May C, Brosseron F, Chartowski P, Schumbrutzki C, Schoenebeck B, Marcus K. Instruments and methods in proteomics. Methods Mol Biol. 2011;696:3–26.
24. Glish GL, Vachet RW. The basics of mass spectrometry in the twenty-first century. Nat Rev Drug Discov. 2003;2(2):140–50.
25. Oliva R, De Mateo S, Castillo J, Azpiazu R, Oriola J, Ballescà JL. Methodological advances in sperm proteomics. Hum Fertil (Camb). 2010;13(4):263–7.
26. Agarwal A, Durairajanayagam D, Halabi J, Peng J, Vazquez-Levin M. Proteomics, oxidative stress and male infertility. Reprod Biomed Online. 2014;29(1):32–58.
27. Zhou T, Zhou ZM, Guo XJ. Bioinformatics for spermatogenesis: annotation of male reproduction based on proteomics. Asian J Androl. 2013;15(5):594–602.
28. Michalski A, Damoc E, Hauschild JP, Lange O, Wieghaus A, Makarov A, et al. Mass spectrometry-based proteomics using Q Exactive, a high-performance benchtop quadrupole Orbitrap mass spectrometer. Mol Cell Proteomics. 2011;10(9):M111.011015.
29. Mahmood T, Yang PC. Western blot: technique, theory, and trouble shooting. N Am J Med Sci. 2012;4(9):429–34.
30. Fedorova M, Bollineni RC, Hoffmann R. Protein carbonylation as a major hallmark of oxidative damage: Update of analytical strategies. 2014;33(2):79–97.
31. Yan L-J. Protein redox modification as a cellular defense mechanism against tissue ischemic injury. Oxidative medicine and cellular longevity. 2014;2014:343154-.
32. Baraibar MA, Ladouce R, Friguet B. Proteomic quantification and identification of carbonylated proteins upon oxidative stress and during cellular aging. Journal of proteomics. 2013;92:63–70.
33. Butterfield DA, Gu L, Di Domenico F, Robinson RA. Mass spectrometry and redox proteomics: applications in disease. Mass spectrometry reviews. 2014;33(4):277–301.
34. Ayaz A, Agarwal A, Sharma R, Arafa M, Elbardisi H, Cui Z. Impact of precise modulation of reactive oxygen species levels on spermatozoa proteins in infertile men. Clinical proteomics. 2015;12(1):4.
35. Marino SM, Gladyshev VN. Cysteine function governs its conservation and degeneration and restricts its utilization on protein surfaces. Journal of molecular biology. 2010;404(5):902–16.
36. Go Y-M, Jones DP. The redox proteome. The Journal of biological chemistry. 2013;288(37):26512–20.
37. Paulech J, Liddy KA, Engholm-Keller K, White MY, Cordwell SJ. Global analysis of myocardial peptides containing cysteines with irreversible sulfinic and sulfonic acid post-translational modifications. Mol Cell Proteomics. 2015;14(3):609–20.
38. Baker AF, Dragovich T, Tate WR, Ramanathan RK, Roe D, Hsu CH, et al. The antitumor thioredoxin-1 inhibitor PX-12 (1-methylpropyl 2-imidazolyl disulfide) decreases thioredoxin-1 and VEGF levels in cancer patient plasma. The Journal of laboratory and clinical medicine. 2006;147(2):83–90.
39. Ishii T, Warabi E, Yanagawa T. Novel roles of peroxiredoxins in inflammation, cancer and innate immunity. Journal of clinical biochemistry and nutrition. 2012;50(2):91–105.
40. Lourenço Dos Santos S, Petropoulos I, Friguet B. The Oxidized Protein Repair Enzymes Methionine Sulfoxide Reductases and Their Roles in Protecting against Oxidative Stress, in Ageing and in Regulating Protein Function. Antioxidants (Basel). 2018;7(12):191.

41. Brock JW, Jenkins AJ, Lyons TJ, Klein RL, Yim E, Lopes-Virella M, et al. Increased methionine sulfoxide content of apoA-I in type 1 diabetes. J Lipid Res. 2008;49(4):847–55.
42. Jones DP. Radical-free biology of oxidative stress. American journal of physiology Cell physiology. 2008;295(4):C849–68.
43. Dinkova-Kostova AT, Holtzclaw WD, Cole RN, Itoh K, Wakabayashi N, Katoh Y, et al. Direct evidence that sulfhydryl groups of Keap1 are the sensors regulating induction of phase 2 enzymes that protect against carcinogens and oxidants. Proc Natl Acad Sci U S A. 2002;99 (18):11908–13.
44. Rossi R, Dalle-Donne I, Milzani A, Giustarini D. Oxidized forms of glutathione in peripheral blood as biomarkers of oxidative stress. Clin Chem. 2006;52(7):1406–14.
45. Ullevig SL, Kim HS, Short JD, Tavakoli S, Weintraub ST, Downs K, et al. Protein S-Glutathionylation Mediates Macrophage Responses to Metabolic Cues from the Extracellular Environment. Antioxid Redox Signal. 2016;25(15):836–51.
46. Mandal A, Woodi M, Sood V, Krishnaswamy P, Rao A, Ballal S, et al. Quantitation and characterization of glutathionyl haemoglobin as an oxidative stress marker in chronic renal failure by mass spectrometry. Clinical biochemistry. 2007;40:986–94.
47. Chen H-JC, Lin W-P, Chiu S-D, Fan C-H. Multistage Mass Spectrometric Analysis of Human Hemoglobin Glutathionylation: Correlation with Cigarette Smoking. Chemical Research in Toxicology. 2014;27(5):864–72.
48. Sampathkumar R, Balasubramanyam M, Sudarslal S, Rema M, Mohan V, Balaram P. Increased glutathionylated hemoglobin (HbSSG) in type 2 diabetes subjects with microangiopathy. Clinical biochemistry. 2005;38(10):892–9.
49. Niwa T, Naito C, Mawjood A, Imai K. Increased Glutathionyl Hemoglobin in Diabetes Mellitus and Hyperlipidemia Demonstrated by Liquid Chromatography/Electrospray Ionization-Mass Spectrometry. Clinical chemistry. 2000;46:82–8.
50. Kim JK, Lee JR, Kang JW, Lee SJ, Shin GC, Yeo W-S, et al. Selective Enrichment and Mass Spectrometric Identification of Nitrated Peptides Using Fluorinated Carbon Tags. Analytical Chemistry. 2011;83(1):157–63.
51. Yeo WS, Kim YJ, Kabir MH, Kang JW, Ahsan-Ul-Bari M, Kim KP. Mass spectrometric analysis of protein tyrosine nitration in aging and neurodegenerative diseases. Mass spectrometry reviews. 2015;34(2):166–83.
52. Bergt C, Pennathur S, Fu X, Byun J, O'Brien K, McDonald TO, et al. The myeloperoxidase product hypochlorous acid oxidizes HDL in the human artery wall and impairs ABCA1-dependent cholesterol transport. 2004;101(35):13032–7.
53. Zheng L, Nukuna B, Brennan M-L, Sun M, Goormastic M, Settle M, et al. Apolipoprotein A-I is a selective target for myeloperoxidase-catalyzed oxidation and functional impairment in subjects with cardiovascular disease. J Clin Invest. 2004;114(4):529–41.
54. Gole MD, Souza JM, Choi I, Hertkorn C, Malcolm S, Foust RF, 3rd, et al. Plasma proteins modified by tyrosine nitration in acute respiratory distress syndrome. Am J Physiol Lung Cell Mol Physiol. 2000;278(5):L961–7.
55. Pignatelli B, Li C-Q, Boffetta P, Chen Q, Ahrens W, Nyberg F, et al. Nitrated and Oxidized Plasma Proteins in Smokers and Lung Cancer Patients. 2001;61(2):778–84.
56. Redondo-Horcajo M, Romero N, Martínez-Acedo P, Martínez-Ruiz A, Quijano C, Lourenço CF, et al. Cyclosporine A-induced nitration of tyrosine 34 MnSOD in endothelial cells: role of mitochondrial superoxide. Cardiovascular Research. 2010;87(2):356–65.
57. Barglow KT, Knutson CG, Wishnok JS, Tannenbaum SR, Marletta MA. Site-specific and redox-controlled S-nitrosation of thioredoxin. 2011;108(35):E600–E6.
58. Sinha V, Wijewickrama GT, Chandrasena REP, Xu H, Edirisinghe PD, Schiefer IT, et al. Proteomic and mass spectroscopic quantitation of protein S-nitrosation differentiates NO-donors. ACS Chem Biol. 2010;5(7):667–80.
59. Havelund JF, Wojdyla K, Davies MJ, Jensen ON, Møller IM, Rogowska-Wrzesinska A. A biotin enrichment strategy identifies novel carbonylated amino acids in proteins from human plasma. Journal of proteomics. 2017;156:40–51.

60. du Plessis SS, Kashou AH, Benjamin DJ, Yadav SP, Agarwal A. Proteomics: a subcellular look at spermatozoa. Reprod Biol Endocrinol. 2011;9:36.
61. Sadowski PG, Dunkley TP, Shadforth IP, Dupree P, Bessant C, Griffin JL, et al. Quantitative proteomic approach to study subcellular localization of membrane proteins. Nat Protoc. 2006;1 (4):1778–89.
62. Amaral A, Castillo J, Estanyol JM, Ballescà JL, Ramalho-Santos J, Oliva R. Human sperm tail proteome suggests new endogenous metabolic pathways. Mol Cell Proteomics. 2013;12(2): 330–42.
63. Gu B, Zhang J, Wu Y, Zhang X, Tan Z, Lin Y, et al. Proteomic analyses reveal common promiscuous patterns of cell surface proteins on human embryonic stem cells and sperms. PLoS One. 2011;6(5):e19386.
64. Naaby-Hansen S, Herr JC. Heat shock proteins on the human sperm surface. J Reprod Immunol. 2010;84(1):32–40.
65. Stein KK, Go JC, Lane WS, Primakoff P, Myles DG. Proteomic analysis of sperm regions that mediate sperm-egg interactions. Proteomics. 2006;6(12):3533–43.
66. Martínez-Heredia J, Estanyol JM, Ballescà JL, Oliva R. Proteomic identification of human sperm proteins. Proteomics. 2006;6(15):4356–69.
67. Nowicka-Bauer K, Ozgo M, Lepczynski A, Kamieniczna M, Malcher A, Skrzypczak W, et al. Human sperm proteins identified by 2-dimensional electrophoresis and mass spectrometry and their relevance to a transcriptomic analysis. Reprod Biol. 2018;18(2):151–60.
68. Cui Z, Sharma R, Agarwal A. Proteomic analysis of mature and immature ejaculated spermatozoa from fertile men. Asian J Androl. 2016;18(5):735–46.
69. Samanta L, Sharma R, Cui Z, Agarwal A. Proteomic analysis reveals dysregulated cell signaling in ejaculated spermatozoa from infertile men. Asian J Androl. 2019;21(2):121–30.
70. Intasqui P, Agarwal A, Sharma R, Samanta L, Bertolla RP. Towards the identification of reliable sperm biomarkers for male infertility: A sperm proteomic approach. Andrologia. 2018;50(3).
71. Baker MA, Naumovski N, Hetherington L, Weinberg A, Velkov T, Aitken RJ. Head and flagella subcompartmental proteomic analysis of human spermatozoa. Proteomics. 2013;13 (1):61–74.
72. de Mateo S, Castillo J, Estanyol JM, Ballescà JL, Oliva R. Proteomic characterization of the human sperm nucleus. Proteomics. 2011;11(13):2714–26.
73. Kim YH, Haidl G, Schaefer M, Egner U, Mandal A, Herr JC. Compartmentalization of a unique ADP/ATP carrier protein SFEC (Sperm Flagellar Energy Carrier, AAC4) with glycolytic enzymes in the fibrous sheath of the human sperm flagellar principal piece. Dev Biol. 2007;302(2):463–76.
74. Panner Selvam MK, Finelli R, Agarwal A, Henkel R. Proteomics and metabolomics - Current and future perspectives in clinical andrology. Andrologia. 2021;53(2):e13711.
75. Zhou M, Philips MR. Nitrogen Cavitation and Differential Centrifugation Allows for Monitoring the Distribution of Peripheral Membrane Proteins in Cultured Cells. J Vis Exp. 2017(126).
76. Nixon B, Mitchell LA, Anderson AL, McLaughlin EA, O'Bryan M K, Aitken RJ. Proteomic and functional analysis of human sperm detergent resistant membranes. J Cell Physiol. 2011;226(10):2651–65.
77. Shetty J, Diekman AB, Jayes FC, Sherman NE, Naaby-Hansen S, Flickinger CJ, et al. Differential extraction and enrichment of human sperm surface proteins in a proteome: identification of immunocontraceptive candidates. Electrophoresis. 2001;22(14):3053–66.
78. Domagala A, Pulido S, Kamieniczna M, Kurpisz M, Herr JC. An isoimmune response to human sperm clathrin in an infertile woman with systemic lupus erythematosus. J Reprod Immunol. 2011;89(1):95–102.
79. Agarwal A, Baskaran S, Panner Selvam MK, Barbăroşie C, Master K. Unraveling the Footsteps of Proteomics in Male Reproductive Research: A Scientometric Approach. Antioxid Redox Signal. 2020;32(8):536–49.

80. Pilch B, Mann M. Large-scale and high-confidence proteomic analysis of human seminal plasma. Genome Biol. 2006;7(5):R40.
81. Milardi D, Grande G, Vincenzoni F, Messana I, Pontecorvi A, De Marinis L, et al. Proteomic approach in the identification of fertility pattern in seminal plasma of fertile men. Fertil Steril. 2012;97(1):67–73.e1.
82. Intasqui P, Camargo M, Antoniassi MP, Cedenho AP, Carvalho VM, Cardozo KHM, et al. Association between the seminal plasma proteome and sperm functional traits. Fertil Steril. 2016;105(3):617–28.
83. Alikhani M, Mirzaei M, Sabbaghian M, Parsamatin P, Karamzadeh R, Adib S, et al. Quantitative proteomic analysis of human testis reveals system-wide molecular and cellular pathways associated with non-obstructive azoospermia. J Proteomics. 2017;162:141–54.
84. Liu M, Hu Z, Qi L, Wang J, Zhou T, Guo Y, et al. Scanning of novel cancer/testis proteins by human testis proteomic analysis. Proteomics. 2013;13(7):1200–10.
85. Agarwal A, Parekh N, Panner Selvam MK, Henkel R, Shah R, Homa ST, et al. Male Oxidative Stress Infertility (MOSI): Proposed Terminology and Clinical Practice Guidelines for Management of Idiopathic Male Infertility. World J Mens Health. 2019;37(3):296–312.
86. Baskaran S, Finelli R, Agarwal A, Henkel R. Reactive oxygen species in male reproduction: A boon or a bane? Andrologia. 2021;53(1):e13577.
87. Panner Selvam MK, Agarwal A. Update on the proteomics of male infertility: A systematic review. Arab J Urol. 2018;16(1):103–12.
88. Aitken RJ, Koppers AJ. Apoptosis and DNA damage in human spermatozoa. Asian J Androl. 2011;13(1):36–42.
89. Dorostghoal M, Kazeminejad SR, Shahbazian N, Pourmehdi M, Jabbari A. Oxidative stress status and sperm DNA fragmentation in fertile and infertile men. Andrologia. 2017;49(10).
90. Novotny J, Aziz N, Rybar R, Brezinova J, Kopecka V, Filipcikova R, et al. Relationship between reactive oxygen species production in human semen and sperm DNA damage assessed by Sperm Chromatin Structure Assay. Biomed Pap Med Fac Univ Palacky Olomouc Czech Repub. 2013;157(4):383–6.
91. Sharma R, Agarwal A, Mohanty G, Du Plessis SS, Gopalan B, Willard B, et al. Proteomic analysis of seminal fluid from men exhibiting oxidative stress. Reprod Biol Endocrinol. 2013;11:85.
92. Intasqui P, Antoniassi MP, Camargo M, Nichi M, Carvalho VM, Cardozo KH, et al. Differences in the seminal plasma proteome are associated with oxidative stress levels in men with normal semen parameters. Fertil Steril. 2015;104(2):292–301.
93. Sharma R, Agarwal A, Mohanty G, Hamada AJ, Gopalan B, Willard B, et al. Proteomic analysis of human spermatozoa proteins with oxidative stress. Reprod Biol Endocrinol. 2013;11:48.
94. Ayaz A, Agarwal A, Sharma R, Kothandaraman N, Cakar Z, Sikka S. Proteomic analysis of sperm proteins in infertile men with high levels of reactive oxygen species. Andrologia. 2018;50(6):e13015.
95. Agarwal A, Ayaz A, Samanta L, Sharma R, Assidi M, Abuzenadah AM, et al. Comparative proteomic network signatures in seminal plasma of infertile men as a function of reactive oxygen species. Clin Proteomics. 2015;12(1):23.
96. Panner Selvam MK, Agarwal A. Sperm and Seminal Plasma Proteomics: Molecular Changes Associated with Varicocele-Mediated Male Infertility. World J Mens Health. 2020;38(4): 472–83.

Chapter 21
Other Techniques for Oxidative Stress Assessment

A. Khelfi

Abstract In this chapter, we describe a battery of techniques used for oxidative stress assessment including Western blot, capillary electrophoresis, COMET assay, PCR, and flow cytometry. Each technique is devoted to the qualitative and quantitative determination of particular metabolic products of oxidative stress (OS). The recently developed ultrasound, nuclear magnetic resonance (NMR) (metabolomics), magnetic resonance imaging (MRI) techniques as well as Raman spectroscopy in order to measure reactive species (RS) and oxidation products are also presented. New topics presented in this chapter include phosphorescence and the exciting new area of OS assessment using nanoparticles (NPs) and photoluminescence.

Keywords Western blot · Capillary electrophoresis · COMET assay · PCR · Flow cytometry · Nuclear magnetic resonance (NMR) · Metabolomics · Magnetic resonance imaging (MRI) · Raman spectroscopy · Phosphorescence

21.1 Nuclear Magnetic Resonance (NMR)-Based Metabolomics/Metabonomics

Metabolomics, also called metabonomics, is defined as the quantitative measurement of the multi-parametric metabolic responses in living systems to genetic modification or pathophysiological stimuli [1]. In this context, high-resolution nuclear ^{1}H-magnetic resonance (NMR) spectroscopy allows for simultaneous detection of tens to hundreds of low-molecular weight molecules within biological materials, resulting in a metabolic profile or the so-called "NMR fingerprint." The generated NMR spectra are analyzed using pattern recognition and multivariate statistical methods to generate models for samples classification [2].

A. Khelfi (✉)
Faculty of Pharmacy, University of Algiers1, Algiers, Algeria

National Centre of Toxicology, Algiers, Algeria

Department of Chemistry and Biomolecular Science, Clarkson University, Potsdam, NY, USA
e-mail: ab.khelfi@univ-alger.dz; akhelfi@clarkson.edu

S. Andreescu et al. (eds.), *Biomarkers of Oxidative Stress*,
https://doi.org/10.1007/978-3-031-60738-7_21

NMR technology has been widely applied to recent studies with a range of biological fluids, such as plasma and urine, in both experimental animals and humans [3–7]. Metobolomic analysis particularly fits with the analysis of other biomaterials from experimental or clinical studies such as EBC (exhaled breath condensate), with the possibility of measuring simultaneously multiple OS markers [8–10].

Analysis of biological samples may lead to the identification of novel biomarkers for wider studies in patient populations. Besides, metabolomics allows for the simultaneous analysis of a wide range of metabolic end products [10]. It may be that these "fingerprints" represent the most valuable characteristic of a particular disease and a more robust and powerful mean by which disease severity and progression can be stratified [11].

21.2 Magnetic Resonance Imaging (MRI)

MRI is a widely used imaging technique providing high-dimension resolution and soft tissue information. MRI is based on contrast, which is largely dependent on differences in the relaxation times (T1 and T2) of protons in water. Contrast agents (CAs) are frequently used to improve image resolution and quality through interactions with water molecules that disturb relaxation times.

Although very few CAs responsive to RS have been developed to date, interest in redox-responsive sensors and non-invasive methods for imaging OS and inflammation processes is expending and becomes an emerging area of research [12–16].

Numerous studies have been successfully using several CAs containing Gd (III) by responding to hydrogen peroxide generation and myeloperoxidase (MPO) activity, a key biomarker for inflammation [17, 18]. Further development of these probes has been shown to increase MRI signals in the presence of H_2O_2 and MPO, effectively enabling the non-invasive tracking of its activity in tissues (Fig. 21.1) [18].

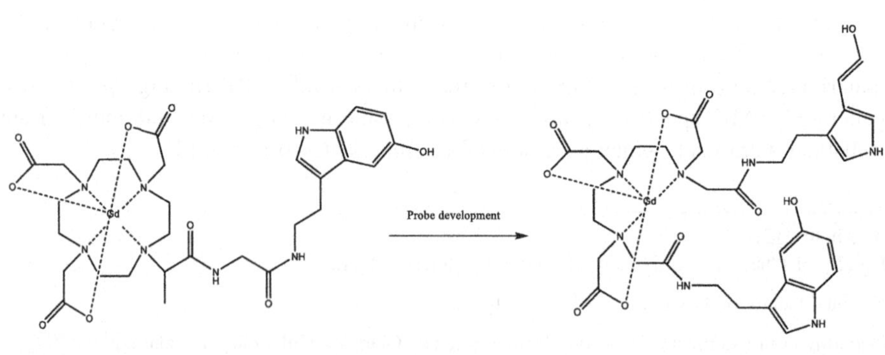

Fig. 21.1 Development of contrast agent based Gd(III) for the detection of MPO and hydrogen peroxide generation

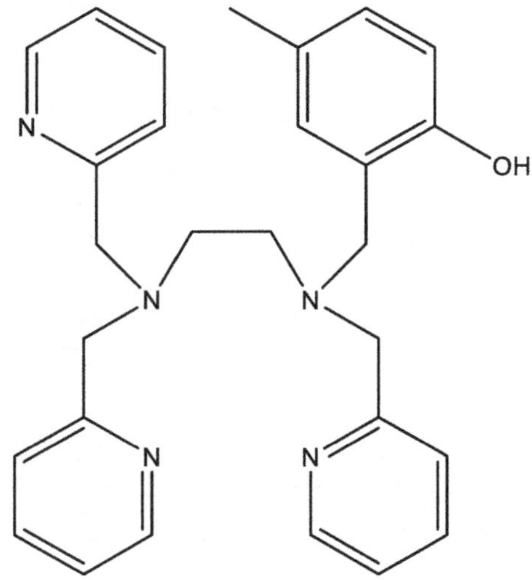

Fig. 21.2 Structure of Hptp1 probe

The new derivatives have shown low toxicity and have been used to track the inflammatory response in vasculitis, ischemic stroke, and atherosclerotic plaques [14, 18–20].

Yu et al. reported the synthesis of a CA in which a redox-active ligand, Hptp1 (N-(2-hydroxy-5- methylbenzyl)-N,N′,N′-tris(2-pyridinylmethyl)-1,2-ethanediamine) (Fig. 21.2) was complexed to Mn(II). The $[Mn(Hptp1)(MeCN)]^{2+}$ MRI CA proved to be responsive to H_2O_2 and has been shown to be independent of MPO [12].

Towner et al. combined immune-spin trapping approach with MRI technique in order to detect in vivo trapped radicals in the spinal cords of SOD1G93A-transgenic mice, a model for amyotrophic lateral sclerosis disease [21]. Using DMPO as a trapping agent, the anti-DMPO probe (Fig. 21.3) was effectively employed to detect DMPO-radical adducts through a significant and sustained increase in MRI signal intensities.

Nitroxides or aminyloxyl radicals have proven to be effective CAs for use in MRI. These probes were found to be useful for sensing the intracellular redox status of tumors [22–24]. For example, Hyodo et al. showed that cell-permeable nitroxides, 3-carbamoyl-2,2,5,5-tetramethylpyrrolidine-1-oxyl (3CP) and TEMPOL (4-Hydroxy-TEMPO, 4-hydroxy-2,2,6,6-tetramethylpiperidin-1-oxyl), showed faster decay in tumor tissue when compared to the cell-impermeable 3CxP (3-carboxy-2,2,5,5-tetramethylpyrrolidine-1-oxyl) (Fig. 21.4) [24]. In a further example, Ustimi et al. were able to isotopically label nitroxide CAs (14N and 15N) in order to simultaneously monitor both reduction and oxidation processes in cells through overhauser enhanced MRI imaging [22].

Prussian blue NPs (PBNs) have also been reported as potential reactive oxygen species (ROS) imaging sensors. These particles have the ability to detect ROS in both MRI and ultrasound imaging [25]. They can be stabilized by a biocompatible

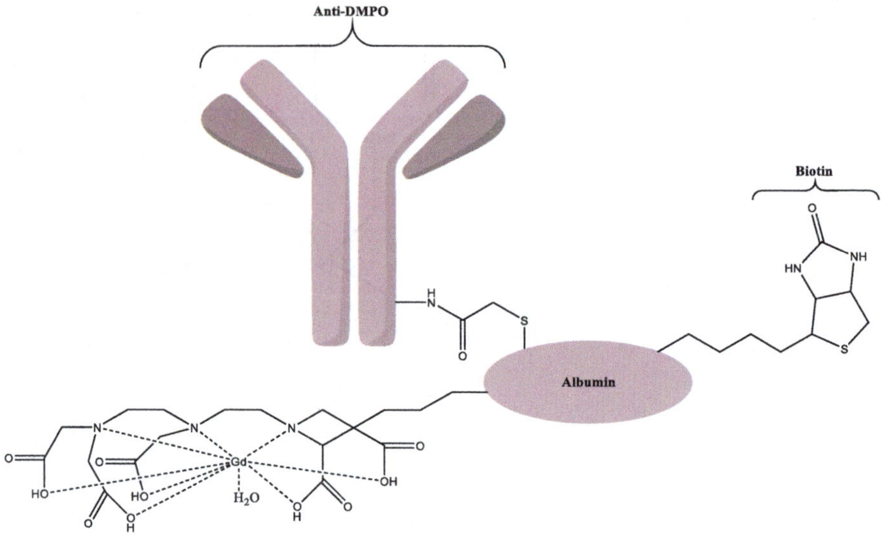

Fig. 21.3 Design of anti-DMPO-Gd probe

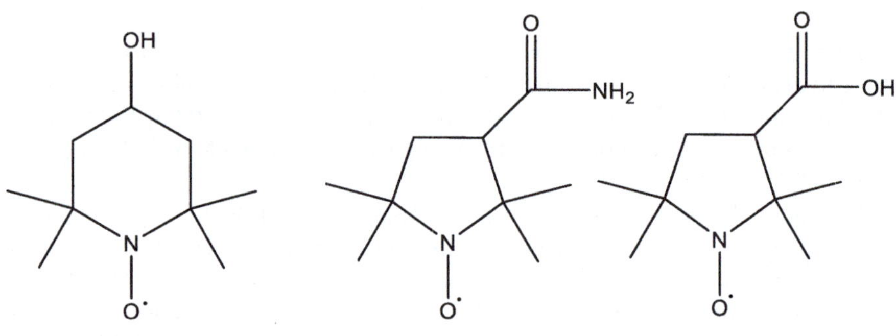

Fig. 21.4 Structure of nitroxide-based CAs

poly(vinylpyrrolidone) polymer coating to ensure decent solubility and particle dispersion [26]. These PBN sensors possess catalase-like ability as they were found to catalyze the breakdown of H_2O_2 to H_2O and molecular oxygen resulting in the generation of microbubbles detectable by ultrasound imaging instruments [27]. The potential of PBNs in MRI has been previously discerned. The relaxivity of these particles arises through five unpaired electrons per unit (Fe^{2+}–CN–Fe^{3+}) and the fact that some of the Fe^{3+} centers are accessible to water enhances T1 relaxation [28]. Increased T1 image enhancement in the presence of both PBN and ROS compared to PBN alone demonstrates that both PBNs and the paramagnetic O_2

microbubbles assist T1 relaxation and, as a consequence, MRI response enhancement in the presence of ROS.

Overall, there is a surge of interest in multimodal imaging technology that combines the complimentary abilities of different techniques [29]. The ability of these probes to act as CAs to assess ROS production using both MRI and ultrasound technology makes them particularly attractive.

21.3 Raman Spectroscopy

The use of NPs in Raman spectroscopy is a growing area of interest. Several NP-based approaches for the direct detection and imaging of ROS production have already been developed. Great technical advances in this field include the ROS-responsive surface-enhanced Raman spectroscopy (SERS) (Fig. 21.5) developed by Jiang et al. [30]. The Raman inactive dihydrorhodamine probe can be oxidized by several ROS including H_2O_2, $^1O_2$, $OH^•$, and HOCl to give a Raman active moiety. A subsequent "lighting up" of the SERS sensor occurs

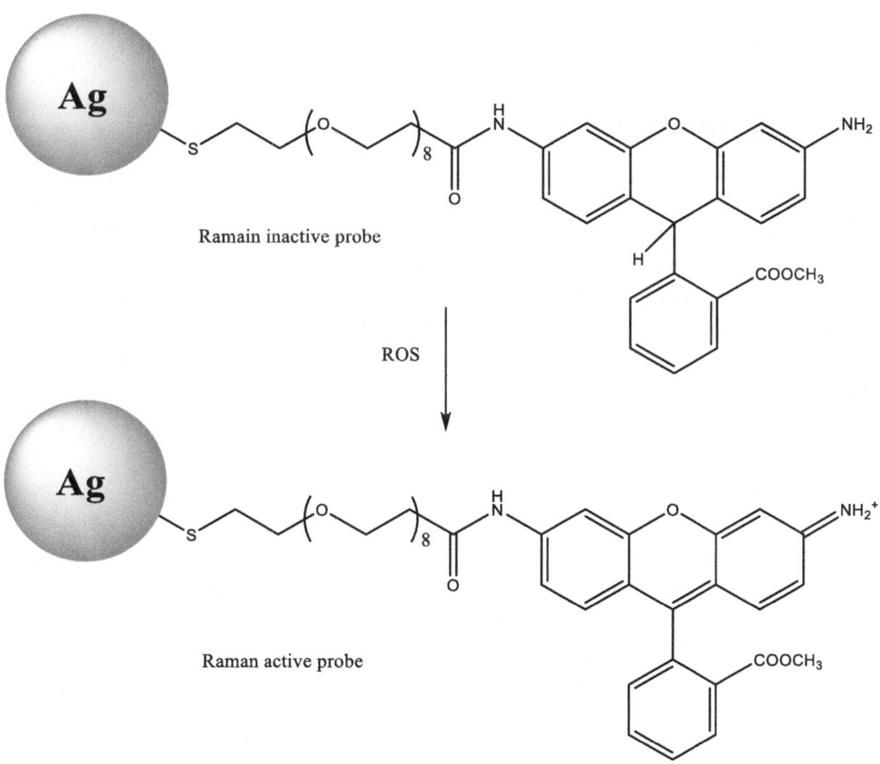

Fig. 21.5 Design of SERS probe and its reaction with ROS

through surface enhancement occurring on silver NPs. The developed probe by Jiang et al. was shown to be biocompatible and was effective in detecting ROS in cells at very low levels (femtomolar concentrations). Additionally, the probe allowed for a direct observation of ROS distribution in cells through high special resolution Raman imaging [30].

21.4 Phosphorescence

Singlet oxygen can be directly monitored via its own property called "phosphorescence" with a maximum around 1268 nm [31]. Even $^1O_2$ microscopy within the visible spectrum has been reported by Snyder et al. [32]. Nevertheless, singlet oxygen detection and real time-dependent profiles in biological materials are difficult [33]. This is due to the very low quantum yield of emitted rays, which ranges from 10^{-7} to 10^{-4} depending on the solvent [34].

For the direct activation of $^1O_2$, the decay time of this species can be used for monitoring, but it is highly dependent on the surrounding conditions with typical values of 70 ns in cells, while some sources report around 200 ns for the lifetime of $^1O_2$ in living cells [35, 36].

21.5 Photoluminescence

Carbon dots (CDs) are a new class of fluorescent NP probes that have attracted growing interest in recent years because of their powerful chemical and photoluminescent properties [37–39]. These particles have decent tunable surface functionality, biocompatibility, cell membrane permeability, water solubility, and excellent photostability [37]. In a study conducted by Wu et al., a CD-based nanosensor for the ratiometric determination of $NO^•$ was developed as illustrated in reaction 21.1 [40].

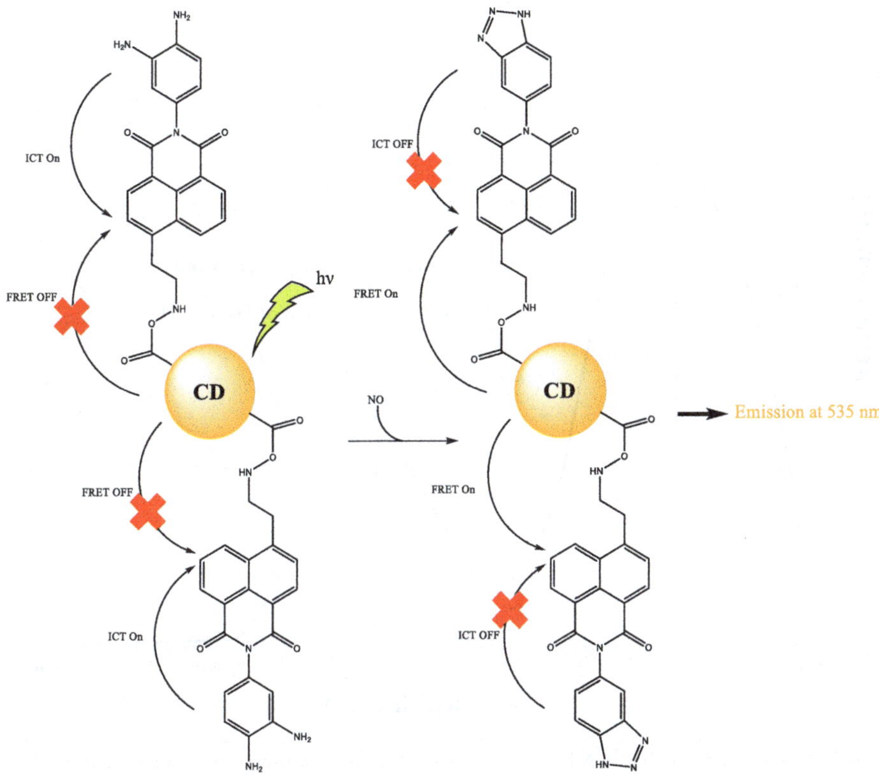

(21.1)

Recent advances in the design of carbon NPs provides a new way for optical sensor fabrication. Some carbon nanotubes (CNTs) are photoluminescent, absorbing radiation, and emitting photons at specific wavelengths (Fig. 21.6) [41, 42]. The photoluminescent properties displayed by SWNTs (single-walled carbon nanotubes) are dependent on integers (*m* and *n*), which means that they can emit specific photon wavelengths. SWNTs have some powerful properties. First, they are easily deliverable to eukaryotic cells and their short-term toxicity is negligible, making them useful in experimental studies [43]. One study highlighted the potential of SWNTs in multimodal analysis of ROS production in cells by reporting levels of H_2O_2, $^1O_2$, and $OH^•$ [44]. A further development of these particles allowed for nanomolar detection of H_2O_2 in cells with a precise temporal and spatial specificity [45]. SWNTs have also been used to image nitric oxide production in vivo, whereby a coating of SWNTs using DAP-dex (3,4-diaminophenyl-functionalized dextran) displayed selectivity for $NO^•$ with a decent sensitivity of 60 μM in mice [46]. Although these proof-of-principle studies are exciting, there are a number of issues that need to be addressed before SWNTs become commonly used optical sensors. Most of these issues are related to the aggregation of SWNTs due to the synthesis of non-homogeneous structures that typically result in poor

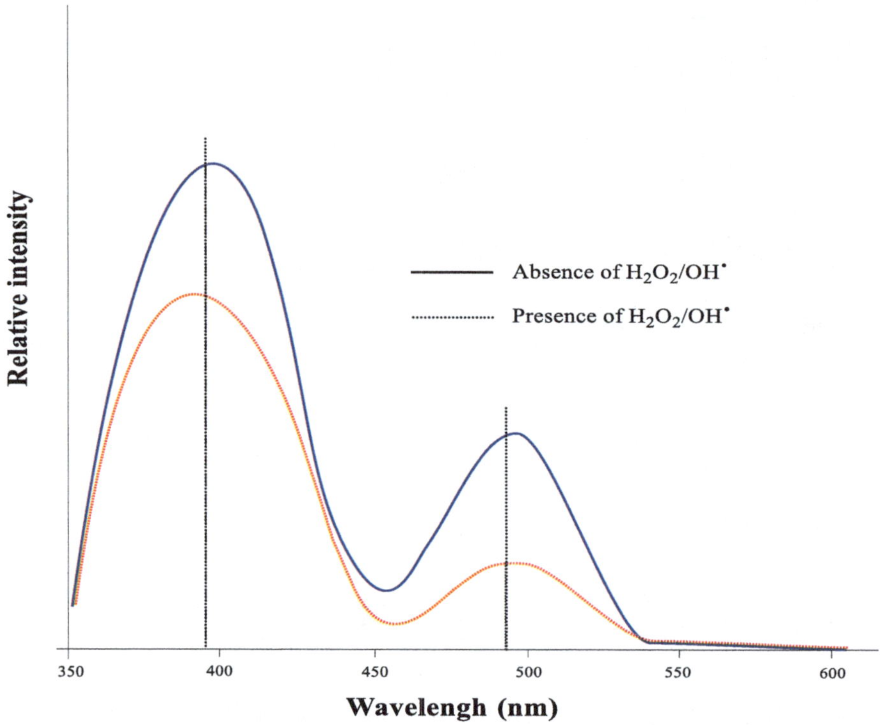

Fig. 21.6 Carbon nanotubes (CNTs) emission spectrum in the presence or absence of ROS Some CNTs can react differentially to specific ROS providing a multiplex analysis. Some ROS (e.g., $H_2O_2/OH^•$) can produce a reduction in emission intensity or a pronounced redshift in the fluorescent profile. This allows the measurement of specific ROS under experimental conditions

photoluminescence and high reactivity, which presents a challenge for specificity [47].

21.6 Ultrasound Contrast Agents

Ultrasound is a commonly used medical imaging technique that uses the reflection of echoes from tissues with different acoustic impedances to provide real-time images. These technologies were used for OS assessment in biological materials such as blood, tissues, and organs. This can be achieved by using microbubble-based ultrasound contrast agents (UCAs), which result in the widespread use of contrast-enhanced ultrasound in echographic imaging. The first microbubble contrast-enhanced ultrasound techniques were based on the introduction of a continuous stream of short-lived gas microbubbles into the bloodstream, while modern techniques employ prefabricated microbubbles consisting of a gas enclosed within proteins, lipids, or polymer shell. The early and the majority of clinical applications focused on blood samples, echohepatology, and echocardiology, whereas the focal point of the latest research applications is the development of site-directed UCAs coated with targeting ligands. These ligands include monoclonal antibodies, peptides, and glycoproteins specific to a particular cell receptor. This results in a site-specific accumulation of the UCAs [48–50].

Some studies exploited the high sensitivity of UCA-based microbubbles to detect ROS at physiologically relevant levels [27, 51]. This cannot be achieved using conventional UCAs. In this context, UCAs are designed such that the ROS generate microbubbles in situ following reaction with a targeted substrate. For instance, liposome-encapsulated allyl hydrazine was used to generate microbubbles detectable by ultrasound imaging, in the presence of $OH^{\bullet}$ [51]. As illustrated in reaction 21.2, oxidation of allyl hydrazine by $OH^{\bullet}$ generates gaseous N_2 and propene. The ability of these liposome-encapsulated allyl hydrazine sensors to detect ROS in vivo was demonstrated using a model of LPS-induced inflammation in mice [51].

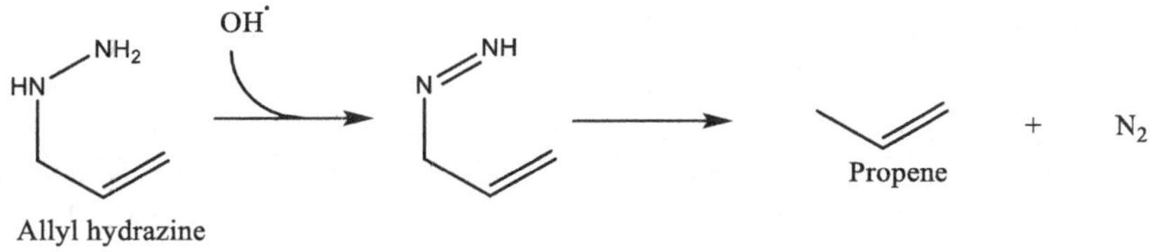

(21.2)

An alternative design for the detection of ROS using ultrasound technology has been tested and was inspired by the ability of tubular micromotor converters (TMMCs) to use H_2O_2 as a fuel for propulsion [27]. These converters are internally coated with either platinum or a dual layer of catalase (CAT) over gold that catalyzes the conversion of H_2O_2 to H_2O and O_2. The produced oxygen microbubbles provide propulsion and a trail for detection. Olson et al. successfully demonstrated the ability of ultrasound rays to detect microbubbles generated from platinum-coated TMMCs, suggesting their potential application in ultrasound imaging. As the motion of these converters was not sufficient for imaging, spherical nanospheres containing concentric shells of CAT and poly(sodium styrenesulfonate) were selected as promising alternatives. Triple layer nanospheres were shown to produce oxygen microbubbles that could be exploited for hydrogen peroxide detection using ultrasound imaging in ex vivo experiments such as activated neutrophils or in vivo models like abscess in rats [27].

21.7 Radio Assays

21.7.1 Protein Thiol/Disulfide Redox Status

Several radiolabeled probes including, N-iodoacetyl(^{125}I)-3-iodotyrosine [52] and (^{3}H) acetyl-4-aminophenylarsine oxide [53], were developed for the assessment of thiol/disulfide redox status of cellular proteins. These probes covalently bind to free thiol groups of a particular protein, which is then detected using ultra-sensitive instruments.

21.7.2 S-Thiolation

S-thiolation is the non-enzymatic linkage of glutathione (GSH) to a sulfhydryl group on proteins. This type of reaction has been shown to be a reversible process during protein lifetime. It is also considered to be an early event in OS conditions [54, 55]. This conjugation usually occurs in cellular proteins after exposure to peroxides and in monocytes undergoing respiratory burst. S-thiolation is believed to occur following GSSG formation. The protein–glutathione bond formed during OS usually disturbs protein function and inactivates some enzymatic activities [56]. Radiolabeled cystine in cultured cells is often used to measure S-thiolation in biological systems [54, 57]. Radiolabeled cystine is introduced into cells and undergoes reduction reaction to form cysteine, which is incorporated into glutathione molecules. Radiolabeled GSH will label proteins if S-thiolation occurs. During the experience, protein synthesis inhibitors (e.g., cycloheximide) should be added to the system to ensure that cysteine is not used for the synthesis of new proteins.

21.8 Western Blot

21.8.1 Protein Carbonyls

Determination of proteins containing carbonyls can be accomplished by Western blotting. In this technique, protein carbonyls are identified by derivatization with DNPH and electrophoretic separation, followed by immunodetection using antibodies specific to DNP [58]. Western blot of the carbonyl-2,4-dinitrophenylhydrazine adduct improves the sensitivity of immunoassays to pmol level and may be used to identify specifically the proteins that are modified. An interesting advantage of the Western blot is that, after detection of carbonyls, the blots can be erased and probed with other antibodies to characterize a protein of interest.

Simple Western blot assays are available as commercial kits for protein carbonyl detection including the Oxyblot kit provided by Intergen, Inc. (Chemicon, CA, USA) and the OxiSelect kit from Cell Biolabs, Inc. (San Diego, CA, USA).

Western blot can be performed after one-dimensional or two-dimensional electrophoretic separation [59]. Nakamura and Goto found more than 25 proteins with high carbonyl contents in two-dimensional immunoblot of tissue-soluble proteins. Rabbit polyclonal antibodies against 2,4-dinitrophenyl were used to study the derivatized proteins with DNPH and very low sensitivity was reached using this assay (1 pmol of carbonyls) [60].

In order to determine protein oxidation in cell membranes and particularly in erythrocyte membranes, a Western blot immunoassay was developed by Dumaswala et al. as a semi-quantitative estimation using chemiluminescence detection [61]. The minimum detection limit of this assay is about 500-fold lower (pmol) than that for carbonyl determinations by spectrophotometric methods.

Quantitative determination of protein carbonyls using immunoblotting (slot blot) has also been developed on cultured cells and small tissue samples. This assay has allowed the detection of approximately 60 pmol carbonyl/mg protein in tissue [62].

21.8.2 Nitrated Proteins

Commercial high-activity rabbit polyclonal antibody for detecting nitrotyrosine-containing proteins and peptides is available for Western blotting (Invitrogen) applications. The assay is useful for identifying the nitrated proteins and determining levels of protein nitrosylation in tissues.

It is worth mentioning that blotting techniques for NO_2Tyr detection lack specificity and sensitivity [63]. By employing two-dimensional electrophoresis (2DE) and Western blot, coupled with mass spectroscopy, targets of protein nitration and hydroxynonenal (HNE) modification have been determined in protein systems [64]. Two-dimensional Western blot has also been used to detect and characterize four nitrated proteins including actin in human pituitary, which participate in neurotransmission as well as cellular immunity, structure, and motility [65].

21.8.3 AGEs/ALEs

A number of methods have been reported to measure advanced glycation end products (AGEs) using immunobloting. For instance, Inoue et al. used the 6D12 antibody (see Chap. 17) demonstrated that CML derived from debris may help distinguish between patients with multiple post-operative ischemic events and those with non-multiple ischemic events who undergo carotid artery stenting [66].

Western blot is also used to detect modified proteins by lipid peroxidation end products. Using this technique, Kim et al. noted the presence of epitopes of lipid peroxide-modified proteins in atherosclerotic plaque and showed that their antibody could also detect lipid peroxide-modified proteins in plasma of normal individuals [67].

21.8.4 Peroxiredoxin (Prx)

The Prx family has a key role in the clearance of H_2O_2 because it is highly abundant and reacts with high efficiency. Oxidized Prx-6 in the cerebrospinal fluid has been evaluated using Western blot. It has been proposed as a biomarker of OS in brain injury and as a good predictor of the outcome [68].

Using Western blot, ELISA, and PCR methods, the importance of the redox state of Prx as a biomarker has been demonstrated, showing that Prx-2 is secreted via a non-classical pathway by inflammatory cells and can be found in serum and synovial fluid of rheumatoid arthritis (RA) patients [69–71]. Bayer et al. have shown that Prx-2 of erythrocytes undergoes a transient overoxidation following lipopolysaccharide exposure. The authors suggested that oxidized erythrocyte Prx-2 could be a sensitive real-time marker of systemic neutrophil activation and an early indicator of inflammation and OS. This marker can be released in the oxidized form and its redox state, not just its levels, can be an early marker of OS as well [72].

21.8.5 Other Compounds

Several other biomarkers of OS have been explored in biological fluids and tissue homogenates using Western blot. For example, analysis of Nrf-2 levels by immunoblot in tumor biopsies is becoming a tool to clinically assess tumor malignancy [73]. Western blot is more commonly used for protein S-glutathiolation determination as well [74]. Although S-glutathionylation can be easily measured by Western blot under non-reducing conditions, the use of more effective approaches is often required for an accurate quantification and confirmation. Finally, NAD(P)H oxidase (NOX) isoforms are currently widely studied using this approach, but there is little data on the specificity of the corresponding antibodies.

21.9 Capillary Electrophoresis

Capillary electrophoresis is currently a very powerful technique for the assessment of OS. Here are the potential applications of capillary electrophoresis with different detectors:

- Procedures based on high-performance capillary electrophoresis have been developed to monitor changes in the electrophoretic behavior of isolated LDL and to separate oxidized from non-oxidized LDL [75, 76]. Since protein migration is monitored at 234 nm, changes occurring in the absorption spectra of lipoproteins, for instance because of the presence of conjugated diene, can be simultaneously recorded.
- Nitric oxide radical is rapidly metabolized into the stable end products NO_2^- and NO_3^- in most body fluids, including plasma. The most commonly used assay for their assessment is based on the Griess diazotization reaction coupled with spectrophotometric detection. Other methods have been developed especially capillary electrophoresis with laser-induced fluorescence detection, which can estimate reactive nitrogen species (RNS) production even in single cells [77, 78].
- Due to the rapid nature of the reduction of GSSG relative to its synthesis or secretion, the ratio of GSH to GSSG is a good indicator of OS within cells. GSH and GSSG levels can be determined by capillary electrophoresis coupled with laser induced fluorescence detection [79, 80].
- Measures of the oxidized bases, 8-oxo-dG and 8-OHdG, in urine, plasma, and digested DNA can be made by capillary electrophoresis with electrochemical detection. Capillary electrophoresis allows greater separation efficiency than HPLC methods for these compounds [81].
- Analysis of malondialdehyde (MDA) in EBC has been made using capillary electrophoresis coupled with laser induced fluorescence detection and resulted in very low limit of quantification (LOQ) allowing the determination of this compound at physiological levels [82].
- It can be noted that quantification of vitamin C and homocysteine in human plasma using capillary electrophoresis with different detectors (UV, electrochemical, and electrochemiluminescence detection) has been described and developed by some researchers [83–86].

21.10 COMET Assay

Single-cell gel electrophoresis-comet assay is a sensitive way for the detection of DNA damage at the individual mammalian cell level [87]. It can be relatively simple by running DNA taken from lymphocytes. This method employs purified endonucleases that specifically recognize the damaged bases in DNA and create instead a DNA break. Formamidopyrimidine DNA glycosylase (FPG) is a specific repair endonuclease that specifically removes 8-OHdG, leaving a base-less sugar and therefore converts 8-OHdG to apurinic site [88]. The latter is hydrolyzed by an associated AP endonuclease, introducing a new strand break at the site of the oxidative damage. DNA loops containing breaks lose supercoiling and unwind. They are released easily from the nucleus and form a comet tail after single-cell electrophoretic assay. Extent of DNA movement from lysed cells (visualized using fluorescent DNA stain) on electrophoresis gel depends on supercoiling changes and strand breaks. DNA breaks can be quantified by image intensification and computer analysis or by visual grading.

Since this method does not require DNA extraction or derivatization, the generation of artifacts is minimized. The 8-OHdG values obtained with the FPG-comet assay are the lowest so far, the highest were reported with GC-MS method, and intermediate with HPLC-EC method [89].

Another variant of the COMET assay has been introduced by Collins et al. In this assay, OGG-1 (8-oxoguanine DNA glycosylase) activity is measured in lymphocyte extracts. The assay creates 8-oxo-G oxidation then measures the activity of repair enzyme in extract of lysed cells [90].

21.11 PCR and Gene Expression

A general caveat for COMET assay is that this method does not tell us whether the damage occurs in inactive genes, active genes, or telomeres. There is a tremendous need to know the location of damage because it will tell the investigator the biological importance and consequences of such alterations. The combined use of analytical chemistry and molecular genetics is likely to be very productive to solve this issue. With redox biology and ROS/RNS species playing such vital roles in living systems, their impact on gene expression has been well studied. Direct signaling by ROS/RNS can impact gene expression, affecting cell health and cell proliferation.

Molecular techniques such as PCR have been used to determine the expression of several specific gene sequences related to OS. Table 21.1 provides a small listing of the popular ROS/RNS-associated genes examined.

Nrf-2 is activated in response to OS and its quantification by RT-PCR may have clinical significance and could, therefore, be used as an indicator of RS generation that exceeds the existing antioxidant defense systems. It has been reported that Nrf-2 activity declines with age and with some degenerative disorders [91]. On the other hand, an increased Nrf-2 activity has been noted in transformed cells [92].

Table 21.1 Selected genes involved in RS metabolism and gene response to OS

ALOX12	ANGPTL7	AOX1	APOE	ATOX1	BNIP3	CAT	CCL5
CCS	CSDE1	CYBA	CYGB	DGKK	DUOX1	DUOX2	DUSP1
EPHX2	EPX	FOXM1	GLRX2	GPR156	GPX1	GPX2	GPX3
GPX5	GPX6	GR	GSS	GTF2I	IPCEF1	KRT1	LPO
MBL2	MPO	MPV17	MSRA	MT3	MTL5	NCF2	NME5
NOS2	NF-κB	NOX1	NOX2	NOX3	NOX4	NOX5	NRF-2
NUDT1	OXR1	OXSR1	PDLIM1	PNKP	PRDX2	PRDX5	PREX1
PRG3	PRNP	RNF7	SCARA3	SEPP1	SFTPD	SGK2	SIRT2
PRX	SOD1	SOD2	SOD3	SRXN1	STK25	TPO	TTN
TXNRD2	XO						

Determination of its levels in tumor samples by qRT-PCR is becoming a tool to clinically assess tumor malignancy. For all these reasons, Nrf-2 can be considered an interesting and valid biomarker for OS.

Nuclear factor-κB (NF-κB) is a transcriptional factor, which stimulates the expression of genes coding for ROS-generating enzymes (COX-2 and iNOS) and pro-inflammatory cytokines. Evaluation of NF-κB expression using quantitative PCR could be used as an indicator of the state of the mucosal immune and oxidative environment.

Scarpa et al. analyzed early events in ulcerative colitis (UC)-related and non-inflammatory carcinogenesis with reference to CD80 expression to clarify whether NF-κB pathway, evaluated using PCR, is involved in its up-regulation in UC patients. They found that NF-κB mRNA expression was directly correlated with CD80 mRNA mucosal levels. This finding is of undoubted practical value as it can help to identify early events of cancerogenesis in UC patients [93].

NADPH oxidases are heme flavoenzymes that produce cellular ROS. PCR is a useful technique to evaluate the expression of this enzyme in OS-related diseases and therapeutic evaluation of antioxidants. This enzyme, in particular NOX 4, which is responsible for the constant production of $O_2^{\bullet -}$ at the endothelium, was evaluated using quantitative PCR in order to evaluate the activity of natural antioxidants [94]. Besides, measurement of NOX4 expression by quantitative real-time PCR can be useful to identify the origin of intestinal fibrosis, which is a severe complication of Crohn's disease (CD). Indeed, Hotta et al. support a function for NOX4 in intestinal fibrogenesis and identified it as a potential therapeutic target in recalcitrant fibrotic disorders of CD patients [95].

Finally, with our growing understanding of gene regulation and RNA metabolism, miRNAs are increasingly being researched for their role in OS and ROS/RNS regulation. Table 21.2 lists some of the miRNAs that are known to touch OS and redox biology.

21.12 Flow Cytometry (FC)

FC is one of the most powerful tools for single-cell analysis. It is routinely used in immune system exploration, diagnosis, and evaluation of blood cancers, as well as detection of human immunodeficiency virus (HIV) infection [96–98]. This technique can be used for the detection of RS in cells with fluorescent probes.

The advantage of FC-based assays is that they represent a rapid and semi-quantitative measure of RS in biological systems. In addition, they offer the advantage of being able to assess the intracellular generation of RS in cells cultured in artificial media and therefore distinguish between the cells showing differential OS status, which can never be distinguished using spectrometric techniques. Quantitative data on the numbers of cells emitting fluorescent probes can be obtained rather than relative fluorescence units.

Table 21.2 miRNA targets associated with OS

MicroRNAs	Association with OS
miR-150	Downregulated
miR-142-5p	Downregulated
miR-122	Downregulated
miR-34c	Upregulated
miR-34-5p	Upregulated
miR-29b	Upregulated
miR-743a	Involved
miR-335	Involved
miR-34a	Involved
miR-200c	Involved
miR-145	Involved
miR-205	Involved
miR-320	Involved
Let-7	Involved
miR-23	Involved
miR-144	Involved
miR-451	Involved

Flow cytometry-based assays are also advantageous in that the thiol heterogeneity of different cell populations can be observed. Indeed, this technique has been used to demonstrate that distribution of GSH is not always homogenous throughout the whole cell population but can be heterogeneous, especially when cultured cells are treated with various agents [99]. However, FC has few disadvantages as cells are required to be in suspension and require either scraping or trypsinization, which can be a source of OS elevation. Control experiments to optimize assay conditions must always be conducted to limit this issue.

In addition to the significant contribution of FC in the evaluation of oxidative burst during chronic granulomatous disease [100], this instrumentation has been used for many years to evaluate oxidative burst in several pathological conditions including asymptomatic HIV^+ individuals [101] and autoimmune neutropenia [102]. It is noteworthy to know that the decrease in the T-cell GSH content in HIV^+ patients was first demonstrated by FC [103].

Flow cytometry is a technique that is based on the separation of cells using hydrodynamic focusing and reading of specific molecule using laser excitation and subsequent detection. Cytometry or fluorimetry can be used for the detection of RS with fluorescent probes. Many probes for the detection of RS have been developed in last years. A large number of fluorescent probes are now commercially available with specific binding to RS but with different degrees of sensitivity. Conventionally, RS are determined using $DCFH_2$ probe and more advanced probes are also being used nowadays.

21.12.1 DCFH and Derivatives

2-7-dichlorodihydrofluorescein (DCFH) is one of the most widely used techniques for directly measuring the redox state of a cell (see Chap. 11). It is a cell-permeable, non-fluorescent precursor that can be used as an intracellular probe for OS. The oxidation of DCFH to the fluorescent compound 2-7-dichlorofluorescein (DCF) was initially proposed to be a relatively specific indicator of hydrogen peroxide formation [104]. It has been used extensively in the detection of oxidants (H_2O_2, $NO^{\bullet}$, $ONOO^{-}$, and $OH^{\bullet}$) produced during the respiratory burst in a variety of intact cells using FC [105–108].

The diacetate form of DCFH (DCFH-DA) is taken up by cells where intracellular esterases hydrolyze the diacetate moiety to yield DCFH. The latter is a relatively polar and cell membrane-impermeable product which has been suggested to remain trapped within the intracellular space. This non-fluorescent molecule accumulates intracellularly and subsequent oxidation yields fluorescence.

Peroxidases (e.g., horseradish peroxidase (HRP) and MPO), heme proteins (e.g., cytochrome c), and reduced iron (Fe^{2+}) have been shown to oxidize $DCFH_2$ to DCF in the presence of H_2O_2. Hence, cellular peroxidase levels and heme protein content are critical variables to consider when interpreting results using fluorescent probes, and the suitability of $DCFH_2$-DA as an indicator of RS production is increasingly being questioned [109].

Fluorescence of DCF in cells is measured with excitation at 495 nm and emission at 522. Fluorescence can be measured using a flow cytometer and is intended to be proportional to the concentration of H_2O_2 in cells.

With the development and widespread use of FC, DCFH oxidation is now routinely used to quantify ROS generation within individual cells or populations of cells. It has many advantages over other techniques. It is very easy to use, inexpensive, extremely sensitive to redox changes in cells and can be used to follow changes in ROS over time. However, there are limitations to the interpretation of DCF fluorescence as a specific marker for quantitative intracellular hydrogen peroxide formation. Other limitations encountered with this probe are undesirable hemolysis when performing FC, the presence of esterase inhibitors or antioxidants, self-propagation of DCF radicals, and being multidrug resistance (MDR) substrates or inducers.

There is another analog called 5-(and-6)-chloromethyl-2′,7′-dichlorodihydrofluorescein diacetate, acetyl ester (CM-DCFH-DA), which is a chloromethyl derivative of DCFH-DA. This probe passively diffuses into cells where acetate groups are cleaved by intracellular esterases and thiol-reactive chloromethyl group reacts with intracellular thiols including GSH. Subsequent oxidation of the probe yields a fluorescent adduct that is trapped inside the cell, therefore facilitating long-term studies.

Other derivatives of DCFH have been developed. For instance, intracellularly converted diacetate derivatives of 4,5-diaminofluorescein diacetate (DAF-2-DA) and 4-amino-5-methylamino-2′,7′-difluorofluorescein diacetate (DAF-FM-DA) have widely been used for ROS/RNS detection [110–114]. Once taken up by cells, these probes are hydrolyzed by intracellular esterases, generating the non-fluorescent and membrane-impermeable DAF-2 and DAF-FM respectively. Subsequent oxidation by ROS/RNS results in the formation of fluorescent probes. These compounds share some limitations with $DCFH_2$-DA as they are sensitive to esterase inhibitors and can be recognized as MDR substrates or inducers in cells.

Finally, carboxymethyl compounds (CM-DCFH-DA) are newer derivatives for H_2O_2, $ONOO^-$, $OH^•$, and $ROO^•$ detection with much popular choice than $DCFH_2$-DA due to long shelf life and higher quantum yield.

21.12.2 Monohalobimanes

Monochlorobimanes (MCB) and monobromobimanes (MBB) (see Chap. 10) are versatile and membrane-permeable probes that bind irreversibly to sulfhydryl groups yielding fluorescent products. These probes are commonly used for thiol detection in FC, which gain fluorescence upon conjugation with thiol-containing compounds. Fluorescence is measured by setting the excitation and emission maxima at 395 and 480 nm, respectively.

Because MCB and MBB react with all thiols, including protein thiols, they can only provide an estimate of cell thiol status and GSH levels. Measuring GSH levels in a cell can often be valuable in determining the overall redox state of the cell.

MCB conjugation with GSH is dependent on glutathion S-transferase (GST) activity, and results from these GSH measurements tend to reflect GST activity rather than actual GSH levels [115, 116]. For this reason, the use of MCB has been heavily criticized. Monobromobimane, on the other hand, is less dependent on GST for its conjugation with thiols and gives a better indication of GSH levels [117, 118].

21.12.3 Hydroethidine

Hydroethidine (dihydroethidium, HE) is a cell-permeant probe that can undergo two-electron oxidation to form a DNA-intercalating fluorophore (see Chaps. 10 and 11). HE is preferentially oxidized by $O_2^{•-}$, with minimal interferences of H_2O_2, $ONOO^-$, and HOCl [119]. The intracellular oxidation of HE by $O_2^{•-}$ results in the generation of ethidium, which can be analyzed with FC. Visualization of cells, tissues, and anatomic regions displaying increased rates of $O_2^{•-}$ production with digital imaging microfluorimetry has been described in vivo as well as in isolated blood vessels and tissue slices [120, 121]. Unlike other intracellular probes for $O_2^{•-}$, there is little artifactual formation of $O_2^{•-}$ by HE due to redox cycling.

21.12.4 DHR 123

Dihydrorhodamine 123 (DHR123) is an uncharged probe that passively diffuses across cell membranes. It is converted upon oxidation by HClO, H_2O_2, or $ONOO^-$ to the fluorescent membrane-impermeant rhodamine 123 (Rho 123) (see Chap. 11), which is predominantly sequestered in the mitochondria. This probe presents some inconvenient including self-propagation of DHR radicals, sensitivity to cellular antioxidants, and being recognized as MDR substrate or inducer in cells.

21.12.5 C11-BODIPY

4,4-difluoro-5-(4-phenyl-1,3-butadienyl)-4-bora-3a,4a-diaza-s-indacene-3-undecanoic acid (C11-BODIPY) is a lipophilic probe used to evaluate RS production in leukocytes and platelets (see Chap. 11) [122, 123].

C11-BODIPY is a 11-carbon fatty acid in which boron dipyrromethene difluoride core (BODIPY) is substituted by a phenyl group via a conjugated diene. The latter is sensitive to oxidation and when oxidized by $ROO^•$, shortening of the conjugated electron resonance structures between the phenyl group and the BODIPY core shifts C11-BODIPY's fluorescence from red to green. Oxidation and also nitration of BODIPY in the presence of $ONOO^-$ result in reduced red fluorescence but not necessarily increased green fluorescence.

One of the major advantages of this probe is neither C11-BODIPY nor its oxidized products are able of spontaneous leak from the lipid bilayer. The ratio of oxidized to non-oxidized C11-BODIPY can be used to normalize probe incorporation in cells of different size (granulocytes lymphocytes, and monocytes) [123, 124]. Only hemolysis and antioxidants, particularly uric acid, could bias the measurement of $ROO^•$ generation using this probe [125].

21.12.6 Other Probes

There are many probes for measuring RS production in cells using FC. Chemiluminescence of luminol and lucigenin (see Chap. 9) and fluorescence (see Chaps. 10 and 11) of several compounds have all been used successfully to detect RS generation. A list of commercially available probes used for identification RS in cells using FC is provided in Table 21.3.

Table 21.3 Commonly used fluorescent/chemiluminescent probes for ROS/RNS detection by flow cytometry

RS	Probes	Localization
Singlet oxygen	Trans-1-(2′-methoxyvinyl)pyrene	Intracellular
Superoxide anion	HE Coelenterazine TEMPO-9-AC MCLA XTT MTT NBT Luminol Lucigenin	Intracellular
Hydrogen peroxide	$DCFH_2$ $DCFH_2$-DA CM-$DCFH_2$-DA Dihydrocalcein AM DHR123 Lucigenin Luminol	Intracellular
Hydroxyl radical	$DCFH_2$ $DCFH_2$-DA CM-$DCFH_2$-DA Proxyl fluorescamine 3′-(*p*-Aminophenyl) fluorescein 3′-(*p*-Hydroxyphenyl) fluorescein	Intracellular
Peroxyl radicals	$DCFH_2$ $DCFH_2$-DA *cis*-Parinaric acid C11-BODIPY CM-$DCFH_2$-DA DPPP Luminol	Membrane
Nitric oxide	DAF-2 DA DAF-FMDA 2,3-Diaminonaphthalene Luminol	Intracellular
Peroxynitrite	$DCFH_2$ $DCFH_2$-DA 3′-(*p*-Aminophenyl) fluorescein 3′-(*p*-Hydroxyphenyl) fluorescein CM-$DCFH_2$-DA Coelenterazine DHR123 Luminol	Intracellular
Hypochlorous acid	Aminophenyl fluorescein DHR123 Luminol	Intracellular
Glutathione	Monochlorobimanes Monobromobimanes CM-$DCFH_2$-DA	Intracellular

References

1. Nicholson JK, Lindon JC, Holmes E. 'Metabonomics': understanding the metabolic responses of living systems to pathophysiological stimuli via multivariate statistical analysis of biological NMR spectroscopic data. Xenobiotica. 1999;29(11):1181–9.
2. el-Deredy W. Pattern recognition approaches in biomedical and clinical magnetic resonance spectroscopy: a review. NMR Biomed. 1997;10(3):99–124.
3. Jasbi P, Nikolich-Žugich J, Patterson J, Knox KS, Jin Y, Weinstock GM, et al. Targeted metabolomics reveals plasma biomarkers and metabolic alterations of the aging process in healthy young and older adults. GeroScience. 2023;45(6):3131–46.
4. Liu L-q, Zhang P, Qi Y-z, Li H, Jiang Y-h, Yang C-h. Quercetin Attenuates Atherosclerosis via Modulating Apelin Signaling Pathway Based on Plasma Metabolomics. Chinese Journal of Integrative Medicine. 2023;29(12):1121–32.
5. Choi B, San José Estépar R, Godbole S, Curtis JL, Wang JM, San José Estépar R, et al. Plasma metabolomics and quantitative interstitial abnormalities in ever-smokers. Respiratory Research. 2023;24(1):265.
6. Ma L, Liu J, Deng M, Zhou L, Zhang Q, Xiao X. Metabolomics analysis of serum and urine in type 1 diabetes patients with different time in range derived from continuous glucose monitoring. Diabetology & metabolic syndrome. 2024;16(1):21.
7. Chen J, Liu J, Cao D. Urine metabolomics for assessing fertility-sparing treatment efficacy in endometrial cancer: a non-invasive approach using ultra-performance liquid chromatography mass spectrometry. BMC Women's Health. 2023;23(1):583.
8. Fernández-Peralbo MA, Calderón Santiago M, Priego-Capote F, Luque de Castro MD. Study of exhaled breath condensate sample preparation for metabolomics analysis by LC–MS/MS in high resolution mode. Talanta. 2015;144:1360–9.
9. Costa CP, Marques J, Silva D, Barbosa C, Oliveira AS, Santos M, et al. Metabolomics profiling of human exhaled breath condensate by SPME/GC × GC-ToFMS: Exploratory study on the use of face masks at the level of lipid peroxidation volatile markers. Microchemical Journal. 2021;171:106830.
10. Bang G, Park JH, Park C, Kim K-j, Kim JK, Lee SY, et al. High-resolution metabolomics-based biomarker discovery using exhaled breath condensate from patients with lung cancer. Journal of Analytical Science and Technology. 2022;13(1):37.
11. Rahman I, Biswas SK. Non-invasive biomarkers of oxidative stress: reproducibility and methodological issues. Redox report : communications in free radical research. 2004;9(3): 125–43.
12. Yu M, Beyers RJ, Gorden JD, Cross JN, Goldsmith CR. A Magnetic Resonance Imaging Contrast Agent Capable of Detecting Hydrogen Peroxide. Inorganic Chemistry. 2012;51(17): 9153–5.
13. Loving GS, Mukherjee S, Caravan P. Redox-activated manganese-based MR contrast agent. J Am Chem Soc. 2013;135(12):4620–3.
14. Su HS, Nahrendorf M, Panizzi P, Breckwoldt MO, Rodriguez E, Iwamoto Y, et al. Vasculitis: Molecular Imaging by Targeting the Inflammatory Enzyme Myeloperoxidase. Radiology. 2012;262(1):181–90.
15. Tsitovich PB, Spernyak JA, Morrow JR. A redox-activated MRI contrast agent that switches between paramagnetic and diamagnetic states. Angew Chem Int Ed Engl. 2013;52(52): 13997–4000.
16. Tsitovich PB, Burns PJ, McKay AM, Morrow JR. Redox-activated MRI contrast agents based on lanthanide and transition metal ions. Journal of inorganic biochemistry. 2014;133:143–54.
17. Querol M, Chen JW, Bogdanov JAA. A paramagnetic contrast agent with myeloperoxidase-sensing properties. Organic & Biomolecular Chemistry. 2006;4(10):1887–95.
18. Querol M, Chen JW, Weissleder R, Bogdanov A, Jr. DTPA-bisamide-based MR sensor agents for peroxidase imaging. Org Lett. 2005;7(9):1719–22.

19. Breckwoldt MO, Chen JW, Stangenberg L, Aikawa E, Rodriguez E, Qiu S, et al. Tracking the inflammatory response in stroke in vivo by sensing the enzyme myeloperoxidase. Proc Natl Acad Sci U S A. 2008;105(47):18584–9.
20. Ronald JA, Chen JW, Chen Y, Hamilton AM, Rodriguez E, Reynolds F, et al. Enzyme-sensitive magnetic resonance imaging targeting myeloperoxidase identifies active inflammation in experimental rabbit atherosclerotic plaques. Circulation. 2009;120(7):592–9.
21. Towner RA, Smith N, Saunders D, Lupu F, Silasi-Mansat R, West M, et al. In vivo detection of free radicals using molecular MRI and immuno-spin trapping in a mouse model for amyotrophic lateral sclerosis. Free Radic Biol Med. 2013;63:351–60.
22. Utsumi H, Yamada K, Ichikawa K, Sakai K, Kinoshita Y, Matsumoto S, et al. Simultaneous molecular imaging of redox reactions monitored by Overhauser-enhanced MRI with 14N- and 15N-labeled nitroxyl radicals. Proc Natl Acad Sci U S A. 2006;103(5):1463–8.
23. Matsumoto K-i, Hyodo F, Matsumoto A, Koretsky AP, Sowers AL, Mitchell JB, et al. High-Resolution Mapping of Tumor Redox Status by Magnetic Resonance Imaging Using Nitroxides as Redox-Sensitive Contrast Agents. Clinical Cancer Research. 2006;12(8): 2455–62.
24. Hyodo F, Matsumoto K, Matsumoto A, Mitchell JB, Krishna MC. Probing the intracellular redox status of tumors with magnetic resonance imaging and redox-sensitive contrast agents. Cancer research. 2006;66(20):9921–8.
25. Yang F, Hu S, Zhang Y, Cai X, Huang Y, Wang F, et al. A Hydrogen Peroxide-Responsive O2 Nanogenerator for Ultrasound and Magnetic-Resonance Dual Modality Imaging. Advanced Materials. 2012;24(38):5205–11.
26. Uemura T, Kitagawa S. Prussian Blue Nanoparticles Protected by Poly(vinylpyrrolidone). Journal of the American Chemical Society. 2003;125(26):7814–5.
27. Olson ES, Orozco J, Wu Z, Malone CD, Yi B, Gao W, et al. Toward in vivo detection of hydrogen peroxide with ultrasound molecular imaging. Biomaterials. 2013;34(35):8918–24.
28. Shokouhimehr M, Soehnlen ES, Hao J, Griswold M, Flask C, Fan X, et al. Dual purpose Prussian blue nanoparticles for cellular imaging and drug delivery: a new generation of T1-weighted MRI contrast and small molecule delivery agents. Journal of Materials Chemistry. 2010;20(25):5251–9.
29. Louie A. Multimodality Imaging Probes: Design and Challenges. Chemical reviews. 2010;110 (5):3146–95.
30. Jiang C, Liu R, Han G, Zhang Z. A chemically reactive Raman probe for ultrasensitively monitoring and imaging the in vivo generation of femtomolar oxidative species as induced by anti-tumor drugs in living cells. Chemical Communications. 2013;49(59):6647–9.
31. Wessels JM, Rodgers MAJ. Detection of the O2(1.DELTA.g) .fwdarw. O2(3.SIGMA.g-) Transition in Aqueous Environments: A Fourier-Transform Near-Infrared Luminescence Study. The Journal of Physical Chemistry. 1995;99(43):15725–7.
32. Snyder JW, Zebger I, Gao Z, Poulsen L, Frederiksen PK, Skovsen E, et al. Singlet Oxygen Microscope: From Phase-Separated Polymers to Single Biological Cells. Acc Chem Res. 2004;37(11):894–901.
33. Li H, Melø TB, Arellano JB, Razi Naqvi K. Temporal profile of the singlet oxygen emission endogenously produced by photosystem II reaction centre in an aqueous buffer. Photosynthesis Research. 2012;112(1):75–9.
34. Schweitzer C, Schmidt R. Physical Mechanisms of Generation and Deactivation of Singlet Oxygen. Chemical reviews. 2003;103(5):1685–758.
35. Mattila H, Khorobrykh S, Havurinne V, Tyystjärvi E. Reactive oxygen species: Reactions and detection from photosynthetic tissues. Journal of Photochemistry and Photobiology B: Biology. 2015;152:176–214.
36. Gorman AA, Rodgers MA. Current perspectives of singlet oxygen detection in biological environments. J Photochem Photobiol B. 1992;14(3):159–76.

37. Yong K-T, Roy I, Swihart MT, Prasad PN. Multifunctional nanoparticles as biocompatible targeted probes for human cancer diagnosis and therapy. Journal of Materials Chemistry. 2009;19(27):4655–72.
38. Sun X, Lei Y. Fluorescent carbon dots and their sensing applications. TrAC Trends in Analytical Chemistry. 2017;89:163–80.
39. Wang B, Lu S. The light of carbon dots: From mechanism to applications. Matter. 2022;5(1): 110–49.
40. Yu C, Wu Y, Zeng F, Wu S. A fluorescent ratiometric nanosensor for detecting NO in aqueous media and imaging exogenous and endogenous NO in live cells. Journal of Materials Chemistry B. 2013;1(33):4152–9.
41. Gao C, Guo Z, Liu J-H, Huang X-J. The new age of carbon nanotubes: An updated review of functionalized carbon nanotubes in electrochemical sensors. Nanoscale. 2012;4(6):1948–63.
42. Hersam MC. Progress towards monodisperse single-walled carbon nanotubes. Nature Nanotechnology. 2008;3(7):387–94.
43. Leeuw TK, Reith RM, Simonette RA, Harden ME, Cherukuri P, Tsyboulski DA, et al. Single-Walled Carbon Nanotubes in the Intact Organism: Near-IR Imaging and Biocompatibility Studies in Drosophila. Nano Letters. 2007;7(9):2650–4.
44. Heller DA, Jin H, Martinez BM, Patel D, Miller BM, Yeung T-K, et al. Multimodal optical sensing and analyte specificity using single-walled carbon nanotubes. Nature Nanotechnology. 2009;4(2):114–20.
45. Kim J-H, Patra CR, Arkalgud JR, Boghossian AA, Zhang J, Han J-H, et al. Single-Molecule Detection of H2O2 Mediating Angiogenic Redox Signaling on Fluorescent Single-Walled Carbon Nanotube Array. ACS Nano. 2011;5(10):7848–57.
46. Kim J-H, Heller DA, Jin H, Barone PW, Song C, Zhang J, et al. The rational design of nitric oxide selectivity in single-walled carbon nanotube near-infrared fluorescence sensors for biological detection. Nature Chemistry. 2009;1(6):473–81.
47. O'Connell MJ, Bachilo SM, Huffman CB, Moore VC, Strano MS, Haroz EH, et al. Band gap fluorescence from individual single-walled carbon nanotubes. Science. 2002;297(5581): 593–6.
48. Unnikrishnan S, Klibanov AL. Microbubbles as ultrasound contrast agents for molecular imaging: preparation and application. AJR Am J Roentgenol. 2012;199(2):292–9.
49. Gessner R, Dayton PA. Advances in molecular imaging with ultrasound. Mol Imaging. 2010;9 (3):117–27.
50. Ten Kate GL, van den Oord SC, Sijbrands EJ, van der Lugt A, de Jong N, Bosch JG, et al. Current status and future developments of contrast-enhanced ultrasound of carotid atherosclerosis. J Vasc Surg. 2013;57(2):539–46.
51. Perng JK, Lee S, Kundu K, Caskey CF, Knight SF, Satir S, et al. Ultrasound imaging of oxidative stress in vivo with chemically-generated gas microbubbles. Ann Biomed Eng. 2012;40(9):2059–68.
52. Gitler C, Mogyoros M, Kalef E. Labeling of protein vicinal dithiols: role of protein-S2 to protein-(SH)2 conversion in metabolic regulation and oxidative stress. Methods in enzymology. 1994;233:403–15.
53. Frost SC, Schwalbe MS. Uptake and binding of radiolabelled phenylarsine oxide in 3T3-L1 adipocytes. The Biochemical journal. 1990;269(3):589–95.
54. Schuppe-Koistinen I, Gerdes R, Moldéus P, Cotgreave IA. Studies on the reversibility of protein S-thiolation in human endothelial cells. Arch Biochem Biophys. 1994;315(2):226–34.
55. Seres T, Ravichandran V, Moriguchi T, Rokutan K, Thomas JA, Johnston RB, Jr. Protein S-thiolation and dethiolation during the respiratory burst in human monocytes. A reversible post-translational modification with potential for buffering the effects of oxidant stress. Journal of immunology (Baltimore, Md : 1950). 1996;156(5):1973–80.
56. Thomas JA, Poland B, Honzatko R. Protein sulfhydryls and their role in the antioxidant function of protein S-thiolation. Arch Biochem Biophys. 1995;319(1):1–9.

57. Ravichandran V, Seres T, Moriguchi T, Thomas JA, Johnston RB. S-thiolation of glyceraldehyde-3-phosphate dehydrogenase induced by the phagocytosis-associated respiratory burst in blood monocytes. Journal of Biological Chemistry. 1994;269(40):25010–5.
58. Alomari Ea, Bruno S, Ronda L, Paredi G, Bettati S, Mozzarelli A. Protein carbonylation detection methods: A comparison. Data in Brief. 2018;19:2215–20.
59. Wehr N, Levine R. Quantification of Protein Carbonylation. Methods Mol Biol. 2013;965: 265–81.
60. Nakamura A, Goto S. Analysis of protein carbonyls with 2,4-dinitrophenyl hydrazine and its antibodies by immunoblot in two-dimensional gel electrophoresis. Journal of biochemistry. 1996;119(4):768–74.
61. Dumaswala UJ, Zhuo L, Jacobsen DW, Jain SK, Sukalski KA. Protein and lipid oxidation of banked human erythrocytes:: Role of glutathione. Free Radical Biology and Medicine. 1999;27(9):1041–9.
62. Robinson CE, Keshavarzian A, Pasco DS, Frommel TO, Winship DH, Holmes EW. Determination of protein carbonyl groups by immunoblotting. Anal Biochem. 1999;266(1):48–57.
63. Duncan MW. A review of approaches to the analysis of 3-nitrotyrosine. Amino Acids. 2003;25(3–4):351–61.
64. Sultana R, Reed T, Butterfield DA. Detection of 4-hydroxy-2-nonenal- and 3-nitrotyrosine-modified proteins using a proteomics approach. Methods Mol Biol. 2009;519:351–61.
65. Zhan X, Desiderio DM. The human pituitary nitroproteome: detection of nitrotyrosyl-proteins with two-dimensional Western blotting, and amino acid sequence determination with mass spectrometry. Biochemical and Biophysical Research Communications. 2004;325(4):1180–6.
66. Inoue R, Sakata N, Nakai K, Aikawa H, Tsutsumi M, Nii K, et al. Nε-(Carboxymethyl)lysine in Debris from Carotid Artery Stenting: Multiple versus Nonmultiple Postoperative Lesions. Journal of Stroke and Cerebrovascular Diseases. 2014;23(10):2827-33.
67. Kim JG, Sabbagh F, Santanam N, Wilcox JN, Medford RM, Parthasarathy S. Generation of a Polyclonal Antibody Against Lipid Peroxide-Modified Proteins. Free Radical Biology and Medicine. 1997;23(2):251–9.
68. Manevich Y, Hutchens S, Halushka PV, Tew KD, Townsend DM, Jauch EC, et al. Peroxiredoxin VI oxidation in cerebrospinal fluid correlates with traumatic brain injury outcome. Free Radical Biology and Medicine. 2014;72:210–21.
69. Szabó-Taylor KÉ, Eggleton P, Turner CAL, Faro MLL, Tarr JM, Tóth S, et al. Lymphocytes from rheumatoid arthritis patients have elevated levels of intracellular peroxiredoxin 2, and a greater frequency of cells with exofacial peroxiredoxin 2, compared with healthy human lymphocytes. The international journal of biochemistry & cell biology. 2012;44(8):1223–31.
70. Salzano S, Checconi P, Hanschmann EM, Lillig CH, Bowler LD, Chan P, et al. Linkage of inflammation and oxidative stress via release of glutathionylated peroxiredoxin-2, which acts as a danger signal. Proc Natl Acad Sci U S A. 2014;111(33):12157–62.
71. Checconi P, Salzano S, Bowler L, Mullen L, Mengozzi M, Hanschmann E-M, et al. Redox Proteomics of the Inflammatory Secretome Identifies a Common Set of Redoxins and Other Glutathionylated Proteins Released in Inflammation, Influenza Virus Infection and Oxidative Stress. PLOS ONE. 2015;10(5):e0127086.
72. Bayer SB, Maghzal G, Stocker R, Hampton MB, Winterbourn CC. Neutrophil-mediated oxidation of erythrocyte peroxiredoxin 2 as a potential marker of oxidative stress in inflammation. 2013;27(8):3315–22.
73. Frijhoff J, Winyard PG, Zarkovic N, Davies SS, Stocker R, Cheng D, et al. Clinical Relevance of Biomarkers of Oxidative Stress. Antioxid Redox Signal. 2015;23(14):1144-70.
74. Stanicka J, Landry W, Cotter TG. Oxidative Stress Biomarkers and ROS Molecular Probes. Oxidative Stress: Diagnostics, Prevention, and Therapy Volume 2. ACS Symposium Series. 1200: American Chemical Society; 2015. p. 353-74.

75. Tanigawa H, Miura S-i, Zhang B, Uehara Y, Matsuo Y, Fujino M, et al. Low-density lipoprotein oxidized to various degrees activates ERK1/2 through Lox-1. Atherosclerosis. 2006;188(2):245–50.
76. Zinellu A, Sotgia S, Galistu F, Lumbau F, Pasciu V, Mario Pes G, et al. Applications on the monitoring of oxidative modification of LDL by capillary electrophoresis: a comparison with spectrophotometer assay. Talanta. 2004;64(2):428–34.
77. Yang Q, Zhang X, Bao X, Lu H, Zhang W, Wu W, et al. Single cell determination of nitric oxide release using capillary electrophoresis with laser-induced fluorescence detection. Journal of Chromatography A. 2008;1201(1):120–7.
78. Floyd PD, Moroz LL, Gillette R, Sweedler JV. Capillary electrophoresis analysis of nitric oxide synthase related metabolites in single identified neurons. Anal Chem. 1998;70(11): 2243–7.
79. Hodáková J, Preisler J, Foret F, Kubáň P. Sensitive determination of glutathione in biological samples by capillary electrophoresis with green (515nm) laser-induced fluorescence detection. Journal of Chromatography A. 2015;1391:102–8.
80. Zhang J, Hu Z, Chen X. Quantification of glutathione and glutathione disulfide in human plasma and tobacco leaves by capillary electrophoresis with laser-induced fluorescence detection. Talanta. 2005;65(4):986–90.
81. Zhang S, Song X, Zhang W, Luo N, Cai L. Determination of low urinary 8-hydroxy-2`-deoxyguanosine excretion with capillary electrophoresis and molecularly imprinted monolith solid phase microextraction. Science of The Total Environment. 2013;450–451:266–70.
82. Lačná J, Foret F, Kubáň P. Sensitive determination of malondialdehyde in exhaled breath condensate and biological fluids by capillary electrophoresis with laser induced fluorescence detection. Talanta. 2017;169:85–90.
83. Ivanov AV, Virus ED, Luzyanin BP, Kubatiev AA. Capillary electrophoresis coupled with 1,1′-thiocarbonyldiimidazole derivatization for the rapid detection of total homocysteine and cysteine in human plasma. Journal of Chromatography B. 2015;1004:30–6.
84. Tao Y, Zhang X, Wang J, Wang X, Yang N. Simultaneous determination of cysteine, ascorbic acid and uric acid by capillary electrophoresis with electrochemiluminescence. Journal of Electroanalytical Chemistry. 2012;674:65–70.
85. Cheng M-L, Liu T-Z, Lu F-J, Chiu DT-Y. Simultaneous detection of vitamin C and uric acid by capillary electrophoresis in plasma of diabetes and in aqueous humor in acute anterior uveitis. Clinical Biochemistry. 1999;32(6):473–6.
86. Wang X, Li L, Li Z, Wang J, Fu H, Chen Z. Determination of ascorbic acid in individual liver cancer cells by capillary electrophoresis with a platinum nanoparticles modified electrode. Journal of Electroanalytical Chemistry. 2014;712:139–45.
87. Khelfi A, Djemili I, Bellouard M-M, Hedhili A, Nebbali M, Hoummani H, et al. CHAPITRE 8 - Toxicologie des autres agents/produits toxiques. In: Khelfi A, Djemili I, Bellouard M-M, Hedhili A, Nebbali M, Hoummani H, et al., editors. Tout-En-un de Toxicologie. Paris: Elsevier Masson; 2023. p. 623–86.
88. Taranenko MV, Volkov EM, Saparbaev MK, Kuznetsova SA. New Noncleavable Analogs of 8-Oxoguanine-DNA Glycosylase Substrates. Molecular Biology. 2004;38(5):728–36.
89. Collins AR, Cadet J, Mőller L, Poulsen HE, Viña J. Are we sure we know how to measure 8-oxo-7,8-dihydroguanine in DNA from human cells? Archives of Biochemistry and Biophysics. 2004;423(1):57–65.
90. Collins AR. Investigating oxidative DNA damage and its repair using the comet assay. Mutation Research/Reviews in Mutation Research. 2009;681(1):24–32.
91. Bahn G, Jo DG. Therapeutic Approaches to Alzheimer's Disease Through Modulation of NRF2. Neuromolecular medicine. 2019;21(1):1–11.
92. Gañán-Gómez I, Wei Y, Yang H, Boyano-Adánez MC, García-Manero G. Oncogenic functions of the transcription factor Nrf2. Free Radical Biology and Medicine. 2013;65:750–64.

93. Scarpa M, Cardin R, Bortolami M, Kotsafti A, Scarpa MC, Pozza A, et al. Mucosal immune environment in colonic carcinogenesis: CD80 expression is associated to oxidative DNA damage and TLR4–NFκB signalling. European Journal of Cancer. 2013;49(1):254–63.
94. Silva V, Genta G, Möller MN, Masner M, Thomson L, Romero N, et al. Antioxidant Activity of Uruguayan Propolis. In Vitro and Cellular Assays. Journal of Agricultural and Food Chemistry. 2011;59(12):6430–7.
95. Hotta Y, Uchiyama K, Takagi T, Kashiwagi S, Nakano T, Mukai R, et al. Transforming growth factor β1-induced collagen production in myofibroblasts is mediated by reactive oxygen species derived from NADPH oxidase 4. Biochemical and Biophysical Research Communications. 2018;506(3):557–62.
96. Bordoni V, Casetti R, Capuano G, De Stefani B, Piselli P, Gioia C, et al. A novel 8-color flow cytometry panel to study activation, maturation and senescence of CD4 and CD8 T lymphocytes in HIV-infected individuals at different stages of disease. International journal of immunopathology and pharmacology. 2012;25(2):415–24.
97. Whitby L, Whitby A, Fletcher M, Barnett D. Current laboratory practices in flow cytometry for the enumeration of CD 4(+) T-lymphocyte subsets. Cytometry Part B, Clinical cytometry. 2015;88(5):305–11.
98. Sharpe C, Davis J, Mason K, Tam C, Ritchie D, Koldej R. Comparison of gene expression and flow cytometry for immune profiling in chronic lymphocytic leukaemia. Journal of Immunological Methods. 2018;463:97–104.
99. Han D, Handelman G, Marcocci L, Sen CK, Roy S, Kobuchi H, et al. Lipoic acid increases de novo synthesis of cellular glutathione by improving cystine utilization. Biofactors. 1997;6(3): 321–38.
100. El Hawary R, Meshaal S, Deswarte C, Galal N, Abdelkawy M, Alkady R, et al. Role of Flow Cytometry in the Diagnosis of Chronic Granulomatous Disease: the Egyptian Experience. Journal of clinical immunology. 2016;36(6):610–8.
101. Spear GT, Kessler HA, Rothberg L, Phair J, Landay AL. Decreased oxidative burst activity of monocytes from asymptomatic HIV-infected individuals. Clinical immunology and immunopathology. 1990;54(2):184–91.
102. Macey MG, Sangster J, Veys PA, Newland AC. Flow cytometric analysis of the functional ability of neutrophils from patients with autoimmune neutropenia. Journal of microscopy. 1990;159(Pt 3):277–83.
103. Roederer M, Staal FJ, Osada H, Herzenberg LA, Herzenberg LA. CD4 and CD8 T cells with high intracellular glutathione levels are selectively lost as the HIV infection progresses. International immunology. 1991;3(9):933–7.
104. Keston AS, Brandt R. The fluorometric analysis of ultramicro quantities of hydrogen peroxide. Analytical Biochemistry. 1965;11(1):1–5.
105. Soto-Mercado V, Mendivil-Perez M, Jimenez-Del-Rio M, Fox JE, Velez-Pardo C. Cannabinoid CP55940 selectively induces apoptosis in Jurkat cells and in ex vivo T-cell acute lymphoblastic leukemia through H2O2 signaling mechanism. Leukemia Research. 2020;95:106389.
106. Wang Y-X, Han F-Y, Duan Z-K, Chang Y, Lin B, Wang X-B, et al. Phenolics from Archidendron clypearia (Jack) I.C.Nielsen protect SH-SY5Y cells against H2O2-induced oxidative stress. Phytochemistry. 2020;176:112414.
107. Huang L, Jiang W, Zhu L, Ma C, Ou Z, Luo C, et al. γ-Oryzanol suppresses cell apoptosis by inhibiting reactive oxygen species-mediated mitochondrial signaling pathway in H2O2-stimulated L02 cells. Biomedicine & Pharmacotherapy. 2020;121:109554.
108. Pan Y, Deng Z-Y, Chen X, Zhang B, Fan Y, Li H. Synergistic antioxidant effects of phenolic acids and carotenes on H2O2-induced H9c2 cells: Role of cell membrane transporters. Food Chemistry. 2021;341:128000.
109. Kalyanaraman B, Darley-Usmar V, Davies KJ, Dennery PA, Forman HJ, Grisham MB, et al. Measuring reactive oxygen and nitrogen species with fluorescent probes: challenges and limitations. Free Radic Biol Med. 2012;52(1):1–6.

110. Sharma R, Bani Hani S, Bayachou M, Sabanegh E, Agarwal A. Intracellular nitric oxide measurement in human sperm using 4, 5-diaminofluorescein-2-diacetate and flow cytometry. Fertility and sterility. 2011;96(3, Supplement):S232–S3.
111. de Andrade CM, Rey FM, Cintra ACO, Sampaio SV, Torqueti MR. Effects of crotoxin, a neurotoxin from Crotalus durissus terrificus snake venom, on human endothelial cells. International Journal of Biological Macromolecules. 2019;134:613–21.
112. Tanoue S, Chang L-Y, Li Y, Kaplan DE. Monocyte-derived dendritic cells from cirrhotic patients retain similar capacity for maturation/activation and antigen presentation as those from healthy subjects. Cellular Immunology. 2015;295(1):36–45.
113. Cohen J, D'Agostino L, Tuzer F, Torres C. HIV antiretroviral therapy drugs induce premature senescence and altered physiology in HUVECs. Mechanisms of Ageing and Development. 2018;175:74–82.
114. Deshane JS, Redden DT, Zeng M, Spell ML, Zmijewski JW, Anderson JT, et al. Subsets of airway myeloid-derived regulatory cells distinguish mild asthma from chronic obstructive pulmonary disease. Journal of Allergy and Clinical Immunology. 2015;135(2):413–24.e15.
115. Ublacker GA, Johnson JA, Siegel FL, Mulcahy RT. Influence of glutathione S-transferases on cellular glutathione determination by flow cytometry using monochlorobimane. Cancer research. 1991;51(7):1783–8.
116. Cook JA, Iype SN, Mitchell JB. Differential specificity of monochlorobimane for isozymes of human and rodent glutathione S-transferases. Cancer research. 1991;51(6):1606–12.
117. Hedley DW, Chow S. Evaluation of methods for measuring cellular glutathione content using flow cytometry. Cytometry. 1994;15(4):349–58.
118. Sen CK, Roy S, Han D, Packer L. Regulation of cellular thiols in human lymphocytes by alpha-lipoic acid: a flow cytometric analysis. Free Radic Biol Med. 1997;22(7):1241–57.
119. Benov L, Sztejnberg L, Fridovich I. Critical evaluation of the use of hydroethidine as a measure of superoxide anion radical. Free Radic Biol Med. 1998;25(7):826–31.
120. Bindokas VP, Jordán J, Lee CC, Miller RJ. Superoxide production in rat hippocampal neurons: selective imaging with hydroethidine. J Neurosci. 1996;16(4):1324–36.
121. Miller FJ, Gutterman DD, Rios CD, Heistad DD, Davidson BL. Superoxide Production in Vascular Smooth Muscle Contributes to Oxidative Stress and Impaired Relaxation in Atherosclerosis. 1998;82(12):1298–305.
122. Becatti M, Fiorillo C, Gori AM, Marcucci R, Paniccia R, Giusti B, et al. Platelet and leukocyte ROS production and lipoperoxidation are associated with high platelet reactivity in Non-ST elevation myocardial infarction (NSTEMI) patients on dual antiplatelet treatment. Atherosclerosis. 2013;231(2):392–400.
123. Peluso I, Manafikhi H, Altieri F, Zanza C, Palmery M. The effect of sample storage on the Peroxidation of Leukocytes Index Ratio (PLIR) measure. Scientific Reports. 2014;4.
124. Drummen GP, van Liebergen LC, Op den Kamp JA, Post JA. C11-BODIPY(581/591), an oxidation-sensitive fluorescent lipid peroxidation probe: (micro)spectroscopic characterization and validation of methodology. Free Radic Biol Med. 2002;33(4):473–90.
125. Peluso I, Manafikhi H, Reggi R, Longhitano Y, Zanza C, Palmery M. Relationship between the Peroxidation of Leukocytes Index Ratio and the Improvement of Postprandial Metabolic Stress by a Functional Food. Oxid Med Cell Longev. 2016;2016:5630985.

Chapter 22
Oxidative Stress Monitoring: Needs and Opportunities

A. Khelfi, A. Mahmoud, S. Perrone, R. Marín, F. Hussain, T. Fulop, A. Khalil, F. Pilo, R. Rehman, S. Roumeliotis, V. Liakopoulos, and R. Henkel

Abstarct Oxidative stress is characterized by the diminished capacity of a biological system to overcome the overproduction of reactive species (RS) and free radicals (FR). This process has been related to the pathogenesis of virtually every disease. Since this process is a leading cause of many diseases (e.g., ischemic stroke, heart failure, hypertension, cardiomyopathy, rheumatoid arthritis, β-thalassemia, and myelodysplastic syndrome), there is an urgent need for its monitoring. Detecting the levels of this oxidation could provide an early warning signal as biochemical markers of oxidative stress are a feasible way of early detection of these diseases. The lack of accessibility of appropriate early predictors of these diseases has determined that vulnerable patients should be identified on the basis oxidative stress status in their body. Therefore, global indicators of oxidative stress represent a convenient tool for patient stratification and can be used to assess novel therapeutic approaches that target oxidative stress. Indeed, several lines of evidence support a beneficial role of antioxidant supplementation in attenuating oxidative events involved in several chronic diseases.

A. Khelfi (✉)
Faculty of Pharmacy, University of Algiers, Algiers, Algeria

National Centre of Toxicology, Algiers, Algeria

Department of Chemistry and Biomolecular Science, Clarkson University, Potsdam, NY, USA
e-mail: ab.khelfi@univ-alger.dz; akhelfi@clarkson.edu

A. Mahmoud
Department of Life Sciences, Faculty of Science and Engineering, Manchester Metropolitan University, Manchester, UK

S. Perrone
University of Parma, Parma, Italy

R. Marín
Center for Biophysics and Biochemistry (CBB), Venezuelan Institute for Scientific Research (IVIC), Caracas, Venezuela

S. Andreescu et al. (eds.), *Biomarkers of Oxidative Stress*,
https://doi.org/10.1007/978-3-031-60738-7_22

In this chapter, clinical trials are summarized for major oxidative stress-related disorders. The efficacy of antioxidant treatments with particular focus on the evaluation of oxidative stress biomarkers is also presented. The chapter also explores opportunities to improve the diagnosis of disease and the design of new therapeutic agents. In addition, it presents the clinical evidence to support the monitoring of the levels of therapeutic antioxidants in patients with oxidative stress-related diseases.

Keywords Ischemic stroke · Heart failure · Hypertension · Cardiomyopathy · Rheumatoid arthritis · β-Thalassemia · Myelodysplastic syndrome · Therapeutic antioxidants

There is growing evidence that redox regulators, related functional mediators, cellular organelles, and surrounding environments are all tied together in intricate networks that affect the whole body. The scientific community has learned a great deal about these networks in development, metabolism, energetic, state of health, and disease.

With the onset and progression of many diseases being increasingly linked with changes in the cellular redox status, accurately examining redox changes has become increasingly important. As such, the number of publications that involve ROS has increased over the last 10 years.

F. Hussain
Clinico-Molecular Biochemistry Laboratory, Department of Biochemistry, University of Agriculture, Faisalabad, Pakistan

T. Fulop · A. Khalil
Department of Medicine, Geriatrics Service, Faculty of Medicine and Health Sciences, University of Sherbrooke, Sherbrooke, QC, Canada

F. Pilo
Hematology and Transplant Center, Ospedale Oncologico di Riferimento Regionale "Armando Businco", Azienda Ospedaliera Brotzu, Cagliari, Italy

R. Rehman
Department of Biological & Biomedical Sciences, Aga Khan University, Karachi, Pakistan

S. Roumeliotis · V. Liakopoulos
2nd Department of Nephrology, Medical School, AHEPA Hospital, Aristotle University of Thessaloniki, Thessaloniki, Greece

R. Henkel
LogixX Pharma Ltd., Theale, Berkshire, UK

Department of Medical Bioscience, University of the Western Cape, Bellville, South Africa

Department of Metabolism, Digestion & Reproduction, Imperial College, London, UK

Over the last two decades, a variety of distinct oxidative biomarkers have been developed, methodologically validated, and applied in clinical populations. Such markers may represent an underlying specific disease mechanism and a means for therapeutic monitoring and outcome prediction. Examining oxidative stress (OS) biomarkers has also the potential to provide greater insight into the characteristics of redox-related disease and may help the setup of new diagnostic approaches. Overall, exploration of OS in clinical settings is urgently needed for the following reasons:

22.1 Involvement of OS in Physiological and Pathophysiological Conditions

Redox signaling is universally integrated with central homeostatic mechanisms at the molecular, cellular, tissue, and organismic levels. Some of RS actions are considered physiological and others are deeply involved in so many pathological situations, especially if they are uncontrolled (Table 22.1). One of the well-known functions that demonstrates the involvement of OS in human homeostasis is the pivotal and apparent contribution of redox balance to gastrointestinal physiology and gut health. This has significantly changed our understanding of the way that oxidants maintain the intestinal barrier and control complex host–flora interactions.

22.2 Involvement of OS in Several Diseases

There are several diseases where OS play a key role in pathogenesis. Even if it remains difficult to demonstrate the mechanistic implication of redox status in numerous pathologies, it is indisputable and collectively admitted that RS play a fundamental role in numerous situations. Moreover, there is growing evidence supporting the role of deficient antioxidant system in severe conditions. Extensive data suggest that oxidative damage highly correlate with a number of human diseases. Some of which, OS has been hypothesized to represent a major risk factor of diseases with very high rates of morbidity and mortality:

- The data are quite robust in showing strong associations of OS with various manifestations of preclinical to advanced cardiovascular (CVD) and/or prediction of CVD events. There may be a case for looking at markers that are related directly to the oxidation of LDL since this is a central event that occurs in atherosclerotic plaques. For instance, intima-media thickness was strongly and positively associated with oxLDL/oxLDL-Ab ratio in carotid atherosclerosis [130]. Moreover, oxLDL was found to be correlated with the extent of atherosclerosis [131]. Recently, the role of these oxidative biomarkers in response to cardiovascular therapeutics is being actively studied [132, 133].

Table 22.1 Oxidative stress-associated pathologies

System/organ	Pathology	Mechanisms	Biomarkers of OS	References
Neurological system	Alzheimer	Aβ aggregation and generation of FR leading to consequent damage to neuronal membrane systems (lipid peroxidation and oxidation of membrane proteins). Interaction between transition metals and Aβ (Aβ:metal complexes promote further aggregation and facilitate the generation of FR) Excitotoxicity resulting from the Aβ-induced release of glutamate from surrounding glial cells and aberrant stimulation of NMDA receptors by glutamate increasing the influx of calcium with the activation of neuronal NOS and excessive formation of toxic $NO^{\bullet}$ Disturbance of calcium homeostasis resulting in mitochondrial dysfunction Generation of RS by mitochondria and redox-active metal ions (aberrant amounts of $O_2^{\bullet -}$ and H_2O_2, which diffuses across biological membranes and decomposes by redox metals, thereby contributing the flux of $OH^{\bullet}$ through Fenton chemistry) Induction of microglia activation and neuroinflammation Generation of ROS and especially $O_2^{\bullet -}$ by the microglia via the NOX catalytic activity Reduction in the activity of several components of the respiratory chain (e.g., cytochrome oxidase) and disruption of mitochondrial energy metabolism Disruption in the homeostasis of redox-inert metals (Zinc) Generation of neurotoxic MDA, 4-hydroxynonenal, isoprostanes, and acrolein in brain Generation of nitrotyrosine in the brain Initiation of Aβ deposition via 4-HNE- or iron-activated g-secretase activity Formation of AGEs in the brain Decrease of GST amount in the brain Decrease of GST amount in the blood	NO, MDA, 4-HNE, isoprostanes, acrolein, AGEs, GST, 8-OH-dG, 5-OH-methylC, and nitrotyrosine in brain 4-HNE, MPO, Homocysteine and GSH in blood Oxidized proteins	[1–11]

	Parkinson's disease	Chronic neuro-inflammation mediated by glial cells via iNOS, mitochondrial complex I deficiency, and excess of iron (correlation between damage to the brain's substantia nigra and iron accumulation) Autoxidation of iron–neuromelanin complexes Increased lipid, DNA, and protein damage in substantia nigra Increase of 4-HNE in the brain Depletion of GSH in the brain α-Synuclein aggregation by the inhibition of mitochondrial function impairing in turn further mitochondrial function in the brain Familial PD: A53T mutation of α-synuclein with impaired vesicular storage of dopamine leading to its accumulation and interaction resulting in high generation of ROS, a process that increases with age	MDA in blood 8-OH-dG, protein carbonyls, AGEs, GSH, and 4-HNE in the brain	[12–15]
	Amyotrophic lateral sclerosis (ALS)	Some inherited ALS: mutations in copper/zinc SOD, which converts the protein from an antioxidant to a pro-oxidant capable of causing oxidative damage to the cells with generation of 4-HNE and isoprostanes and depletion of GSH	4-HNE, MDA, SOD, and GSH in blood Protein carbonyls, 4-HNE, nitrotyrosine, isoprostanes, and 8-OH-dGua in spinal cords	[16]
Cardiovascular system	Atherosclerosis	Oxidation and deposition of LDL particles by ROS, forming ox-LDL, which are internalized by macrophages and monocytes (possible sources of ROS is NOX in phagocytic cells and auto-oxidation of homocysteine). Transformation of the latter to foam cells, which form fibrous dome-shaped plaques Overgeneration of $ONOO^-$ due to the reaction of $O_2^{\bullet -}$ with endothelium-generated $NO^{\bullet}$, which induces a "second round" of OS and LDL peroxidation oxLDL can induce several effects: The stimulation of the production of pro-inflammatory cytokines and growth factors leading to the activation of additional accumulation of phagocytes Upregulation of the expression of cellular adhesion molecules that	oxLDL, MPO, 3-chlorotyrosine, nitrotyrosine, MDA, 4-HNE, isoprostanes, homocysteine, and AGEs in blood	[17–21]

(continued)

Table 22.1 (continued)

System/organ	Pathology	Mechanisms	Biomarkers of OS	References
		facilitate leukocyte binding Induction of the release of factors that stimulate smooth muscle cell proliferation Stimulation of NOX and overgeneration of $O_2^{\bullet -}$ Parallel oxidation of hemoglobin and release of heme and iron, promoting further oxidation of lipids and vascular cell cytotoxicity Generation of HOCl by phagocyte-derived MPO with increased production of 3-chlorotyrosine High generation of nitrotyrosine, MDA, 4-HNE, and isoprostanes Glycation of protein and generation of AGEs, especially relevant in diabetic patients associated with atherosclerosis		
	Ischemic stroke	Reduced state of the respiratory chain due to the low amount of oxygen resulting in ATP depletion and Na^+/K^+-ATPase pump inhibition Reperfusion event: Overgeneration of ROS due to the quick influx of oxygen and the fully respiratory chain available, leading to enhanced formation of $O_2^{\bullet -}$ and H_2O_2 Overproduction of ROS overwhelms existing antioxidant capacities and causes damage to macromolecules (DNA, proteins, and lipids) leading to severe dysfunctions in the heart/brain Post-ischemic peroxidation of phospholipids in ER and high increase in cytoplasmic calcium concentration leading to uncontrolled activation NOS with an increase of the bioavailability of $NO^{\bullet}$ and $ONOO^{-}$ as well as the opening of the mitochondrial permeability transition pore, which results in diffusion of molecules including ROS and cytochrome c from the mitochondria to the cytoplasm Activation of XO (producing further $O_2^{\bullet -}$, which can be converted to more hazardous RS such as $OH^{\bullet}$ via metal-mediated Fenton reaction),	Isoprostanes, oxLDL, protein carbonyl, homocysteine, and GSH in blood	[22–25]

		COX and NOX in activated phagocytes Auto-oxidation of homocysteine		
	Heart failure	Excess ROS negatively affects the cardiomyocytes through stimulation of hypertrophy, alteration of Ca^{2+} regulation, induction of fibrosis and apoptosis, and activation of inflammatory response. ROS can directly activate G-protein coupled receptor (GPCR) and upon activation, GPCR leads to ROS generation. Activation of GPCR and hypertrophic signaling pathways in the myocardium are induced by α-adrenergic stimulation, Ang II, endothelin-1 (ET-1), and bradykinin. GPCR activation-mediated NADPH oxidase-dependent CaMKII activation promotes apoptosis via the activation of p38-MAPK. ROS affect the function of ryanodine receptors (RyRs) by oxidizing thiol (SH) groups, and the abnormal Ca^{2+} leaking may provoke inward Na^{+} current, leading to delayed afterdepolarizations. ROS can alter sarcoplasmic reticulum (SR) Ca^{2+}-ATPase (SERCA) activity and increase L-type Ca^{2+} channels opening resulting in early afterdepolarizations (EADs). ROS can activate transforming growth factor (TGF)-β that promotes the transdifferentiation of fibroblasts into myofibroblasts, and JNK that is involved in differentiation and proliferation of cardiac fibroblasts. ROS can also activate matrix metalloproteinase (MMPs) leading to the proliferation of cardiac fibroblasts, matrix remodeling, and fibrosis.	MDA, GSH, SOD, uric acid, GPx, $NO^{•}$	[26–32]
	Hypertension	Mitochondrial dysfunction, NADPH oxidase activation, eNOS uncoupling, and endoplasmic reticulum stress are the main sources of ROS implicated in OS in hypertension. Excess ROS increase intracellular Ca^{2+}, resulting in vasoconstriction and the development of hypertension. Superoxide radicals react with $NO^{•}$ producing the potent oxidant ONOO−, leading to vasoconstriction due to decreased nitric oxide bioavailability. Ang-II-induced hypertension involves the generation of ROS via	MDA, GSH, SOD, GPx, $NO^{•}$, and homocysteine	[33–38]

(continued)

Table 22.1 (continued)

System/organ	Pathology	Mechanisms	Biomarkers of OS	References
		activation of NOXs and all NOXs are regulated by Ang II. Mitochondrial dysfunction associated with inactivation of sirtuin 3 (SIRT3) and SOD2 contributes to the development and progression of hypertension. Excessive ROS generated by the dysfunctional mitochondria cause OS and vascular dysfunction along with hypertension in pre-clinical studies. ER stress is involved in vascular cell dedifferentiation, phenotypic switching, apoptosis, and calcification. It also contributes to vascular remodeling and the dysfunction of the endothelial lining in the pathogenesis of hypertension and atherosclerosis.		
	Cardiomyopathy	OS in the myocardium results in altered signaling pathways and oxidative damage to major cell molecules and subsequently cell death via apoptosis or necrosis. Excess ROS negatively affect cardiomyocytes by stimulating hypertrophy, disrupting Ca^{2+} handling, and provoking inflammation, fibrosis and apoptosis.	MDA, GSH, SOD, GPx, and $NO^{\bullet}$	[39]
	Atrial fibrillation	Surplus ROS generation can cause re-entry and focal activity by modifying several ionic currents in the myocardium and OS is therefore arrhythmogenic. Excess ROS are associated with altered Ca^{2+} homeostasis and downregulation of L-type calcium current mediated via S-nitrosylation. Mitochondrial-derived ROS oxidizes atrial RyR2 resulting in increased leak of intracellular Ca^{2+} and dysregulated Ca^{2+} homeostasis and AF development. ROS reduce total Na^{+} channels delaying cardiac conduction and facilitating re-entry, prolong the duration of action potential by enhancing late Na^{+} current, prolong cardiac conduction by interfering with gap junctions, and activate inflammatory and fibrotic processes. Increased Ang II and TGF-β1 and subsequent activation of Rac1 lead	Nitrotyrosine, protein carbonyls, MDA, GSH, SOD, and MPO	[40–45]

		to NOX activation, elevated ROS, and fibrosis in AF. Xanthine oxidase-derived superoxide generation has been demonstrated in atrial appendages in the setting of atrial-tachypacing. Myeloperoxidase generates HOCl that elicits signaling cascades leading to collagen deposition and arrhythmias. In patients with AF, high circulating MPO activity and large burden in right atrium have been reported.		
Respiratory system	Asthma	Release of ROS ($O_2^{\bullet-}$ and H_2O_2) by inflammatory cells in the airways (NOX of macrophages, neutrophils, and eosinophils), which can result in lung injury through direct oxidative cell damage to epithelial cells Contribution of ROS (particularly $O_2^{\bullet-}$, H_2O_2, and $ONOO^-$) to the induction of allergic airway inflammation Induction of NF-κB activator and generation of pro-inflammatory cytokines by ROS Induction of bronchial hyperreactivity, contraction of airway smooth muscle, histamine release from mast cells, and mucous production from epithelial cells by ROS Increased production of $NO^{\bullet}$ in patients with severe asthma Increased generation of 4-HNE and protein carbonyl in lung Impaired SOD activity associated with airflow obstruction (possible damage by ROS causing tyrosine nitration of SOD) Oxidative damage to proteins and lipids	H_2O_2 in breath condensate Isoprostanes (8-iso-PGF2α) in plasma and BAL (markers of severity) Exhaled $NO^{\bullet}$ (marker of treatment) Exhaled ethane 4-HNE and protein carbonyl in lung Nitrotyrosine and chlorotyrosine in plasma 3-bromotyrosine in BAL proteins	[46–48]
	COPD	Presence of numerous RS and FR in cigarette smoke resulting in COPD (e.g., $NO^{\bullet}$, tar-semiquinones) Sequestration of immune cells in the pulmonary microvasculature leading to the alteration of the mitochondrial respiration and high production of $O_2^{\bullet-}$ and H_2O_2 Formation of MDA and 4-HNE-protein adducts resulting from exacerbated lipid peroxidation Stimulation of smooth muscle contraction by 8-isoprostanes High levels of nitrotyrosine and iNOS in lung tissue Decrease of enzymatic antioxidant system (e.g. SOD and CAT) and	Exhaled H_2O_2 Isoprostanes and hydrocarbons (ethane and pentane) in plasma and exhaled air condensate Isoprostane $F_{2\alpha}$–III in urine MDA and 4-HNE in in plasma and exhaled breath Protein carbonyls,	[49–51]

(continued)

Table 22.1 (continued)

System/organ	Pathology	Mechanisms	Biomarkers of OS	References
		GSH Increase of biomarker levels of oxidative damage to proteins, lipids, and DNA	3-chlorotyrosine, MPO GSH, SOD, and CAT in blood Nitrotyrosine and iNOS in lung tissue	
Gastrointestinal	IBD	Presence of phagocytes-derived MPO in inflamed intestine resulting in a prolonged and intense mucosal exposure to RHS. RHS non-specifically reacts with sulfhydryls, polyunsaturated fatty acids, DNA, and aromatic amino acids Production of ROS by XO, AO, and amine oxidase as well as the NOX in the macrophages of mucosal lamina propria Activation of NF-kB signaling pathway by AGE and NOX leading to the generation of pro-inflammatory cytokines and increased expression of caspase 3 and ICAM Generation of ROS and $NO^{\bullet}$ by highly produced cytokines such as TNF-α and IFN-γ Considerable production of H_2O_2 in the lamina propria due to the accumulation of phagocytic leukocytes Increase of $O_2^{\bullet-}$ synthesis leading to vascular endothelium permeability Modification of mucosal vasodilatation, increased epithelial permeability, and altered motility by $NO^{\bullet}$ produced by macrophages, IECs, and myofibroblasts iNOS-derived $NO^{\bullet}$ stimulates TNF-α production in the middle and distal colon, which promotes the infiltration of neutrophils through stimulation of synthesis of intracellular adhesion molecule (ICAM) and P-selectin Both direct and indirect measurements of RS levels Simultaneous increase of $O_2^{\bullet-}$ and $NO^{\bullet}$ leading to the production of peroxynitrite stress that can cause damage to enzymes and DNA	AGEs content in plasma NO, NOS, HOCl, MPO, H_2O_2, 3-nitrotyrosine, malondialdehyde, 4-hydroxynonenal, protein carbonyls, SOD, GPx, CAT, GSH in the intestines	[52, 53]

		Excessive production of ROS and RNS can lead to tissue damage of the host via oxidation of lipids, proteins, and DNA Slightly decreased levels of Cu/Zn-SOD and extracellular SOD but markedly upregulated expression of manganese (Mn)-SOD in inflamed mucosa Decrease of CAT and GPX of the antioxidant system Depletion of small anti-oxidative agents (GSH, α- and β-carotene, lycopene, ascorbic acid, and β-cryptoxanthin)		
Eyes	Macular degeneration disease AMD	Direct light exposure of the macula and generation of $^1O_2$ Plasma membranes of cells in the macula have the highest polyinsatured fatty acid amount of any body tissue Retinal oxygen turnover is very high and cellular mitochondria are abundant → The macula strongly sensitive to OS and ROS overgeneration	NA	[54–56]
	Cataract	Formation of $^1O_2$ and H_2O_2 in the nucleus Accumulation of oxidized lens components and decreased efficiency of repair mechanisms and antioxidant systems (e.g., pyruvate and GSH) Lipid peroxidation and loss of protein sulfhydryl groups with oxidation of cysteine and methionine residues	NA	[57, 58]
Endocrine/metabolic system Multiorgan	Diabetes	Enhanced production of $O_2^{\bullet-}$ and H_2O_2 through activation of NOX and dysfunctional cellular antioxidant response Due to NOX activation, there are elevated levels of MDA in diabetic individuals ROS damage PARP-1 (Poly [ADP-ribose] polymerase 1) is released and blocks G3PDH (glyceraldehyde 3-phosphate dehydrogenase) activity that leads to excess of glycolytic intermediates Defects in both insulin secretion, action, and signaling by chronic OS (OS-mediated insulin resistance, e.g., modification of IRS-1 and serine/threonine kinases, activation of activates redox-sensitive signaling pathways including NF-κB, p38MAPK, JNK/SAPK, and PKC isoforms)	Isoprostane and AGEs in blood MDA, AGEs, $NO^{\bullet}$, CAT, and GSH	[59–67]

(continued)

Table 22.1 (continued)

System/organ	Pathology	Mechanisms	Biomarkers of OS	References
		Increase in ROS levels by insulin resistance (vicious circle) and excessive flux of glucose and free fatty acids into cells stimulating the hexosamine pathway, leading to impaired sensitivity to insulin signaling Increased metabolism in mitochondria as a result of high glucose levels and subsequent high electron leak with ROS overproduction Higher degree of lipid peroxidation (isoprostanes) in skeletal muscle of obese insulin-resistant patients High concentrations of protein carbonylation, as well as o- and m-tyrosine β-Cell dysfunction due to increased mitochondrial activity with increased ROS production induced by high glucose and fatty acid levels Low expression of antioxidant enzymes (SOD, CAT, and GPX) in β-cells Reduced glutathione levels contribute to oxidative damage and cellular dysfunction Diabetes complications: OS and glucose auto-oxidation induce non-enzymatic glycation of proteins and increase the formation of AGEs AGEs, including pentosidine and CML, upon interaction with the receptor RAGE, induce tissue damage, elevate ROS production via NOX activation and prompt inflammatory responses through NF activation, contributing to diabetic complications. AGEs activate OS through NOX and NF-κB, contributing to inflammation and exacerbating diabetic complications. Protein kinase C (PKC) pathway, stimulated by elevated glucose-induced diacylglycerol production, intricately influences gene expression, promoting oxidant and AGE synthesis through the NOX complex AGEs enhance LDL-induced endothelial cell toxicity via redox-		

		sensitive MAPK activation LDL modification by endogenous AGEs contributes to cardiovascular issues ROS generation and lipid peroxidation may play a vital role in the development of ocular injury in diabetic patients AO-derived ROS may play a critical role in diabetic cardiovascular complications		
Reproductive system	Male infertility	Release of high amount of ROS by immature sperm cells with flawed morphology and diminished fertility potential High production of H_2O_2 molecules in pathophysiological conditions such as infection and inflammation resulting in the generation of $OH^{\bullet}$ in the presence of transition metals via Fenton reaction High generation of ROS results in infertility by: Damage the sperm lipidic membrane, which results in decreased sperm motility and ability to fuse with the oocyte Alteration of the sperm DNA (modification of all bases, formation of base-free sites, frameshifts, deletions, DNA cross-links, chromosomal arrangements, single- and double-strand breaks and fragmentation of DNA) Induction of high inflammation process Reduced antioxidative ability of spermatozoa Lipid peroxidation induced by ROS generation causes axonemal damage, low sperm viability and increased defects in the midpiece of the sperms with diminished sperm motility Mutations in nDNA and mtDNA cause defect in respiratory chain and mitochondrial energy metabolism Caspase-mediated apoptosis by ROS	ROS, MDA, lipid peroxidation, and ORP	[68, 69]
	Preeclampsia	There is an alteration in the development of the placental vasculature during the first weeks of pregnancy. This leads to low perfusion and increased hypoxia, with a concomitant release of antiangiogenic factors into the maternal circulation, together with a modification of the maternal systemic endothelial function, which can explain the	TBARS and conjugated dienes peroxidation in red blood cells, blood plasma, and placental tissue SODF2-IsoPs and protein	[70–78]

(continued)

Table 22.1 (continued)

System/organ	Pathology	Mechanisms	Biomarkers of OS	References
		hypertension that is characteristic of this clinical syndrome There is an increased lipid peroxidation, vasoconstriction, vascular resistance, $NO^{\bullet}$ metabolic changes, and coagulation abnormalities There is also a reduction of the $NO^{\bullet}$ availability, prostacyclin synthesis, and a rise of the expression of cell adhesion molecules, prothrombotic factors, and several cytokines. Mitochondria and the ER produce byproducts of auto-oxidation of ETC components including ROS. NOX is another significant source of ROS Vascular endothelial damage is leading to OS with lipid peroxidation which in turn results in endothelial dysfunction	carbonyl mRNA expressions of HO-1 and HO-2, SOD, GPx, and CAT PMCA activity in red blood cells, myometrial, and placental plasma membranes	
	Endometriosis	Iron, apoptotic endometrial tissue, and desquamated menstrual cells are transported into the peritoneal cavity after retrograde menstruation and can induce a chronic inflammatory response with cytokines activating granulocytes and macrophages. In turn these leukocytes produce large amounts of ROS. As a result, important molecular pathways are modified	MDA, AOPP, NOS 2, TAC in follicular fluid can be measured. Toll-like receptors in the ovarian cortex	[78–82]
	PCOS	Obesity, androgen excess, visceral adiposity, and insulin resistance form a vicious cycle that leads to increased production of ROS and OS	8-OHdG, TAC in the follicular fluid	[83, 84]
Multiorgan	Sepsis	Systemic inflammatory response, with neutrophil infiltration, activation, and respiratory burst generated by NOX High production of $O_2^{\bullet-}$ by endothelial cells via XO catalytic activation by neutrophils High generation of HOCI (generated by MPO) and $NO^{\bullet}$ (generated by iNOS), which further reaction with $O_2^{\bullet-}$ to form $ONOO^-$	NA	[85]
Immune system	Systemic lupus erythematosus	Increase of oxidized amino acids and correlation with disease activity (methionine sulfoxide, 3-nitrotyrosine, and protein carbonyls) Decrease of protein thiols Elevated serum levels of MDA-, 4-HNE-modified proteins, and AGEs	Protein carbonyls, 3-nitrotyrosine, AGE, oxLDL, MDA, 4-HNE, F2-isoprostane, and GSH	[86–88]

		Presence of autoantibodies to MDA-modified epitopes High levels of oxLDL and F2-isoprostane linked to high disease activity Depletion of GSH	in plasma Autoantibodies to MDA-modified epitopes in plasma	
	Acute leukemia	Acute myeloid leukemia (AML): Excessive ROS production and oxidative DNA damage Altered ROS levels stimulate leukemogenesis through regulation of redox-sensitive transcription factors (e.g., Nrf2, Bach1, NF-κB, AP-1, and HIF1α) Acute lymphoblastic leukemia (ALL): Depletion of GSH and vitamin C Impairment of antioxidant enzymes CAT and SOD Increase in the production of H_2O_2 by the cancer cells	CAT, SOD, GSH, and vitamin C in blood	[89–91]
Joint system	Rheumatoid arthritis	High production of ROS ($O_2^{\bullet -}$, H_2O_2, lipid hydroperoxides, hypohalous acids) by activated neutrophils and macrophages in the inflamed rheumatoid joint Chronic inflammation and gradual degeneration of the joints by ROS High activity of MPO and production of TNF-α Enhanced OS status in synovial cells Possible release of ferrous ion and generation of $OH^{\bullet}$ in synovial fluid Excessive production of RS and damage to protein (e.g., protein carbonyls chlorotyrosine, and nitrotyrosine), carbohydrates, lipids (e.g., oxLDL), nucleic acids (8-OHdGua), lipids (isoprostanes), and matrix components (e.g., hyaluronic acid and collagen) in the synovial fluid Signaling functions of ROS amplifying the synovial inflammatory response (T-cells, exposed to increased OS, become hyporesponsive and refractory to growth and death stimuli) Decline in the levels of cellular antioxidant sulfhydryls	Protein carbonyls, MPO, chlorotyrosine, nitrotyrosine, oxLDL, and OH-dG in the synovial fluid Isoprostanes and GSH in blood	[92–94]
Blood	Beta-thalassemia	Increase of free iron levels, lipid peroxides, and RS intermediates Subsequent production of $OH^{\bullet}$ via Fenton reaction Decrease in TAC	NA	[95–98]

(continued)

Table 22.1 (continued)

System/organ	Pathology	Mechanisms	Biomarkers of OS	References
	Myelodysplastic syndrome	Iron overload markedly decreased the ratio of immature HSCs and injured the clonogenic HSPC's capacity through the NOX4/ROS/P38 MAPK signaling pathways OS is also involved in the pathogenesis of the bone loss during iron excess and may affect the relationship between hemopoietic cells and microenvironment in MDS	NA	[99–101]
Kidney	Urolithiasis	Increase of RS production in stone-forming conditions Initiation of lipid peroxidation by high concentrations of oxalate	MDA in serum	[102–104]
	Chronic kidney disease	CKD-related heightened inflammatory environment disrupts redox balance. Inactivation of $NO^{\bullet}$ by $O_2^{\bullet-}$ impairs vasodilation and causes overproduction of $ONOO^-$ and generated vasoconstriction molecules disrupting kidney hemodynamics, endothelium malfunction, and salt retention. As CKD progresses, reduced clearance of oxidants and accumulation of uremic toxins further augment OS. OS-related comorbidities like diabetes, hypertension, and aging usually coexist with CKD. Dietary restrictions aiming to prevent hyperkalemia, (fruits and vegetables, natural sources of anti-oxidants) limit the restoration of anti-oxidant mechanisms In dialysis, both hemodialysis and peritoneal dialysis induce oxidant production and antioxidants' loss	MDA, ox-LDL, AOPPs, AGEs, F2-isoprostanes, and GSH-Px in blood	[105–107]
Skin	Melanoma	Pathophysiological processes including DNA damage and lipid peroxidation	NA	[108–110]
Bone	Osteoporosis, Paget's disease, periodontal disease, osteosar-coma, and cancer bone metastasis	ROS-mediated production by NOX2 and mitochondria resulting in activation of redox-sensitive signaling pathways (e.g., NF-κB, MAPK, PI3K/Akt) and RANKL-induced osteoclastogenesis	NA	[111, 112]

Muscle system	Rhabdomyolysis	Generation of free iron by heme oxygenase (HO) from the heme moiety Release of heme iron from myoglobin and formation of OH·via Fenton/Haber–Weiss reactions with heme iron conversion to iron-based FR, namely ferryl or perferryl radicals	NA	[113, 114]
	Extensive physical exercise	During moderate physical exercise, increase of the oxygen consumption (10- to 15-fold above resting levels) resulting in a temporary increase in the rate of production of mitochondrial ROS During high-intensity exercise, overgeneration of potentially harmful FR that may potentially lead to skeletal muscle damage and inflammation with the onset of an inflammatory signaling cascade Lipid and protein oxidation induced by high-intensity exercise	MDA in blood	[116]
Fat system	Obesity	Elevation of superoxide production by intracellular triglycerides within the ETC and subsequent inhibition of the mitochondrial adenosine nucleotide transporter Increase of OS resulting from the generation of pro-inflammatory cytokines from preadipocytes and adipocytes, which stimulate the production of ROS/RNS by macrophages and monocytes through the upregulation of the activity of NOX, iNOS, and MPO Induction of adipogenesis by ROS through activation of redox-sensitive signaling pathways (e.g., insulin/IGF, and ERK, by inhibiting PTP1B activities), and transcription factors (e.g., C/EBPβ, and PPARγ) Involvement of Nrf2 and OS in adipocyte differentiation	NA	[115]
–	Aging	Decline of mitochondrial integrity and function with the onset of OS as a function of age Increase of FR which promote the processes of aging Direct relationship between OS and telomere shortening is the major cause of senescence Expression of OS-inducible genes is enhanced with aging Elevations of endothelial cell markers of OS (nitrotyrosine) and	Nitrotyrosine in blood	[117–120]

(continued)

Table 22.1 (continued)

System/organ	Pathology	Mechanisms	Biomarkers of OS	References
		correlation with the elevation of p65 (component of the redox sensitive NF-κB transcription factor complex)		
–	Cancer	Production of ROS by proliferating cancer cells Resistance of tumor cells to OS (enhanced antioxidative capacity) Initiation: Induction of cancerogenesis by the action of ROS on nucleic acids (direct oxidation and indirect action by end products of lipid peroxidation such as MDA) leading to DNA damage and mutations Increase of the sensitivity to mutagenic agents by ROS Activation of oncogenes and inhibition of tumor suppressor genes by RS Promotion: Regulation of cellular signaling linked to the mitosis-apoptosis balance by RS (e.g., organic peroxides) leading to an increase in cell proliferation, migration, and survival Modulation of the endogenous antioxidant systems by some tumor promoters High generation of RS during inflammation leading to the activation of nuclear hormone receptors Metastasis: Regulation of the levels of tumor-secreted angiogenic growth factors by ROS Induction of the stability of hypoxia-inducible factor 1 by ROS, a transcription factor that induces vascular endothelial growth factor expression, an important player in angiogenesis Modulation of the expression and activity of matrix metalloproteinases by ROS	8-oxo-dGua/8-OHdGua in blood and urine MDA, 4-HNE, and GSH in blood	[121–123]
Neonatal diseases	Brain injury	Iron release from red blood cell following hypoxia and increased free radical generation through Fenton reaction	Non-protein bound iron in blood Advanced oxidative	[124–126]

		Neuronal injury from FR White matter injury from FR	protein products in blood Isoprostanes in blood	
	Kidney damage	Renal tubule damage mediated by oxygen FR, following hypoxia	Advanced oxidative protein products and total hydroperoxides in blood	[127]
	Necrotizing enterocolitis	Perinatal oxidative insults in intestinal epithelial cells	Advanced oxidative protein products, total hydroperoxides, and non-protein bound iron in blood	[128]
	Down syndrome	Oxidative stress early in pregnancy: increased SOD-1 activity may not be compensated by a rise in GPx	Isoprostanes in amniotic fluid	[129]
Other pathologies and health conditions	Pathologies: Hypertension, myocardial infarction, HIV infection, B and C hepatitis, pancreatitis, acute renal failure, glomerulonephritis, uremia, drug-induced nephrotoxicity (cyclosporine, tacrolimus, and gentamycin), retinal degeneration, emphysema, Keshan disease, porphyria, hyperhomocysteinemia, cystic fibrosis, scleroderma, G6PD deficiency, systemic lupus erythematous, metabolic syndrome, acute respiratory distress syndrome (ARDS), phenylketonuria Health conditions: Kidney transplantation, hemodialysis, nutrition deficiency with decreased GSH and natural antioxidants (vitamin C, vitamin E, carotenoids, coenzyme Q, flavonoids)			

- Basic research has demonstrated the important etiological role of OS in neurodegenerative diseases. Data are emerging that implicate aberrant redox interactions with key proteins in AD, PD, and ALS as well as the subsequent induction of OS leading to neurodegeneration. Vast works has been done to disclose the role of Cu/Zn-SOD in ALS and lipid peroxidation end products with several neurodegenerative diseases [134, 135]. Delineation of lipid peroxidation as a major pathway in most degenerative diseases would depend on adequate standardization and control of measurement conditions.
- All the available evidence suggests that RS-mediated events are important in both primary and downstream pathophysiological mechanisms underlying intestinal inflammation. As such, it has been suggested that ROS play a key role in the initiation of the progression of IBD [136]. Attenuating OS in IBD patients has already been a therapeutic strategy for nearly 70 years. Commonly used drugs, in particular sulfasalazine and its active moiety 5-aminosalicylic acid, are potent RS scavengers [137].
- There is now growing evidence that overgeneration of ROS plays a major pathophysiological role in COPD and asthma, as well as the severity of these diseases [138, 139].
- OS is a pathogenic hallmark in RA because ROS production in monocytes and whole blood increases by fivefold in RA patients as compared to healthy ones. In addition, the generation of ROS in neutrophils is positively correlated with the severity of RA [140–144].
- There is now reasonable evidence that ROS play a co-factor role in several allegedly diseases. When persistent and inappropriate levels occur, OS can impair cells and accelerate various physiological processes including aging.
- Large number of studies has demonstrated that FR contribute to the initiation and progression of cancer. It has been confirmed that excessive ROS production may lead to overexpression of oncogenes and formation or mutagen compounds.
- OS plays a major role in the outcome of critically ill patients, making biomarkers of OS a major area of research to predict the outcome of patients in the ICU. Multiple studies have been conducted to quantify the effects of these markers in patients, especially in sepsis and ARDS [145, 146]. The potential use of antioxidant therapy requires further elucidation and attention for the managements of such critically illness.
- An expanding body of evidence supports a role of OS as a significant cause of male infertility [147]. Treatment that would minimize OS in the male reproductive system is highly recommended by some groups of research.
- Numerous recent studies evaluated the diagnostic, prognostic, and therapeutic values of OS biomarkers in pathological conditions and found some success in many of them. Table 22.2 illustrates some pertinent examples with related diseases.

The borderline between OS and pathophysiological conditions is highly context dependent and needs to be better characterized. Certainly, the list of diseases

Table 22.2 Diagnostic, prognostic, and therapeutic values of OS biomarkers

Medical field	Biomarker	Pathological condition	Value	Nature of study	Reference
Nephrology	Anti-oxLDL ab	Hemodialysis	Therapeutic	Comparative	[148]
	AOPP	Steroid-sensitive nephrotic syndrome	Prognostic	Comparative	[149]
	MDA	Hemodialysis	Prognostic	Comparative	[150]
	AOPP, TH	Tubular damage	Diagnostic	Comparative	[127]
Cardiology	TOS	Left ventricular remodeling	Prognostic	Retrospective	[151]
	NO	Coronary artery disease	Prognostic	Comparative	[152]
	NO	Cardiovascular disease	Prognostic	Prospective	[153]
	NO	Acute myocardial infarction	Prognostic	Comparative	[154]
Gastroenterology	TAS and TOS	Acute appendicitis	Diagnostic	Comparative	[155]
	Thiol	Acute appendicitis	Diagnostic	Comparative	[156]
	BAP test	Inflammatory bowel disease	Therapeutic	Randomized double-blinded placebo-control	[157]
	MDA, GPx, and GR	Acute pancreatitis	Therapeutic	Retrospective	[158]
	GPx	Hepatic encephalopathy	Diagnostic	Comparative	[159]
	MDA and MPO	Ulcerative colitis	Diagnostic and prognostic	Comparative	[160]
	ROS	Cholestasis	Diagnostic	Comparative	[161]
	NPBI, AOPP, TH	Necrotizing enterocolitis	Prognostic	Comparative	[162]
Cancerology	8-OHdG, MDA, vitamin C, and vitamin E	Oral cancer	Diagnostic	Comparative	[163]
	TOS and TAS	Breast cancer	Prognostic	Comparative	[164]
	MDA, ROS, and 8-OHdG	Myelodysplasia Acute leukemia	Prognostic	Explorative	[165]
Others	MDA, GSH, and SOD	Exposure to chrome	Prognostic	Comparative	[166]
	TAS and TOS	Embolism	Diagnostic	Comparative	[167]

(continued)

Table 22.2 (continued)

Medical field	Biomarker	Pathological condition	Value	Nature of study	Reference
	Protein carbonyls	Schizophrenia	Prognostic	Comparative	[168]
	SOD and TBARS	Macular degeneration	Diagnostic	Comparative	[169]
	8-IsoPs and SOD	Parapneumonic pleural effusions	Prognostic	Retrospective	[170]
	8-OHdG	Chronic periodontitis	Prognostic	Comparative	[171]
	MDA	Post-surgical aneurysmal subarachnoid hemorrhage	Prognostic	Comparative	[172]
	HO-1	Preeclampsia	Diagnostic	Comparative	[70]
	AOPP and TAS	Type 1 diabetes	Diagnostic	Comparative	[173]
	MDA, AGEs, NO˙, CAT, and GSH	Type 2 diabetes	Diagnostic and prognostic	Comparative	[63–65, 174]
	TAS and GSH	Congenital abnormalities	Diagnostic	Prospective	[175]
	SOD and H_2O_2	Fetal distress	Diagnostic	Comparative	[176]
	Isoprostanes, AOPP	Brain injury	Diagnostic	Retrospective	[125, 126]

associated with OS status will grow in the future. Many diseases, for which an indirect association with OS has been suggested, need further research to establish a firm correlation between these diseases and the role of OS in their pathophysiology.

22.3 Stratification of Patients

The presence of OS at different levels in human beings and its involvement in most biological interactions opens several possibilities in clinical practice. Turn to view on this process as a cause of numerous diseases will considerably change therapeutic approaches and will provide a solid platform to better manage such diseases. In order to maintain defense against the destructive RS, an appropriate balance of enzymatic and non-enzymatic antioxidative defenses might be one of the best ways for the management of OS related diseases. Therefore, OS biomarkers may help identifying patient populations that need antioxidant treatments, allowing patient stratification based on pathogenic mechanisms rather than just disease severity.

22.4 Monitoring of Antioxidant Treatment Effects on OS Status

Protection from the influence of RS has become the center of attraction for scientists. Several research groups in recent years tried to understand the mechanism of action of various antioxidants. Antioxidants, as class of compounds able to counteract and mitigate RS effects on health, gained enormous attention from the biomedical research community. This is due to the fact that these compounds not only showed a good degree of efficacy in terms of prevention and treatment of some diseases but also because of the general perception that most of them are free from important side effects.

As ROS are involved in the pathophysiology of many diseases, it is an obvious step to use treatments that reduce or protect against OS. It is possible that such approach could be complementary to classical treatments in the prevention and treatment of RS-associated chronic diseases. At present, several natural antioxidants have been investigated to evaluate their possible therapeutic effect in conditions linked to OS. In general, the strategies in use are the administration of antioxidants and RS scavengers, as well as boosting of repair processes.

First observations underscored the importance of controlled tests for the evaluation of benefits and risks of antioxidant therapies. Currently, the effect of antioxidant treatments as a defense against several pathologies is being widely explored in clinical trials. Herein, we review selected recent trials of antioxidants in human diseases and the use of OS biomarkers as indicators and criteria of therapeutic success.

22.4.1 Evaluation of OS in Diabetic Patients Treated with Antioxidants

Vitamins E and C are possibly the most widely used antioxidant. The effect of vitamin C and/or E supplementations on T2DM has been evaluated in a single-blinded randomized controlled clinical trial involving 40 patients. Among several markers, serum MDA and GSH, GR, GPx, and GST were estimated. The authors reported an improvement of GSH levels after 90 days of treatment [177].

The role of long-term use of vitamin C (more than 2 years) in gestational diabetes mellitus has been evaluated in a randomized controlled trial ($n = 200$) using oxidant and antioxidant biomarkers (GSH, SOD, CAT, GPx, and MDA). These markers were measured in different fluids including placental tissue homogenates, maternal blood, and neonatal blood. Based on these biomarkers, the authors concluded that the use of vitamin C markedly reverses OS in women with gestational diabetes mellitus with marked improvement on neonatal outcome [178].

Ponzo et al. evaluated the effects of 6 months' treatment of resveratrol in patients with T2DM ($n = 192$), using a double-blinded randomized controlled trial. At baseline and at the trial end (6 months), PTX3 and TAS values were determined as biomarker and global index of OS status as well as indicators of therapeutic success. They found that resveratrol supplementation increased PTX3 and TAS levels in a dose-dependent manner with potential clinical implications for DM management [179].

Curcumin, an active polyphenol of an Indian spice, was used in treatment of DM for their scavenging activity of FR. Antioxidant effects of curcuminoids (especially curcumin) in patients with T2DM have been studied in several randomized controlled trials. For instance, Panahi et al. reported antioxidant effects of curcuminoids supplementation (8 weeks) in patients with T2DM ($n = 118$) based on the measurement of serum TAC, SOD activities, and MDA levels [180]. Besides, ROS scavenging compounds such as α-lipoic acid have also been studied. A clinical trial (3 months) about the impact of α-lipoic acid on OS in type 2 diabetic patients ($n = 105$) has been undertaken by Derosa et al. Several markers were monitored and an improvement of OS markers was reported by the authors (increase of SOD and GPx and a decrease of MDA) in the group receiving the supplement containing α-lipoic acid [181].

The benefits of antioxidant therapy in diabetic complications are also reported in the literature. Neuropathic and nephropathic complications of T2DM are often explored using OS parameters in placebo-controlled clinical trials. In some of these trials, different antioxidants were used (e.g., curcumin, carnosine, vitamin E) and some markers of lipid peroxidation were explored. Overall, the trials reported an improvement of OS as well as amelioration of neurologic and renal functions in patients [182–184].

22.4.2 Evaluation of OS in Hemodialysis Patients Treated with Antioxidants

Dietary antioxidants have been extensively studied for their vast antioxidant, biological, and anti-inflammatory properties. For instance, manipulation of these antioxidants was shown to modify OS and clinical outcomes of hemodialysis. F_2-IsoPs and $NO^{\bullet}$ were measured in two separate studies to evaluate the therapeutic effect of coenzyme Q10 and ω-3 fatty acid associated with α-tocopherol in hemodialysis patients [185, 186]. The results showed improved levels of these parameters with enhanced renal functions.

22.4.3 Evaluation of OS in Treated Patients with Neurodegenerative Diseases

Concerning neurodegenerative diseases, recent trial encompassing 79 patients explored the therapeutic effects of selenium and melatonin for the management of Alzheimer's ($n = 79$; duration = 12 weeks) and Parkinson's ($n = 60$; duration = 12 weeks) diseases, respectively [187, 188]. Overall, these randomized, double-blind, controlled clinical trials used antioxidant parameters and particularly GSH as criteria of therapeutic success and reported favorable effects of the tested drugs on these diseases with improved cognitive functions.

22.4.4 Evaluation of OS in ICU Patients Treated with Antioxidants

Biomarkers of OS can be used to monitor treatments given to critically ill patients and to manage emergency situations. Indeed, several recent studies have focused on this possibility and used a wide range of biomarkers for this purpose. For instance, Yazdi et al. used this approach in a clinical trial to assess the effect of N-acetylcysteine (NAC) infusion on redox status of the body in patients with sepsis [189]. The authors found that NAC considerably increased TAC levels in treated groups compared to placebo. Likewise, NAC treatment decreased MDA levels in treated patients. According to authors, NAC treatment in patients with sepsis could have protective effects and is consistent with the hypothesis that antioxidant therapy could be of importance in sepsis based on the analysis of OS parameters as indicators. Glutathione prodrug, NAC, represents a promising antioxidant category that has received only limited use in clinical settings (with the exception of acetaminophen overdose to overcome liver damage). It is a direct ROS scavenger and a provider of cysteine, which is the precursor for the rate-limiting step of the synthesis of GSH. It is, by far, the most tested molecule because of its good antioxidant properties and its lack of severe side effects.

One study has reported that high doses of NAC added to GSH significantly decreased the peroxidative stress in patients with septic shock [190]. The protection against peroxidative stress was evaluated by measuring expired ethane and plasma MDA. Clinical scores were also significantly improved in this group.

Vitamins seem also to be effective in emergency situations. Supplemental administration of vitamins C and E decreased mortality (45.7% vs. 67.5%) in ICU patients according to Derosa et al. [191]. The authors also demonstrated that plasma TBARS and IsoPs decreased in treated patients and antioxidants improved LDL resistance to OS by approximately 30%.

Overall, antioxidant supplementation seems to be a promising therapy for acute diseases. The assessment of oxidative biomarkers will allow the determination of the dose and route of antioxidant supplementation through clinical outcome and level of OS. Currently, more randomized and controlled trials are granted for this purpose so that further recommendations can be made.

Antioxidant supplements have now been shown in randomized placebo-controlled studies to protect sperm from oxidative-related DNA damage. It may, therefore, be the right time to consider the use of antioxidants in infertile men exhibiting OS, surely after extensive evaluation of the corresponding markers.

22.4.5 *Evaluation of OS in Atherosclerosis Patients Treated with Antioxidants*

Antioxidants have been used widely in the treatment of atherosclerosis. Using biomarkers of oxidative damage, Carpenter et al. demonstrated that oral α-tocopherol supplementation decreased lipid oxidation in established human atherosclerotic lesions [192]. In this study (placebo-controlled double-blind trial (3 months)), the authors have found in treated patients with carotid endarterectomy ($n = 53$) increased levels of α-tocopherol in plasma and lower susceptibility of circulating LDL to ex vivo oxidation. Moreover, there was a significant inverse correlation in lesions between cholesterol-standardized levels of α-tocopherol and 7β-hydroxycholesterol, a free radical oxidation product of cholesterol.

22.4.6 *Evaluation of OS in Treated Patients with Respiratory Diseases*

Pilot studies demonstrated that some antioxidant compounds are effective in modulating of OS in asthma. For instance, inhaled apocynin was shown to decrease ROS concentration in EBC in mild asthmatics. In a completed clinical trial, effect of inhaled apocynin on ROS and NOS generation was demonstrated in 14 non-smoking mild asthmatics. In comparison to placebo, H_2O_2 and NO_2 were shown to reduce in EBC of asthmatic subjects in response to nebulized apocynin and showed no adverse side effects [193].

In addition, one early study demonstrated that selenium supplementation in smokers resulted in reduced $O_2^{\bullet-}$ release from leukocytes [194].

Randomized double-blind placebo-controlled Phase II trials of a 6–12 month oral dose of NAC (600 mg, once or twice a day) showed reduction in various plasma and BAL fluid oxidative biomarkers in smokers and COPD patients (H_2O_2, TBARs, lipophilic DNA adducts, and 8-oxo-dG) [195, 196].

Similarly, two additional studies showed a clear benefit of NAC in reducing oxidant burden in COPD patients [197, 198]. Given orally, 600 mg of NAC led to a transient increase in lung GSH levels after daily dosing for 2 weeks [199].

More clinical trials of existing and newly developed antioxidants (alone or in combination with anti-inflammatory agents) are urgently needed to validate these compounds as therapeutic agents and to provide the proof of concept that OS is highly involved in the pathogenesis of COPD. To successfully use redox modulation as a therapeutic strategy, it is of crucial importance to characterize redox balance in different conditions and contexts of COPD.

In a study conducted by Vasu et al., enzymatic producers of RS were explored. The findings of this work revealed that MPO activity and its reactive products represent useful indicators of thiol antioxidant doses (NAC and GSH) required to ameliorate airway OS and inflammation in CF patients and provide mechanistic insight into the antioxidative mechanisms of action of GSH and NAC when administered into the CF lung [200].

22.4.7 Evaluation of OS in RA Patients Treated with Antioxidants

Biomarkers of OS were also evaluated in blood, synovial fluid, and synovial tissue to assess the effect of vitamin E on severe knee osteoarthritis in a randomized controlled study (2 months) conducted by Tantavisut et al. [201]. The results indicated that vitamin E can improve OS status in patients with late-stage knee osteoarthritis ($n = 72$) as demonstrated by the levels of MDA, α-tocopherol, and TAC.

22.4.8 Evaluation of OS in IBD Patients Treated with Antioxidants

Clinical studies with existing and potential anti-IBD drugs show promising outcomes for IBD. For instance, it was demonstrated that sulfasalazine decreased ROS concentration in several forms of IBD [202]. In patients with moderate UC treated with sulfasalazine (2–4 g/day for 8–36 weeks), a decreased activity of NF-κB accompanied by reduced expression of pro-inflammatory cytokines (IL-1β and IL-8) was observed using Western blot and immunohistochemical staining

[203]. Indeed, stimulation of NF-κB signaling pathway activates genes for pro-inflammatory cytokines production. Therefore, down-regulation of NF-κB activity under sulfasalazine treatment is desired for IBD treatment.

Mesalazine (5-ASA) is widely used for the treatment of UC. At the cellular level, 5-ASA reduces OS by inhibiting superoxide and hydrogen peroxide production, as well as preventing mucosal GAPDH inhibition [204]. Moreover, 5-ASA potently scavenges $ONOO^-$, curbs peroxynitrite-mediated DNA strand breakage, and affects peroxynitrite-mediated radical formation responsible in part for 5-ASA anti-inflammatory effects [205].

Clinical trials indicated that 4-week treatment 5-ASA (2.4 g/day) plus NAC (0.8 g/day) in patients with UC not only improved clinical response but also correlated with decreased blood concentrations of IL-6, IL-8, and TNF-α, as well as improved GSH content. 5-ASA administrated alone also improved clinical outcome but with low effect on IL-6 and IL-8 content and with no influence on GSH and TNF-α concentration [206].

Using HPLC method, Romagnoli et al. found that NAC prevents TNF-α-induced depletion of GSH/GSSG ratio in intestinal subepithelial myofibroblasts isolated from patients with active CD [207]. The improvement of cell redox status negatively correlated with secreted MMP-2, which is involved in the dysfunction of epithelial barrier in CD patients.

Biomarkers of OS were also used to assess the efficiency of other types of drugs, mainly known for their anti-inflammatory effects. A study conducted by Makitalo et al. indicated that glucocorticoid therapy effectively inhibited neutrophil activity, reflected by decreased MPO and elastase serum contents in pediatric IBD [208]. The anti-oxidative and anti-inflammatory action of glucocorticoids can be explained by their influence on NF-κB. It has been demonstrated that prednisolone (0.75 mg/kg/day for 3 weeks) strongly inhibits intestinal NF-κB activation by stabilizing the cytosolic IkBα activation in tissue from patients with colitis ($n = 13$) [209]. Ljung et al. demonstrated a significant decrease of NOS mucosal level and rectal nitric oxide production in patients with UC ($n = 22$) and CD ($n = 24$) treated with prednisolone (0.5–1 mg/kg orally for 1 month) [210].

22.4.9 Evaluation of OS in Treated Patients with Other Diseases

Antioxidant supplementation has been proposed to be a useful complementary therapy in other diseases based on the determination of OS biomarkers. These markers were evaluated in several large studies exploring various diseases and conditions including coronary artery disease, post-myocardial infarction patients, methotrexate-induced hepatotoxicity, cirrhosis, and alcohol rehabilitation [211–215].

With respect to cancer, concern is focused on whether modulation of OS by antioxidant supplementation is helpful. According to Poprac et al., 30–70% of cancer patients, depending on the type of cancer, take antioxidant supplements. However, it remains controversial whether OS modulation by antioxidants in cancer treatment is beneficial [216]. In this context, OS monitoring and molecular understanding of the interactions between OS and transcriptional pathways have already facilitated and will continue to provide a driving force for developing effective therapeutics against RS-associated cancers.

22.4.10 Overview of Clinical Trials

Taken together, the results of recently conducted research studying the molecular and cellular involvement of OS in several diseases and monitoring of its biomarkers in clinical trials and possibly in practice offer promising venues as they propose a new way of evaluating clinical success of novel potential therapeutic agents for RS-linked diseases.

Antioxidant supplementation has been shown to improve antioxidant capacity, but whether this actually produces a clinical benefit still needs further studies as the results from clinical trials have been mixed in some diseases.

Initial pilot studies are promising but need largest randomized study. Large prospective placebo-controlled trials are needed because they will enhance the statistical significance of the tests used and evaluate more precisely the efficacy of antioxidant therapy in patients.

The role of antioxidant therapy as an adjunct to standard therapies requires further investigation with appropriate biomarkers. Indeed, an important challenge facing investigators is lack of a panel of OS markers for particular diseases. In fact, the limited benefit of antioxidant therapeutic agents used so far in several clinical trials might be the result of untargeted approach of OS monitoring and more importantly the possible evaluation of non-related redox regulators. Attention to the timing of therapy and monitoring as well as the types of antioxidants used against OS damage will be critical as well.

In many cases, the determination of oxidative end products and antioxidants still remains the common approach to assess OS, but in the near future immunological determination of oxidative damage will certainly become the more precise way to investigate OS in humans. As mentioned earlier, the progression of oxidative damage in the human body may result in the generation of aAb with the capacity to react even after sampling. Thus, the determination of OS adducts along with antibodies against oxidized derivatives would be a method of choice.

22.5 Monitoring of Therapeutic Antioxidants

It is generally assumed that an increase in RS burden is reflected in a decrease in the antioxidant levels and vice versa [217]. Some antioxidant-based treatments can be used to alleviate antioxidant deficiencies especially if they meet the depleted antioxidant(s). In this case, therapeutic drug monitoring finds its interest by using the appropriate method for measurement. Such approach is currently used in several therapies dedicated for several diseases. Analytical methods for antioxidant monitoring should be coupled with careful observation of the progress of clinical symptoms, and the effects of any antioxidant intervention therapies should be studied from both aspects. Only in this way can the clinical symptoms be reliably correlated with the levels of RS generation/repair and an understanding of their association be reached.

Recently, monitoring of antioxidant therapy has been used in two-center study in patients with COPD and DM. Plasma concentrations of coenzyme Q10 were monitored in COPD patients who received supplementation with Qter® (coenzyme Q10). In this study, increased bioavailability of this drug was confirmed by the authors [218]. Vitamin E, along with other OS indicators, has been evaluated in type 2 diabetic patients ($n = 83$) in a clinical trial conducted by Rafraf et al. The authors reported that vitamin E supplementation (8 weeks) improves serum paraoxonase-1 activity as well as vitamin E level and total antioxidant status (TAS) in patients [219]. A study of high doses of vitamin E in hepatitis C refractory to α-interferon therapy demonstrated initial positive results with increased levels of α-tocopherol in blood [220].

Overall, studies that use antioxidant monitoring seem to be much more effective in treatment of diseases rather than studies that use only clinical symptoms and functional parameters as indicators of therapeutic success. In the absence of therapeutic antioxidant monitoring, the "antioxidant dose" and its effect on various "risk" populations are unclear. There are few data, which characterize dose–response relationships of putative antioxidant strategies. Oddly, large-scale clinical trials of antioxidants have yet to incorporate such measures either in patient selection or in determination of drug response. It is hardly surprising that clinical trials have failed to yield a clear picture in any disease to date.

Several explanations can be made for the failure of convincing evidence from antioxidant trials. One of them is related to the use of insufficient amount of antioxidant without an effective monitoring of its levels in blood. In addition, it is plausible that appropriate evaluation of additional OS markers will help to follow disease progression hence evaluate better therapeutic outcomes. This highlights an important clinical benefit of using at the same time antioxidant drug and OS monitoring. Further, the potential of exogenous antioxidants in terms of relative specificity and efficiency to reduce each RS could be different. We further emphasize that injury causing oxidants must be intelligently identified and monitored.

22.6 Clinical Practice: Need to Use Biomarkers of OS in Antioxidant Therapy

We propose that the most important points to be addressed in regard to the clinical use of antioxidants center on optimizing the most effective dose of the "drug" by monitoring OS biomarkers, and its early administration, so as to optimize the overall effectiveness of the antioxidant therapy. We can confirm that OS, as phenomenon, although being one of the major harms to individuals' wellness and health, can also be exploited as a treatment tool when and if we will be able to operate a fine tuning of this process inside human organism.

Toxicological investigations to promptly detect the danger of an oxidative imbalance and to restore antioxidant defense before triggering irreversible damage cascade are thus mandatory and should be performed in the future as routine exams through standard evaluations, and repeated along the treatment plan, to monitor the efficacy of antioxidant treatments.

We cannot exclude, in the future, that an oxidative-status check-up followed by appropriate targeted treatment will achieve better social health perspectives, preventing major illnesses and complications as well as heavier social costs burden. Evaluation of OS has markedly improved in basic medicine, but there is much room left for its clinical application and development.

The measurements of the antioxidants themselves may represent at least an indirect relevance of OS. Antioxidants can be detected by a wide range of reliable methods. They can be estimated separately or as a total global antioxidant capacity of tissues. In parallel with the exciting developments of analytical methods and clinical indications of biomarkers, antioxidant and patterns that correlate best with treatment efficacy will eventually advance the field of OS and will help in the management of several diseases. It is important to well-establish antioxidant drug monitoring to promote reasonable spread of anti-oxidative interventions.

References

1. Bhatia V, Sharma S. Role of mitochondrial dysfunction, oxidative stress and autophagy in progression of Alzheimer's disease. Journal of the Neurological Sciences. 2020:117253.
2. García-Blanco A, Baquero M, Vento M, Gil E, Bataller L, Cháfer-Pericás C. Potential oxidative stress biomarkers of mild cognitive impairment due to Alzheimer disease. Journal of the Neurological Sciences. 2017;373:295–302.
3. Qu Z, Sun J, Zhang W, Yu J, Zhuang C. Transcription factor NRF2 as a promising therapeutic target for Alzheimer's disease. Free Radical Biology and Medicine. 2020;159:87–102.
4. Forster MJ, Dubey A, Dawson KM, Stutts WA, Lal H, Sohal RS. Age-related losses of cognitive function and motor skills in mice are associated with oxidative protein damage in the brain. Proceedings of the National Academy of Sciences of the United States of America. 1996;93(10):4765–9.
5. Lovell MA, Markesbery WR. Oxidative DNA damage in mild cognitive impairment and late-stage Alzheimer's disease. Nucleic acids research. 2007;35(22):7497–504.

6. Gella A, Durany N. Oxidative stress in Alzheimer disease. Cell adhesion & migration. 2009;3 (1):88–93.
7. Verri M, Pastoris O, Dossena M, Aquilani R, Guerriero F, Cuzzoni G, et al. Mitochondrial alterations, oxidative stress and neuroinflammation in Alzheimer's disease. International journal of immunopathology and pharmacology. 2012;25(2):345–53.
8. Tönnies E, Trushina E. Oxidative Stress, Synaptic Dysfunction, and Alzheimer's Disease. Journal of Alzheimer's disease: JAD. 2017;57(4):1105–21.
9. Tamagno E, Bardini P, Guglielmotto M, Danni O, Tabaton M. The various aggregation states of beta-amyloid 1-42 mediate different effects on oxidative stress, neurodegeneration, and BACE-1 expression. Free radical biology & medicine. 2006;41(2):202–12.
10. Loh KP, Huang SH, De Silva R, Tan BK, Zhu YZ. Oxidative stress: apoptosis in neuronal injury. Current Alzheimer research. 2006;3(4):327–37.
11. Melov S, Adlard PA, Morten K, Johnson F, Golden TR, Hinerfeld D, et al. Mitochondrial oxidative stress causes hyperphosphorylation of tau. PloS one. 2007;2(6):e536.
12. Ren X, Butterfield DA. Fidelity of the PINK1 knockout rat to oxidative stress and other characteristics of Parkinson disease. Free Radical Biology and Medicine. 2021;163:88–101.
13. Sharma A, Weber D, Raupbach J, Dakal TC, Fließbach K, Ramirez A, et al. Advanced glycation end products and protein carbonyl levels in plasma reveal sex-specific differences in Parkinson's and Alzheimer's disease. Redox Biol. 2020;34:101546.
14. Colamartino M, Duranti G, Ceci R, Sabatini S, Testa A, Cozzi R. A multi-biomarker analysis of the antioxidant efficacy of Parkinson's disease therapy. Toxicology in Vitro. 2018;47:1–7.
15. Vural G, Gumusyayla S, Bektas H, Deniz O, Alisik M, Erel O. Impairment of dynamic thiol–disulphide homeostasis in patients with idiopathic Parkinson's disease and its relationship with clinical stage of disease. Clinical Neurology and Neurosurgery. 2017;153:50–5.
16. D'Amico E, Factor-Litvak P, Santella RM, Mitsumoto H. Clinical perspective on oxidative stress in sporadic amyotrophic lateral sclerosis. Free Radical Biology and Medicine. 2013;65: 509–27.
17. Li H, Horke S, Förstermann U. Vascular oxidative stress, nitric oxide and atherosclerosis. Atherosclerosis. 2014;237(1):208–19.
18. Rosenson RS, Stafforini DM. Modulation of oxidative stress, inflammation, and atherosclerosis by lipoprotein-associated phospholipase A2. J Lipid Res. 2012;53(9):1767–82.
19. Lönn ME, Dennis JM, Stocker R. Actions of "antioxidants" in the protection against atherosclerosis. Free Radical Biology and Medicine. 2012;53(4):863–84.
20. Moschonas IC, Tselepis AD. The pathway of neutrophil extracellular traps towards atherosclerosis and thrombosis. Atherosclerosis. 2019;288:9–16.
21. Prashar Y, Ritu, Bais S, Gill NS. Emerging role of various signaling pathways in the pathogenesis and therapeutics of atherosclerosis. Reviews in Vascular Medicine. 2017;10–11:1–12.
22. Frijhoff J, Winyard PG, Zarkovic N, Davies SS, Stocker R, Cheng D, et al. Clinical Relevance of Biomarkers of Oxidative Stress. Antioxid Redox Signal. 2015;23(14):1144–70.
23. Doeppner TR, Hermann DM. Free radical scavengers and spin traps – therapeutic implications for ischemic stroke. Best Practice & Research Clinical Anaesthesiology. 2010;24(4):511–20.
24. Wu J, Li R, Li W, Ren M, Thangthaeng N, Sumien N, et al. Administration of 5-methoxyindole-2-carboxylic acid that potentially targets mitochondrial dihydrolipoamide dehydrogenase confers cerebral preconditioning against ischemic stroke injury. Free Radical Biology and Medicine. 2017;113:244–54.
25. Narne P, Pandey V, Phanithi PB. Interplay between mitochondrial metabolism and oxidative stress in ischemic stroke: An epigenetic connection. Molecular and Cellular Neuroscience. 2017;82:176–94.
26. Nishida M, Schey KL, Takagahara S, Kontani K, Katada T, Urano Y, et al. Activation mechanism of Gi and Go by reactive oxygen species. J Biol Chem. 2002;277(11):9036–42.
27. Sabri A, Hughie HH, Lucchesi PA. Regulation of hypertrophic and apoptotic signaling pathways by reactive oxygen species in cardiac myocytes. Antioxid Redox Signal. 2003;5 (6):731–40.

28. Johnston AS, Lehnart SE, Burgoyne JR. Ca(2+) signaling in the myocardium by (redox) regulation of PKA/CaMKII. Front Pharmacol. 2015;6:166.
29. Karagueuzian HS, Nguyen TP, Qu Z, Weiss JN. Oxidative stress, fibrosis, and early afterdepolarization-mediated cardiac arrhythmias. Front Physiol. 2013;4:19.
30. Liu RM, Desai LP. Reciprocal regulation of TGF-β and reactive oxygen species: A perverse cycle for fibrosis. Redox Biol. 2015;6:565–77.
31. Siwik DA, Colucci WS. Regulation of matrix metalloproteinases by cytokines and reactive oxygen/nitrogen species in the myocardium. Heart Fail Rev. 2004;9(1):43–51.
32. Palomeque J, Rueda OV, Sapia L, Valverde CA, Salas M, Petroff MV, et al. Angiotensin II-induced oxidative stress resets the Ca2+ dependence of Ca2+-calmodulin protein kinase II and promotes a death pathway conserved across different species. Circ Res. 2009;105(12): 1204–12.
33. Griendling KK, Camargo LL, Rios FJ, Alves-Lopes R, Montezano AC, Touyz RM. Oxidative Stress and Hypertension. Circulation Research. 2021;128(7):993–1020.
34. Bedard K, Krause KH. The NOX family of ROS-generating NADPH oxidases: physiology and pathophysiology. Physiol Rev. 2007;87(1):245–313.
35. Santillo M, Colantuoni A, Mondola P, Guida B, Damiano S. NOX signaling in molecular cardiovascular mechanisms involved in the blood pressure homeostasis. Front Physiol. 2015;6:194.
36. Dikalova AE, Itani HA, Nazarewicz RR, McMaster WG, Flynn CR, Uzhachenko R, et al. Sirt3 Impairment and SOD2 Hyperacetylation in Vascular Oxidative Stress and Hypertension. Circ Res. 2017;121(5):564–74.
37. Su H, Zeng H, Liu B, Chen JX. Sirtuin 3 is essential for hypertension-induced cardiac fibrosis via mediating pericyte transition. J Cell Mol Med. 2020;24(14):8057–68.
38. Ochoa CD, Wu RF, Terada LS. ROS signaling and ER stress in cardiovascular disease. Mol Aspects Med. 2018;63:18–29.
39. Burgoyne JR, Mongue-Din H, Eaton P, Shah AM. Redox signaling in cardiac physiology and pathology. Circ Res. 2012;111(8):1091–106.
40. Carnes CA, Janssen PM, Ruehr ML, Nakayama H, Nakayama T, Haase H, et al. Atrial glutathione content, calcium current, and contractility. J Biol Chem. 2007;282(38):28063–73.
41. Chelu MG, Sarma S, Sood S, Wang S, van Oort RJ, Skapura DG, et al. Calmodulin kinase II-mediated sarcoplasmic reticulum Ca2+ leak promotes atrial fibrillation in mice. J Clin Invest. 2009;119(7):1940–51.
42. Korantzopoulos P, Letsas K, Fragakis N, Tse G, Liu T. Oxidative stress and atrial fibrillation: an update. Free Radic Res. 2018;52(11–12):1199–209.
43. Adam O, Lavall D, Theobald K, Hohl M, Grube M, Ameling S, et al. Rac1-induced connective tissue growth factor regulates connexin 43 and N-cadherin expression in atrial fibrillation. J Am Coll Cardiol. 2010;55(5):469–80.
44. Dudley SC, Jr., Hoch NE, McCann LA, Honeycutt C, Diamandopoulos L, Fukai T, et al. Atrial fibrillation increases production of superoxide by the left atrium and left atrial appendage: role of the NADPH and xanthine oxidases. Circulation. 2005;112(9):1266–73.
45. Rudolph V, Andrié RP, Rudolph TK, Friedrichs K, Klinke A, Hirsch-Hoffmann B, et al. Myeloperoxidase acts as a profibrotic mediator of atrial fibrillation. Nat Med. 2010;16(4): 470–4.
46. Teng Y, Sun P, Zhang J, Yu R, Bai J, Yao X, et al. Hydrogen Peroxide in Exhaled Breath Condensate in Patients with Asthma: A Promising Biomarker? Chest. 2011;140(1):108–16.
47. Fitzpatrick AM, Teague WG, Holguin F, Yeh M, Brown LA. Airway glutathione homeostasis is altered in children with severe asthma: evidence for oxidant stress. The Journal of allergy and clinical immunology. 2009;123(1):146–52.e8.
48. Misso NL, Thompson PJ. Oxidative stress and antioxidant deficiencies in asthma: potential modification by diet. Redox report: communications in free radical research. 2005;10(5): 247–55.

49. Bikov A, Lazar Z. Chapter 19 – Chronic obstructive pulmonary disease and smokers. In: Beauchamp J, Davis C, Pleil J, editors. Breathborne Biomarkers and the Human Volatilome (Second Edition). Boston: Elsevier; 2020. p. 305–20.
50. Marushchak M, Maksiv K, Krynytska I, Dutchak O, Behosh N. The Severity of Oxidative Stress in Comorbid Chronic Obstructive Pulmonary Disease (COPD) and Hypertension: Does it Depend On ACE and AGT Gene Polymorphisms? J Med Life. 2019;12(4):426–34.
51. Antczak A, Ciebiada M, Pietras T, Piotrowski WJ, Kurmanowska Z, Górski P. Exhaled eicosanoids and biomarkers of oxidative stress in exacerbation of chronic obstructive pulmonary disease. Archives of medical science: AMS. 2012;8(2):277–85.
52. Moura FA, de Andrade KQ, dos Santos JCF, Araújo ORP, Goulart MOF. Antioxidant therapy for treatment of inflammatory bowel disease: Does it work? Redox Biol. 2015;6:617–39.
53. Chami B, Martin NJJ, Dennis JM, Witting PK. Myeloperoxidase in the inflamed colon: A novel target for treating inflammatory bowel disease. Archives of Biochemistry and Biophysics. 2018;645:61–71.
54. Nowak JZ. Oxidative stress, polyunsaturated fatty acidsderived oxidation products and bisretinoids as potential inducers of CNS diseases: focus on age-related macular degeneration. Pharmacological Reports. 2013;65(2):288–304.
55. Ardeljan D, Chan C-C. Aging is not a disease: Distinguishing age-related macular degeneration from aging. Progress in Retinal and Eye Research. 2013;37:68–89.
56. Fu Y, Dong Y, Gao Q. Age-related cataract and macular degeneration: Oxygen receptor dysfunction diseases. Medical Hypotheses. 2015;85(3):272–5.
57. Taylor A, Nowell T. Oxidative Stress and Antioxidant Function in Relation to Risk for Cataract. In: Sies H, editor. Advances in Pharmacology. 38: Academic Press; 1996. p. 515–36.
58. Goutham G, Manikandan R, Beulaja M, Thiagarajan R, Arulvasu C, Arumugam M, et al. A focus on resveratrol and ocular problems, especially cataract: From chemistry to medical uses and clinical relevance. Biomedicine & Pharmacotherapy. 2017;86:232–41.
59. Wojcik M, Krawczyk M, Zieleniak A, Mac Marcjanek K, Wozniak LA. Chapter 14 – Associations of high blood sugar with oxidative stress and inflammation in patients with type 2 diabetes. In: Preuss HG, Bagchi D, editors. Dietary Sugar, Salt and Fat in Human Health: Academic Press; 2020. p. 305–23.
60. Yuan T, Yang T, Chen H, Fu D, Hu Y, Wang J, et al. New insights into oxidative stress and inflammation during diabetes mellitus-accelerated atherosclerosis. Redox Biol. 2019;20:247–60.
61. Masenga SK, Kabwe LS, Chakulya M, Kirabo A. Mechanisms of Oxidative Stress in Metabolic Syndrome. International Journal of Molecular Sciences. 2023;24(9):7898.
62. He M, Pan H, Xiao C, Pu M. Roles for Redox Signaling by NADPH Oxidase in Hyperglycemia-Induced Heme Oxygenase-1 Expression in the Diabetic Retina. Investigative Ophthalmology & Visual Science. 2013;54(6):4092–101.
63. Tiwari BK, Pandey KB, Abidi A, Rizvi SI. Markers of oxidative stress during diabetes mellitus. Journal of biomarkers. 2013;2013.
64. Nandi A, Yan L-J, Jana CK, Das N. Role of catalase in oxidative stress-and age-associated degenerative diseases. Oxidative medicine and cellular longevity. 2019;2019.
65. Luc K, Schramm-Luc A, Guzik T, Mikolajczyk T. Oxidative stress and inflammatory markers in prediabetes and diabetes. Journal of Physiology & Pharmacology. 2019;70(6).
66. Yan Lj. Redox imbalance stress in diabetes mellitus: Role of the polyol pathway. Animal models and experimental medicine. 2018;1(1):7–13.
67. Li J-s, Ji T, Su S-l, Zhu Y, Chen X-l, Shang E-x, et al. Mulberry leaves ameliorate diabetes via regulating metabolic profiling and AGEs/RAGE and p38 MAPK/NF-κB pathway. Journal of Ethnopharmacology. 2022;283:114713.
68. Ihsan AU, Khan FU, Khongorzul P, Ahmad KA, Naveed M, Yasmeen S, et al. Role of oxidative stress in pathology of chronic prostatitis/chronic pelvic pain syndrome and male infertility and antioxidants function in ameliorating oxidative stress. Biomedicine & Pharmacotherapy. 2018;106:714–23.

69. Asa E, Ahmadi R, Mahmoodi M, Mohammadniya A. Supplementation of freezing media with alpha lipoic acid preserves the structural and functional characteristics of sperm against cryodamage in infertile men with asthenoteratozoospermia. Cryobiology. 2020;96:166–74.
70. Erdemli HK, Yıldırımlar P, Alper TY, Kocabaş R, Salis O, Bedir A. Increased serum heme oxygenase-1 levels as a diagnostic marker of oxidative stress in preeclampsia. Hypertension in pregnancy. 2014;33(4):488–97.
71. Aouache R, Biquard L, Vaiman D, Miralles F. Oxidative Stress in Preeclampsia and Placental Diseases. International journal of molecular sciences. 2018;19(5).
72. Chiarello DI, Abad C, Rojas D, Toledo F, Vázquez CM, Mate A, et al. Oxidative stress: Normal pregnancy versus preeclampsia. Biochim Biophys Acta Mol Basis Dis. 2020;1866(2): 165354.
73. Zeeman GG, Dekker GA, van Geijn HP, Kraayenbrink AA. Endothelial function in normal and pre-eclamptic pregnancy: a hypothesis. Eur J Obstet Gynecol Reprod Biol. 1992;43(2): 113–22.
74. Burton GJ, Redman CW, Roberts JM, Moffett A. Pre-eclampsia: pathophysiology and clinical implications. Bmj. 2019;366:l2381.
75. Huppertz B. The Critical Role of Abnormal Trophoblast Development in the Etiology of Preeclampsia. Curr Pharm Biotechnol. 2018;19(10):771–80.
76. Haram K, Mortensen JH, Myking O, Magann EF, Morrison JC. The Role of Oxidative Stress, Adhesion Molecules and Antioxidants in Preeclampsia. Curr Hypertens Rev. 2019;15(2): 105–12.
77. Abad C, Vallejos C, De Gregorio N, Díaz P, Chiarello DI, Mendoza M, et al. Na^+, K^+-ATPase and Ca^{2+}-ATPase activities in basal and microvillous syncytiotrophoblast membranes from preeclamptic human term placenta. Hypertension in pregnancy. 2015;34(1):65–79.
78. Touhouche S, Guenifed A, Yaker ND, Khelfi A. Phtalates et fonctions reproductrices de la femme: effets réels ou mythes ? Toxicologie Analytique et Clinique. 2023.
79. Sopasi F, Spyropoulou I, Kourti M, Vasileiadis S, Tripsianis G, Galazios G, et al. Oxidative stress and female infertility: the role of follicular fluid soluble receptor of advanced glycation end-products (sRAGE) in women with endometriosis. Hum Fertil (Camb). 2023;26(6): 1400–7.
80. Lambrinoudaki IV, Augoulea A, Christodoulakos GE, Economou EV, Kaparos G, Kontoravdis A, et al. Measurable serum markers of oxidative stress response in women with endometriosis. Fertility and sterility. 2009;91(1):46–50.
81. Singh AK, Chattopadhyay R, Chakravarty B, Chaudhury K. Markers of oxidative stress in follicular fluid of women with endometriosis and tubal infertility undergoing IVF. Reproductive toxicology (Elmsford, NY). 2013;42:116–24.
82. Carvalho LF, Samadder AN, Agarwal A, Fernandes LF, Abrão MS. Oxidative stress biomarkers in patients with endometriosis: systematic review. Arch Gynecol Obstet. 2012;286(4): 1033–40.
83. Murri M, Luque-Ramírez M, Insenser M, Ojeda-Ojeda M, Escobar-Morreale HF. Circulating markers of oxidative stress and polycystic ovary syndrome (PCOS): a systematic review and meta-analysis. Human reproduction update. 2013;19(3):268–88.
84. Onal M, Agar M. Follicular fluid redox balance and DNA damage according to PCOS phenotype. European review for medical and pharmacological sciences. 2023;27(12):5583–9.
85. Taylor DE, Piantadosi CA. Oxidative metabolism in sepsis and sepsis syndrome. Journal of Critical Care. 1995;10(3):122–35.
86. Arif Z, Tarannum A, Arfat MY, Arif B, Shahab S, Arif M, et al. Impact of endogenous stress on albumin structure in systemic lupus erythematosus (SLE) patients. International Journal of Biological Macromolecules. 2020;151:891–900.
87. Elloumi N, Ben Mansour R, Marzouk S, Mseddi M, Fakhfakh R, Gargouri B, et al. Differential reactive oxygen species production of neutrophils and their oxidative damage in patients with active and inactive systemic lupus erythematosus. Immunology Letters. 2017;184:1–6.

88. Ferreira HB, Pereira AM, Melo T, Paiva A, Domingues MR. Lipidomics in autoimmune diseases with main focus on systemic lupus erythematosus. Journal of Pharmaceutical and Biomedical Analysis. 2019;174:386–95.
89. Rasool M, Farooq S, Malik A, Shaukat A, Manan A, Asif M, et al. Assessment of circulating biochemical markers and antioxidative status in acute lymphoblastic leukemia (ALL) and acute myeloid leukemia (AML) patients. Saudi Journal of Biological Sciences. 2015;22(1): 106–11.
90. Battisti V, Maders LDK, Bagatini MD, Santos KF, Spanevello RM, Maldonado PA, et al. Measurement of oxidative stress and antioxidant status in acute lymphoblastic leukemia patients. Clinical Biochemistry. 2008;41(7):511–8.
91. Er T-K, Tsai S-M, Wu S-H, Chiang W, Lin H-C, Lin S-F, et al. Antioxidant status and superoxide anion radical generation in acute myeloid leukemia. Clinical Biochemistry. 2007;40(13):1015–9.
92. Yamada T, Steinz MM, Kenne E, Lanner JT. Muscle Weakness in Rheumatoid Arthritis: The Role of Ca2+ and Free Radical Signaling. EBioMedicine. 2017;23:12–9.
93. Huang H-H, Chen L-Y, Chen K-Y, Lee Y-c, Tsai C-Y, Chen C-Y. Increased monocyte chemoattractant protein-1 and nitrotyrosine are associated with increased body weight in patients with rheumatoid arthritis after etanercept therapy. Neuropeptides. 2020;84:102100.
94. Phull A-R, Nasir B, Haq Iu, Kim SJ. Oxidative stress, consequences and ROS mediated cellular signaling in rheumatoid arthritis. Chemico-Biological Interactions. 2018;281:121–36.
95. El-Shanshory M, Hablas NM, Aboonq MS, Fakhreldin AR, Attia M, Arafa W, et al. Nigella sativa improves anemia, enhances immunity and relieves iron overload-induced oxidative stress as a novel promising treatment in children having beta-thalassemia major. Journal of Herbal Medicine. 2019;16:100245.
96. Sajadi Hezaveh Z, Azarkeivan A, Janani L, Hosseini S, Shidfar F. The effect of quercetin on iron overload and inflammation in β-thalassemia major patients: A double-blind randomized clinical trial. Complementary Therapies in Medicine. 2019;46:24–8.
97. Banerjee M, Kumar A, Batra HS, Bandyopadhyay S, Kapoor R. Are thalassemia patients oxidatively challenged? Medical Journal Armed Forces India. 2019;75(4):383–8.
98. Manafikhi H, Drummen G, Palmery M, Peluso I. Total Antioxidant Capacity in beta-thalassemia: A systematic review and meta-analysis of case-control studies. Critical Reviews in Oncology/Hematology. 2017;110:35–42.
99. Chai X, Li D, Cao X, Zhang Y, Mu J, Lu W, et al. ROS-mediated iron overload injures the hematopoiesis of bone marrow by damaging hematopoietic stem/progenitor cells in mice. Sci Rep. 2015;5:10181.
100. Jin X, He X, Cao X, Xu P, Xing Y, Sui S, et al. Iron overload impairs normal hematopoietic stem and progenitor cells through reactive oxygen species and shortens survival in myelodysplastic syndrome mice. Haematologica. 2018;103(10):1627–34.
101. Bulycheva E, Rauner M, Medyouf H, Theurl I, Bornhäuser M, Hofbauer LC, et al. Myelodysplasia is in the niche: novel concepts and emerging therapies. Leukemia. 2015;29 (2):259–68.
102. Tahri A, Lahyani A, Kallel R, Ayadi F, Boudawara T, Sahnoun Z. Exposure to imipenem/cilastatin causes nephrotoxicity and even urolithiasis in Wistar rats. Toxicology. 2018;404–405:59–67.
103. Zhou J, Jin J, Li X, Zhao Z, Zhang L, Wang Q, et al. Total flavonoids of Desmodium styracifolium attenuates the formation of hydroxy-L-proline-induced calcium oxalate urolithiasis in rats. Urolithiasis. 2018;46(3):231–41.
104. Golovanov SA, Ianenko EK, Dzeranov NK, Drozhzheva VV, Beĭshliev DA, Kon'kova TA. [Lipid peroxidation and the antioxidant system of patients with urolithiasis after extracorporeal lithotripsy]. Urologiia i nefrologiia. 1998(2):14–6.
105. Roumeliotis S, Liakopoulos V, Veljkovic A, Dounousi E. Redox Systems Biology in Chronic Kidney Disease. Hindawi; 2023.

106. Roumeliotis S, Liakopoulos V, Dounousi E, Mark PB. Oxidative Stress in End-Stage Renal Disease: Pathophysiology and Potential Interventions. Hindawi; 2023.
107. Wojtaszek E, Oldakowska-Jedynak U, Kwiatkowska M, Glogowski T, Malyszko J. Uremic toxins, oxidative stress, atherosclerosis in chronic kidney disease, and kidney transplantation. oxidative medicine and cellular longevity. 2021;2021.
108. Pasqual-Melo G, Bernardes SS, Souza-Neto FP, Carrara IM, Ramalho LNZ, Marinello PC, et al. The progression of metastatic melanoma augments a pro-oxidative milieu locally but not systemically. Pathology – Research and Practice. 2020;216(11):153218.
109. Arslanbaeva LR, Santoro MM. Adaptive redox homeostasis in cutaneous melanoma. Redox Biol. 2020;37:101753.
110. Cannavò SP, Tonacci A, Bertino L, Casciaro M, Borgia F, Gangemi S. The role of oxidative stress in the biology of melanoma: A systematic review. Pathology – Research and Practice. 2019;215(1):21–8.
111. Xin Z, Wu X, Yu Z, Shang J, Xu B, Jiang S, et al. Mechanisms explaining the efficacy of psoralidin in cancer and osteoporosis, a review. Pharmacological Research. 2019;147:104334.
112. Wimalawansa SJ. Chapter 59 – Skeletal Effects of Nitric Oxide: Novel Agent for Osteoporosis. In: Bilezikian JP, Raisz LG, Martin TJ, editors. Principles of Bone Biology (Third Edition). San Diego: Academic Press; 2008. p. 1273–310.
113. Zorova LD, Pevzner IB, Chupyrkina AA, Zorov SD, Silachev DN, Plotnikov EY, et al. The role of myoglobin degradation in nephrotoxicity after rhabdomyolysis. Chemico-Biological Interactions. 2016;256:64–70.
114. Boutaud O, Roberts LJ. Mechanism-based therapeutic approaches to rhabdomyolysis-induced renal failure. Free Radical Biology and Medicine. 2011;51(5):1062–7.
115. Ruiz-Ojeda FJ, Olza J, Gil Á, Aguilera CM. Chapter 1 – Oxidative Stress and Inflammation in Obesity and Metabolic Syndrome. In: del Moral AM, Aguilera García CM, editors. Obesity: Academic Press; 2018. p. 1–15.
116. Feter N, Spanevello RM, Soares MSP, Spohr L, Pedra NS, Bona NP, et al. How does physical activity and different models of exercise training affect oxidative parameters and memory? Physiology & Behavior. 2019;201:42–52.
117. Bokov A, Chaudhuri A, Richardson A. The role of oxidative damage and stress in aging. Mechanisms of Ageing and Development. 2004;125(10):811–26.
118. Fleenor BS, Sindler AL, Marvi NK, Howell KL, Zigler ML, Yoshizawa M, et al. Curcumin ameliorates arterial dysfunction and oxidative stress with aging. Experimental Gerontology. 2013;48(2):269–76.
119. Yu BP, Chung HY. Oxidative stress and vascular aging. Diabetes Research and Clinical Practice. 2001;54:S73–S80.
120. Abdelmegeed MA, Choi Y, Ha S-K, Song B-J. Cytochrome P450-2E1 is involved in aging-related kidney damage in mice through increased nitroxidative stress. Food and Chemical Toxicology. 2017;109:48–59.
121. Chiou C-C, Chang P-Y, Chan E-C, Wu T-L, Tsao K-C, Wu JT. Urinary 8-hydroxydeoxyguanosine and its analogs as DNA marker of oxidative stress: development of an ELISA and measurement in both bladder and prostate cancers. Clinica Chimica Acta. 2003;334(1):87–94.
122. Abudawood M, Tabassum H, Almaarik B, Aljohi A. Interrelationship between oxidative stress, DNA damage and cancer risk in diabetes (Type 2) in Riyadh, KSA. Saudi Journal of Biological Sciences. 2020;27(1):177–83.
123. Guo C, Chen Q, Chen J, Yu J, Hu Y, Zhang S, et al. 8-Hydroxyguanosine as a possible RNA oxidative modification marker in urine from colorectal cancer patients: Evaluation by ultra performance liquid chromatography-tandem mass spectrometry. Journal of Chromatography B. 2020;1136:121931.
124. Buonocore G, Perrone S, Longini M, Paffetti P, Vezzosi P, Gatti MG, et al. Non protein bound iron as early predictive marker of neonatal brain damage. Brain. 2003;126(Pt 5):1224–30.

125. Negro S, Benders M, Tataranno ML, Coviello C, de Vries LS, van Bel F, et al. Early Prediction of Hypoxic-Ischemic Brain Injury by a New Panel of Biomarkers in a Population of Term Newborns. Oxid Med Cell Longev. 2018;2018:7608108.
126. Coviello C, Perrone S, Buonocore G, Negro S, Longini M, Dani C, et al. Isoprostanes as Biomarker for White Matter Injury in Extremely Preterm Infants. Front Pediatr. 2020;8: 618622.
127. Perrone S, Mussap M, Longini M, Fanos V, Bellieni CV, Proietti F, et al. Oxidative kidney damage in preterm newborns during perinatal period. Clin Biochem. 2007;40(9–10):656–60.
128. Perrone S, Tataranno ML, Negro S, Cornacchione S, Longini M, Proietti F, et al. May oxidative stress biomarkers in cord blood predict the occurrence of necrotizing enterocolitis in preterm infants? The journal of maternal-fetal & neonatal medicine: the official journal of the European Association of Perinatal Medicine, the Federation of Asia and Oceania Perinatal Societies, the International Society of Perinatal Obstet. 2012;25 Suppl 1:128–31.
129. Perrone S, Longini M, Bellieni CV, Centini G, Kenanidis A, De Marco L, et al. Early oxidative stress in amniotic fluid of pregnancies with Down syndrome. Clin Biochem. 2007;40(3–4): 177–80.
130. Pawlak K, Mysliwiec M, Pawlak D. Oxidized LDL to autoantibodies against oxLDL ratio – The new biomarker associated with carotid atherosclerosis and cardiovascular complications in dialyzed patients. Atherosclerosis. 2012;224(1):252–7.
131. Sevinc Ok E, Kircelli F, Asci G, Altunel E, Ertilav M, Sipahi S, et al. Neither oxidized nor anti-oxidized low-density lipoprotein level is associated with atherosclerosis or mortality in hemodialysis patients. Hemodialysis international International Symposium on Home Hemodialysis. 2012;16(3):334–41.
132. Choi SH, Chae A, Miller E, Messig M, Ntanios F, DeMaria AN, et al. Relationship Between Biomarkers of Oxidized Low-Density Lipoprotein, Statin Therapy, Quantitative Coronary Angiography, and Atheroma Volume: Observations From the REVERSAL (Reversal of Atherosclerosis with Aggressive Lipid Lowering) Study. Journal of the American College of Cardiology. 2008;52(1):24–32.
133. Xu RX, Zhang Y, Zhang Y, Wu YR, Li XL, Guo YL, et al. Effects of Pitavastatin on Lipoprotein Subfractions and Oxidized Low-density Lipoprotein in Patients with Atherosclerosis. Current medical science. 2020;40(5):879–84.
134. Rumfeldt JAO, Lepock JR, Meiering EM. Unfolding and Folding Kinetics of Amyotrophic Lateral Sclerosis-Associated Mutant Cu,Zn Superoxide Dismutases. Journal of Molecular Biology. 2009;385(1):278–98.
135. Bocci T, Pecori C, Giorli E, Briscese L, Tognazzi S, Caleo M, et al. Differential motor neuron impairment and axonal regeneration in sporadic and familiar amyotrophic lateral sclerosis with SOD-1 mutations: lessons from neurophysiology. International journal of molecular sciences. 2011;12(12):9203–15.
136. Zhang Q, Tao H, Lin Y, Hu Y, An H, Zhang D, et al. A superoxide dismutase/catalase mimetic nanomedicine for targeted therapy of inflammatory bowel disease. Biomaterials. 2016;105: 206–21.
137. Machado APdF, Geraldi MV, do Nascimento RdP, Moya AMTM, Vezza T, Diez-Echave P, et al. Polyphenols from food by-products: An alternative or complementary therapy to IBD conventional treatments. Food Research International. 2021;140:110018.
138. Elmasry SA, Al-Azzawi MA, Ghoneim AH, Nasr MY, AboZaid MMN. Role of oxidant–antioxidant imbalance in the pathogenesis of chronic obstructive pulmonary disease. Egyptian Journal of Chest Diseases and Tuberculosis. 2015;64(4):813–20.
139. Kwon H-S, Bae Y-J, Moon K-A, Lee YS, Lee T, Lee K-Y, et al. Hyperoxidized peroxiredoxins in peripheral blood mononuclear cells of asthma patients is associated with asthma severity. Life Sciences. 2012;90(13):502–8.
140. Pradhan A, Bagchi A, De S, Mitra S, Mukherjee S, Ghosh P, et al. Role of redox imbalance and cytokines in mediating oxidative damage and disease progression of patients with rheumatoid arthritis. Free radical research. 2019;53(7):768–79.

141. Gorlino CV, Dave MN, Blas R, Crespo MI, Lavanchy A, Tamashiro H, et al. Association between levels of synovial anti-citrullinated peptide antibodies and neutrophil response in patients with rheumatoid arthritis. European journal of immunology. 2018;48(9):1563–72.
142. Datta S, Kundu S, Ghosh P, De S, Ghosh A, Chatterjee M. Correlation of oxidant status with oxidative tissue damage in patients with rheumatoid arthritis. Clinical rheumatology. 2014;33 (11):1557–64.
143. Kundu S, Ghosh P, Datta S, Ghosh A, Chattopadhyay S, Chatterjee M. Oxidative stress as a potential biomarker for determining disease activity in patients with rheumatoid arthritis. Free radical research. 2012;46(12):1482–9.
144. Martelli-Palomino G, Paoliello-Paschoalato AB, Crispim JC, Rassi DM, Oliveira RD, Louzada P, et al. DNA damage increase in peripheral neutrophils from patients with rheumatoid arthritis is associated with the disease activity and the presence of shared epitope. Clinical and experimental rheumatology. 2017;35(2):247–54.
145. Nagasawa R, Hara Y, Murohashi K, Aoki A, Kobayashi N, Takagi S, et al. Serum heme oxygenase-1 measurement is useful for evaluating disease activity and outcomes in patients with acute respiratory distress syndrome and acute exacerbation of interstitial lung disease. BMC pulmonary medicine. 2020;20(1):310.
146. Scigliano G, Scigliano GA. Acute respiratory distress syndrome from Covid-19: A perfect storm from free radicals? Proposal for a new treatment. Medical Hypotheses. 2020;144: 110120.
147. Agarwal A, Qiu E, Sharma R. Laboratory assessment of oxidative stress in semen. Arab J Urol. 2017;16(1):77–86.
148. Cengiz N, Baskin E, Sezgin N, Agras P, Haberal M. Oxidative stress in children on hemodialysis: value of autoantibodies against oxidized low-density lipoprotein. Pediatric nephrology (Berlin, Germany). 2009;24(2):387–93.
149. Fan A, Jiang X, Mo Y, Tan H, Jiang M, Li J. Plasma levels of oxidative stress in children with steroid-sensitive nephrotic syndrome and their predictive value for relapse frequency. Pediatric nephrology (Berlin, Germany). 2016;31(1):83–8.
150. Wang Z, Yu C, Li XH, Deng BQ. The prognostic value of oxidative stress and inflammation in Chinese hemodialysis patients. Renal failure. 2017;39(1):54–8.
151. Türkoğlu C, Gür M, Şeker T, Selek Ş, Koçyiğit A. The predictive value of M30 and oxidative stress for left ventricular remodeling in patients with anterior ST-segment elevation myocardial infarction treated with primary percutaneous coronary intervention. Coronary artery disease. 2016;27(8):690–5.
152. Adam LN, Oraha AY, Shekha MS, Al-Habib OAM. Exploring nitric oxide as a crucial prognostic biomarker of coronary artery disease. Prostaglandins & Other Lipid Mediators. 2023;165:106717.
153. Bahadoran Z, Mirmiran P, Tahmasebinejad Z, Azizi F, Ghasemi A. Serum nitric oxide metabolites and hard clinical endpoints: a population-based prospective study. Scand Cardiovasc J. 2019;53(4):176–82.
154. Gumanova NG, Bogdanova NL, Metelskaya VA, Tarasov VI, Kots AY, Kutsenko VA, et al. Serum biomarkers, including nitric oxide metabolites (NOx), for prognosis of cardiovascular death and acute myocardial infarction in an ESSE-RF case-control cohort with 6.5-year follow up. Sci Rep. 2022;12(1):18177.
155. Kavakli H, Erel O, Becel S. Oxidative stress in diagnosis of acute appendicitis patients. Scientific Research and Essays. 2011;6.
156. Elmas B, Yildiz T, Yazar H, İlçe Z, Bal C, Özbek B, et al. New Oxidative Stress Markers Useful in the Diagnosis of Acute Appendicitis in Children: Thiol/Disulfide Homeostasis and the Asymmetric Dimethylarginine Level. Pediatric emergency care. 2020;36(8):362–7.
157. Ballini A, Santacroce L, Cantore S, Bottalico L, Dipalma G, Topi S, et al. Probiotics Efficacy on Oxidative Stress Values in Inflammatory Bowel Disease: A Randomized Double-Blinded Placebo-Controlled Pilot Study. Endocrine, metabolic & immune disorders drug targets. 2019;19(3):373–81.

158. Stolyarchuk OV, Sergiychuk OV, Tytarenko NV, Sergiychuk OL. [DIAGNOSTIC MARKERS AS PREDICTORS OF NECESSITY FOR THE OPERATIVE INTERVENTIONS PERFORMANCE IN PATIENTS FOR CORRECTION OF OXIDATIVE STRESS IN AN ACUTE PANCREATITIS]. Klinichna khirurhiia. 2016(12):23–5.
159. Sangeetha K, Krishnasamy N, Padma K, Rajendran DK. Evaluation of Oxidative Stress in Liver Cirrhosis Patients to Early Diagnosis of Minimal Hepatic Encephalopathy. International Neuropsychiatric Disease Journal. 2016;5:1–9.
160. P389. The value of oxidative stress parameters to predict histological severity in ulcerative colitis. Journal of Crohn's and Colitis. 2016;10(suppl_1):S294-S.
161. Masubuchi N, Sugihara M, Sugita T, Amano K, Nakano M, Matsuura T. Oxidative stress markers, secondary bile acids and sulfated bile acids classify the clinical liver injury type: Promising diagnostic biomarkers for cholestasis. Chem Biol Interact. 2016;255:83–91.
162. Perrone S, Cremonini I, Marinelli F, Monaco S, Nicoletti L, Giordano M, et al. New Strategies for Necrotizing Enterocolitis Diagnosis and Prevention in Newborns. Curr Pediatr Rev. 2021;17(3):191–200.
163. Kaur J, Politis C, Jacobs R. Salivary 8-hydroxy-2-deoxyguanosine, malondialdehyde, vitamin C, and vitamin E in oral pre-cancer and cancer: diagnostic value and free radical mechanism of action. Clinical oral investigations. 2016;20(2):315–9.
164. Feng JF, Lu L, Dai CM, Wang D, Yang YH, Yang YW, et al. Analysis of the diagnostic efficiency of serum oxidative stress parameters in patients with breast cancer at various clinical stages. Clin Biochem. 2016;49(9):692–8.
165. Gonçalves AC, Cortesão E, Oliveiros B, Alves V, Espadana AI, Rito L, et al. Oxidative stress and mitochondrial dysfunction play a role in myelodysplastic syndrome development, diagnosis, and prognosis: A pilot study. Free radical research. 2015;49(9):1081–94.
166. Khan FH, Ambreen K, Fatima G, Kumar S. Assessment of health risks with reference to oxidative stress and DNA damage in chromium exposed population. The Science of the total environment. 2012;430:68–74.
167. Stresin O, Önemi P, Pulmoner V, Hastaların E, Değeri T, Kavaklı H, et al. Importance of Oxidative Stress in Pathogenesis and its Value in Diagnosis of Pulmonary Embolism Patients. Journal of Academic Emergency Medicine. 2012;11.
168. Verbenko VA, Malev AL, Zakharova AN. P-783 – The diagnostic role of carbonyl derivatives, as a marker of oxidative stress in schizophrenia and alcoholic psychiatric disorders. European Psychiatry. 2012;27:1.
169. Dănulescu R, Costin D. Use of blood markers in early diagnosis of oxidative stress in age related macular degeneration. Revista medico-chirurgicala a Societatii de Medici si Naturalisti din Iasi. 2012;116(4):1136–42.
170. Tsilioni I, Kostikas K, Kalomenidis I, Oikonomidi S, Tsolaki V, Minas M, et al. Diagnostic accuracy of biomarkers of oxidative stress in parapneumonic pleural effusions. Eur J Clin Invest. 2011;41(4):349–56.
171. Villa-Correa YA, Isaza-Guzmán DM, Tobón-Arroyave SI. Prognostic Value of 8-Hydroxy-2′-Deoxyguanosine and Human Neutrophil Elastase/α1-Proteinase Inhibitor Complex as Salivary Biomarkers of Oxidative Stress in Chronic Periodontitis. Journal of periodontology. 2015;86(11):1260–7.
172. Kaneda K, Fujita M, Yamashita S, Kaneko T, Kawamura Y, Izumi T, et al. Prognostic value of biochemical markers of brain damage and oxidative stress in post-surgical aneurysmal subarachnoid hemorrhage patients. Brain research bulletin. 2010;81(1):173–7.
173. Krzystek-Korpacka M, Salmonowicz B, Boehm D, Berdowska I, Zielinski B, Patryn E, et al. Diagnostic potential of oxidative stress markers in children and adolescents with type 1 diabetes. Clinical Biochemistry. 2008;41(1):48–55.
174. Song H, Ma H, Shi J, Liu Y, Kan C, Hou N, et al. Optimizing glycation control in diabetes: An integrated approach for inhibiting nonenzymatic glycation reactions of biological macromolecules. International Journal of Biological Macromolecules. 2023:125148.

175. Pietryga M, Dydowicz P, Toboła K, Napierała M, Miechowicz I, Gąsiorowska A, et al. Selected oxidative stress biomarkers in antenatal diagnosis as 11-14 gestational weeks. Free Radic Biol Med. 2017;108:517–23.
176. Raicevic S, Cubrilo D, Arsenijevic S, Vukcevic G, Živkovic V, Vuletic M, et al. Oxidative Stress in Fetal Distress: Potential Prospects for Diagnosis. Oxidative Medicine and Cellular Longevity. 2010;3:870507.
177. El-Aal AA, El-Ghffar EAA, Ghali AA, Zughbur MR, Sirdah MM. The effect of vitamin C and/or E supplementations on type 2 diabetic adult males under metformin treatment: A single-blinded randomized controlled clinical trial. Diabetes & metabolic syndrome. 2018;12(4): 483–9.
178. Maged AM, Torky H, Fouad MA, GadAllah SH, Waked NM, Gayed AS, et al. Role of antioxidants in gestational diabetes mellitus and relation to fetal outcome: a randomized controlled trial. The journal of maternal-fetal & neonatal medicine: the official journal of the European Association of Perinatal Medicine, the Federation of Asia and Oceania Perinatal Societies, the International Society of Perinatal Obstet. 2016;29(24):4049–54.
179. Bo S, Ponzo V, Evangelista A, Ciccone G, Goitre I, Saba F, et al. Effects of 6 months of resveratrol versus placebo on pentraxin 3 in patients with type 2 diabetes mellitus: a double-blind randomized controlled trial. Acta diabetologica. 2017;54(5):499–507.
180. Panahi Y, Khalili N, Sahebi E, Namazi S, Karimian MS, Majeed M, et al. Antioxidant effects of curcuminoids in patients with type 2 diabetes mellitus: a randomized controlled trial. Inflammopharmacology. 2017;25(1):25–31.
181. Derosa G, D'Angelo A, Romano D, Maffioli P. A Clinical Trial about a Food Supplement Containing α-Lipoic Acid on Oxidative Stress Markers in Type 2 Diabetic Patients. International journal of molecular sciences. 2016;17(11).
182. Jiménez-Osorio AS, García-Niño WR, González-Reyes S, Álvarez-Mejía AE, Guerra-León S, Salazar-Segovia J, et al. The Effect of Dietary Supplementation With Curcumin on Redox Status and Nrf2 Activation in Patients With Nondiabetic or Diabetic Proteinuric Chronic Kidney Disease: A Pilot Study. Journal of Renal Nutrition. 2016;26(4):237–44.
183. Elbarbary NS, Ismail EAR, El-Naggar AR, Hamouda MH, El-Hamamsy M. The effect of 12 weeks carnosine supplementation on renal functional integrity and oxidative stress in pediatric patients with diabetic nephropathy: a randomized placebo-controlled trial. 2018;19 (3):470–7.
184. Ng YT, Phang SCW, Tan GCJ, Ng EY, Botross Henien NP, UD MP, et al. The Effects of Tocotrienol-Rich Vitamin E (Tocovid) on Diabetic Neuropathy: A Phase II Randomized Controlled Trial. Nutrients. 2020;12(5).
185. Rivara MB, Yeung CK, Robinson-Cohen C, Phillips BR, Ruzinski J, Rock D, et al. Effect of Coenzyme Q10 on Biomarkers of Oxidative Stress and Cardiac Function in Hemodialysis Patients: The CoQ10 Biomarker Trial. American Journal of Kidney Diseases. 2017;69(3): 389–99.
186. Asemi Z, Soleimani A, Shakeri H, Mazroii N, Esmaillzadeh A. Effects of omega-3 fatty acid plus alpha-tocopherol supplementation on malnutrition-inflammation score, biomarkers of inflammation and oxidative stress in chronic hemodialysis patients. International urology and nephrology. 2016;48(11):1887–95.
187. Tamtaji OR, Heidari-Soureshjani R, Mirhosseini N, Kouchaki E, Bahmani F, Aghadavod E, et al. Probiotic and selenium co-supplementation, and the effects on clinical, metabolic and genetic status in Alzheimer's disease: A randomized, double-blind, controlled trial. Clinical nutrition (Edinburgh, Scotland). 2019;38(6):2569–75.
188. Daneshvar Kakhaki R, Ostadmohammadi V, Kouchaki E, Aghadavod E, Bahmani F, Tamtaji OR, et al. Melatonin supplementation and the effects on clinical and metabolic status in Parkinson's disease: A randomized, double-blind, placebo-controlled trial. Clin Neurol Neurosurg. 2020;195:105878.
189. Peivandi Yazdi A, Razavi M, Sheikh S, Boroumand N, Salehi M, Hashemy SI. Clinical Trial Assessment of Intermittent and Continuous Infusion Dose of N-Acetylcysteine on Redox

Status of the Body in Patients with Sepsis Admitted to the ICU. Journal of intensive care medicine. 2020;35(12):1383–8.
190. Ortolani O, Conti A, De Gaudio AR, Moraldi E, Cantini Q, Novelli G. The effect of glutathione and N-acetylcysteine on lipoperoxidative damage in patients with early septic shock. American journal of respiratory and critical care medicine. 2000;161(6):1907–11.
191. Crimi E, Liguori A, Condorelli M, Cioffi M, Astuto M, Bontempo P, et al. The Beneficial Effects of Antioxidant Supplementation in Enteral Feeding in Critically Ill Patients: A Prospective, Randomized, Double-Blind, Placebo-Controlled Trial. Anesthesia & Analgesia. 2004;99(3).
192. Carpenter KLH, Kirkpatrick PJ, Weissberg PL, Challis IR, Dennis IF, Freeman MA, et al. Oral a-tocopherol Supplementation Inhibits Lipid Oxidation in Established Human Atherosclerotic Lesions. Free radical research. 2003;37(11):1235–44.
193. Wieczfinska J, Sarniak A, Wlodarczyk A, Sokolowska M, Pniewska E, Doniec Z, et al. Apocynin reduces reactive oxygen species concentrations in exhaled breath condensate in asthmatics. Experimental lung research. 2012;38:90–9.
194. Clausen J. The influence of selenium and vitamin E on the enhanced respiratory burst reaction in smokers. Biological trace element research. 1991;31(3):281–91.
195. Van Schooten FJ, Besaratinia A, De Flora S, D'Agostini F, Izzotti A, Camoirano A, et al. Effects of oral administration of N-acetyl-L-cysteine: a multi-biomarker study in smokers. Cancer epidemiology, biomarkers & prevention: a publication of the American Association for Cancer Research, cosponsored by the American Society of Preventive Oncology. 2002;11(2): 167–75.
196. Kasielski M, Nowak D. Long-term administration of N-acetylcysteine decreases hydrogen peroxide exhalation in subjects with chronic obstructive pulmonary disease. Respiratory medicine. 2001;95(6):448–56.
197. De Benedetto F, Aceto A, Dragani B, Spacone A, Formisano S, Pela R, et al. Long-term oral n-acetylcysteine reduces exhaled hydrogen peroxide in stable COPD. Pulmonary pharmacology & therapeutics. 2005;18(1):41–7.
198. Sadowska AM, van Overveld FJ, Górecka D, Zdral A, Filewska M, Demkow UA, et al. The interrelationship between markers of inflammation and oxidative stress in chronic obstructive pulmonary disease: modulation by inhaled steroids and antioxidant. Respiratory medicine. 2005;99(2):241–9.
199. Bridgeman MM, Marsden M, Selby C, Morrison D, MacNee W. Effect of N-acetyl cysteine on the concentrations of thiols in plasma, bronchoalveolar lavage fluid, and lung tissue. Thorax. 1994;49(7):670–5.
200. Vasu VT, de Cruz SJ, Houghton JS, Hayakawa KA, Morrissey BM, Cross CE, et al. Evaluation of thiol-based antioxidant therapeutics in cystic fibrosis sputum: Focus on myeloperoxidase. Free radical research. 2011;45(2):165–76.
201. Tantavisut S, Tanavalee A, Honsawek S, Suantawee T, Ngarmukos S, Adisakwatana S, et al. Effect of vitamin E on oxidative stress level in blood, synovial fluid, and synovial tissue in severe knee osteoarthritis: a randomized controlled study. BMC musculoskeletal disorders. 2017;18(1):281.
202. Couto D, Ribeiro D, Freitas M, Gomes A, Lima JL, Fernandes E. Scavenging of reactive oxygen and nitrogen species by the prodrug sulfasalazine and its metabolites 5-aminosalicylic acid and sulfapyridine. Redox report: communications in free radical research. 2010;15(6): 259–67.
203. Gan HT, Chen YQ, Ouyang Q. Sulfasalazine inhibits activation of nuclear factor-kappaB in patients with ulcerative colitis. Journal of gastroenterology and hepatology. 2005;20(7): 1016–24.
204. Campregher C, Luciani MG, Biesenbach P, Evstatiev R, Lyakhovich A, Gasche C. The position of the amino group on the benzene ring is critical for mesalamine's improvement of replication fidelity. Inflammatory bowel diseases. 2010;16(4):576–82.

205. Graham PM, Li JZ, Dou X, Zhu H, Misra HP, Jia Z, et al. Protection against peroxynitrite-induced DNA damage by mesalamine: implications for anti-inflammation and anti-cancer activity. Molecular and cellular biochemistry. 2013;378(1–2):291–8.
206. Guijarro LG, Mate J, Gisbert JP, Perez-Calle JL, Marin-Jimenez I, Arriaza E, et al. N-acetyl-L-cysteine combined with mesalamine in the treatment of ulcerative colitis: randomized, placebo-controlled pilot study. World journal of gastroenterology. 2008;14(18):2851–7.
207. Romagnoli C, Marcucci T, Picariello L, Tonelli F, Vincenzini MT, Iantomasi T. Role of N-acetylcysteine and GSH redox system on total and active MMP-2 in intestinal myofibroblasts of Crohn's disease patients. International journal of colorectal disease. 2013;28(7):915–24.
208. Mäkitalo L, Rintamäki H, Tervahartiala T, Sorsa T, Kolho KL. Serum MMPs 7-9 and their inhibitors during glucocorticoid and anti-TNF-α therapy in pediatric inflammatory bowel disease. Scandinavian journal of gastroenterology. 2012;47(7):785–94.
209. Ardite E, Panés J, Miranda M, Salas A, Elizalde JI, Sans M, et al. Effects of steroid treatment on activation of nuclear factor kappaB in patients with inflammatory bowel disease. Br J Pharmacol. 1998;124(3):431–3.
210. Ljung T, Lundberg S, Varsanyi M, Johansson C, Schmidt PT, Herulf M, et al. Rectal nitric oxide as biomarker in the treatment of inflammatory bowel disease: responders versus non-responders. World journal of gastroenterology. 2006;12(21):3386–92.
211. Dehghani F, Sezavar Seyedi Jandaghi SH, Janani L, Sarebanhassanabadi M, Emamat H, Vafa M. Effects of quercetin supplementation on inflammatory factors and quality of life in post-myocardial infarction patients: A double blind, placebo-controlled, randomized clinical trial.n/a(n/a).
212. Nachvak SM, Shabanpur M, Mostafai R, Heidari Moghaddam R, Moludi J. L-Carnitine supplementation reduces biomarkers of inflammatory and oxidative stress in patients with coronary artery disease: a randomised controlled trial. Archives of Physiology and Biochemistry. 2020:1–8.
213. Gueguen S, Pirollet P, Leroy P, Guilland J-C, Arnaud J, Paille F, et al. Changes in Serum Retinol, α-Tocopherol, Vitamin C, Carotenoids, Zinc and Selenium after Micronutrient Supplementation during Alcohol Rehabilitation. Journal of the American College of Nutrition. 2003;22(4):303–10.
214. Lai CY, Cheng SB, Lee TY, Hsiao YF, Liu HT, Huang YC. Impact of Glutathione and Vitamin B-6 in Cirrhosis Patients: A Randomized Controlled Trial and Follow-Up Study. Nutrients. 2020;12(7).
215. Elbarbary NS, Ismail EAR, Farahat RK, El-Hamamsy M. ω-3 fatty acids as an adjuvant therapy ameliorates methotrexate-induced hepatotoxicity in children and adolescents with acute lymphoblastic leukemia: A randomized placebo-controlled study. Nutrition. 2016;32(1):41–7.
216. Poprac P, Jomova K, Simunkova M, Kollar V, Rhodes CJ, Valko M. Targeting Free Radicals in Oxidative Stress-Related Human Diseases. Trends in pharmacological sciences. 2017;38(7):592–607.
217. Dalle-Donne I, Rossi R, Colombo R, Giustarini D, Milzani A. Biomarkers of oxidative damage in human disease. Clin Chem. 2006;52(4):601–23.
218. De Benedetto F, Pastorelli R, Ferrario M, de Blasio F, Marinari S, Brunelli L, et al. Supplementation with Qter® and Creatine improves functional performance in COPD patients on long term oxygen therapy. Respiratory medicine. 2018;142:86–93.
219. Rafraf M, Bazyun B, Sarabchian MA, Safaeiyan A, Gargari BP. Vitamin E Improves Serum Paraoxonase-1 Activity and Some Metabolic Factors in Patients with Type 2 Diabetes: No Effects on Nitrite/Nitrate Levels. Journal of the American College of Nutrition. 2016;35(6):521–8.
220. von Herbay A, Stahl W, Niederau C, Sies H. Vitamin E improves the aminotransferase status of patients suffering from viral hepatitis C: a randomized, double-blind, placebo-controlled study. Free radical research. 1997;27(6):599–605.

Chapter 23
Future Directions, Pitfalls, and Solutions in Oxidative Stress Assessment

A. Khelfi

Abstract Increased evidence links reactive species (RS) to the etiology of various diseases. Despite the impressive developments taking place in oxidative stress (OS) monitoring, many important aspects of free radical (FR) assessment are open for investigation. The available methods for RS determination in biological systems are challenging as they have rapidly evolved over the few last decades. Therefore, the knowledge of the gaps that face any investigator will undoubtedly lead to a better assessment of oxidative stress to prevent, control, or diagnose oxidative stress related diseases. In all cases, RS levels and the redox state of a biological system (e.g., cells, tissues, and whole organism) should be measured precisely by analytical methods that offer greater specificity and sensitivity for each and every RS. The use of global indicators in a valuable alternative has gained considerable interest for their capacity to assess the overall redox state in cells.

This chapter begins with a brief introduction about the very short half-lives of oxidative stress markers and gives an in-depth account of the present knowledge and current problems in the field of oxidative assessment in biological systems. The critical discussion of the methodologies is especially helpful. In order to minimize the variability of results, optimal conditions for sampling and storage are described. New approaches are also presented like the use of the combination of several biomarkers for optimal results. In this chapter, developing an index for summarizing oxidative stress status is recommended. Future perspectives are highlighted such as standardization of protocols, inter-laboratory comparisons, technical validation, and consensus establishment.

Keywords Sampling · Storage · Techniques · Biomarker · Standardization · Inter-laboratory · Consensus

---

A. Khelfi (✉)
Faculty of Pharmacy, University of Algiers, Algiers, Algeria

National Centre of Toxicology, Algiers, Algeria

Department of Chemistry and Biomolecular Science, Clarkson University, Potsdam, NY, USA
e-mail: ab.khelfi@univ-alger.dz; akhelfi@clarkson.edu

S. Andreescu et al. (eds.), *Biomarkers of Oxidative Stress*,
https://doi.org/10.1007/978-3-031-60738-7_23

Measurement of RS and oxidative products provides a fingerprint of oxidative damage and gives the strongest evidence of OS in biological systems. Biomarkers should be useful to monitor the real redox potential of RS compounds in vivo in order to better evaluate their potential effects, but also whether their beneficial or detrimental effects are directly linked to several diseases. Examining a panel of OS biomarkers has the potential to provide greater insight into the characteristics of redox-related disease as well as help identify new therapeutic and diagnostic methods. Indeed, OS likely plays a role in several diseases, yet very few OS markers have made it into routine clinical use, which may have several reasons. Although there are many parameters to quantify during an oxidative insult, the measurement of OS in biological systems has proven to be a difficult task.

The main problems in assessing OS concern the complexity of some techniques requiring special equipment, as well as their respective sensitivity and specificity for markers. Among other factors that make OS assessment difficult to perform include the short lifetime of RS, the low levels of oxidized biomolecules, sampling, artificial oxidation during sample preparation, and the choice of the best markers demonstrating the special implication of OS in a particular pathological condition.

23.1 Short Half-Life of Oxidative Stress Markers

Generally, authentic RS, such as $O_2^{\bullet -}$, $NO^{\bullet}$, and $ONOO^{-}$ are highly reactive and have a too short half-life, making their detection extremely difficult. Also, the production of RS is not a regular and constant mechanism but is often a fast and pulsed phenomenon. A high concentration of cellular antioxidant enzymes and the considerable kinetic rates of their reactions with RS result in molecular competition between the probes used for RS detection and antioxidant enzymes. With these conditions, and because they are produced mainly in some tissues inside the body and less in plasma, it is difficult to assess the primary RS and FR from a patient blood sample. Overcoming these difficulties remains a significant challenge in the analytical field. Fortunately, primary radicals generated by the activation of oxygen will form secondary and more stable radicals when encountering an organic compound. Therefore, ongoing oxidative damage is generally analyzed by measurement of secondary products, including oxidized derivatives of amino acids, nucleic bases, and lipids.

Molecules modified by RS, such as carbonyls, 4-HNE, and 8-oxo-dG, are derivatives of proteins, lipids, and nucleic acids, respectively, and are rather stable and easier to determine. The amounts of these products are proportional to OS status, and they are detected at the sites where free radical attacks occur.

23.2 Sampling and Storage

Since the specific effects of OS are manifested within cells and inner tissues, access to these biomaterials in patients would be difficult. Assessment of OS in readily accessible biological fluids is the only option, but this in itself may lead to some difficulties in interpretation. Indeed, the use of biological fluids assumes that the products resulting from OS have been released into the circulation and that they were not completely removed by other organs as part of repair mechanisms and antioxidant defenses. Additionally, the markers may also represent the extent of OS in the whole body and not simply that of a specific organ or tissue. At the same time, the mass of the targeted tissues is usually small compared to that of the body, and therefore, significant amounts would have to exit the tissue for any change to be measurable. Some biological samples are non-invasive and can overcome these analytical limits. For instance, the examination of exhaled breath condensate (EBC) samples being simple and non-invasive could be routinely used for prognostic purposes to follow the course of diseases and to monitor a selected therapy, especially if an imbalance of the redox status is located in the respiratory system. EBC assays can satisfy the increasing demand to measure biomarkers as a means to assess the efficacy of new treatments in an easily accessible and cost-effective manner. This type of sample is already used for the assessment of H_2O_2, 3-NO_2Tyr, MDA, 4-HNE, isoprostanes (IsoPs), ethane, pentane, and GSH (Table 7.2).

It is worth mentioning that some biomarkers such as 15-F_{2t}-IsoP may increase upon storage time so that the given value at the time point of analysis may differ greatly. Under such circumstances, analysis may provide apparently increased levels of biomarkers, for instance, in the control group, even if the analytical procedure is similar to the tested group, thus leading to wrong conclusions.

23.3 Sample Preparation and Analysis

The choice of the assay for OS assessment is a compromise between ease, ability to collect real-time data, need for spatial information, time of sample preparation, and the availability of instruments. At present, real-time measures of OS parameters are limited and there are no truly non-invasive methods. Otherwise, investigators have to deal with a long preparation process and heavy methods in order to closely assess highly reactive compounds.

One of the greatest problems in assessing OS is the generation of artifacts during sample processing. The fact that the atmosphere contains 20% molecular oxygen creates a possibility of spontaneous oxidation of antioxidants, lipids, proteins, and DNA during sample analysis.

Even if a very small fraction of the biomarker precursor is oxidized during sample handling, the artifactual formation rate may be considerable. It may reach and even by far exceed the basal level of the biomarker. Artifactual formation of OS biomarkers is very widespread and has been reported for numerous compounds including GSH, and IsoPs. Other well-known examples of the artificial generation of biomarkers concern the oxidative modification of guanine to 8-OHdG and tyrosine to 3-NO_2Tyr. It has been shown for both these derivatives that their concentrations can be artifactually increased several times compared to the respective basal levels during sample treatment. The molar ratio of guanine to 8-OHdG and of tyrosine to 3-NO_2Tyr in proteins is estimated to be much higher than usually reported, i.e., 1 to 10×10^5:1 [1, 6].

Oxidation of biomarkers of OS may occur during extractions, sample processing, derivatization, and clean-up steps unless special precautions are taken. These precautions include the chelation of iron in solution, the bubbling of nitrogen or argon to remove oxygen from solutions, the storage of samples at −80 °C until the analysis, and eventually addition of antioxidants to prevent samples from atmospheric oxidation. Some precautions protect from oxidation more than others, which may result in a discrepancy in the measurement of OS parameters.

Some of the assays are prone to numerous artifacts resulting from the analytical step itself [30, 31]. Assays requiring sample derivatization are inherently problematic because of the reactive environment in which such a process is conducted. For some derivatization protocols, samples that usually would be kept at subzero temperatures, are heated to higher temperatures at which oxidation occurs at a high rate. For instance, some procedures for the derivatization of GSH or DNA base pairs resulted in oxidation of samples and yielded artificially high values [8]. For such cases, it is recommended as possible to separate the reduced from the oxidized form before being derivatized.

Finally, molecular products formed from the reaction of RS with various biomolecules are generally more stable than RS themselves and can be more easily assessed and prepared for analysis. Indeed, in most cases, RS is commonly traced by measuring their oxidized targets (e.g., 3-NO_2Tyr and protein carbonyls), MDA, or anti-oxidants (e.g., anti-oxidant enzymes).

23.4 Analytical Techniques

As mentioned several times above, two analytical factors are decisive for biomarkers of OS, the artifactual generation of biomarkers during sample treatment and the lack of the performance of the analytical method. Indeed, in vivo determination of OS biomarkers has been difficult to accomplish because among other reasons the lack of sensitive, specific, and selective assays. While the oxidation of lipids, protein, and nucleic acids generates unique products, these products remain structurally similar to many other molecules that are also present in the cell and much higher amounts. In the past three decades, many methodologies and various approaches have been developed, which have enabled scientists and medical physicians to measure and quantify with a great degree of accuracy the levels of different oxidants, antioxidants, and oxidized biomolecules. It is apparent that no single method will accurately reflect quantitative rates of formation of RS under all circumstances and no single method best assesses OS state. In most cases, many different methodologies must be implemented before an accurate evaluation of OS can be made.

Although some RS can be directly detected, currently available methods are not generally applicable to clinical examination. One of the current challenges facing the application of OS assessment is that most of the proposed methods have not yet been tested in relevant clinical trials, and that the optimal choice of an analytical method is still dependent on the pathology being tested. Moreover, most of the analytical procedures used to measure some biomarkers require expensive instruments (e.g., LC-MS/MS and GC-MS/MS) and skilled technical personnel. Overall, what is urgently needed is an assembly of simple and precise assays for OS assessment and disease risk stratification. Analytical methods with the specificity and sensitivity to detect a small number of oxidized molecules among many structurally similar compounds are difficult to develop in clinical laboratories. Conversely, some biomarkers with high physiological values can be measured by means of rather simple and inexpensive procedures. The only feature required is to check the specificity of these methods in biological systems. Indeed, commercially available immunoassay kits are increasingly used by many groups in basic and clinical research without adequate adaptation and evaluation. For instance, a number of comparison studies have shown unsatisfactory agreement between EIA and LC-MS/MS or GC-MS/MS results [3, 9, 10]. While mass spectrometry provides sensitivity and specificity and has become more accessible, antibody-based methods remain, for now, the clinical standard. As we have seen earlier, some EIA methods fall short of specificity to the point where antibodies specific for oxLDL or any new antibody-based tracer require rigorous verification of specificity and sensitivity.

Immunoassays using antibodies to oxidize molecules are more convenient in clinical practice than chromatographic methods. They are especially useful when samples from different groups are compared and when absolute quantitation is not required. Although variations in the specificity and avidity as well as the possible cross-reactivity of antibodies lead to variability between laboratories, these limitations are often sufficient in comparative studies, for example, to determine whether levels of AGEs, ALEs, or lipoxidation products are higher or lower in patients, compared to control groups, or in response to therapy.

Another major limitation of immunoassays is the relatively low sensibility that needs solubilization of tested molecules, especially for the detection of cross-link structures. By contrast, CML is recognized with over 1000-fold higher avidity in the protein-bound form, compared to the free amino acid. This variability might result from differences in the size of the antigenic epitope on protein, and therefore, major haptens must be independently evaluated [5].

Other antibody-based assays, such as immunohistochemistry, have proven to be an exceptional tool for detection of oxidation products in situ. However, there are limitations to the identification of specific products because, among other reasons, of the frequent and possible cross-reactivity of antibodies. For instance, *N*-(carboxymethyl)ethanolamine, derived from either glycoxidation or lipoxidation of phosphatidylethanolamine, can react with CML-antibodies prepared against AGEs (carbohydrate-oxidized proteins) [11].

Second, concerns have been issued in several studies about the cross-reactivity and limited specificity related to polyclonal antibodies that contain a heterogeneous population of antibodies against a wide range of epitopes, even if a single hapten is used for immunization. These inconveniences primarily lead to differential affinity to OS products especially if these compounds contain several epitopes that are heterogeneously expressed.

A third concern with immunohistochemistry is that the staining with antibodies does not necessarily mean that the detected epitopes are necessarily the same as the primary tested specificity of the antibody. A minor cross-reactive fraction of antibody may "light up" the tissue section.

It is helpful to isolate the epitope-specific fraction of antibody cocktail by immunoaffinity chromatography using immobilized hapten and demonstrate their specificity by competitive and known haptens before concluding that a specific epitope is present in the tissue. Even in this case, the immunizing epitope might be bound more tightly by the antibody than its analogous form in tissues, and a determinant with low cross-reactivity may still result in a positive reaction.

Fourth, when it comes to quantitative applications of immunohistochemistry some issues may arise, yielding misleading interpretations. Immunohistochemistry has been used in a semi-quantitative fashion to detect oxidative changes in several diseases. However, a possible limitation may emerge, as quantitative methods are intrinsically related to the assumption that epitopes are equally exposed on the surface of the modified molecules. This may not always be true, especially if metabolic turnovers and repair mechanisms in tissues are activated by the disease process. In some cases, tissue sections are treated by enzymatic digestion to favor the exposure of epitopes and enhance the sensitivity for immunostaining. This procedure may differently affect control and diseased tissues and can lead to misinterpretations. The manifold limitations to rigorous interpretations of immunohistochemical experiments highlight the importance of confirming immunological observations by chemical analysis.

Fifth, immunohistochemistry techniques are usually performed on inaccessible tissues. Less use of this type of assays and utilization of newer and easier strategies, such as real-time visualization techniques that provide spatial and temporal understanding of RS in cells, will furnish a more complete picture and precise understanding of OS. Circulating cell harvesting methods may provide an additional solution to this issue.

Finally, from the point of view of clinical application, the methods for OS determination have to respond to some criteria of quality:

- First, good precision and accuracy are usually not achieved when determining RS and enzymes, i.e., some published techniques give poor performances in terms of quantitative analysis.

- Second, whatever method is used to assess RS or to measure oxidative damage, it is necessary to think carefully about how the method works, what is likely to confound it and how quantitative it can be.

23.5 Low Concentrations of Biomarkers

Other difficulties that can be faced by analysts include the lack of sensitive assays for RS determination. Oxidized lipids, proteins, and DNA are difficult to measure due to their low abundance in cells and blood circulation. Similarly, the labile nature of some modifications, such as cysteine oxidation, poses significant challenges to translate them into a high-throughput, cost-effective clinical diagnostic.

In the basal state, many biomarkers of OS occur in most biological systems at very low concentrations. For instance, the basal plasma concentrations of 3-NO_2Tyr are 1.4 nmol/L. The reported mean urinary o,o′-dityrosine concentration is around 0.08 μmol/L in urine. Normal values of MDA depend on the technique, ranging from more than 4 μmol/L for TBA-reactants by colorimetric tests to 2.5 μmol/L by fluorometry and to 1 μmol/L by HPLC. The 4-HNE normal values are around 0.12 μmol/L in plasma [32–34]. The 15-F_{2t}-IsoP is present in human blood as free acids at about 20 pmol/L and as esters of lipids at about 400 pmol/L. In human urine, the concentration of 15-F_2t-IsoP and its metabolites is higher, on the order of 1 nmol/L (or 250 pmol/mmol creatinine), although these concentrations are difficult to be detected with accessible instruments. For other products of lipoxidation, normal values of ethane and pentane range from 0.3 to 1.6 nmol/L and from 0.2 to 4 nmol/L without lung washout and 24 pmol/L and 4 to 11 pmol/L using lung washout. Besides, it is noteworthy to know that plasma NO_3^- concentrations are around 20–50 μmol/L, whereas plasma or serum nitrite concentrations are much lower, i.e., 0.20–4.8 μmol/L. Such extremely low concentrations represent a great challenge to most analytical methods as very low LOD is highly required.

23.6 Time of Oxidative Stress Assessment

A high steady-state level of oxidized molecules rarely occurs in vivo during times of OS. Oxidized targets of RS disrupt cell function and are then rapidly removed by antioxidants and cellular repair systems. For instance, cells contain GPx that removes lipid peroxides, proteases that recognize and remove oxidized proteins, and DNA repair enzymes that remove oxidized bases. Determination of oxidized/reduced glutathione ratio is a good index of fast and transient oxidation, as found after ischemia-reperfusion episodes. This is because this ratio can increase rapidly during the first minute after stress; however, it recovers to a normal level within a few minutes. To monitor chronic stress such as in atheroma or aging, MDA remains the most popular indicator despite its deficiencies and controversy.

23.7 Cause–Effect Relationship

Redox signaling is universally integrated with central homeostatic systems at the molecular, cellular, and organismic levels. The scientific community has learned a great deal about these systems in health and disease. This led to the emergence of a new field of redox therapy that explores the possibility of targeting the production of oxidants as a medical strategy. However, because of the large pleiotropy of RS and their role as physiological signaling and toxic agents, it remains difficult to explicitly define their involvement in terms of a simple cause–effect relationship. This is the reason behind the complex problem of making the link between OS and pathologies. This link may be complicated by the existence of other diseases that may have OS as part of their etiology. For instance, LDL oxidation is not unique to atherosclerotic plaques and may also occur in biological fluids (e.g., synovial fluid or plasma) of systemic sclerosis, ischemic stroke, systemic lupus erythematosus, or RA patients (Table 22.1) [12, 13].

Another issue may arise when changes in the OS status occur as a result of aging process or diet habits. This makes measurement of generalized OS difficult to interpret as an indicator of a specific disease. Paradoxically, an interesting value of OS biomarkers may come from being indicators of a disease mechanism common to several pathologies rather than diagnostic for a specific disease.

An additional difficulty in examining the putative relationship between OS and diseases is to define the existence of a specific oxidative modification in the pathogenesis of a particular disease. While studies of oxidative modification and diseases have been reasonably well validated, it is still unusual to find both deployed together in a characterized way. Whether OS is merely a by-product of pathophysiological conditions or if it plays an active role in the progression of these pathologies is another difficult task and has yet to be decided. Of note, it is also difficult to know whether OS is the origin or the consequence of diseases to which it is bound. Some successful compelling data suggest that OS plays either an initiating and sustaining role in many types of liver disease, inflammatory bowel disease (IBD), and pancreatitis [14–16].

Although a general pattern for OS exists, a specific pattern of OS often varies from one biological system to another. There are numerous potential cellular sources of RS and many endogenous compounds with which RS may react. For these reasons, a set outcome of oxidative damage is difficult to predict for a specific disease. In some cases, an increase in RS can occur without a significant decrease in certain antioxidants; or lipid peroxidation may occur without DNA oxidation.

23.8 Panels and Global Indicators

Generally, simultaneous determination of biomarkers of OS using different methods is certainly recommended. Only recently, an emphasis on the formation of a panel of biomarkers for specific disease states and environmental exposures has been

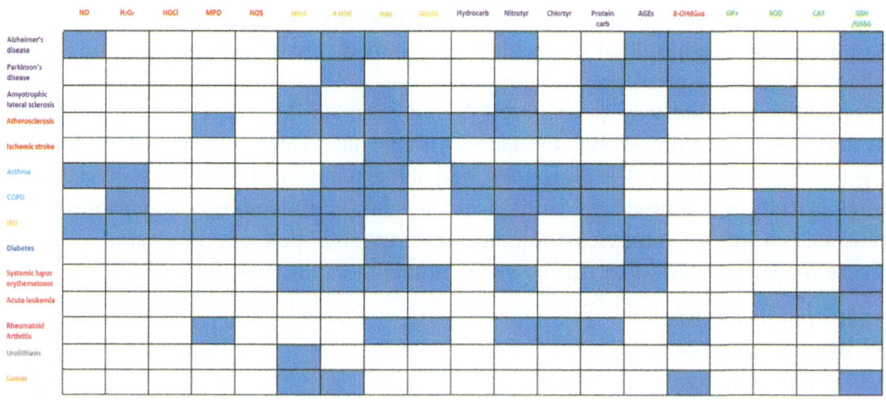

Fig. 23.1 Relevance of a battery of oxidative stress biomarkers in clinical practice. Lipid (orange), protein (purple), and DNA (pink) biomarkers were analyzed in various diseases listed across the columns. Pro-oxidant (red) and anti-oxidant (green) systems are represented as well. Color coding of disease illustrates the type of pathogenesis as illustrated across the rows

proposed. As many of the markers have been measured in similar diseases, a combination of them in large-scale panels and profile analysis could provide an additional approach to assess disease progression and therapeutic outcomes (Fig. 23.1).

Measurement of larger panels of markers will help give a more comprehensive picture of their significance. Basically, as many as possible independent biomarkers should be measured in biological systems. The use of a panel of biomarkers would enhance the positive predictive value of a test and minimize the proportion of false-positive and false-negative results. There is a great chance of making this panel as long as biomarkers of damage to several different biomolecules frequently rise in parallel in cells subjected to severe OS [30].

From a practical point of view, a good strategy for OS assessment would be to measure simultaneously typical biomarkers representing damage to cellular components such as oxidized proteins (carbonyls or AOPP) in plasma, oxidized bases (8-OHdG) in urine, and oxidized lipids (IsoPs) in urine. For instance, many biomarkers of OS have been extensively and simultaneously explored in cigarette smokers [17, 18]. Many of these compounds were explored in cancers, pulmonary diseases, and CVD, to name a few. A panel of biomarkers that range from oxidized bases to IsoPs has provided a tool to better understand different patterns of disease progression and biological outcomes.

Because no single method exists to measure all of these breakdown products at once, an accurate assessment of total oxidative damage is almost impossible to obtain in everyday practice. Few oxidized compounds are preferentially chosen as indicators of the severity of oxidative damage and are monitored as biomarkers of OS.

In most of the studies summarized in this work, one to three biomarkers were used to explore the magnitude of OS, with a small percentage of articles reporting the use of four or more biomarkers.

Another possible way to explore OS and estimate the redox state in biological systems is to explore global indicators of the antioxidant system using in vitro assays. These indicators are grouped under the name of TAC. They measure the total activity of antioxidant substances present in the biological samples (e.g., plasma, serum, erythrocytes) able to scavenge RS that are generated in vitro. There are several tests that can be differentiated according to the:

- RS producer system
- Biological target
- Detection system

On the basis of the chemical reactions involved, TAC assays can be mainly divided into two categories: hydrogen atom transfer (HAT)-based methods and single electron transfer (SET)-based methods. A list of commercially available assays is provided in Table 23.1.

Overall, the benefit of using these assays is to readily evaluate the radical scavenging capacity and measure the cumulative reducing ability of all antioxidants in a given system (erythrocytes, blood, and seminal plasma). However, the main inconvenience is the inability of these assays to give a specific identification of the level of a particular antioxidant. Moreover, these assays measure non-enzymatic antioxidants only and the predictive value has yet to be established.

Other global indicators have been widely used in research including total oxidant status (TOS), ROS-TAC score, and oxidation-reduction potential (ORP).

Erel's TOS method is based on the oxidation of ferrous ions (e.g., Fe^{+2} o-dianisidine complex) to ferric ion in the presence of oxidant molecules (e.g., RS) within plasma samples and on the formation of a colored complex with the formed ferric ions and xylenol orange in an acidic environment. The color intensity, which is proportional to the amount of oxidants in the sample, is measured spectrophotometrically.

ROS-TAC score is a better predictor compared to TOS and TAC assessment alone by minimizing variability. However, it requires statistical modeling and provides a cumulative prediction of OS in a sample but not a direct measure of ROS nor TAC.

With regard to ORP, this assay measures the potential of movement of electrons from one chemical species to another. It offers several advantages:

- Measure of all known and unknown oxidants and antioxidants
- Real-time assessment of redox balance
- Quick and relatively simple test with ready-to-use sensor and analyzer
- Applicable to various types of samples including seminal ejaculate or seminal plasma, both fresh and frozen
- Requires small volume of semen for analysis
- Test results are not affected by semen age for up to 2 h

Table 23.1 Assessment of total antioxidant capacity of plasma and erythrocytes

Type of assays	Assay	Sample	Principle	Limitations
Hydrogen atom transfer (HAT) $X^{\cdot} + AH \rightarrow XH + A^{\cdot}$	HAT-ORAC	Blood plasma Seminal plasma	Fluorimetric assay: AAPH-induced: B-phycoerythrin (red) or fluorescein (green) Ability of compounds to protect these probes from free radical damage resulting in fluorescence decay competitive reaction kinetic, AUC Sensitive and covers a wide range of antioxidants	AAPH used as a free radical source undergoes spontaneous decay and it is sensitive to temperature Lipophilic antioxidants not included Proteins
	HAT-TRAP	Blood plasma Seminal plasma	Fluorimetric assay: AAPH—Induced: B-phycoerythrin fluorescence decay (red) $DCFH_2 \rightarrow DCF$ fluorescence increase (green) Competitive reaction kinetic, lag phase Gives an idea of the rate of free radical formation	AAPH used as a free radical source undergoes spontaneous decay and it is sensitive to temperature Lipophilic antioxidants not included Not all the antioxidants give a lag phase Self-propagation of DGF radicals
	HAT-CB	Blood plasma Seminal plasma	Spectrophotometric assay: AAPH- or AMVN-induced absorbance decay (450 nm) Competitive reaction kinetic, stern-Volmer-like relation	Bilirubin and carotenoids that absorb at the wavelength of determination
	HAT-TAP	Blood plasma Seminal plasma	Fluorimetric assay: MeO-AMVN induced C11-BODIPY fluorescence increase (green) Competitive reaction kinetic, AUC	–
Single–electron transfer (SET) $X^{\cdot} + AH \rightarrow X^{-} + AH^{+\cdot}$	$ABTS^{\cdot+}$	Blood plasma Seminal plasma	Spectrophotometric assay: Capacity of antioxidant compounds to quench and decolorize $ABTS^{\cdot+}$ Absorbance decay (734 nm) Measurement is compared to an internal standard (Trolox equivalent antioxidant capacity (TEAC), a vitamin E analog)	Results vary with sample dilution; antioxidant used may interact with solvent molecules; specificity varies

(continued)

Table 23.1 (continued)

Type of assays	Assay	Sample	Principle	Limitations
			Non-competitive, endpoint, TEAC Extremely fast and simple	
	DPPH$^{•}$	Blood plasma Seminal plasma	Spectrophotometric assay: Ability of antioxidant compounds to quench DPPH$^{•}$ leading to absorbance decay (515 nm) Non-competitive, endpoint, EC50	Carotenoids that absorb at the wavelength of determination
	FRAP	Blood plasma Seminal plasma	Spectrophotometric assay: Ability of compounds to affect ferric to ferrous reduction Absorbance increase (593 nm) noncompetitive, endpoint	Fe may generate FR Not every FR reduces Fe; GSH is not measured Bilirubin absorbs at the wavelength of determination
	CUPRAC	Blood plasma Seminal plasma	Spectrophotometric assay: Neocuproine absorbance increase (450 nm) noncompetitive, endpoint	Bilirubin and carotenoids that absorb at the wave-length of determination
	TAS	Blood plasma Seminal plasma	Spectrophotometric assay: Capacity of antioxidant compounds to curb Fenton reaction-formation of the colored dianisidyl radical (reaction between O-dianisidine and OH$^{•}$). This results in the decrease of absorbance at 444 nm. Competitive, endpoint Results are expressed in μmol Trolox Eq/L.	–
–	BAP	Plasma and erythrocytes	Spectrophotometric assay: Ability of antioxidants to reduce a colored thiocyanate-derived substrate, which contains bounded Fe^{3+} ions, to Fe^{2+} form and the capacity to decrease the absorption following this conversion The absorbance is assessed by measuring the absor-bance at 505 nm Extremely fast, simple and covers a wide variety of antioxidants	Hyperlipemic samples can underestimate results

–	Hemolysis assay	Erythrocytes	In vitro system generating ROS into contact with the RBCs. The rate of hemolysis is measured as a marker of oxidative damage. The latency time before the onset of hemolysis is directly proportional to the amount of antioxidants present in the erythrocytes Kirial laboratories (KRL) test: Time needed to hemolyze 50% of whole RBCs Measures mainly membrane resistance to oxidants, therefore very linked to vitamin E Requires fresh, unfrozen blood	–

AAPH 2,2′-azobis(2-methylpropionamidine) dihydrochloride, *ABTS* 2,2′-azino-bis(3-ethylbenzothiazoline-6-sulfonic acid), *AMVN* 2,2′-azobis 2,4-dimethylnaleronitrile, *AUC* area under the curve, *BAP* Biological antioxidant potential, *CB* Crocin bleaching, *CUPRAC* copper-reducing assay, *DCFH*$_2$ 2′,7-dichlorodihydrofluorescein, *DPPH* 2,2-diphenyl-l-picrylhydrazyl, *EC50* efficient concentration 50, i.e., the amount of antioxidant necessary to decrease by 50%, *FRAP* ferric reducing antioxidant power, *MeO-AMVN* 2,2′-azobis(4-methoxy-2,4-dimethylvaleronitrile), *ORAC* oxygen radical antioxidant capacity, *TAP* total antioxidant performance, *TAS* total antioxidant status, *TEAC* Trolox equivalent antioxidant capacity, *TRAP* total radical antioxidant potential

The following points are the reported limitations of the assay:

- Affected by viscosity of the sample
- Does not provide information on individual oxidant and antioxidant that lead to the imbalance of the redox state

An interesting application of global assessment of RS using ORP (MiOXSYS) in male infertility has been described by Agarwal et al. [7]. Using semen samples, the assay gives a direct and global estimation of RS by reacting them with Luminol/ lucigenin probes and monitoring the signal using a luminometer. This assay presents high sensitivity and specificity and reference values have been reported (cutoff 1.36 mv/10^6 SPZ). However, test results are affected by several factors including semen age and volume, repeated centrifugation, and background luminescence. Moreover, it is time-consuming assay and requires expensive equipment.

An alternative has been described, which is the nitroblue tetrazolium test coupled with optic microscopy. This test estimates ROS generation in semen samples and allows the differentiation between both sources, i.e., spermatozoa and leukocytes. It is highly sensitive and inexpensive. However, scoring of relative density of formazan granules inside spermatozoa and leukocytes requires experienced personnel. Inter-observer variability may occur and a quantitative measure of ROS is not possible in this test.

23.9 How to Choose the Most Appropriate Biomarker?

When investigating the OS in relation to disease state, three questions should be addressed:

- What are the most appropriate biomarkers for OS?
- What is the best strategy to measure them?
- How does an investigator decide to measure a specific biomarker instead of another to assess OS?

Although there are many biomarkers of OS to choose from, admittedly, no currently used biomarker has yet fulfilled the key requirement of the "gold" biomarker that it be predictive of the initial and later development of disease. It is therefore evident that the choice of the most appropriate biomarker is left to the "free interpretation" of investigators. No consensus in this sense has been reached in this area, although some recommendations have been issued for analytical methods (see Sect. 23.11). With regard to this point, there is a clear need for 'integrative' scheme to allow the evaluation of the role and significance of specific molecular biomarkers (Fig. 23.2).

A good and reliable biomarker should satisfy a series of specific criteria in order to function as a suitable indicator of oxidative damage in relation to disease:

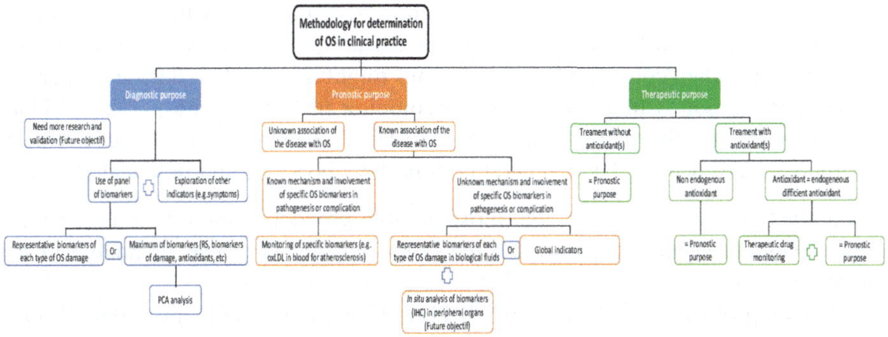

Fig. 23.2 Methodologies for OS assessment in the medical field

- It should be chemically unique, stable, accurately quantified, and reflect specific oxidation pathways (e.g., protein carbonyls, 3-NO_2Tyr, some lipid oxidation products, DNA/RNA oxidation products).
- Directly implicated in the onset and/or progression of disease.
- Its concentrations in biological samples should change during times of OS and correlate with disease severity so that it can be used both as a prognostic tool and a monitor to follow the efficiency of a particular therapy.
- It should not be affected by cellular processes (e.g., cell cycle, metabolism, etc.).
- It has sensitive and specific methods for its determination.
- Requires noninvasive methods of sampling and accessible techniques for level determination.
- Low intra- and inter-variability in humans (e.g., circadian rhythm).
- Accurate, precise, sensitive, specific, interference-free, and validated assays for quantitative analysis.
- Consensus on reference intervals and physiological values.

Among the numerous compounds appearing during OS, only a few can really be determined in clinical situations. In order to get the most out of the presently available biomarkers, it's important that they be used intelligently. On top of the given criteria, some guidance can help to address this issue. As a general rule, the choice of an indicator depends on the type of OS. Indeed, some biomarkers may be better suited than others according to their origin. The choice must be appropriate for the putative mechanism of damage:

- Some indicators like DNA damage products are preferentially interesting to analyze when the OS is related to nucleus (irradiation) or when the disease involves a DNA modification (cancer). If DNA damage is implicated, measurement of lipid peroxidation and protein oxidation may be irrelevant.
- Lipid peroxidation is a better choice for an OS occurring close to the plasma membrane or during the activation of immune cells.
- Protein oxidation is supposed to better reflect stress in cytosol or extracellular fluid.

- Another possibility is to associate biomarkers of damage with defense system measurements. Among these measurements, antioxidant enzyme activities and small antioxidant compound concentrations might be useful, to decide on starting supplementation with antioxidants or to monitor the effectiveness of antioxidant medication (see Sects. 22.4 and 22.5). Most of these markers have recently been proposed and need to be evaluated both in the laboratory and practical clinic.

For each chosen biomarker, the most suitable procedure should be used after a critical evaluation and adaptation of the proposed methodologies available in the literature.

Another problem may emerge when high-throughput results are required. One should distinguish between clinical investigators who need an easy-to-run, inexpensive, and useful technique, and fundamental scientists probing RS at the origin of OS using various selective and sensitive techniques. Investigators have to be pragmatic and realize that sophisticated and expensive instruments do not always offer more information than easy-to-run assays. When making diagnosis in an individual patient, physicians need a precise and discriminative method. Such a method is not so required in epidemiological studies where a great number of samples and a statistical evaluation compensate for the lack of precision. In the latter case, investigators need easy-to-run methods and highly stable parameters because a great number of samples are collected far from the laboratory.

23.10 Technical Validation

Whereas a panel of more than 70 biomarkers of OS has been developed, mainly with respect to lipid peroxidation, there is still a lack of validated indicators to study the outcomes of OS and its status, which is often reversible and precedes oxidation in many pathological conditions. In addition, too many molecules have been proposed as biomarkers of OS before they have been adequately and technically validated. The application of any new technical procedure in investigative studies or in routine practice requires validation in terms of sensitivity, specificity, precision, accuracy, reproducibility, and correlation of the measurements with disease status. Such new validated procedures should also allow the investigation of the real and true role of OS during the progression of diseases, as well as the evaluation of the potential benefit of therapeutic interventions aimed at fighting it.

Usually, most spectrophotometric methods suffer from a lack of selectivity and specificity. The use of additional chromatographic steps may represent a good solution as it greatly improves the selectivity and specificity by separating potential interferences prior to detection.

The highest selectivity and specificity is provided by mass spectrometry-based methods, notably GC-MS/MS and LC-MS/MS, due to their "inherent" accuracy. However, more simple instruments such as GC-MS and LC-MS, may suffer from insufficient specificity/selectivity for some biomarkers (e.g., IsoPs).

Unlike other analytical methods, mass spectrometry-based methods use isotope-labeled analogs of the biomarkers as internal standards in quantitative analyses. These analogs and their biological counterparts behave identically during extraction and all subsequent procedures. Therefore, they minimize all possible errors that may emerge from the system, which leads to an enhancement of the analytical performances of mass spectrometric methods. The lack of commercially available stable isotope-labelled analogs for some biomarkers is a limiting factor in mass spectrometry-based techniques.

Finally, there is still a desperate need for validated biomarkers in non-invasive samples to study the oxidative status that precedes oxidation in most pathological conditions. The validity of EBC as a reliable tool for the assessment of OS in airways is still questionable owing to limitations in the reproducibility of analyzed oxidative biomarkers in gaseous matrices, with respect to both intra- and inter-individual variability [19].

23.11 Need for Consensus

In order to overcome the evident difficulties in OS assessment, the creation of a consortium of laboratories aimed at finding consensus and rules regarding the clinical determination of biomarkers of OS is needed. This consortium should address pertinent questions including:

- Validation of analytical methods dedicated to the quantification of OS markers.
- Appropriate pre-analytical and analytical procedures are to be used to measure a specific biomarker or a group of markers.
- Choice of biomarkers to be measured and the respective biological system to assess.
- The determination of reference values and intervals for suitable biomarkers according to biological system.
- Standard methods for the determination of cause–effect relationship. The correct panel to perform and reference materials to use represent other issues to be addressed.

Regarding the first point, analytical methodologies used for the determination of biomarkers of OS should be treated in the same way as those used for the measurement of other endogenous compounds and metabolites. This means that the methods should be thoroughly validated in detail, preferably according to standardized protocols (e.g., ICH protocols). Briefly, validation should include at least the following indicators: stability of the biomarker and extraction yield, accuracy and precision, selectivity and specificity, reproducibility and robustness, linearity, and finally limits of detection and quantification.

The use of validated and proven methods will provide accurate results regarding the physiological concentrations with the potential to serve as reference values and intervals for certain biomarkers in relevant biological systems as pointed out in the fourth purpose of the consortium.

With respect to the second point, the consortium should deal with the appropriate methods to use with a specific biomarker and therefore define reference methods with standardized protocols. For example, the ESCODD was set up to solve problems in the measurement of oxidized bases (see Sect. 15.6) [2]. Thus, in addition to providing reference analytical methodologies, comparative studies like those undertaken by ESCODD may also reveal accurate concentrations that may serve as reference values and intervals for particular biomarkers and biological systems and will help to ensure further achievement of the fourth point. Moreover, the validity of biomarkers and the analytical methodology used for their quantification in clinical practice as well as the determination of basal concentrations may form the basis for an independent and unbiased evaluation of the given results and their outcome.

Concerning the third point, prior to the application of an analytical method in clinical studies and explorations, the suitability of an analyte to function as a reliable biomarker should be assessed by evaluating its ability to:

- Specifically reflect changes in the OS status in a particular disease. This can be best attained by comparative studies, for instance, through the active collaboration of different laboratories working on the same biological system (e.g., blood), and using different biomarkers that best reflect oxidative changes in this system with a particular disease.
- Assess the therapeutic effect of an antioxidant in analogy with the administered doses.

An interesting example of how to measure the suitability of a potential biomarker and reference analytical methods for OS assessment is represented by the nationwide BOSS, organized by the National Institute of Environmental Health Sciences.

This work represents the first comprehensive comparative study that examined several biomarkers of OS in the same model system to determine which ones are the most specific, sensitive, and selective [20, 21].

23.12 Interferences with Diet

Another problem with the development of OS in clinical exploration and research is that food contains already some oxidant/antioxidant substances and more than a few of them have large amounts of degradable compounds (e.g., lipids), which makes biomarker identification questionable. For instance, a significant fraction of NO_3^- in blood and urine is of dietary origin. Therefore, NO_3^- in these biological samples may be of little or even no utility for the measurement of endogenous nitric oxide production [22]. It is noteworthy to mention that there is a possibility that the OS assessment may be confounded by high exposure to dietary fat. Indeed, lipid hydroperoxides and aldehydes such as MDA can be absorbed from the diet and then excreted in urine. Moreover, it has been demonstrated that post-prandial levels of 15-F_{2t}-IsoP may rise following a fatty meal [4].

Achievement of clinical exploration under standardized conditions, for example with low NO_2^-, NO_3^-, and fat diet may help reduce considerably the impact of exogenous impact on analysis [23]. Measurements of these biomarkers in plasma and urine can be confounded by diet and should not be used as an index of whole-body lipoxidation until diet is strictly monitored.

23.13 Inter-Laboratory Variations in Biomarker Assessment

Major problem with OS analysis has been the tremendous variation in the absolute values of biomarkers given by different laboratories. In this work, the occurrence of OS was investigated by measuring more than 50 different biomarkers (e.g., TBARS, MDA, thiols, 8-OHdG, F_2-isoPs, protein carbonyls, NO_2Tyr, antioxidant enzymes, etc.), for many of which mean values reportedly span over one to two orders of magnitude even in healthy controls (Table 7.2). Major discrepancies resulted from different methods used for OS assessment. For instance, the wide variety of methods has created a certain amount of discrepancies about the values of GSH and GSSG in biological fluids.

As pointed out by Collins et al., the published values of some biomarkers in normal human cells vary considerably. When different research groups undertook similar analyses, physiological concentrations of 8-OHdG in human cells were found to vary up to three orders of magnitude, depending on the analytical technique used [6].

While a comparison can be made between normal and pathological samples measured by the same laboratory, comparison between different laboratories is not reliable even when the same method is employed. Therefore, the clinical results cannot be easily compared from one laboratory to another. The variations in the reproducibility between laboratories could be due to methodological issues regarding sample collection, storage, sample preparation, and analysis.

Another important defect is the lack of reference materials for quality control operations. Currently, we do not have any standardized biological materials for MDA, 4-HNE, oxLDL, protein carbonyl, AGEs, GPx, or MPO. Fortunately, some collaborative works are in progress such as those done in the 90th by the European FLAIR group on total blood certified for GPx [24].

Overall, the use of biomarkers to examine OS has continued to grow and as a result, much attention has been placed on increasing accuracy through improvement of all aspects of analysis and standardization of measurement procedures.

23.14 Variability in Results

An examination of the literature reveals that different laboratories have published conflicting results on whether a physiological condition or disease involves OS. For example, some studies have reported that extensive exercise causes lipid peroxidation, while others have reported no lipid peroxidation during this condition [25–27]. There are also conflicting results on whether protein oxidation increases with age [28, 29]. These inconsistencies may be related to technical issues that occur in measuring OS in biological systems. As mentioned earlier, major discrepancies have been noted between the various measurement methods used, and standardized protocols are urgently needed for determination of OS biomarkers.

23.15 Standardization of Protocols

Addressing the physiological role of RS and their involvement in diseases is certainly methodologically complex and not easily applicable in medical field because of the intrinsic properties of RS and the technical tools that have been developed to this point, which are most of the time neither standardized nor optimized for daily use. Therefore, one of the most important needs in the field of OS exploration is the standardization of analytical techniques. This will allow, among other goals, a better design and evaluation of antioxidant therapeutic strategies. Technical work and standardizations are necessary before OS assessment can provide precise measurements of the concentration of a given biomarker and hence be used reliably in the clinical field. It should be pointed out that few methods have been already standardized and optimized by the standardization committees of scientific societies.

References

1. Tsikas D, Caidahl K. Recent methodological advances in the mass spectrometric analysis of free and protein-associated 3-nitrotyrosine in human plasma. J Chromatogr B Analyt Technol Biomed Life Sci. 2005;814(1):1–9.
2. Measurement of DNA oxidation in human cells by chromatographic and enzymic methods. Free Radic Biol Med. 2003;34(8):1089–99.
3. Proudfoot J, Barden A, Mori TA, Burke V, Croft KD, Beilin LJ, et al. Measurement of urinary F (2)-isoprostanes as markers of in vivo lipid peroxidation-A comparison of enzyme immunoassay with gas chromatography/mass spectrometry. Anal Biochem. 1999;272(2):209–15.
4. Gopaul NK, Zacharowski K, Halliwell B, Änggård EE. Evaluation of the postprandial effects of a fast-food meal on human plasma F2-isoprostane levels. Free Radical Biology and Medicine. 2000;28(5):806–14.

5. Reddy S, Bichler J, Wells-Knecht KJ, Thorpe SR, Baynes JW. N epsilon-(carboxymethyl)-lysine is a dominant advanced glycation end product (AGE) antigen in tissue proteins. Biochemistry. 1995;34(34):10872–8.
6. Collins AR, Cadet J, Mőller L, Poulsen HE, Viña J. Are we sure we know how to measure 8-oxo-7,8-dihydroguanine in DNA from human cells? Archives of Biochemistry and Biophysics. 2004;423(1):57–65.
7. Agarwal A, Qiu E, Sharma R. Laboratory assessment of oxidative stress in semen. Arab J Urol. 2017;16(1):77–86.
8. Nuhu F, Gordon A, Sturmey R, Seymour A-M, Bhandari S. Measurement of Glutathione as a Tool for Oxidative Stress Studies by High Performance Liquid Chromatography. Molecules. 2020;25(18):4196
9. Devaraj S, Hirany SV, Burk RF, Jialal I. Divergence between LDL oxidative susceptibility and urinary F(2)-isoprostanes as measures of oxidative stress in type 2 diabetes. Clin Chem. 2001;47(11):1974–9.
10. Tsikas D, Schwedhelm E, Suchy M-T, Niemann J, Gutzki F-M, Erpenbeck VJ, et al. Divergence in urinary 8-iso-PGF2α (iPF2α-III, 15-F2t-IsoP) levels from gas chromatography–tandem mass spectrometry quantification after thin-layer chromatography and immunoaffinity column chromatography reveals heterogeneity of 8-iso-PGF2α: Possible methodological, mechanistic and clinical implications. Journal of Chromatography B. 2003;794(2):237–55.
11. Requena JR, Ahmed MU, Fountain CW, Degenhardt TP, Reddy S, Perez C, et al. Carboxymethylethanolamine, a biomarker of phospholipid modification during the maillard reaction in vivo. J Biol Chem. 1997;272(28):17473–9.
12. James MJ, van Reyk D, Rye KA, Dean RT, Cleland LG, Barter PJ, et al. Low density lipoprotein of synovial fluid in inflammatory joint disease is mildly oxidized. Lipids. 1998;33 (11):1115–21.
13. Bruckdorfer KR, Hillary JB, Bunce T, Vancheeswaran R, Black CM. Increased susceptibility to oxidation of low-density lipoproteins isolated from patients with systemic sclerosis. 1995;38(8): 1060–7.
14. Dryden GW, Deaciuc I, Arteel G, McClain CJ. Clinical implications of oxidative stress and antioxidant therapy. Current Gastroenterology Reports. 2005;7(4):308–16.
15. Balmus IM, Ciobica A, Trifan A, Stanciu C. The implications of oxidative stress and antioxidant therapies in Inflammatory Bowel Disease: Clinical aspects and animal models. Saudi J Gastroenterol. 2016;22(1):3–17.
16. Takaki A, Kawano S, Uchida D, Takahara M, Hiraoka S, Okada H. Paradoxical Roles of Oxidative Stress Response in the Digestive System before and after Carcinogenesis. Cancers (Basel). 2019;11(2):213.
17. Mehta S, Vijayvergiya R, Dhawan V. Activation of NLRP3 inflammasome assembly is associated with smoking status of patients with coronary artery disease. International Immunopharmacology. 2020;87:106820.
18. Mateu-Jiménez M, Sánchez-Font A, Rodríguez-Fuster A, Aguiló R, Pijuan L, Fermoselle C, et al. Redox Imbalance in Lung Cancer of Patients with Underlying Chronic Respiratory Conditions. Molecular medicine (Cambridge, Mass). 2016;22:85–98.
19. Rahman I. Reproducibility of oxidative stress biomarkers in breath condensate: are they reliable? 2004;23(2):183–4.
20. Kadiiska MB, Gladen BC, Baird DD, Germolec D, Graham LB, Parker CE, et al. Biomarkers of oxidative stress study II: are oxidation products of lipids, proteins, and DNA markers of CCl4 poisoning? Free Radic Biol Med. 2005;38(6):698–710.
21. Kadiiska MB, Gladen BC, Baird DD, Graham LB, Parker CE, Ames BN, et al. Biomarkers of oxidative stress study: III. Effects of the nonsteroidal anti-inflammatory agents indomethacin and meclofenamic acid on measurements of oxidative products of lipids in CCl4 poisoning. Free Radical Biology and Medicine. 2005;38(6):711–8.

22. Lauer T, Preik M, Rassaf T, Strauer BE, Deussen A, Feelisch M, et al. Plasma nitrite rather than nitrate reflects regional endothelial nitric oxide synthase activity but lacks intrinsic vasodilator action. Proc Natl Acad Sci U S A. 2001;98(22):12814–9.
23. Tsikas D. Analysis of nitrite and nitrate in biological fluids by assays based on the Griess reaction: appraisal of the Griess reaction in the L-arginine/nitric oxide area of research. J Chromatogr B Analyt Technol Biomed Life Sci. 2007;851(1-2):51–70.
24. Belsten JL, Wright AJ. European Community--FLAIR common assay for whole-blood glutathione peroxidase (GSH-Px); results of an inter-laboratory trial. Eur J Clin Nutr. 1995;49(12): 921–7.
25. Viitala PE, Newhouse IJ, LaVoie N, Gottardo C. The effects of antioxidant vitamin supplementation on resistance exercise induced lipid peroxidation in trained and untrained participants. Lipids Health Dis. 2004;3(1):14.
26. Fogarty M, Hughes C, Burke G, Brown J, Trinick T, Duly E, et al. Exercise-Induced Lipid Peroxidation: Implications for Deoxyribonucleic Acid Damage and Systemic Free Radical Generation. Environmental and molecular mutagenesis. 2011;52:35–42.
27. Kawamura T, Muraoka I. Exercise-Induced Oxidative Stress and the Effects of Antioxidant Intake from a Physiological Viewpoint. Antioxidants (Basel). 2018;7(9):119.
28. Reeg S, Grune T. Protein Oxidation in Aging: Does It Play a Role in Aging Progression? Antioxid Redox Signal. 2015;23(3):239–55.
29. Goto S, Radak Z. Implications of oxidative damage to proteins and DNA in aging and its intervention by caloric restriction and exercise. Journal of Sport and Health Science. 2013;2(2): 75–80.
30. Halliwell B, Whiteman M. Measuring reactive species and oxidative damage in vivo and in cell culture: how should you do it and what do the results mean? Br J Pharmacol. 2004;142(2): 231–55.
31. Shulaev V, Oliver DJ. Metabolic and Proteomic Markers for Oxidative Stress. New Tools for Reactive Oxygen Species Research. 2006;141(2):367–72.
32. Orhan H, Coolen S, Meerman J. Quantification of urinary o,o′-dityrosine, a biomarker for oxidative damage to proteins, by high performance liquid chromatography with triple quadrupole tandem mass spectrometry: A comparison with ion-trap tandem mass spectrometry. Journal of chromatography B, Analytical technologies in the biomedical and life sciences. 2005;827:104–8.
33. Tsikas D, Schwedhelm E, Suchy MT, Niemann J, Gutzki FM, Erpenbeck VJ, et al. Divergence in urinary 8-iso-PGF(2alpha) (iPF(2alpha)-III, 15-F(2t)-IsoP) levels from gas chromatography-tandem mass spectrometry quantification after thin-layer chromatography and immunoaffinity column chromatography reveals heterogeneity of 8-iso-PGF(2alpha). Possible methodological, mechanistic and clinical implications. Journal of chromatography B, Analytical technologies in the biomedical and life sciences. 2003;794(2):237–55.
34. Ryberg H, Caidahl K. Chromatographic and mass spectrometric methods for quantitative determination of 3-nitrotyrosine in biological samples and their application to human samples. Journal of Chromatography B. 2007;851(1):160–71.

Index

S. Andreescu et al. (eds.), *Biomarkers of Oxidative Stress*,
https://doi.org/10.1007/978-3-031-60738-7

SPRINGER NATURE

GPSR Compliance

The European Union's (EU) General Product Safety Regulation (GPSR) is a set of rules that requires consumer products to be safe and our obligations to ensure this.

If you have any concerns about our products, you can contact us on ProductSafety@springernature.com

In case Publisher is established outside the EU, the EU authorized representative is:

Springer Nature Customer Service Center GmbH
Europaplatz 3
69115 Heidelberg, Germany

The manufacturer's authorised representative in the EU is Springer
Nature Customer Service Centre GmbH, Europaplatz 3, 69115 Heidelberg,
Germany. If you have any concerns regarding our products, please
contact ProductSafety@springernature.com

Printed and bound by CPI Group (UK) Ltd, Croydon, CR0 4YY
28/09/2026
02224056-0017